D0771332

INDUSTRIAL and RESIDENTIAL

ELECTRICITY

SECOND EDITION

INDUSTRIAL and RESIDENTIAL ELECTRICITY

REX MILLER

State University College
Buffalo, New York

Glencoe
A Division of Macmillan Publishing Company
Mission Hills, California

Send all inquiries to:
Glencoe Publishing Company
15319 Chatsworth Street
P.O. Box 9509
Mission Hills, California 91345-9509

Printed in the United States of America

ISBN 0-02-676360-5 (Text)
ISBN 0-02-676370-2 (Teacher's Resource Guide)
ISBN 0-02-676380-1 (Student Workbook)

2 3 4 5 6 93 92 91 90 89

Preface

This new edition of a popular text contains much new material. For example, it introduces information on alternate energy sources, such as microhydro and solar energy systems. It discusses the use of voltage spike protectors and the use of PVC conduit systems. Information throughout the text has been updated to reflect changes in the National Electrical Code. Procedures for estimating a job have been included.

The value of this new edition has been enhanced through the inclusion of the following:

- Learning objectives at the beginning of each chapter.
- A terms list at the end of each chapter.
- A glossary, defining over 150 key terms in electricity.

The purpose of this book is to aid you in your study of electricity. *Industrial and Residential Electricity* presents information that will enable you to understand and work with electricity in the industrial, commercial, rural, and residential areas of the field.

This book has been designed for students in vocational classes, as well as for electricians enrolled in apprenticeship courses who desire training in residential, commercial, and industrial wiring. Although it covers some theory of the principles of electricity, the main emphasis of the text is on the practical, everyday application of those principles.

Many illustrations are included which show a variety of parts and techniques found in present day practice in the field. Obviously, not all related problems can be presented here since there is a great deal of ingenuity required by the worker on the job. For standard procedure, however, the National Electrical Code handbook does give a guide to the number of wires and the types of equipment safe to install in a given location. This handbook should be a part of your tool kit.

To familiarize the reader with the metric measurement system, symbols approved by the International Organization for Standardization are used where applicable throughout this text. They belong to the SI—the International System of Units—and are given in parentheses, following the customary units.

It is not possible to learn all there is to know about electrical work by reading about it. You must be willing to get some practical experience and devote some time to the development of skills related to the job. The information presented in this text can provide a foundation for a rewarding career in the field of electricity.

Rex Miller

Acknowledgments

As with many books, people have given generously of their time in assisting the author of *Industrial and Residential Electricity*. Their efforts and suggestions have been of great value. I would like to take this opportunity to thank each of them for his or her contributions.

Special appreciation is expressed to Dr. Thomas H. Arcy of the Department of Technical Education, University of Houston, for his review of the manuscript.

Many manufacturers of electrical parts and equipment have provided illustrations and information to this book. The following manufacturers and agencies are but a few of the many who make the generation, installation, and use of electricity as efficient and safe as possible. In addition, the technical data furnished by these companies and agencies are much appreciated. Without their assistance, this text could have neither appeal nor objectivity.

Allen-Bradley Company, Milwaukee, Wisconsin

Aluminum Association, New York, New York

Amerace Corporation, Hackettstown, New Jersey

Appleton Electric Company, Chicago, Illinois

Automatic Switch Company, Florham Park, New Jersey

Bernzomatic Corporation, Rochester, New York

Biddle Instruments, Inc., Blue Bell, Pennsylvania

Black & Decker Manufacturing Company, Towson, Maryland

Bodine Electric Company, Chicago, Illinois

Brown & Sharpe Manufacturing Company, Providence, Rhode Island

Bryant Electric Company, Bridgeport, Connecticut

Buchanan Electric Products Division, Union, New Jersey

Burndy Corporation, Norwalk, Connecticut

Canadian Standards Association, Rexdale, Ontario, Canada

Carlon Electrical Sciences, Inc., Cleveland, Ohio

Central Illinois Light Company (CILCO), Peoria, Illinois

Chromalox Comfort Conditioning, St. Louis, Missouri

Chrysler Corporation, Detroit, Michigan

Circle F Industries, Trenton, New Jersey

Combustion Engineering, Windsor, Connecticut

Conduflor Corporation, Comstock Park, Michigan

Crescent Tool Company, Jamestown, New York

Crescent Wire and Cable Company, Trenton, New Jersey

Crouse-Hinds, Syracuse, New York

Daniel Woodhead, Inc., Northbrook, Illinois

Delco, Muncie, Indiana

Eastman Kodak Co., Rochester, New York

Edison Electric Institute, New York, New York

EICO, Long Island City, New York

Eller Mfg. Co., Brooklyn, New York

Emerson Electric Company, St. Louis, Missouri

General Electric Company, Providence, Rhode Island

Greenlee Tool Company, Rockford, Illinois

Harvey Hubbell, Inc., Bridgeport, Connecticut

Heath Company, Benton Harbor, Michigan

Honeywell, Minneapolis, Minnesota

Ideal Industries, Inc., Sycamore, Illinois

Kellems, Div. of Harvey Hubbell, Inc., Stonington, Connecticut

Lennox Industries, Inc., Portland, Oregon

Lincoln Electric Company, Cleveland, Ohio

Lockwood Corporation, Gering, New Brunswick

P.R. Mallory Company, Indianapolis, Indiana

Midwest Electric Products, Inc., Mankato, Minnesota

Misener Manufacturing Company, Inc., Syracuse, New York

National Fire Protection Association, Boston, Massachusetts

National Safety Council, Chicago, Illinois

New York Board of Fire Underwriters, New York, New York

New York State Power Authority, Niagara Falls, New York

Niagara Mohawk Power Corporation, Buffalo, New York

NuTone Division of Scovill, Cincinnati, Ohio

Ohmite, Skokie, Illinois

Onan, Minneapolis, Minnesota

Pass & Seymour, Inc., Syracuse, New York

Philips Drill Co., Inc., Michigan City, Indiana

Raco, Inc., South Bend, Indiana

Ramset Fastening Systems, Branford, Connecticut

Rawlplug Company, Inc., New Rochelle, New York

Rural Electrification Authority (REA), Washington, D.C.

Sangamo Electric Company, Springfield, Illinois

Sears, Roebuck & Company, Chicago, Illinois

Seatek Company, Riverside, Connecticut

H.B. Sherman Manufacturing Company, Battle Creek, Michigan

Sierra Electric Company, Gardenia, California

Slater Electric, Inc., Glen Cove, New York

Spaulding Fibre Company, Tonawanda, New York

Square D Company, Lexington, Kentucky

Star Expansion Company, Mountainville, New York

Superior Electric Company, Bristol, Connecticut

Thomas & Betts Company, Elizabeth, New Jersey

3M, Electro-Products Division, St. Paul, Minnesota

Underwriters' Laboratories, Inc., Chicago, Illinois

Union Insulating Company, Parkersburg, West Virginia

Wagner Electric Corporation, St. Louis, Missouri

Westinghouse Electric Company, Pittsburgh, Pennsylvania

Weston Electrical Instruments Company, Newark, New Jersey

Wiremold Company, West Hartford, Connecticut

Contents

SECTION ONE GENERATING and DISTRIBUTING ELECTRICITY

CHAPTER 1

NATURE OF ELECTRICITY

Objectives

After studying this chapter, you will be able to:

- Give a simple definition of current electricity.
- List seven ways in which electricity may be generated.
- Identify the four factors that determine resistance.
- Identify the various types of circuits.
- State Ohm's law.
- Discuss the ways in which electricity is measured.
- Identify the switches used to control electricity.
- Read a resistor color code.

WHAT IS ELECTRICITY?

Though you cannot see electricity, you are aware of it every day. You see it used in countless ways. You cannot taste or smell electricity, but you can feel it. You can taste food cooked with its energy. You can smell the gas (ozone) that forms when lightning passes through the air.

Basically there are two kinds of electricity—static (stationary) and current (moving). This book is chiefly about current electricity because that is the kind commonly put to use.

Current electricity can be simply defined as the *flow of electrons along a conductor*. To understand that definition, you must know something about chemical elements and atoms.

ELEMENTS AND ATOMS

Elements are the most basic materials in the universe. Ninety-four elements, such as iron, copper, and nitrogen, have been found in nature. Scientists have made eleven others in laboratories. Every known substance—solid, liquid, or gas—is composed of elements.

It is very rare for an element to exist in a pure state. Nearly always the elements are found in combinations called *compounds*. Even such a common substance as water is a compound rather than an element. Fig. 1-1.

An *atom* is the smallest particle of an element that retains all the properties of that element. Each element has its own kind of atom. That is, all hydrogen atoms are alike, and they are different from the atoms of all other elements. However, all atoms have certain things in common. They all have an inner part, the *nucleus*. This is composed of tiny particles called *protons* and *neutrons*. An atom also has an outer part. It consists of other tiny particles, called *electrons*, which orbit around the nucleus. Figs. 1-2 & 1-3.

Neutrons have no electrical charge, but protons are positively charged. Electrons have a negative charge. Because of these charges, protons and electrons are particles of energy. That is, these charges form an electric field of force within the atom. Stated very simply, these charges are always pulling and pushing each other, which makes energy in the form of movement.

The atoms of each element have a definite number of electrons, and they have the same number of protons. A hydrogen atom has one electron and one proton. An aluminum atom has thirteen of each. The opposite charges—negative electrons and positive protons—at-

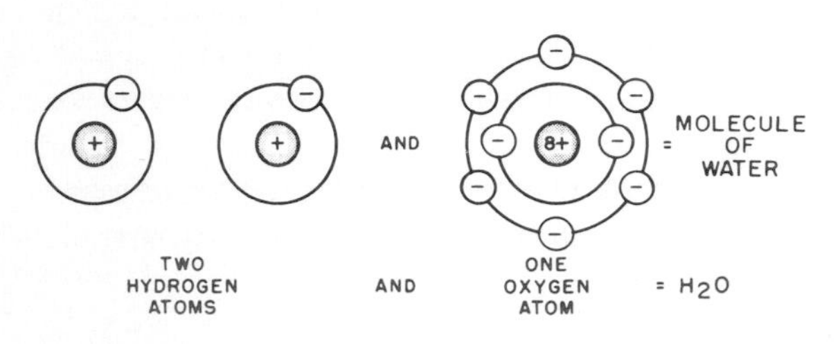

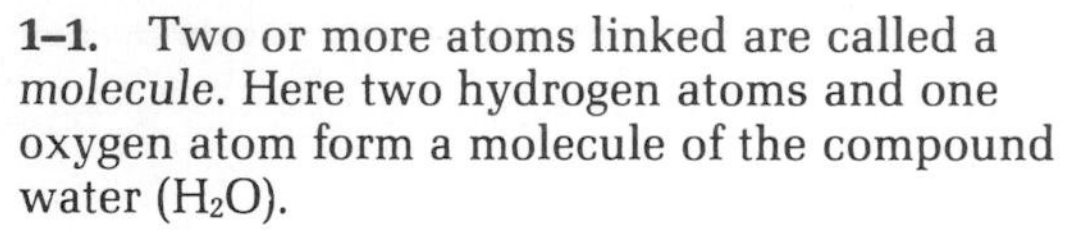

1–1. Two or more atoms linked are called a *molecule*. Here two hydrogen atoms and one oxygen atom form a molecule of the compound water (H_2O).

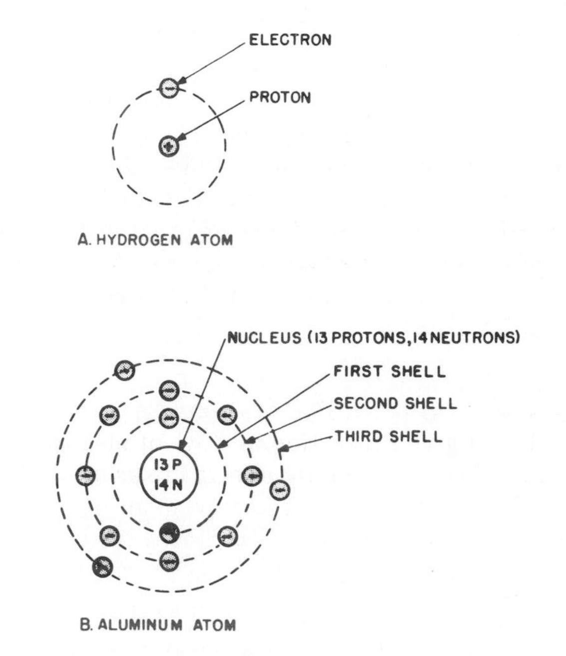

1–2. Atoms contain protons, neutrons, and electrons.

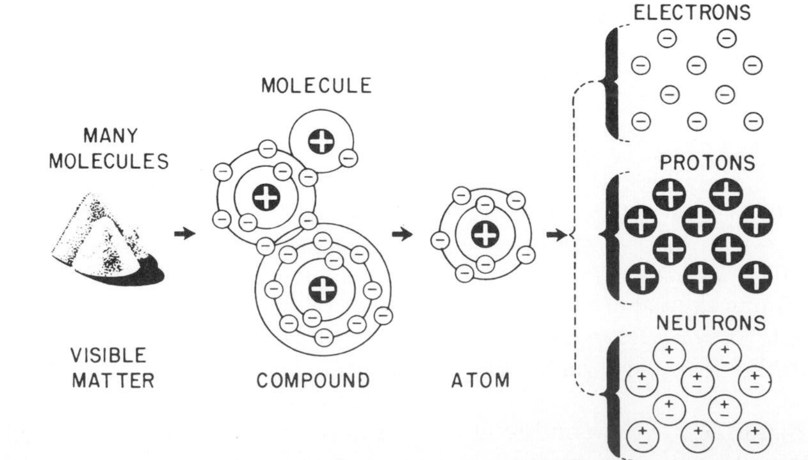

1–3. Molecular structure.

tract each other and tend to hold electrons in orbit. As long as this arrangement is not changed, an atom is electrically balanced.

However, the electrons of some atoms are easily pushed or pulled out of their orbits. This ability of electrons to move or flow is the basis of current electricity.

Free Electrons

In some materials, heat causes electrons to be forced loose from their atoms. In other materials, such as copper, electrons may be easily forced to drift, even at room temperatures. When electrons leave their orbits, they may move from atom to atom at random, drifting in no particular direction. Electrons that move in such a way are referred to as *free electrons.* However, a force can be applied to direct them in a definite path.

Current Flow

If the movement of free electrons is channeled in a given direction, a flow of electrons occurs. This is commonly referred to as *current flow.* Thus you see that the movement of electrons is related to current electricity.

Energy

Electrons are incredibly small. The diameter of an electron is about 0.00000000000022 inch. You may wonder how anything so small can be a source of energy. Much of the answer lies in the fact that electrons move at nearly the speed of light. Also, billions of them can move at once through a wire. Speed and concentration produce great energy.

ELECTRICAL MATERIALS

■ *Conductors.* A *conductor* is a material through which electrons move. Actually, all metals and most other materials are conductors to some extent. Some, however, are better than others. Thus the term *conductor* is usually used to mean a material through which electrons move freely.

What makes one material a better conductor than another? A material which has many free electrons tends to be a good conductor. For practical purposes, however, there are other points which must be considered when choosing a material to use as a conductor.

For example, gold, silver, aluminum, and copper are all good conductors. However, the cost of gold and silver limits their use. Copper, because of its superior strength in both hot and cold weather, is preferred over aluminum for many uses.

■ *Insulators.* An *insulator* is a substance that restricts the flow of electrons. Such materials have a very limited number of free electrons. Thus you see that the movement of free electrons classifies a material as either a conductor or an insulator. No material is known to be a perfect insulator—that is, entirely void of free electrons. However, there are materials which are such poor conductors that for all practical purposes they are placed in the insulator class.

Wood, glass, mica, and polystyrene are insulators. Fig. 1-4. They have varying degrees of resistance to the movement of their electrons. The higher the line on the chart in Fig. 1-4, the better are the insulating qualities of the material.

■ *Semiconductors.* You have heard the word *semiconductor* in relation to transistors and diodes used in electronic equipment. Materials used in the manufacture of transistors and diodes have a conductivity between that of a good conductor and a good insulator. Therefore, the name semiconductor is given them. Germanium and silicon are the two most commonly known semiconductor materials. Through the introduction of small amounts of other elements these nearly pure (99.999999%) elements become limited conductors. The manufacture of semiconductors is a fascinating process. However, it would take too long to go into details at this time. You may wish to research the topic on your own by checking out a book from your library.

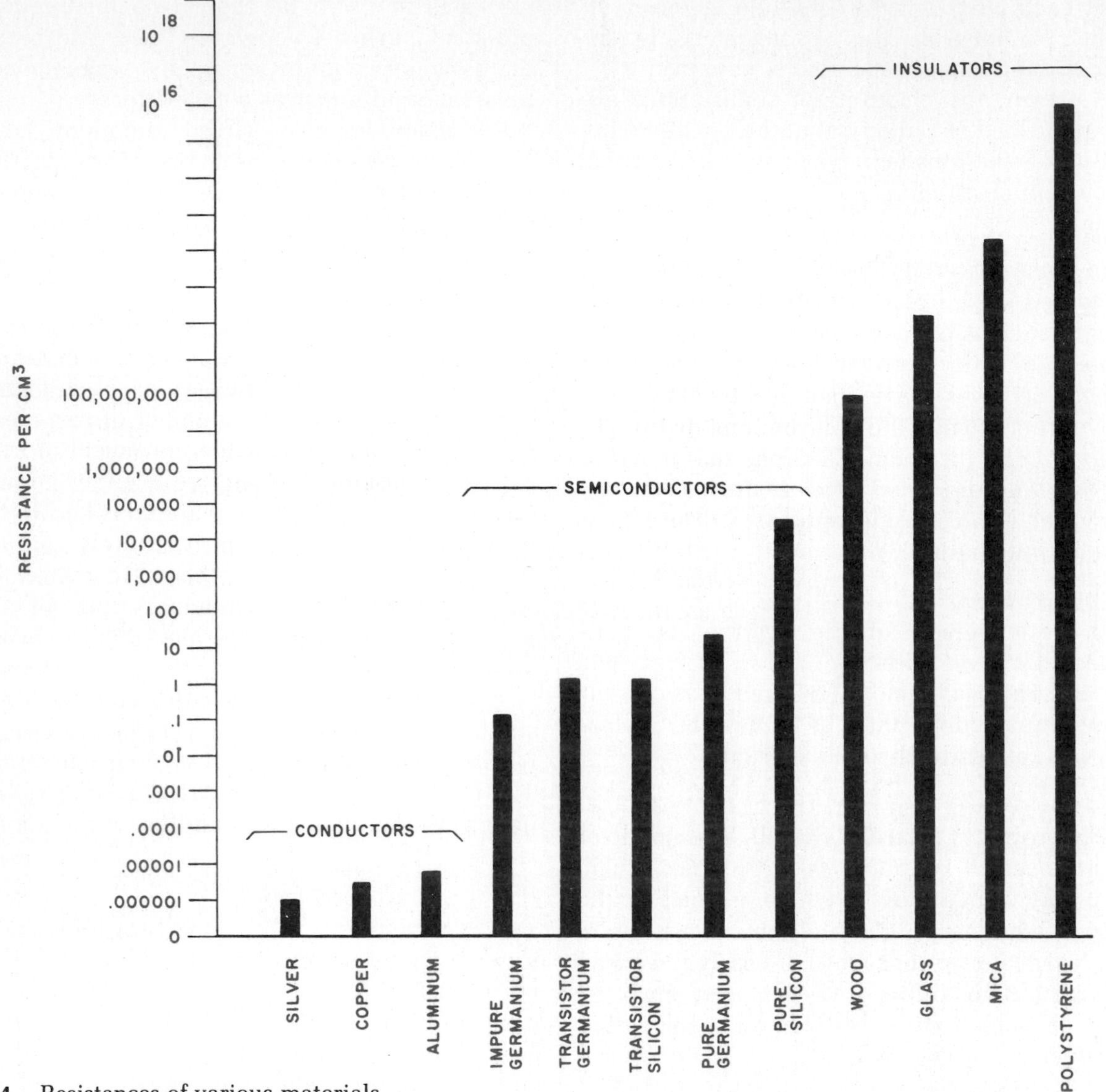

1–4. Resistances of various materials.

GENERATING ELECTRICITY

There are several ways to produce electricity. Remember: *Electricity is the flow of electrons along a conductor*. Friction, pressure, heat, light, chemical action, and magnetism are among the more practical methods used to make electrons move along a conductor. Other methods (sometimes called "exotic") are used to generate electricity for special purposes. For instance, experimental cells developed for the space program are termed "exotic."

■ *Friction*. Electricity is produced when two materials are rubbed together. The movement of your shoes against the carpet can cause static electricity. Some practical applications of static electricity are in the manufacture of sandpaper and in the cleaning of polluted air. Fig. 1-5.

■ *Pressure.* Electricity is produced when pressure is applied to a crystal. The crystals are usually Rochelle salts or quartz. The special properties of crystals are utilized in the crystal microphone. Fig. 1-6. Here, bending of the crystal produces a small electrical output. This is known as the *piezoelectrical* effect. This small voltage can be amplified to drive a speaker. In fact, crystal pickups are used in inexpensive record players and for some industrial jobs.

■ *Heat.* Electricity is produced when heat is applied to the junction of two dissimilar metals. This junction is usually referred to as a *thermocouple.* The thermocouple is used to measure temperatures in industrial applications. This is especially true in checking the temperature of kilns for ceramic work. Fig. 1-7.

■ *Light.* Electricity is produced when light strikes a photosensitive material. (The word *photo* means light). Photoelectric cells are used in cameras, spacecraft, and in radios. Fig. 1-8.

■ *Chemical action.* Electricity is produced when a chemical action takes place between two metals in a cell. A single unit is called a cell. Connecting two or more cells together produces a battery. Batteries are used in flashlights, radios, hearing aids, and in calculators. The automobile uses a lead-acid cell combination. You could not start today's cars without a battery. Many types of cells are available today. Fig. 1-9.

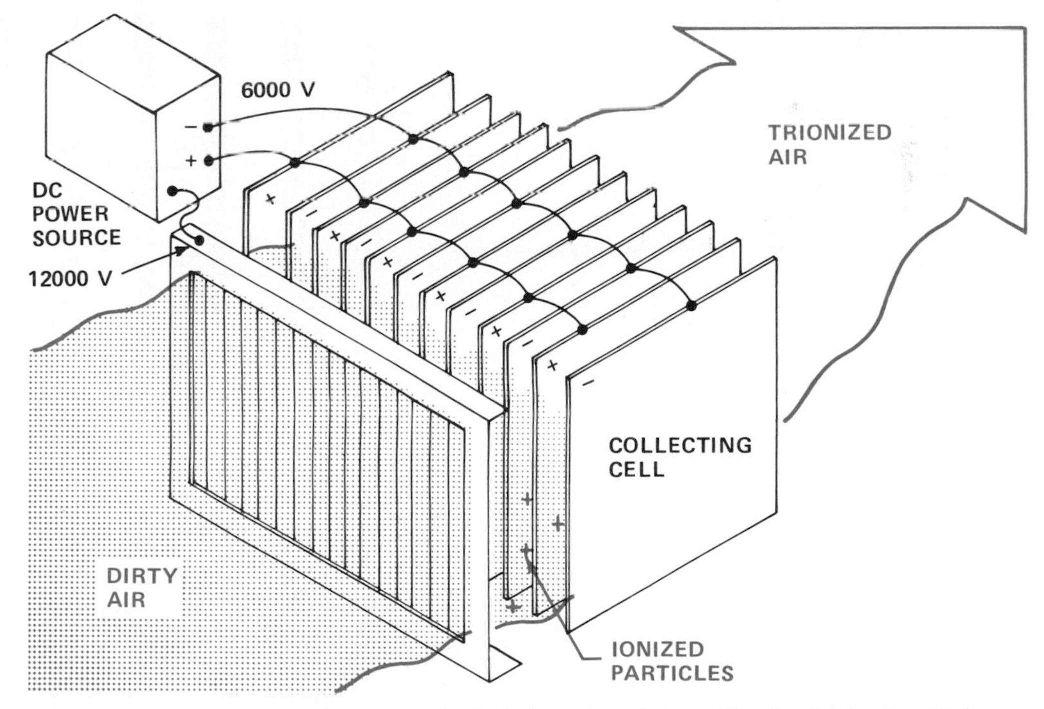

1–5. Electrostatic precipitator uses a two-stage method of cleansing air by collecting ionized particles on charged plates. Ionizing wires of tungsten are charged with 12 000 volts DC (+). All particles are then electrically charged by ionization (+). The positively charged particles are attracted to the negatively charged plates with 6000 volts (−). The negatively charged plate is the collecting cell. Clean air is exhausted from the precipitator.

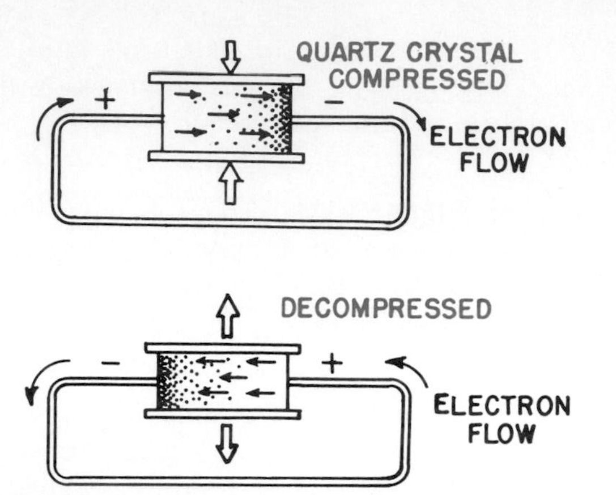

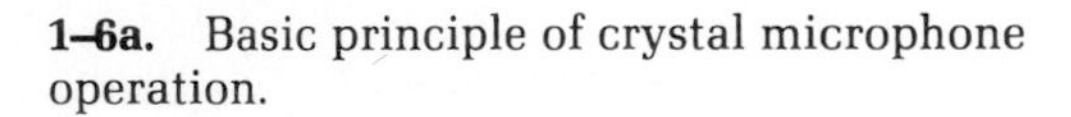
1–6a. Basic principle of crystal microphone operation.

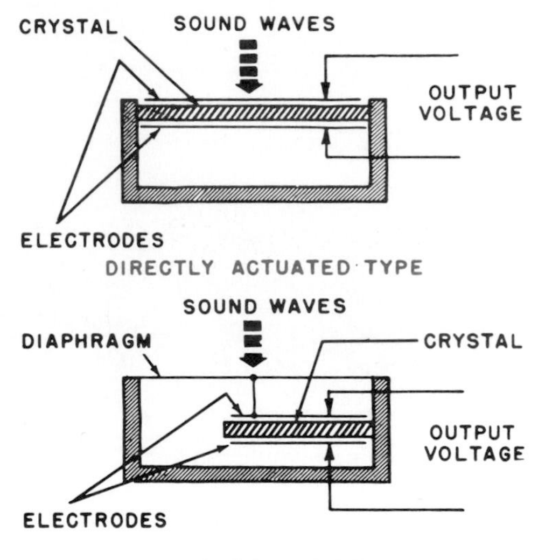

1–6b. Crystal microphones: directly actuated type and diaphragm type.

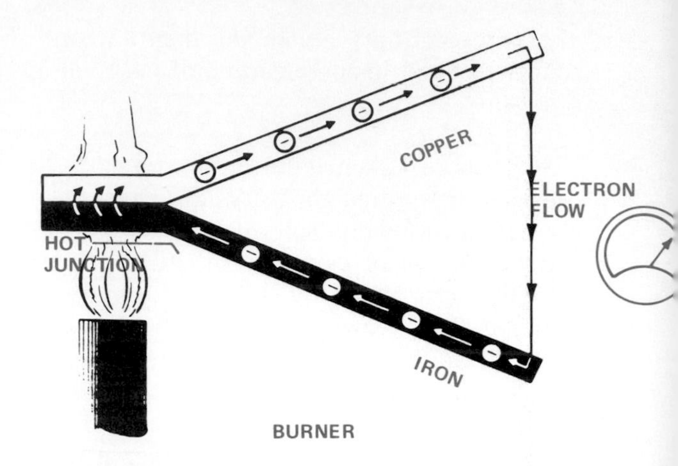

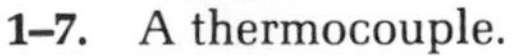
1–7. A thermocouple.

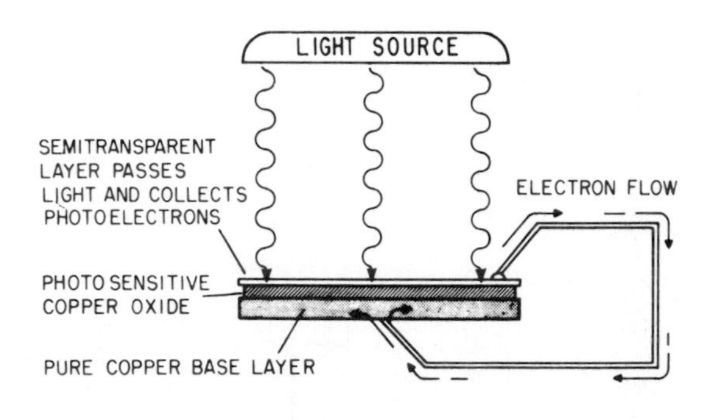

1–8. Photoelectric cell.

■ *Magnetism.* Electricity is produced when a magnet is moved past a piece of wire. Or, a piece of wire can be moved through a magnetic field. The result is the same. Motion, a magnetic field, and a piece of wire are needed to produce electricity. To date, magnetism is the most inexpensive way of producing electrical power. We use magnetism to produce electricity for homes and cars. An electric generator is found under the hood of every automobile. This device can produce great amounts of electrical energy. It is called an *alternator* because it generates alternating current. Alternating current (AC) flows first in one direction and then in the other. Direct current (DC) flows in one direction only. Fig. 1-10.

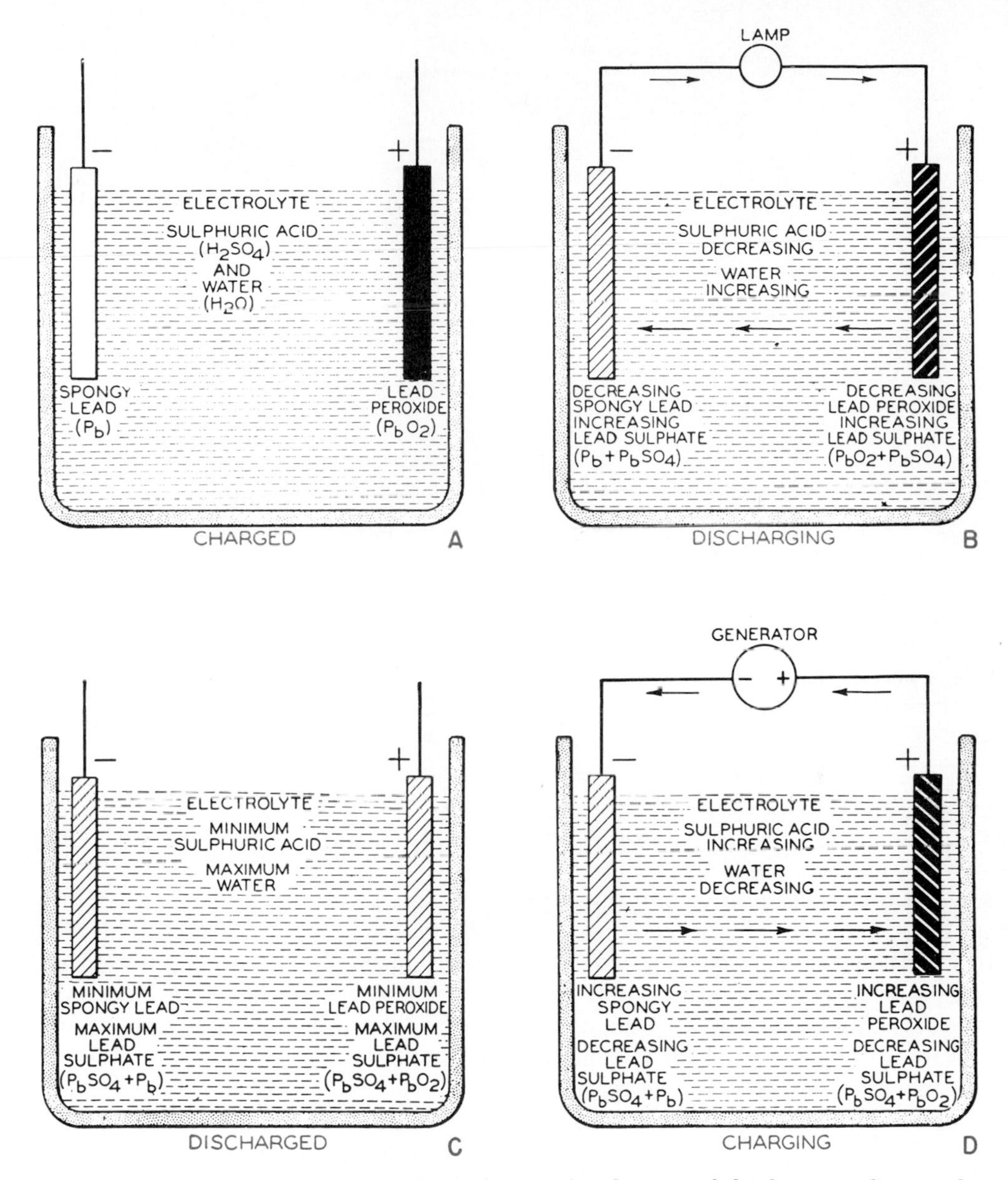

1–9a. A secondary cell is one that can be recharged. Here the charge and discharge cycles are shown.

■ *Exotic generators.* The fuel cell is one of the latest developments for the production of electricity. The *oxygen-concentration cell* includes an electrolyte. The electrolyte conducts an electric charge in the form of oxygen ions, but acts as an insulator to electrons. The electrolyte is located between two electrodes. (The electrolyte is wet—electrodes are usually metal rods or sheets.) By causing oxygen of different concentrations to pass by the electrodes, it is possible to produce electricity.

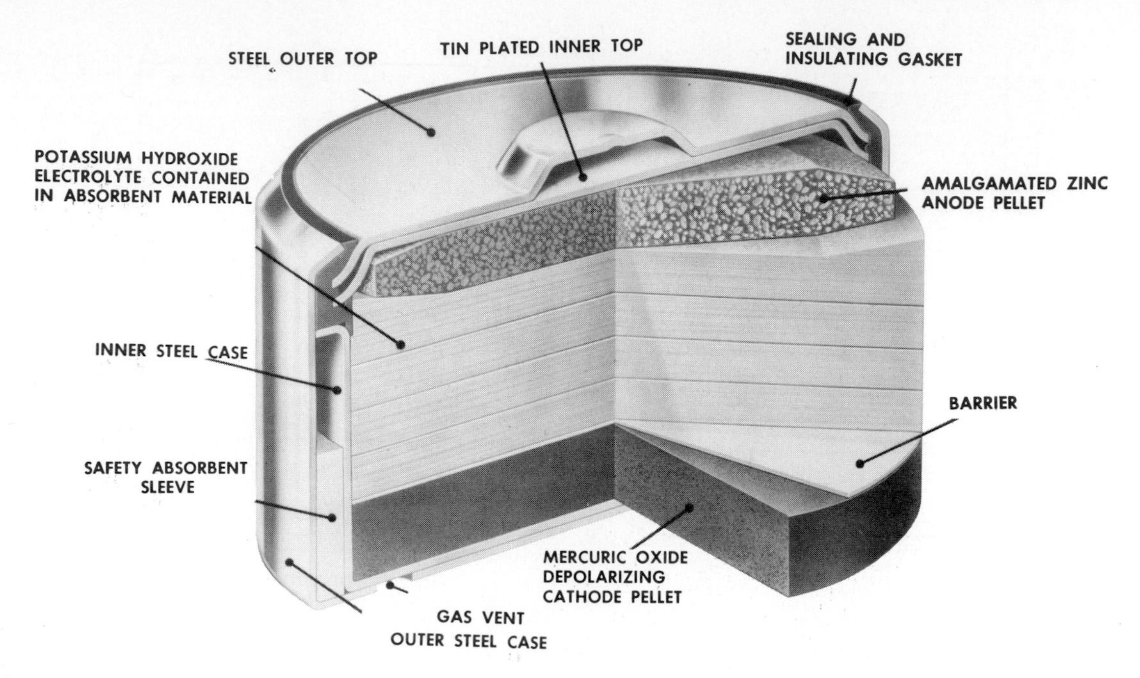

1–9b. Mercury cell: cutaway view.

The hydrogen-oxygen cell produces water and electricity. Such a cell was used on one of the spaceflights to supply both drinking water and electricity in a very small space. Other exotic cells—not all of them perfected yet—are the redox fuel cell, the hydrocarbon fuel cell, the ion-exchange membrane, and MHD (magnetohydrodynamic). Fig. 1-11. In the MHD generator, hot plasma is generated and seeded in a burner similar to a rocket engine. It then travels through a magnetic field which is applied at right angles to the flow, and past electrodes which are exposed to this stream of gas. Electrons in the gas are deflected by the field. Between collisions with the particles in the gas, they make their way to one of the electrodes. Electricity flows as the electrons move from the cathode, through the load, to the anode, and back again to the gas stream. There are thousands of other methods of producing electricity.

VOLTAGE AND CURRENT

So far you have become aware of what electricity is. You have learned some of the ways it is produced. Now it is time to learn how electrical energy is measured. The units of measurement most frequently used are *voltage* and *current.*

■ *Volts.* We measure the difference in potential between two plates in a battery in terms of volts. It is actually *electrical pressure* exerted on electrons in a circuit. (A *circuit* is a pathway for the movement of electrons.) An external force exerted on electrons to make them flow through a conductor is known as *electromotive force*, or EMF. It is measured in volts. Electrical pressure, potential difference, and EMF mean the same thing. The words *voltage drop* and *potential drop* can be interchanged.

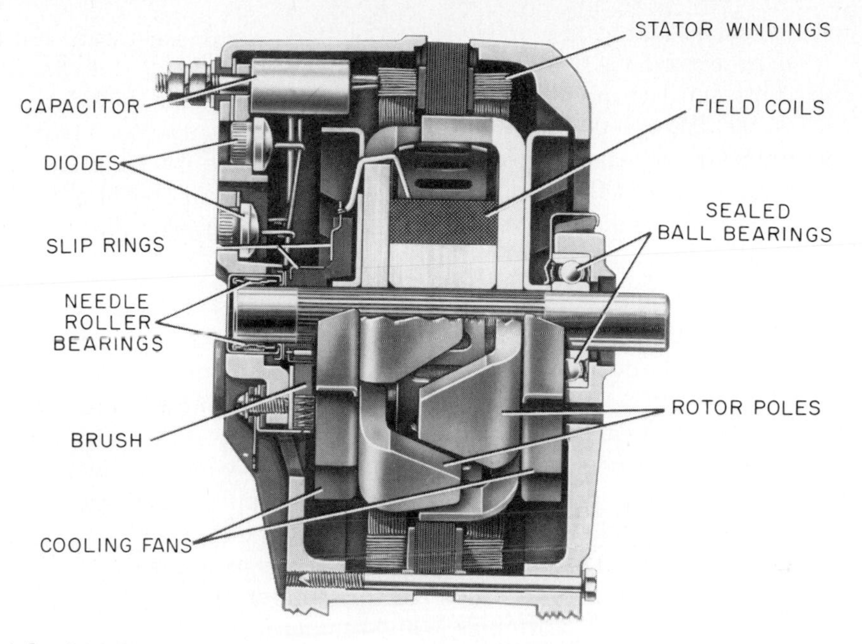

1–10. Automobile alternator.

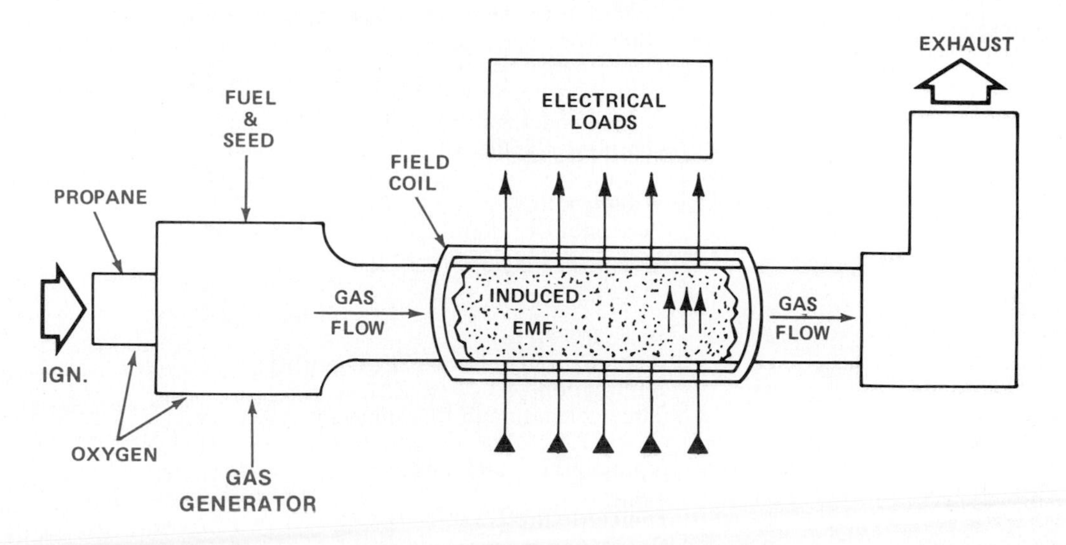

1–11. One exotic power source is the MHD generator.

■ *Current*. For electrons to move in a particular direction, it is necessary for a potential difference to exist between two points of the EMF source. If 6 280 000 000 000 000 000 electrons pass a given point in one second, there is said to be one *ampere* of current flowing. The symbol for ampere is A. The same number of electrons stored on an object (static charge) and *not moving* is called a *coulomb* (symbol C).

Current is assumed to flow from negative (−) to positive (+) terminals of a battery or generator.

Current is measured in amperes. In electronics it is sometimes necessary to use smaller units of measurement. The *milliampere* (mA) is used to indicate $\frac{1}{1000}$ of an ampere (0.001 A). If an even smaller unit is needed, it is usually the *microampere* (μA). The microampere is one millionth of an ampere. This may be written as 0.000001 A. The Greek letter mu (μ) is used to indicate *micro*. (Table 1-A lists the Greek alphabet and the terms they designate.)

A voltmeter is used to measure voltage. An ammeter is used to measure current in amperes. A microammeter or a milliammeter may be used to measure smaller units of current.

Resistance

The movement of electrons along a conductor meets with some opposition. This opposition is *resistance*. Resistance is useful in electrical and electronics work. Resistance makes it possible to generate heat, control electron flow, and supply the correct voltage to a device.

Resistance in a conductor depends on four factors: material, length, cross-sectional area, and temperature.

1-A. Greek Alphabet

NAME	CAPITAL	SMALL	USED TO DESIGNATE
alpha	Α	α	Angles, area, coefficients, and attentuation constant
beta	Β	β	Angles and coefficients
gamma	Γ	γ	Electrical conductivity and propagation constant
delta	Δ	δ	Angles, increment, decrement, and determinants
epsilon	Ε	ε	Dielectric constant, permittivity, and base of natural logarithms
zeta	Ζ	ζ	Coordinates
eta	Η	η	Efficiency, hysteresis, and coordinates
theta	Θ	ϑ θ	Angles and angular phase displacement
iota	Ι	ι	Coupling coefficient
kappa	Κ	κ	
lambda	Λ	λ	Wavelength
mu	Μ	μ	Permeability, amplification factor, and prefix micro
nu	Ν	ν	
xi	Ξ	ξ	
omicron	Ο	ο	
pi	Π	π	Pi = 3.1416 . . .
rho	Ρ	ρ	Restivity and volume charge density
sigma	Σ	σ s	Summation
tau	Τ	τ	Time constant and time-phase displacement
upsilon	Υ	υ	
phi	Φ	ϕ φ	Magnetic flux and angles
chi	Χ	χ	Angles
psi	Ψ	ψ	Dielectric flux
omega	Ω	ω	Resistance in ohms and angular velocity

■ *Material.* Some materials offer more resistance than others. It depends upon the number of free electrons present in the material.

■ *Length.* The longer the wire or conductor, the more resistance it has. The resistance is said to vary directly with the length of the wire.

■ *Cross-sectional area.* Resistance varies inversely with the size of the conductor in cross section. In other words, the larger the wire, the smaller the resistance per foot of length.

■ *Temperature.* For most materials, the higher the temperature, the higher the resistance. However, there are some exceptions to this in devices known as *thermistors.* Thermistors change resistance with temperature. They decrease in resistance as the temperature increases. Thermistors are used in meters and as temperature indicators.

Resistance is measured by a unit called the *ohm.* The Greek letter omega (Ω) is used as the symbol for electrical resistance.

Wire Size

As you become more familiar with electricity and its circuits and with some of the requirements for wiring a house or building, you will become more aware of the current-carrying abilities of wire. Size of the wire is given in numbers. This size usually ranges from 0000 (referred to as four-ought) to No. 40. The *larger the wire, the smaller its number.*

For instance, No. 32 wire is smaller than No. 14. Table 1-B shows the resistance (in ohms per 1000 feet) in relation to the cross-sectional area. Note how the temperature affects the resistance at 77°F and at 149°F (25°C and 65°C). Temperature can make quite a difference in resistance for long wires. Long wires pick up heat when exposed to summer weather and expand.

Copper vs. Aluminum Wire

Although silver is the best conductor, its use is limited because of high cost. Two commonly used conductors are aluminum and copper. Each has advantages and disadvantages. For instance, copper has high conductivity and is more ductile (can be drawn out thinner). It is relatively high in tensile strength and can be soldered easily. But it is more expensive than aluminum.

Aluminum has only about 60% of the conductivity of copper. It is used in high-voltage transmission lines and sometimes in home wiring. Its use increased in recent years. However, most electricians will NOT use it to wire a house today. There are a number of reasons for this, which will become apparent as we progress through the book.

If copper and aluminum are twisted together, as in a wire nut connection, it is possible for moisture to get to the open metals over a period of time. Corrosion will take place, causing a high-resistance joint. This can result in a dimmer light or a malfunctioning motor.

Circuits

■ *Complete circuit.* A complete circuit is necessary for the controlled flow or movement of electrons along a conductor. Fig. 1-12a. *A complete circuit is made up of a source of electricity, a conductor, and a consuming device.* The flow of electrons through the consuming device produces heat, light, or work.

In order to form a complete circuit, these rules must be followed:

1. Connect one side of the power source to one side of the consuming device (A to B).
2. Connect the other side of the power source to one side of the control device, usually a switch (C to D).
3. Connect the other side of the control device to the consuming device which it is supposed to control (E to F).

1-B. Standard Annealed Solid Copper Wire

(American wire gage—B & S)

GAGE NUMBER	DIAMETER (MILS)	CROSS SECTION		OHMS PER 1000 FEET		OHMS PER MILE 25°C (= 77°F)	POUNDS PER 1000 FEET
		CIRCULAR MILS	SQUARE INCHES	25°C (= 77°F)	65°C (= 149°F)		
0000	460.0	212 000.0	0.166	0.0500	0.0577	0.264	641.0
000	410.0	168 000.0	0.132	0.0630	0.0727	0.333	508.0
00	365.0	133 000.0	0.105	0.0795	0.0917	0.420	403.0
0	325.0	106 000.0	0.829	0.100	0.116	0.528	319.0
1	289.0	83 700.0	0.0657	0.126	0.146	0.665	253.0
2	258.0	66 400.0	0.0521	0.159	0.184	0.839	201.0
3	229.0	52 600.0	0.0413	0.201	0.232	1.061	159.0
4	204.0	41 700.0	0.0328	0.253	0.292	1.335	126.0
5	182.0	33 100.0	0.0260	0.319	0.369	1.685	100.0
6	162.0	26 300.0	0.0206	0.403	0.465	2.13	79.5
7	144.0	20 800.0	0.0164	0.508	0.586	2.68	63.0
8	128.0	16 500.0	0.0130	0.641	0.739	3.38	50.0
9	114.0	13 100.0	0.0103	0.808	0.932	4.27	39.6
10	102.0	10 400.0	0.00815	1.02	1.18	5.38	31.4
11	91.0	8 230.0	0.00647	1.28	1.48	6.75	24.9
12	81.0	6 530.0	0.00513	1.62	1.87	8.55	19.8
13	72.0	5 180.0	0.00407	2.04	2.36	10.77	15.7
14	64.0	4 110.0	0.00323	2.58	2.97	13.62	12.4
15	57.0	3 260.0	0.00256	3.25	3.75	17.16	9.86
16	51.0	2 580.0	0.00203	4.09	4.73	21.6	7.82
17	45.0	2 050.0	0.00161	5.16	5.96	27.2	6.20
18	40.0	1 620.0	0.00128	6.51	7.51	34.4	4.92
19	36.0	1 290.0	0.00101	8.21	9.48	43.3	3.90
20	32.0	1 020.0	0.000802	10.4	11.9	54.9	3.09
21	28.5	810.0	0.000636	13.1	15.1	69.1	2.45
22	25.3	642.0	0.000505	16.5	19.0	87.1	1.94
23	22.6	509.0	0.000400	20.8	24.0	109.8	1.54
24	20.1	404.0	0.000317	26.2	30.2	138.3	1.22
25	17.9	320.0	0.000252	33.0	38.1	174.1	0.970
26	15.9	254.0	0.000200	41.6	48.0	220.0	0.769
27	14.2	202.0	0.000158	52.5	60.6	277.0	0.610
28	12.6	160.0	0.000126	66.2	76.4	350.0	0.484
29	11.3	127.0	0.0000995	83.4	96.3	440.0	0.384
30	10.0	101.0	0.0000789	105.0	121.0	554.0	0.304
31	8.9	79.7	0.0000626	133.0	153.0	702.0	0.241
32	8.0	63.2	0.0000496	167.0	193.0	882.0	0.191
33	7.1	50.1	0.0000394	211.0	243.0	1,114.0	0.152
34	6.3	39.8	0.0000312	266.0	307.0	1,404.0	0.120
35	5.6	31.5	0.0000248	335.0	387.0	1,769.0	0.0954
36	5.0	25.0	0.0000196	423.0	488.0	2,230.0	0.0757
37	4.5	19.8	0.0000156	533.0	616.0	2,810.0	0.0600
38	4.0	15.7	0.0000123	673.0	776.0	3,550.0	0.0476
39	3.5	12.5	0.0000098	848.0	979.0	4,480.0	0.0377
40	3.1	9.9	0.0000078	1,070.0	1,230.0	5,650.0	0.0299

This method is used to make a complete path for electrons to flow from one terminal of the battery or power source containing an excess of electrons, to the terminal which has a deficiency of electrons. The movement of the electrons along the completed path provides energy. Of course, in order for the path to be complete, the switch must be closed.

If the circuit is so arranged that the electrons have only one path, the circuit is called a *series circuit*. If there are two or more paths for electrons, the circuit is called a *parallel circuit*.

■ *Series circuit*. Fig. 1-12b shows three resistors connected in series. The current flows through each of them before returning to the positive terminal of the battery.

Kirchoff's Law of Voltages states that the sum of all voltages across resistors or loads is equal to the applied voltage. Voltage drop is considered across the resistor. In Fig. 1-12b the current flows through three resistors. The voltage drop across R_1 is 5 volts. Across R_2 it is 10 volts, and across R_3 it is 15 volts. The sum of the individual voltage drops is equal to the total or applied voltage, 30 volts.

To find the total resistance in a series circuit, just add the individual resistances ($R_T = R_1 + R_2 + R_3 + \ldots$).

■ *Parallel circuit*. In a parallel circuit each load (resistance) is connected directly across the voltage source. There are as many separate paths for current flow as there are branches. Fig. 1-12c.

The voltage across all branches of a parallel circuit is the same. This is because all branches are connected across the voltage source. Current in a parallel circuit depends on the resistance of the branch. Ohm's Law (discussed later) can be used to determine the current in each branch. You can find the total current for a parallel circuit by adding the individual currents. As a formula, this reads:

$$I_T = I_1 + I_2 + I_3 + \ldots$$

The total resistance of a parallel circuit cannot be found by adding the resistor values. Two formulas are used for finding parallel resistances. If there are only two resistors in parallel, a simple formula can be used:

$$R_T = \frac{R_1 \times R_2}{R_1 + R_2}$$

If there are more than two resistors in parallel, you can use the following formula. (You can also use this formula if there are only two resistors.)

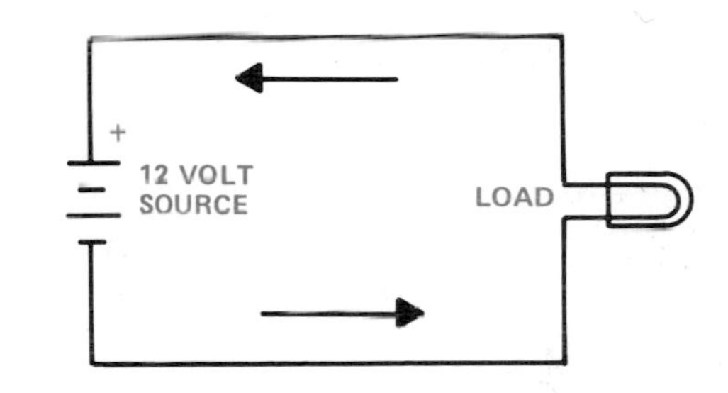

1–12a. A simple circuit.

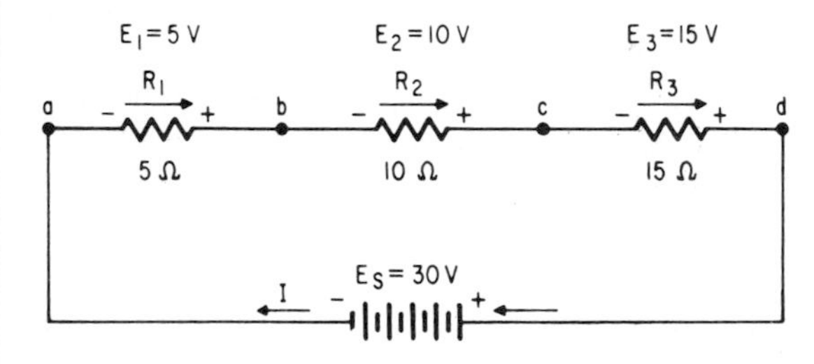

1–12b. Series circuit.

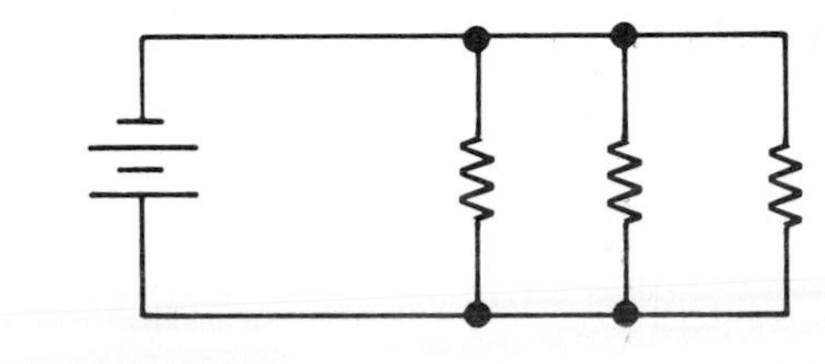

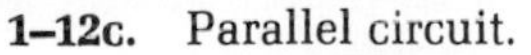

1–12c. Parallel circuit.

$$\frac{1}{R_T} = \frac{1}{R_1} + \frac{1}{R_2} + \frac{1}{R_3} + \ldots$$

One thing should be kept in mind in parallel resistances: *The total resistance is always less than the smallest resistance.*

■ *Series-parallel circuits.* Series-parallel circuits are a combination of the two circuits. Fig. 1-12d shows a series-parallel resistance circuit.

■ *Open circuit.* An open circuit is one which does not have a complete path for electrons to follow. Such an incomplete path is usually brought about by a loose connection or the opening of a switch. Fig. 1-13.

■ *Short circuit.* A short circuit is one which has a path of low resistance to electron flow. It is usually created when a low-resistance wire is placed across a consuming device. The greater number of electrons will flow through the path of least resistance rather than through the consuming device. A short usually generates an excess current flow which results in overheating, possibly causing a fire or other damage. Fig. 1-14.

It is easy to compute the amount of current flowing in a circuit if the voltage and the resistance are known. The relationship between voltage, current, and resistance in any circuit is shown by *Ohm's Law.*

OHM'S LAW

There are three basic quantities of electricity, and each has a relationship to the other two. A physicist named Georg S. Ohm discovered the relationship in 1827. He found that *in any circuit where the only opposition to the flow of electrons is resistance, there is a relationship between the values of voltage, current, and resistance.* The strength or intensity of the current is directly proportional to the voltage and inversely proportional to the resistance.

It is easier to work with Ohm's Law when it is expressed in a formula. In the formula, **E** represents EMF, or voltage. **I** is the current, or the intensity of electron flow. **R** stands for resistance. The formula is $E = I \times R$. This is the formula to use in order to find the EMF (voltage) when the current and resistance are known.

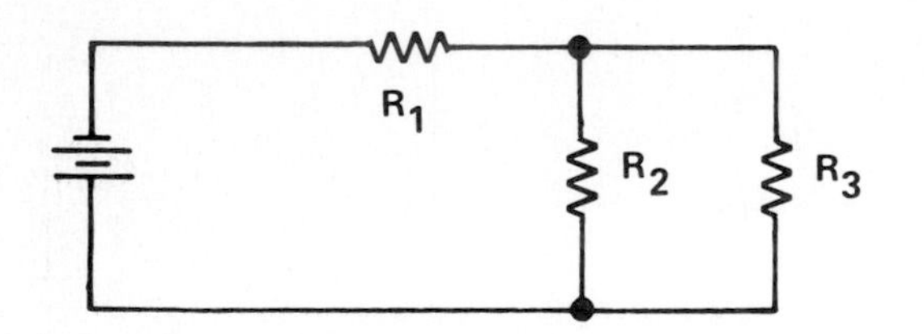

1–12d. Series-parallel circuit.

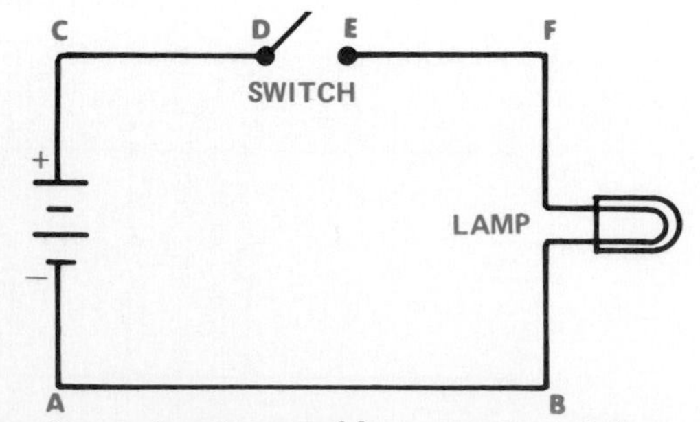

1–13. Open circuit caused by an open switch.

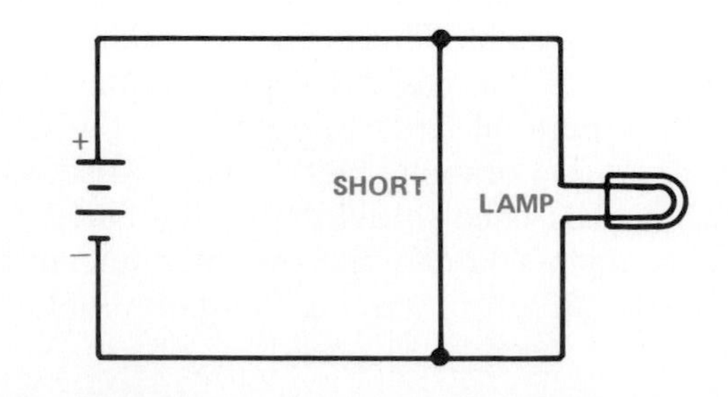

1–14. A short circuit. The wire has less resistance than the lamp.

To find the current when the voltage and resistance are known, the formula to use is:

$$I = \frac{E}{R}$$

To find the resistance when the voltage and current are known, the formula to use is:

$$R = \frac{E}{I}$$

Using Ohm's Law

There are many times in electrical work when you will need to know Ohm's Law; for example, to determine wire size in a particular circuit or to find the resistance in a circuit.

The best way to become accustomed to solving problems is to start with something simple, such as:

1. If the voltage is given as 100 volts and the resistance is known to be 50 ohms, it is a simple problem and a practical application of Ohm's Law to find the current in the circuit.

$$I = \frac{E}{R}$$

$$I = \frac{100 \text{ volts}}{50 \text{ ohms}}$$

$$I = 2 \text{ amperes}$$

2. If the current is given as 4 amperes (shown on an ammeter) and the voltage (read from a voltmeter) is 100 volts, it is easy to find the resistance.

$$R = \frac{E}{I}$$

$$R = \frac{100 \text{ volts}}{4 \text{ amperes}}$$

$$R = 25 \text{ ohms}$$

3. If the current is known to be 5 amperes, and the resistance is measured (before current is applied to the circuit) and found to be 75 ohms, it is then possible to determine how much voltage is needed to cause the circuit to function properly.

$$E = I \times R$$

$$E = 5 \text{ amperes} \times 75 \text{ ohms}$$

$$E = 375 \text{ volts}$$

Fig. 1-15 illustrates the way the formula works.

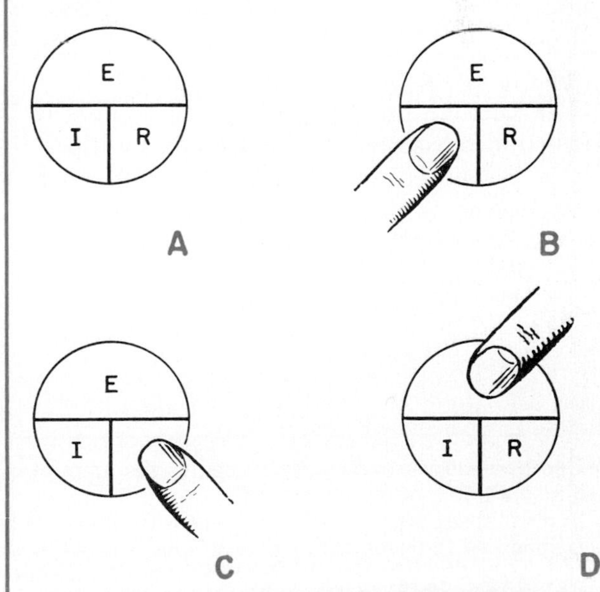

1–15. Ohm's Law. Place finger on the unknown value. The remaining two letters will give the formula to use for finding the unknown value.

POWER

Power is defined as the rate at which work is done. It is expressed in metric measurement terms of watts, for power, and in joules for energy or work. (For some common metric conversions, see Appendix A.) A *watt* is the power which gives rise to the production of energy at the rate of one joule per second. ($W = J/s$). A joule is the work done when the point of application of a force of one newton is displaced a distance of one meter in the direction of the force. ($J = N \times m$).

It has long been the practice in this country to measure work in terms of *horsepower*. Electric motors are still rated in horsepower and probably will be for some time.

Power can be electrical or mechanical. When a mechanical force is used to lift a weight, work is done. The rate at which the weight is moved is called power. Horsepower is defined in terms of moving a certain weight over a certain distance in one minute. Energy is consumed in moving a weight, or work is done. The findings in this field have been equated with the same amount of work done by electrical energy. *It takes 746 watts of electrical power to equal one horsepower*. Table 1-C.

The horsepower rating of electric motors is arrived at by taking the voltage and multiplying it by the current drawn under full load. This power is measured in watts. In other words, one volt times one ampere equals one watt. When put into a formula:

$$\text{Power} = \text{Volts} \times \text{Amperes, or } P = E \times I$$

Kilowatts

The prefix *kilo* means one thousand. Thus one thousand watts equal one kilowatt. The abbreviation for kilowatt is kW. There is a unit known as the kilowatt-hour also. It is abbreviated kWh and is equivalent to one thousand watts used for one hour. Electric bills are figured in kilowatt-hours. Usage for an entire month is computed on an hourly basis and then read in the kWh unit.

Power formulas are sometimes needed to figure the wattage of a circuit. Here are the three most commonly used formulas:

$$P = E \times I$$

$$P = \frac{E^2}{R}$$

$$P = I^2 \times R$$

This means that if any one of the three—voltage, current, or resistance—is missing, it is possible to find the missing quantity by using the relationship of the two known quantities. In later chapters you will encounter the problem of the *I^2R losses* and some other terminology related to the formulas just shown.

1-C. Horsepower

One horsepower is usually defined as the amount of work required to move a 550-pound weight a distance of one foot in one second.
In most cases the modern way to measure power is in kilowatts rather than horsepower. In case a motor is specified in terms of horsepower, but is rated in watts or kilowatts, the conversion is simple:

1 horsepower = 746 watts

Divide the number of watts or kilowatts by 746 or 0.746, respectively, to find the horsepower rating.

The *milliwatt* (mW) is sometimes used in referring to electrical equipment. For instance, the rating of the speaker in a transistor radio may be given as 100 mW or 300 mW. This means a 0.1 watt or 0.3 watt rating, since the prefix *milli* means one-thousandth. Transistor circuits are designated in milliwatts but power-line electrical power is usually in kilowatts.

MEASURING ELECTRICITY

Electricity must be measured if it is to be sold, or if it is to be fully utilized. There are a number of ways to measure electricity. It can be measured in volts, amperes, or watts. The kilowatt hour meter is the device most commonly used to measure power.

Meters

In order to measure anything, there must be a basic unit in which to measure. In electricity, the current (flow of electrons) is measured in a basic unit called the ampere. The current is usually measured with a permanent magnet and an electromagnet arranged to indicate the amperes. Such a device is necessary since we are unable to see an electron—even with the most powerful microscopes. Obviously, counting the number of electrons passing a given point in a second is impossible when there are no visible particles to count. Therefore a magnetic field is used to measure the effect of the electrons.

The D'Arsonval meter movement uses a permanent magnet as a base over which a wire or electromagnet is pivoted and allowed to move freely. When current flows through the coil, a magnetic field is set up. Fig. 1-16 a & b. The strength of the magnetic field determines how far the coil will be deflected. The polarity of the moving coil is the same as that of the permanent magnet. A repelling action results. This is in proportion to the strength of the magnetic field generated by the current flowing through the coil. The number of turns in the coil, times the current through the coil, determines the strength of the magnetic field. Since the meter coil is pivoted on jeweled bearings to reduce any friction, the movement is calibrated against a known source of current or against another meter. The scale on a new device is calibrated to read in amperes, milliamperes, or microamperes, Fig. 1-16c.

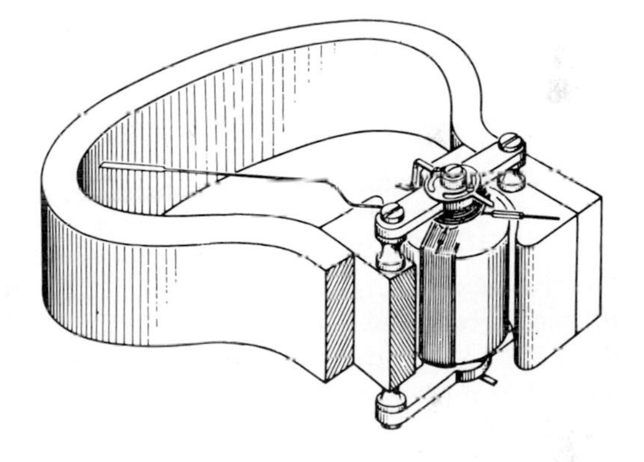

1–16a. D'Arsonval meter movement.

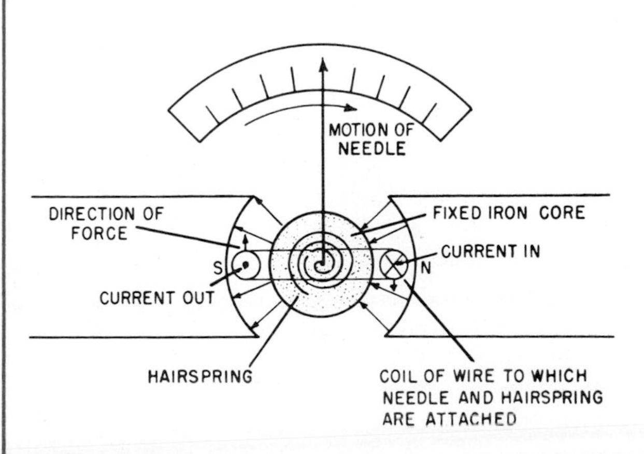

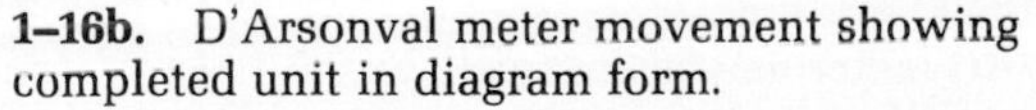

1–16b. D'Arsonval meter movement showing completed unit in diagram form.

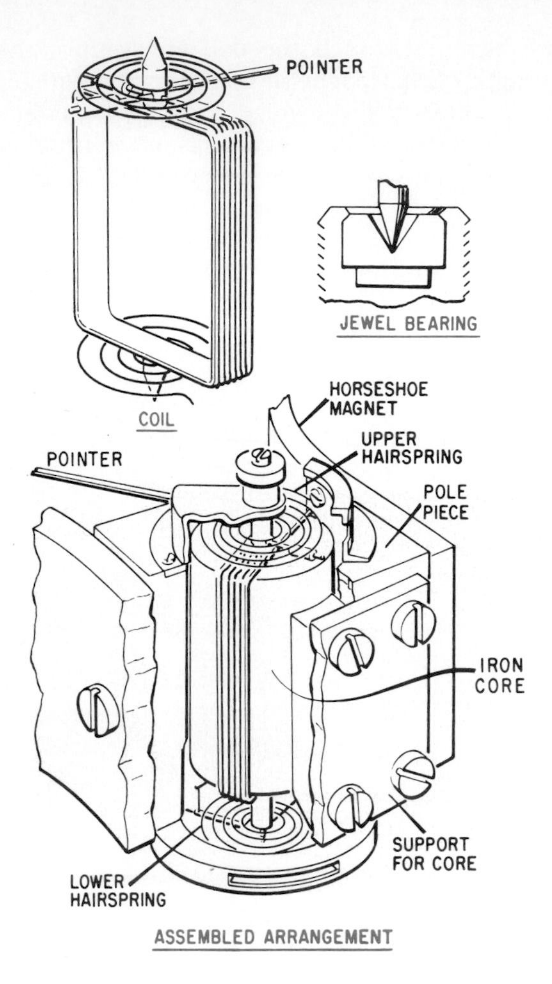

1–16c. Assembled arrangement of the D'Arsonval meter movement.

■ *AC ammeter.* If alternating current is to be measured by a DC meter movement, a rectifier is inserted in the circuit (meter circuit) to change the AC to DC. It can then be measured by the meter movement. Otherwise the alternating current will make the needle on the meter vibrate rapidly. This vibration means there is little or no movement from zero. Fig. 1-17.

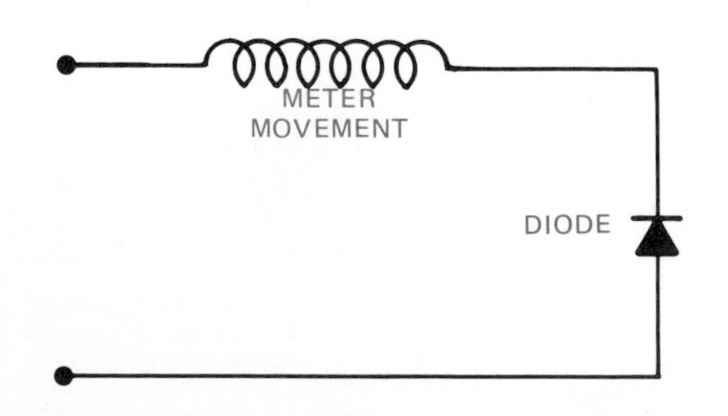

1–17. AC meter movement made by adding a diode to a DC meter movement.

■ *Shunts.* A *shunt* is a means of bypassing current around a meter movement. A resistor of the proper size is inserted across the meter movement to bypass the current around the movement. Most of the current is bypassed with only the necessary amount left to cause the meter to deflect at its designed limit. The meter is calibrated to read on its scale the full amount of current flowing in the circuit. Fig. 1-18.

A number of shunts can be placed in a meter case and switched. A different resistor (shunt) can be switched in for each range needed. A meter with more than one range is called a *multimeter.*

A multimeter (one which can measure volts and ohms as well) is shown in Fig. 1-19. It can measure from 0 to 1 mA, 0 to 10 mA, and from 0.1 A to 1 A. This is a 1 mA meter movement and needs no shunt when used to measure as high as 1 mA. A shunt is switched in, however, to measure a range of 0 to 10 mA and 0.1 A to 1 A. A switch on the meter can also add a diode to help with measuring AC.

Common and *positive* holes have test leads inserted to attach to the circuit being measured. Common is negative (−) or black, and the positive is (+) or red. However, polarity isn't necessary to measure AC. Either lead can be used at any terminal in an AC circuit. *Ammeters are always placed in series in a circuit.* This usually means the circuit has to be broken and the meter inserted in the line.

■ *Voltmeter.* The *voltmeter* measures electrical pressure, or volts. It is nothing more than an ammeter with a resistor added in the meter circuit. The high resistance of the voltmeter makes it possible to place it across a power source (in parallel). Fig. 1-20. A number of resistors, called *multipliers*, can be switched into a meter circuit to increase its range or make it capable of measuring higher voltages. The voltmeter in Fig. 1-21 is capable of measuring from 0 to 150, 0 to 300, and 0 to 750 volts by placing the proper multiplier into the meter circuit. Note how the terminals on top of the meter allow for placing test probes in a number of different positions for the purpose of varying the range of measurement.

■ *Ohmmeter.* The basic unit for measuring resistance is the *ohm*. An *ohmmeter* is a device used to measure resistance or ohms (Ω). It is an ammeter (or milliammeter or micrometer) movement, modified to measure resistance. Fig. 1-22.

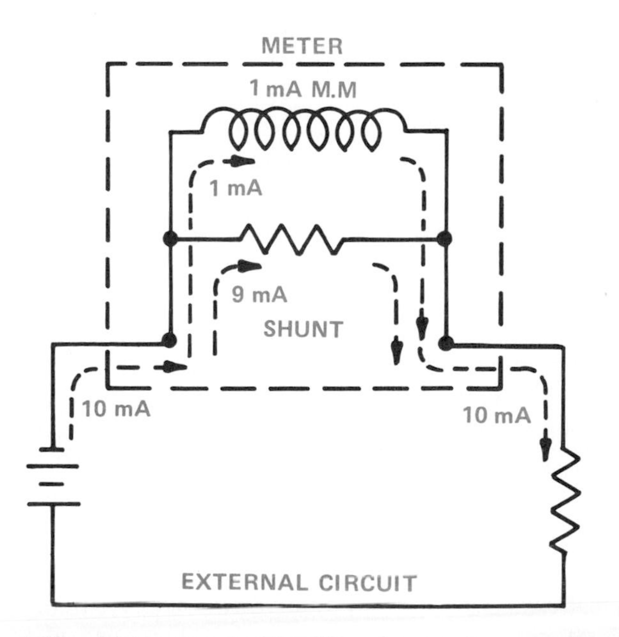

1–18. Meter movement with a shunt to increase its range to 10 mA.

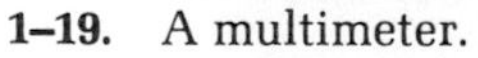

1–19. A multimeter.

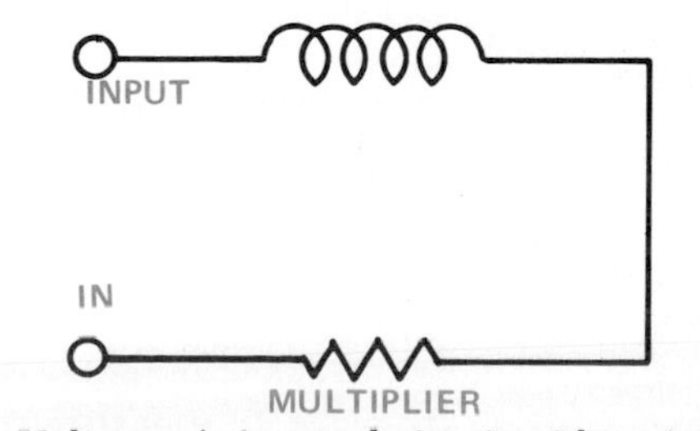

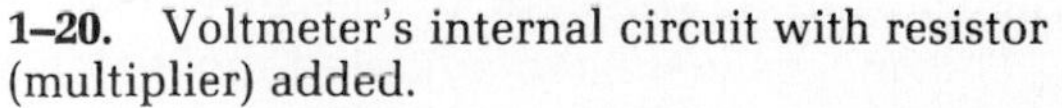

1–20. Voltmeter's internal circuit with resistor (multiplier) added.

1–21. DC voltmeter.

Fig. 1-22 shows a multimeter capable of measuring ohms with three different ranges: R × 1, R × 100, and R × 10K. This means it can measure from 0 to 200 ohms on the R × 1 scale and 0 to 200 000 ohms on the R × 100 range. Within the R × 10K or R × 10 000 range it is capable of measuring from 0 to 20 000 000 ohms or 0 to 20 megohms (*mega* means *one million*). The meter scale has to be multiplied by the 100 or 10 000 number in order to have it read the proper value. By changing resistors, it is possible to vary the resistance measuring range of an ohmmeter. Fig. 1-23 shows a basic ohmmeter using a 1 mA meter movement and its necessary parts. Note how the battery serves as the power source. This makes it necessary to turn off power whenever you read the resistance of a circuit. THE OHMMETER HAS ITS OWN POWER SOURCE. Do not connect it to a LIVE CIRCUIT or one with the power on. To do so will result in the destruction of the meter movement.

Adjust the ohmmeter so that the meter reads zero before starting to use it to measure resistance. This means you have adjusted the meter

1–22. Multimeter used to measure ohms, volts, and milliamperes as well as microamperes.

circuit to compensate for the battery voltage changes. Battery voltages decrease with shelf life. It doesn't matter whether or not the battery is used. It will, in time, lose its voltage.

Some meters are not portable. They need external sources of power. Electronic circuits using vacuum tubes or transistors are used to improve the meter capabilities. Fig. 1-24 shows a digital voltmeter. The voltmeter is set to read DC volts.

■ *Digital meters.* The digital meter is entirely electronic. It uses printed circuits and integrated circuit chips to measure and calculate voltage, resistance, or current. There are no coils or magnets.

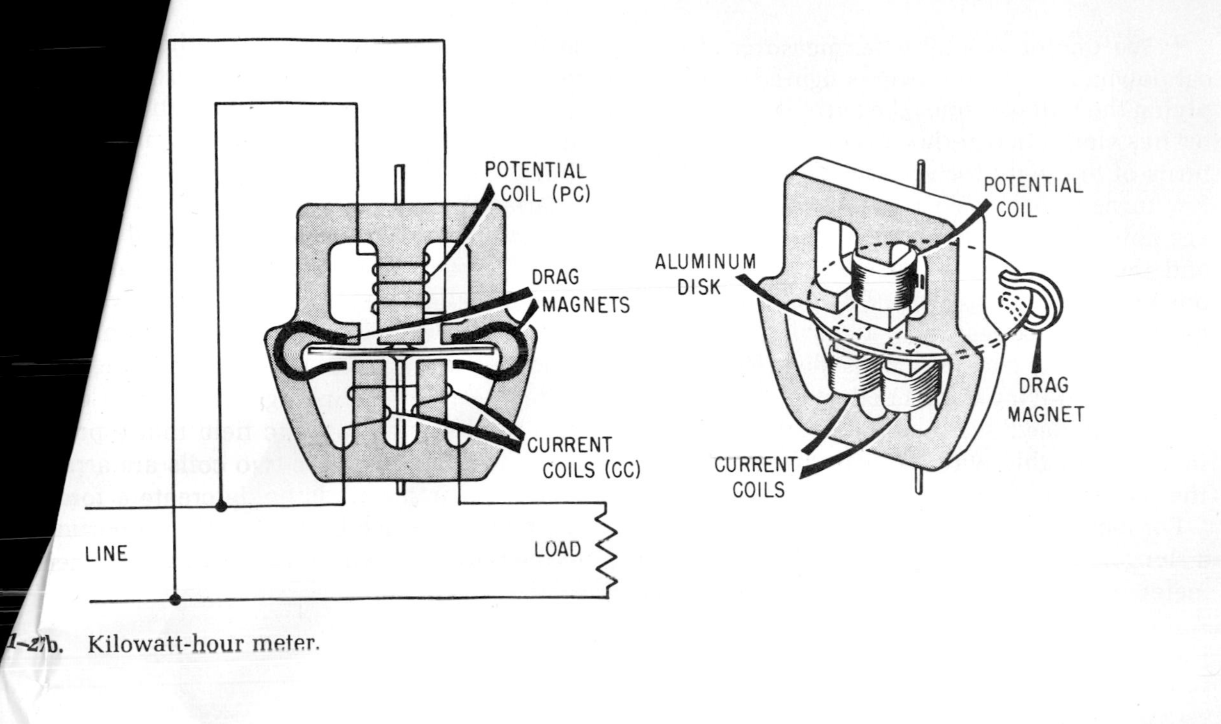

1-27b. Kilowatt-hour meter.

Permanent magnets are used to introduce a retarding, or braking force, which is proportionate to the speed of the disc. The magnetic strength of these retarding magnets regulates the disc speed for any given load so that each revolution of the disc always measures the same quantity of energy or watt-hours. Disc revolutions are converted to kilowatt-hours on the meter register.

Most meters are inserted into a socket on the wall of a structure. Removing the meter interrupts, or terminates, power without handling of dangerous high voltage wires. Three-phase and single-phase power, to be discussed in a later chapter, each require different watt-hour meters. Kilowatt-hour meters are tested by computers in the service centers of power companies. Fig. 1-28.

Other Types of Meters

There are other types of meters used to measure voltage and current. The D'Arsonval movement is only one of many types used today. The taut band type is basically the same as the D'Arsonval except that a tightly stretched and twisted band is used to hold the coil and needle in place between the permanent magnet poles. In addition, no moving points touch the meter case; so jeweled bearings are unnecessary. The band is twisted when it is inserted into the meter frame so that it will cause the coil to spring back to its original resting place upon interruption of current through the coil.

■ *Electrodynamometer.* The electrodynamometer type of meter uses no permanent magnet. Two fixed coils produce the magnetic field. The meter also uses two moving coils. This meter can be used as a voltmeter or an ammeter. Fig. 1-29. It is not as sensitive as the D'Arsonval meter movement.

■ *Inclined-coil iron-vane meter.* The inclined-coil, iron-vane meter is used for measuring AC or DC where large amounts of current are present. This meter can be used as a voltmeter or ammeter. Fig. 1-30.

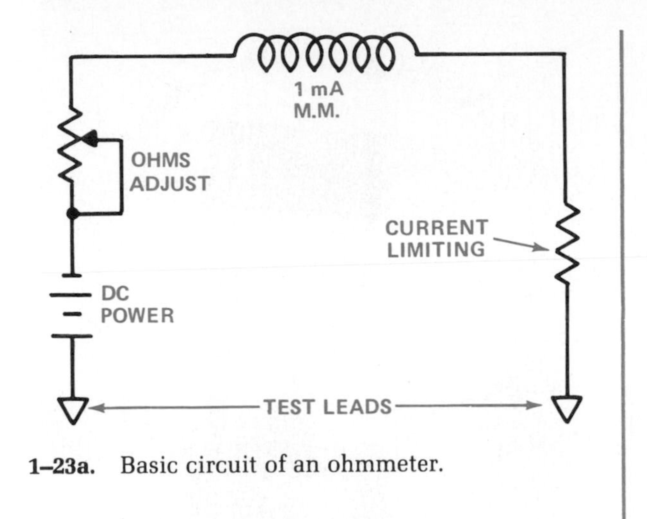

1-23a. Basic circuit of an ohmmeter.

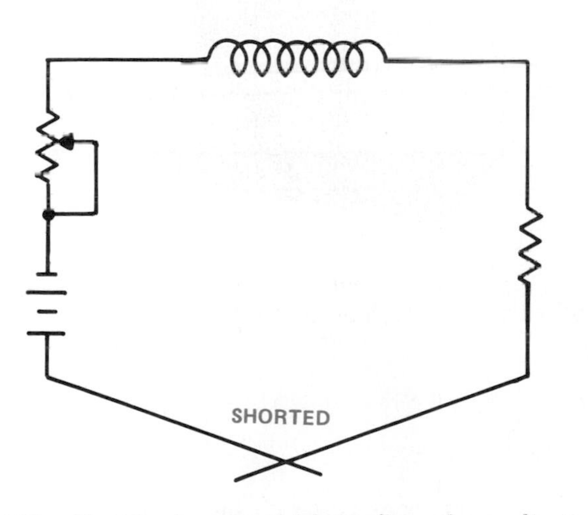

1-23b. Test leads crossed (shorted) so ohms adjust resistor can be adjusted to make the meter read zero.

The meter shown in Fig. 1-24 is a digital type. It is portable and self-contained. The meter indicates the reading on a *liquid crystal display* (LCD). This type of digital readout is found on many instruments, clocks, and watches.

Most of these meters have a number of voltage ranges. These must be selected each time you measure a circuit. Other meters are made with auto-ranging. They select the proper range and measure the voltage without the

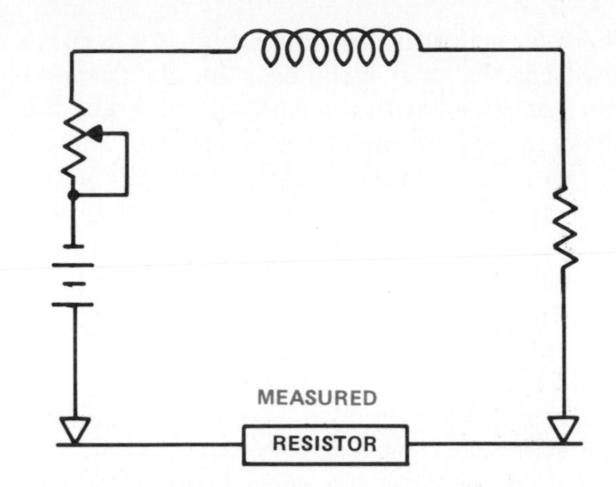

1-23c. Test leads are touching the leads of a resistor to measure it.

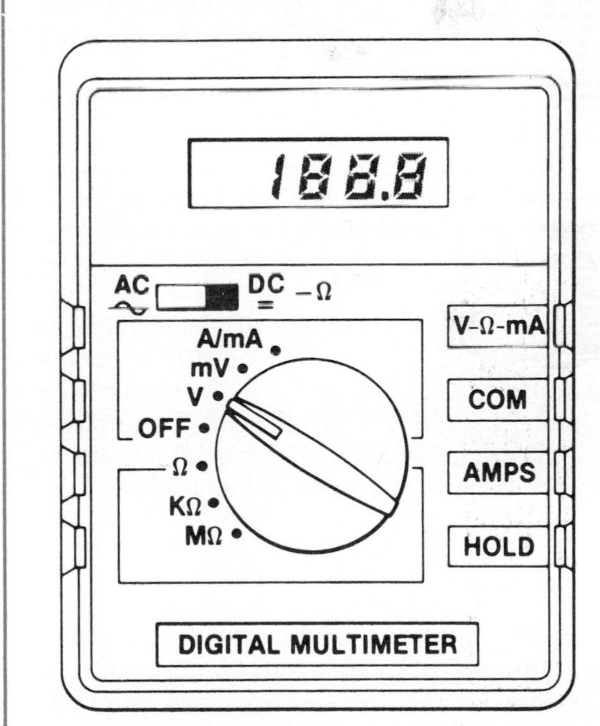

1-24. A digital voltmeter, set to read DC volts.

need for such preselection. The operator must, though, indicate if volts, ohms, or amps are to be measured.

This type of meter will sample the circuit about 5 times per second. It will then display the average of its 5 samples for very accurate readings. In most instances, the electrician is not concerned with the 0.01 volt accuracy that these meters are capable of reaching.

The meter is highly accurate and easy to use. You simply turn on the meter to the needed function. Then select ohms and use the probes across the resistor to measure resistance. If you use this unit as a milliammeter, you have to insert it properly into the circuit being measured. If a circuit is being measured for resistance, make sure there is no power turned on when the meter is connected.

This type of meter is rugged. It can be damaged, however, if it is set for measuring ohms when volts are, instead, being measured. The display can be damaged permanently when left on for too long or when the meter is dropped rather hard. To extend the battery life, turn the meter off when not in use.

Prices of digital meters are dropping rapidly. In time, the digital meter can be expected to replace all other types. However, for some purposes the digital meter can be outperformed by the D'Arsonval movement. For example, some meter indicators simply need a deflection of the needle to show proper operation. In the digital, you have to wait for the numbers to be counted up or down. It takes concentration on the part of the user. The D'Arsonval movement simply shows a deflection of the needle.

■ *AC clamp-on meter.* Figs. 1-25 & 1-26 show two different styles of an AC clamp-on meter. They are inserted over a wire carrying alternating current. The magnetic field around the wire induces a small amount of current in the meters. The scale is calibrated to read amperes or volts. Because the wire is run through the large loop extending past the meter movement, it is possible to read the AC voltage, or current, without removing the insulation from the wire. These meters are very useful when working with AC motors.

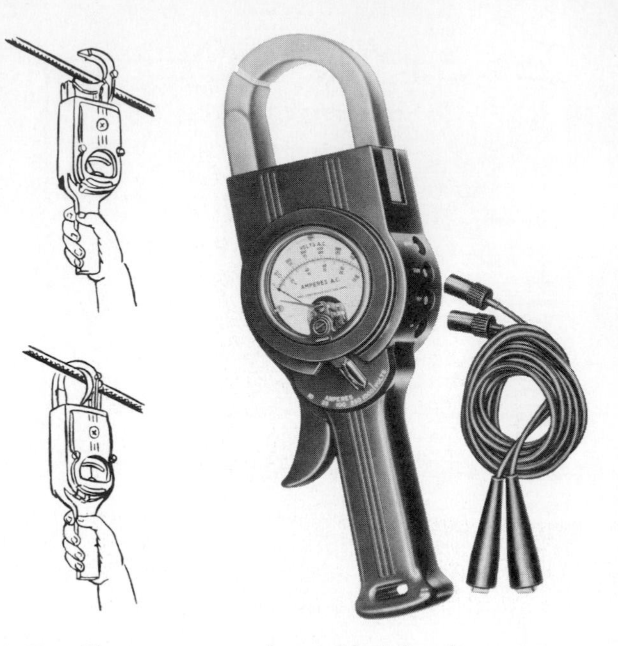

1–25. Clamp-on type of portable AC volt-ammeter.

1–26. Clamp-on type of AC volt-ammeter.

■ *Wattmeter.* A wattmeter measures electrical power. Electrical power is figured by multiplying the voltage times the current. A wattmeter has electromagnetic coils (a coil with many turns of fine wire for voltage and a coil with a few turns of heavy wire for current). The voltage coil is connected across the incoming line, and the current coil is inserted in series with one of the incoming wires. Two coils are stationary and in series with a moving coil. The strength of the magnetic fields determines how much the moving coil is deflected. The deflection of the needle is read on a scale calibrated in watts. In this way the wattmeter measures the power consumed in one second. Fig. 1-27.

For measuring the electrical power used over a longer period of time, the kilowatt-hour meter was designed. The kWh meter is often seen on
measure
riod, such
measures t
thousands of
a certain rate

The kilowatt-
motor. Meter torq
magnet called a st
windings. One wind
produces a magnetic
voltage. Another windi
coil, produces a magneti
the load current. These tw
so that their magnetic field
the meter disc which is direc
to the power, or watts, drawn b
load.

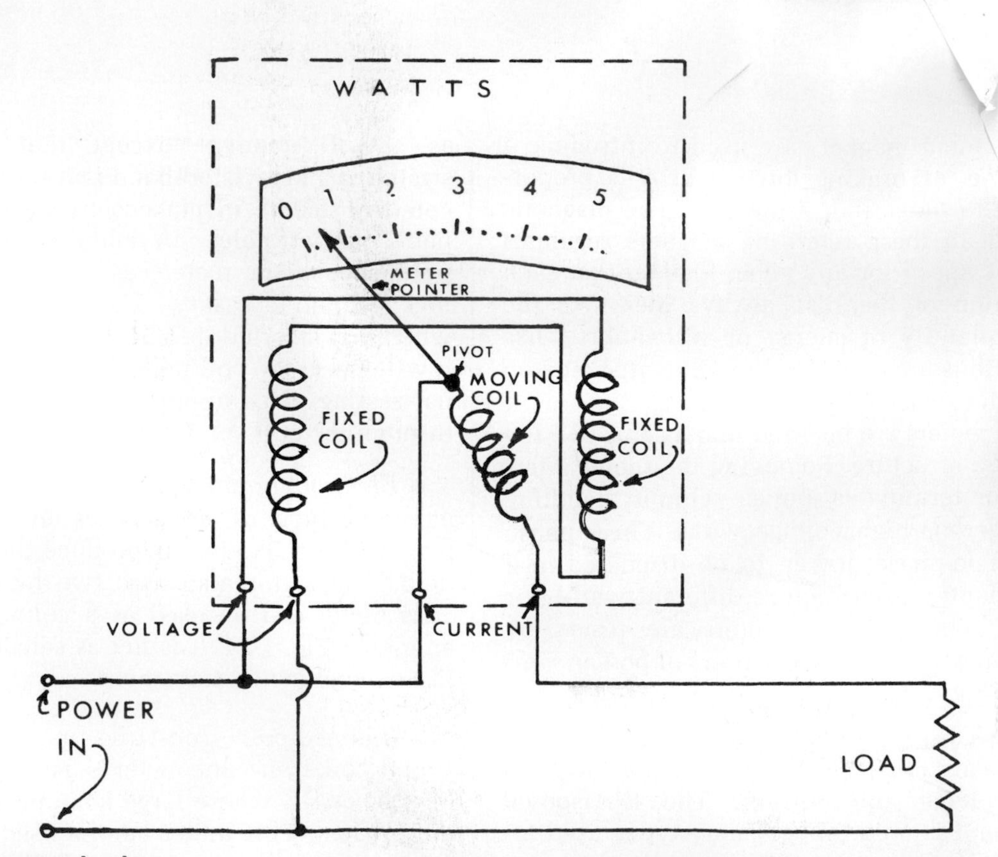

1–27a. Wattmeter hookup.

24 □ Sectio

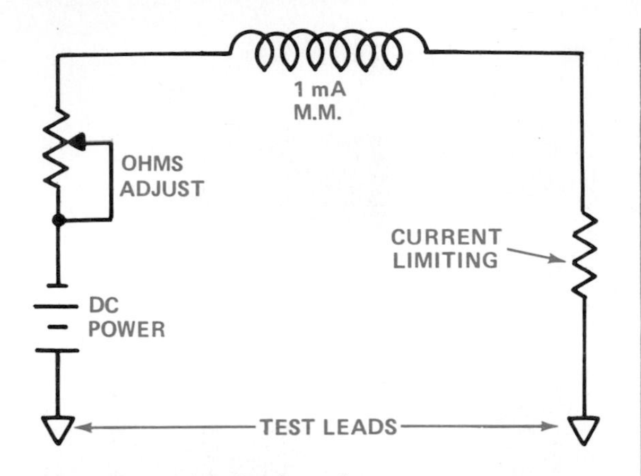

1–23a. Basic circuit of an ohmmeter.

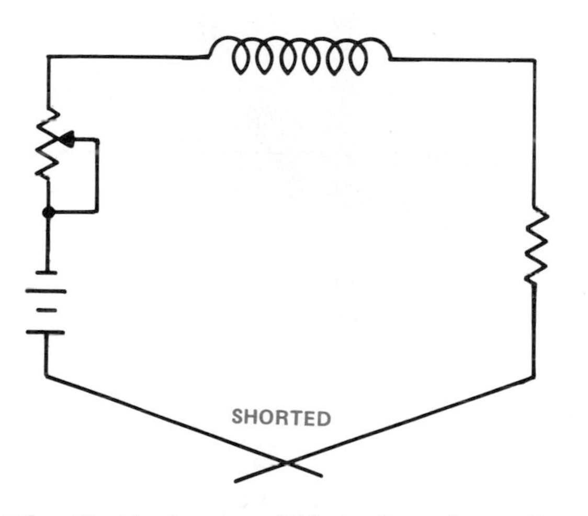

1–23b. Test leads crossed (shorted) so ohms adjust resistor can be adjusted to make the meter read zero.

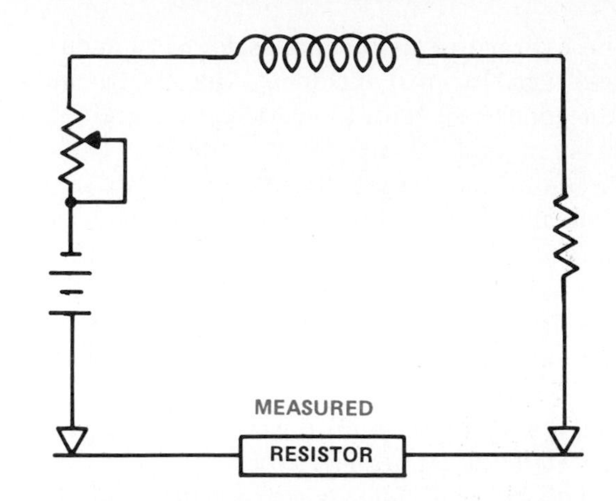

1–23c. Test leads are touching the leads of a resistor to measure it.

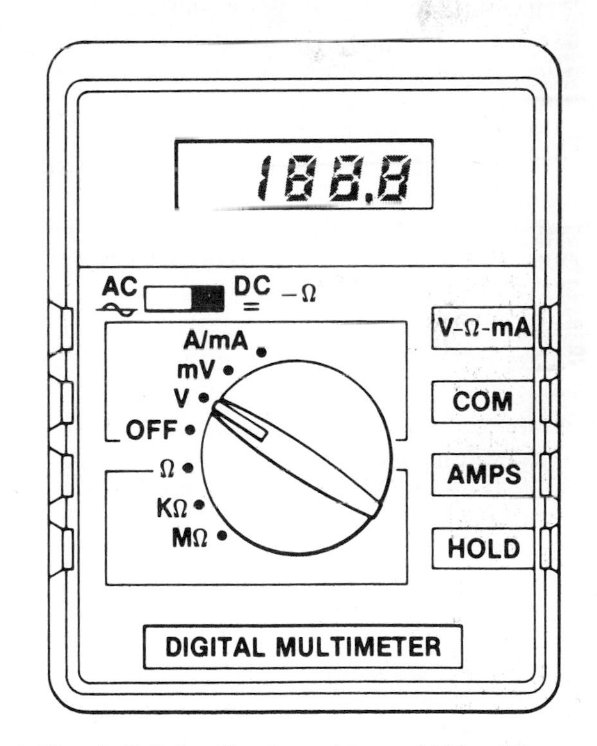

1–24. A digital voltmeter, set to read DC volts.

The meter shown in Fig. 1-24 is a digital type. It is portable and self-contained. The meter indicates the reading on a *liquid crystal display* (LCD). This type of digital readout is found on many instruments, clocks, and watches.

Most of these meters have a number of voltage ranges. These must be selected each time you measure a circuit. Other meters are made with auto-ranging. They select the proper range and measure the voltage without the need for such preselection. The operator must, though, indicate if volts, ohms, or amps are to be measured.

This type of meter will sample the circuit about 5 times per second. It will then display the average of its 5 samples for very accurate readings. In most instances, the electrician is not concerned with the 0.01 volt accuracy that these meters are capable of reaching.

The meter is highly accurate and easy to use. You simply turn on the meter to the needed function. Then select ohms and use the probes across the resistor to measure resistance. If you use this unit as a milliammeter, you have to insert it properly into the circuit being measured. If a circuit is being measured for resistance, make sure there is no power turned on when the meter is connected.

This type of meter is rugged. It can be damaged, however, if it is set for measuring ohms when volts are, instead, being measured. The display can be damaged permanently when left on for too long or when the meter is dropped rather hard. To extend the battery life, turn the meter off when not in use.

Prices of digital meters are dropping rapidly. In time, the digital meter can be expected to replace all other types. However, for some purposes the digital meter can be outperformed by the D'Arsonval movement. For example, some meter indicators simply need a deflection of the needle to show proper operation. In the digital, you have to wait for the numbers to be counted up or down. It takes concentration on the part of the user. The D'Arsonval movement simply shows a deflection of the needle.

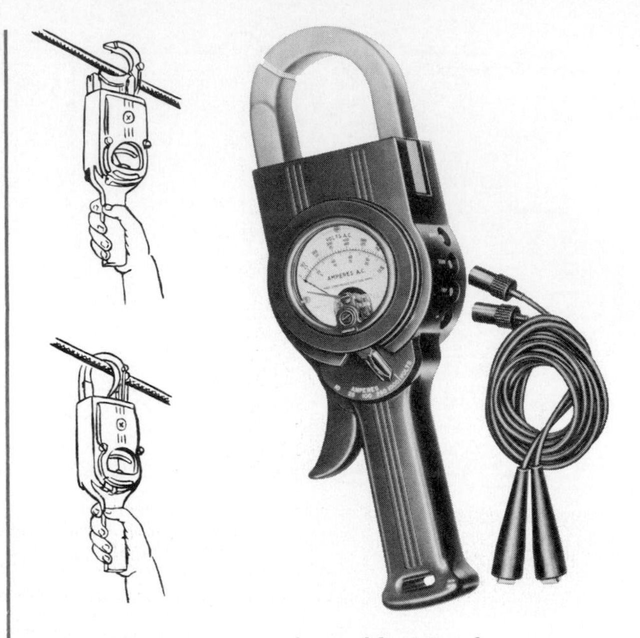

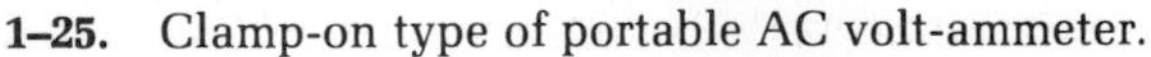

1–25. Clamp-on type of portable AC volt-ammeter.

■ *AC clamp-on meter.* Figs. 1-25 & 1-26 show two different styles of an AC clamp-on meter. They are inserted over a wire carrying alternating current. The magnetic field around the wire induces a small amount of current in the meters. The scale is calibrated to read amperes or volts. Because the wire is run through the large loop extending past the meter movement, it is possible to read the AC voltage, or current, without removing the insulation from the wire. These meters are very useful when working with AC motors.

1–26. Clamp-on type of AC volt-ammeter.

■ *Wattmeter.* A wattmeter measures electrical power. Electrical power is figured by multiplying the voltage times the current. A wattmeter has electromagnetic coils (a coil with many turns of fine wire for voltage and a coil with a few turns of heavy wire for current). The voltage coil is connected across the incoming line, and the current coil is inserted in series with one of the incoming wires. Two coils are stationary and in series with a moving coil. The strength of the magnetic fields determines how much the moving coil is deflected. The deflection of the needle is read on a scale calibrated in watts. In this way the wattmeter measures the power consumed in one second. Fig. 1-27.

For measuring the electrical power used over a longer period of time, the kilowatt-hour meter was designed. The kWh meter is often seen on the side of a house or other building. It measures power used over a certain time period, such as a month. A kilowatt-hour meter measures the power consumed in terms of thousands of watts. Electric power is priced at a certain rate per kWh.

The kilowatt-hour meter is a small induction motor. Meter torque is produced by an electromagnet called a stator, which has two sets of windings. One winding, called a potential coil, produces a magnetic field representing circuit voltage. Another winding, known as a current coil, produces a magnetic field that represents the load current. These two coils are arranged so that their magnetic fields create a force on the meter disc which is directly proportionate to the power, or watts, drawn by the connected load.

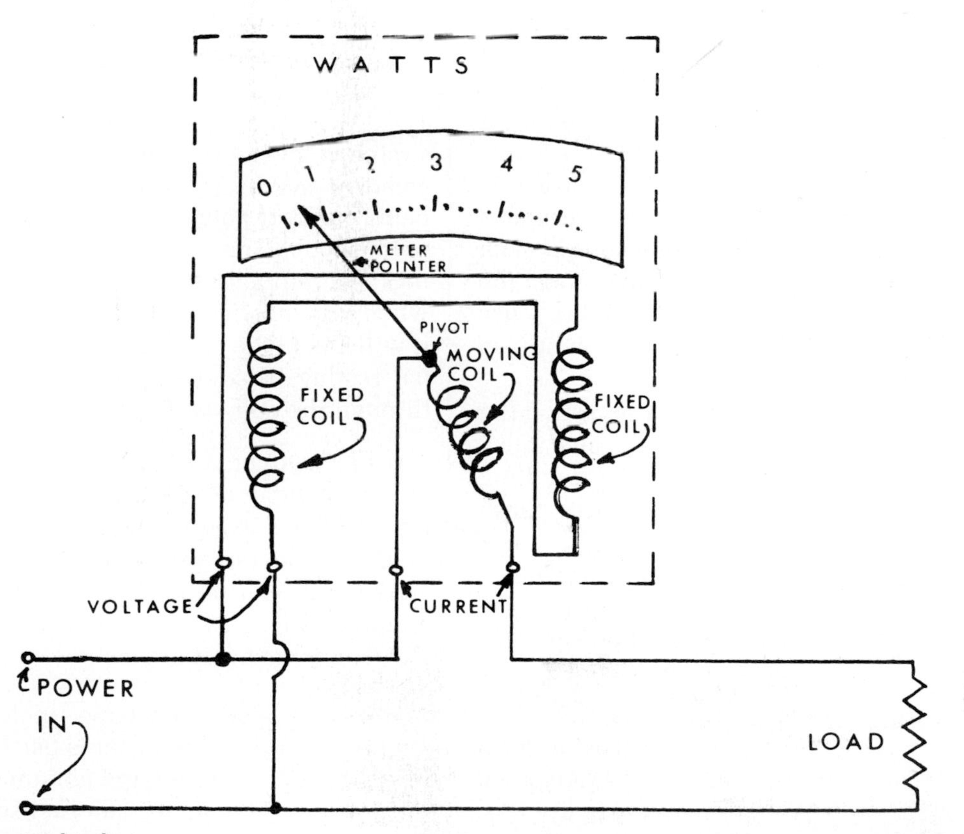

1–27a. Wattmeter hookup.

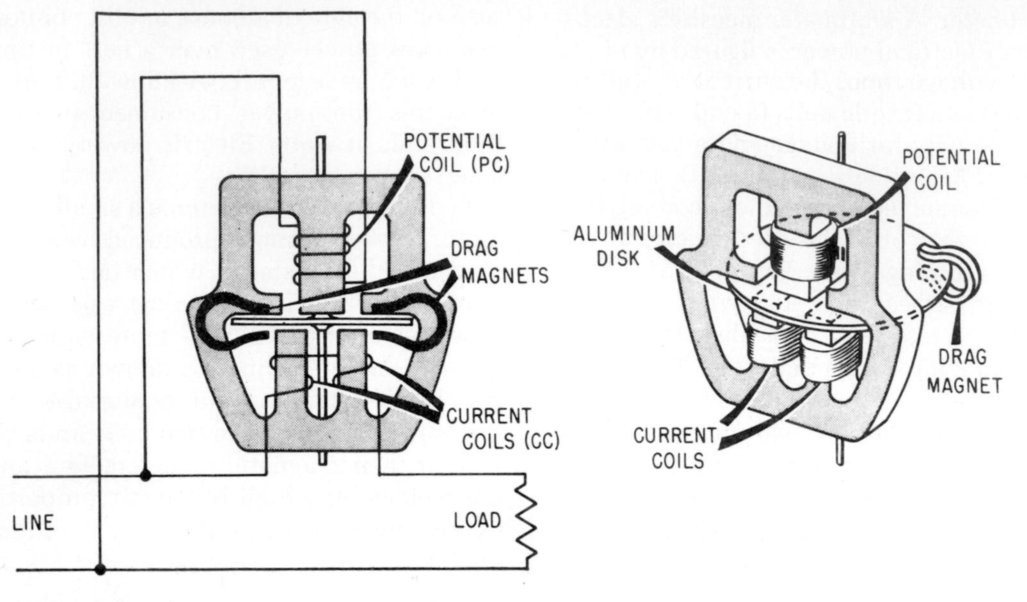

1–27b. Kilowatt-hour meter.

Permanent magnets are used to introduce a retarding, or braking force, which is proportionate to the speed of the disc. The magnetic strength of these retarding magnets regulates the disc speed for any given load so that each revolution of the disc always measures the same quantity of energy or watt-hours. Disc revolutions are converted to kilowatt-hours on the meter register.

Most meters are inserted into a socket on the wall of a structure. Removing the meter interrupts, or terminates, power without handling of dangerous high voltage wires. Three-phase and single-phase power, to be discussed in a later chapter, each require different watt-hour meters. Kilowatt-hour meters are tested by computers in the service centers of power companies. Fig. 1-28.

Other Types of Meters

There are other types of meters used to measure voltage and current. The D'Arsonval movement is only one of many types used today. The taut band type is basically the same as the D'Arsonval except that a tightly stretched and twisted band is used to hold the coil and needle in place between the permanent magnet poles. In addition, no moving points touch the meter case; so jeweled bearings are unnecessary. The band is twisted when it is inserted into the meter frame so that it will cause the coil to spring back to its original resting place upon interruption of current through the coil.

■ *Electrodynamometer.* The electrodynamometer type of meter uses no permanent magnet. Two fixed coils produce the magnetic field. The meter also uses two moving coils. This meter can be used as a voltmeter or an ammeter. Fig. 1-29. It is not as sensitive as the D'Arsonval meter movement.

■ *Inclined-coil iron-vane meter.* The inclined-coil, iron-vane meter is used for measuring AC or DC where large amounts of current are present. This meter can be used as a voltmeter or ammeter. Fig. 1-30.

1–28. Computer testing of kilowatt-hour meters.

CONTROLLING ELECTRICITY

In order to make electricity useful it is necessary to control it. You want it in the proper place at the proper time. Otherwise, it can do great damage—even kill. Electricity can be controlled by using switches, relays, or diodes. These devices are used to direct the current to the place where it will work for you. Each device is carefully chosen to do a specific job. For example, the relay is used for remote-control work, and a diode is used to control large and small amounts of current in electrical as well as electronic equipment. A diode is a device which allows current to flow in one direction only. It can be used to change AC to DC.

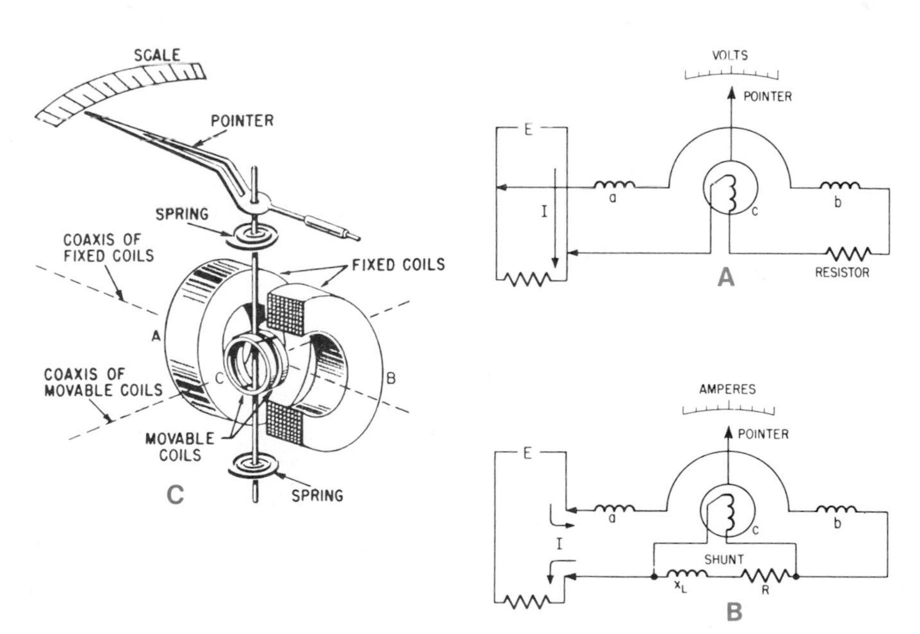

1–29. Electrodynamometer: (A) Circuit when used as voltmeter. (B) Circuit when used as ammeter. (C) Internal construction.

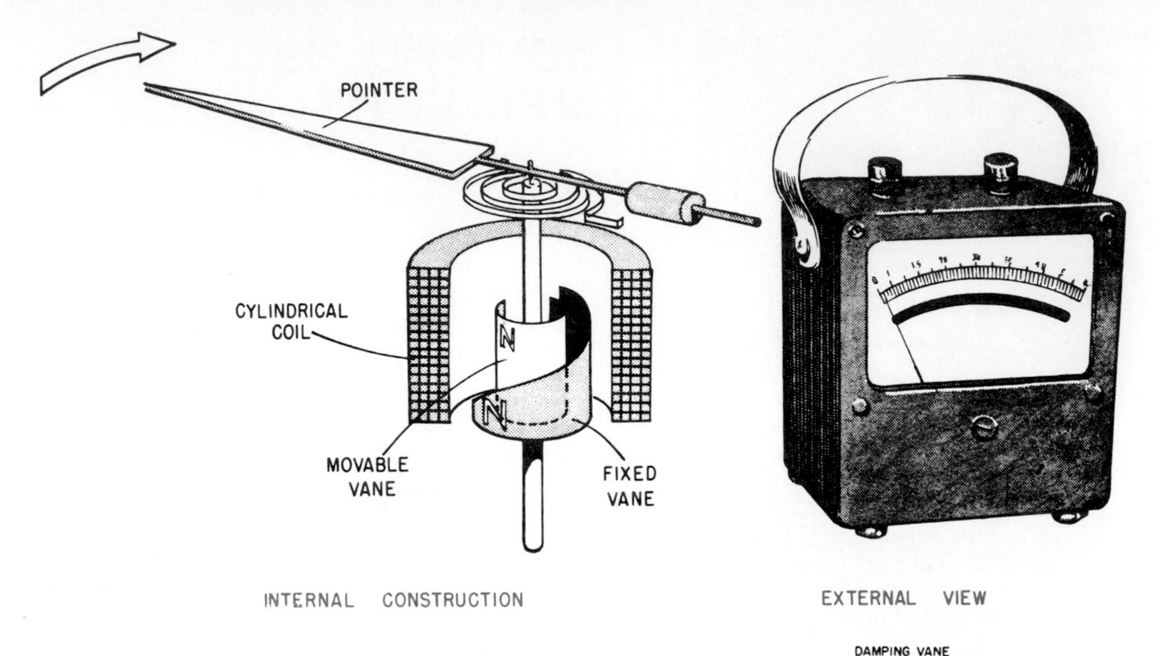

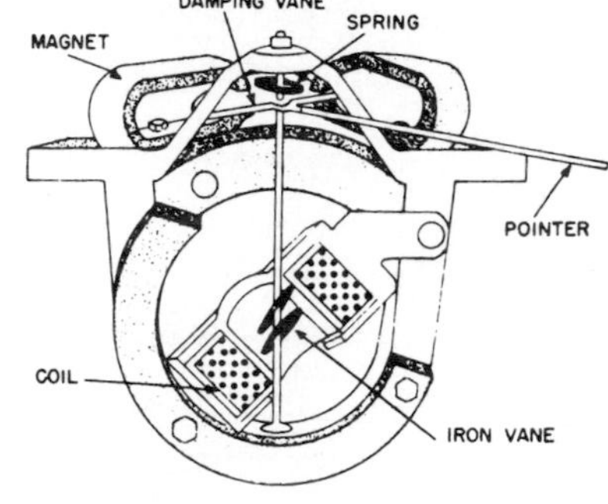

1–30. Inclined-coil iron-vane meter.

Switches

There are a number of switches used for controlling electricity. Each switch has a different name. This helps designate it according to the job it performs. For instance, the single-pole, single-throw switch (SPST) is just that—a single pole which is moved either to make connection between two points or to not make connection. In the off position the contacts are not touching, and the flow of electrons is interrupted. Fig. 1-31.

The double-pole, double-throw switch (DPDT) can be used to control more than one circuit at the same time. It can be used to reverse the direction of rotation of a DC motor by reversing polarity. Fig. 1-32.

Fig. 1-33 shows a single-pole, double-throw (SPDT) switch. These are open switches and

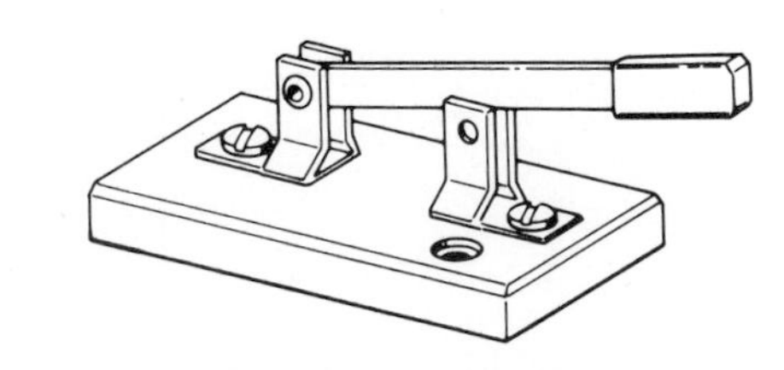

1–31. Knife switch. Single-pole, single-throw.

should not be used with more than 12 volts. They are usually referred to as radio or battery switches and are used here to show the simple operation of a switch. Switches used for higher voltages are totally enclosed and protected from body contact. This prevents shocks or injuries from electrical shorts or contact with high voltages.

The double-pole, single-throw switch (DPST) is used to control two circuits at the same time. It can be used as a simple on-off switch for two circuits. When it is open it interrupts the current in the two circuits. When it is closed it completes the circuits for proper operation. This, too, is an open switch and meant to be used only on low voltages where the danger of shock is very much reduced. Fig. 1-34 illustrates this type of switch.

The doorbell switch or door chime switch is extremely simple. It completes a circuit from the low-voltage transformer to the chime or bell. Fig. 1-35. When the button is pressed, it completes the circuit to the bell or chime, causing current to flow from the transformer to the chime. Fig. 1-36.

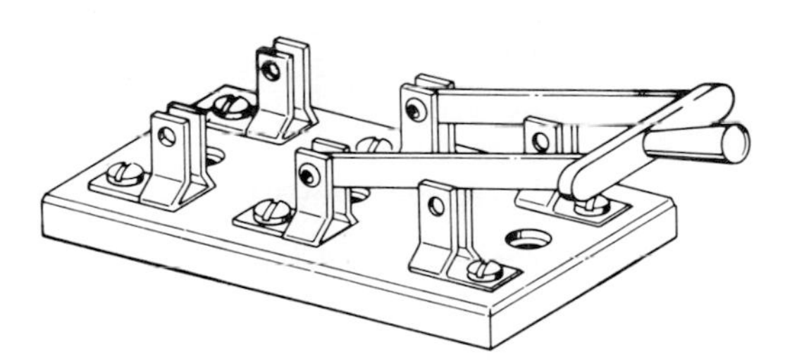

1–32. Knife switch. Double-pole, double-throw.

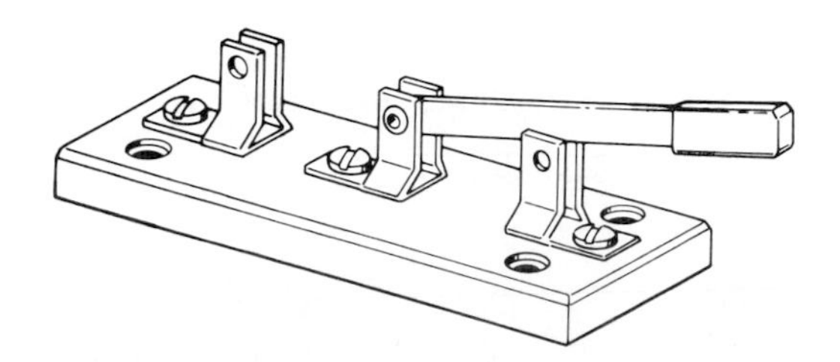

1–33. Knife switch. Single-pole, double-throw.

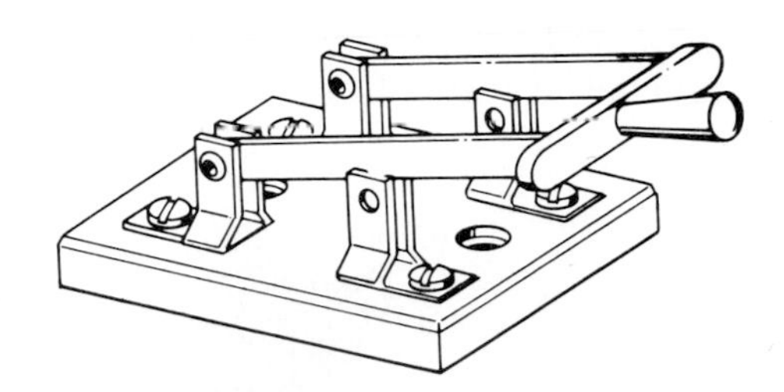

1–34. Knife switch. Double-pole, single-throw.

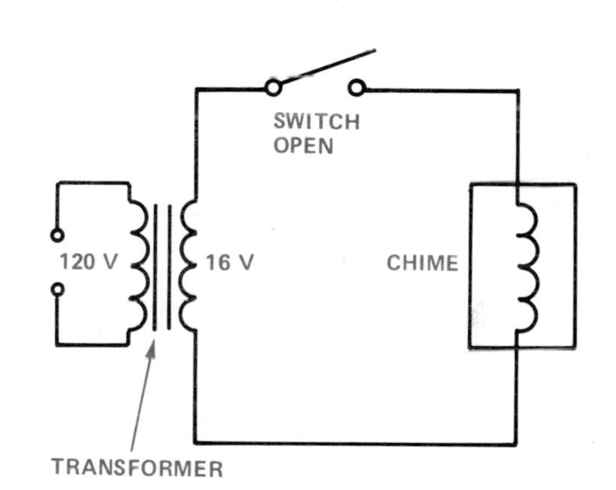

1–35. Door chime circuit. Switch open.

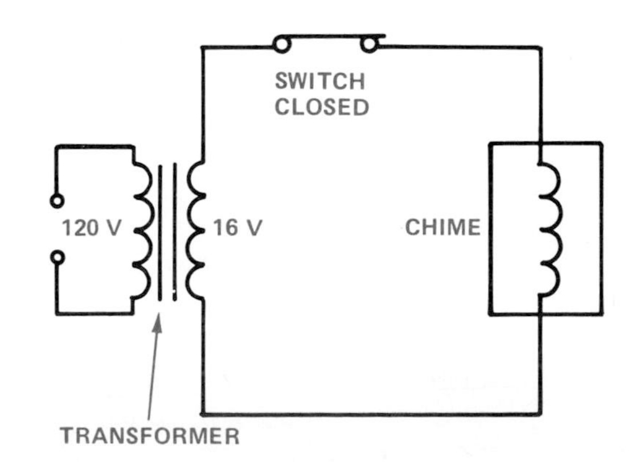

1–36. Door chime circuit. Switch closed.

■ *Toggle switches.* Toggle switches are used to turn various devices on and off, or to switch from one device to another. They are made in a number of configurations to aid in selection for a particular job. Fig. 1-37. These switches usually have a metal handle and are mounted through a round hole. Screw terminals are usually provided for attaching wires. However, some may have wire leads furnished. A *wire nut* is used to attach the switch leads to the circuit wires. A wire nut is a device which makes a connection between two pieces of wire by twisting ends together. It insulates the connection with a plastic coating.

Note the AC rocker switch in Fig. 1-37. It is mounted with screws to attach to the switch's

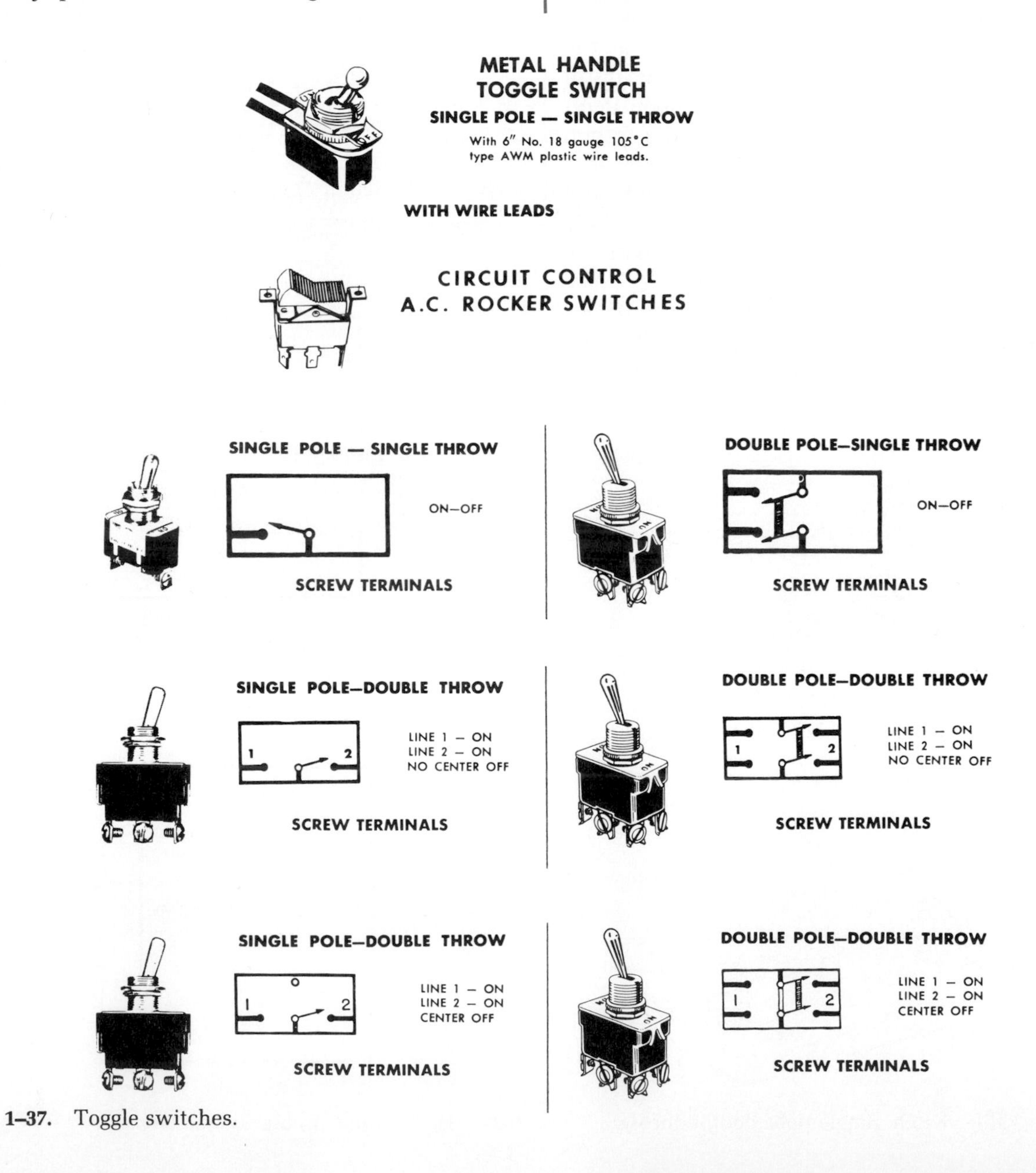

1-37. Toggle switches.

steel bracket. Holes are tapped for a #6-32 screw on $1\frac{5}{8}''$ centers. These switches are usually rated for use on 120 volts AC, with special attention to current. The switches are capable of switching 240 volts. Current is usually doubled in a 120-volt rating as compared to a 240-volt rating. This means a switch rated at 120 volts and 6 amperes must be derated to 3 amperes for 240-volt circuits.

■ Residential toggle switches. Various shapes are encountered when switches are needed for use in business, industry, or the home. For instance, in Fig. 1-38 you will find examples of some of those used to switch 120 volts AC and 240 volts AC in common circuits used in lighting and small motors. In Fig. 1-38, A is a residential toggle switch, rated at 10 amperes on a 125-volt line. Note the absence of "plaster ears" near the long screws. Switch B is like A except that it has wide plaster ears that can be removed easily if not required to hold the switch rigid in its box. The plaster ears are scored (marked) so that they can be bent easily and removed. Switch C is more expensive than either A or B. It has *specification grade* (top of the line) quality with wide plaster ears. Note that the screws for attaching circuit wiring are located topside, instead of along the side of the switch.

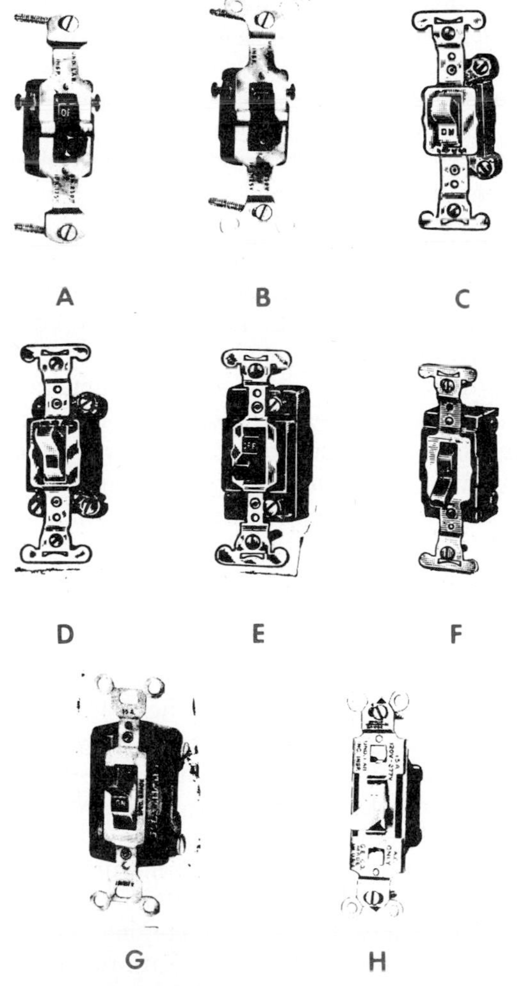

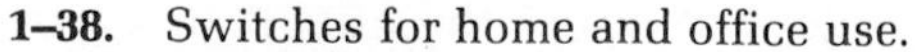

1–38. Switches for home and office use.

Switch D is a high-capacity, heavy-duty, industrial type, rated for 20 amperes at 125 volts. Switch E is an extra heavy-duty industrial type. It is a more expensive switch which minimizes the arcing of contacts when turned on and off. The arc, which occurs each time a switch is turned on or off, creates high heat. The heat can cause the contacts of the switch to become pitted and make a high-resistance contact. The contacts in switch E have an extended life, made possible by the use of arc snuffers.

Switch F is a "no-klik," or quiet, heavy-duty switch. It has eliminated the noise associated with the on-off operation of a switch. Switch G is also heavy-duty, quiet, of high capacity, specification grade, and good for 15 amperes at 120 or 277 volts. Switch H is a quiet 15-ampere AC, side- or back-wired with binding screw, and pressure, or screwless, terminals. Some switches have a *grounding strap* designed for use with nonmetallic systems which use Bakelite boxes and for bonding between the device strap and steel boxes. These newer switches feature a green grounding-screw terminal. They are available in either ivory or brown.

■ *Three-way switches.* Three-way switches are used where you need to control a light or device from more than one location. These switches have three terminals instead of two and do not have the words *on* and *off* on the handles. More discussion of this type of switch and its circuitry will follow in a later chapter. The three-way switch looks like a regular switch with the exception of the three terminals for wiring into a circuit.

■ *Four-way switches.* Four-way switches are double pole, and are used where a light or device needs to be controlled from three or more locations. If three controls are preferred, you need two three-way switches and one four-way switch. The four-way resembles the three-way, but has four terminals for connection into a circuit. It does not have *on* or *off* on its handle, since either up or down may be on or off, as you will see later. More information concerning the four-way switch will be given in a later chapter.

Other types of switches are available for different jobs of current control. They will be shown later, as they are introduced in connection with a specific job.

Switches are used to turn the flow of electricity on or off, thereby causing a device to operate or cease operation. Switches can be used to reverse polarity and, as in the case of electric motors, the direction of rotation can be reversed by this action. Switches, as you have already seen, come in many shapes and sizes. The important thing to remember is to use a switch with proper voltage and current rating for the job to be done. A careful study of the types presented in this chapter will help in the proper selection of a switch for a particular job.

Solenoids

Solenoids are devices that turn electricity, gas, oil, or water on and off. Solenoids can be used, for example, to turn the cold water on, and the hot water off, to get a proper mix of warm water in a washing machine. To control the hot water solenoid, a thermostat is inserted in the circuit.

Fig. 1-39 shows a solenoid for controlling natural-gas flow in a hot-air furnace. Note how the coil is wound around the plunger. The plunger is the core of the solenoid. It has a tendency to be sucked into the coil whenever the coil is energized by current flowing through it. The electro-magnetic effect causes the plunger to be attracted upward into the coil area. When the plunger is moved upward by the pull of the electromagnet, the soft disc (10) is also pulled upward, allowing gas to flow through the valve. This basic technique is used to control water, oil, gasoline, or any other liquid or gas.

The starter solenoid on an automobile uses a similar procedure, except the plunger has electrical contacts on the end that complete the circuit from the battery to the starter. The solenoid uses low voltage (12 volts) and low current to energize the coil. The coil in turn sucks the plunger upward. The plunger then touches heavy-duty contacts which are designed to handle the 300 amperes needed to start a cold engine. In this way, low voltage and low current are used, from a remote location, to control low voltage and high current.

Solenoids are *electromagnets*. An electromagnet is composed of a coil of wire wound around a core of soft iron. When current flows through the coil, the core will become magnetized.

The magnetized core can be used to attract an armature and act as a magnetic circuit breaker. Fig. 1-40. (A circuit breaker, like a fuse, protects a circuit against short circuits and overloads.) In Fig. 1-40, the magnetic circuit breaker is connected in series with both the load circuit to be protected and with the switch contact points. When excessive current flows in the circuit, a strong magnetic field in the electromagnet causes the armature to be attracted to the core. A spring attached to the armatures causes the switch contacts to open and break the circuit. The circuit breaker must be reset by hand to allow the circuit to again

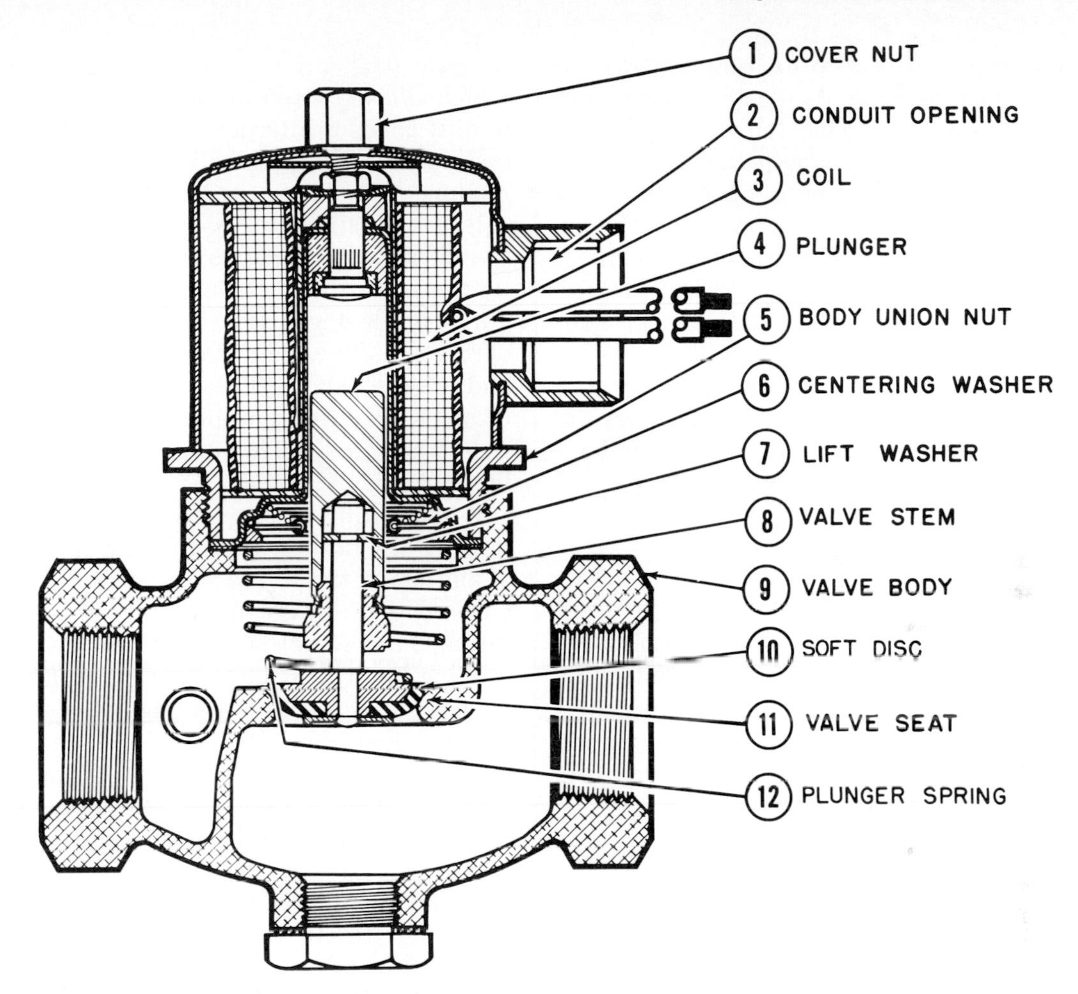

1–39. Solenoid for controlling natural gas flow to a hot-air furnace.

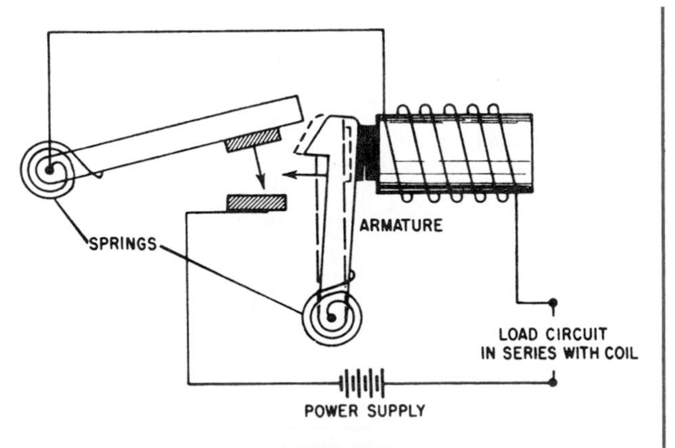

1–40. Magnetic circuit breaker.

operate properly. If the overload is still present, the circuit breaker will "trip" again. It will continue to do so until the cause of the short circuit or overload is found and corrected.

Relays

A *relay* is a device that can control current from a remote position through use of a separate circuit for its own power. Fig. 1-41 shows a simple relay circuit.

When a switch is closed, current flows through the electromagnet, or coil, and energizes it. The pull of the electromagnet causes the soft iron armature to be attracted toward the electromagnet core. As the armature moves toward the coil, it touches the contacts of the other circuits, thereby completing the circuit for the load. When a switch opens, the relay coil *de-energizes*, and the spring pulls the armature back. This action breaks the contact and removes the load from the 12-volt battery. Relays are remote switches that can be controlled from almost any distance if the coil is properly wired to its power source.

Many types of relays are available. They are used in telephone circuits and in almost all automated, electrical machinery.

Diodes

Diodes are semiconductor devices that allow current to flow only in one direction. By properly placing a diode, or diodes, in a circuit, it is possible to control current flow by controlling the direction of the current. Fig. 1-42a. Alternating current (AC) flows first in one direction and then the other, 120 times per second in a 60-hertz (Hz) circuit. The diode is a *rectifier* which allows current to flow in only one direction. It changes alternating current (AC) to direct current (DC), which flows in only one direction.

It is possible to use four diodes, two switches, and two wires to control two lamps located some distance away from the switches. For example, Fig. 1-42b shows how switch A is depressed, allowing current to flow through it and diode No. 1. The current then passes through the wire to diode No. 2, through lamp A, and back to the 120-volt source. Lamp A will light.

When switch B is pressed, current flows through the bottom wire to lamp B, diode No. 3, and along the top wire to diode No. 4. The current then flows to switch B, and back through the wire, to the 120-volt AC source, causing lamp B to light. Whenever lamp B is lit, lamp A is out because current cannot flow through diode No. 2. However, if both switches are depressed at the same time, both lamps will light. This is due to the alternating current taking turns, flowing first in one direction, for instance in lamp A, and then the other, in lamp B. There are pulses of current, or direct current pulsating (PDC), through each lamp. At 60 hertz (60 Hz) it is not possible for the human eye to see the on-off condition of the lamps. The pulses are so fast that the eye cannot re-

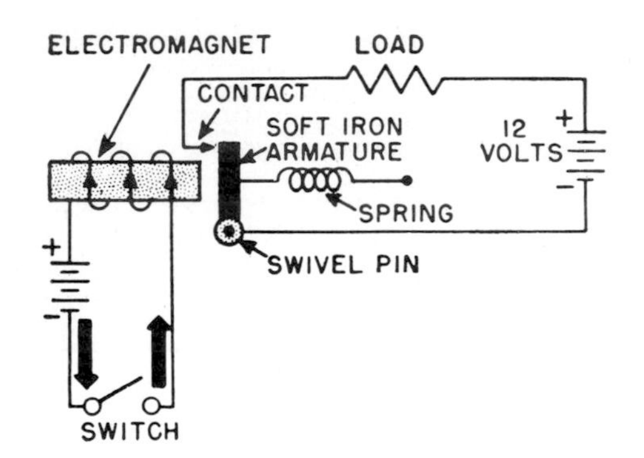

1–41. Simple relay circuit.

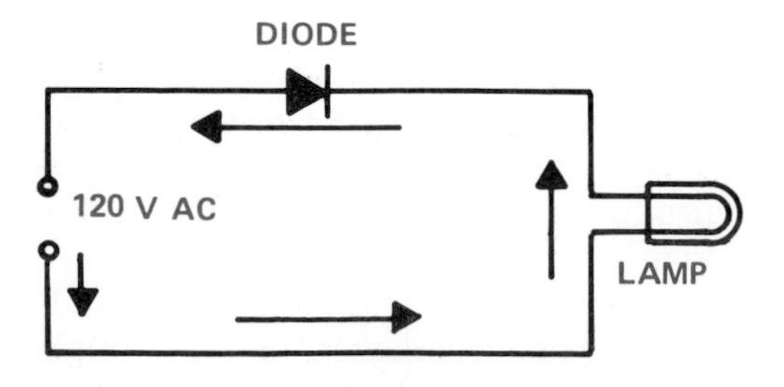

1–42a. Diode control of current.

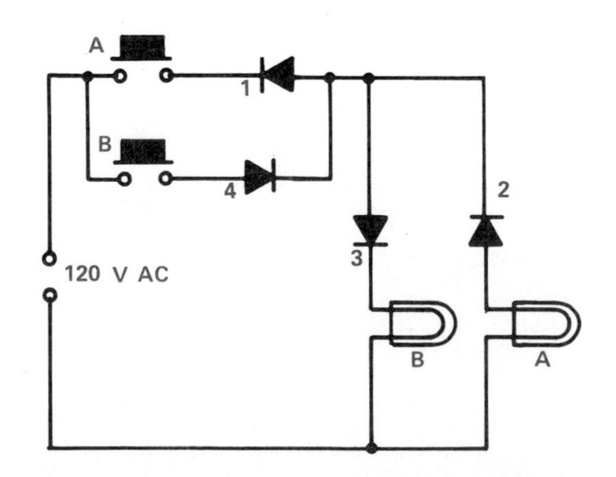

1–42b. Diode control of current for two lamps.

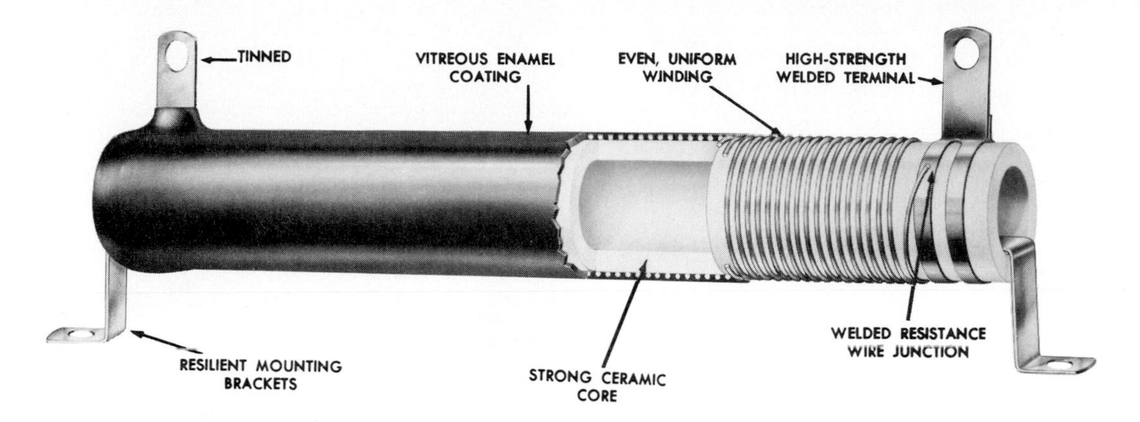

1-43. Wire-wound resistor.

1-44. Variable wire-wound resistor.

spond, thus causing the lamps to appear to be glowing continuously.

You will find applications of diodes in computers and in other less complicated electrical devices. Just remember that the diode allows current to flow in only one direction.

RESISTORS

A *resistor* is a device used to provide a definite, required amount of opposition to current in a circuit. Resistance is the basis for the generation of heat. It is used in circuits to control

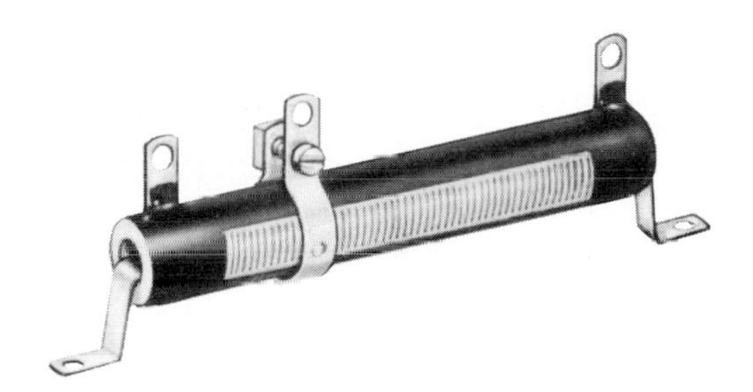

1-45. Adjustable wire-wound resistor.

the flow of electrons and to ensure that the proper voltage reaches a particular device.

Resistors are usually classified as either wire-wound or carbon-composition. The symbol for a resistor of either type is —\/\/\/—.

■ *Wirewound resistors* are used to provide sufficient opposition to current flow to dissipate power of five watts or more. They are made of resistance wire. Fig. 1-43. Variable wire-wound resistors are also available for use in circuits where voltage is changed at various times. Fig. 1-44. Some variable resistors have the ability to be varied but also adjusted for a particular setting. Fig. 1-45.

High wattage, wire-wound resistors may be purchased in many sizes and shapes. Fig. 1-46.

■ *Carbon-composition resistors* are usually found in electronic circuits of low wattage, since they are not made in sizes beyond 2 watts. They can be identified by three- or four-color bands around them. Their resistance can

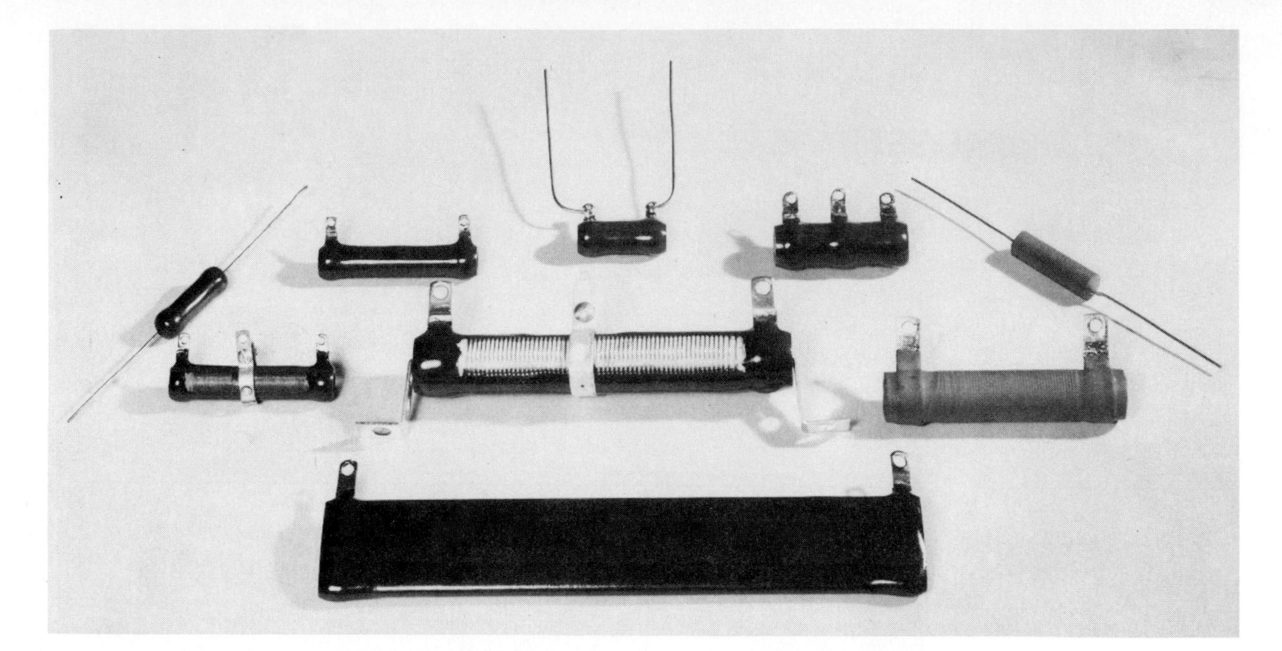

1–46. Various sizes and shapes of wire-wound resistors.

be determined by reading the color bands and checking the resistor color code.

The wattage ratings of carbon-composition resistors are determined by physical size. They come in $\frac{1}{4}$-watt, $\frac{1}{2}$-watt, 1-watt, and 2-watt sizes. By examining them and becoming familiar with them through use, you should be able to identify wattage rating by sight. The larger the physical size of the resistor, the larger the wattage rating. Fig. 1-47.

Resistor Color Code

Take a close look at a carbon-composition resistor. The bands should be to your left. Fig. 1-48. Read from left to right. The first band gives the first number according to the color code. In this case it is *yellow*, or 4. The second band gives the next number, which is *violet*, or 7. The third band represents the *multiplier* or *divisor*. If the third band is a color in the 0 to 9 range in the color code, it states the number of zeros to be added to the first two numbers. *Orange* is 3; so the resistor in Fig. 1-48 has a value of 47 000 ohms resistance.

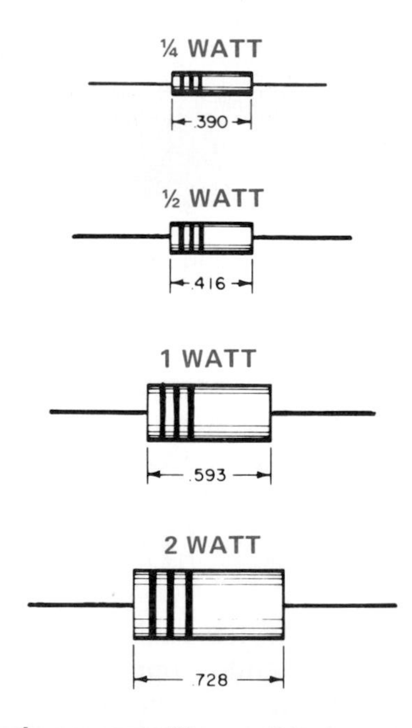

1–47. Carbon-composition resistors.

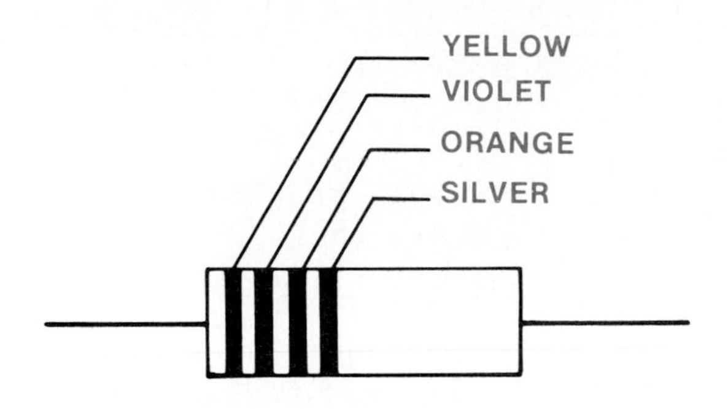

1–48. 47 000-ohm resistor.

If there is no fourth band, the resistor has a tolerance rating of ±20% (± means "plus or minus.") If the fourth band is silver, the resistor has a tolerance of ±10%. If the fourth band is gold, the tolerance is ±5%.

Silver and gold may also be used for the third band, in which case, according to the color code, the first two numbers (obtained from the first two color bands) must be divided by 10 or 100. Silver means divide by 100; gold means divide by 10. For example, if the bands on a resistor are red, yellow, gold, and silver, the resistance would be 24 divided by 10, or 2.4 ohms ±10%. Fig. 1-49.

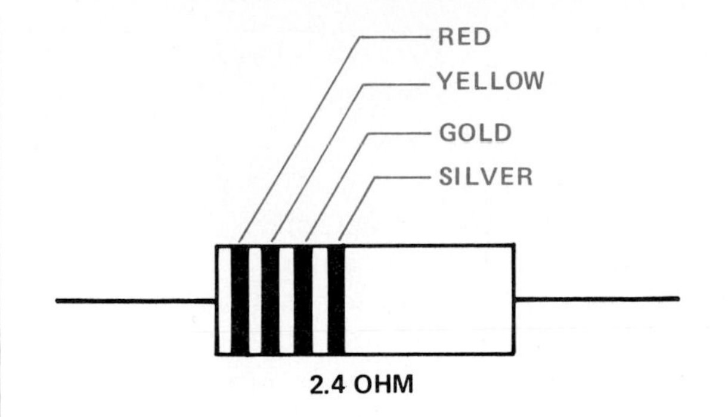

1–49. 2.4-ohm resistor.

Resistors are available in hundreds of sizes and shapes. Once familiar with electronics and electrical circuits, you will be able to identify various components by their shape, size, or markings. Products for such circuits are constantly changing with new items being marketed almost every day. To stay informed about these products, it is necessary to read the literature written about the industry. Each area of electrical energy has its own magazines to keep those on the job informed and up-to-date in their special fields of interest.

QUESTIONS

1. What is electricity?
2. What is an atom?
3. What are elements?
4. What is the difference between static and current electricity?
5. What are free electrons?
6. What is a conductor?
7. What is an insulator?
8. What is a semiconductor?
9. Name six methods used to generate electricity.
10. What is an exotic generator?
11. In what unit of measurement is electrical current measured?
12. In what unit of measurement is Voltage measured?
13. In what unit of measurement is electrical resistance measured?
14. What is the relationship of wire size to its numbering system?
15. What is a complete circuit?
16. What is an open circuit?
17. What is a short circuit?
18. State Ohm's Law.
19. Define kilowatt.
20. What is a kilowatt-hour?
21. How is a meter shunt used?
22. What has to be done to a DC meter to convert it to measure AC?
23. What is the difference between an ohmmeter and a voltmeter?
24. Where is an inclined-coil iron-vane meter used?
25. What do these letters represent: SPST, DPDT, SPDT, and DPST?
26. What is a toggle switch?
27. What is the difference between a three-way and a four-way switch?
28. What is the definition of a solenoid? A relay?
29. What is a diode?
30. What is a resistor?

KEY TERMS

atom
circuit
conductor
current
current flow
diode
element
horsepower
insulator
multimeter
neutron
ohm
ohmmeter
Ohm's law
proton
relay
resistance
resistor
shunt
solenoid
thermistor
voltage
voltmeter
watt

CHAPTER 2

GENERATING ELECTRICITY

Objectives

After studying this chapter, you will be able to:

- Identify the three general groups of electric plants.
- Identify three mechanical energy sources.
- Discuss the economical considerations of various engines.
- Identify the primary components in nuclear fuel.
- Identify three fossil fuels used to power generators.
- Discuss the troubleshooting of a gasoline-engine driven generator.

As discussed in Chapter 1, there are a number of ways to produce electricity, some of which are commercially feasible. The use of magnetism is the most common method of generating electricity in large quantities for businesses, homes, industry, hospitals, and other institutions.

Cells, or batteries, produce direct current (DC). A more economical way of producing DC, however, is with a mechanically driven generator. Mechanical force is used to rotate a wire loop in a magnetic field to generate electricity. The magnetic field is generated by a current-carrying wire, looped around a core. Fig. 2-1.

Since the current from the battery supply is DC with a negative (−) and a positive (+) polarity, the north (N) and south (S) fields are fixed in the positions indicated. If a conductor is moved in an upward direction through the magnetic field between the N and S poles (Fig. 2-1, A), a current will flow in the conductor in the direction indicated by the arrows. But if the conductor moves downward (as in Fig. 2-1, B), then current flows in the opposite direction. The voltage generated will depend upon the following: the intensity of the magnetic field, the number of turns of the wire, and the speed at which the wire passes through the magnetic field.

A simple generator is shown in Fig. 2-2. A loop of wire is wrapped around an iron core called an *armature*. Copper segments, called the *commutator*, which are attached to the ends of the loop of wire, are insulated from the core and from each other. *Brushes* (conductors that make sliding contact) are so placed that they contact the commutator and carry any

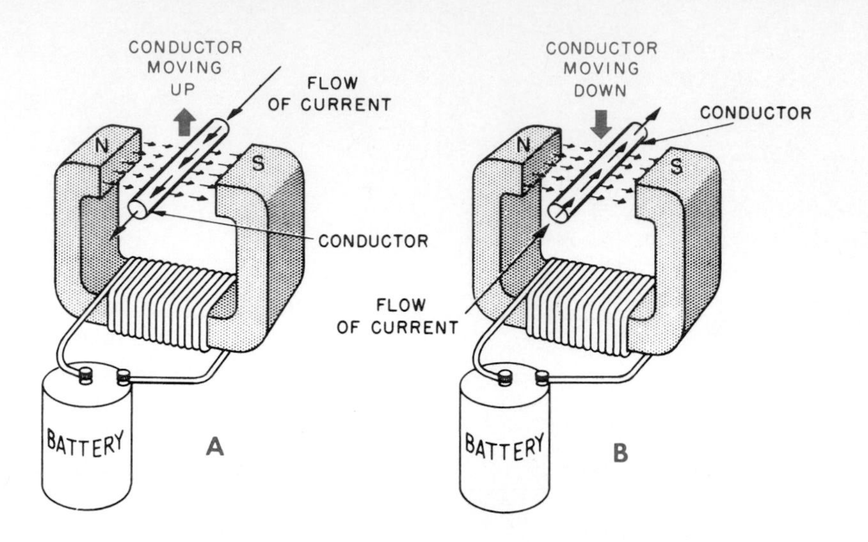

2–1. Magnetism principle of electrical generation.

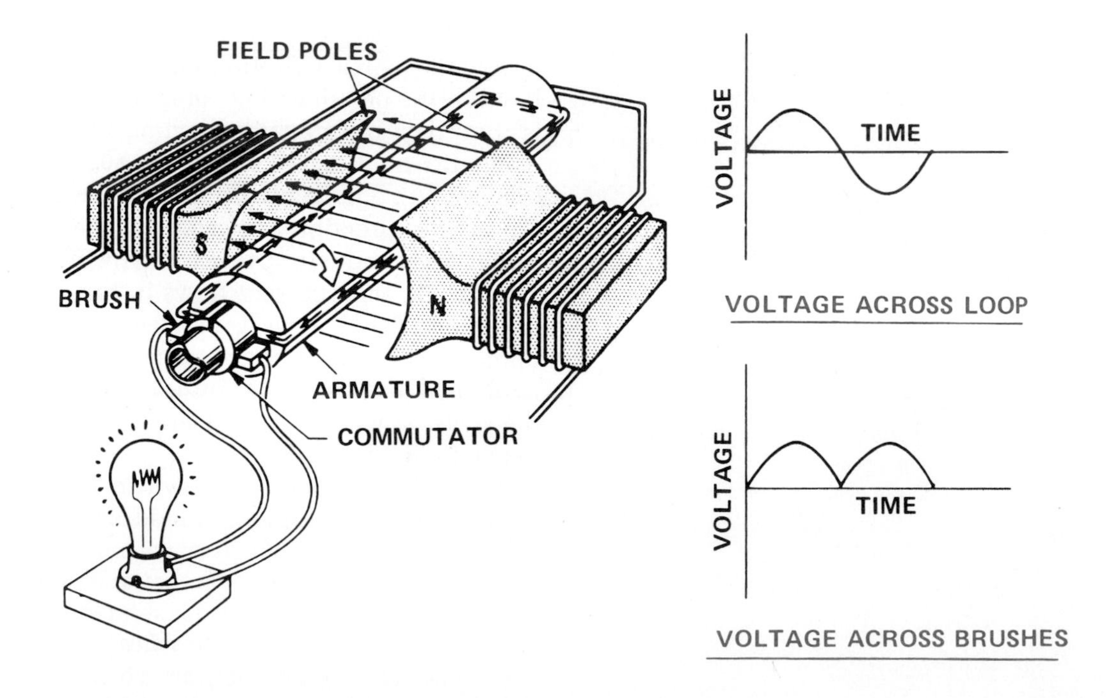

2–2. Simplified DC generator with output voltages shown.

generated electricity to the load, or consuming device. To produce electricity the armature must be mounted between *field coils* so that the magnetic force generated by the electromagnet will be cut by the rotating armature. Field coils form an electromagnet when they are wrapped around soft-iron cores, known as field poles.

The position of the armature in Fig. 2-2 represents the point at which the armature loop cuts directly across the magnetic field. At this point it is generating its maximum output. When the armature has rotated another one-quarter turn, it will be moving parallel with the magnetic field, and no output will be obtained from the generator. During the time that the armature rotates 360° (one revolution), it generates a maximum and a minimum twice—maximum when passing across the S pole, and again when crossing the N pole.

Voltage across the loop is represented as alternating current (AC) in Fig. 2-2, while voltage across the brushes is shown as pulsating direct current (PDC). The commutator acts as a reversing switch as the armature rotates in the different fields. As a result of the switching action, the current output is a series of maximums and minimums with current flowing in only one direction.

■ *AC generator.* An AC generator is usually referred to as an *alternator*. Alternators generate most of the electrical power used today. Large generating plants produce the electricity demanded by a world constantly increasing its need for energy.

Alternating current (AC) generators operate on the basic principle illustrated in Fig. 2-3. Atomic- and steam-powered generators use the same basic idea of electric power generation.

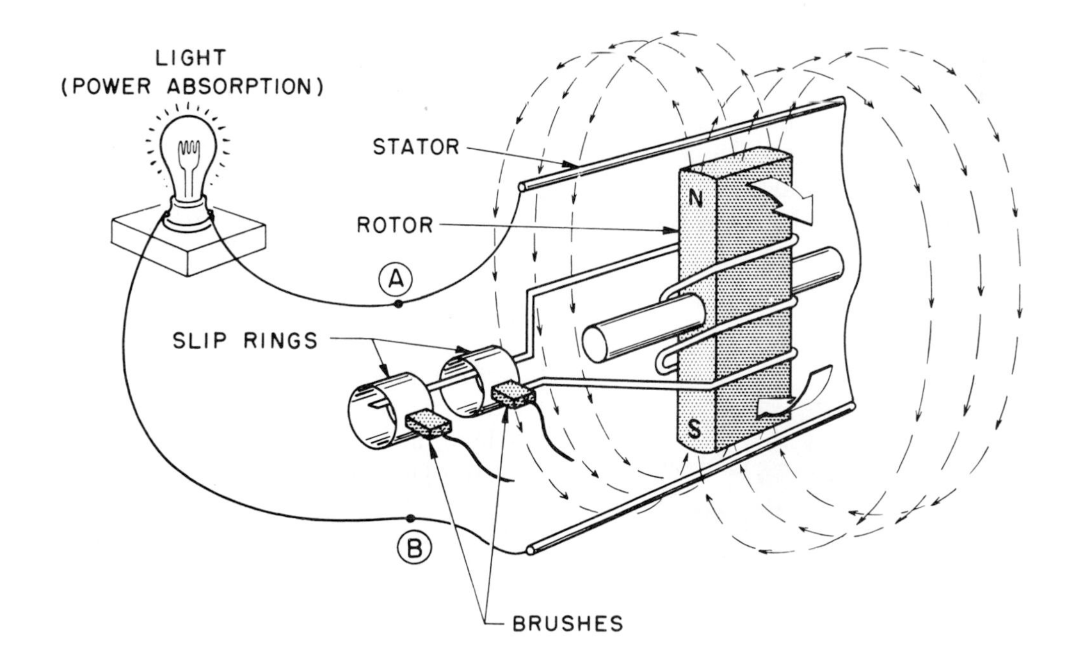

2-3. Simplified alternator (AC generator)

POWER PLANT OPERATIONS

Development of Electric Power Plants

In the early stages of development of the electric plant, most engines were heavy, clumsy, and slow speed. Fig. 2-4. Small engines had not yet reached a stage where they would run for long periods of time without frequent attention and expensive repair. Therefore it was necessary to prolong the life of the engine by using it as little as possible. To do this with an electric plant, it was necessary to use a storage battery. Small electrical loads were operated directly from the battery. The plant engine was used only one or two days each week to recharge the battery. In those early days of electricity, electric plants were known as "light plants"; today's electric plants are more appropriately known as "power plants."

Another problem, at that time, was that the engines could not be made to run dependably at constant speeds. This often resulted in the flickering of lights when electricity was supplied directly from the plant. Batteries had to be used to provide the constant voltage which is necessary for practical lighting. This situation lasted for many years and resulted in the use of many battery-charging plants. Most of these were of the 32-volt type, because the 112-volt, 56-cell battery system was extremely costly.

2-4. Early engine-driven generator.

Gradually, lamp bulbs were manufactured in larger wattages, a variety of appliances became common, and electric motors came into regular use. The increased electrical load required larger batteries and more frequent running of the engine each week, even though the plants were larger. Even for small loads, a battery system of this type was not very efficient. In fact, the continuous increases in load made this inefficiency very apparent. In addition, an intensive plant maintenance system was required. The growing need for a better system resulted in a process of power production directly from the generator.

When plants had to run many more hours each week to operate extra electrical load, the question seemed to be: Why not use electricity directly from the generator? With this method, batteries could be eliminated, and only the amount of electricity actually used would be generated. A more efficient system resulted.

By the middle 1920s, it became practical to do this. Small engines were now running at higher speeds and had attained greater durability, requiring less servicing. In addition, they could run at more constant speeds due to the availability of good engine governors. For the more efficient method of direct production, generators did not need to differ greatly from those used in battery-type plants. Fig. 2-5.

The need for direct power production from generators resulted in the 115-volt, battery-less, direct current (DC) electric plant, which came into use and enjoyed wide popularity for many years. However, this plant did have many limitations. Although satisfactory power was produced for lights and small hand tools, motors which had commonly been AC now had to be of the DC type. These motors were more expensive and harder to obtain than the AC type. Also, the popular AC radio, new on the market at that time, could not be used on DC. Again the power need was reviewed.

With an increasing number of AC appliances, radios, and motors available, the decision was made to construct an electric plant to

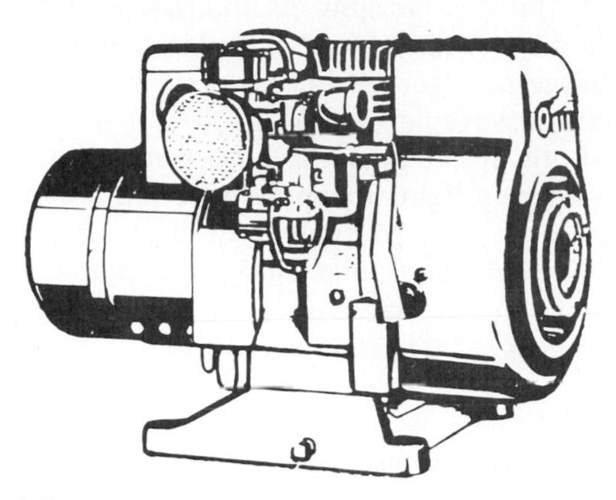

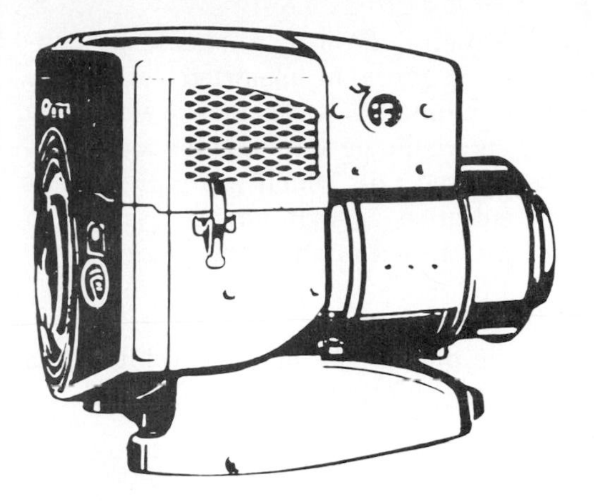

2–5. AC generators.

provide AC. About 1928, the first 350- to 10 000-watt plants were built. AC generators used in these plants were crude, inefficient, rather large and had poor voltage regulation. Gradually, improvements have been made so that today these AC generators are compact, efficient, and have good regulation. They are designed to form an integral part of a single-unit electric plant. These AC electric plants have replaced their dependable DC predecessors. In fact, hundreds of thousands of AC appliances now are being used in homes, farms, manufacturing plants, and industry in general. Fig. 2-6.

2–6. Gas or gasoline-powered 5.5 HP single cylinder 4-cycle engine used to produce 2500 watts @ 120 volts, 60 Hz, 1 Ø (phase), 2-wire AC. It rotates at 3600 RPM and weighs 139 lbs.

Although the widespread acceptance of AC plants had indicated a trend in power usage, it did not necessarily follow that AC plants should be used for all purposes. Both the old battery-charging and DC plants have been improved simultaneously with the development of the newer AC plants. Such a variety of plants can provide service according to the demands of many different types of consuming devices.

Private Engine-Driven Power Plants vs. Public Utility Power Plants

Generally, a private electric plant should not be considered for the primary power source if power is available from a commercial generator.

In some instances there is an advantage in the use of an individual electric plant as the primary power source. Expensive fuels, plus the cost of power distribution, often force public utilities in some states to charge high rates for their power. In such cases an individual plant, using a fuel which is plentiful in that particular location (natural gas, for example) could perhaps prove more economical than public utility power. Generally, however, a small individual plant cannot generate electricity at a cost as low per kilowatt-hour as can a large commercial power company.

An individual electric plant should be considered as a source of electricity if utility power is not available. It can also be a standby source for emergency use in case the main utility power source fails. Such use represents a form of insurance against power loss which cannot be purchased in any other manner.

Types of Plants

All electric plants may be divided into three general groups as follows:

1. Alternating current (AC)
2. Direct current (DC)
3. Battery-charging (DC)

■ *Alternating current plants.* AC plants feed electricity directly to the line to which the lights, motors, and appliances are connected. No battery is used between the plant and the line, although in many cases, a battery may be used for engine-starting purposes only. The plant is run as electricity is needed. The AC output is the same as that commonly used in homes, offices, and factories. Plants have these standard voltages, available in either 50 or 60 hertz (Hz):

Single-phase:
120 volts, 2-wire
120/240 volts, 3-wire
240 volts, 2-wire

Three-phase:
240 volts, 3-wire
120/208 volts, 4-wire
120/240 volts, 4-wire
480 volts, 3-wire
220/380 volts, 4-wire
277/480 volts, 4-wire
600 volts, 3-wire

■ *Direct current plants.* DC plants feed electricity directly to the line to which the lights, motors, and appliances are connected. As in AC generation, no battery is used between the plant and the line, although sometimes a battery may be used for engine-starting purposes only. The plants are run each time electricity is needed. They are generally available in two standard voltages: 115 or 230 volts.

■ *Battery-charging plants.* Battery-charging plants are direct current plants designed to charge the storage batteries with which they must always be used. With this system, electrical power is drawn directly from the batteries to operate lights or appliances. The plants are operated only to charge the batteries or provide additional electricity for heavy loads. They are available in four standard battery voltages: 12, 24, 32, and 110 volts.

Power Plant Operations—AC

An alternating current (battery-less) plant supplies electricity directly to the line to which electrical consuming devices are connected. It must be operated whenever any part, or all of the equipment, is to be used.

This type of plant supplies primary, or portable, power for field or construction work. It will operate electric lights, hand tools with universal motors, and other equipment and motors designed for AC. Alternating current power has numerous other uses, especially for the home, farm, factory, industry, hospital, and communications fields. Its vital function in these fields is that of a reliable source of emergency electricity for those occasions when there is an interruption in regular power service. Fig. 2-7.

2–7. Diesel, water-cooled, 6-cylinder, 4-cycle engine. Also available with 12 cylinders. Uses 24 volts, DC for remote-control system. Larger sizes produce up to 500 kW or 625 kVA @ 277/480 volts, 3 Ø, 4-wire AC. The 12-cylinder model weighs 12 020 lbs. and produces 750 HP to drive a generator.

Since much of the equipment and most motors built today require alternating current, the AC Plant is more widely used than any other type of generating plant.

An AC plant has other advantages. It is available in the widest range of sizes and with more types of controls (ranging from manual-start to full automatic). It will also operate a greater selection of low-cost equipment. Many appliances and some lighting, such as fluorescent, can be operated only on AC. Wiring distribution of the alternating current system is probably the most flexible, efficient, and least expensive to operate.

The AC plant is ideal as a primary source of power. It is capable of operating a variety of large-capacity equipment on regular high-line power (overhead wires) when a conversion in power is made (transformers step down voltage). The plant may then be kept in reserve for emergency standby, or other service. Such versatility makes the AC type of power plant unique.

An engine-driven AC plant with a DC starting system can supply a limited amount of DC power as well. It would be of 12-volt or 32-volt design, depending on the starting characteristics of the plant.

Because a plant must be operated whenever power is needed, there is an interruption in power when the plant needs repair or service. Such a problem is usually resolved through proper care and maintenance.

Power Plant Operations—DC

Because a direct current (battery-less) plant supplies electricity directly to the line to which equipment is connected, it must be operated when part or all of the equipment is to be used. In principle, it is similar to the AC plant, except that it generates a direct current instead of an alternating current.

This type of plant is suitable for field work, construction jobs, and other operations requiring extensive lighting and use of tools, where high-line power is not available. Most hand tools have universal motors which operate on either AC or DC. Yet, DC operation probably affords more efficient use. Tools requiring DC power could be electric drills, saws, shears, and hedge trimmers. Moreover, some construction equipment, such as lifting magnets for cranes, might operate only on DC. Typical users of DC motors would be contractors, fire departments, repair crews, and turf-maintenance crews.

A DC plant has several advantages. Because no main battery is needed, original as well as maintenance costs are reduced. Only the actual amount of electrical load demanded is generated, and thus efficiency is increased. Installation and wiring costs are lower than for a battery system. In addition, more usable power per pound of weight can be developed in a DC plant because operation at full continuous load is rare. Thus there is less fuel consumption and wear on the engine.

There are limitations to the DC plant. It will operate only ordinary lamp bulbs, not fluorescent ones, and motors and appliances that are designed for DC, and motors of the universal type. It cannot operate any equipment designed for AC and therefore is not usable as a standby for emergencies. Because electric-starting systems are expensive, most DC plants are of the manual-starting type. They are not usually available in a complete range of sizes. Because a plant must be operated whenever power is needed, no power is available during periods when the plant is being serviced. Low-cost, high-speed AC plants are generally replacing this type of DC plant. (Higher RPMs are needed to produce 60 Hz AC. Therefore they are considered high-speed.)

Power Plant Operations—Battery-Charging

A battery-charging plant produces only direct current. Storage batteries are always used with this type of plant. The plant is operated for the purpose of generating electricity. This electricity is supplied to the storage battery in the form of a battery-charging current. With this system, electrical power is drawn directly from the batteries to operate lights, appliances, and other devices.

After a plant has charged the battery, it is shut off until all energy has been drawn from the battery and it is discharged. The battery must then be recharged by running the plant.

Electricity is available, even when the plant is running. In such cases, a larger load, with a maximum no greater than the combined capacity of the plant and battery, may be used. If the battery is discharged, the load can be no greater than the capacity of the plant itself. Using the plant with discharged batteries and at full load will prolong the charging time of the battery.

There are several uses for a battery-charging plant system: for example, a situation in which continuous functioning is undesirable due to the noise of operation, as aboard a boat. Another application of battery systems is in communications systems, where a small but continuous amount of direct current is required. In addition, there is often a need in the field for power to charge batteries used in cars, trucks, and tractors and for assistance in starting these vehicles.

Advantages of the system are found in the continuous source of power available from the battery and in the relatively limited plant operation required to charge a battery.

There are several disadvantages. Initial cost is high for both plant and battery. Available appliances and lights which operate on a battery plant are limited and expensive, especially for low-voltage systems. The battery must be periodically charged and discharged, regardless of usage. In addition, it must be replaced with a new one from time to time. Except for marine applications, normal use of this system is generally limited, either because it is insufficient in capacity, too expensive, or too cumbersome.

Mechanical Energy Sources. The previous discussion of electrical power plants covered their electrical-generator component. Now let us review the various types of generator driving power that are often used. The prime source of mechanical power for driving the generator is usually one of these three types: gasoline, diesel, or gas engine. Fig. 2-8.

A high degree of efficiency has been reached in the design and production of these internal combustion engines, which ensures that any one of the three types will produce a dependable source of power.

All engines must maintain a specific generator speed to develop the desired frequency. For example, 60 hertz (Hz) requires 3600 RPM, or some fraction ($\frac{1}{2}$ or $\frac{1}{3}$) of 3600, which would be 1800 or 1200. For 50 hertz, this must be 3000 RPM or 1500 or 1000.

■ *Gasoline engine.* The gasoline engine is probably best known, since its basic design is the same as that of the engine used in automobiles. Sometimes, however, due to the low requirement for power in some plants, a gasoline engine may need only one or two cylinders, instead of four or more, as required in an automobile.

Most of the engines in power plants are of the four-cycle type that operates either at 1800 or 3600 RPM. Some are two-cycle engines, operating at 3600 RPM. These are used for extremely lightweight, portable units. Because these engines present certain service problems,

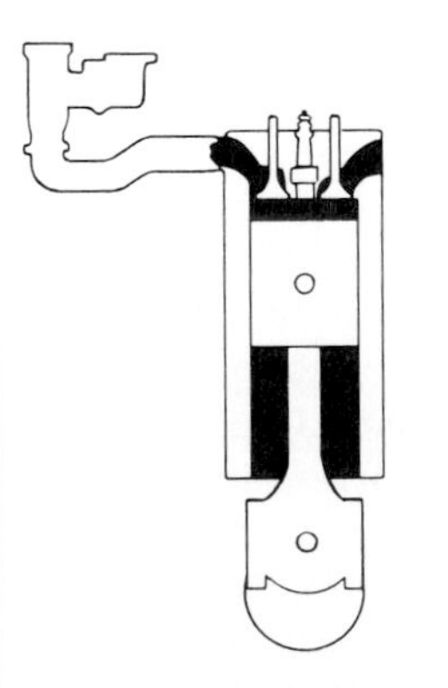

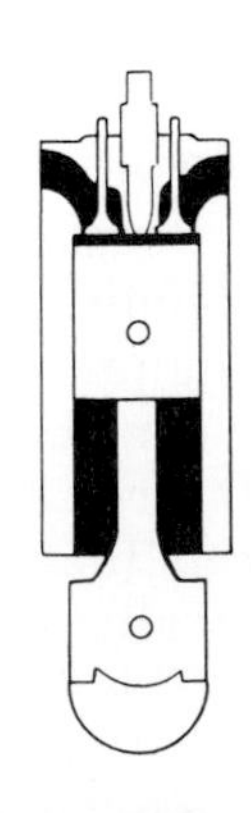

2–8. Gas and gasoline-type engines' intake on left; diesel engine on right.

the trend in design is toward four-cycle engines which run at moderately high speeds, to obtain more generating capacity with less plant weight. Older engines ran at 300 or 450 RPM. Modern engine design has increased the speeds to the point where 1800 RPM is established as standard, and 3600 RPM is universally accepted. Fig. 2-9.

All plant engines have compression ratios designed to take full advantage of high-octane gasoline, which gives greater operating efficiency.

■ *Diesel engine.* The diesel engine is also of the internal combustion type. It uses the heat of compressed air to ignite the fuel, instead of an electric spark as the gasoline and gas engines do. Production problems, very high initial costs, and limited use prevented an early customer acceptance of the diesel. Today diesels are in demand. An important reason for their widespread use is their ability to run on comparatively low-cost fuel, their low operating costs, and their long life.

Two combustion cycles are available in diesel engines: two-cycle and four-cycle. Both types are practical. However, four-cycle engines are sometimes more economical because they use less fuel than two-cycle engines.

Because of extremely high combustion pressures, all components making up this engine must be larger than those of gasoline engines and must be made of very strong material. For example, the injection pump, which times and measures the amount of fuel for each cylinder, must have very close tolerances for good starting and efficient operation. This adds to the expense of the engine. Fig. 2-10.

■ *Gas engine.* Natural, manufactured, a combination of both, and bottled, or LP (liquified petroleum), gas are widely used as engine fuels. A standard gasoline engine, equipped with special carburetion, permits the use of such fuel. In theory, combustion is the same, except the fuel is already in a gaseous state. Thus the process of gasoline vaporization is eliminated. Use of gas fuels allows the engine to run cleaner, with lower maintenance costs as a result. See Fig. 2-6.

Natural gas has a high enough Btu content (British thermal unit; SL: joule, J) to permit the

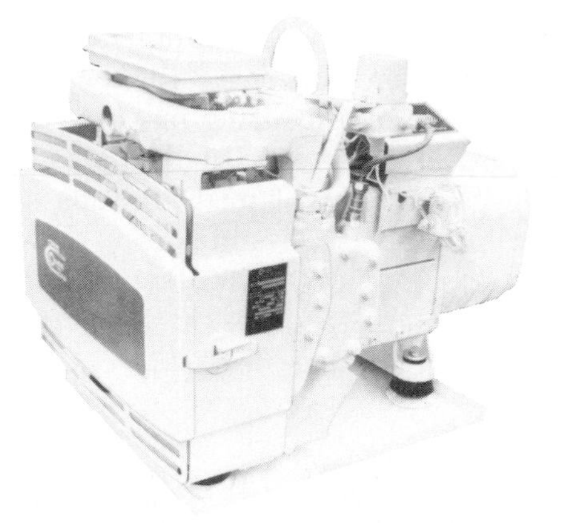

2–9. This gasoline engine produces 13 HP at 1800 RPM. It has remote start-stop capability with 12 volts, DC. This is a marine-type engine, and may be used on ships of various sizes. The generator output is 120 volts, 1 Ø, 60 Hz, 4-wire at 4 kW. It is quiet running, rubber mounted, and water cooled.

2–10. Diesel engine, single cylinder 4-cycle, 5.7 HP, air cooled. Unit weighs 348 lbs. Generator produces 3 kW, 120/240 volts, 60 Hz, 1 Ø, 4-wire, with a 4-pole revolving armature, sealed ball bearings and inherent voltage regulation.

engine to develop near-rated horsepower. Manufactured gas, or a mixture of it and natural gas, both have a reduced Btu content which lowers the output of the engine and plant. The plant may require derating if the fuel used has a Btu content of 1000 or less. Derating means lowering the plant output because the fuel doesn't contain enough energy to meet the standard fuel value.

Bottled, or LP, gas is usually butane or propane or a mixture of both. It is supplied according to local and climatic conditions and is stored in tanks in liquid form under pressure. When pressure is reduced, the liquid becomes a gas and operates the same as natural gas. In all cases, the Btu content is high, and the engine output is comparable to that obtained from gasoline.

Economical Considerations of Various Engines. Because of the apparent lower operating costs of diesel engines using oil for fuel, the person responsible for ordering plant machinery might be inclined to specify that type of engine. There are, however, several factors to consider.

■ *Original cost.* Diesel plants cost quite a lot more than either gasoline or gas engines, for several reasons. There is not the high mass production, which would help reduce manufacturing costs. More expensive components are required for diesel engines. Because they are heavier than other engines, transportation costs are higher for them. Gas plants are slightly higher in cost than gasoline plants.

■ *Installation cost.* For an average job, installation costs are about the same for the three types of plants. But when a gasoline plant must meet local fire codes for a building installation, the cost of gasoline plants may be the highest of the three plant types. Strict regulatory codes exist because volatile fuel is being stored within the building and adequate protection must be provided. Installation of gas engine plants is probably the least expensive, with the diesel rating next in economy. The same comparison would probably apply to insurance on a building. Check insurance policies to make sure, however.

■ *Starting methods.* Gasoline and gas engine plants are easier to start and are manufactured with more types of starting systems than diesel plants. Fig. 2-11. Diesels can be hand cranked, generally in sizes up to 3 kW while other types can be hand-cranked in all sizes. All plants can be electrically started. Smaller gasoline and gas plants may be started through the exciter winding of the generator; larger sizes by an automotive-type starter. Only small diesel plants can be exciter cranked. All diesels can be electrically cranked, however. The ambient temperature (air temperature surrounding the plant) must be above zero Fahrenheit to assure starting. This may be accomplished with glow plugs, air heaters, or room and oil base heaters.

There are many types of starting methods and variations of those methods. Let us consider the five most common methods:

1. *Manual.* This plant, using magneto (permanent magnets) ignition, is manually started by hand crank or rope. It is stopped by a push button. This method is used to start portable, mobile, intermittent, and sometimes, emergency-service units. Fig.2-11.
2. *Electric.* This plant, using magneto or battery ignition, is electrically started or stopped by means of push buttons on the unit. Stopping is possible from any push button located up to 250 feet (80 meters) from the plant. Storage batteries furnish current for cranking. Fig. 2-11.

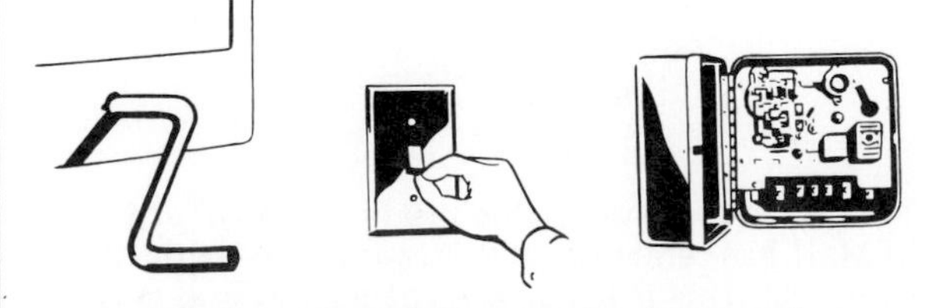

2–11. Starting methods for power plants. Left to right: hand crank, switched remote-control, and automatic load transfer.

3. *Remote control.* This type of plant is electrically started or stopped by push buttons mounted on the plant, or from any number of push-button stations located up to 250 feet (80 meters) from the plant. Fig. 2-11. A set of automotive-type batteries supplies the electric power for cranking. These batteries are kept charged by a special circuit, controlled at the plant. Depending on the voltage and capacity of the batteries, the starting circuit may also be used to supply a small amount of DC current for lighting or other purposes.

   The remote control is practical for many installations. When used with many start-stop stations, it provides almost the same service as an automatic plant, except that its operations are controlled. This eliminates haphazard operation and results in greater efficiency. A water pump cooling system may also be run with this method, but automatic heating or refrigeration requires a full-automatic starting method. To convert from remote control, all that is necessary for most AC remote control plants is the addition of a full-automatic control panel designed for this purpose.

4. *Full automatic.* This plant starts automatically whenever an electric-lighting load of about 75 watts (resistance load much higher) is turned on. It continues to run until the load is turned off. This starting method will operate a refrigerator, or any properly connected automatic equipment, without any individual attention. Basically, this method uses a standard remote control plant, plus an automatic control which makes the plant fully automatic in starting. Thus, a remote-control plant may be converted at any time to full-automatic starting by the addition of the panel.

   A plant with this automatic-starting method is likely to run more frequently than a plant with other types of starting. Any minimum electrical load will start it.

   Special consideration must be given to the size and use of a full-automatic plant. If several automatic devices are to be used, with a possibility that they may all operate at one time, the plant must be large enough to handle the total starting and running load. A plant using the other starting methods may be smaller in size because its load and usage can be controlled.

5. *Automatic load transfer.* This starting method is designed for emergency standby plants. Fig. 2-11. Whenever there is a power interruption on a commercial line, an automatic transfer disconnects the load from the line, starts a standby engine, and keeps the plant running. It connects the load to the plant, then reverses these procedures when normal power is resumed. It ordinarily supplies a trickle charge to the starting battery when the plant is not running.

   An automatic load transfer may be installed and connected to any AC, remote-control plant to provide this type of starting. The plant and transfer mechanism may be connected in several ways: one method is to allow the combination of the two to handle the entire load. Another solution provides that the plant and transfer handle only a certain portion of the load on a preselected circuit. The size of both plant and transfer is selected according to electrical requirements.

   Certain relays may be added to the transfer to service special needs as required. These include voltage-sensitive relays and time-delay relays. Voltage-sensitive relays will start the plant when line voltage varies above or below a predetermined value. Time-delay relays delay starting and stopping so that the plant does not start on momentary power outages nor shut down before high-line power is securely restored. Where the starting load is larger than plant capacity, time-delay relays are necessary to transfer the load to the plant in stages so that motors can be started one at a time.

■ *Operation.* The main point to consider is the cost and availability of the fuels. In almost all cases, diesel fuel costs are the lowest, with gasoline and gas being about the same for similar Btu value. Prices, of course, are subject to change. Exceptions in fuel economy occur where a plant has access to a free gas well, which would obviously give it the lowest fuel cost when compared to other types of plants. However, if this plant operates only a few hours per year, the savings on fuel costs would not offset higher initial equipment costs.

Availability of fuels is of prime importance. While a standard gasoline at somewhat stable prices is available almost everywhere, the other fuels, gas and diesel, cannot always be secured. This should be a determining factor in selection of the type of engine for a plant.

Lubricating oil used in the engine crankcase may be slightly more expensive per quart for the diesel plant, but the small amount required, as compared to the operating fuel, does not make it an important price factor.

Gasoline and gas plants require about equal expense for engine maintenance. Such costs may be lower than for the diesel engine, even though the latter has fewer points to service. An operator of a gasoline, gas, or diesel plant can, in most cases, take care of his own maintenance. However, some diesel engines may require special training for servicing and maintenance personnel. This type of specialized service has become more readily available, although it does add to the overall cost of diesel plant operation.

■ *Overhaul.* All engines require a periodic overhaul, with carbon removal and valve grind. In addition, certain parts require replacement, depending on their type and length of service. A complete overhaul for gasoline and gas plants will probably cost less than for diesel plants. (Such an overhaul, however, will be required more often than for the diesel plants.) As a result, total overhaul costs for the life of a diesel plant may be lower than for other types because most service to a diesel plant is usually for accessory equipment.

■ *Engine cooling.* Two basic methods are used for the job of cooling an engine which develops a large amount of heat in its operation: air and water. Fig. 2-12.

In the early history of power plants, most engines were water cooled, but as improvements were made, air cooling was developed for small engines. Further research resulted in engines of hundreds of horsepower using air cooling. There has been some reluctance on the part of users to this change from water cooling, especially if they had difficulty with one of the earlier air-cooled engines. However, plant experience with millions of engines has proven the reliability and ease of service of an air-cooled system.

Water-cooled engines have radiators which are supplied by municipal water, usually raising costs higher than for air-cooled systems. The single-duct radiator heat exhausting system has been equalled by new air-cooled systems.

The power plants discussed here should give you some idea how electricity is obtained in different locations in the world. It may be necessary someday for you to operate, or at least identify, these power plants. The information presented should be sufficient for you to detect advantages and disadvantages, possibly enabling you to recommend a particular unit to a person who is interested in buying such equipment, either for emergency power or for use in the field.

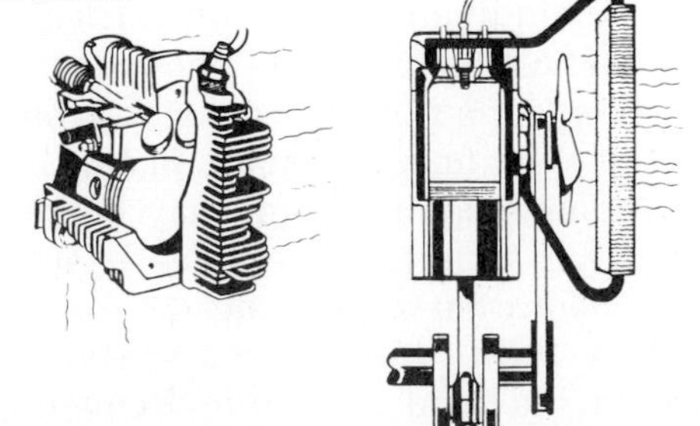

2–12. Air-cooled (left) and water-cooled (right) engines.

FALLING-WATER GENERATORS

The principle of electricity generation is the same, whether it uses mechanical energy produced by gasoline or diesel engines, or falling-water or nuclear energy. In this part of the chapter, you will learn about the production of electricity by falling water. This method uses the energy released by the force of water falling over a distance to turn a generator and produce electricity.

Niagara River Project

The Niagara Power Project is an example of falling water used to generate huge amounts of electrical energy. It is one of the world's largest facilities of this type. As the description of the operation and utilization of the plant progresses, you will probably realize the similarity of this type of power generation to that of fossil-fuel and nuclear power plants. Although these different types of generators of electricity are the same, the method for generating mechanical energy is different. (Mechanical energy is needed to turn the shaft upon which a coil of wire is revolved in a magnetic field.) Initial cost of building a power plant of this magnitude is extremely high, but the system is comparatively maintenance free for about one hundred years.

Niagara is one of the largest hydroelectric developments in the western world. Total installed power at Niagara's two generating plants (Robert Moses and Lewiston) is 2 190 000 kilowatts. Fig. 2-13. There are 340 miles (544 km) of transmission lines to interconnect with the Niagara Project and the dam at St. Lawrence.

Generating Equipment. The generating section at the Robert Moses plant consists of 13 unit blocks. Each of these unit blocks contains a Francis-type hydraulic (water-driven) turbine rated at 200 000 horsepower at 300 feet net head. Fig. 2-13. The 150 000 kW generators operate at 120 RPM, 3-phase, 60 hertz. They are mounted below the deck under removable hatch covers. A 630-ton (571.5 metric ton) capacity traveling crane can be positioned over the generator pits for handling the rotating parts of the turbine and generator assemblies. This crane is the largest of its type ever built. It is 70 feet (21 m) high and weighs more than 900 tons (816.5 metric tons).

Transformers are located on the deck behind the generators. Connection to the switchyard, located south of the power canal, is by high-voltage oil-type cable, housed in a reinforced-concrete, underground power and control tunnel. Power is produced at 13 900 volts and stepped up by seven transformers at the power plant to 115 000 volts. Six additional transformers provide 230 000 volts to meet various transmission requirements. Moreover, transformers at the switchyard increase some of the power to 345 000 volts.

Pump/Generating Plant. The Lewiston Pump/Generating plant was constructed about one mile (1.6 km) from the Niagara Gorge, east of the Robert Moses plant. Fig. 2-13. The plant's dual function is indicated in its name. During the night, from April 1 to October 31, and during periods of low demand, surplus power from the Robert Moses plant is used by units at the Lewiston plant to pump water into the storage reservoir. During daytime peak power-demand periods, the stored water is released to flow through the Lewiston plant units, which, in reverse rotation, operate as turbine-driven generators. After producing 240 000 kilowatts of power at the Lewiston plant, the water then flows down the power canal to be reused through the generating units at the Moses plant.

Pump/Generating Equipment. A reversible pump/generating installation, such as that at the Lewiston plant, consists of 12 motor-generators directly connected to 12 hydraulic pump-turbines. Each electrical unit is rated at 37 500

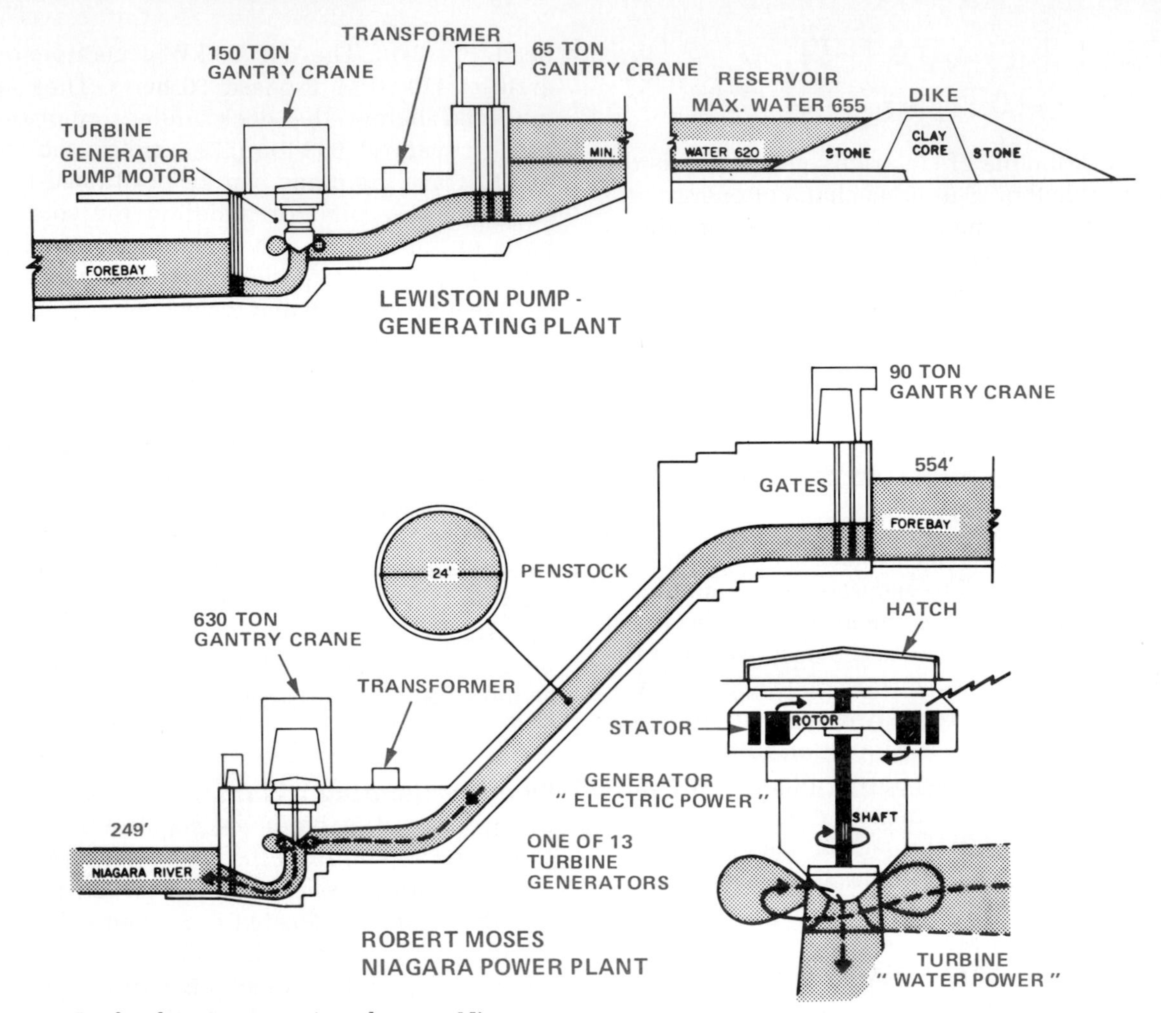

2–13a. Hydroelectric generating plants at Niagara.

HP as a motor, and 20 000 kW as a generator, operating at 112.5 RPM in either capacity. Each hydraulic unit is rated to discharge 3400 cubic feet (95.8 m^3) of water per second, at 85 feet (26 m) net head as a pump, and to develop 28 000 HP at 75 feet (23 m) net head as a turbine. The complete pump/generating units are mounted below the downstream, or lower deck of the plant, under removable hatches. Steel penstocks, 18 to 24 feet (5.5 to 7.3 m) in diameter, are embedded in the mass concrete. A 150-ton (136-metric ton) traveling crane on the lower deck handles the motor generator and pump-turbine assemblies, as well as trash racks and gates.

Switchyard. The Niagara switchyard is situated on a 35 acre (140 m^2) site south of the power canal, halfway between the Robert Moses and the Lewiston power plants. Purpose of the switchyard is to collect and meter power from the generators and send it out over transmission lines. The switchyard has three voltage sections: 115 kV, 230 kV, and 345 kV. Power enters the switchyard from the Robert Moses plant through seven 115 kV cable circuits, and

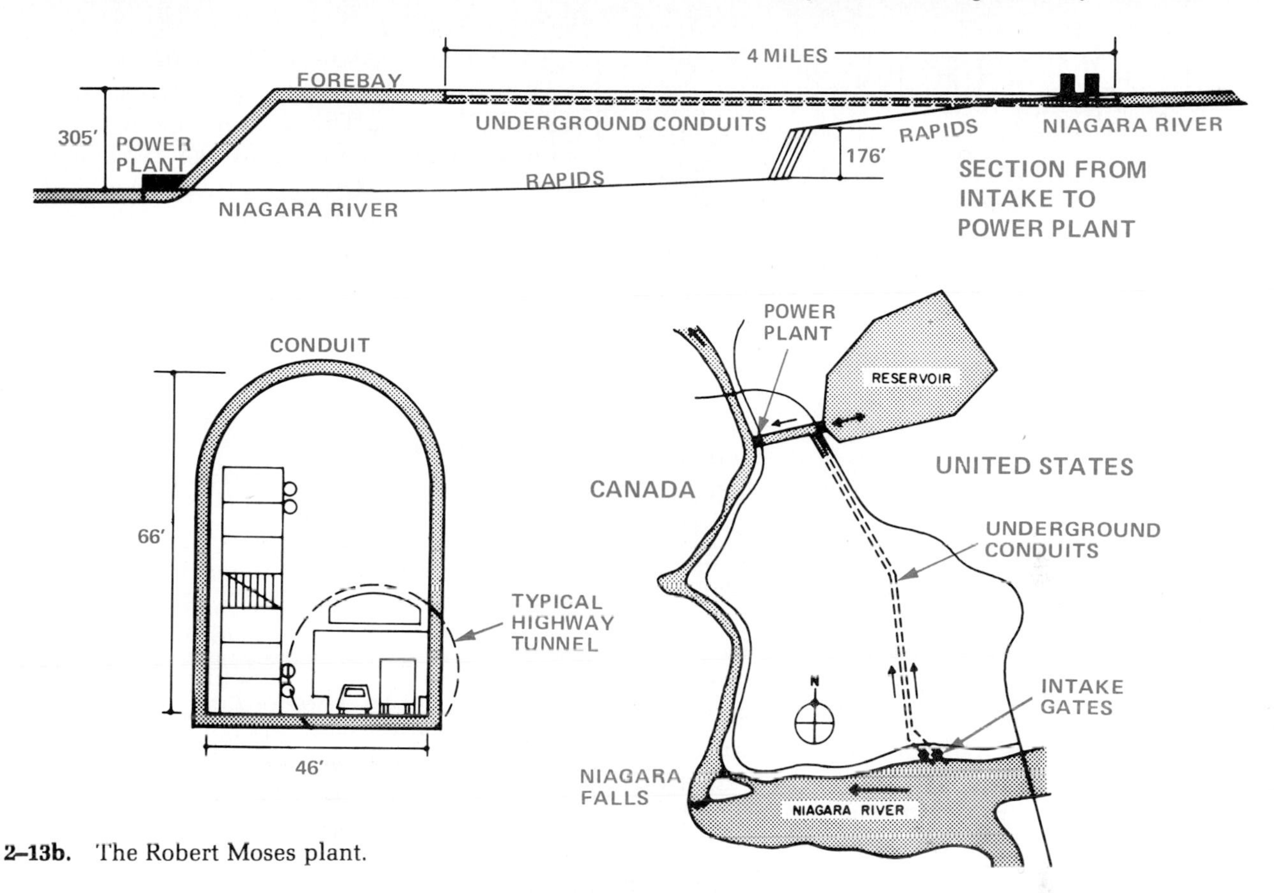

2–13b. The Robert Moses plant.

six 230 kV cable circuits; and from the Lewiston plant by means of four 230 kV circuits. The cables are installed in underground power tunnels. These cable circuits, 61 620 feet in length (18 781 m), are filled with a total of 138 000 gallons (627.3 kl) of a special insulating oil.

The 115 kV section contains 692 foundation structures to support steel towers and electrical equipment. Among these are 78 disconnecting switches and 33 oil circuit breakers, each capable of safely interrupting 10 million kVA of power under short-circuit conditions. There are 368 foundations in the 230 kV switchyard to support the 68 disconnecting switches; and 27 oil circuit breakers, each capable of interrupting 20 million kVA under short-circuit conditions.

The extra high voltage or 345 kV section contains about 120 foundations for 15 disconnecting switches and five oil circuit breakers rated at 25 million kVA interrupting capacity.

The 115 kV and 230 kV sections are tied together by two 200 000 kVA autotransformers. The 230 kV and 345 kV sections are tied together with two 400 000 kVA autotransformers and one 800 000 kVA autotransformer. The latter unit, incidentally, weighs 982 600 pounds (445697.5 kg) and was, at the time of construction, the highest rated autotransformer built in the United States.

Underground, the switchyard is laced with about 12 miles (19.2 km) of various lines, for water drainage, for fire protection, and for oil piping. In addition, there are about 30 miles (48 km) of electrical conduit, which is used for the 248 miles (396.8 km) of various size control and miscellaneous power cables.

NUCLEAR POWER PLANTS

A nuclear power plant is in many aspects similar to a conventional fossil-fuel burning plant. The chief difference is in the way the heat is generated, controlled, and used to produce steam to turn the turbine generator. Fig. 2-14.

In a nuclear power plant, the furnace for burning coal, oil, or gas is replaced by a reactor which contains a core of nuclear fuel. Energy is produced in the reactor by a process called fission. This fission process splits the center, or nucleus, of certain atoms when they are struck by a subatomic particle called a neutron. The resulting fragments, or fission products, then fly apart at great speed, generating heat as they collide with surrounding matter.

The splitting of an atomic nucleus into parts is accompanied by the emission of high-energy electromagnetic radiation and the release of additional neutrons. The released neutrons may in turn strike other fissionable nuclei in the fuel, causing further fissions.

A nuclear reactor is thus a device for starting and controlling a self-sustaining fission reaction. The nuclear core of the reactor generally consists of fuel elements in a chemical form of uranium and thorium, or plutonium, depending on the type of reactor. Heat energy is produced by the fissioning of the nuclear fuel. A coolant is then used to remove this heat energy from the reactor core so that it can be used in producing electricity.

Fuel elements for water-cooled reactors are metal tubes containing small cylindrical pellets of uranium oxide. Also under evaluation for commercial power production is the gas-cooled reactor, in which fuel elements are fabricated basically of a uranium-carbide compound and of graphite, the latter acting as the structural material, as well as a protective enclosure for the fuel material. The protective enclosure, whether of graphite or metal, is called cladding.

To control the rate at which fission occurs, most reactors regulate the "population" of neutrons in the core. This is done mainly by rods which, when inserted into the core, absorb neutrons and retard the fission process. If a plant operator wishes to increase the power level or reaction rate, the regulating rods are withdrawn; to shut down a reactor, the rods are fully inserted.

Neutrons liberated in the fissioning process travel at very high speeds. This is not desirable in some reactor systems, where slow-moving neutrons are more effective in triggering fission than are high-velocity neutrons. To attain the desired slow-moving neutrons, a material called a moderator is used. It checks the speed of the neutrons, but its tendency to absorb them is minimal. Materials used for this purpose may be graphite or ordinary water, the latter also serving as a coolant. Most power reactors presently in operation, or under construction, utilize slow-moving neutrons. They are called thermal reactors.

In addition to water, other materials, such as gas or molten sodium, may act as coolants. A coolant transfers heat from the reactor to produce steam. In some types of reactor plants which use water as a coolant, the water is allowed to boil in the reactor, with the resultant steam used directly in the turbine.

Nuclear Fuel, the Heart of a Reactor

Nuclear fuel is a mixture of what are called fissionable and fertile materials.

- *Fissionable material* is that capable of capturing a neutron and splitting into other particles, causing a great release of kinetic energy. The most abundant natural fissionable material is uranium-235, an isotope of uranium.†

†The nuclei of a particular element always have the same number of protons. However, the number of neutrons may vary. Each variation in the neutron number is known as an isotope, and many elements have naturally occurring isotopes. Some uranium isotopes are: U^{234}, U^{235}, and U^{238}.

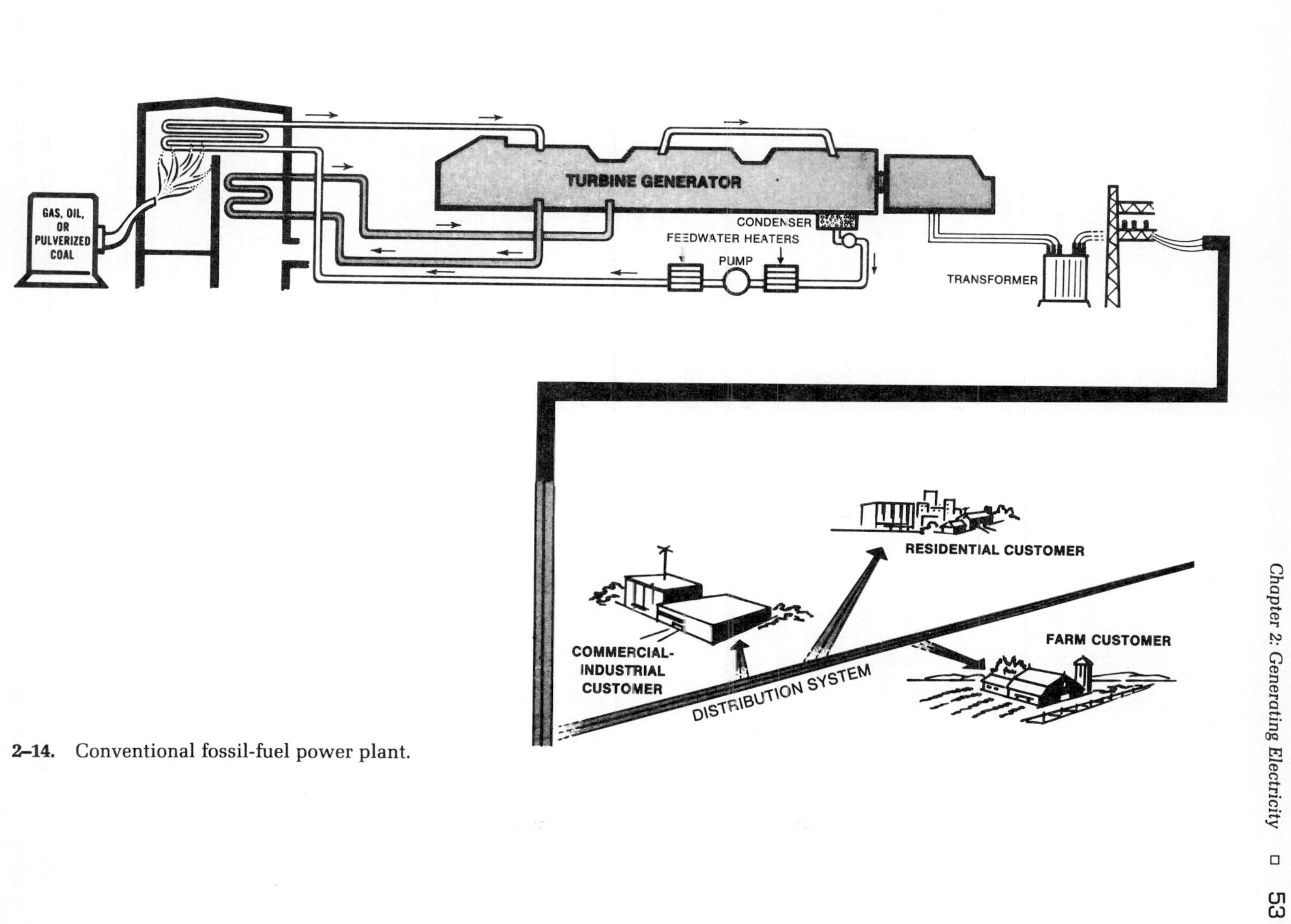

2–14. Conventional fossil-fuel power plant.

■ *Fertile material* is that which is capable of being transformed into fissionable material by capturing a neutron. Another isotope, uranium-238, is a fertile material. It is transformed into plutonium when it captures a neutron. Another naturally occurring fertile material is the element thorium. When it captures a neutron, it is changed into uranium-233.

As has been explained, a reactor is a device for starting and controlling a self-sustaining fission chain reaction. In order to assure a chain reaction, a certain amount of fissionable material must be present. This amount is called the "critical mass." A uranium-fueled reactor must have a critical mass of uranium-235. This can be accomplished in two ways: (1) by building a reactor with a large core of natural uranium† or (2) by changing the "mix" of uranium-235 and uranium-238, through a process called enrichment, to ensure a greater percentage of fissionable isotope uranium-235 in the core of the reactor.

In the second instance, the reactor is said to use enriched uranium fuel, while in the first, the reactor is said to use natural uranium fuel. In the United States, many scientists and engineers have selected enriched fuels as the basis for developing nuclear power. In Canada and Great Britain, the emphasis is placed on natural-uranium-fueled reactors.

No matter what type of nuclear fuel is used, there are certain production steps that must be taken before the fuel can be "burned" in a reactor. The first step is to mine the uranium ore. In this country, ore is found primarily in New Mexico, Wyoming, Colorado, and Utah. Canada and South Africa are also significant producers of uranium ore.

After the ore is mined, the uranium must be extracted. This process, called milling, involves several chemical steps. The end product is an ore concentrate, called "yellow cake" because of its brilliant yellow appearance. This concentrate is then further refined to prevent impurities from absorbing neutrons unproductively, thereby detracting from the efficiency of the system.

If enriched fuel is to be used, the next step in the process is to enrich the refined or purified uranium, increasing the proportion of the fissionable isotope uranium-235. The enriched uranium fuels used in this country generally have about 2 to 3% uranium-235. Natural uranium has only 0.7% of uranium-235. Enrichment involves a complex process called gaseous diffusion which increases the ratio of uranium-235 atoms to uranium-238 atoms.

There are three gaseous-diffusion plants in the United States, all of them owned by the federal government. The enrichment process is the only step in the nuclear-fuel cycle which must be performed in government-owned facilities. Industry, however, is discussing with the U.S. Energy Research and Development Administration (formerly the Atomic Energy Commission) possible conditions and dates when industry can participate more fully in gaseous-diffusion operations.

During the enrichment stage, uranium is in a gaseous state—uranium hexafluoride (UF_6). Before it can be made into fuel elements, it must be converted to the solid state. This requires a chemical process. Generally, most of the fuel that is made for the reactors now operating, or being built today, is in the form of uranium dioxide (UO_2).

Water-Cooled Reactor

In a *water-cooled reactor*, the nuclear fuel, which has been compacted into uniform pellets, is placed into tubes or fuel elements. These fuel elements are then sealed at the top and bottom and arranged by spacer devices into bundles called fuel assemblies. The spacer devices separate the fuel elements to permit coolant to flow around all of the elements, a process necessary to remove heat produced by

†The mixture which occurs in nature is almost entirely uranium-238 (about 99%) so that a large amount of it must be used to get enough uranium-235.

the fissioning uranium atoms. Scores of fuel assemblies, precisely arranged, are required to make up the core of a reactor. Fig. 2-15.

This geometric arrangement is necessary for several reasons. Nuclear fuel, unlike fossil fuel, has a very high energy density—that is, tremendous quantities of heat are produced by a small amount of fuel. Therefore, the fuel must be arranged as stated above to permit the coolant to carry away the heat. To accomplish this, the fuel must be dispersed, rather than lumped together in a large mass.

Chemical reactions between the fuel and the coolant must be avoided; so, as a safety precaution, radioactive materials produced must be enclosed. For these reasons, the fuel is contained in individual tubes, or fuel elements. The cladding material, from which the tubes are made, must meet rigid specifications. It must have good heat-transfer characteristics. It must not react chemically with either the fuel or the coolant. Also, it must not absorb the neutrons produced in the fissioning of the fuel at a rate which could be to the detriment of the chain reaction. The cladding material generally used for this purpose is thin-walled stainless steel. An alloy of the element zirconium can also be used.

Gas-Cooled Reactor

A *gas-cooled reactor*, which utilizes the inert gas helium as the coolant, has a different core structure than the water-cooled reactor. The fuel elements are made of graphite. Graphite acts as the structural material and the neutron moderator. It also serves as the cladding material. Nuclear fuel consists of both uranium and thorium. It is formed into the center of the fuel element. Because the helium is inert, the graphite serves as cladding for the nuclear fuels. (An inert gas will not react with nor corrode the graphite or any other structural material.)

Physically, the fuel elements for a gas-cooled reactor are much larger in size than those of a water-cooled reactor. They are not bundled into individually arranged fuel assemblies and spaced for the circulation of coolant as are those in water-cooled systems. Despite their larger size, however, several hundred fuel elements are required to make up the core of a reactor. Fig. 2-16.

Nuclear fuel, whether in the form of fuel assemblies for a water-cooled reactor or fuel elements for a gas-cooled reactor, is placed in the reactor to produce heat. The heat in turn is converted into electricity. After several years of operation, a nuclear core must be replaced. With the passage of time, the absorption of neutrons by the accumulated fission products is so great that there are too few neutrons remaining to maintain a chain reaction. Fig. 2-17.

After fuel assemblies, or elements, are removed from the reactor, they retain some material. The material must be reclaimed because of its economic value. In fact, only about 1 to 3% of the uranium has been "burned" in nuclear reaction. The remaining 97 to 99% is locked in the hundreds of fuel elements of the "spent" core.

The procedure by which the nuclear fuel is reclaimed is called *reprocessing*. The first industry-owned reprocessing plant began operation in 1966.

Reprocessing is a complex chemical operation. It must be performed by remote control due to the radioactive nature of the fissionable products. The process includes removing the cladding (metal or graphite) of each fuel element and separating the fission products. Uranium and other valuable materials, including fissionable plutonium, or fissionable U^{233}, are returned for enriching and refabricating. Waste fission products are stored in underground facilities. These facilities must meet strict specifications of the U. S. Energy Research and Development Administration.

Advanced Reactors

Much research remains to be done before the full benefits of nuclear power are obtained. One of the most promising types of nuclear re-

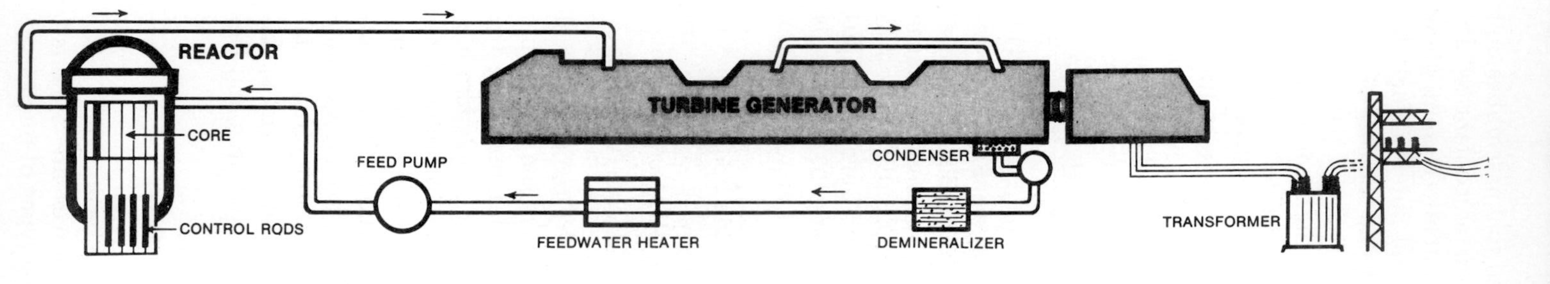

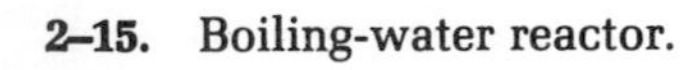
2–15. Boiling-water reactor.

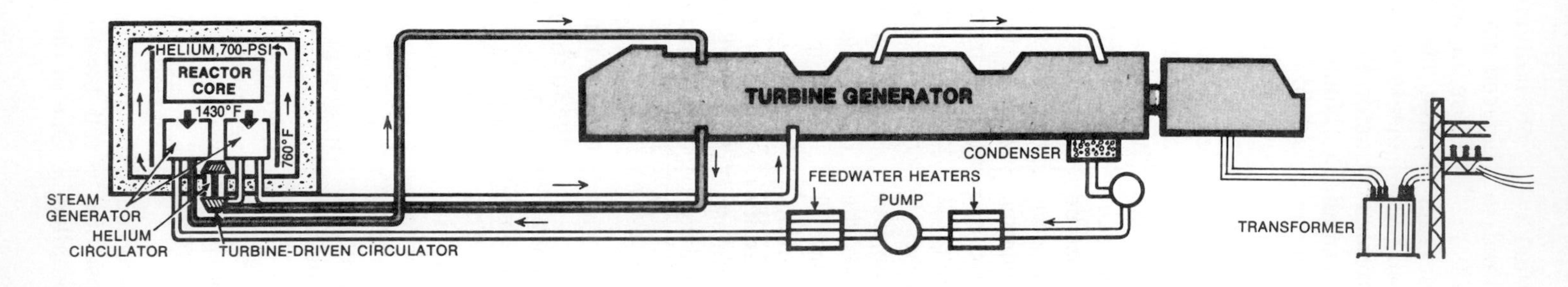

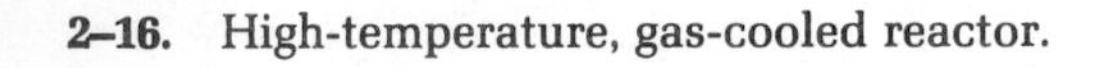
2–16. High-temperature, gas-cooled reactor.

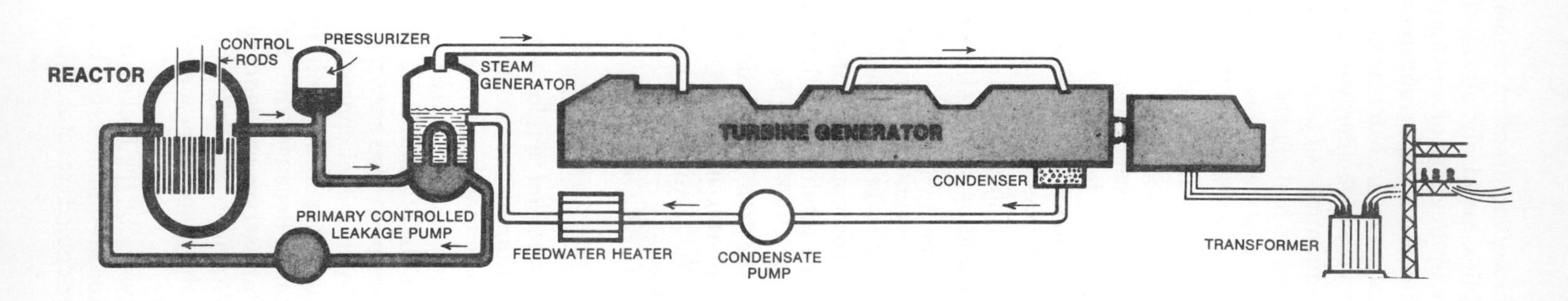

2–17. Pressurized-water reactor.

actors is the *breeder reactor*—one which produces more fissionable material than it consumes. This particular process may increase the nation's energy resources by producing nuclear fuel. It also shows promise of producing less expensive electric power. Electric power industries are participating in the development of such a reactor. Fig. 2-18.

All reactors that are fueled with uranium produce plutonium. *Plutonium* is a man-made fissionable material. It is produced when a fertile U^{238} isotope captures a neutron. Some of this plutonium is burned, or fissioned, in the reactor in which it is produced. This happens in the same fashion as the fissioning of the U^{235} atom described earlier. Yet most of the plutonium is not utilized by the time the nuclear core must be removed from the reactor for reprocessing. Most experts agree that plutonium will best be utilized in a fast-breeder reactor. It may be several years, however, before economical fast-breeder reactors will be developed and built in enough quantity to require large amounts of plutonium fuel. Meanwhile, large amounts of plutonium will have accumulated. Rather than stockpile it for later use in breeder reactors, industry and government are engaged in research to determine ways to reuse plutonium in the thermal reactors which now produce it. Fig. 2-19.

Water-cooled reactors presently operating or planned for the near future produce plutonium. Production is minimal compared to the amount of uranium fuel they consume. These reactors are said to have low conversion ratios. There are other types of reactors that have higher conversion ratios. In fact, they approach "breeding"; that is, producing more fissionable material than they consume. Some experts believe that in the period before breeder reactors are fully developed, the so-called "advanced-converter" reactors should be developed to help conserve nuclear resources.

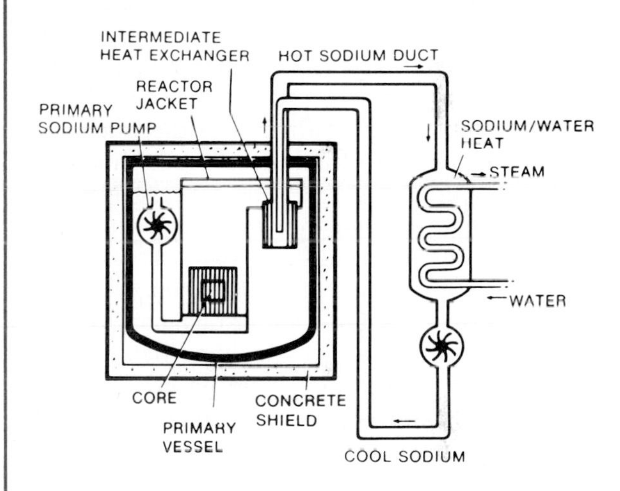

2–19. Liquid-metal, fast-breeder reactor.

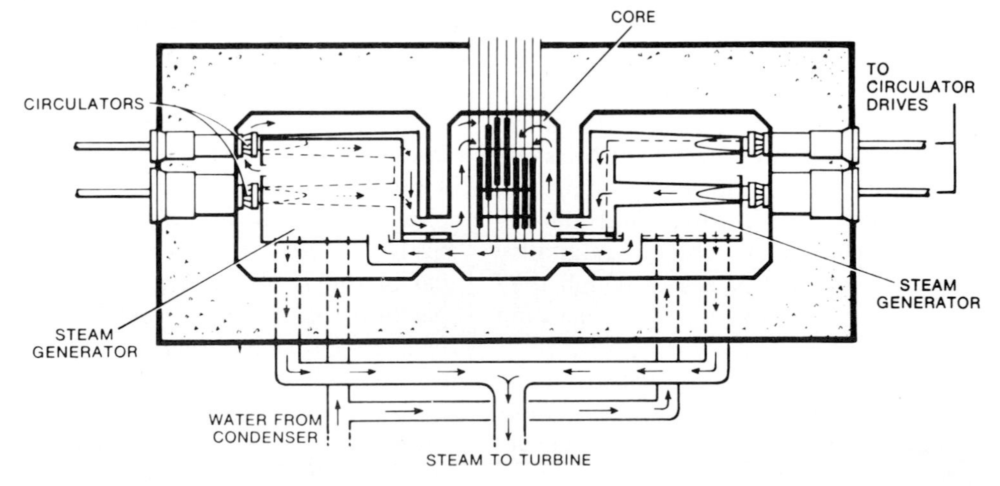

2–18. Gas-cooled, fast-breeder reactor.

A prototype advanced converter of the high-temperature, gas-cooled type is in operation on the system of one electric utility company. Another still larger prototype is being built by a utility in cooperation with the U. S. Energy Research and Development Administration and an equipment manufacturer. This type of reactor will, in addition to uranium, utilize a fertile material, thorium, which is convertible to fissionable uranium-233. Such experimentation may offer another nuclear-fuel resource.

Fusion

Many companies are investing in research on the fusion process. In this process, light elements are fused together, with a resulting release of heat energy. *Fusion* is the opposite of fission, which is the splitting of the nuclei of heavy elements. Problems in producing and controlling the fusion process are staggering, a result of the fact that the reaction takes place at temperatures of millions of degrees. Such high temperatures make containment of materials impossible. Here, electromagnetic fields are used. But, although the production problems are enormous, potential benefits stagger the imagination. A source of fuel for the fusion process could be the vast oceans of water, which cover so much of the earth.

Safety

Reactors are designed to be reliable and safe in operation. Their reliability in the production of electricity is an economic requirement to the electric utility company. Their safe operation is a prime consideration of electric utilities, reactor manufacturers, and the government.

Utility companies have excellent performance records. They have achieved reliability and safety during the 20 years in which they have been planning, building, and operating power reactors.

Reactor safety starts with the design of the reactor core and control system. It continues with the engineering of the reactor installation. A vital principle for safe plant operations is to anticipate possible human errors and electromechanical failures and to make provisions for them in the design. For example, safety control rods that can shut down a reactor are designed on such a fail-safe basis. The rods are automatically inserted into a reactor if operation becomes abnormal, or even if important measuring instruments fail. Once in the reactor, these "nuclear brakes" halt the fission process quickly by absorbing neutrons.

Another step in assuring the safety of the public is to surround power reactors with gas-tight enclosures. These shells, which give a domed appearance to many nuclear power plants, are designed to withstand the maximum amount of pressure that might be generated in the event that a rupture should occur in the reactor cooling system. Moreover, power reactors of the water type have a self-regulating characteristic assuring that, as the fission process increases and temperatures rise, an inherent response of the reactor is to slow down the reaction.

The operation of nuclear power plants is supervised and performed by people who have received thorough general training in all aspects of nuclear reactors. There is a particular emphasis placed on the specific reactor they are to supervise or operate. They are required to pass written, oral, and operating tests. The tests are given by the U. S. Energy Research and Development Administration.

Despite extensive educational efforts by utility companies, reactor manufacturers, and government agencies, some people still believe that a nuclear reactor is something resembling a bomb in a box. Actually, the design principles of a reactor and an atomic bomb are entirely different. Concentrating and fabricating fissionable material in such a way as to make it perform like a bomb is extremely complex. In fact, after first learning to produce a chain reaction of the type used in reactors, several additional years of intense research were required by the nation's scientists and engineers to develop a nuclear explosive device. An atomic bomb encloses two pieces of almost pure fissionable material that are brought quickly together. They are held long enough to generate a very large explosive force. The reactor holds fissionable material which has been diluted with structural and control material, coolant, moderator, and non-fissionable uranium or thorium. Moreover, construction materials enclose the fissionable material in a uniform, widely dispersed arrangement.

Federal law prohibits building or operating a nuclear power plant without a construction permit and an operating license from the U. S. Energy Research and Development Administration. The licensing procedure involves a thorough analysis of the safety of the proposed plant. The analysis is not only by the Administration's own qualified regulatory staff, but also by outside experts. The procedure provides opportunity for state and local authorities and the public to keep fully informed on the progress of license applications. Each can participate in hearings held before action is taken to grant or deny the licenses.

Future of Atomic Energy

Yesterday, nuclear power was an exotic subject of the scientific world. Today it is a competitive element in the electric-power production market. The stage has been reached where engineers speak of conventional nuclear power plants to distinguish them from the more advanced reactor concepts that are the object of research and development programs. It is possible that nuclear power will play an important role in supplying future electricity.

FOSSIL-FUEL POWER GENERATORS

The steam necessary for driving turbines (which in turn drive electric generators) must be produced by heat. The method of heat production often becomes a rather difficult engineering problem. With the development of some dependable sources of heat from fossil fuels, former design problems have been simplified. Almost any substance may be used as a fuel. If it can be pulverized and fed into furnaces with extremely high temperatures, it will burn.

Coal, Natural Gas, and Oil

Coal has been one of the most abundant sources of energy. It is used frequently to produce electricity by providing steam in those power plants designed for the use of coal. It is not, however, the only fossil fuel used to produce steam for power plants. Natural gas and oil are also used. There are some problems associated with these fuels. The waste products can pollute the atmosphere. Combustion products discharged into the atmosphere from a steam generator must meet EPA (Environmental Protection Agency) emission standards for particulate, sulfur oxides, and nitrogen oxides. Clean fuels for a steam generator may be defined as those fuels whose combustion products do not require stack-gas cleanup in order to meet the limits set by antipollution agencies.

In Table 2-A coal, gas, and oil are shown with their federal performance standards noted; SO_2 is sulfur dioxide, NO_x is nitrogen

2-A. Federal Performance Standards for Fossil-Fuel Fired Boilers

	SO_2		NO_x		Particulate
	lb/10^6 Btu	ppm	lb/10^6 Btu	ppm	lb/10^6 Btu
Coal	1.20	520	0.70	525	0.10
Oil	0.80	550	0.30	227	0.10
Gas	—	—	0.20	165	—

oxide, and ppm is parts-per-million. Btu is, of course, British thermal units.

Pollution Problems. Table 2-B shows some of the products that require monitoring at the stack for excessive atmospheric pollution by use of a particular fuel. In this chart, low Btu and medium Btu gas, coke oven gas, and natural gas are analyzed. Fuels of the type shown in the chart may be used in steam generators without cleanup. They contain little or no sulfur and essentially no ash. Therefore they can comply with sulfur oxide and particulate-emission standards, as shown in Table 2-A. To prevent emissions of NO_x, it has been demonstrated that tangentially fired steam generators can comply with the limits in Table 2-A when burning fuel oil, natural gas, or coal.

Steam Generator Design

Operating conditions for the cycle selected are shown in Table 2-C for a nominal 550-megawatt unit (MW). A natural gas unit for this capacity is shown in Fig. 2-20. The furnace width is 51 feet, 8 inches; furnace depth is 41 feet; and volume is 163 000 cubic feet. At the full load rating, this unit has a furnace combustion rate of 31 000 Btu/hr/cu ft and a furnace heat release rate of 280 000 Btu/hr/sq ft. This is a Controlled Circulation®, radiant reheat unit which is tangentially fired. It has a two-section super heater, single-section reheater, spiral fin economizer and one Ljungstrom® air heater. Reheat steam temperature control is accomplished through the use of tilting tangential fuel nozzles. The primary steam temperature is controlled by spray desuperheating. Ignition and warm-up fuels are natural gas.

2-B. Fuel Analysis

	VOLUMETRIC PERCENT				WEIGHT PERCENT
	LOW Btu	MEDIUM Btu	COKE OVEN GAS	NATURAL GAS	OIL
CO_2	10.33	6.01	0.75	0.2	—
CO	21.50	55.90	6.00	—	—
H_2	16.40	37.39	53.00	—	11.0
N_2	51.50	0.70	12.10	0.6	0.2
CH_4	0.27	—	28.15	99.2	—
O_2	—	—	—	—	0.5
C	—	—	—	—	87.8
S	—	—	—	—	0.5
HHV, Btu/cu ft	127	301	475	1006	—
HHV, Btu/lb	1857	5916	16 431	23 518	18 500
STOICHIOMETRIC AIR, lb/10^6 Btu	576	568	658	722	738
ATMOSPHERIC STOICHIOMETRIC AIR, lb/10^6 Btu	584	575	667	731	747
FUEL IN PRODUCTS, lb/10^6 Btu	538	169	61	42.5	54
EXCESS AIR USED, %	20	20	10	8	5
TOTAL AIR COMBUSTION, lb/10^6 Btu	701	690	734	790	787
TOTAL COMBUSTION PRODUCTS, lb/10^6 Btu	1239	859	795	832.5	841
SCF @ 60° F					

2-C. Steam Generator Design Conditions

	CONTROL LOAD	CONTINUOUS RATING
EVAPORATION, lb/hr	2 200 000	3 800 000
REHEAT FLOW, lb/hr	1 970 000	2 380 000
SUPERHEATER STEAM TEMP., °F	1005	1005
REHEAT STEAM TEMP., °F	1005	1005
FEEDWATER TEMP., °F	430	486
SUPERHEATER OUTLET PRESSURE, psig	2550	2600
DESIGN PRESSURE, psig		2950

This unit is arranged for pressurized firing. At full load, the uncorrected exit-gas temperature from the Ljungstrom® air heater is 245°F, (118°C) when operating at 8% excess air, with a thermal efficiency of 85.63%.

Application of Low Btu Gas to Existing Units

Some operators of steam generators are interested in what can be done to use LBG (low Btu gas) in existing plants. This is particularly true of those owners of natural gas units. The supply of natural gas has become unreliable, and these steam generators cannot operate on fuel oil. In such cases, there is the possibility that a coal or oil LBG gasification plant could be constructed adjacent to the existing steam generator.

Converting an existing fossil-fuel fired unit to LBG would result in a shift in the heat absorption pattern from that in the original design. The pressure level of the cycle (1800, 2400, or 3500 psig) and the changes from original design fuel would have considerable effect on performance. It could possibly require heating-surface revisions to the unit.

Conversion of Fuel Sources. Conversion to LBG firing would be in the following order of increasing difficulty for steam generators designed to fire:

- Oil, with gas recirculation.
- Coal.
- Natural gas.

As in the design of new units, the difficulty in converting existing units to firing LBG results from a large increase in combustion gas weight and the change in furnace outlet temperature.

Combustion Engineering, Inc. has completed the first phase of a four-phase program to develop a coal-gasification system which will produce a clean, low Btu gas from coal for use as an electric utility fuel. The work was done under an Office of Coal Research contract.

Phase 1 of the program consisted of system and component design studies to provide a basis for selection of a gasification system to meet the clean-fuel needs of United States utilities. The Phase 1 study determined that an atmospheric-pressure, entrainment-type gasification system would be best suited for such requirements. The study proved it to have the greatest potential for scale-up and handling the load changes required in power plant operation. In addition, the system can use any type of coal, producing a gaseous fuel with a heating value ranging from 127 to 285 Btu/scf.

Phases 2 and 3 will include the design and construction of a 5-ton per hour coal-gasification process development unit. The 5-ton per hour size was selected because it is suitable for direct scale-up to plants in the 200- to 500-MW range.

Phase 3 will also involve design of a pilot unit to serve a 200-MW electric utility generating plant and selection of a site for this installation. Phase 4 will include erection of the

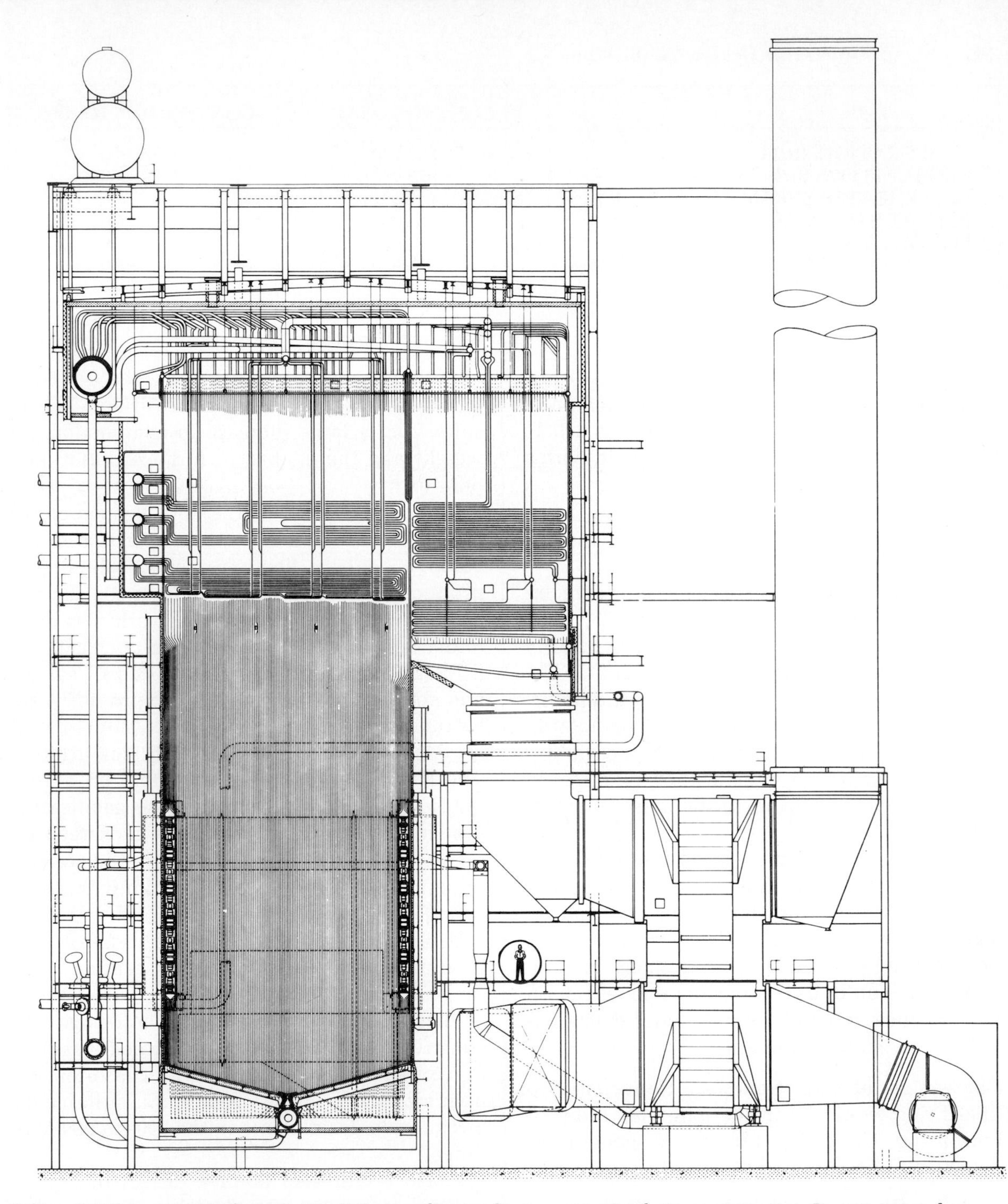

2–20. A 550 megawatt, forced-circulation, radiant-reheat, pressurized steam generator. It uses natural gas for fuel, and a tangential firing system.

200-MW pilot plant, with operation and testing to demonstrate system capability.

Generally, the efficiencies of the gasifiers are high enough to transfer 75 to 80% of the heating value of the coal to the gas. Coupled with the 35 to 40% efficiencies realizable from electrical plants, the overall efficiencies of coal to electricity become between 26 and 32%. Improvements in both processes would certainly be of interest to the power industry.

Synthetic Gas. Steam generation from synthetic low or medium Btu gas is a development probably several years in the future. This is due to the lack of availability of a commercial gasifier system at this time. As indicated, the design of LBG-steam generators presents problems, none of which is insurmountable. Conversions of existing units to gas must be handled on an individual basis, to ensure operation within original design limitations.

In Figs. 2-20 through 2-24 (pages 64-67) you will see different methods of producing steam for driving generators to produce a very large amount of electrical power.

Coal-Fired Power Plants. Fig. 2-25 (page 68) shows a Niagara Mohawk Power Corporation plant located on Lake Erie in Dunkirk, New York. It is near an adequate supply of water for the production of steam.

Fig. 2-26 (page 68) shows another Niagara Mohawk power plant in the town of Tonawanda, New York, located on the Niagara River. A huge pile of coal is needed to fire the furnaces. The river here also provides an ample supply of water to the plant.

Fig. 2-27 (page 69) shows the inside of the power plant where the turbines, fed by steam, rotate electrical generators to produce the power needed.

Fig. 2-28 (page 69) shows the power control center in a large nuclear power plant. Operation is by remote control, with monitoring charts checked constantly for performance and demand.

Of great concern for power plants today is the use of inexpensive fuels that will pollute the atmosphere as little as possible. New designs and more research help assure a solution advantageous to everyone who uses electricity.

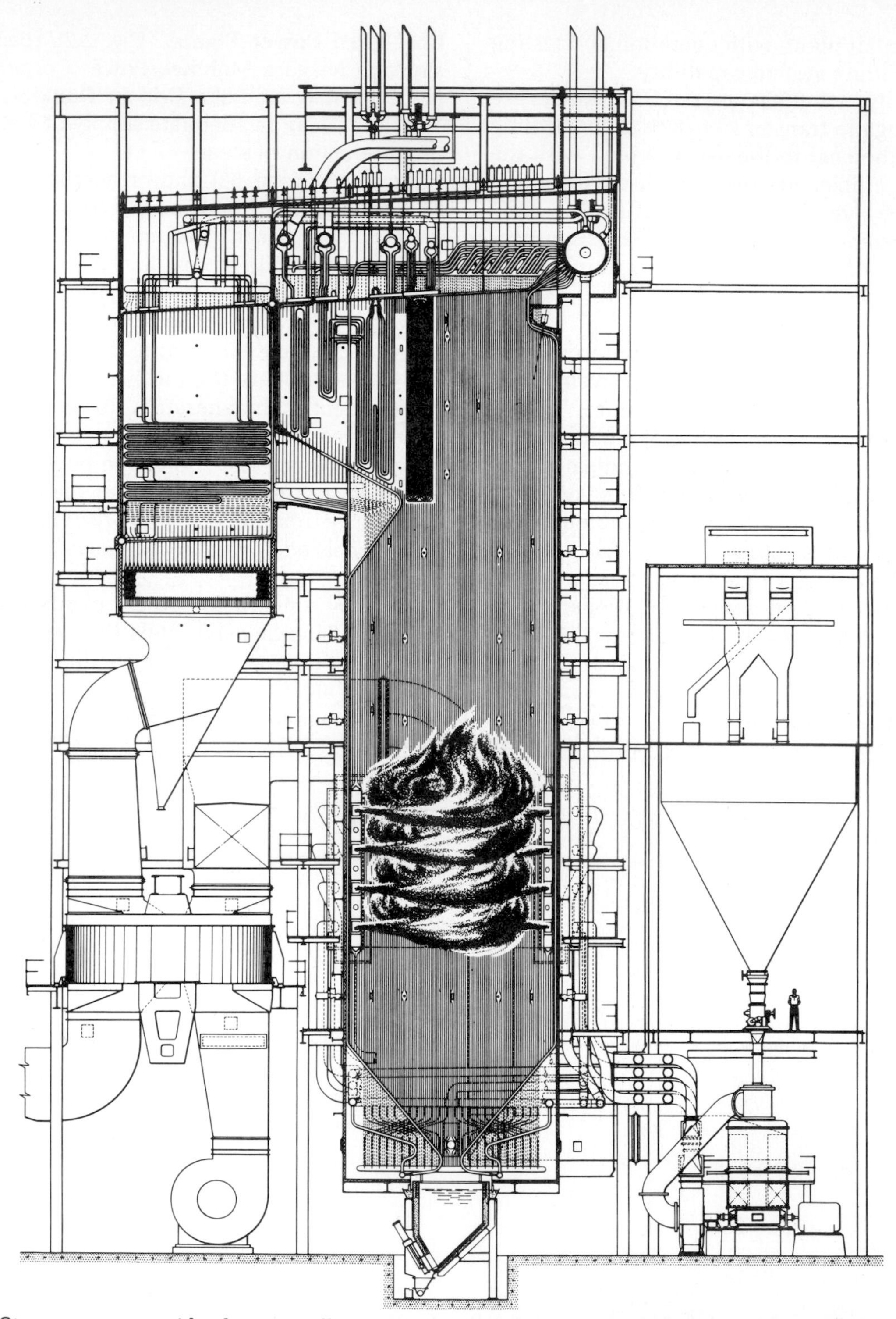

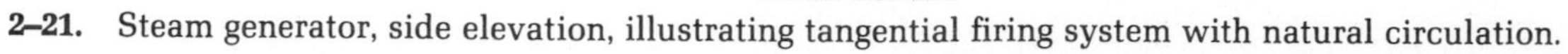

2–21. Steam generator, side elevation, illustrating tangential firing system with natural circulation.

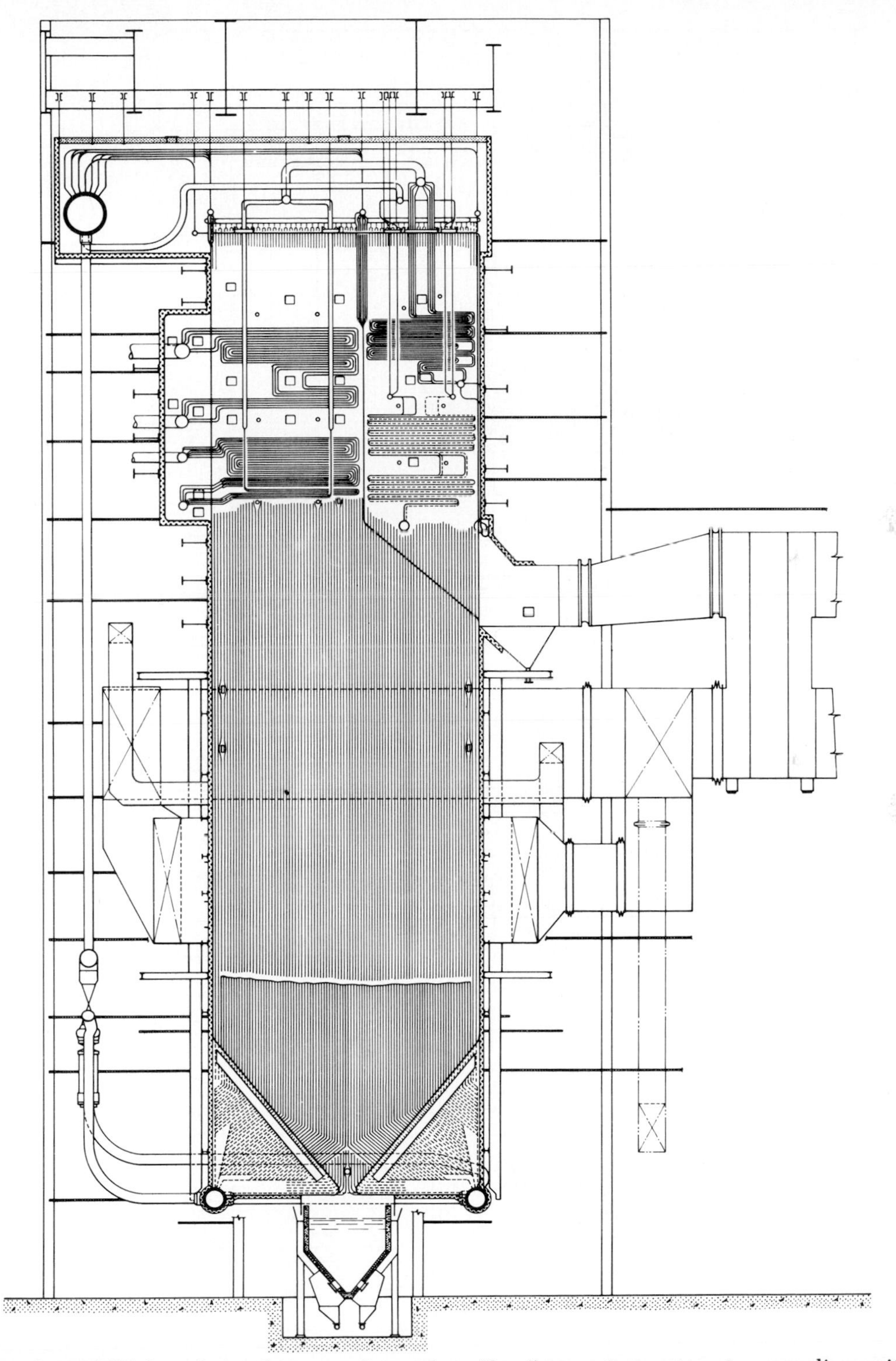

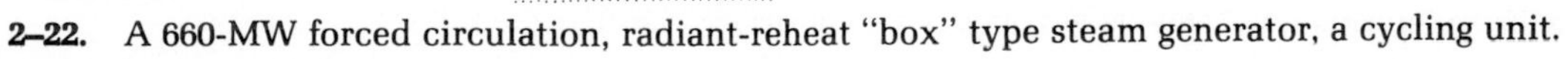

2–22. A 660-MW forced circulation, radiant-reheat "box" type steam generator, a cycling unit.

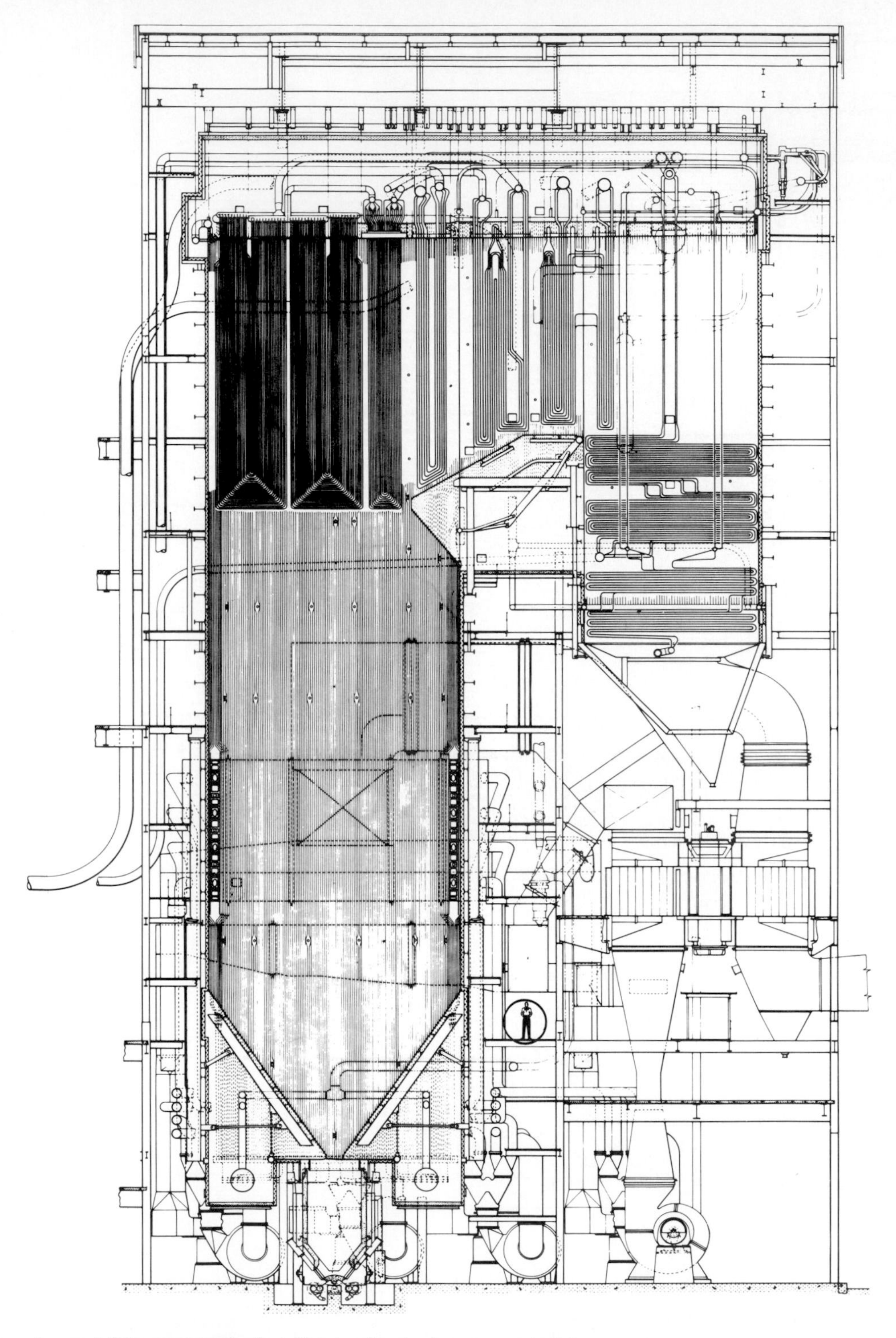

2–23. A 350-MW supercritical radiant-reheat, steam generator.

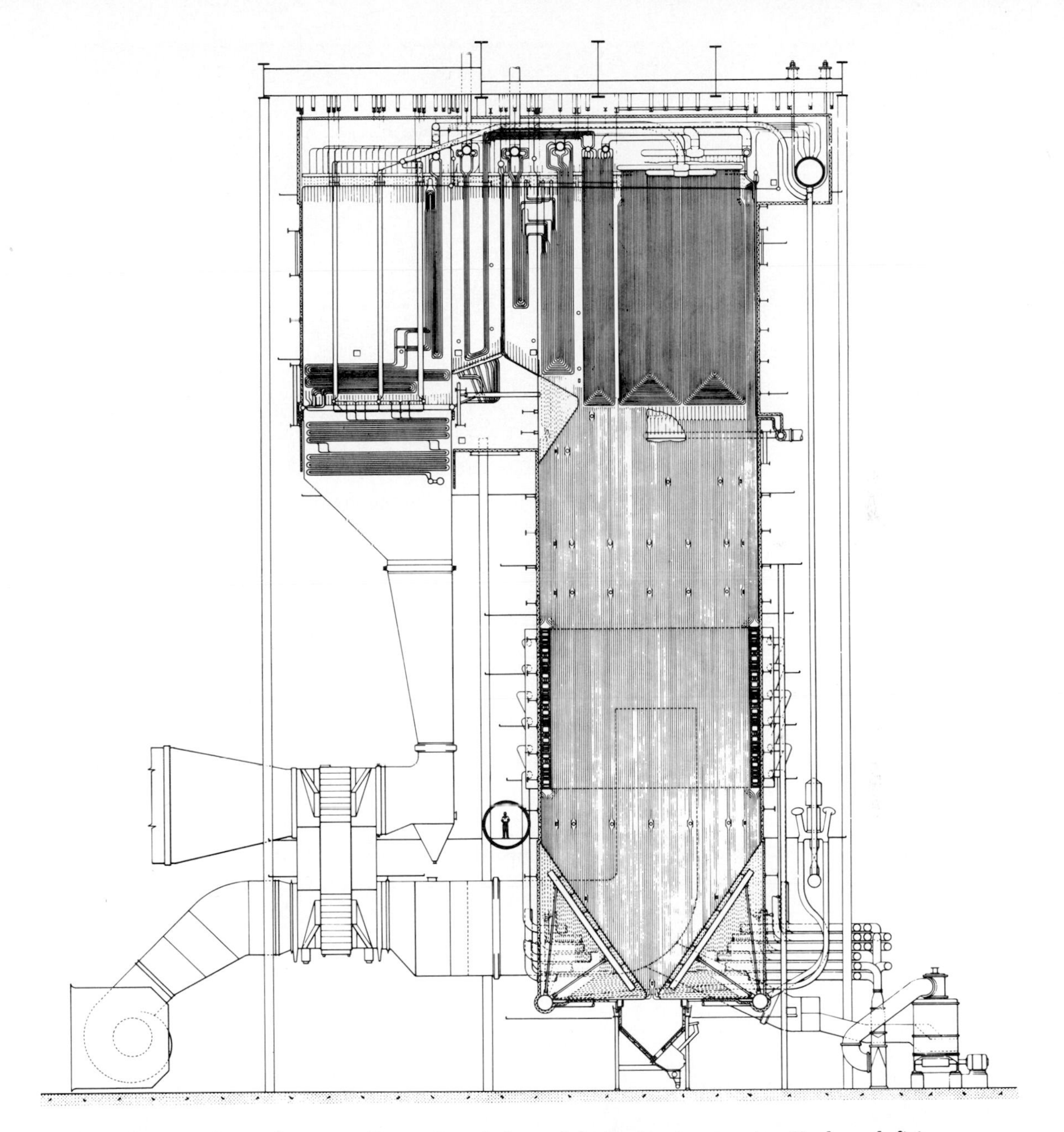

2–24. A 600-MW forced-circulation, radiant-reheat balanced draft, steam generator. Fuel: coal; firing system: tangential.

2–25. Niagara Mohawk's Dunkirk, N.Y., coal-fired power generating plant.

2–26. Niagara Mohawk's coal-fired power plant located on the Niagara River supplies power for Buffalo, N.Y.

2–27. Generator room, where steam turbines drive alternators to produce electrical energy.

2–28. The power control center in a large nuclear power plant.

WIND, WATER, AND SOLAR POWER SYSTEMS

Many people generate their own electricity. The U.S. government has made grants available for those who want to experiment with alternate energy sources. However, most are finding that having their own generator is not less expensive than using the power company's electricity. Producing electricity by using wind machines, microhydro-systems, and solar cells can require a large commitment of time and money.

Windmills

Windmills were used by many rural families for electricity before the establishment of the REA (Rural Electrification Administration) in 1935. The windmill produced enough power to charge a number of batteries. These batteries were used to light the house and power some electrical appliances in the house and barn. With the advent of the REA, inexpensive electricity became available to rural subscribers. The windmill stayed, but in most cases it was used to pump water for cattle.

To produce electricity, a windmill must be properly located. In some areas the wind is very strong and blows for long periods each day. In other locations it is not strong enough to turn a system. The wind must be evaluated as a resource at each location being considered. Collection of data for five years is recommended before the installation of a system. Keep in mind that the power of the wind is directly proportional to the wind speed.

A system that has proven useful and practical is shown in Fig. 2-29. This system has been designed to use batteries to store energy that is not immediately needed. It connects to the public utility so power can be fed to the utility's lines when it is not needed by the owner. The unit shown here was built for $21,019, which does not include labor.

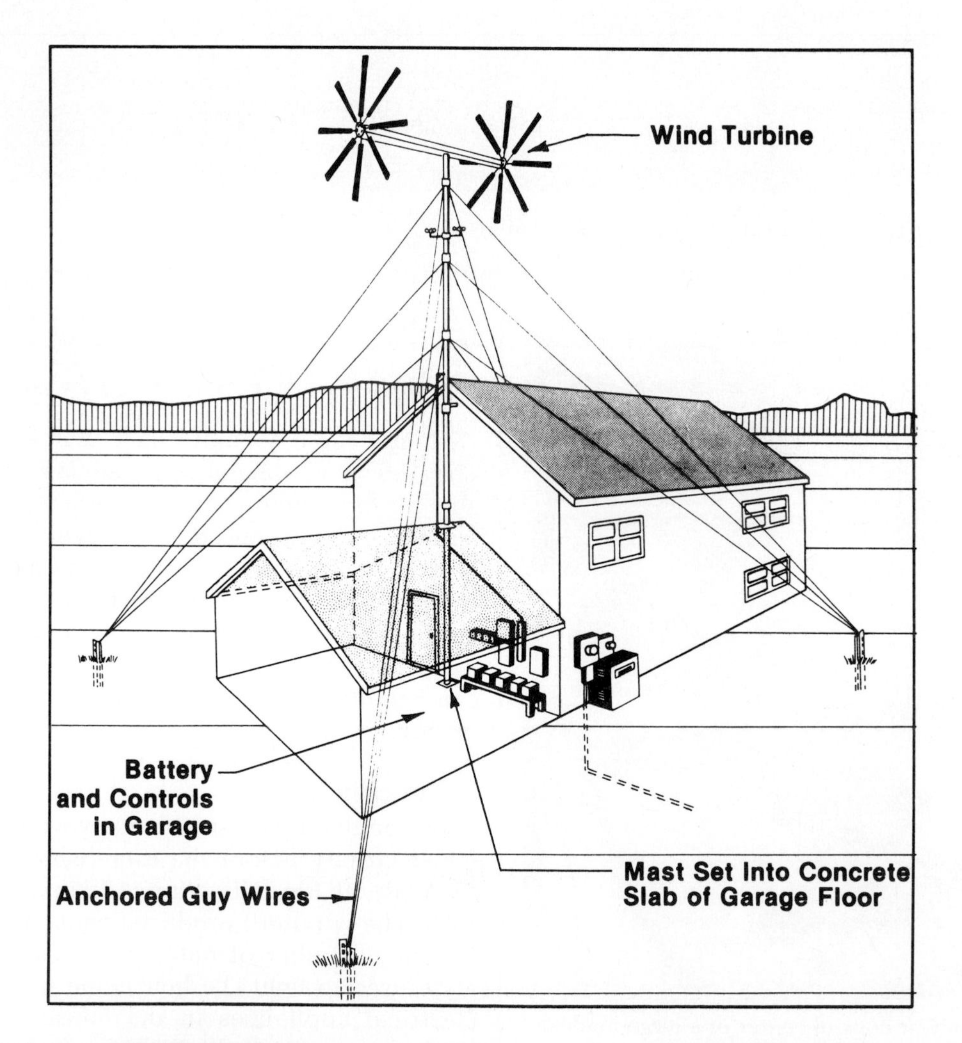

2–29. Wind-power electrical system.

MICROHYDROS

More than half of the electricity generated in the United States at the turn of the century was generated by flowing and falling water.

Microhydro systems produce up to 100 kilowatts of electricity. They can be designed for a variety of water-flow, from a creek to a small river.

The system shown in Fig. 2-30 consists of a 4.75 inch Pelton wheel turbine, directly mounted to a 1500 watt induction generator. Both are housed in an 8- by 12-foot wooden shed. Water is fed into the system through a 4-inch diameter PVC pipe, or penstock, which is buried 2 to 3 feet underground. The water flows a distance of 1700 feet from a first intake. A second intake was added 1700 feet from the first intake to test the difference in power production. Both intakes are covered with wire mesh to keep debris out of the system. When the water reaches the turbine, it flows into two 2-inch nozzle feed pipes from where it is directed to the $\frac{1}{2}$-inch nozzles. Depending on the brook's rate of flow, the water can be directed to one nozzle only or it can be split and directed to two nozzles. This nearly doubles the system's production. It took about 2 years to construct this system, which cost $13,905, plus labor.

SOLAR SYSTEMS

Solar cells, or *photovoltaics*, were first used by the space program to generate electricity for spacecraft communications and for satellites. There are a number of advantages to a system consisting of solar cells. They are clean, reliable, and nonpolluting. They are almost maintenance-free. They have no moving parts.

A solar system must conform to state and local building and electrical codes. For example, if the system is mounted on a roof, it will have to pass the local building codes. Planning and zoning boards may also need to approve the system.

Make sure that nearby buildings do not cast a shadow on the unit, which would greatly reduce power output. Also look at the present location of trees, especially those that may grow to block the sun.

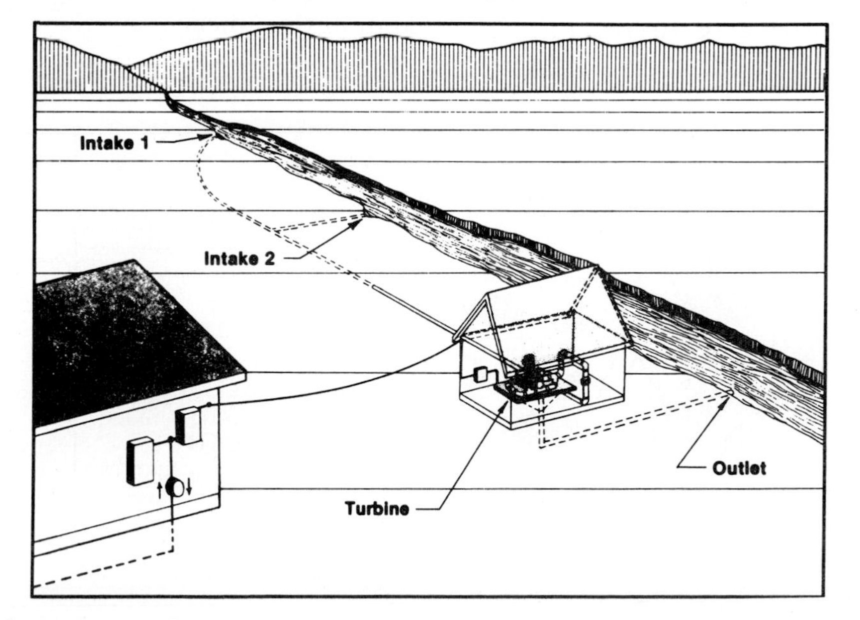

2–30. Microhydro system used to produce electricity.

Keep in mind that photovoltaics produce only DC, which has to be converted to AC to run most appliances. Maintenance involves keeping the panels clean, plus checking the batteries and connections at least once a year.

The system shown in Fig. 2-31 consists of 24 modules, four batteries connected in parallel, and a $\frac{1}{2}$-horsepower DC motor that drives the water pump. Water is pumped into a 3300 gallon storage tank that feeds a water trough by gravity flow. In addition, there is a charge controller that disconnects the photovoltaic panels from the batteries when they are fully charged, and a blocking diode to prevent discharge of the four batteries at night. It is estimated that the system can easily pump over 1000 gallons of water a day.

Many of these units are used by farmers and ranchers to pump water for cattle. Solar-powered units can operate for long periods without attention. For example, one system was used on I-70 in Utah to power a flashing sign that warned truckers of a dangerous section of road. To have extended power lines to the site would have cost $125,000. The photovoltaic system cost approximately $12,000. The system shown in Fig. 2-31 cost $14,934. That does not include labor or installation costs.

GASOLINE-ENGINE DRIVEN GENERATORS

Small gasoline engines are used to drive AC generators for emergency power and for use in the field. They provide dependable service with very little maintenance. Most alternators have a self-excited revolving field (rotor). The rotor connects directly to the engine crankshaft with a tapered fit. A steel stud passes through the hollow center of the shaft and secures the rotor to the crankshaft. The stationary armature (stator) contains two 125-volt AC windings to supply AC power to the load. The stator contains a separate exciting winding. AC voltage of the winding is rectified to DC by a solid state rectifier bridge. DC passes to the rotor through slip rings and low current brushes thereby enabling the alternator to produce its own excitation current.

Fig. 2-32 shows how the alternator is grounded. This should be done before attempting to start the engine. One of the first things to do before attempting to use the alternator is to check the power requirements of the appliance or tool to which you are going to supply power. Most of these items have the wattage require-

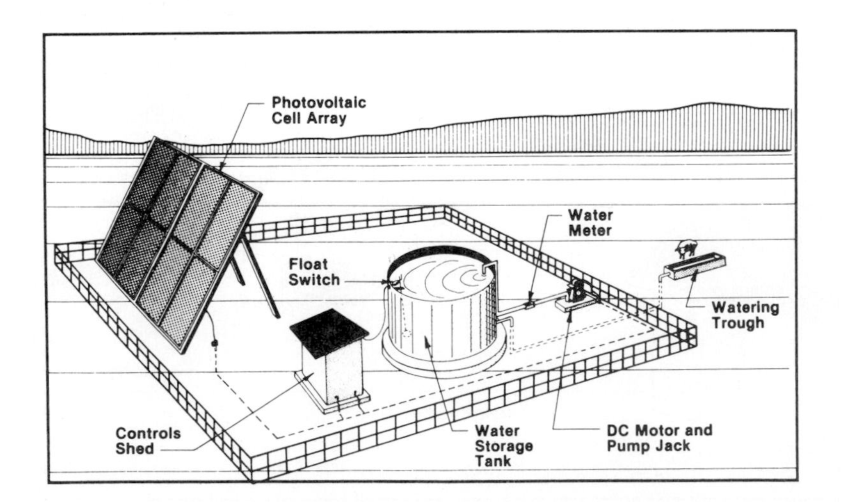

2–31. Photovoltaic power system used to produce electricity for powering a water pump.

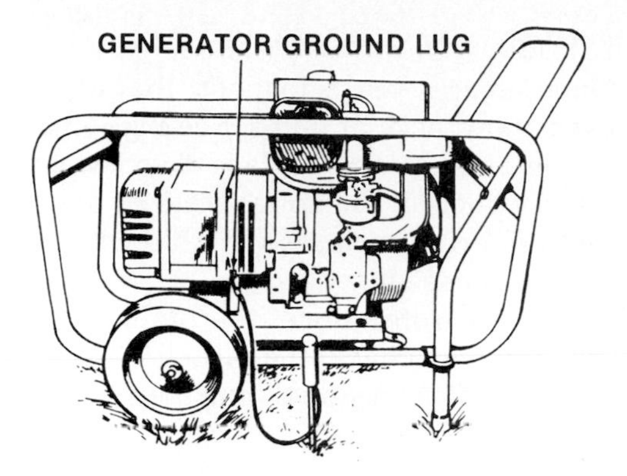

2–32. The portable gasoline engine drive alternator properly grounded with a rod driven into the earth alongside the unit.

ments imprinted on their nameplate. Use Table 2-D as a guide. If the wattage requirements are not listed on the equipment, check to see if the amperage and voltage are shown. If shown, multiply the number of amps times the number of volts to obtain the wattage requirements.

■

Do not exceed the wattage output of the alternator. Overloading may cause the fuses to blow or the engine to slow down. This will

2-D. Power Requirements for Appliances

APPLIANCE OR TOOL	APPROXIMATE RUNNING WATTAGE
Air Conditioner	800-4000
Attic Fan	375
Battory Chargor	Up to 800
Broiler	1325
Clothes Dryer	4500
Clothes Washer	250-1000
Coffee Perculator	550-700
Dishwasher (conventional)	300
Dishwasher (heating element)	1150
Electric Blanket	50-200
Electric Broom	200-500
Electric Drill	250-750
Electric Frying Pan	1000-1350
Electric Iron	500-2000
Electric Saw	400-1500
Electric Stove (per element)	350-1000
Electric Water Heater	1000-1500
Electric Water Pump	500-600
Freezer	300-1000
Furnace Fan	225
Garbage Disposal Unit	325
Hair Dryer	350-500
Space Heater	1000-1500
Microwave Oven	700-1500
Oil Burner	250
Radio	50-200
Refrigerator	600-1000
Sump Pump	250-500
Television	200-600
Vacuum Cleaner	500-1500
Well Water Pump	250-1000

produce a lower voltage than required for normal operation of some devices.

■

Starting

Adjust the carburetor choke as necessary for temperature conditions. The choke controls the amount of air-to-fuel mixture into the carburetor. Cold starting requires a full choke. Be sure the stop switch is not against the end of the spark plug. Pull the starting rope with a fast, steady pull to crank the engine. Do not jerk the rope or let it snap back into the rewind mechanism. As the unit warms up, gradually open the choke. Some models, of course, will be equipped with a battery and starter for electrical starting.

Applying the Load

Do not apply the load until the unit has reached normal operating temperature. Continuous overloading may shorten the life of the unit. Connect the load by inserting the load wire plugs into the proper output receptacle. See Fig. 2-33. Receptacles are located in the alternator end bell or in the optional receptacle box.

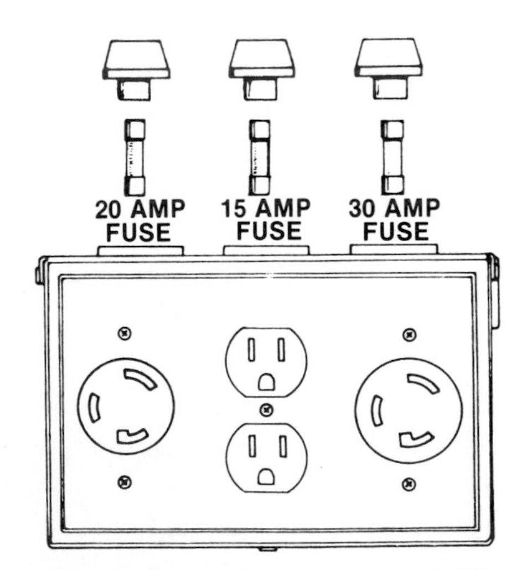

2–33. Receptacles on the generator unit. Note the 125-volt receptacles in the middle and the 240-volt receptacles on the sides. Also check the location of fuses in those units that contain fuses.

Alternator Maintenance

Once every year, inspect the alternator brushes for cracks or chips. See Fig. 2-34. Measure the brushes and replace them when they wear to $\frac{5}{16}$-inch. If the collector rings are rough, smooth the ring surfaces with No. 240 sandpaper. Do not use emery paper.

If the generator does not build up to the rated voltage, check the brushes, diodes or bridge rectifier, residual magnetism, and capacitor.

1. *Brushes.* Check for cracks or excessive wear.
2. *Diodes or bridge rectifier.* See Fig. 2-35 for location inside the end bell housing.

   The best way to check the diodes is with an ohmmeter. Using an ohmmeter, connect one lead to each end of the diode and observe the resistance reading. Reverse the ohmmeter leads and again observe the resistance readings. A good diode should have a high reading in one direction and a low reading in the other direction when the probes of the ohmmeter are reversed. If both readings are high or low, the diode is defective. It should be replaced with a new one that has an identical number.

   The rectifier bridge is made up of four diodes. However, in most instances, only three wires come from the unit. Fig. 2-36. Remove the two small yellow and one black lead from the bridge rectifier. Connect one lead of an accurate ohmmeter to

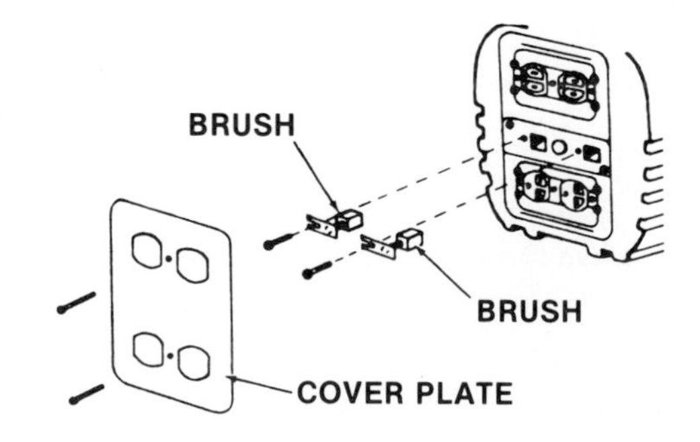

2–34. Removal of brushes for checking and/or replacement.

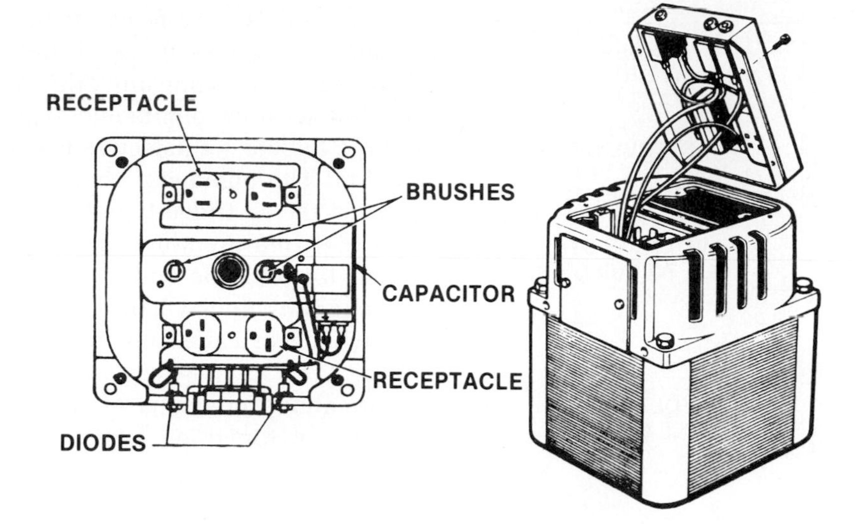

2–35. Location of brushes, capacitor, diodes, and receptacle on an end bell (inside view).

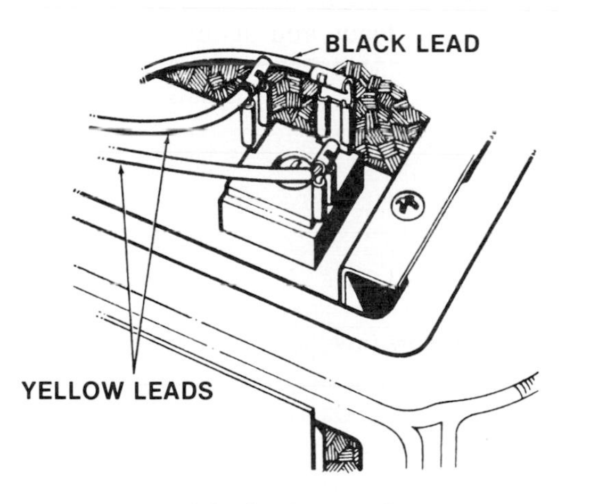

2–36. Location of the bridge rectifier.

one of the AC terminals and the ohmmeter lead to ground. Observe the reading. Next reverse the ohmmmeter leads and again observe the meter. (Check both AC terminals to positive [+] in this manner.) A good rectifier will have a higher reading in one direction than the other. If both readings are high or low, the rectifier unit is defective. Replace with one that is identical. This unit is used to produce the DC from the generated AC so the rotor can be excited to produce more power.

3. *Loss of residual magnetism.* This can be remedied with a 6-volt battery and a couple of clip leads. Fig. 2-37. The process is called *flashing the field.* Connect a 6-volt battery as shown in Fig. 2-37. Start the unit with *no load* connected to the generator. Momentarily touch the positive brush with the + lead of the lantern battery while grounding the − lead to the generator frame. Remove the + lead as soon as voltage starts to build up. Use a plug-in voltmeter in one of the receptacles if the unit is not equipped with a built-in voltmeter.
4. *Bad capacitor.* If it is necessary to flash the field daily or each time the generator is used, the capacitor is probably defective.

Fig. 2-35 shows the location of the capacitor. Discharge the capacitor first with a screwdriver across its terminals. This does, of course, assume that the engine is not running. Remove the capacitor and replace it with one of the same type. Make sure it is identical in voltage rating and frequency.

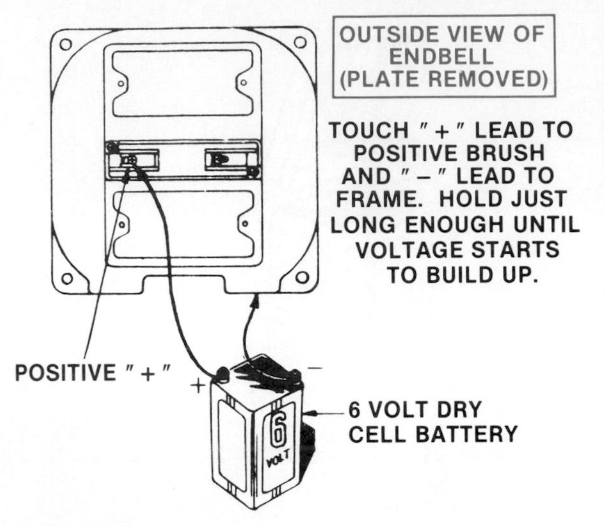

2–37. Flashing the field with a lantern battery.

Some models have fuses mounted in the receptacle box. Check to see if they are in good condition before doing any other troubleshooting. Test with an ohmmeter. It is not always possible to detect a defective fuse by visual inspection. The ohmmeter should read continuity when the cartridge-type fuse is checked.

Normal maintenance for the engine is needed for long life and carefree operation of the generator unit. Change the oil and spark plug whenever called for in the owner's manual.

More about the theory of operation of generators can be found in Chapter 14.

2-E. Metric Measurement Conversions

SYMBOLS	
Hertz	Hz
kilohertz	kHz
megahertz	MHz
RPM, rpm	r/min

CONVERSIONS	
1 mile	= 1.609344 km
1 cubic foot (cu ft)	= 0.028316846592 cubic meters (m^3)
1 ton (short)	= 0.90718474 metric ton
1 foot	= 0.3048 meter

Cycles per second, cps. or ~, has been changed to *hertz* per second. The time interval of the *second* is understood to be there. So it is referred to simply as hertz without the *per second* added. Instead of 60 cps, 60 cycles per second, or 60 ~, it is now 60 hertz or 60 Hz.

Revolutions per minute, rpm, or RPM, has been changed to r/min. This will probably be showing up later as the switch to metric becomes more noticeable.

QUESTIONS

1. What determines the voltage generated by a mechanical generator?
2. What are the parts of a simple mechanical generator?
3. What is another name for an AC generator?
4. Where are portable power plants used?
5. What are the three general groups into which all electric power plants can be divided?
6. Where do battery-charging plants serve most effectively?
7. What are the mechanical energy sources used to drive portable electric generating plants?
8. How would you go about deciding which type of engine to purchase for a power plant?
9. Why are falling-water power plants desirable today?
10. What is a pump-generating plant?
11. What is a switchyard?
12. What is meant by fusion?
13. What are fossil-fuel power generators?
14. What causes pollution in fossil-fuel generators?
15. What is LBG?
16. How do coal-fired generators turn coal into electricity?

KEY TERMS

alternator
armature
breeder reactor
brush
field coil
fusion
gas-cooled reactor
microhydro systems
nuclear power plant
photovoltaics
plutonium
solar cells
water-cooled reactor

CHAPTER 3

DISTRIBUTING ELECTRIC POWER

Objectives

After studying this chapter, you will be able to:

- Discuss design trends in underground distribution systems.
- Identify the two basic types of faults in underground cable.
- Discuss the fourfold role of the federal government in the distribution of electric power.
- Discuss those provisions of the National Electrical Code that concern irrigation machines.
- Identify the service equipment in common use in mobile home parks.
- Discuss the function of an automatic transfer switch in providing emergency power.
- Identify the two types of uninterruptible power systems.

URBAN AND SUBURBAN DISTRIBUTION

Electricity may be generated by fossil-fuel power generators, by atomic-powered generators, or by the use of falling water to drive generators. It is also possible to generate power by engine-driven generators. In Alaska, for example, where long distribution lines are impractical due to ice conditions which would result in line damage during storms, engine-driven generators are common.

Once electricity is generated in sufficient amounts for consumption in large quantities, the second necessary step is to get the energy to the consumer. Herein lies a distribution problem: that of stringing and maintaining long lines.

This chapter will be concerned with some of the problems and methods of distributing electricity after it is generated by public utilities.

Fig. 3-1 diagrams the process of getting power from the generating plant along high-voltage transmission lines, to voltage step-down substations, and then to office buildings, stores, apartments, and large factories. Further reductions are necessary in voltage to reduce the power to proper voltages (120-240) for home use.

A main concern of most utility companies is to get the greatest possible amount of usable energy to their customers. Voltage losses on the way from the plant to consumers are minimized by proper use of step-up and step-down transformers, and the correct size of wire.

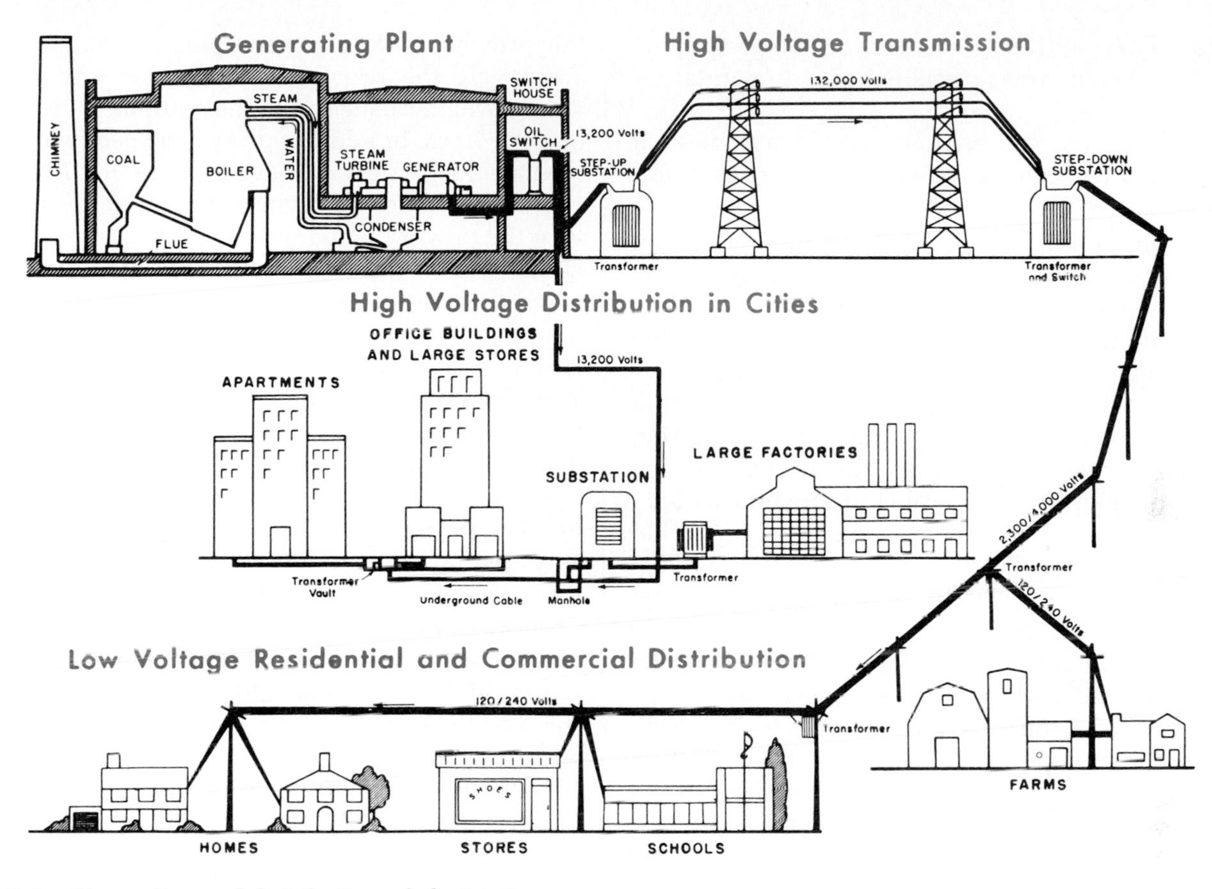

3–1. Generation and distribution of electricity.

Electrical Power

Most electrical power is generated as three-phase (3∅). (See Chapter 13.) It is stepped up to 132 000 volts, 238 000 volts, or even to 750 000 volts. The frequency is 60 hertz. Sometimes, however, 25 hertz AC is generated for use by some consumers who still may have older equipment that uses this frequency. Motors, and other equipment using this lower frequency, would be too expensive to replace. However, most 25-hertz equipment, when worn out, is replaced with 60-Hz equipment. A separate distribution system is necessary to send 25 Hz power to its destination, an operation which is expensive.

United States Power Companies. In the United States, power is generated by shareholder-owned electric companies, by government-owned facilities such as TVA (Tennessee Valley Authority), and by privately owned companies. The most popular system in the United States is the shareholder-owned. In this system an electric company is owned by people who buy shares in the company on the stock exchange, as they would in any other company.

Shareholder-owned electric companies differ in two major respects from commercial and manufacturing companies:

- Their activities are confined to geographic service areas established by regulatory agencies.
- They operate free of competition from other electric companies as long as the best interests of the public are served.

Competition by suppliers of other forms of energy, such as gas, coal, and oil is intensive, however.

Electric utilities are granted exclusive franchise areas because of the extremely large investment required of them to supply electricity, compared to the investment necessary from other types of businesses. For instance, an average manufacturer is able to function with an investment in buildings, machinery, and other facilities of about 50 cents for each dollar of annual sales. An electric company, on the other hand, needs an investment of about 4 dollars for each dollar of annual revenue. In other words, for each dollar of annual sales, electric companies must have eight times as much invested in facilities as a manufacturer. This in turn means an annual payment of eight times the amount of property taxes, eight times the depreciation expense, and eight times the amount of yearly interest and dividends.

With this investment (approximately 200 million dollars) required to supply electric service, it is economically unsound for two or more companies to set up duplicate facilities to compete for the customer's dollar. In recognition of this, electric utilities are allowed to operate without competition from other electric companies, but are placed under strict federal and state regulations. A utilities commission grants an electric company permission to do business in a designated area or areas as long as the commission is satisfied that the public is receiving adequate service at reasonable rates. In this way, the interests of the public are protected without resorting to expensive competition among power companies or to government ownership.

Canadian Power Companies. In Canada the power companies are owned and controlled by the provinces. Because equipment standards are set by the provinces, companies can be very particular concerning the equipment used on their lines. In addition they can specify and regulate inspection of devices hooked to the power lines (such as home appliances). In this way the power companies, along with the Canadian Standards Association, are able to demand and get safer manufactured devices in homes and industry. Their regulations can specify whether equipment is safe to operate. The Canadian companies can also designate the type of wiring for a house, business, or industry.

There are, as you can see, some advantages to this type of operation. There is, of course, always the possibility that such regulations will make devices too expensive for many people. A question here lies in determining just where an individual becomes responsible for the proper and safe use of equipment, and where the responsibility of the regulatory agency ends. This question could be debated for some time, with no definite answer. It is mentioned here to stimulate thought on the topic because it does affect everyone who uses electricity.

High Voltage Transmission. Figs. 3-2 through 3-5 illustrate some of the devices and equipment used to get the high voltage from the generator to the customer.

3-2. Distribution network for a large generating plant.

3-3. Control room of a large generating plant.

3-4. Installing a new 238 000-volt transmission line.

3-5. Note the size of the transformers in comparison to the man nearby. This is a high-voltage substation.

Local Distribution

Figs. 3-6 through 3-16 (pages 82-84) are concerned with transferring power from the substation to the local user. Most of these photos show overhead installations of power distribution systems. There are also a number of underground installations. Large cities and most suburban areas are now demanding removal of unsightly wires and poles. Underground wiring, although more expensive, is often used.

3–6. Cross-sectional view of a high voltage underground cable. Note the various materials used for insulation.

3–7. Stringing new cable for local service.

3–8. Providing power to an apartment complex.

Underground Systems. A prime objective of underground systems for distributing electrical energy has been to reduce the cost of these systems so that they could be used in areas previously served only by overhead systems.

Actually, underground distribution is as old as the electric power industry itself. In 1882, when Thomas A. Edison built the first central power station on Pearl Street, New York City, he installed feeders and mains under the city streets to distribute the power generated in the station to its customers. These were pipe-type cables, with a pair of copper conductors which were separated from each other, and from the enclosing pipe, by specially designed washers. The pipe was then filled with an insulating compound, consisting of asphaltum boiled in oxidized linseed oil, with paraffin and a little beeswax added.

Development of lead-sheathed insulated cables which could be pulled into previously installed ducts greatly simplified construction of underground distribution systems, but the high costs of such systems remained. As a re-

3–9. Laying wires in a trench for an underground distribution system. These wires will be used to provide underground power circuits for homes in a subdivision.

3–10. A buried transformer being spliced into service. Notice how it is buried in an enclosure.

sult, for many years underground distribution was limited largely to the central portions of cities, where load densities are high and where the congestion that would result from overhead systems is highly undesirable. Fig. 3-6.

3–11. Aboveground distribution system for homes. The transformer is connected to the top wire for high-voltage input; the three wires out to the lower wires are 240 volts, single-phase.

3–12. Three conductors, one red, one black, and one uninsulated ground wire, bring power from the transformer on the pole to the house for distribution throughout the home.

Today, costs of underground distribution have been reduced to levels that are economically practical for service to light- and medium-density load areas. Underground systems now are being installed in most new residential areas. Home buyers in these areas generally agree that the improved appearance and enhanced value of their property more than justify the added cost of underground service. Figs. 3-9 & 3-10.

3–13. Wires from the pole are spliced to a larger three-wire Romex cable to carry power to the meter and then down to the distribution panel in the basement.

3–14. Close-up of a box containing a meter. The kilowatt-hour meter can be removed from the plug-in socket if necessary.

3–15. This is a five-in-line clock dial, single-phase, 240-volt, 3-wire kilowatt-hour meter.

3–16. Mercury-vapor streetlights are part of a local distribution system. Maintenance and installation are both easily accomplished with the hydraulic lift mounted on a truck.

Some utilities are replacing existing overhead systems with underground systems in residential areas, particularly where load growth requires added capacity. Even some rural systems are being installed underground today.

Reduced costs of underground systems have been achieved through the development of efficient new materials and low-cost system components, and by progress made in simplifying construction techniques and installation methods. One of the most important developments has been new synthetic insulating material, notably thermosetting, or cross-linked, polyethylene. Cables with this type of insulation may be buried directly in the earth without expensive extra duct work. The development of improved cable plowing and trenching equipment has also helped to reduce installation costs.

Cables account for a major portion of the cost of an underground system. Initially, aluminum cable was used mostly in secondary portions of underground systems, but today it is widely used on both primary and secondary portions. It is also being employed increasingly for primary feeders. Figs. 3-17 through 3-21.

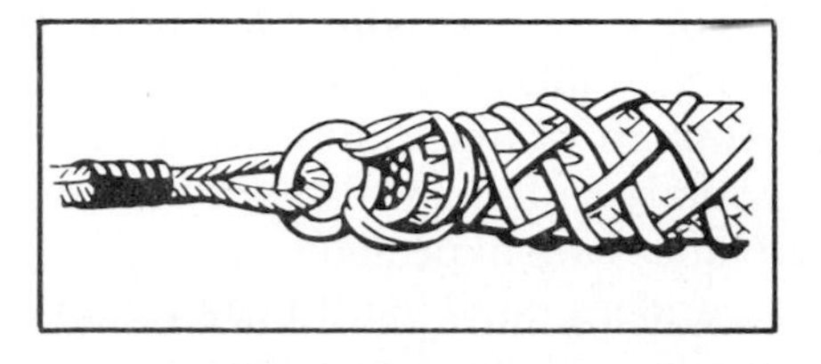

3–17. Method of attaching a line to a wire for pulling it through a conduit. Use an insulated basket grip to attach pull line to the conductors. Bare steel grip should not be used as it will score the inside of the conduit.

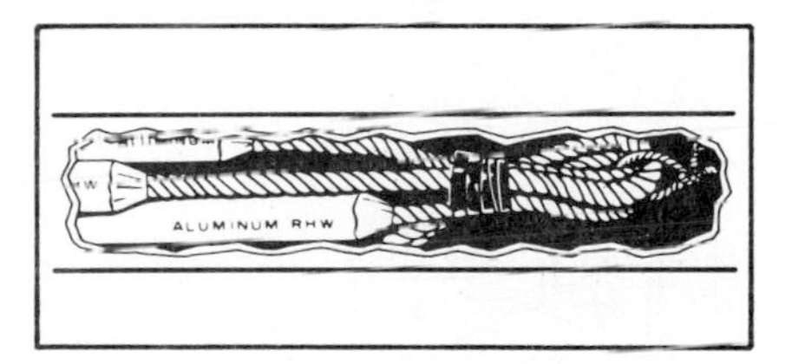

3–18. When a rope pull has to be used, skin the cable ends and stagger them, after locking with tape. This will hold tie to a minimum cross-section.

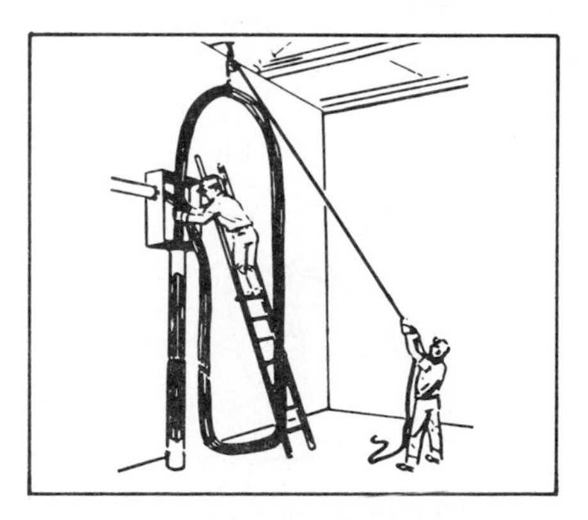

3–19. Plan pulls so that cable moves downward. This will speed work and reduce strain on cables and pull line.

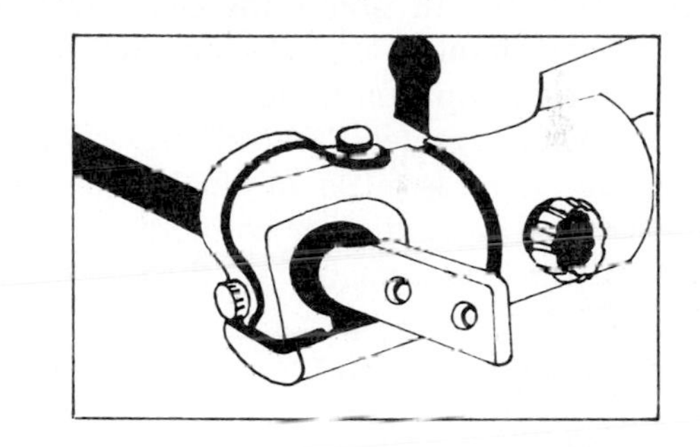

3–20. Crimping tool must be firmly closed. Failure to close crimping tool will lead to an unsatisfactory and weak joint.

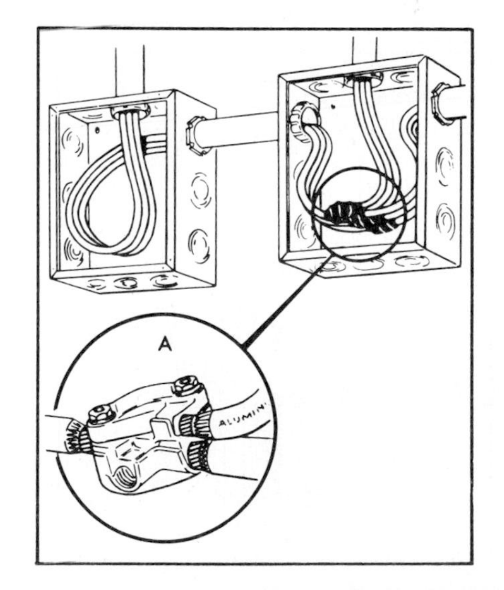

3–21. Use proper connectors, sufficient insulation, and leave enough slack for possible wiring changes and cable movement. Connector allows tap without cutting the feeder (A).

■ *Design trends in underground systems.* The revolution in underground distribution has brought significant changes in system design, mostly toward simplification. This is particularly true of systems for residential and light-load areas. Although design details vary, some trends are noticeable.

Residential areas are served almost entirely by single-phase systems. Some utilities use radial primary systems, while others employ loop primaries. Some companies have both, the choice for a given area depending primarily on load density. In some instances, radial primaries are changed over to loop when the load reaches an appropriate level.

Some utilities prefer "rear-of-lot" installation, where there is little or no interference with water and gas mains or with other underground utilities. Others install their systems along the more readily accessible street side of the property. Pad-mounted transformers, service pedestals, and switching equipment have generally been used above ground, but an increasing number of utilities are installing their entire systems underground, using submersible equipment. Fig. 3-10.

Manufacturers of transformers have developed compact units for pad mounting which require smaller and thus less conspicuous enclosures. Figs. 3-22 & 3-23. At the same time improvements in submersible units are being made, notably in their resistance to corrosion. Such advances are expected to stimulate an increase in underground vault installations. Dry-type transformers for direct burial also are in use, which, by eliminating the need for vaults, reduce the cost of installing a system entirely underground.

3–22. Transformer on a pad. Power transformer terminals, if copper, must employ short copper stub, spliced to the aluminum cable. Connections are illustrated to show the crimped terminal.

3–23. Copper terminations on switch connections may be replaced with aluminum by using smaller, multiple aluminum conductors to reduce voltage drop and increase current-carrying capacity.

Primary voltages today are mostly in the 15 kV range. Tables 3-A & 3-B. Some utilities still use lower voltages, mostly in the 5 kV range, and many utilize both 5 kV and 15 kV on different portions of their systems. There is a trend today, however, toward higher primary voltages, with many utilities going to 25 kV or 35 kV. Some are installing 35 kV cable on systems now operating at 15 kV. Upgrading such a system to higher voltage will not require an expensive change of cable when later load growth demands it. A few are using still higher voltages—up to 69 kV. Adopting higher primary voltages of course reduces the number of substations required, which is an important consideration with today's generally rising levels of power consumption.

A maximum number of twelve homes are being served from a single transformer on some systems. There is a trend, however, toward fewer homes per transformer, especially in developments where homes are built with all-electric utilities and appliances. Location preferences for service connections show no strong trend; yet connections at the transformer compartment, and in junction boxes below grade, are methods preferred by utility companies. The use of pedestal and dome type connections appears to be diminishing. Some companies are experimenting with a single transformer per house, with the service connection extending directly from a transformer that is either pad or wall mounted, or installed in an underground vault.

Considerable diversity is noted in lightning protection and in switching. A majority of utilities install lightning arresters on riser poles; many install arresters both on the poles and at the open point of a loop. Transformer primary protection is mostly by internal weak-link or fuses, but some companies use both. Various types of transformer switches are being employed, but the use of load-break elbows for both transformer switching and sectionalizing is increasing.

■ *Types of aluminum underground cables.* Most widely used on primary circuits is *concentric-type aluminum underground cable,* with bare neutral, consisting of equally spaced wires wound helically around an insulated conductor. The insulation is either high molec-

3-A. 15 000 Volt Copper Central Conductor Concentric Neutral U.D. Cable

SIZE AWG OR MCM	STRANDING NO. & SIZE OF WIRES, INCHES	INSULATION THICKNESS, INCHES	DIAMETER OVER INSUL., ± .030″	SHIELD/JACKET THICKNESS, INCHES	DIAMETER OVER JKT., ± .050″	ELECTRICAL REQUIREMENTS AC TEST, kV	ELECTRICAL REQUIREMENTS DC TEST, kV	ELECTRICAL REQUIREMENTS MINIMUM PARTIAL DISCHARGE EXTINCTION LEVEL, kV	NEUTRAL NO. & SIZE OF WIRES	APPROX. WT., lb/1000 ft
2	7 × 0.0974	0.175	0.700	0.030	0.780	27.0	70.0	13	6 × #14	443
1	19 × 0.0664	0.175	0.740	0.030	0.820	27.0	70.0	13	7 × #14	515
1/0	19 × 0.0745	0.175	0.785	0.030	0.865	27.0	70.0	13	9 × #14	642
2/0	19 × 0.0837	0.175	0.830	0.030	0.910	27.0	70.0	13	11 × #14	738
3/0	19 × 0.0940	0.175	0.880	0.030	0.960	27.0	70.0	13	14 × #14	905
4/0	19 × 0.1055	0.175	0.940	0.030	1.020	27.0	70.0	13	18 × #14	1121
250	37 × 0.0822	0.175	0.995	0.030	1.075	27.0	70.0	13	13 × #12	1280
350	37 × 0.0937	0.175	1.100	0.050	1.220	27.0	70.0	13	18 × #12	1768
500	37 × 0.1162	0.175	1.235	0.050	1.355	27.0	70.0	13	17 × #10	2456

3-B. 15 000 Volt Aluminum (Compressed Stranding) Central Conductor Concentric Neutral U.D. Cable

						ELECTRICAL REQUIREMENTS				
SIZE AWG OR MCM	STRANDING NO. & SIZE OF WIRES, INCHES	INSULATION THICKNESS, INCHES	DIAMETER OVER INSUL., ± .030″	SHIELD JACKET THICKNESS, INCHES	DIAMETER OVER JKT., ± .050″	AC TEST, kV	DC TEST, kV	MINIMUM PARTIAL DISCHARGE EXTINCTION LEVEL, kV	NEUTRAL NO. & SIZE OF WIRES	APPROX, WT., lb/1000 ft
2	7 × 0.0974	0.175	0.695	0.030	0.775	27.0	70.0	13	6 × #14	293
1	19 × 0.0664	0.175	0.730	0.030	0.810	27.0	70.0	13	6 × #14	319
1/0	19 × 0.0745	0.175	0.770	0.030	0.850	27.0	70.0	13	6 × #14	352
2/0	19 × 0.0837	0.175	0.815	0.030	0.895	27.0	70.0	13	7 × #14	405
3/0	19 × 0.0940	0.175	0.865	0.030	0.945	27.0	70.0	13	9 × #14	480
4/0	19 × 0.1055	0.175	0.920	0.030	1.000	27.0	70.0	13	11 × #14	558
250	37 × 0.0822	0.175	0.980	0.030	1.060	27.0	70.0	13	13 × #14	634
350	37 × 0.0937	0.175	1.080	0.050	1.200	27.0	70.0	13	18 × #14	860
500	37 × 0.1162	0.175	1.210	0.050	1.330	27.0	70.0	13	16 × #12	1136
750	61 × 0.1109	0.175	1.400	0.050	1.520	27.0	70.0	13	15 × #10	1568

ular weight polyethylene or cross-linked polyethylene, its thickness depending on the voltage rating of the cable. To provide electrical shielding, a thin layer of extruded, semiconducting polyethylene is applied to the conductor under the insulation and over the outside surface of the insulation. Neutral wires are wound around the outer shielding layer. Fig. 3-24. Coated copper wires generally are used for bare neutral. These should be of sufficient size and number to assure that the neutral has the same conductivity as the conductor. The neutral wires usually are round, but cables with flat neutral wires, for greater mechanical protection, also are being manufactured. In some locations, construction utilizes flat aluminum wires, with an overall plastic jacket.

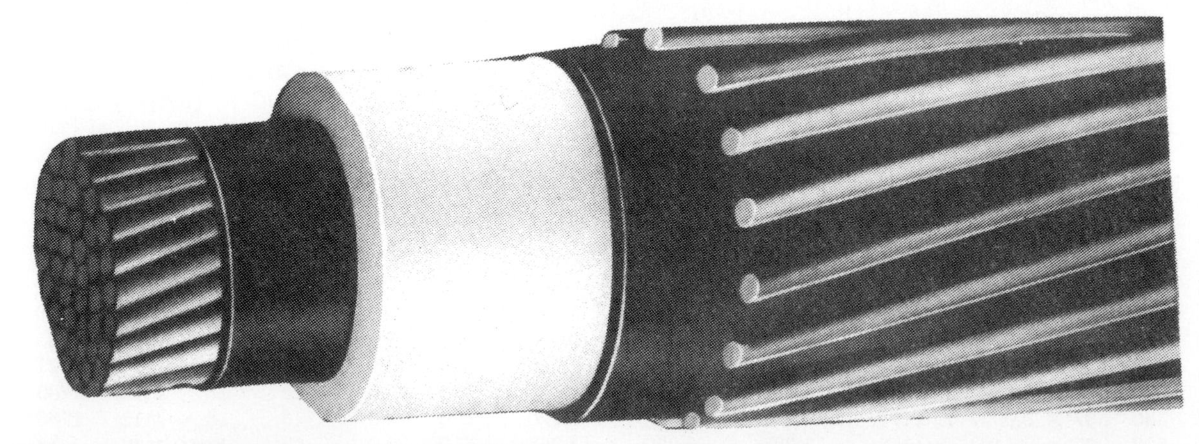

3–24. Primary URD and concentric neutral UD cable.

Secondary and service cable of aluminum most commonly used is insulated with a single pass of polyethylene or cross-linked polyethylene. Fig. 3-25. The latter is more widely used, despite its higher cost, because it withstands a higher conductor-operating temperature; therefore cables with this type of insulation have higher ampacities (ampere capacities). Table 3-C.

For use on the usual three-wire secondary system, two-phase conductors, plus a reduced-size, insulated neutral are commonly triplexed, either twisted or parallel, for convenience in installation. The neutral conductor is color coded for ease of identification. A construction consisting of two insulated conductors in parallel configuration, with bare neutral wires wound helically around the conductors, is also used on secondary systems and for underground service entrances.

Although the prevailing practice today is direct burial of both primary and secondary cables, some utilities still prefer installation in ducts. To meet the needs of these companies, cable, preassembled in plastic pipe, is available. Both concentric-neutral primary cable and triplexed secondary cables are available in this type of construction. The pipe may be either plain or corrugated for greater flexibility.

Where underground primaries are fed from overhead lines, ampacities for this cable will be limited by the riser portions, because of higher ambient temperatures to which those lengths are exposed and because of poor heat dissipation in air. Fig. 3-26 & Table 3-D.

■ *Charging current,* which is negligible on secondary cables, also may be neglected on primary cables of short laterals. On long, single-phase primary circuits, however, the cable-charging current may be sufficient to account for an appreciable voltage drop.

3-C. Comparison of Cross-Linked Polyethylene and Butyl-Neoprene Cables.

(See Fig. 3–25.)

CROSS-LINKED POLYETHYLENE CABLES

INSULATION THICKNESS	B	MINIMUM NO. OF TAPE HALF-LAPPED LAYERS)
0.055″	$2\frac{1}{2}$″	2
0.065″	$2\frac{3}{4}$″	2
0.075″	$2\frac{3}{4}$″	3
0.085″	3″	3
0.095″	3″	4
0.105″	$3\frac{3}{4}$″	4
0.110″	$3\frac{3}{4}$″	4

BUTYL-NEOPRENE CABLES

INSULATION THICKNESS	JACKET TKNS.	B	MINIMUM NO. OF TAPE (HALF-LAPPED LAYERS)
$\frac{4}{64}$″	0.030″	$3\frac{1}{4}$″	4
$\frac{5}{64}$″	0.045″	$3\frac{3}{4}$″	5
$\frac{6}{64}$″	0.065″	4″	5

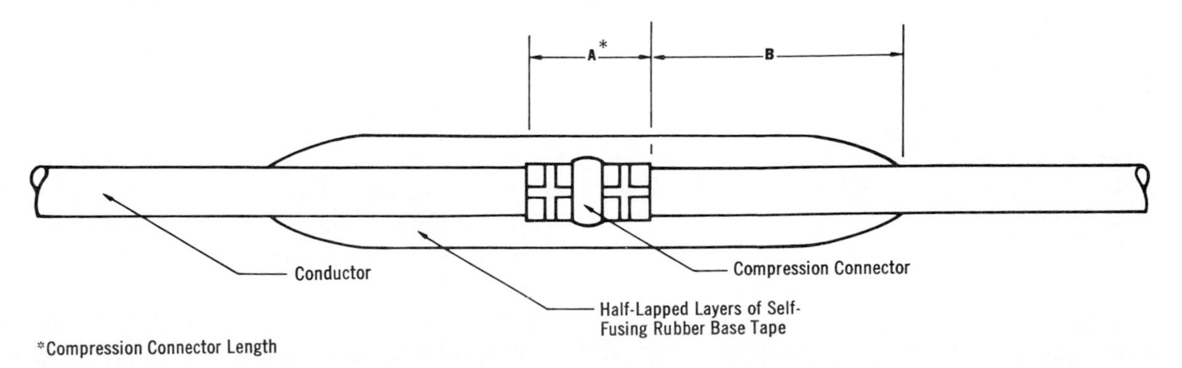

3–25. Splicing a 600-volt secondary cable. Single conductor, two-way straight splice (cross-linked polyethylene or butyl-neoprene insulated aluminum).

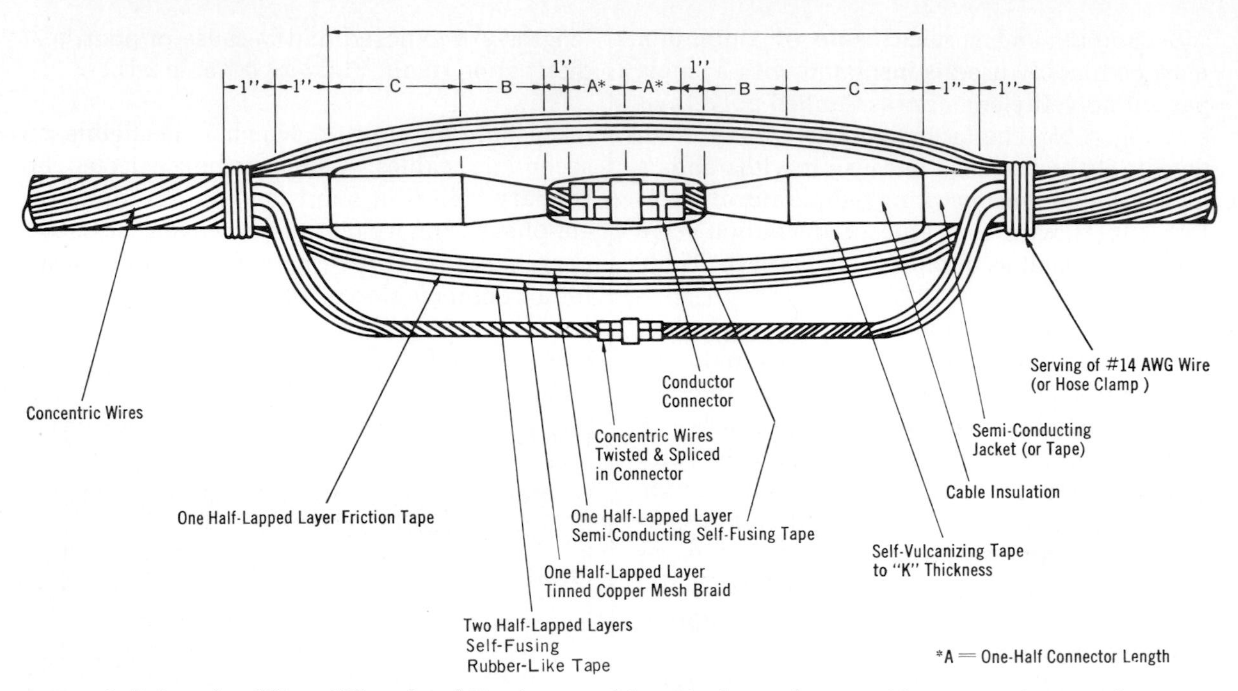

3–26. Splicing of 15 kV, 25 kV, and 35 kV primary cables. Single conductor with concentric neutral, straight splice (conventional or cross-linked polyethylene insulated, solid or stranded) for grounded neutral service. See also Table 3-D.

3-D. This Table Accompanies Fig. 3–26.

INSULATION THICKNESS	A	B	C	D	K
			15 kV		
0.175" or 0.220"	One-half Connector Length	2"	4¼"	2A + 14½"	3/8"
			25 kV		
0.260"	One-half Connector Length	2½"	5¼"	2A + 17½"	7/16"
			35 kV		
0.345"	One-half Connector Length	3"	7"	2A + 22"	9/16"

Charging current is higher with cross-linked polyethylene insulation than with conventional polyethylene because of the higher dielectric constant of the former. This is a factor to be considered in choosing cable for long, single-phase primaries, in addition to the lower cost of conventional polyethylene. Increasing the insulation thickness will reduce the charging current. This increases the diameter of the cable over the insulation and thus reduces the cable capacitance. Charging current increases with voltage.

On long rural underground primary circuits, shunt reactors may be used to reduce the charging current. These may be installed at sectionalizing points within the same enclosures as the sectionalizing equipment.

■ *Installation practices.* Installation of underground distribution cables is done either by trenching or plowing. Trenching, the older of the two methods, is still used by many utilities.

Fig. 3-9. With the development of improved plowing equipment, however, installation by that method is increasing. Whichever method is used, cables are almost universally installed in conduit where they pass under roadways, sidewalks, or other paved areas.

Depth of conduit burial ranges from about 30 to 48 inches (762 to 1219 mm) for primary cable, and from about 24 to 42 inches (610 to 1067 mm) for secondary cables when buried separately from primary. On many systems, both primary and secondary cables are buried in the same trench with no separation. In many areas, the trenches are shared jointly with other utilities, notably communication cables—both telephone and CATV (community antenna television). Joint use of trenches requires close collaboration on installation schedules, but offers substantial economies to the cooperating utilities.

Formerly, a separation of one foot was required between primary-power and communication cables, a standard still held by many companies. A recent amendment to the National Electrical Code, however, permits random-lay (no deliberate separation) installation of certain communication and power cables in the same trenches. These shared cables must have grounded wye power systems, operating at voltages not in excess of 22 kV to ground; or delta systems, operating at voltages not in excess of 5.3 kV, phase to phase. There is a growing trend toward joint use on systems operating within these voltage ranges because the distances found on most systems are nominal. Joint use with long, single-phase primary circuits is not recommended, however, because of inductive pickup of harmonics from power cables by the communication cables.

Care in handling cable during installation will help avoid trouble later. Damage sustained by the cable during installation has proved to be a major cause of subsequent cable failure.

■ *Splices, connections, and terminations.* The revolution in underground distribution system design has also included devices and methods for making splices, connections, and terminations. The objective in design here has been to reduce the amount of skill and time required in the field, thereby reducing installation costs. Prefabricated system components are produced under factory-controlled conditions. The need for heating and pouring of insulating compounds, or for extensive taping in the field, has been greatly reduced. Aluminum cable is available in a variety of lengths. Splices seldom are required.

Aluminum connections and terminating devices should be used with aluminum conductors. This avoids the differential expansion and contraction from heating and cooling that could result from the combination of connectors with dissimilar metals. Compression-type connectors and lugs, applied with a tool and die, are widely used. Figs. 3-27 through 3-29.

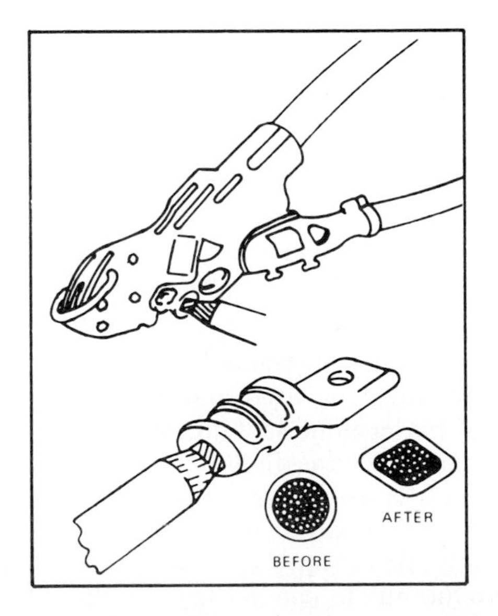

3–27. Typical terminals with compression lugs.

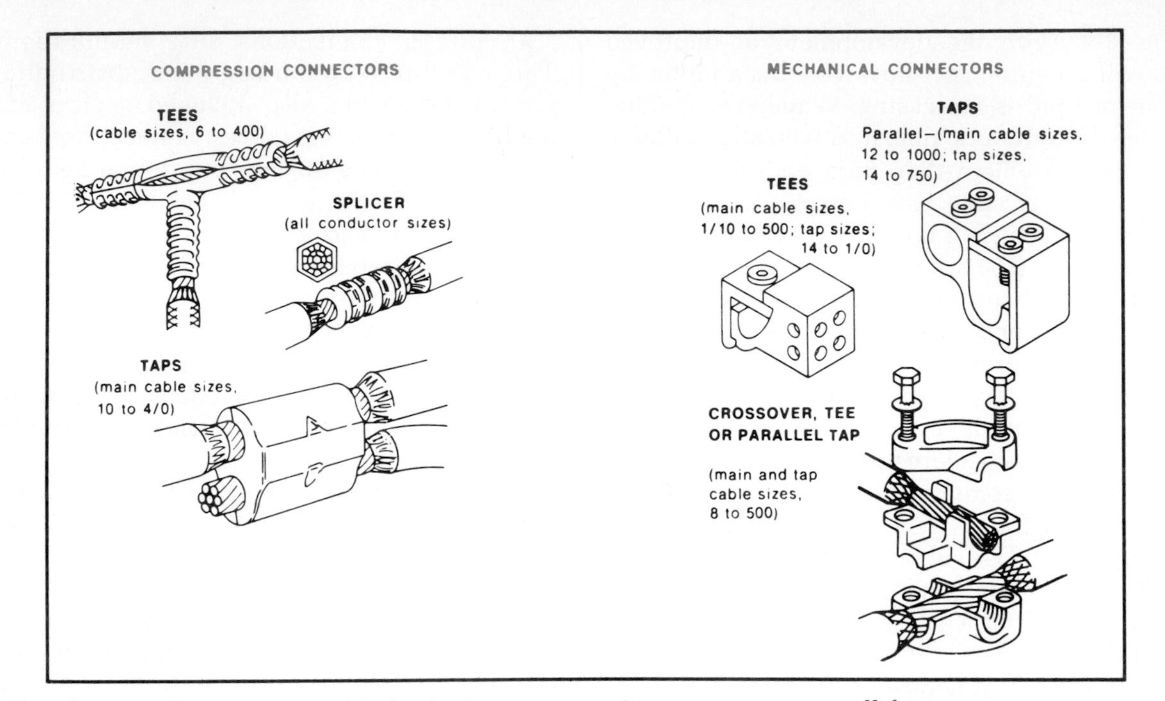

3–28. Compression connectors. Mechanical connectors. Crossover, tee, or parallel tap.

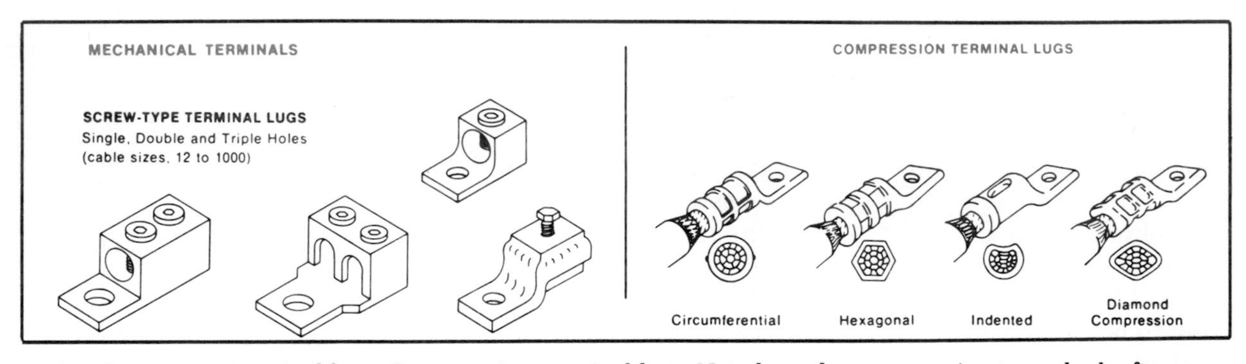

3–29. Screw-type terminal lugs. Compression terminal lugs. Note how the compression types look after a crimping tool has been used on them. Different crimping tools make the various patterns.

When installing these devices, dies of the correct size must be used and full pressure applied. This is important for obtaining permanently sound connections.

When it is necessary to attach an aluminum terminal to a copper bus, via a copper or steel stud, a Belleville spring washer should be used under the nut to compensate for the different rates of expansion and contraction of the dissimilar metals. Fig. 3-30.

Termination of primary underground cables requires some type of stress cone. Initially, these were made up by taping. Taping is a most tedious and time-consuming job for the field worker. Today, most utilities use some form of preshaped or prefabricated stress cone. The cone can be installed in a fraction of the time required to tape.

Typical terminations for primary cables are molded, precut tape, and porcelain types for

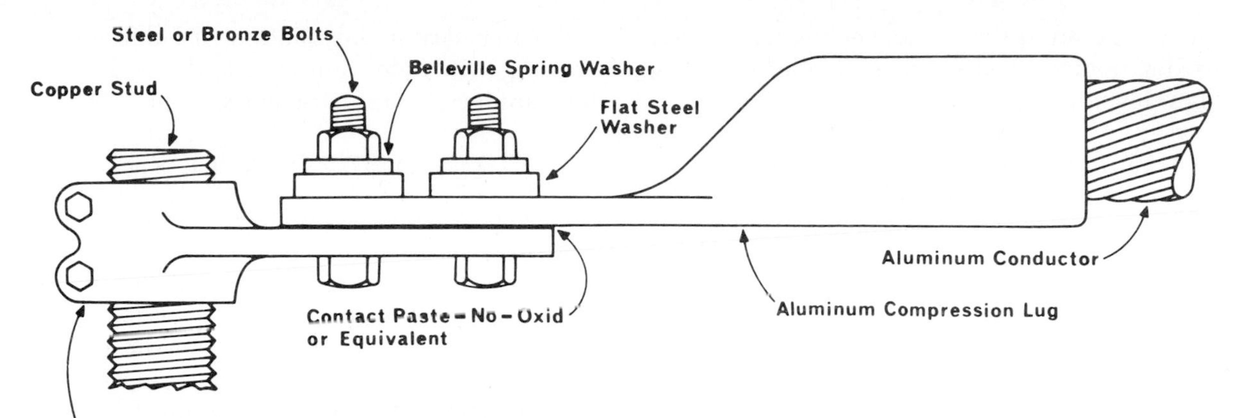

3–30. Aluminum and copper do not connect directly without trouble later. Note the use of the Belleville spring washer and its location in the aluminum conductor, and compression lug connection to a copper or bronze pad on a copper stud.

use indoors. Outdoors, porcelain units are most often used. One of the most significant developments for primary cable terminations has been the introduction of plug-in connectors for joining cables to equipment, or to other cables. With these devices, it is almost as easy to connect a primary cable as to plug in or remove an appliance cord from a convenience outlet. Figs. 3-31 & 3-32.

It is important that preformed splices and terminals be ordered to fit actual cable diameters. Otherwise, the fit may not be precise. Improper fit could result in later failure of service. Or, it could be impossible to assemble for initial installation.

Connector manufacturers have made important advances, too, in devices for secondary circuits. There are far too many different types to be described here. Suffice it to say that connectors are available to meet the expanding needs of the industry. Much of the recent developmental work on secondary and service cable connections has concentrated on finding better and cheaper ways to insulate and waterproof connections. This is especially important in view of the trend toward totally underground systems. Heat-shrinkable boots (plastic boots filled with insulating compound), epoxy, or urethane encapsulating materials are among the insulating materials in wide use. Continu-

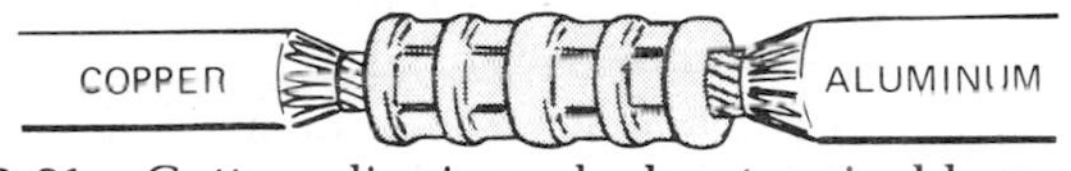

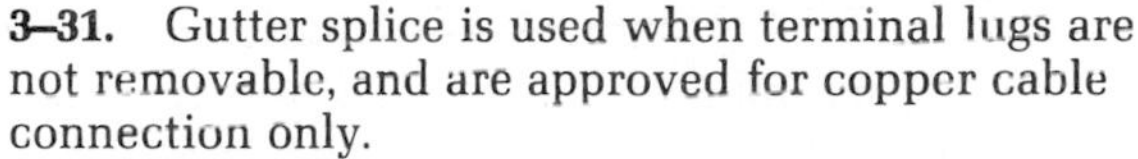

3–31. Gutter splice is used when terminal lugs are not removable, and are approved for copper cable connection only.

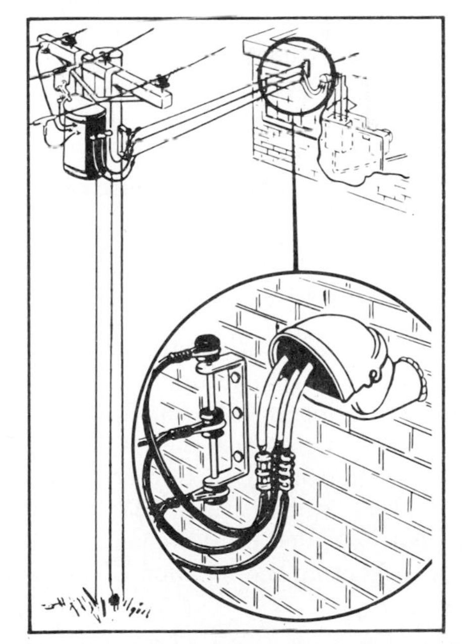

3–32. Overhead distribution system from pole to house. Copper service-drop cables are quickly field-spliced to aluminum service entrance with variety of compression connectors. Drip loop should be provided, with aluminum higher than copper in all cases.

ous work on splicing, connecting, and terminating devices is still proceeding at the design and development levels.

■ *Maintenance of underground systems.* Experience has demonstrated conclusively that the reliability of underground distribution is superior to that of overhead systems. The underground systems are subject to fewer hazards. When a fault does occur, however, it is essential that it be located promptly. Repairs must be made and service restored with little delay.

The greatest single cause of faults on direct buried cables has been "dig-ins," or disturbance by construction equipment. Faults from this cause are easiest to locate because they can usually be spotted visually. The second most common cause has been damage to the insulation during installation. Failures from this cause usually do not show up until some time after the installation has been completed and is in service.

Whatever their cause, faults in underground cable are of two basic types: (1) open circuit or series faults, in which the conductor has burned open or pulled out of the connector and (2) parallel faults, involving failure of the insulation.

■ *Fault location.* Present methods of locating faults may be divided into two groups:

1. Those utilizing an instrument which is connected to one terminal of the faulted cable, to determine the location of the fault by means of electrical measurements.
2. Those which apply a tracer signal to one end of the faulted cable. The circuit route is then patrolled until a change in the signal, indicating the location of the fault, is detected.

There are some modifications of these two methods which produce results for different users. Each type of tracing method has to be perfected for use by a particular individual, or by a group of individuals working as a team.

Greater care in installation of cable and in splicing will help eliminate faults and therefore produce a more efficient system.

RURAL ELECTRICITY

The industrial revolution of the nineteenth century, which had transformed living styles in cities the world over, scarcely touched life on the farms of the United States. As a result, the American farmer, at the dawn of the present century, was earning his living in a way that had changed little from that of the first colonists who settled along the Atlantic seaboard. The tools he used were simple and ancient: the wheel, the lever, the block and tackle, the plow. For most tasks, the farmer could draw only on his own strength, or that of horses and other animals. The children studied by the dim light of a kerosene lamp; the women were slaves to the wood stove and washboard.

City versus Country Power Development

For people in cities and towns, life was different. Electricity for power and lights was available to them. It was probably one of the attractions which began to pull people away from farms and into the cities. Some towns and cities were served by municipally owned electric power plants. Typically, however, the American city was served—and still is served—by electric power companies. Organized as stock corporations to build generating plants and distribution systems, these companies served urban areas, where concentration of consumers assured profits.

In almost all cases, operation of a power company is based on a charter granted by the city. The distribution of electric power is almost by its very nature a monopoly. In recognition of the industry's monopoly status, and of its nature as an indispensable public utility, electric companies are subject to varying degrees of regulation. They are regulated by public commission in almost all of the fifty states. Although differing from state to state, controls usually regulate rates a company may charge for service. They also set standards of service, territory to be served, and sometimes jurisdiction over financing and capital investments.

Federal Agencies

In the United States, the federal government does not provide retail electric service over distribution systems, either to urban or to rural people. The role of government in the electric power industry is fourfold:

1. The Federal Power Commission exercises licensing authority over the utilization of hydroelectric sites on navigable rivers of the country. It also maintains certain controls over the interstate transmission and sale at wholesale of electric energy generated by electric companies.
2. The Bureau of Reclamation, the Army Corps of Engineers, and the Tennessee Valley Authority build and operate some of the nation's hydroelectric generating plants. The TVA also builds steam plants.
3. The Bureau of Reclamation, TVA, the Bonneville Power Administration, and similar agencies build and operate transmission lines for marketing power wholesale.
4. The Rural Electrification Administration (REA) makes loans for rural electrification, including generation, transmission, and distribution systems.

This, then, is some of the background information which may prove helpful when studying the development of rural electrification.

Electricity for Everyone

The notion that electricity generated at a central station could be distributed to every farm in the United States took hold of people's minds slowly. Electric service was theoretically within the reach of rural families when the discovery was made that alternating-current voltage can be "stepped up" for transmission and "stepped down" for utilization. This of course, means that power can be delivered economically to areas that are distant from the generating station.

Of course, technological theory was not sufficient to bring electrical power to rural areas. Financing on a large scale had to be provided. In the United States, farmers live on the land they cultivate. Farmhouses, therefore, are widely scattered across the countryside. In some ranch areas of the western states, houses are many miles apart. In such thinly populated rural areas, electric companies could see little prospect for profit. Farmers usually were required to pay construction costs of individual line extensions to provide service, and rates were high. Most of them could not afford this and therefore remained without electricity.

REA

In 1935, President Franklin D. Roosevelt created the *Rural Electrification Administration* as an emergency relief program. In May, 1936, the Congress of the United States passed the Rural Electrification Act, which established REA as a lending agency of the federal government. It had the responsibility of developing a program for rural electrification. The act authorized and empowered REA to make self-liquidating loans to companies, cooperatives, municipalities, and public power districts. These loans were to finance the construction and operation of generating plants, transmission and distribution lines, and related facilities for the purpose of furnishing electricity to persons in rural areas.

In addition to making loans, REA furnishes technical assistance to its borrowers in engineering, management, accounting, public relations, power use, and legal matters. REA, however, does not construct, own, or operate electric facilities.

Electricity Works for the Farmer. A single kilowatt-hour of electricity, billed to the farmer at perhaps $0.025, will perform any of the following:

- Light a 100-watt bulb for reading or working for 10 hours.

- Pump 500 gallons (1893 L) of water from a well.
- Grind 400 pounds (181.4 kg) of feed—enough to feed 3 pigs for a month. Fig. 3-33.
- Protect food by running a freezer for 12 hours.
- Milk a cow twice daily for 15 days. Fig. 3-34.

3-33. Farmers use electricity to do many chores around the farm. This farmer is using electricity to grind feed for the animals.

3-34. Dairy farmers use electric cow milkers.

- Hatch 5 chicks in an incubator.
- Run a TV set for 4 hours.
- Operate a washing machine for 3 hours.

Farmers soon learned that a 1-horsepower motor could do as much work in an hour as a man could do in a day. Often, repetitious tasks had an element of drudgery connected with them. Such jobs were accomplished quickly and efficiently by the electric motor.

Rural Distribution of Power. A typical rural distribution system serves a relatively large geographical area, usually with a uniform, light-load demand. An average system serves 5000 consumers with 1500 miles (2400 km) of distribution line. However, some rural systems serve over 25 000 customers. Here, farms are quite large, generally averaging 300 acres (121.4 ha) or more. Typically, about three farms are served by each mile of distribution line. Loads are predominately single-phase, although a few three-phase installations may be located within the area. Towns, villages, and cities are served by separate distribution facilities not financed by REA. The typical rural system load is growing rapidly and requires good voltage regulation and a high level of service continuity. Fig. 3-35.

These factors have a major influence on design and construction of a rural distribution system. The preponderance of single-phase loads has led to systems which are essentially

3-35. Today's farmer often has use of a complete workshop due to the availability of electrical power.

single-phase in construction and operation. Approximately 80% of the lines being used are single-phase. The system which has evolved for rural service is typically an overhead multi-grounded, neutral, radial system on wood poles, using relatively small conductors which are usually aluminum with steel cores (ACSR).

The most common voltage is 7200 V phase-to-ground, or 12 470 V phase-to-phase. In sparsely settled areas which are remote from power sources but where the individual farm loads are heavy (and occasionally under other conditions), it has been more economical to use 14.4/24.9 kV as a distribution voltage.

Compared to other types of systems, the multi-grounded neutral system offers a number of distinct advantages for serving the remote area just described. Long-span, single-phase branches may be constructed most economically in vertical configuration. This is the basic, line-supporting structure used in United States rural electrification.

With addition of a crossarm, this structure permits conversion of a single-phase line to a two-phase circuit, or to a three-phase circuit, when such changes are required. These structures are shown in Fig. 3-36. On single-phase, multi-grounded, neutral circuits, the neutral carries only 30 to 40% of the phase current. In effect, the earth becomes a conductor in parallel with the neutral, which reduces voltage drop and conductor losses. Since 80% of the rural lines in the United States are single-phase, the economic advantage of the multi-grounded neutral system is obvious.

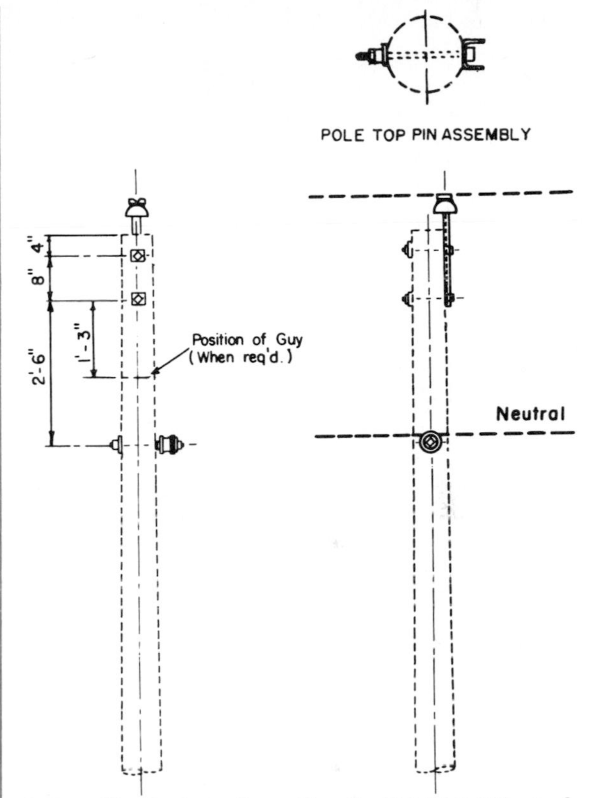

3–36. Vertical configuration for 7.2/12.5 kV, single-phase line supporting structure.

■ *Economical equipment.* Transformers with only one high-voltage bushing are satisfactory for rural distribution systems. Moreover, it is necessary to use only one primary fuse cutout and one primary lightning arrester. Since a transformer is usually required for each rural consumer, these reduced requirements result in significant economies. An installation of this type is shown in Fig. 3-37.

The capability of a distribution system generally is limited by voltage drop, rather than permissible current-carrying capacity. Good voltage levels are obtained by limiting the voltage drop on the primary to a maximum of 7% (or 8 volts on a 120-volt base). Voltage regulators of the step type are used at the substation to maintain the feeder voltage within satisfactory limits. Voltage drop on the grid is held within design limits by limiting circuit lengths, by selecting an adequate size conductor, and by the use of line voltage regulators as necessary. However, the system is essentially designed for use without line regulators. Omission of such equipment initially is a method generally relied upon to defer investment for growth until a time when it will be more economical to make such an investment.

■ *Sectionalizing.* Service reliability in rural areas is obtained principally by a very effective system of *sectionalizing.* Automatic circuit rec-

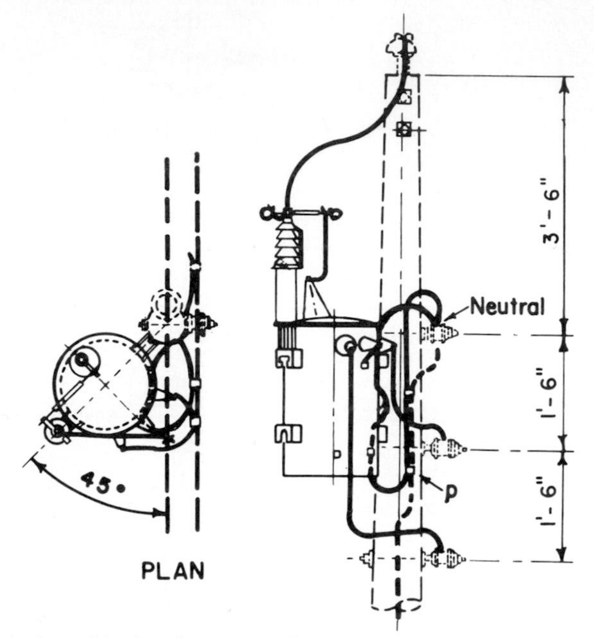

3–37. Single-phase transformer.

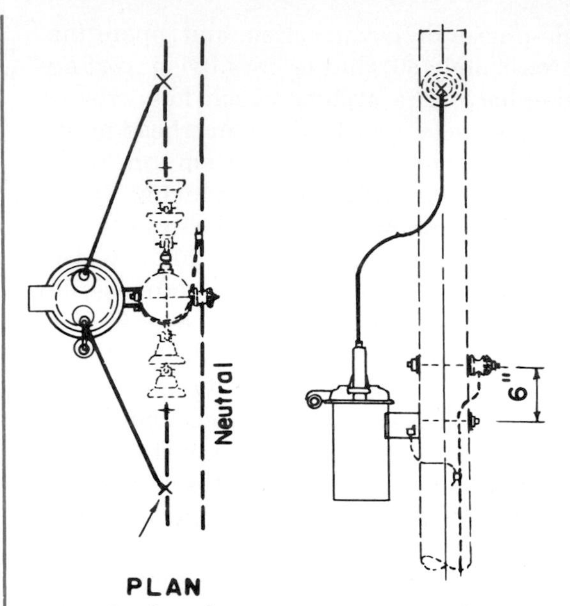

3–38. Single-pole automatic circuit recloser.

losers, automatic line sectionalizers, and (to a limited extent) fuses are used at frequent intervals to isolate any fault which may occur. Sectionalizing devices trip for temporary faults, then reclose rapidly to restore service. Since temporary faults account for more than 90% of all faults, a very high degree of reliability can be obtained through the use of sectionalizing. If the fault persists, the device closest to the fault will lock open. It then becomes necessary to locate and remove the fault and reset the sectionalizing device.

Single-pole sectionalizing devices are most commonly used. They have a higher degree of service reliability than three-pole devices, inasmuch as a fault on one phase does not interrupt service on the other phases. The single-pole, automatic circuit recloser was developed to meet the specific needs of rural distribution systems. These systems required a sectionalizing device that would operate for long periods without servicing, yet would be less expensive than the traditional type of circuit breaker. Fig. 3-38 shows the installation of a single-pole automatic circuit recloser.

■ *Grounding.* A system is adequately grounded if there are four ground connections in each mile of line. REA specifications require a driven ground (a copper rod driven into the ground) at each equipment installation, at secondary dead ends, and (including all driven grounds normally associated with the primary) at intervals of 1500 feet or less along the line. On many systems, pole-protection grounds are added to protect the pole from lightning. Although they do not meet U.S. National Electrical Code requirements as system or equipment ground connections, these grounds contribute to the effectiveness of the grounding of the neutral. Common primary and secondary neutrals are usually employed, and all other grounds, including guy wires, are bonded to the neutral. As a rule, no attempt is made to reduce the resistance of the individual grounding electrode. Even in areas of very high resistivity, experience has shown that the overall resistance of the multi-grounded neutral is very low—usually less than three ohms.

■ *Telephone and power lines.* The experience of rural distribution systems operating in close proximity to telephone or other communication systems has been satisfactory. The two systems frequently parallel each other at roadside separation, and in addition a large amount of line is built with the telephone and electric systems sharing the same poles. No special construction is required for the electric distribution system, but transformers used are built to a specification which controls the harmonic content of the excitation current. Telephone systems in the United States have determined that the most satisfactory method of operating is to adopt construction standards which are characterized by a very low susceptibility to power line influence. These standards make it unnecessary for telephone companies to request special construction from power systems for the purpose of reducing power line influence on their communication cables.

Single-Phase Power. The principal reason for favoring single-phase rural distribution lines is that most rural loads are single-phase. An average farm has no three-phase requirements. Most farm motors, including those used for feed grinding, are below 5 horsepower. Larger and deeper irrigation wells, however, frequently require pump motors of 25 horsepower or more, for which three-phase service is provided. In the past, $7\frac{1}{2}$ horsepower generally was regarded as the largest practical single-phase motor, and the cost of this size motor was about double that of a comparable three-phase motor.

Several hundred dollars for an occasional motor seems preferable to thousands of dollars for three-phase lines. Meanwhile, manufacturers are becoming interested in larger single-phase motors which will perform as well as, yet cost no more than, three-phase motors. It also should be noted that many rural systems have the capacity to serve 30-horsepower, single-phase motors without undue flicker problems. In fact, single-phase loads no longer are limited to any specific size; the present practice is to allow the capability of the system, at the time of installation, to determine the maximum size of the single-phase load that can be served.

Rural Grid. A typical *rural grid* (distribution network) consists of radial feeders, which leave the substation as three-phase circuits, with each of the phases fanning out to serve the countryside as a single-phase circuit. Since most rural loads can be served satisfactorily by single-phase circuits, this grid arrangement has resulted in worthwhile economies. Fig. 3-39. A three-phase line in the United States costs approximately one and one-half times as much as a single-phase line costs.

IRRIGATION SYSTEMS

Irrigation of farm land has become more common. Irrigation systems must be properly wired to make sure the person operating or moving them is not shocked. The ground is almost always wet around a system that is to be towed. Thus, it is extremely important that the proper electrical safety precautions be taken both in operating the equipment and in the original installation.

Center-pivoted systems are used to irrigate large acreages. Fig. 3-40. These use large electric motors to run the water pumps. In some cases; the power is generated on-site with engine-driven generators. In other instances the machines utilize the power available from REA or whatever service is nearby.

Most of the controls for the system are located on the center pivot column or tower. A unit being towed to a new location by tractor-power is shown in Fig. 3-41.

As you can see from the cover on the control box in Fig. 3-42 the 480 volts inside makes it necessary to exercise caution around the exposed connections that may appear inside the box.

The National Electrical Code deals with Electrically Driven or Controlled Irrigation Machines in Article 675. Electric motors used to supply water to the irrigation unit are covered by the general requirements of the Code.

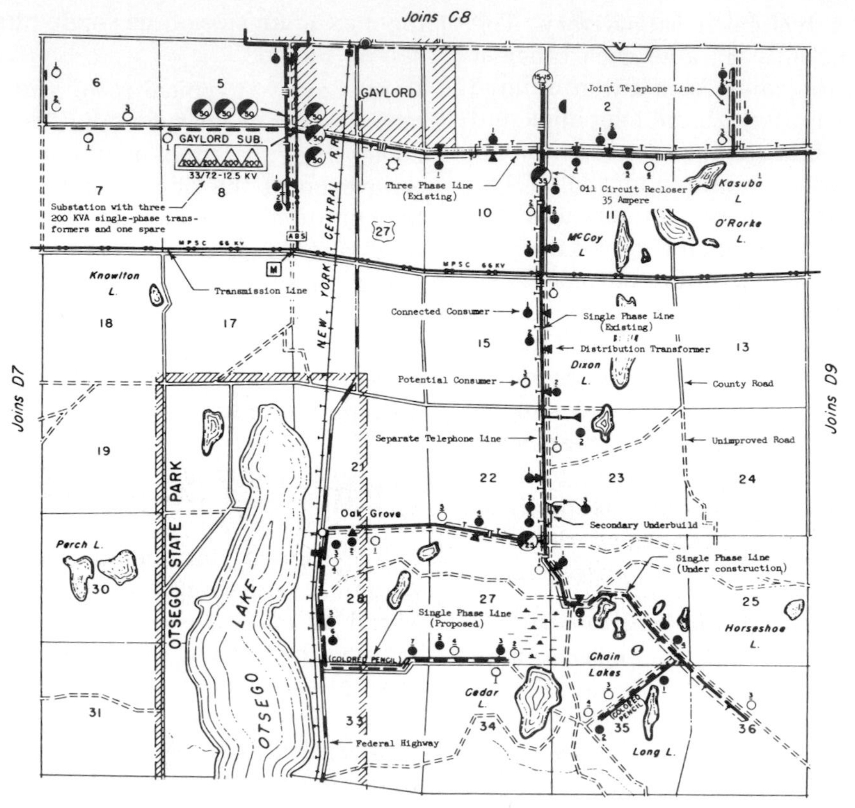

3–39. Sample detail map used for rural electrification.

3–40. An irrigation system designed for center feed and a minimum of spray loss due to wind.

3–41. Modified center tow model designed for easy towing from field to field.

3–42. Control panel from which the electricity is controlled and fed to the various parts of the irrigation system.

Of special interest in wiring the irrigation unit is the wire to be used. Because it will be used in a wet location, the cable used for wiring should be of stranded, insulated conductors with nonhygroscopic and nonwicking filler in a core of moisture and flame-resistant, nonmetallic material. This should be overlaid with a metallic covering and jacketed with a moisture-, corrosion-, and sunlight-resistant nonmetallic material.

The Code permits the main disconnecting means to be up to 50 feet from the machine if readily accessible and capable of being locked in the open position. If the circuit originates at the motor control panel for the irrigation pump and the panel is within 50 feet of the center pivot machine, one set of overcurrent protective devices and one disconnecting means can be eliminated.

The Code should be checked more closely when more than one motor is on a branch circuit. Also check the Code exceptions to grounding of the equipment. Lightning may be a factor to consider in some locations. Then, it is usually necessary to provide a grounding electrode, a rod driven into the ground. Lightning protection is covered in Sections 250-81 and 250-83 of the Code.

INSTALLATION OF SERVICE BY POWER COMPANY

You have, no doubt, noticed that there are some standards for the transmission of electrical power over distances. The towers, poles, and supports of various types are standardized as to size and treatment to endure years of exposure. There are some very definite minimums and maximums for the crossing of roads and various other obstacles on the way from the generator to the user. Fig. 3-43. shows the minimum clearance of service drops below 600 volts.

Fig. 3-44 shows the attention given to the service entrance riser support on a low building. This work is usually done by the electrician, and at least 24″ service conductors are left for connection by the power company when it brings power up to the house or building.

Most new suburbs are demanding that electrical service be brought in from the line to the house by underground cables. Fig. 3-45 (page 104) shows the requirements for such an installation. Note how far the meter is mounted above ground. This service shows the pole in the rear of the house. Some localities also require that the entire electrical service to an area be underground. This eliminates poles in the rear of the house.

Not all farms in the United States are serviced by the Rural Electrification Administration (REA). Some of them are serviced by local power companies. The standards for a farm meter pole by a local power company are shown in Fig. 3-46 (page 105). Note the division of ownership. Also note the $\frac{5}{8}''$ × 8′ ground rods at least 3′ from the pole. This is a single-phase,

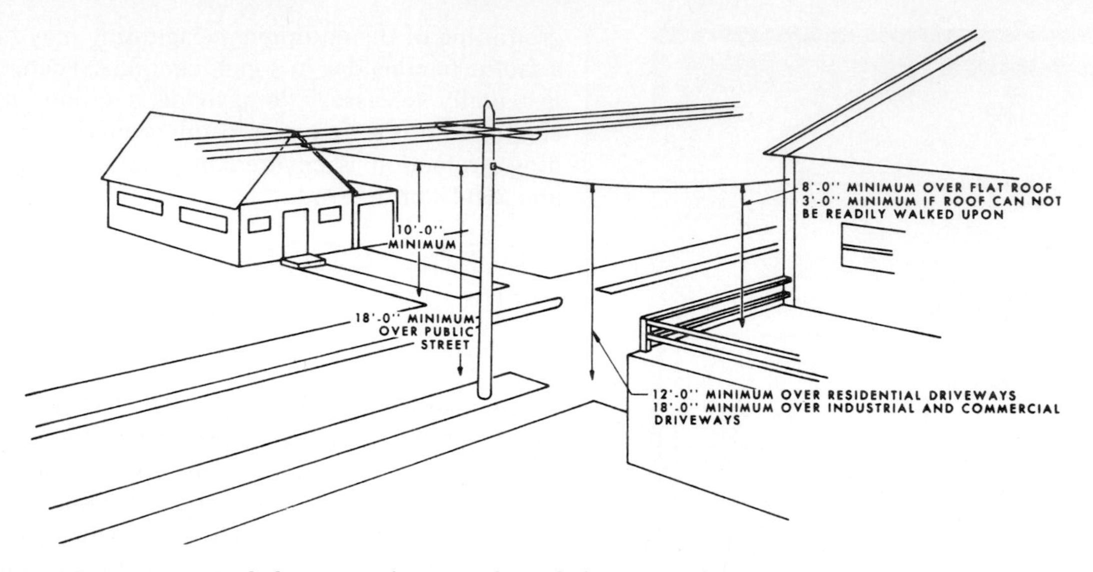

3–43. Minimum vertical clearance of service drops below 600 volts.

3-wire 120/240-volt installation for loads exceeding 40 kW demand. This would be for a pretty large farm.

Meter Installations

The electrician is usually required to place the meter socket trough. It is usually supplied by the power company and installed by the customer. Fig. 3-47 (page 106) shows a typical one-meter installation with a socket for plugging in the meter once the service has been connected. In the case of apartment houses it will be necessary in some cases to install two or more meters. Fig. 3-48 (page 107) shows the proper installation of two to six meters for single-phase, 3-wire, 120/240-volt, 150-ampere minimum service entrance.

Note in Fig. 3-49 (page 108) that six or more meters can be installed indoors in a single-phase, 3-wire, 120/240-volt service or single-phase, 3-wire, 120/208-volt service. Notice the distance the meters should be above the floor. These are the plug-in type meter sockets. Note also the materials furnished by the power company and installed by the electrician. The meter board is usually painted black.

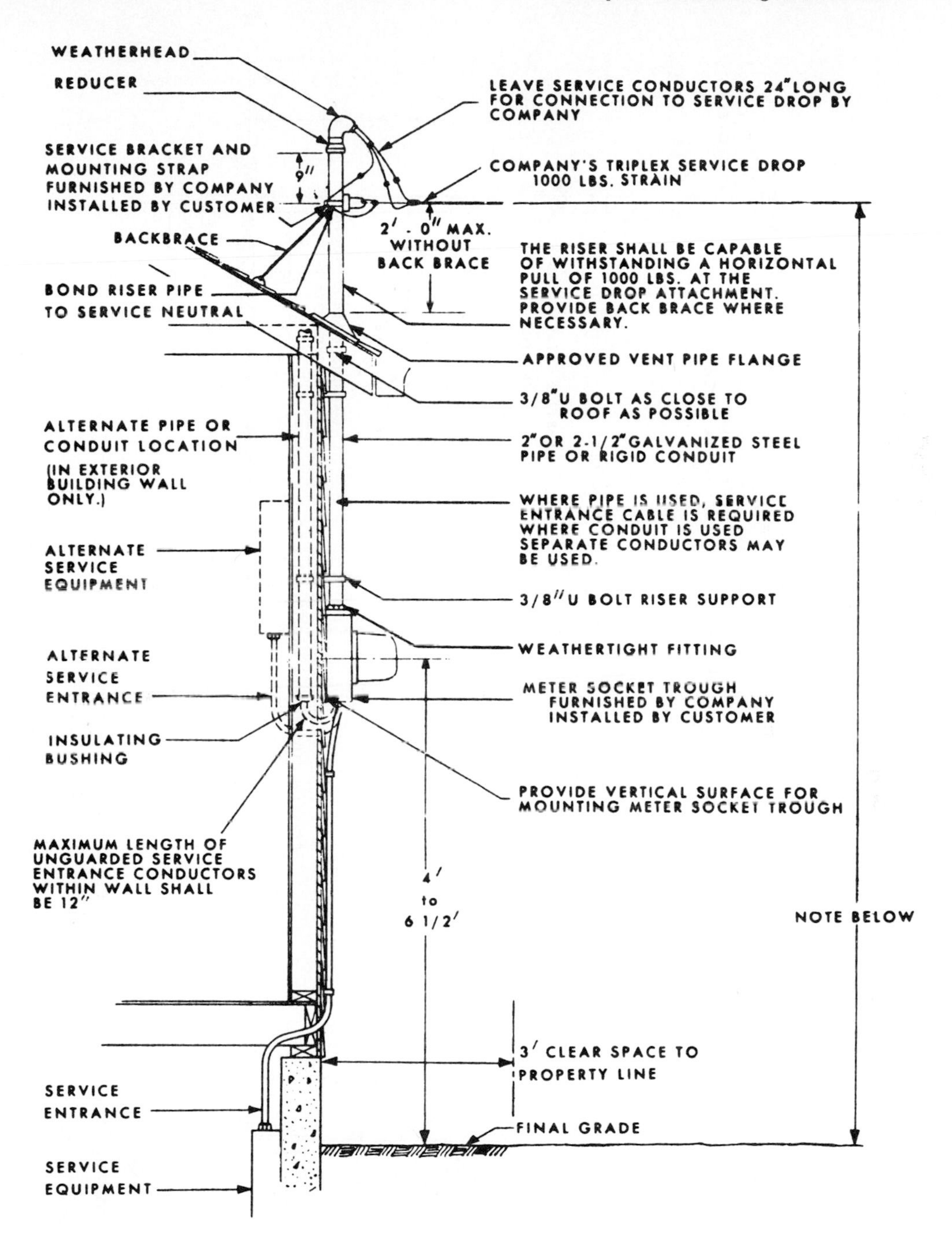

3-44. Service entrance riser support on low building.

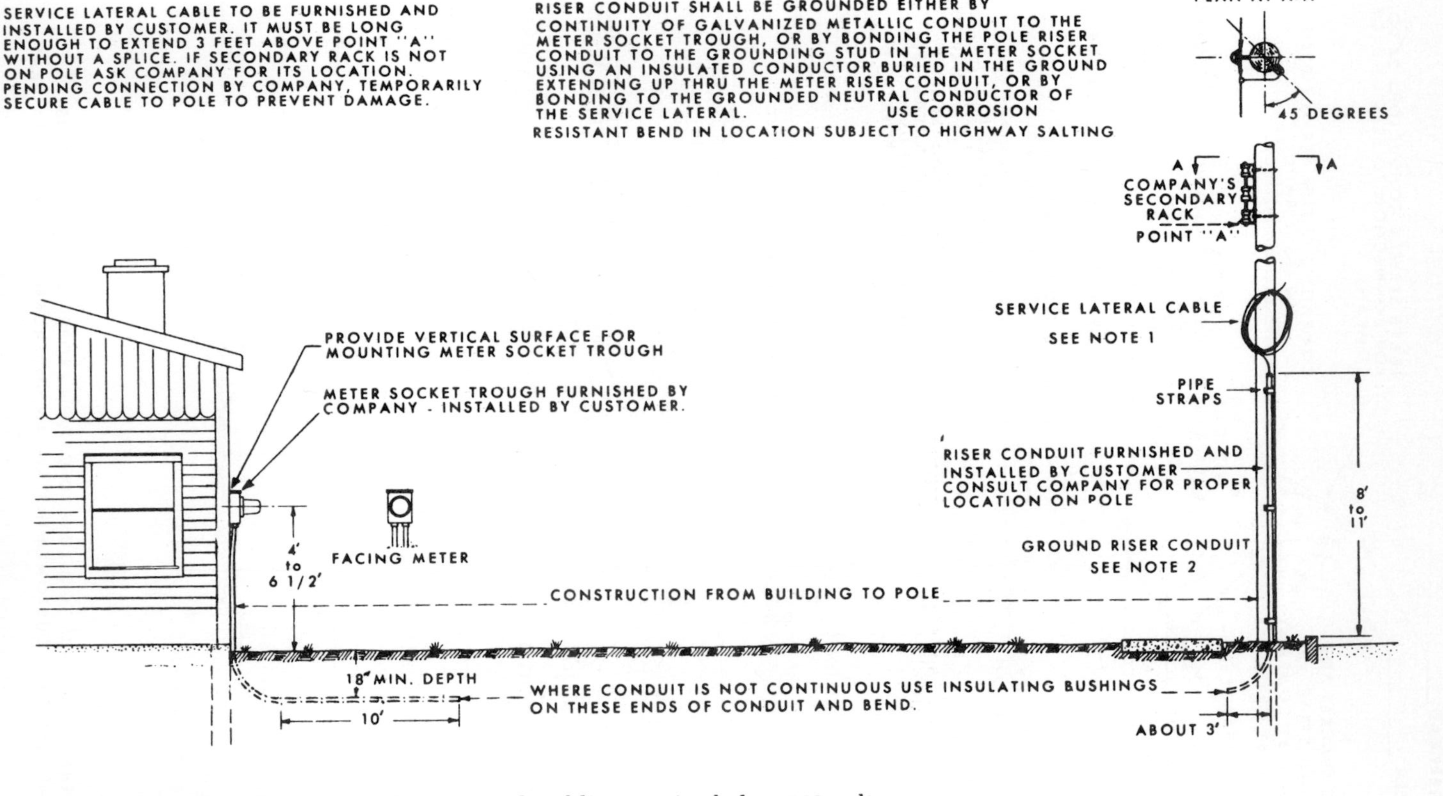

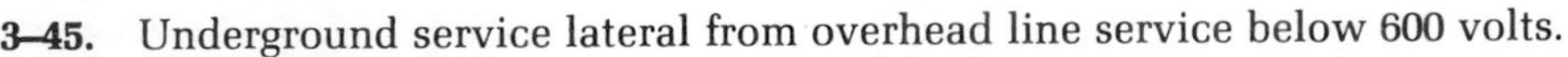
3–45. Underground service lateral from overhead line service below 600 volts.

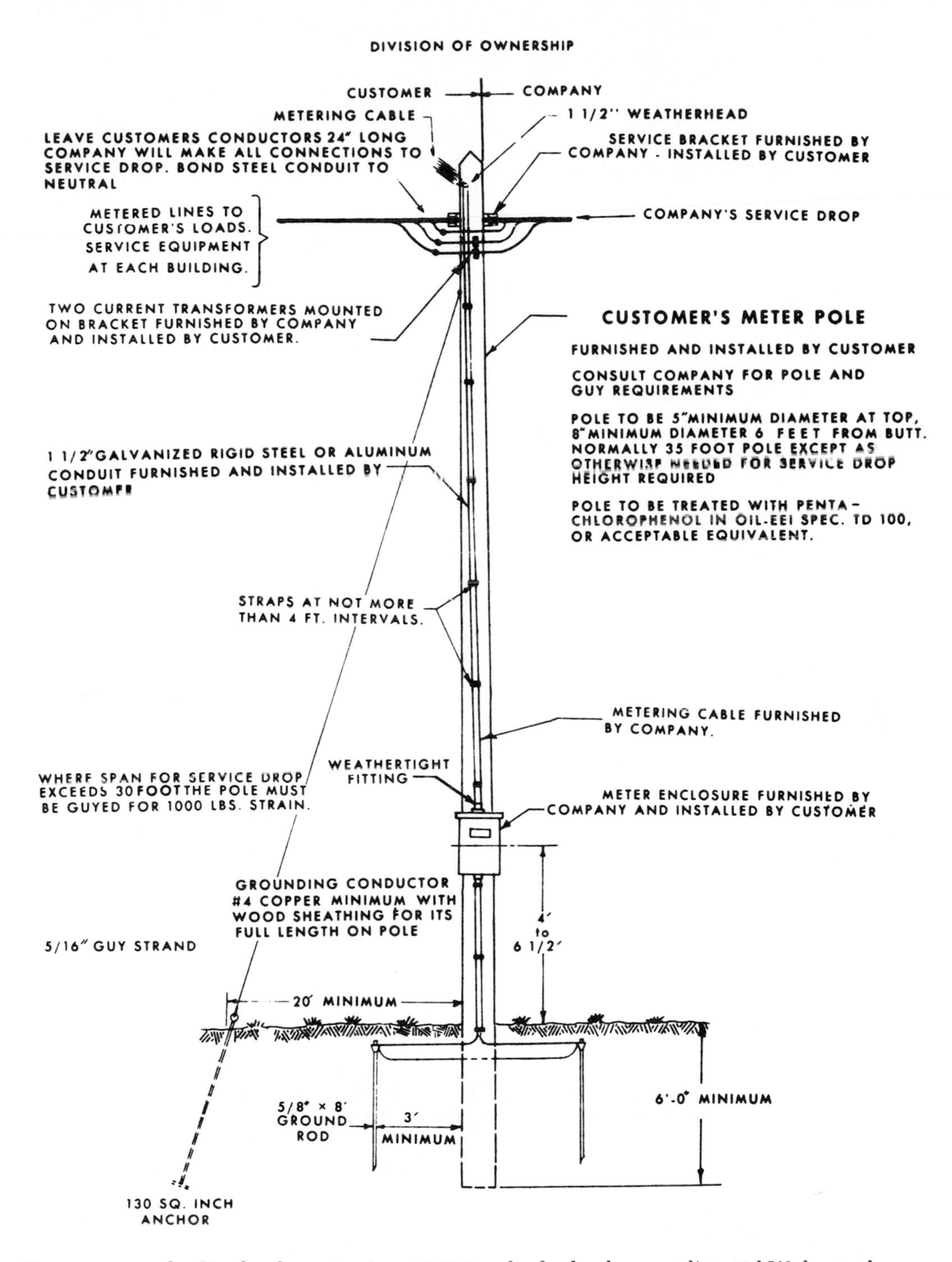

3–46. Farm meter pole. Single-phase, 3-wire, 120/240 volts for loads exceeding 40 kW demand.

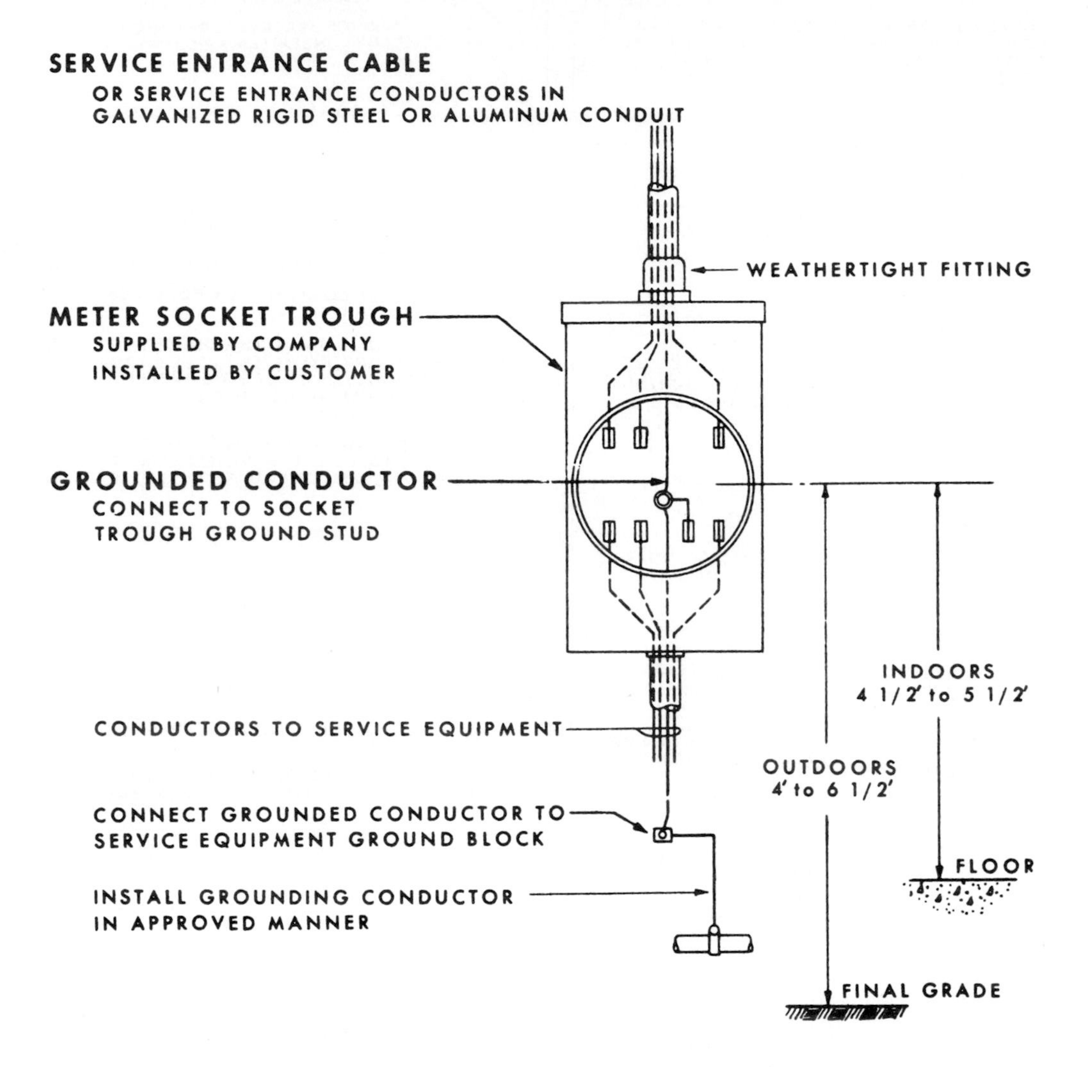

3–47. One-meter installation, socket type. Three-phase, 4-wire 208 wye/120 volts, 200-A maximum service entrance. Three-phase, 4-wire, 480 wye/277 volts, 200-A maximum service entrance.

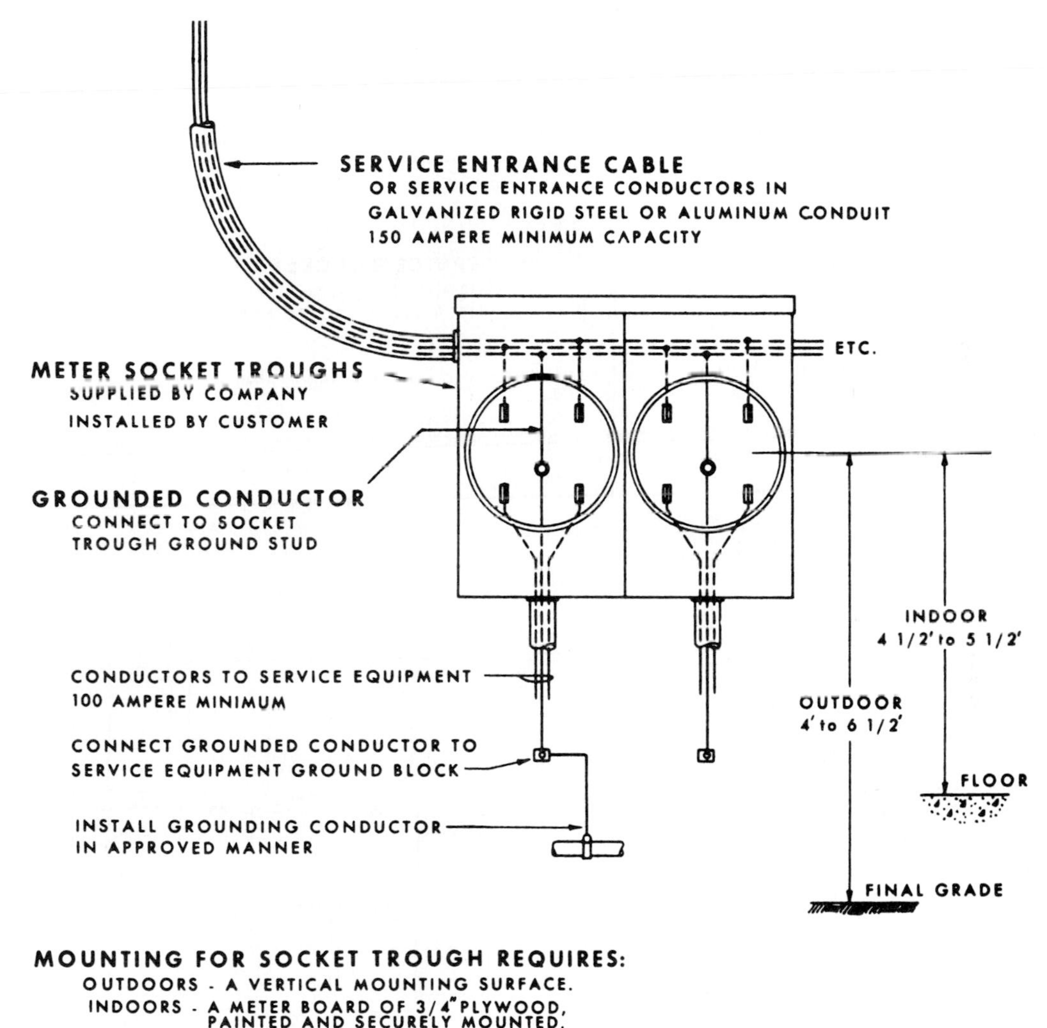

3–48. Two to six meter installation, socket type. Single-phase, 3-wire, 120/240 volts, 150-A minimum service entrance.

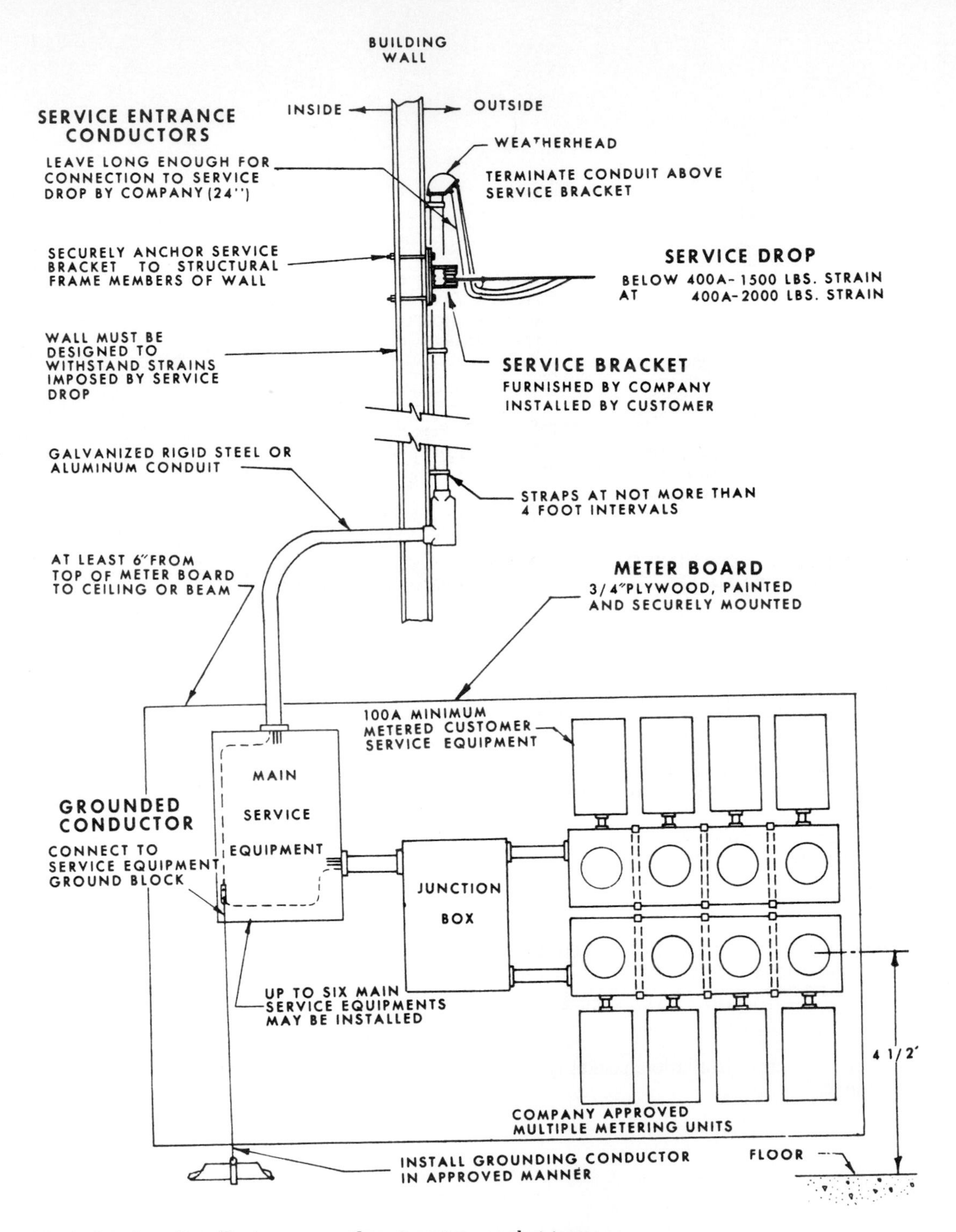

3–49. Typical indoor installation—more than 6 meters, socket type.

Mobile Home Installation

Mobile homes have been getting an increasing amount of attention from the National Electrical Code and other agencies. They are becoming a permanent dwelling for many people. These homes have particular needs which must be met to keep them electrically safe. Fig. 3-50 shows a typical installation for mobile home parks.

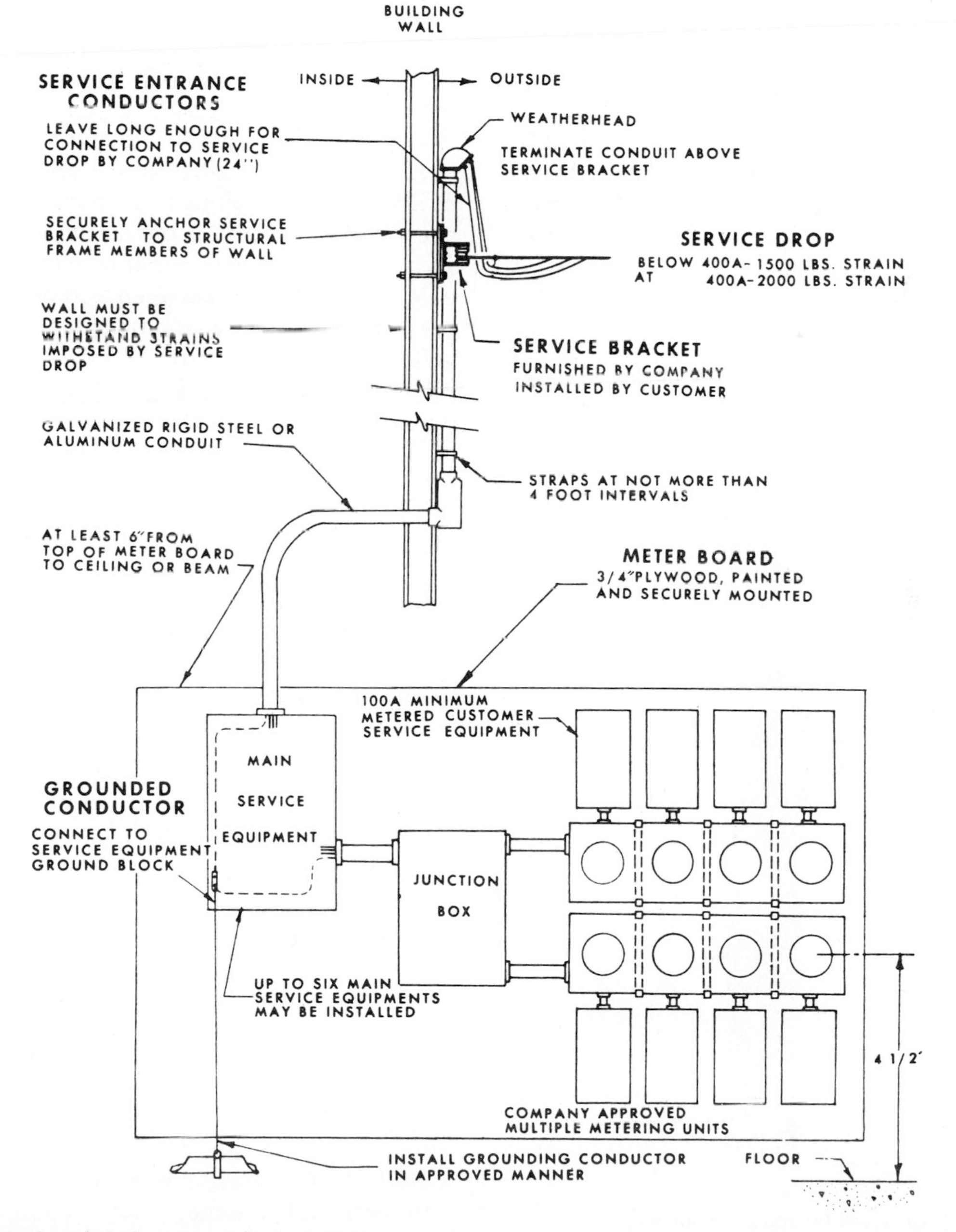

3–50. Typical mobile homes parks installation.

Where electrical power enters a building or other structure, the National Electrical Code requires service equipment. Such service equipment usually consists of a circuit breaker (or switch and fuses) and accessories for use as a main disconnect and to provide overcurrent protection. Fig. 3-51 shows a post made for use in mobile home parks and for recreational vehicles. This type of service equipment allows for easy disconnection of the service and for the use of electrical plug-in equipment near the post. Note that the unit includes a main disconnect consisting of one or more circuit breakers or, in some cases, fusible pullers in a single enclosure suitable for outdoor installation. Fig. 3-52 shows how the posts are wired and mounted either pad post style or embedded in concrete. They may also be used to service meters back to back.

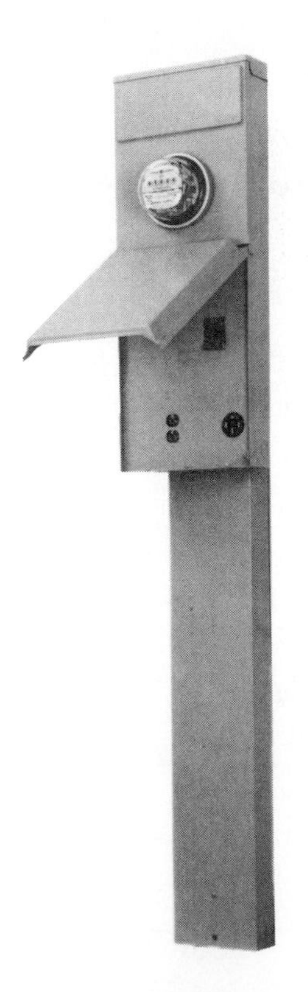

3–51. A post made for mobile homes or recreational vehicle parks.

Mobile home service equipment must be located in sight from the mobile home. The equipment may not be more than 30 feet from any point on the exterior wall of the mobile home. The NEC (Article 551) recognizes the use of feeder raceways external to the mobile home.

EMERGENCY POWER

Manufacturing plants, computer installations, airports, high-rise office and apartment buildings and hospitals rely upon a continuous supply of electricity. A power outage can have serious consequences. What is needed is a reliable way to supply emergency power to these places automatically when needed.

Starting the engine-driven emergency generator is but one of the jobs needed when there is a power failure. If there is a power failure the engine generator can supply power during the emergency. In some systems, controls will monitor the normal power source. When it fails, the engine generator will start automatically. Then, when the generator reaches proper voltage and frequency, the controls will transfer the emergency loads from the dead utility source to the generator to resupply the emergency loads with power. Fig. 3-53.

Automatic Transfer Switches

An *automatic transfer switch* can monitor the normal power source. Fig. 3-54. If there is a power outage, the switch signals the engine generator to start. When the generator reaches proper voltage and frequency, it transfers selected loads to it. When the normal source is restored, the switch retransfers the load to it. It then shuts down the engine after a cool-down period.

At times a transfer switch handles more than its normal or continuous current rating. A reliable transfer switch must be able to handle all situations with no harmful effects on the switch. *Motor starting current* is one of these situations. When a motor starts up, it can draw as much as six times its running current. It can

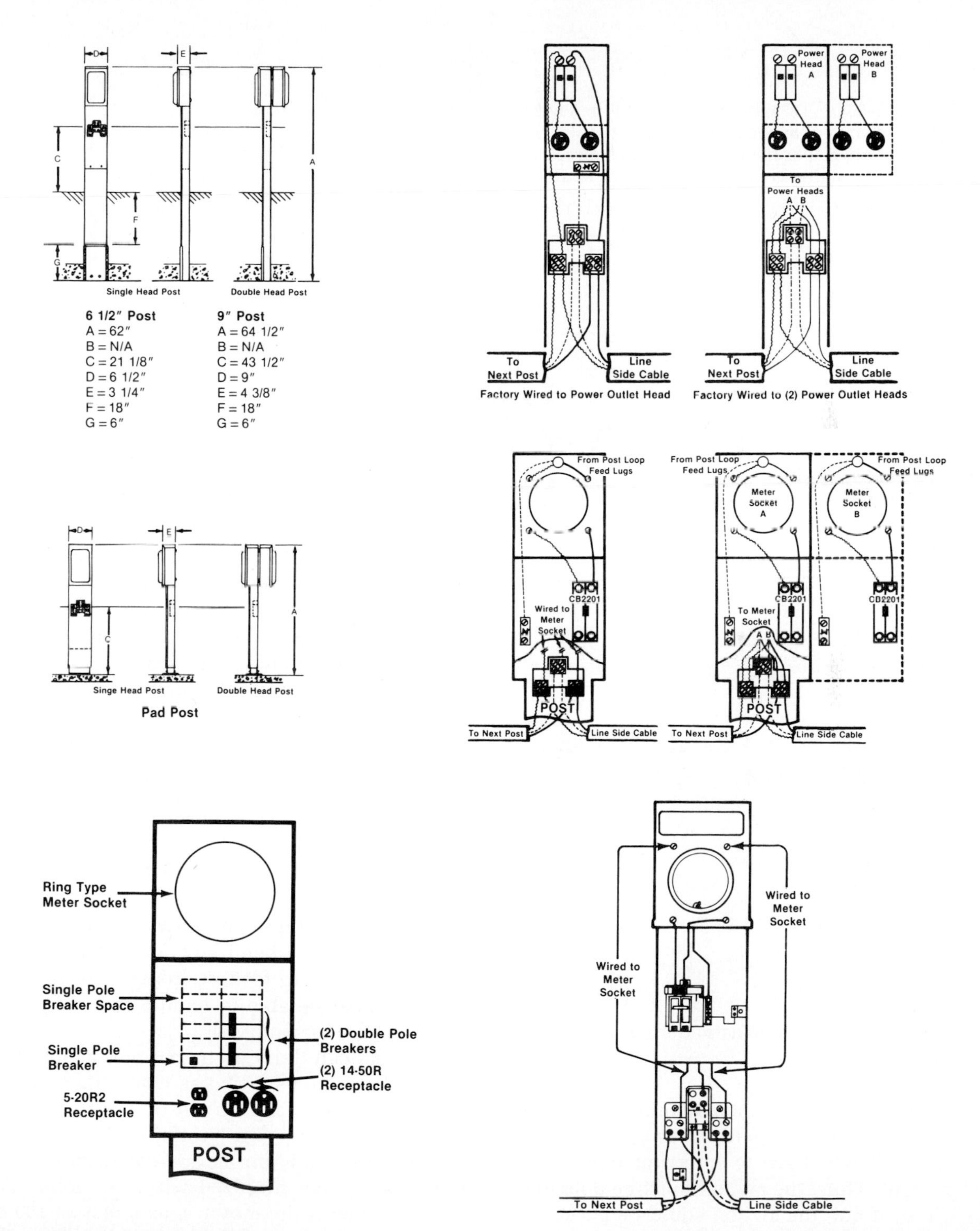

3–52. Wiring of the post shown in Fig. 3–51.

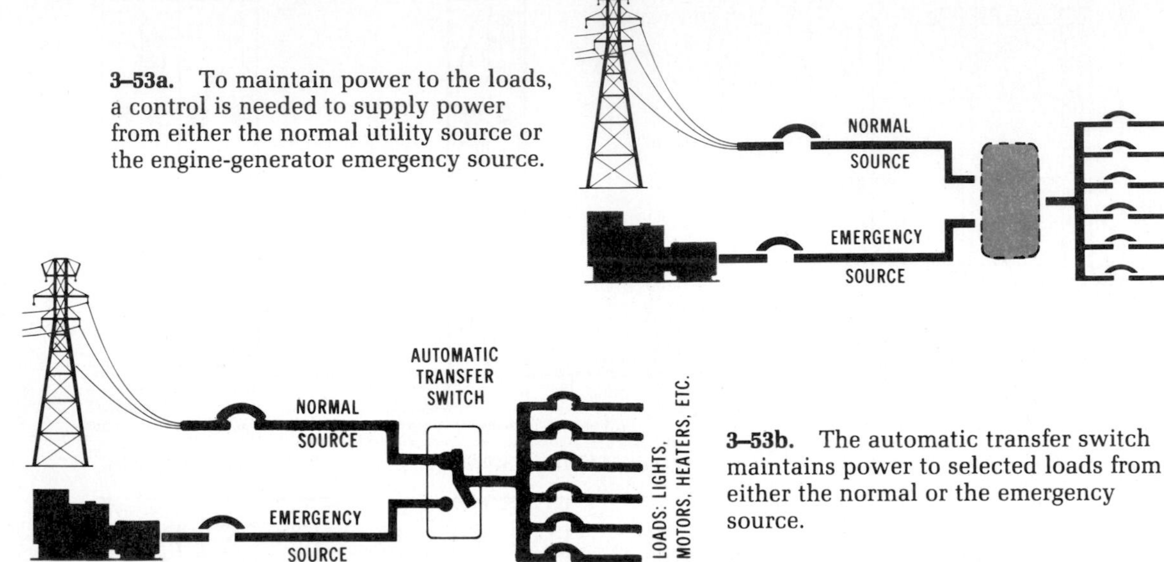

3–53a. To maintain power to the loads, a control is needed to supply power from either the normal utility source or the engine-generator emergency source.

3–53b. The automatic transfer switch maintains power to selected loads from either the normal or the emergency source.

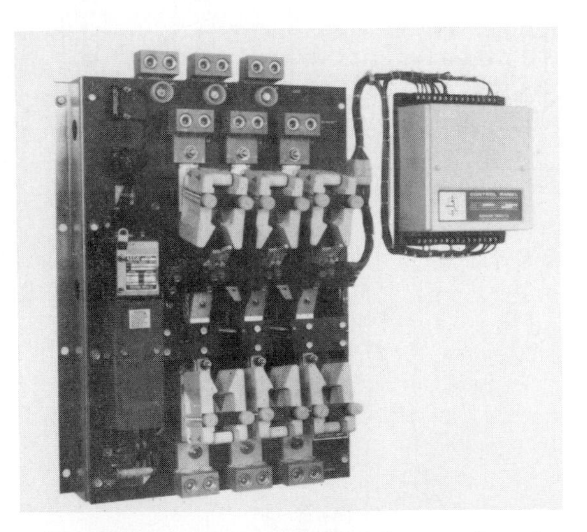

3–54. A 600-amp automatic transfer switch and control panel.

draw that much if it stalls while running. If the transfer switch operates at either of these times, it will have to interrupt that six-times current. Thus, the switch must be able to handle it; if not, the switch could be permanently damaged. Fig. 3-55.

All current goes through the transfer switch. Thus, a short circuit on the load side will cause the maximum available current to go through the transfer switch until the circuit breaker opens the line. Therefore, the transfer switch must also withstand short circuit loads.

Magnetic forces are so great during a short circuit that they can cause the contacts of an inadequately designed switch to open. Thus, the switch must be able to lock its contacts closed until the circuit breaker operates.

Tungsten lamps can draw up to 16 times more current when they are cold. Thus, a transfer switch feeding power to tungsten lamps must be able to handle this above-normal *inrush current*.

Motors can draw up to 15 times as much inrush current on starting as when they are running normally. The inrush of 15 times results when a motor is still running and is reconnected to a new normal or emergency source that is not synchronized with the motor. This can happen if the motor is as much as 180 degrees out-of-phase with the new source.

3–55. Contractor installing a 150-amp automatic transfer switch.

Electronic Loads

Almost every power system has some electronic load. Increased automation and the economies of solid-state control suggest that ever larger percentages of power system loads will be electronic in the future. Typical examples of electronic loads include data processing equipment, intensive care unit monitoring equipment, numerical control machinery, and security equipment.

There is virtually no commercial or industrial facility in which electronic loads are not becoming a critical part of the overall load profile.

Uninterruptible Power Systems (UPS)

Electronic loads are very susceptible to voltage and frequency variations. As a typical example, real-time access computers require an AC power system that does not deviate from nominal voltage by more than +8% or less than −10%. Allowable frequency deviations are typically ±0.5 Hz. In addition to the deviation stated, some data processing loads are affected by rate of change of deviation as well. To protect computer systems, uninterruptible power systems (UPS) have been developed.

Essentially there are two types of UPS systems. *The motor generator flywheel set* and the *static inverter with battery backup*. Due to advances in solid-state equipment, the more commonly used UPS is now the static inverter with battery backup. This is primarily due to the lower cost per kVA and the elimination of high starting kVA requirements.

A solid-state UPS is made up of three major sections. The first section is the rectifier that converts the AC input into DC. The second section is a DC bus on which floats a battery system. The third section is the static inverter that converts the DC back to a clean AC sine wave. The principal function of the UPS is to isolate the protected electronic load from power deviations that would affect the operation of the connected electronic load. Any short-term outage or transient that occurs on the AC power bus is filtered out or overridden by the battery in the DC bus section providing constant power into the inverter.

Sometimes the battery supply of the UPS is required to carry the output of the inverter for extended periods of time (in excess of 3 to 5 minutes). Upon reconnecton of the AC power source to the rectifier, the UPS draws power not only to carry the output load, but also to recharge the batteries.

QUESTIONS

1. Why are transformers needed in a distribution system?
2. Is most electrical power generated as three-phase?
3. What is a shareholder-owned electric company?
4. What are the advantages and disadvantages of an underground electrical distribution system?
5. What are the noticeable trends in underground wiring?
6. Which voltages are used in distributing power underground in residential areas?
7. What type of aluminum cable is used most widely on primary circuits?
8. What are some methods used in installing underground distribution cables?
9. How deep are cables usually buried?
10. What is the greatest cause of trouble in underground cables?
11. For what is a Belleville washer used?
12. Why are underground distribution systems more reliable than overhead?
13. What are the two basic types of faults in underground cables?
14. How are faults in underground cables found?
15. What was the reason for the long delay in obtaining electrical service for rural areas of the United States?
16. What federal agency is involved in regulating electrical-power generation and distribution in the United States?
17. What is the REA? When did it start? Why?
18. What is the phase of the power used in rural distribution lines?
19. What voltages are used in rural distribution lines?
20. What type of grounding system do rural distribution lines use? Why can they operate this way?
21. Identify the two types of UPS systems.

KEY TERMS

automatic transfer switch
inrush current
motor starting current
Rural Electrification Administration
rural grid
sectionalizing
shareholder-owned electric companies
underground cable
uninterruptible power

SECTION TWO WIRING CODES and ELECTRICAL SAFETY

CHAPTER 4

WIRING CODES AND REGULATING AGENCIES

Objectives

After studying this chapter, you will be able to:

- State the purpose of the National Electrical Code.
- State the objectives of Underwriters' Laboratories.

Electricity hasn't been around too long in the relative span of time. Still, it is old enough to be considered an energy source no one would want to be without.

This wasn't always the case, however. Thomas Edison met with resistance from those who preferred gas lighting. To this day, we have some streets lighted by gas. But open flame, with its constant threat of uncontrolled fires, has always been a danger to people and their wooden houses and buildings.

Some examples of the destruction that fire has wrought: Property worth more than $15 million was destroyed in New York City when 13 acres burned in 1835. Much of the town of Charleston, South Carolina, burned to the ground in 1861, with $10 million damage. In Portland, Maine, following the July 4th celebration in 1866, a $10 million blaze destroyed 1500 buildings. At today's prices this would be billions instead of millions.

Because this series of fires meant heavy losses to fire insurance companies, they began a cooperative effort to prevent fires. The New York Board of Fire Underwriters was formed in May of 1867. The function of this organization was to arrange for inspection of premises. It was also to make recommendations for the removal of possible fire hazards. In 1882, the Committee of Surveys drew up a set of safeguards for arc and incandescent lighting. These safeguards were the forerunners of the present National Electrical Code.

A letter from Thomas Edison to the New York Board of Fire Underwriters, in May of 1881, shows how firmly he believed that his electric lighting company was "... free from any possible danger from fire." He believed his system of lighting was far superior to natural gas, that it provided protection from fire, and was safer in all regards. Fig. 4-1.

As you know, electricity can cause shock and fire when improperly used or installed. This is why there are a number of codes written for the protection of those who generate and use electricity.

NATIONAL ELECTRICAL CODE (NEC)

The *National Electrical Code*, first developed in 1911, is sponsored by the National Fire Protection Association, an organization established in 1896.

The Code is reprinted every three years. There are a number of changes each time a new edition is issued. Additions or changes are agreed upon by a committee which, after studying recommendations from many sources, passes on their final form and wording before presenting them to the Association for publishing.

"This Code is purely advisory as far as the National Fire Protection Association and National Standards Institute are concerned, but it is offered for use in law and for regulatory purposes in the interest of life and protection of property."†

The National Electrical Code serves as a basis on which local governmental authorities can write ordinances that deal with protection of the lives of people working with or using electricity or electrical devices. Local laws almost always refer to the National Electrical Code as the "standard minimum," sometimes making additions to it to meet local conditions.

For example, burying of metal in some types of soils can cause deterioration of the metal used for grounding, or for conduits, more quickly than for most other locations. This unique soil condition must therefore be pointed out in a local modification to the National Electrical Code.

The Code is concerned with the how of electrical installation. It states the approved quantity and sizes of wire that can safely be used in a conduit as well as the number of wires in a box, and whether the wire should be aluminum or copper.

For example, No. 12 AWG aluminum wire, also copper-clad, can carry 15 amperes safely, if the types of wire are RUW, T, TW and if the temperature in which the wire is installed for operation does not exceed 60°C (140°F). This assumes that there are not more than three conductors in a raceway (conduit) or buried cable.

RUW means the insulation is made of moisture-resistant latex rubber and is suitable for dry and wet locations with a maximum operating temperature of 60°C (140°F).

T means the insulation is thermoplastic, with the same maximum temperature, but for use in dry locations only.

TW means the insulation is moisture-resistant thermoplastic, for the same maximum temperatures, and can be used in both dry and wet locations.

The National Electrical Code may be bought in paperback or as a hardcover book. It contains much of the information needed by an electrician on the job. It has definitions of electrical terms and chapters on wiring design and protection; wiring methods and materials; equipment for general use; special occupancies, equipment, and conditions; sections on communications; and tables, an index, and an appendix. A membership list of the National Electrical Code (NEC) Committee is included stating titles held, and addresses of individual members. The members come from all parts of the country. This gives greater input to Code revisions, keeping it current and, as nearly as possible, applicable to the entire country. The 1981 Code, for example, contains a very much revised article on swimming pools and hot tubs (Article 680). The use of ground-fault interrupt-

†National Fire Protection Association, National Electrical Code, 1984.

The Edison Electric Light Company,

65 Fifth Avenue,

Norvin Green, Pres.
S. B. Eaton, Vice-Pres.
E. P. Fabbri, Treas.
C. Goddard, Sec'y.

New York, May 6th. 1881

To the New York Board of Fire Underwriters.

No. 115 Broadway, New York City.

Gentlemen:-

Referring to a publication in the New York Evening Post of yesterday, stating that your Board has passed a resolution requesting persons interested in Electric Lighting to furnish any facts to the Board bearing upon the subject of danger of fire from electric wires, I beg to say that the system of Electric Lighting of the Edison Electric Light Company is absolutely free from any possible danger from fire, even in connection with the most inflamable material: and that it is the intention of the Edison Company, before actually furnishing light, to the public, to invite your Board to give a most critical test of the absolute safety of the system, by the aid of such experts as you may select. From the outset I have had especially in view this subject of protection from fire, and I have succeeded in perfecting a system, which, in that respect is not only safer than Gas, but is, as I will show to your Board, absolutely secure under all and every condition.

Very respectfully
Your obedient servant,

Thomas A Edison

4–1. Letter from Thomas Edison.

ers is called for here to protect those around the pool as well as in the water. Grounding methods for electrical devices are stressed in the handbook.

No electrician should be without the latest edition of the Code book in his tool box. It has most answers for problems in the wiring of electrical installations.

The Code book may be obtained from the:

National Fire Protection Association
Battery March Park
Quincy, Massachusetts 02269

UNDERWRITERS' LABORATORIES

Underwriters' Laboratories, Inc., was founded in 1894 by William Henry Merrill. He came to Chicago to test the installation of Thomas Edison's new incandescent electric light at the Columbian Exposition. He later started the UL for insurance companies to test products for electric and fire hazards. It continued as a testing laboratory for insurance underwriters until 1917. It then became an independent, self-supporting, safety-testing laboratory. The National Board of Fire Underwriters (now American Insurance Association) continued as sponsors of UL until 1968. At that time, sponsorship and membership were broadened to include representatives of consumer interests, governmental bodies or agencies, education, public safety bodies, public utilities, and the insurance industry, in addition to safety and standardization experts.

UL has expanded its testing services to more than 13,000 manufacturers throughout the world. Over one billion UL labels are used each year on products listed by Underwriters' Laboratories.

UL is chartered as a not-for-profit organization without capital stock, under the laws of the State of Delaware, to establish, maintain, and operate laboratories for the examination and testing of devices, systems, and materials.

Its objectives, as stated by UL, are: "By scientific investigation, study, experiments and tests, to determine the relation of various materials, devices, products, equipment, constructions, methods and systems to hazards appurtenant thereto or to the use thereof, affecting life and property, and to ascertain, define and publish standards, classifications and specifications for materials, devices, equipment, construction, methods and systems affecting such hazards, and other information tending to reduce and prevent loss of life and property from such hazards."

The corporate headquarters, together with one of the testing laboratories, is located at 207 East Ohio Street, Chicago, Illinois. Other offices and laboratories are located at Northbrook, Illinois; at Melville, New York; at Santa Clara, California; and at Tampa, Florida.

Underwriters' Laboratories, Inc., has a total staff of over 2000 employees. More than 700 persons are engaged in engineering work, and of this number, approximately 425 are graduate engineers. Fig. 4-2. Supplementing the engineering staff are more than 500 factory inspectors.

Engineering functions of Underwriters' Laboratories are divided among these six departments:

- Burglary protection and signaling. Fig. 4-3.
- Casualty and chemical hazards.
- Electricity.
- Fire protection. Fig. 4-4.
- Heating, air conditioning, and refrigeration.
- Marine equipment. Fig. 4-5.

The electrical is the largest of the six engineering departments. Safety evaluations are made on hundreds of different types of appliances for use in homes, commercial buildings, schools, and factories. The scope of the work in this department includes electrical construction materials which are used in buildings to distribute electrical power from the meter location to the electrical outlet. Fig. 4-6.

4–2. Men testing furnaces and air-conditioning equipment at the Underwriters' Laboratories.

4–3. Firing a weapon at glass to see if it is bulletproof. This is one of UL's testing procedures.

4–4. Testing fire extinguishers at UL.

4–5. This tank is used to check buoyancy of jackets, vests, rings, and cushions for compliance with Underwriters' Laboratories and U.S. Coast Guard requirements. Services which comply with all requirements are listed by UL and approved by the USCG.

4–6. This man is checking electrical equipment at the Underwriters' Laboratories' test facility.

Product Listing Service

A person or business desiring an investigation and report on a product may write the UL, giving a brief description of the product. From this information, the Laboratories determine the nature and extent of the necessary examination and test. An application form is then sent to the submitter. Information is requested concerning the kind and number of samples to be furnished, and the cost limit the applicant will agree to. A description of the work to be performed is included, along with the inspection service to be established, if and when the product is found acceptable.

Reports. At the completion of the investigation, a report is given to the applicant. Any objectionable features disclosed by the investigation of the product must be corrected. Later, revised samples must be found acceptable before a listing can be published. With some new products, a report, including recommendations of the staff, may be submitted to one or more of the five engineering councils of Underwriters' Laboratories. These councils are composed of prominent persons who have outstanding knowledge and field experience in areas of public safety. Final listing of an appliance or material may depend on the concurrence of the council concerned. This provides a countercheck of the findings of the staff with the judgment of others who have wide field experience.

Procedures. A manufacturer whose product meets the Laboratories' requirements is provided with a procedure prepared by the engineers. It describes and illustrates the product in detail, particularly its construction and test performance. The procedure becomes the manufacturer's guide for future production and is used by the Laboratories' inspectors for periodic review of the listed product. Procedures are kept up to date by the manufacturer, who informs the Laboratories of any changes in the product or additions to the line.

Follow-up Services. Of equal importance with examination and test work is the follow-up program by UL in the factories of clients and in the marketplace. Follow-up service is designed to serve as a formal check on the supervision which the manufacturer exercises. The follow-up determines compliance of the product with the applicable requirements of the Laboratories.

A manufacturer must conduct specified examinations and tests to be certain the product is in compliance with UL requirements. A Laboratories' representative makes frequent visits to the factory in which the products are manufactured for the purpose of checking the efficiency of the manufacturer's own inspection program.

Should examination or testing by the Laboratories' representative disclose features not in compliance with requirements, the manufacturer must either correct such items or remove any UL identification from the product. In some cases, examinations and tests are conducted by the Laboratories on products purchased on the open market, a procedure which serves as a countercheck on factory inspections.

Inspection centers and representatives are located in more than 200 cities in the United States and Puerto Rico, and in 28 foreign countries, including Canada, Mexico, Japan, Yugoslavia, and several in Western Europe. Fig. 4-7.

More than 200 000 factory follow-up inspections are conducted on listed products each year by the UL inspection force.

Special Services. Underwriters' Laboratories publishes annual lists of manufacturers whose products have met UL safety requirements. These lists are kept up to date by quarterly supplements. There are eleven lists published each year.

The UL presently publishes more than 300 *Standards for Safety* for materials, devices, constructions, and methods. Copies of these are available to interested persons, and a catalog is available free.

4–7. Inspection centers maintained by UL.

Research

UL undertakes numerous research projects in areas of safety. More than 60 research bulletins have been published.

For Your Own Safety

For your own safety, the products you are using to wire a house, building, or installation of any kind should be marked with the "UL." Fig. 4-8.

CANADIAN STANDARDS ASSOCIATION

The parallel organization to the UL in Canada is the *Canadian Standards Association*, or CSA. However, in comparison to the UL, CSA has more authority to remove from the market products which do not meet standards. The UL program is strictly voluntary. If an electrical (or in some cases other type) product used in Canada is connected in any way with the consumption of power from the electrical power sources owned by the provinces, that product must have CSA approval. This is a measure in the interest of public safety. Fig. 4-9.

4–8. UL stickers and labels.

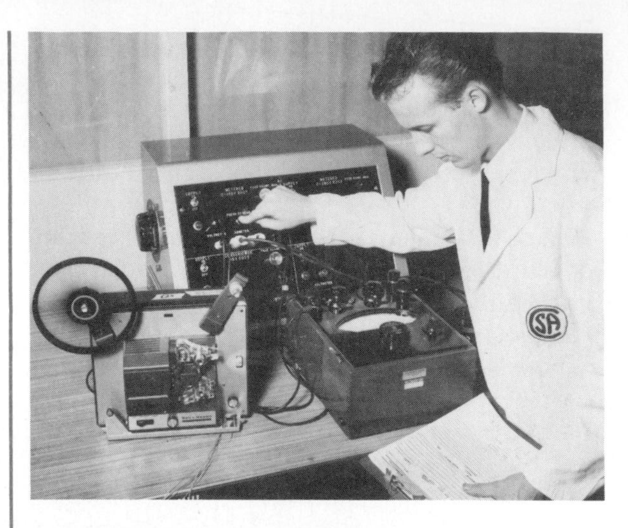

4–9. This technician reads a meter for the Canadian Standards Organization to determine power consumed by an 8mm movie projector.

History of the CSA

As the Canadian economy expanded after the turn of the century, it became apparent that technical standardization was essential if Canada was to compete in the world markets. The original standards association was incorporated under the Dominion Companies Act in 1919, with the name Canadian Engineering Standards Association, a nonprofit, independent, voluntary organization.

By Supplementary Letters Patent, in 1944, this association became the Canadian Standards Association and extended its activities to a broader field of standardization.

Purpose of the Association

The basic objectives of the CSA are:

- To develop voluntary national standards.
- To provide certification services for national standards.
- To represent Canada in international standards activities.

Location and Operation of the Association

CSA headquarters and major testing facilities are located in Rexdale, a suburb of Toronto, in Ontario, Canada. These facilities house a staff of over 400 which includes professional engineers and skilled laboratory technicians, in addition to management and secretarial personnel. Figs. 4-10 & 4-11.

Regional offices and test facilities in Montreal, Winnipeg, and Vancouver are maintained for the convenience of clients across Canada. There is a branch office in Edmonton, Alberta, and executive offices are located in the national capital of Ottawa, Ontario. Fig. 4-12.

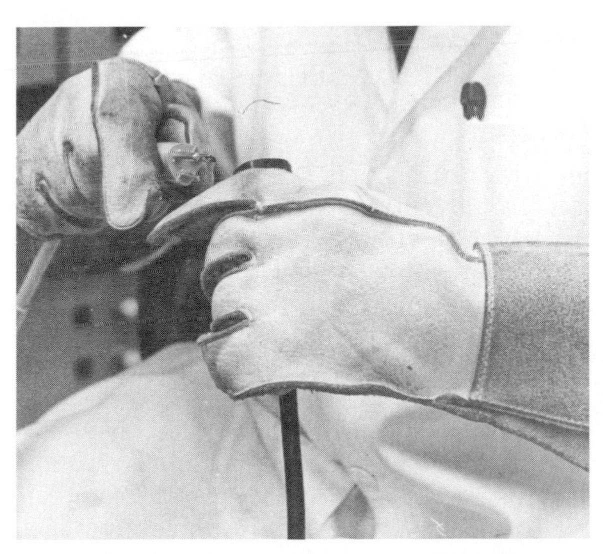

4–10. A technician performs an overload test on a cord-connector body. The cord-connector body, rated at 15 A, 125 V, is tested by manually inserting and withdrawing a cap or plug 50 times, at a rate not exceeding 10 cycles of operation per minute. The test current is 150% of the rated current of the connector body. The test is conducted on DC.

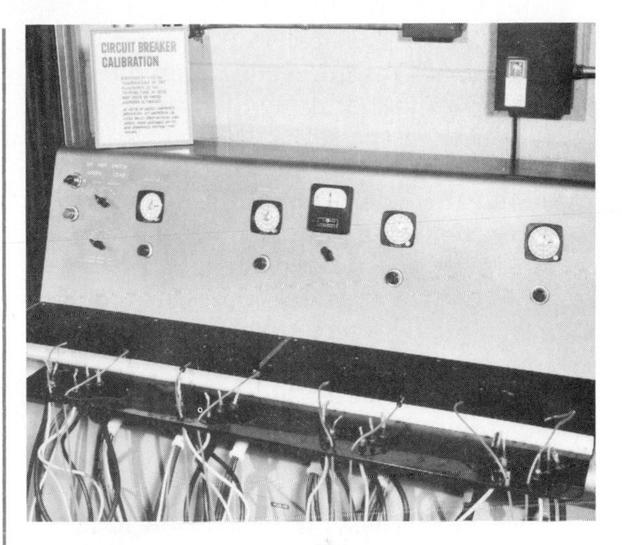

4–11. This circuit breaker calibration console made by CSA technicians is used to check the time it takes for circuit breakers to open at overloads of 125% and 200% of their current ratings. Samples of a manufacturer's product are continually checked.

4–12. Electrical properties of samples of wires and cables are measured in water baths at elevated temperatures.

Who Writes the Standards?

CSA committees are made up of more than 3000 representatives from industry, government, consumer and safety associations, and of scientific and educational institutions who donate their time and technical skills to the development of standards. Producer and consumer interests, and geographical areas, are taken into account in this representation. The CSA staff provides secretarial services for most committees.

The Standards Policy Board is the final authority, prior to publication, within the Association for approval of technical provisions for completed proposed specifications.

There are ten fields in which CSA has standards:

- Construction and building industry.
- Mechanical engineering.
- Electrical engineering.
- Automotive industry.
- Metallurgy.
- Chemical and petrochemical industry.
- Packaging industry.
- Data processing industry.
- Industrial safety codes, products, and equipment.
- General standards.

How Is a Standard Developed? Although development and revision of a CSA standard is the responsibility of the Standards Committee, it must also be acceptable to the appropriate Sectional Committee and the Standards Policy Board. When the Board has been satisfied that a proposed standard will have suitable application and support, it authorizes the establishment of the project and assigns it to the appropriate Sectional Committee. A Standards Committee, composed of representatives of producers concerned with the item to be standardized, and representatives of consumer and general interests, then meets at intervals until agreement is reached. Its recommendations are then placed before the Sectional Committee and, if accepted, will be presented before the CSA Standards Policy Board for final decision and adoption as a CSA standard.

CSA standards are voluntary and recommendatory. They became mandatory only when adopted by an authority (municipal, provincial, national, etc.) having jurisdiction. All standards are subject to periodic review and revision, as conditions warrant. Established on the basis of minimum requirements, they do not restrict design and development. Approximately 1200 standards have been published to date.

How Does the Certification Progran Operate? A manufacturer, who may be from any part of the world, files an application with CSA. CSA engineers and technicians inspect and test the product for compliance with an applicable standard. If the product meets the standard (modifications may be required), the manufacturer will apply a CSA mark to the product, indicating certification. Fig. 4-13.

CSA provides testing, examination, and certification services covering a wide range of electrical equipment and components, including the industrial, commercial, scientific, and domestic. CSA is the testing authority concerned with safety which is accepted by inspection authorities in Canada.

More than 10 000 companies are served by the CSA certification program. Over 100 million marks are applied annually to CSA certified products. More than 9000 new products are submitted for CSA certification each year, and approximately 16 000 factory inspections are carried out in 25 countries.

4-13. The CSA trademark.

The CSA Mark. Products certified by CSA are eligible to bear the CSA certification mark. Misuse of the mark may result in suspension or cancellation of certification. CSA may resort to legal action to protect its registered trademark in the event of abuse. Fig. 4-13. In addition, CSA information tags and other markings are also made available for certified products and their containers, to supplement the CSA mark.

For your own safety, the CSA symbol on electrical equipment you purchase is an assurance that such equipment has passed rigid inspection.

QUESTIONS

1. What is the National Electrical Code?
2. When was the National Electrical Code first developed?
3. How often is the Code revised?
4. What is a ground-fault interrupter?
5. What is Underwriters' Laboratories? Why was it formed?
6. What does the UL do?
7. Why is the UL product listing service important to you?
8. What are the follow-up services of UL?
9. What is the CSA?
10. What does CSA do?
11. Who writes standards for CSA?
12. How does a CSA standard come about?
13. What is the CSA mark? What does it mean on a piece of equipment?

KEY TERMS

Canadian Standards Association (CSA)

National Electrical Code (NEC)

Underwriters' Laboratories (UL)

CHAPTER 5

ELECTRICAL SAFETY

Objectives

After studying this chapter, you will be able to:

- List the general safety precautions that must be observed when working with electrical equipment.
- List the important steps in first aid for victims of electrical shock.
- Distinguish a grounded conductor from a grounding conductor.
- Implement proper grounding and bonding procedures.
- Describe one technique for limiting EMI.
- Know the NEC requirements for grounding conductors.
- Identify the two main types of GFI.
- Identify the three classes of hazardous locations.

There are a number of devices designed to protect people from their carelessness and from electrical shock. If you work around electricity for long you will probably receive a shock; most likely it will be caused by your own negligence. Such a shock could be fatal. There is very little difference between the experience of a slight tingle and a fatal shock. Eight milliamperes is 0.008 amperes and 50 milliamperes is 0.050 amperes. There is then, only a 0.042-ampere difference between the shock you feel as a tingle and the amount of current that will kill you.

There have been cases where people have been shocked or burned by high voltages. Voltage does not necessarily kill—it is the *amount of current* through the body which does the damage.

These facts and figures are presented, not to scare you, but to make you aware of the dangers of being careless with the power in wires. This chapter will deal with some of the aspects of safety in electrical work. There are ways to be safe and live a long life. There are careless practices which are bound to mean a shorter life. Read these topics with the idea that the life saved could easily be your own.

FATAL CURRENT

Strange as it seems, most fatal electric shocks happen to people who should know better. Here are some "electromedical" facts that should make you think twice before taking chances.

It's the Current That Kills

Offhand, it would seem that a shock of 10 000 volts would be more deadly than a shock of 100 volts. But this is not so. Individuals have been electrocuted by appliances using ordinary house current of 110 volts and by electrical apparatus in industry using as little as 42 volts, direct current (DC). The real measure of a shock's intensity lies in the amount of current (milliamperes) forced through the body, and not in the voltage. Any electrical device used on a house wiring circuit can, under certain conditions, transmit a fatal current.

Although any amount of current over 10 milliamperes (0.01 A) is capable of producing painful to severe shock, currents between 100 and 200 mA (0.1 to 0.2 A) are lethal.

From a practical viewpoint, after a person is knocked out by electric shock, it is impossible to tell how much current passed through the vital organs of the body. Artificial respiration must be applied immediately if breathing has stopped.

The Physiological Effects of Electric Shock. Table 5-A shows the physiological effects of various current densities. Note that voltage is not a consideration. Although it takes voltage to make current flow, the amount of shock-current will vary, depending on the body resistance between the points of contact.

As shown in Table 5-A, shock is more severe as the current rises. At values as low as 20 mA, breathing becomes labored, finally ceasing completely even at values below 75 mA.

As current approaches 100 mA, ventricular fibrillation of the heart occurs—an uncoordinated twitching of the walls of the heart's ventricles.

■

A heart that is in fibrillation cannot be restricted by closed-chest cardiac massage. A special device called a defibrillator is available in some medical facilities and by ambulance services.

■

With more than 200 mA, muscular contractions are so severe that the heart is forcibly clamped during the shock. This clamping prevents the heart from going into ventricular fibrillation, making the victim's chances for survival better.

5-A. Physiological Effects of Electric Currents*

READINGS		EFFECTS
SAFE CURRENT VALUES	1 mA or less	Causes no sensation—not felt.
	1-8 mA	Sensation of shock, not painful; individual can let go at will since muscular control is not lost.
UNSAFE CURRENT VALUES	8-15 mA	Painful shock: individual can let go at will since muscular control is not lost.
	15-20 mA	Painful shock; control of adjacent muscles lost; victim cannot let go.
	20-50 mA	Painful, severe muscular contractions; breathing difficult.
	50-100 mA	Ventricular fibrillation—a heart condition that can result in instant death—is possible.
	100-200 mA	Ventricular fibrillation occurs.
	200 mA and over	Severe burns, severe muscular contractions—so severe that chest muscles clamp the heart and stop it for the duration of the shock. (This prevents ventricular fibrillation.)

Danger—Low Voltage. It is common knowledge that victims of high-voltage shock usually respond to artificial respiration more readily than the victims of low-voltage shock. The reason may be the above-mentioned clamping of the heart, due to the high current densities associated with high voltages. However, to prevent a misinterpretation of those details, remember that 75 volts are just as lethal as 750 volts.

The actual resistance of the body varies, depending upon the points of contact, and whether the skin is moist or dry. The area from one ear to the other, for example, has an internal resistance (which is lower than skin resistance) of only 100 ohms; from hand to foot it is nearer 500 ohms. Skin resistance may vary from 1000 ohms for wet skin, to more than 500 000 ohms for dry skin.

GENERAL SAFETY PRECAUTIONS

■ *When working around electrical equipment, move slowly.* Make sure your feet are firmly placed for good balance. Don't lunge after falling tools. Kill all power and ground all high-voltage points before touching wiring. Make sure that power cannot be accidentally restored. Do not work on ungrounded equipment.

■ *Do not examine live equipment when you are mentally or physically fatigued.* Keep one hand in your pocket while investigating live electrical equipment. Most important, do not touch electrical equipment while standing on metal floors, damp concrete, or other well-grounded surfaces. Do not handle electrical equipment while wearing damp clothing (particularly wet shoes), or while skin surfaces are damp.

■ *Do not work alone!* Remember, the more you know about electrical equipment, the more heedless you're apt to become. Don't take unnecessary risks.

WHAT TO DO FOR VICTIMS

■ *Cut voltage and/or remove victim from contact as quickly as possible—but without endangering your own safety.* Use a length of dry wood, rope, blanket, or similar device to pry or pull the victim loose. Don't waste valuable time looking for the power switch. The resistance of the victim's contact decreases with passage of time. The fatal 100 mA to 200 mA level may be reached if action is delayed.

■ *Start artificial respiration at once if the victim is unconscious and has stopped breathing.* Do not stop resuscitation until a medical authority pronounces the victim beyond help. It may take as long as eight hours to revive a patient. There may be no pulse; a condition similar to rigor mortis may be present. However, these are the manifestations of shock, not necessarily indications that the victim has died.

Table 5-B gives some of the resistances the human body presents to an applied voltage. If you want to know the amount of current which would pass through the body, use Ohm's Law formulas and the applied voltage. Will it be enough to produce a fatal shock?

$$\text{Current through body} = \frac{\text{Voltage applied to body}}{\text{Resistance of body and contacts}}$$

5-B. Human Resistance to Electrical Current

BODY AREA	RESISTANCE, OHMS
Dry skin	100 000-600 000
Wet skin	1000
Internal body—hand to foot	400-600
Ear to ear	(about) 100

Shock severity depends upon the following factors:

- Becomes more severe with increased voltage.
- Increases with the amount of moisture on contact surfaces.
- Increases with an increase in pressure of contact.
- Increases with an increase in area of body contact.
- Resistance of body portions.

Three factors involved in electric shock:

- Voltage (analogous to pressure forcing water in pipe).
- Current (analogous to water in pipe flowing).
- Resistance (analogous to something in pipe tending to hold back water flow).

Practical Problem

A worker with wet clothes, or wet with perspiration, comes in contact with a defective 120 V light cord and establishes a good ground:

From preceding table:

Wet-skin resistance	1000 ohms
Body resistance	500 ohms
Internal resistance	1500 ohms

From formula:

$$\text{Current through body} = \frac{120 \text{ volts}}{1500 \text{ ohms}} = 80 \text{ mA}$$

Note that muscular control is lost and the victim cannot break contact. As the contact is continued, the skin resistance is reduced. If the skin is punctured, skin resistance may then be disregarded. Then for practical purposes, the total resistance of the worker may be in the neighborhood of 600 ohms.

$$\text{Current through body} = \frac{120 \text{ volts}}{600 \text{ ohms}} = 20 \text{ mA (certain death)}$$

There is always a danger of an electrical wire with high voltages leaking to ground through another path (if the ground wire is broken), and you may become the path of least resistance if you come too close. Table 5-C.

In order to be safe, keep away from high-voltage wires. However, if you work with them or on them, make sure you use the proper safety procedures and clothing.

5-C. Safe Distances from Live Circuits in Air

Do not approach live conductors closer than the following distances

751	to	3 500 V	1 ft
3 501	to	10 000 V	2 ft
10 001	to	50 000 V	3 ft
50 001	to	100 000 V	5 ft
100 001	to	250 000 V	10 ft

TYPICAL ELECTRICAL SHOCK HAZARDS IN INDUSTRY

It is impractical to list here all possible electrical shock hazards which might be encountered in industry, but the following list may be helpful in locating hazardous situations. These are divided into two broad categories of unsafe physical conditions and unsafe actions which might lead to injury. Control measures are suggested. Table 5-D.

Extension Cords

A proper extension cord can make a difference in the safe operation of a piece of equipment. Make sure it has the proper capacity to handle the current needed. See Table 5-E.

5-D. Common Electrical Shock Hazards in Industry

UNSAFE PHYSICAL CONDITIONS	CONTROL MEASURES
Worn insulation on extension and drop cords. Splices on cords.	Install a system of inspection and preventive maintenance to uncover dangerous conditions, and to correct them. Use UL approved materials only. Spliced cords should be removed from service.
Exposed conductors at rear of switchboard.	Enclose rear of switchboard to prevent exposure of unauthorized persons. Provide rubber mats for workers who must enter the enclosure.
Open switches and control apparatus on panel and switchboards. Location of machine switches.	Provide enclosed safety switches. Insulate with rubber mats in front of switch and control equipment. Locate machine switches so as not to create hazard to the operator.
Unsafe wiring practices, such as using wires too small for the current being carried; open wiring not in conduit; temporary wiring; wiring improperly located.	Comply with recognized electrical code. Remove temporary wiring as soon as it has served its purpose.
The accidental energizing of noncurrent-carrying parts of machines and tools by means of short circuits, breaks in insulation, etc.	Properly ground all noncurrent-carrying parts of machines, tools, and frames of control equipment.
UNSAFE ACTIONS	**CONTROL MEASURES**
Working on "live" low voltage circuits in the belief that they are not hazardous.	Educate and train workers in the hazards of low voltage circuits.
Working on "live" circuits which are thought to be "dead."	Require that switches on all circuits being worked on be locked open and properly tagged. Use protective equipment such as rubber gloves, blankets, etc.
Replacing fuses by hand on "live" circuits,	Open switch before replacing fuses; use fuse pullers.
Using 120-volt lighting circuits for work in boiler or other similar enclosures.	Use low voltage circuits: 6 volts for lighting, not over 30 volts for power.
Overloading circuits beyond their capacity.	Lock fuse boxes to prevent bridging or replacing with heavier fuse.
Abusing electrical equipment, and poor housekeeping about electrical equipment.	Institute safe work practices, with inspection and preventive maintenance of equipment. Improve housekeeping practices.

5-E. Flexible Cord Ampacities*

	TYPE S, SO, ST	TYPE SJ, SJO, SJT
AWG	AMPERES	
18	7, 10*	7, 10*
16	10, 13*	10, 13*
14	15, 18*	
12	20	
10	25	
8	35	
6	45	
4	60	

Wire size and insulation of extension cords must be considered to assure proper operation with a piece of equipment. Take a look at Table 5-F for an explanation of the letters on a cord body. This chart will help you determine the type of service for which a specific cord is recommended.

Length of an extension cord is important in the sense that it must have the correct size wire to allow full voltage to reach the consuming device. For instance, a 25-foot cord with No. 18 wire size is good for 2 amperes. If the distance

5-F. Types of Flexible Cord

TRADE NAME	TYPE LETTER	SIZE AWG	NO. OF CONDUCTORS	INSULATION BRAID ON	BRAID ON EACH CONDUCTOR	OUTER COVERING	USE		
Junior Hard Service Cord	SJ SJO SJT	18, 14	2, 3 or 4	Rubber Thermo plastic or rubber	None	Rubber Oil resist compound Thermoplastic	Pendant or portable	Damp places	Hard usage
Hard Service Cord	S SO ST	6 18, to 10 incl.	2 or more	Rubber Thermoplastic or rubber	None	Rubber Oil resist compound Thermoplastic	Pendant or portable	Damp places	Extra hard usage

is increased to 50 feet, you're still safe with No. 18 and 2 amperes. However, for a distance of 200 feet, the size of the wire must be increased to No. 16, to carry the 2 amperes without dropping the voltage along the cord and therefore producing a low voltage at the consuming device at the end of the line. See Table 5-G.

Inspection and Preventive Maintenance. All drop or extension cords and other electrical equipment should be the proper size for the purpose intended. Wire size and insulation must be adequate for the service to be expected in industrial use. Only equipment and cords listed by Underwriters' Laboratories, Inc. for the intended purpose should be used. Cords which are perfectly safe when purchased may be dangerous a short time later, unless they are given careful use. They should be kept free of oil and water, and protected from abrasion or other mechanical damage.

A system of preventive maintenance is suggested as a means of control over portable electrical equipment. Preventive maintenance involves inspection at regular intervals, the keeping of records of inspections, and making repairs when necessary. Splicing of extension cords should never be permitted.

Electric Plugs and Receptacles

Electric plugs and receptacles come in a wide variety of shapes and arrangements. Because there must be some regulation to ensure that plugs and receptacles are used properly, the National Electrical Manufacturers Association (NEMA) has developed a standard for manufacturing these devices. Table 5-H will show how different voltages and current combinations are specified for a particular plug or receptacle. Note the wiring diagram, shown for three-phase, single-phase, wye and delta connections, and wye and delta with various grounds. You have to be very careful when wiring these plugs to make sure the proper terminal is connected.

5-G. Size of Extension Cords for Portable Electric Tools

THIS TABLE FOR 115-VOLT TOOLS.						
	FULL-LOAD AMPERE RATING OF TOOL					
	0-2.0 A	2.10-3.4 A	3.5-5.0 A	5.1-7.0 A	7.1-12.0 A	12.1-16.0 A
LENGTH OF CORD, FEET	WIRE SIZE (AWG)					
25	18	18	18	16	14	14
50	18	18	18	16	14	12
75	18	18	16	14	12	10
100	18	16	14	12	10	8
200	16	14	12	10	8	6
300	14	12	10	8	6	4
400	12	10	8	6	4	4
500	12	10	8	6	4	2
600	10	8	6	4	2	2
800	10	8	6	4	2	1
1000	8	6	4	2	1	0

Grounded and Grounding Conductors

One of the major difficulties in utilizing the proper configuration for its proper voltage is a misunderstanding, or lack of knowledge, in distinguishing between the terms *grounded conductor* (neutral wire) and *grounding conductor*. Although these two terms appear to be quite similar, the similarity ends with the appearance, as the following will show. Should these two terms be misunderstood in a circuit, the results could easily prove fatal.

Systems grounds (more commonly called neutral wires) are grounded conductors; they complete the circuit and normally carry current, although at ground potential. (Many people who use polarized, two-wire plug caps have the mistaken impression that only the ungrounded conductor, or "hot wire" carries current.)

Equipment grounds are referred to here as grounding conductors; they carry current only if they happen to be energized by an electrical fault (contact between the hot wire and the equipment enclosure).

■ A *grounded conductor* has a white-colored jacket in a two-or three-wire cable (neutral wire). It is terminated to the white, or silver-colored terminal in a plug cap or connector; and it is terminated at the neutral bar in the distribution box.

When there is an electrical fault that allows the hot line to contact the metal housing of electrical equipment, (in a typical two-wire system) or some other ungrounded conductor, any person who touches that equipment or conductor will receive a shock. The person completes the circuit from the hot line to the ground, and current passes through the body.

5-H. Voltage and Current Combinations for Plugs and Receptacles

WIRING DIAGRAM	NEMA *ANSI*	RECEP-TACLE CONFIG-URATION	RATING	
	5-15 C73.11		15A 125V	2 POLE 3 WIRE
	5-20 C73.12		20A 125V	2 POLE 3 WIRE
	5-30 C73.45		300A 125V	2 POLE 3 WIRE
	5-50 C73-46		50A 125V	2 POLE 3 WIRE
	6-15 C37.20		15A 250V	2 POLE 3 WIRE
	6-20 C73.51		20A 250V	2 POLE 3 WIRE
	6-30 C73.52		30A 250V	2 POLE 3 WIRE
	6-50 C73.53		50A 250V	2 POLE 3 WIRE
	7-15 C73.28		15A 277V	2 POLE 3 WIRE
	7-20 C73.63		20A 277V	2 POLE 3 WIRE
	7-30 C73.64		30A 277V	2 POLE 3 WIRE
	7-50 C73.65		50A 277V	2 POLE 3 WIRE
	10-20 C73.23		20A 125/250V	3 POLE 3 WIRE
	10-30 C73.24		30A 125/250V	3 POLE 3 WIRE
	10-50 C73.25		50A 125/250V	3 POLE 3 WIRE
	11-15 C73.54		15A 3ϕ 250V	3 POLE 3 WIRE
	11-20 C73.55		20A 3ϕ 250V	3 POLE 3 WIRE

	RATING	RECEP-TACLE CONFIG-URATION	NEMA *ANSI*	WIRING DIAGRAM
3P3W	30A 3ϕ 250V		11-30 C73-56	
3P3W	50A 3ϕ 250V		11-50 C73.47	
3 POLE 4 WIRE	15A 125/250V		14-15 C73.49	
3 POLE 4 WIRE	20A 125/250V		14-20 C73.50	
3 POLE 4 WIRE	30A 125/250V		14-30 C73.16	
3 POLE 4 WIRE	50A 125/250V		14-50 C73.17	
3 POLE 4 WIRE	60A 125/250V		14-60 C73.18	
3 POLE 4 WIRE	15A 3ϕ 250V		15-15 C73.58	
3 POLE 4 WIRE	20A 3ϕ 250V		15-20 C73.59	
3 POLE 4 WIRE	30A 3ϕ 250V		15-30 C73.60	
3 POLE 4 WIRE	50A 3ϕ 250V		15-50 C73.61	
3 POLE 4 WIRE	60A 3ϕ 250V		15-60 C73-62	
4 POLE 4 WIRE	15A 3ϕY 120/208V		18-15 C73.15	
4 POLE 4 WIRE	20A 3ϕY 120/208V		18-20 C73.26	
4 POLE 4 WIRE	30A 3ϕY 120/208V		18-30 C73.47	
4 POLE 4 WIRE	50A 3ϕY 120/208V		18-50 C73.48	
4 POLE 4 WIRE	60A 3ϕY 120/208V		18-60 C73.27	

Continued on Next Page

5-H. Voltage and Current Combinations for Plugs and Receptacles (Continued)

RATING		RECEPTACLE CONFIGURATION	NEMA *ANSI*	WIRING DIAGRAM
3P3W	30A 3φ 480V		L12-30 C73.102	
3P3W	30A 3φ 600V		L13-30 C73.103	
3 POLE 4 WIRE	20A 125/250V		L14-20 C73.83	
3 POLE 4 WIRE	30A 125.250V		L14-30 C73.84	
3 POLE 4 WIRE	20A 3φ 250V		L15-20 C73.85	
3 POLE 4 WIRE	30A 3φ 250V		L15-30 C73.86	
3 POLE 4 WIRE	20A 3φ 480V		L16-20 C73.87	
3 POLE 4 WIRE	30A 3φ 480V		L16-30 C73.88	
3 POLE 4 WIRE	30A 3φ 600V		L17-30 C73.89	
4 POLE 4 WIRE	20A 3φY 120/208V		L18-20 C73.104	
4 POLE 4 WIRE	30A 3φY 120/208V		L18-30 C73.105	
4 POLE 4 WIRE	20A 3φY 277/480V		L19-20 C73.106	
4 POLE 4 WIRE	30A 3φY 277/480V		L19-30 C73.107	
4 POLE 4 WIRE	20A 3φY 347/600V		L20-20 C73.108	
4 POLE 4 WIRE	30A 3φY 347/600V		L20-30 C73.109	
4P5W	20A 3φY 120/208V		L21-20 C73.90	
4P5W	30A 3φY 120/208V		L21-30 C73.91	

WIRING DIAGRAM	NEMA *ANSI*	RECEPTACLE CONFIGURATION	RATING	
	ML2 C73.44		15A 125V	2 POLE 3 WIRE
	L5-15 C73.42		15A 125V	2 POLE 3 WIRE
	L5-20 C73.72		20A 125V	2 POLE 3 WIRE
	L5-30 C73.73		30A 125V	2 POLE 3 WIRE
	L6-15 C73.74		15A 250V	2 POLE 3 WIRE
	L6-20 C73.75		20A 250V	2 POLE 3 WIRE
	L6-30 C73.76		30A 250V	2 POLE 3 WIRE
	L7-15 C73.43		15A 277V	2 POLE 3 WIRE
	L7-20 C73.77		20A 277V	2 POLE 3 WIRE
	L7-30 C73.78		30A 277V	2 POLE 3 WIRE
	L8-20 C73.79		20A 480V	2 POLE 3 WIRE
	L8-30 C73.80		30A 480V	2 POLE 3 WIRE
	L8-20 C73.81		20A 600V	2 POLE 3 WIRE
	L9-30 C73.82		30A 600V	2 POLE 3 WIRE
	ML3 C73.30		15A 125/250V	3P3W
	L10-20 C73.96		20A 125/250V	3P3W
	L10-30 C37.97		30A 125/250V	3P3W

Continued on Next Page

5-H. Voltage and Current Combinations for Plugs and Receptacles (Continued)

	RATING	RECEPTACLE CONFIGURATION	NEMA *ANSI*	WIRING DIAGRAM
4 POLE 5 WIRE	20A 3φY 277/480V		L22-20 C73.92	
	30A 3φY 277/480V		L22-30 C73.93	
	20A 3φY 347/600V		L23-20 C73.94	
	30A 3φY 347/600C		L23-30 C73.95	

WIRING DIAGRAM	NEMA *ANSI*	RECEPTACLE CONFIGURATION	RATING	
	L11-15 C73.98		15A 3φ 350V	3 POLE 3 WIRE
	L11-20 C73.99		20A 3φ 250V	
	L11-30 C73.100		30A 3φ 250V	
	L12-20 C73.101		20A 3φ 480V	

Because a body is not a good conductor, the current is not high enough to blow the fuse; it continues to pass through the body as long as the body remains in contact with the equipment. Fig. 5-1.

■ A *grounding conductor,* or equipment ground, is a wire attached to the housing, or other conductive parts of electrical equipment that are not normally energized, to carry current from them to the ground. Thus, if a person touches a part that is accidentally energized, there will be no shock, because the grounding line furnishes a much lower resistance path to the ground. Fig. 5-2. Moreover, the high current passing through the wire conductor blows the fuse and stops the current. In normal operation, a grounding conductor does not carry current.

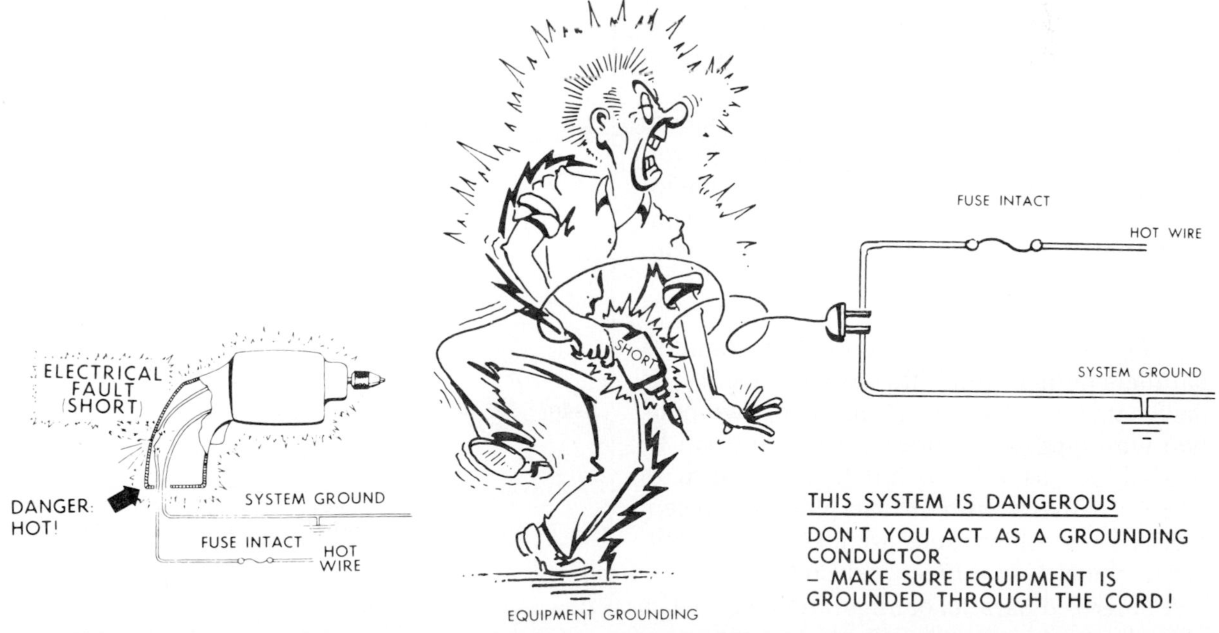

5–1. Because white-jacketed system grounds cannot conduct electricity from short circuits to the ground, they do not prevent the housing of faulty equipment from becoming charged. Therefore, a person who contacts the charged housing becomes the conductor in a short circuit to the ground.

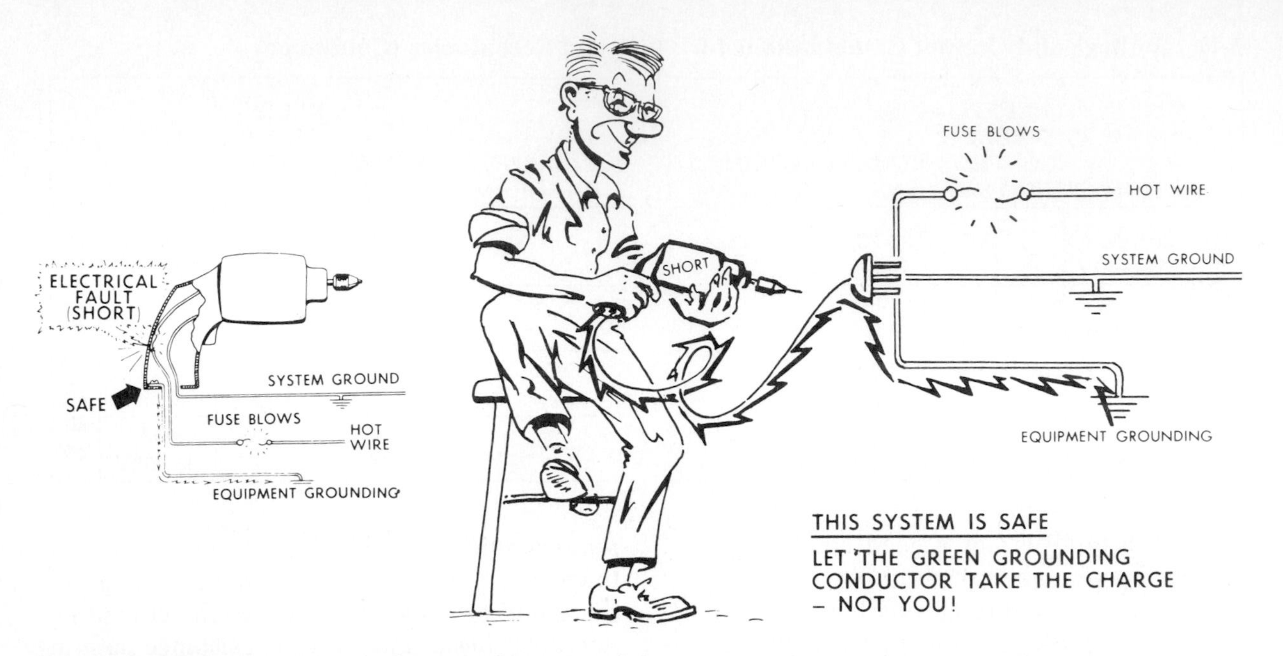

5–2. Equipment stays at ground potential in spite of short circuit, if circuit has a grounding conductor. Internal electrical faults cause current to short-circuit harmlessly to equipment ground in systems with green-jacketed grounding conductors.

The grounding conductor in a threewire conductor cable has a green jacket; it is always terminated at the green-colored hexhead screw on the cap or connector. It utilizes either a green-colored conductor or a metallic conduit as its path to ground. In Canada, this conductor is referred to as the earthing conductor, which is somewhat more descriptive and helpful in distinguishing between grounding conductors and neutral wires, or grounded conductors.

■ *Two-wire configurations.* The standard parallel configuration, illustrated in Fig. 5-3, is designated a 125-volt, 15 ampere, two-pole, two-wire type at the present time, but will be phased out as old wiring is replaced with newer, two-wire with ground plugs and receptacles. These older types are no longer permitted by the National Electrical Code.

A polarized parallel configuration, Fig. 5-4, is also designated a 125-volt, 15-ampere, two-pole, two-wire type. It differs from the standard parallel only in that one of the blades is

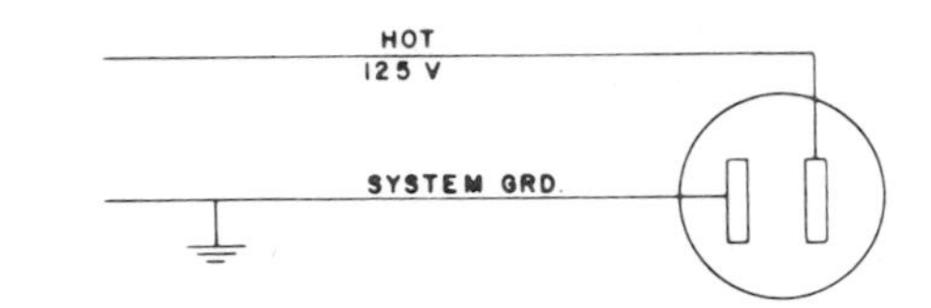

5–3. 125-V, 15-A, standard parallel configuration.

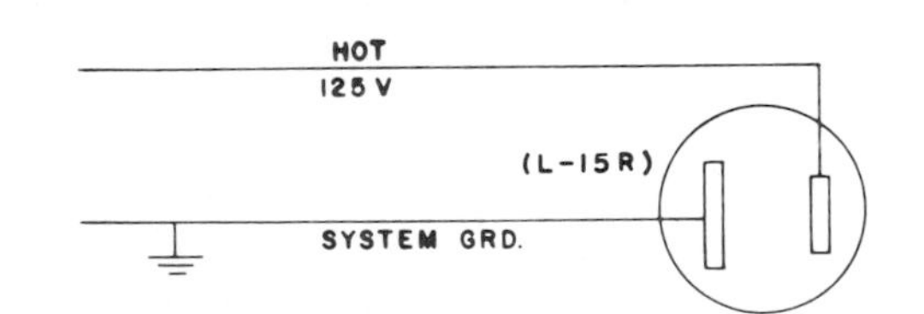

5–4a. 125-V, 15-A, polarized parallel configuration.

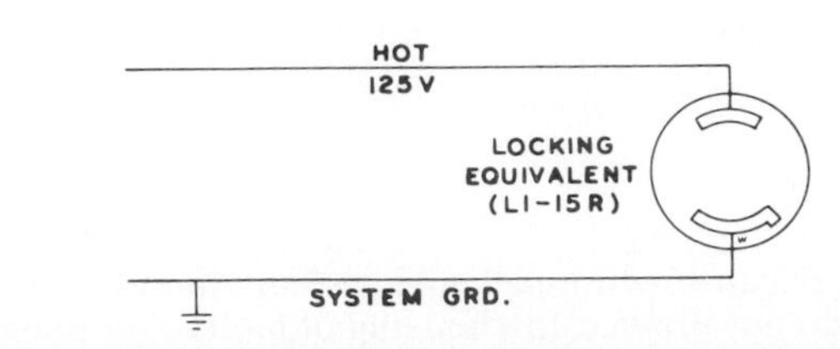

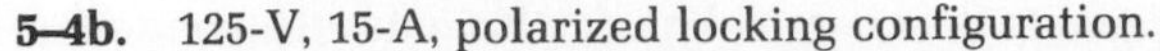
5–4b. 125-V, 15-A, polarized locking configuration.

about $\frac{1}{16}''$ wider than the other. This wider blade is intended for connection to the neutral wire. Cord connectors and receptacles in the configuration also have one wide slot, assuring that, when properly wired, the neutral conductor will be continuous throughout the entire system. Connectors have one wide slot, enabling them to accommodate both the standard and polarized parallel caps. This type of plug may be noticed on some previous models of television sets, where a "hot chassis" was used, making it important to connect the ground wire to the chassis.

Another old type of plug (cap) you may see in some locations is the tandem-position blades (now banned). Fig. 5-5. This is a 250-volt, 15-ampere, two-pole, two-wire type that is used for 230-volt, single-phase circuits where there is no separate equipment ground. The most frequent applications will be found on fractional-horsepower, single-phase motors with 230 volts, where a separate equipment ground in the supply cord is not required. In some cases, grounding is accomplished by a separate flexible wire or strap.

■ *Three-wire configurations.* The U-blade configuration, Fig. 5-6, is most widely used and is designated a 125-volt, 15-ampere, two-pole, three-wire, grounding type. The grounding blade on the plug is slightly longer than the two line blades. The configuration is used for 115-volt, single-phase applications that require a separate equipment ground. It is growing in popularity, and will probably replace the two-pole, two-wire configuration as the older type is phased out by replacements.

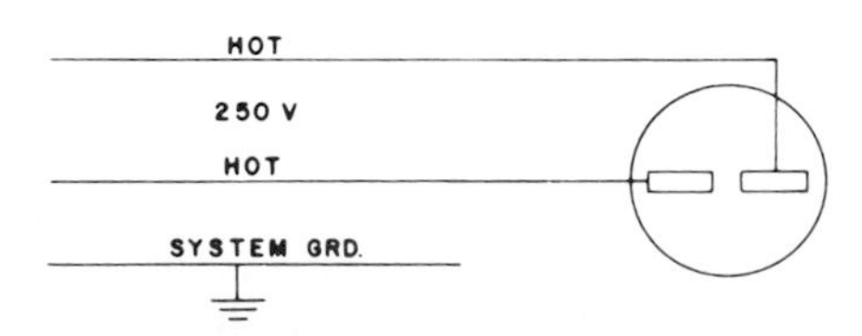

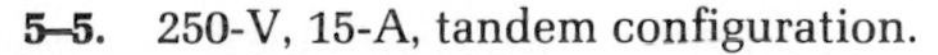

5–5. 250-V, 15-A, tandem configuration.

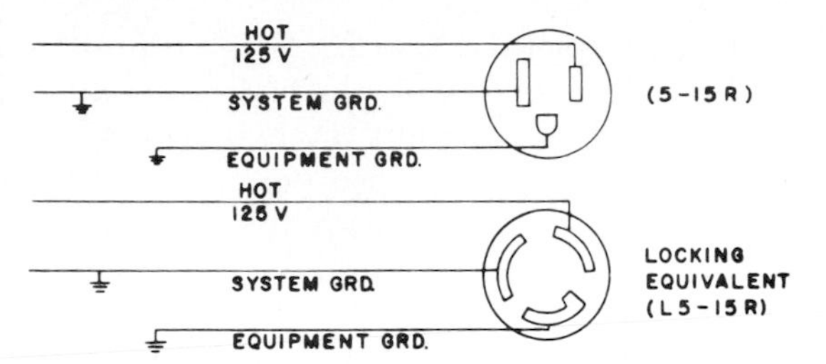

5–6. NEMA 125-V, 15-A, U-blade configuration.

Proper Grounding and Bonding

Grounding must be taken into account wherever electrical current flows. It can never be stressed too strongly that proper grounding and bonding must be correctly applied if the system, the equipment, and the people that come in contact with them are to be protected. The National Electrical Code specifies in Article 250, and in 35 additional sections, the requirements for grounding and bonding. Testing agencies, such as Underwriters' Laboratories, test equipment to make sure it can safely handle conditions which may appear in the design or manufacture of products.

■ *Grounding.* Effective grounding means that the path to ground is permanent and continuous; that the path has an ample current-carrying capacity, and low impedance to permit all current-carrying devices on the circuit to work properly.

■ *Bonding.* Effective bonding means that the electrical continuity of the grounding circuit is assured by proper connections between service raceways, service cable armor, all service equipment enclosures containing service entrance conductors, and any conduit or armor that forms part of the grounding conductor to the service raceway.

■ *A system ground.* A system ground is the grounding of the neutral conductor, or ground leg, of the circuit to prevent lightning or other high voltages from exceeding the design limits of the circuit. It also limits the maximum potential to ground due to normal voltage.

- An equipment ground. An equipment ground is the grounding of exposed conductive materials, such as conduit, switch boxes, or meter frames, which enclose conductors and equipment. This is to prevent the equipment from exceeding ground potential.

Essentials of Grounding. One of the greatest problems in any electrical system is to see that it is properly grounded. Effective grounding limits the voltage on exposed metal parts, allowing circuit-protecting devices to operate when a ground fault occurs.

Fig. 5-7 illustrates the ground-fault return path occurring when one of the phase conductors or metal conductors at phase potential comes in contact with the metal enclosure.

This type of fault differs from the more common short circuit, in which the phase conductors come in contact with one another (phase-to-phase fault), or when a phase conductor contacts the grounded neutral conductor. In either case, the return path is through the actual conductors, which, by design, have a low impedance.

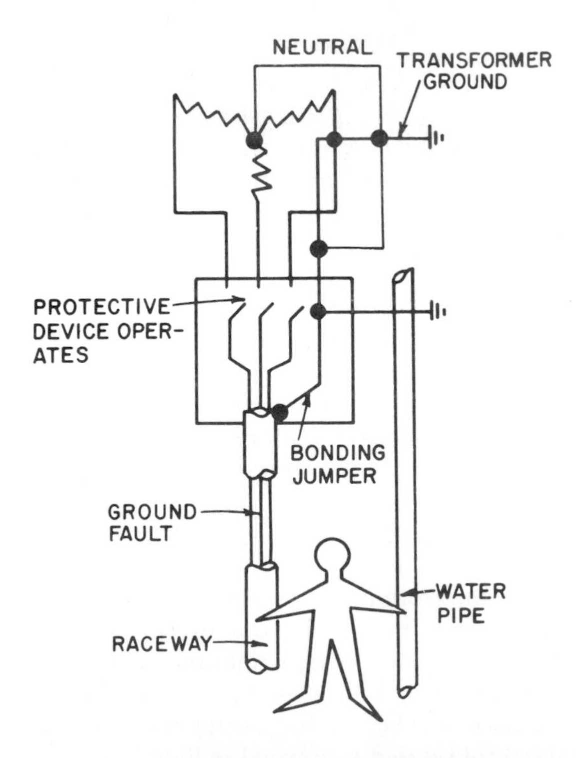

5–7. Low-impedance ground system.

When a phase conductor touches the metal enclosure, the path to ground—which is necessary for the circuit protective device to function—is through the metal enclosure, or raceway.

Sufficient current must flow through this ground return path to operate the circuit-protective device. Unless all of the elements of this system are properly connected, the protective device will not function. The ground-fault return path must have a low impedance and have ample current carrying capacity to function safely. It must maintain this condition under normal and overload operating conditions, where vibration, corrosion, wet or damp locations, and excess heat or cold may be important factors.

The impedance (resistance or opposition) of the ground-fault return path, therefore, is a function of many variables. Improper grounding, or the use of inadequate ground fittings, can cause the path to ground impedance to be high, which prevents proper functioning of circuit-protective devices when a ground fault occurs.

Fig. 5-8 illustrates what can happen if the basic requirements of grounding are not considered in the design of the system. Here, the impedance of the path to ground is too high to allow the circuit-protective equipment to operate.

Fig. 5-8 also illustrates the advantage of connecting a system ground conductor, or neutral, to the equipment-ground electrode to protect from lightning. The tie to ground at this point helps prevent high voltages from entering the electrical system beyond the service.

Isolated Grounding Duplex Receptacles. With the advent of the home computer and many sophisticated solid-state entertainment devices in the home, it is necessary to make sure that the home as well as industry be protected from EMI. EMI is electrical noise.

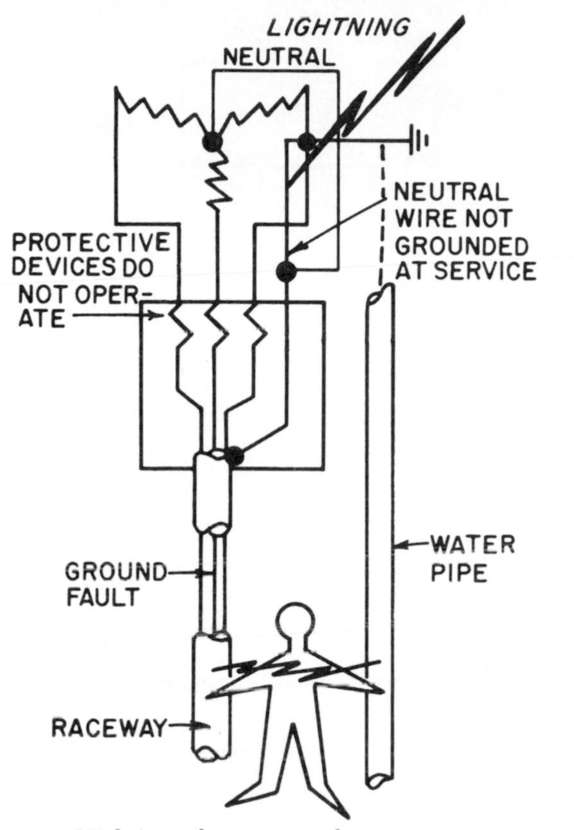

5–8. High-impedance ground system.

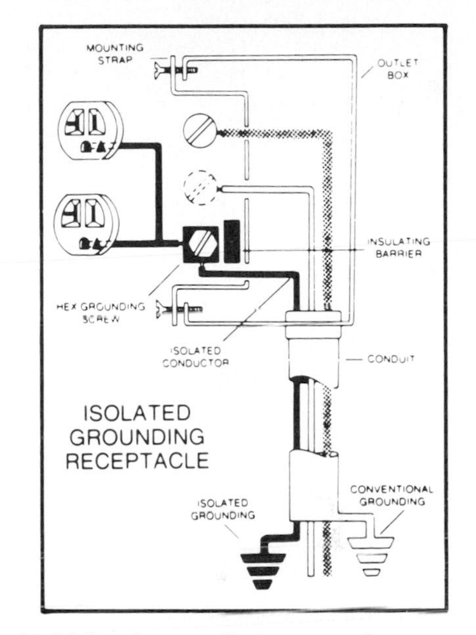

5–9. An isolated grounding receptacle.

When grounding through the normal bonded grounding system, sensitive electronic equipment is often affected by transient electrical noise signals that can cause malfunctions or improper readings. To eliminate this problem, receptacles are required that provide a separate pure grounding path completely insulated from the normal bonded grounding circuits. See Fig. 5-9.

Reduction of electromagnetic interference (EMI) in the electrical grounding system is essential for proper operation of many types of electronic equipment. In conventional grounding circuits, the conduit system is used as the grounding path. The conduit system is grounded at the service entrance and connected to intervening sub-panels, structural steel, and other grounded equipment. Although this will provide personnel and equipment protection, the entire conduit system becomes a large antenna. This collects EMI and transmits it throughout the grounding system. Whenever that occurs, the signals may adversely affect the performance of small computers, electronic testing and calibrating equipment, and solid-state cash registers.

By running an insulated equipment grounding conductor directly to the neutral at the service entrance and installing an isolated ground receptacle, you effectively isolate the grounding circuit from the conduit grounding system. Fig. 5-10. This significantly reduces the size of the antenna and limits the amount of EMI that can be picked up in the isolated grounding system. This provides a relatively noise-free grounding path and improves the operation of electronic equipment.

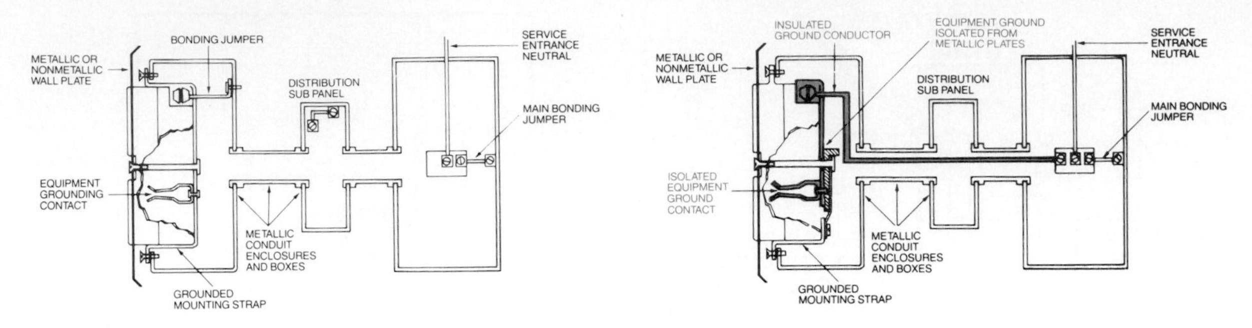

5–10. Isolating the grounding circuit from the conduit grounding system.

Installation of an isolated grounding receptacle is not complete until the green hex-headed grounding screw is connected to the isolated grounding wire.

Isolated ground wall plates are available for this type of receptacle. The receptacles are usually orange or will have "isolated ground" marked on the side of the unit.

Portable Equipment. Fig. 5-11 illustrates a hazard which could occur if the exposed metal frame of a portable appliance is not grounded to the equipment ground. Portable equipment presents such a problem because the connection to ground is disconnectable, not permanent. In addition, even if the metal raceway is well grounded, a hazard still may exist. If the portable equipment is not attached to the raceway system ground, a phase conductor accidentally touching the tool frame will create a path to ground through the operator's body.

Special Ground-Fault Protective Devices. Fuses and circuit breakers cannot distinguish between fault current and normal load current. A ground-fault interrupter, however, is insensitive to load current and responds only to a ground-fault current.

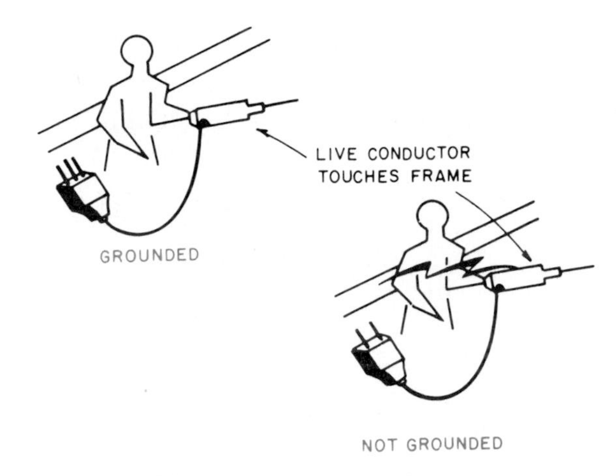

5–11. Portable equipment.

An unbalance between a phase wire and neutral, or between phase wires of as little as 3 milliamperes, will cause a ground-fault interrupter to operate in less than 50 milliseconds. Portable and permanently installed devices are available that will provide safe, portable-tool installation, independent of the ground return path.

The National Electrical Code recognizes the importance of a path to ground of continuous low impedance. No other problem is dealt with more thoroughly. The Code requires certain basic practices to be followed to make the equipment-grounding system safe. It recognizes the importance of Underwriters' Laboratories listings of various connectors used in a system.

The following are some basic factors to be considered in grounding:

1. The type of fault at the point where a phase conductor contacts the metal frame, or raceway.
2. The basic design of the electrical system. Consider wherever the system is grounded or ungrounded and at what point the enclosure is tied to ground.
3. The conductivity of the return path:
   a. Size of the raceway.
   b. Type of raceway.
   c. Type of enclosures, boxes, motor harnesses, housings, etc.
   d. Length of raceway.
4. Number of auxiliary paths to ground.
5. The point at which the equipment ground path is attached to the earth.
6. Type of grounding conductor which joins the metal raceway or equipment ground to the grounding electrode.
7. Type of grounding electrode.
8. Resistivity of the earth at the point of ground.
9. Surrounding conditions, such as wet or dry location, corrosive atmosphere, or possible mechanical damage.
10. Possibly the most important consideration is the use of proper connectors and fittings for the various joints in the equipment grounding system.

Table 5-I shows the size of grounding conductor for the amount of current in a circuit wiring system. Sizes of copper and aluminum wire are given, along with correct sizes of the grounding-electrode conductor.

Code Requirements for Grounding Conductors. The National Electrical Code requires that a system grounding conductor be connected to any local metallic waterpiping system available on the premises, provided that the length of the buried water piping is a minimum of 10 feet. If the system is less than 10 feet long, or if its electrical continuity is broken by either disconnection or nonmetallic fittings, then it should be supplemented by the use of an additional electrode of a type specified by Section 250-81 or 250-83.

Figs. 5-12 through 5-14 illustrate the various types of connectors for grounding or bonding.

■ *Fittings.* Whatever the type of grounding electrode, a system grounding conductor should be connected through any one of a

5-I. Grounding Conductor Sizes

RATING OR SETTING OF AUTOMATIC OVERCURRENT DEVICE IN CIRCUIT AHEAD OF EQUIPMENT, CONDUIT, ETC., NOT EXCEEDING (AMPERES)	SIZE OF GROUNDING CONDUCTOR		SIZE OF GROUNDING ELECTRODE CONDUCTOR	
	COPPER WIRE NO.	ALUMINUM OR COPPER-CLAD ALUMINUM WIRE NO.	CONDUIT OR PIPE (INCHES)	EMT (INCHES)
15	14	12	$\frac{1}{2}$	$\frac{1}{2}$
20	12	10	$\frac{1}{2}$	$\frac{1}{2}$
30	10	8	$\frac{1}{2}$	$\frac{1}{2}$
40	10	8	$\frac{1}{2}$	$\frac{1}{2}$
60	10	8	$\frac{1}{2}$	$\frac{1}{2}$
100	8	6	$\frac{1}{2}$	$\frac{1}{2}$
200	6	4	$\frac{1}{2}$	1
400	3	1	$\frac{3}{4}$	$1\frac{1}{4}$
600	1	2/0	$\frac{3}{4}$	$1\frac{1}{4}$
800	0	3/0	1	2
1000	2/0	4/0	1	2

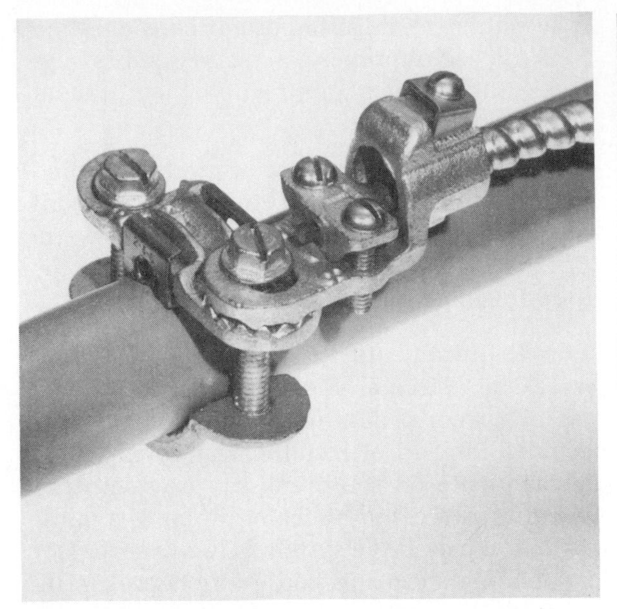

5–12. This is a typical residential ground using armored cable to positively ground conductor from service enclosure to the grounding electrode water pipe.

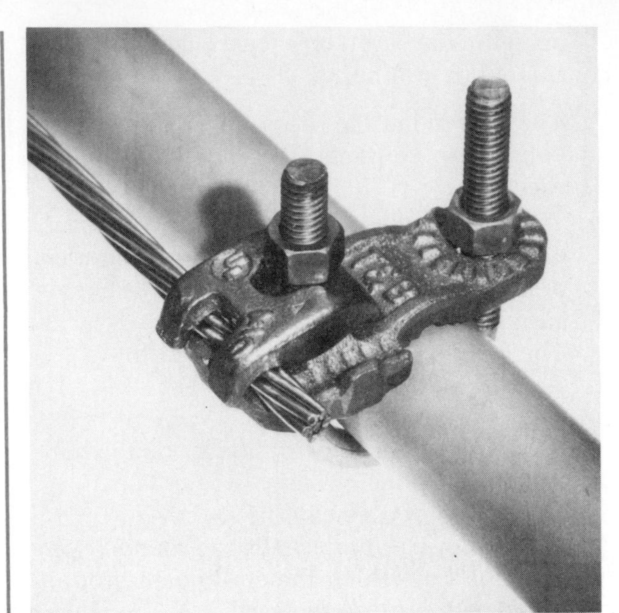

5–14. A heavy-duty ground clamp especially suited for continuous bonding along one pipe or for several parallel pipes. Designed for all handlers of volatile liquids where good grounding is absolutely essential.

5–13. This clamp bounds jumpers around meters or other breaks. No additional hubs are needed.

number of UL-listed clamps. Grounding wire may be either bare, insulated, or armored conductor, or it may be in EMT (thinwall conduit) or rigid conduit. Selection of the proper fitting depends upon the type of conductor and conduit, and the type and diameter of the grounding electrode.

An approved clamp should be made of cast bronze or brass, or of plain or malleable cast iron. In attaching the ground clamp, care should be taken to make sure that any nonconductive protective coating, such as paint or enamel, is removed and that the surfaces are clean, to insure a good electrical connection.

■ *Bonding jumpers.* Wherever electrical continuity of the grounding circuit may be interrupted, either by insulation or by disconnection, a *bonding jumper* is required by the Code. If a grounding attachment to metallic underground water piping is not made on the street side of a water meter, a bonding jumper must be used around the meter, to ensure an uninterrupted path to ground. Bonding jumpers must also be used to span other items likely to become disconnected, such as valves and service unions. In addition, they must be used to span insulating sections, such as plastic pipe, certain water softeners, or other insulating links. Bonding jumpers require the same size conductor as that of the grounding conductor which is a run to water pipe, or other grounding electrode.

Grid Grounding System. Sometimes a *grid grounding* system is necessary to ensure that equipment in a given location is safe to touch. Such installations, where high currents are carried or generated, are capable of creating shock hazards to persons close to the equipment. Large industrial facilities, such as substations, utility generating plants, refineries, chemical plants, steel mills—any distribution areas with large currents—are capable of producing shock hazards. To prevent shock hazards in such locations, a grid system may be installed. Constructed of standard copper cable, the grid is buried underground, or imbedded in concrete and connected to above-ground equipment by copper leads, and to driven ground rods made of copper-clad steel. Grounding grids are necessary in installations with high currents because of induced currents in machinery and wiring.

Grid systems must be able to handle ground-fault currents of up to 50 000 amperes in less than 4 seconds. The grids are connected to one another by compression connections, or by exothermic welding. Fig. 5-15 shows the installation of a grid system using compression connections.

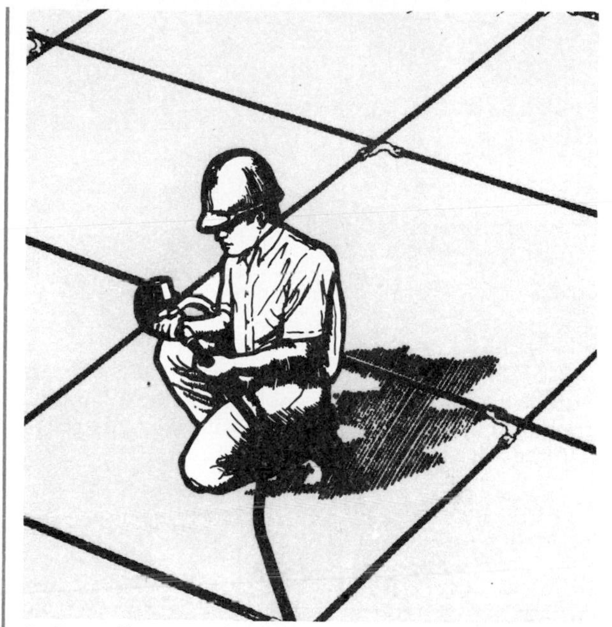

5–15a. Compression-type, copper grounding-grid connector.

5–15b. Close-up of job in 5–15a.

Service Equipment. Continuity at the service equipment is essential and must be assured. Fittings shown in Figs. 5-16 & 5-17, when properly applied, will help assure the continuity of ground. Connections of the fittings are made at the metallic enclosures by means of grounding wedges, locknuts, and bushings. They are installed so as to ensure continuity of ground. Loose or insecure fittings do not make good electrical continuity; so care must be exercised to select fittings which will maintain good con-

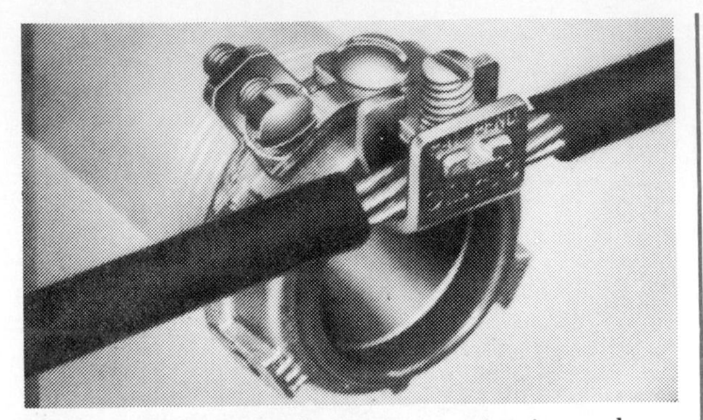

5–16. Grounding bushings must maintain good electrical contact with the enclosure. The lay-in lug provides unusual flexibility in positioning and fastening the grounding conductor.

5–17. This grounding wedge works equally well on existing or new installations. This design reduces installing time while assuring continuity of ground.

tact. Bonding locknuts should be used if the raceway does not use the largest concentric knockout.

Grounding bushings may have an insulated throat to protect the insulation of the current-carrying wire. However, this does not affect their function as a grounding connection. Good electrical contact is maintained with the metallic enclosure by means of the ground wire attached to the screw, and good electrical contact is also maintained with the grounding conduit. All threaded fittings must be made up wrench-tight, as specified by the Code.

The system grounding conductor is attached to the white wire (the grounded circuit conductor) within the service equipment, and should run without joint or splice to the grounding electrode. Generally, attachment of the grounding conductor is made to a built-in terminal on the neutral bar. In the event the neutral bar terminal will not accommodate the size of grounding conductor required, a pressure terminal of the correct size must be bolted on and used for the system ground attachment.

■ *Bonding.* In order to ensure electrical continuity of the grounding circuit, bonding is required at all conduit connections in service equipment, and at points where any nonconductive coating exists which might impair such continuity.

■ *Where bonding is required.* Bonding is required at connections between service raceways, service cable armor, at all service equipment enclosures containing service entrance conductors (including meter fittings and boxes), and at any conduit or armor which forms part of the grounding conductor to the service raceway.

When meter housings are mounted outdoors, separate from the service equipment, the neutral wire in the meter should run to ground.

A meter located inside a building should have the neutral wire bonded to the meter housing and attached to the grounded service conductor.

On rigid conduit, good ground connections will be assured at the point where conduit meets the tapped hub or boss on the enclosure. If service-entrance cable is used, it is recommended that watertight connectors be installed, to protect the joint from moisture.

An iron-bodied, universal-type entrance cap is recommended where service conductors from supply enter the building. Such a fitting works equally well on rigid conduit or on thinwall conduit with no threading. These caps are of weather-proof design, and function to protect, as well as to fasten conductors to the structure. These caps are also available in aluminum.

Grounding and Bonding Raceways. Copper wire, aluminum wire, conduit, EMT or other metal raceway, or cable enclosure can be used to ground equipment. This metal grounding circuit must run from the service equipment enclosure to the termination of all circuits.

Some flexible raceways have a relatively high resistance to ground, and unless the raceway and connector are UL listed for this purpose, a separate conductor should be run around or through the raceway to supplement the normal equipment ground-fault return path. This conductor, which corresponds to the green conductor in a portable cord, must be properly attached to the equipment frame, using an approved connector. The National Electrical Code allows flexible metal conduit to be used for grounding, if the length of the conduit is six feet or less.

Recheck Table 5-I for the size of conductors required for raceway grounds. In nonmetallic sheath cable, the equipment grounding conductor is carried in the cable. It must be connected to all metal enclosures and connected from one cable directly to the other so that "continuity of the branch circuit grounding conductor to the box is maintained when switch or receptacle is removed" (from NEC Handbook).

Conduit and EMT raceway should be attached to all enclosures by an approved means, not by ordinary locknuts or bushings. This usually means a bonding locknut, or a tapped hub, or boss on the enclosures.

Maintaining Continuity of Ground at Equipment Connections. The National Electrical Code specifies repeatedly that the path to ground shall be electrically continuous and of low resistance. To ensure this, many fittings originally designed for use on current-carrying conductors may be used in making connections with ground wires. Selection may be made on the basis of conductor size and the number of conductors to be handled. While the screw-on connector makes a good and acceptable connection, the compression type makes a superior connection. The screw-on connector may be removed if necessary; the compression type is a permanent installation and should be favored for installations of limited accessibility, or where long-term installation without change is expected. Fig. 5-18.

Tapered, threaded openings for rigid conduit fittings in a box make excellent low resistance paths and are watertight and explosion proof. The addition of grease-metallic thread compound aids in sealing the joint, preventing galling of aluminum conduit, and helps give a low resistance path.

■ *Bonding jumpers.* Bonding jumpers must be utilized to assure continuity around concentric or eccentric knockouts which are punched or otherwise formed in such a manner that would impair electrical current flow. The current must flow freely through the reduced cross section of metal that bridges the enclosure wall and punched ring of the knockout. Fig. 5-19.

An emphatic warning is repeated throughout the National Electrical Code: NEVER USE SOLDER FOR GROUNDING CONNECTIONS.

Grounding Electrodes. The National Electrical Code (NEC) calls for proper grounding by using electrodes in some instances. Section 250-84 deals with the resistance of made electrodes. These electrodes are of concern if you do not have a good underground metallic piping system that can be used for grounding purposes. The Code accepts a one-rod ground if

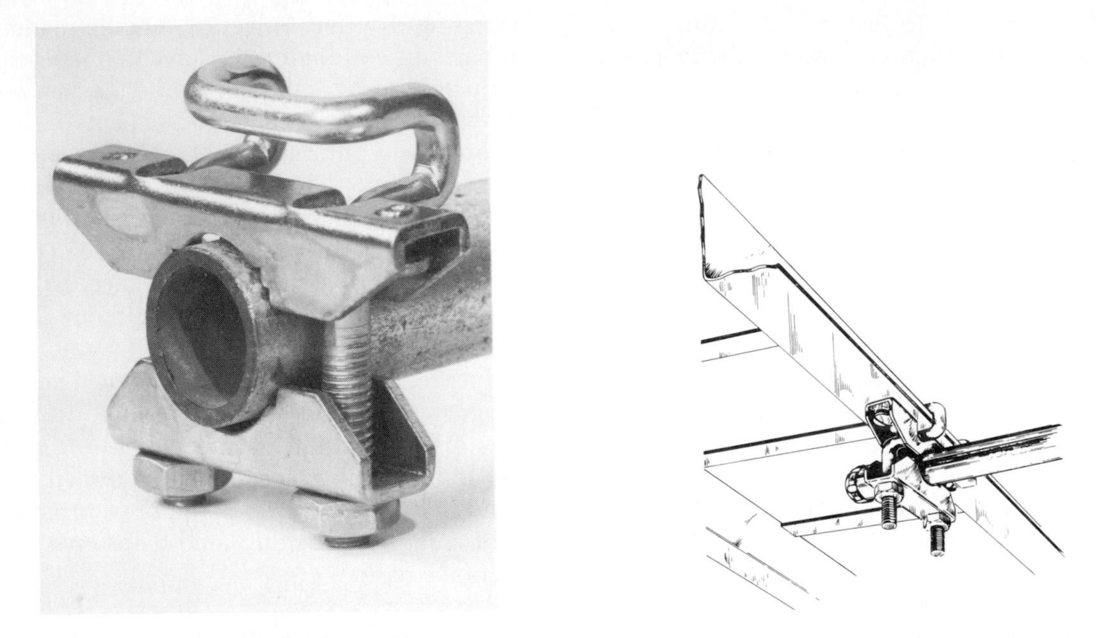

5–18. Cable tray clamp design actually eliminates need for bonding jumper, since it has two fasteners which create a clamping and bonding action on the tray.

5–19. A bonding locknut such as this helps assure low impedance path to ground. Designed with a deep thread that fits tightly over conduit thread, and a bonding setscrew to bite into the wall of the enclosure.

the rod measures less than 25-ohms resistance to ground. Fig. 5-20. If the rod measures more than 25 ohms to ground, you must add another rod in parallel. The other rod should not be less than 6 feet from the original rod. You do not have to measure the resistance if the second rod is installed properly.

In dealing with electrodes, you will find that there is a wide variety of resistance. You should also keep in mind that some types of soil may create a high rate of corrosion. This produces a need for periodic replacement of the grounding electrodes. You should also note that the intimate contact of two dissimilar metals, such as iron and copper, can result in electrolytic corrosion when located in wet soil.

To measure the resistance from the electrode to ground, the megger can be used. Fig. 5-21. This is a battery-operated earth tester that does not require the use of a hand-cranked generator to produce the electricity needed for the reading. This direct-reading instrument is used by electrical contractors, plant electricians,

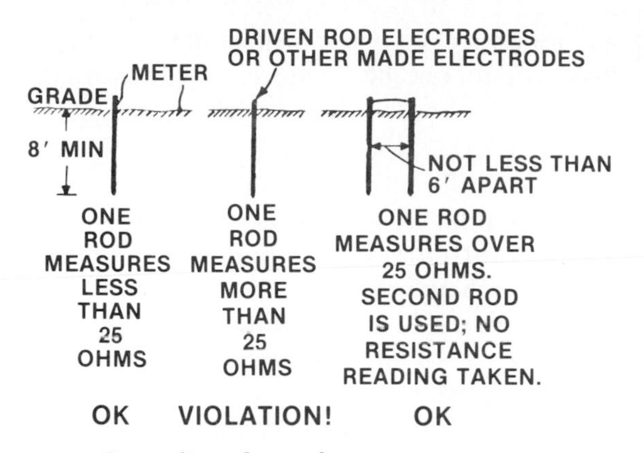

5–20. Grounding electrodes.

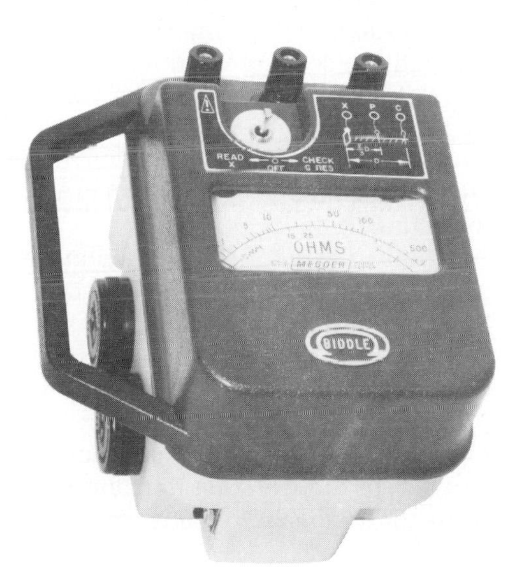

5–21. A battery-operated megger.

telephone, power utility, and CATV technicians to make routine ground tests specified by the NEC and OSHA. The tester is very easy to use and readings are easy to understand. There are three terminals. It can be used to measure fall-of-potential earth-resistance. Operation is simple. After the probes and leads are connected, push the selector switch to the left. Read the earth resistance of the ground under test. If a check of probe resistance and test lead continuity is desired, push the selector switch to the right and use the probe resistance subscale. The direct-reading meter reads from 0.5 to 500 ohms. There is no need for range changing, multipliers, or calibrating adjustments. A separate pushbutton is provided for the battery check. It can be used to test distribution transformers, service entrances, portable generators, telephone and CATV systems, and computers. Fig. 5-22 shows how easy it is to operate.

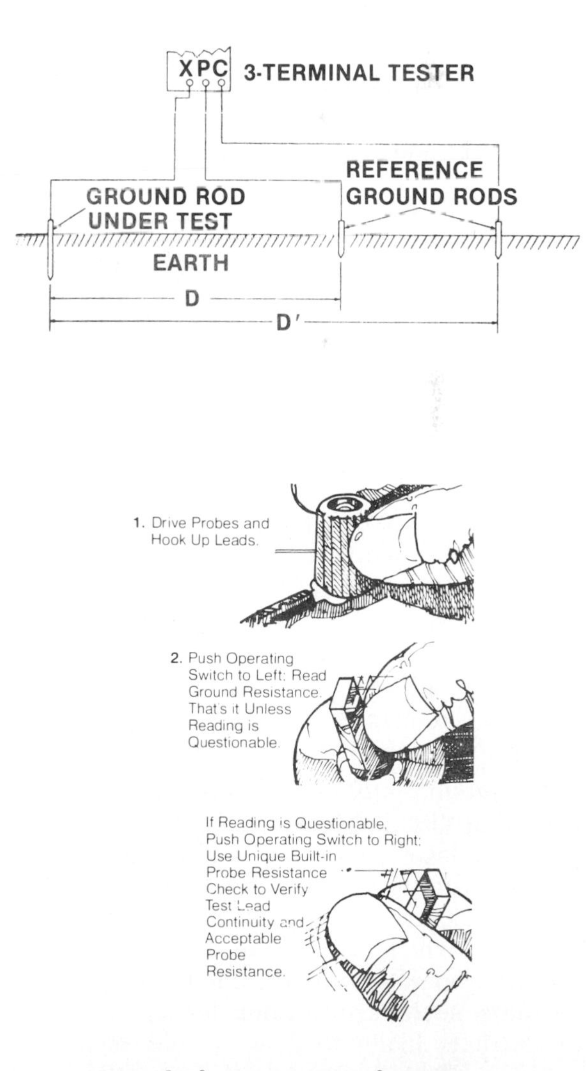

5–22. Using the battery-operated megger.

SAFETY DEVICES

As a power source, electricity can create conditions almost certain to result in bodily harm, property damage, or both. It is important for workers to understand the hazards involved when they are working around electrical power tools, maintaining electrical equipment, or installing equipment for electrical operation.

■ *Overcurrent devices* should be installed in every circuit. They should be of the size and type to interrupt current flow when it exceeds the capacity of the conductor. Proper selection takes into account not only the capacity of the conductor, but also the rating of the power supply and potential short circuits.

■ *Fuses* must be chosen according to type and capacity to fill a specific need. It is recommended that a switch be placed in the circuit so that fuses can be deenergized before they are handled. Insulated fuse pullers should also be used. Blown fuses should be replaced by others of the same type. Fuses should never be inserted in a live circuit.

■ *Circuit breakers* have long been used in high-voltage circuits with large current capacities. Recently they have become more common in other types of circuits. Circuit breakers should be selected for a specific installation by qualified engineers, and checked regularly by experienced maintenance personnel.

■ A *ground-fault circuit interrupter* is a fast-operating circuit breaker that is sensitive to very low levels of current leakage to ground. The interrupter is designed to limit electric shock to a current- and time-duration value below that which can produce serious injury. The unit operates only on line-to-ground fault currents, such as insulation leakage currents, or currents likely to flow during accidental contact with a "hot" wire of a 120-volt circuit and ground. It does not protect in the event of a line-to-line contact. There are two main types of GFIs:

1. The differential ground-fault interrupter, available with various modifications, has current-carrying conductors passing through the circular iron core of a doughnut-shaped differential transformer. As long as all the electricity passes through the transformer, the differential transformer is not affected, and will not trigger the sensing circuit. If a portion of the current flows to ground and through the fault-detector line, however, the flow of electricity through the sensing windings of the differential transformer causes the sensing circuit to open the circuit breaker. These devices can be arranged to interrupt a circuit for currents of as little as 5 mA flowing to ground.
2. Another design is the isolation-type ground-fault interrupter. This unit combines the safety of an isolation system with the response of an electronic sensing circuit. In this setup, an isolating transformer provides an inductive coupling between load and line; both hot and neutral wires connect to the isolating transformer. There is no continuous wire between.

In the latter type of interrupter, a ground fault must pass through the electronic sensing circuit, which has sufficient resistance to limit current flow to as low as 2 mA—well below the level of human perception.

The ground-fault interrupters by Hubbell operate on the differential transformer principle. They monitor load currents flowing in the protected circuit, comparing current flowing to the load with current flowing from it. Leakage to ground (ground fault) appears as the difference between these currents. When this exceeds a preset level, the GFI interrupts the circuit.

Fig. 5-23 is a mini-portable plug-in type of GFI with a 120-volt, 15-ampere capability. Just

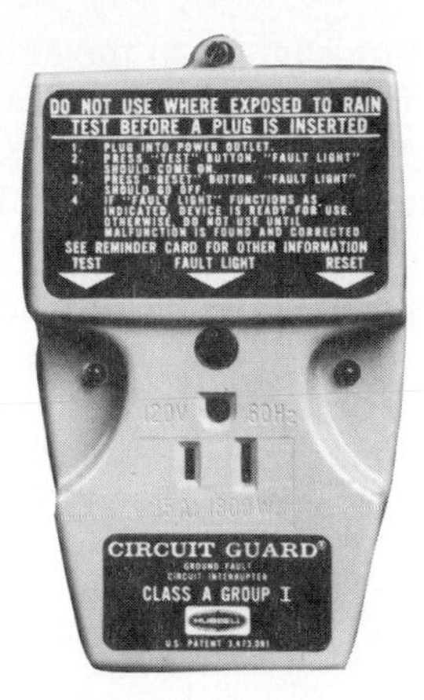

5–23. Mini-portable, plug-in 120-V, 15-A ground fault interrupter.

plug it into the socket and then plug the device to be used on 120 volts and under 15 amperes into the GFI. It guards against shock hazard. It is used with pumps, extension cords and hand tools as well as many other applications in the home, office or factory. Safety minded electricians use this unit for on-the-job protection.

This device has a built-in test circuit which imposes an artificial ground fault on the load circuit to assure that the ground-fault protection is functioning properly. This particular device trips at 5mA.

The circuit guard in Fig. 5-24 wires into an existing receptacle box, providing a GFI-protected receptacle. Cord-connected equipment plugs into it. Weather-proof lift cover encloses the single grounding receptacle, two test switches, and the reset actuator.

Fig. 5-25 shows a surface-mounted GFI, to be installed close to the load, in situations where the length of wire from the load center exceeds the limit set by UL for reliable operation. It is used for general purposes and for swimming pool equipment. It is also used in motor-control circuits, where a model with overcurrent protection would cause coordination problems with thermal overloads. It may be purchased in 15-, 20-, or 30-ampere models for indoor or outdoor use. Trip level is 5 mA.

5–24. Raintight circuit guard, 120-V, GFI 15-A.

5–25. Class A, Group I GFI for general use and swimming pool equipment. In 15-, 20-, and 30-ampere models.

Some GFIs may be used in conjunction with underwater lights in older pools (built prior to 1965) which required less sensitive equipment, or one that has a 20 mA sensitivity. Some less sensitive types may also be used in industrial applications where moisture and chemicals preclude the use of 5 mA type.

Fig. 5-26 shows the wiring of a GFI for use with a swimming pool. Pools require either a Class A interrupter, which calls for a sensitivity of 5 mA to ground, or a Class B model, which has a sensitivity of 20 mA to ground.

Larger GFI models are available for industrial applications. They have a dual-voltage Edison system and are used to monitor a distribution panel with branch circuits. Each circuit is protected. The design provides ground-fault protection to all three wires and any load condition, as shown in Fig. 5-27. It provides protection, even when there is a loss of power or continuity on either of the hot load wires. This device protects the entire load center, protecting all branch circuits fed from it, including single-phase combinations of 120/240 volts; 3-wire systems; or 2-wire, 120-volt circuits derived from these systems.

If a circuit is from a 120/208-volt, 3-phase source (resulting in a 115/208-volt output), the unit should be derated approximately 10%, due to the circulating current in the neutral wire. In the event the fault to ground appears anywhere in the load center, it will cut off power to the entire load. Sensitivity of this GFI is 5 mA, and the unit will take 30 or 50 A.

These units are available for all types of climates. For instance, most of them will operate at 150°F (66°C) or at −31°F (−35°C). The nominal time for tripping is about 25 thousandths of a second (0.025 second). You should specify when ordering whether they are to be used indoors or outdoors.

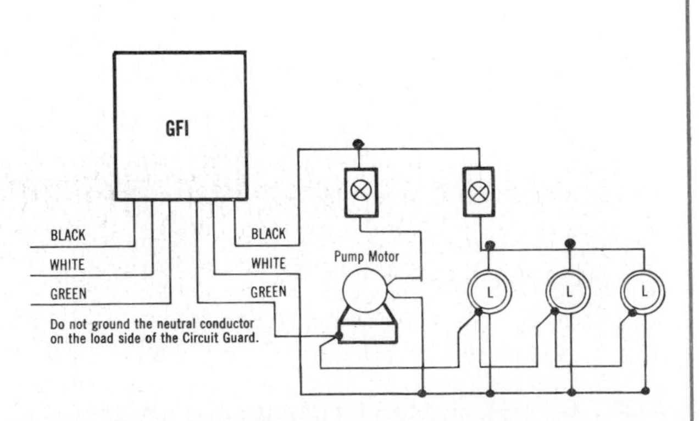

5–26. Wiring diagram for GFI designed for wet locations and outdoor installation.

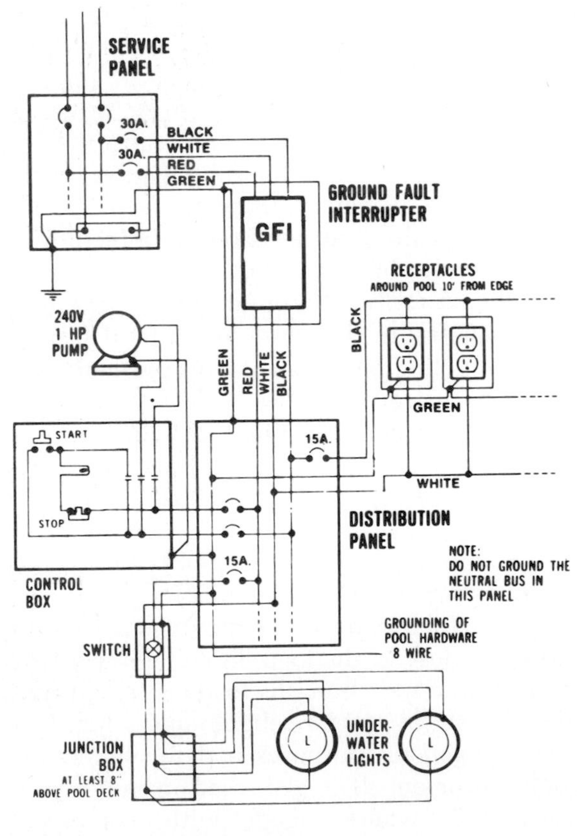

5–27. Dual-voltage commercial pool application GFI wiring diagram.

Fig. 5-28 has three wiring diagrams which indicate how GFIs are wired into various circuits. Note how different devices are taken from the overcurrent distribution panel.

The Occupational Safety and Health Act (OSHA) makes the National Electrical Code of 1971 the basic standard for grounding equipment. It requires equipment grounding in both new construction and existing installations where:

- Exposed, non-current-carrying parts are likely to become energized under abnormal conditions.
- Where a person may bridge the gap between metal and ground.
- In damp or wet locations.
- In hazardous locations.
- In every instance where equipment operates at 150 volts or more to ground.
- On all major applications, and all portable hand tools, unless they are of the double-insulated type.

As stated before, one of the greatest problems in any electrical system is to see that it is properly grounded.

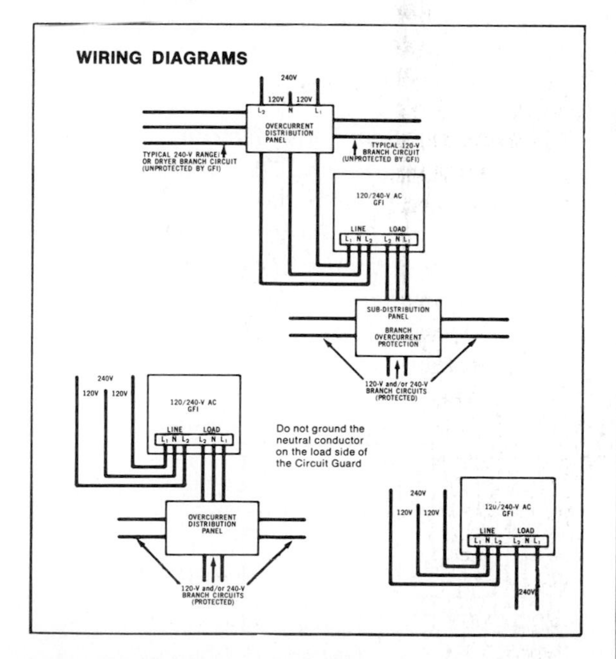

5–28. Wiring diagrams for GFIs.

Effective grounding limits voltage on exposed metal parts and allows circuit protection devices to operate when a ground fault occurs.

Three-Phase Protection

Fig. 5-29 is an oil-tight GFI, designed for use with 25 amperes at 120 /208 volts, 60 Hz, 4-wire wye, and with 138/240 volts, 60 Hz, 4-wire wye connections. It is designed specifically for industrial installations where moisture and chemicals rule out the use of Class A interrupters. It may be used to eliminate the uncertainty of "panic button" safety measures in research and development laboratories, protecting personnel at all times. This GFI unit detects ground-fault current on the load circuitry. When the ground-fault current exceeds 20 milliamperes, the three-pole circuit interrupter trips, shutting off the power. It will, however, continue to operate even with the loss of 2 phases, but does not trip automatically on a lost phase condition. A test button applies an artificial ground fault, to assure that the ground-fault protection is functioning prop-

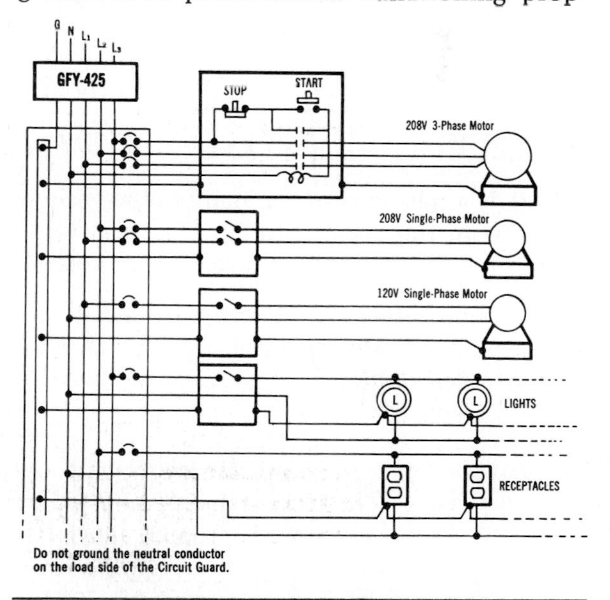

Similar circuitry may be used industrially

5–29. Three-phase protection for indoor or outdoor location of a GFI. It takes 120/208 V, 25-A, 60 Hz, 4-wire wye; and 138/240 V, 4-wire wye.

erly. Inspect the wiring diagram in Fig. 5-29 for the circuitry adaptable to this type of protection. Fig. 5-30 shows a connection diagram for three-phase hookup of the GFI. Note how it fits into the cicuit wiring.

Portable Single-Phase GFI

A lightweight, compact GFI is available. It has four outlets. One automatically trips when the supply voltage is interrupted, or when the line voltage drops too low. It may be plugged into any 120-volt, 60-hertz line which has the proper overcurrent protection. The construction industry uses it where frequent damage to electrical equipment develops ground faults. Use of the portable GFI can help protect an employer by preventing liability cases or claims.

In addition it can be used in industrial plants where temporary wiring is used. If a ground fault occurs when personnel are working in wet areas or corrosive environments, or in metal tanks or enclosures, this GFI will trip at 5 mA in 0.025 seconds. It is rated at 20 amperes or 15 amperes, and has a 6-foot cord.

GFI-RELATED CODE REQUIREMENTS

This is a paraphrase summary of the various Sections of the 1984 National Electrical Code that relate to ground-fault circuit interrupters:

- *Section 210-8(a):* This requires that all 15- and 20-ampere, single-phase, 120-volt receptacles installed outdoors, in garages, and in bathrooms of all residential occupancies are to be protected with ground-fault, circuit-interrupters.

- *Section 305-4(a):* This requirement makes it mandatory for all 15- and 20-ampere, 120-volt, single-phase receptacles on construction sites which are not part of the permanent wiring of the building or structure to be protected by GFIs for personnel. There is an exception to this basic requirement that says that receptacles on portable generators 5 kW or less need not be protected.

- *Section 210-52:* This basic section tells you how many and where you have to put receptacle outlets. Of specific interest is the fact that it requires a receptacle outlet in the bathroom adjacent to the basin location. Further, it requires that there be at least one receptacle outlet installed outdoors on all one-family dwellings. Finally, this section indicates that the receptacle outlets that are required by the section, such as that in the bathroom, must be in addition to any receptacles that are part of a lighting fixture or appliance or that are located within cabinets or cupboards.

- *Section 115-9:* This section states that feeders supplying 15- and 20-ampere receptacle branch circuits shall be permitted to be protected by a ground-fault circuit-interrupter. This is in lieu of the provisions for such interrupters as specified in Section 210-8 and Article 305 of the National Electrical Code.

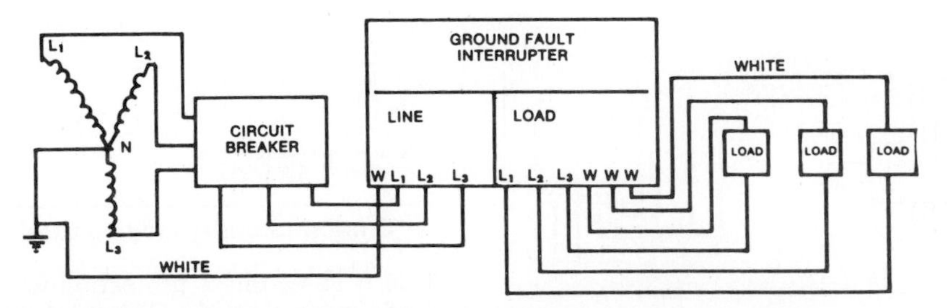

5-30. GFI connection diagram for three-phase lines with wye and center ground.

■ *Section 517-90(c):* This section deals specifically with patient care areas in health care facilities. It requires that all 120-volt, single-phase, 15- or 20-ampere receptacles supplying power to locations which are commonly subject to wet conditions be protected by GFIs, if the interruption of power under fault conditions can be tolerated without causing trouble. If a power interruption cannot be tolerated, then those receptacles must be supplied by an isolated power system. Specifically, this is aimed at receptacles supplying power in hydrotherapy areas.

■ *Section 550-6(b):* Article 550 deals with mobile homes. The ground-fault circuit-interrupter requirements in Section 210-8(a) which apply to outdoor and bathroom receptacles in residences are also applicable to outdoor and bathroom receptacles in mobile homes.

■ *Section 550-23(c):* This basic requirement concerns the receptacles in mobile home service equipment (pedestals). It indicates that, where receptacles are required in this power outlet to supply the mobile home, they shall be 50-ampere receptacles. If there are additional receptacles in this power outlet of the 15- or 20-ampere, 120-volt, single-phase type, these additional receptacles shall be protected with GFIs.

Swimming Pools

■ *Section 680-6(a):* Article 680 deals with swimming pools. This particular requirement says that no receptacle on the property shall be located within 10 feet of the inside walls of the pool, and that any receptacle located 10 to 15 feet from the inside walls shall be protected with a GFI. Fig. 5-31.

■ *Section 680-6(b)(1):* This section pertains to lighting fixtures installed around swimming pools. It requires that any lighting fixture 5 feet from the inside walls of the pool be protected by a GFI.

■ *Section 680-6(b)(2):* When a new swimming pool is installed next to an existing building with a lighting fixture on the building, it's possible for that existing lighting fixture to be within 5 feet of the inside edge of the pool. If this is the case, then the existing lighting fixture must be protected by a GFI.

■ *Section 680-6(b)(3):* This section requires GFI protection of any lighting fixture where the fixture or its supporting means is located within 16 feet of any point on the water surface.

■ *Section 680-31:* This requirement mandates the use of GFIs to protect all electrical equipment, including the power supply cords used with storable (above ground) swimming pools.

■ *Section 680-50:* This requirement applies to fountains. It requires that all electrical equipment such as lighting fixtures, submersible pumps, etc., be protected by GFIs except where that equipment is operated at 15 volts or less and supplied by a transformer complying with Section 680-5(a).

5-31. This GFI provides protection for swimming pools, hot tubs, and industrial controls. It is designed for installation within 10 feet of a swimming pool. Class A GFIs trip when the current to ground has a value in the range of 4 to 6 milliamperes. Swimming pool circuits installed before local adoption of the 1965 Code may include sufficient leakage current to cause a Class A GFI to trip.

ble substance, displacing the oxygen until the percentage of oxygen in the mixture is too low to allow combustion.

To be effective, the inert gas must reduce the amount of oxygen in the air from the normal 21% to a point below the ignition point of the substance involved. The actual amount of reduction depends on the gases involved.

For instance, an inert gas such as carbon dioxide must reduce the oxygen in the air to 6% to prevent combustion of carbon monoxide. Other examples are: 14% for gasoline and 15% for cotton dust.

Here again, as in the process of purging, consideration must be given to the safe disposal of vented flammable gases. Also to be considered is the fact that inert gas in a confined space can lead to an oxygen-deficient atmosphere. Therefore, before a person enters the space, it should be tested to see if the atmosphere will support life. If it will not, an individual entering the space should be equipped with approved respiratory-protective equipment and a harness and lifeline. The individual should be working with another person at all times.

Some lesser-known methods to prevent sparks in hazardous locations include oil immersion, sealing, and encapsulation.

Oil immersion. Although the technique of oil immersion is not likely to be used on a complete instrument, it is of interest, along with sealing and encapsulation, because the three processes may be applied to components or subassemblies.

According to the NEC, any equipment that is oil immersed to reduce explosion hazard must be specifically approved. Some UL requirements for such apparatuses include:

1. The enclosure must be of substantial metallic construction, designed to preclude the possibility of sparks being produced above the oil level.
2. If there is a drain hole, it must be provided with a pipe plug with five full threads engaged.
3. Normal oil level must extend six inches above any electrical joints, or arcing parts.
4. Oil level must be indicated by a visible-level indicator with graduations to show minimum, normal, and maximum.
5. Ordinary fuses are not approved for use within the enclosure.

■ *Sealing.* The NEC says equipment must be "hermetically" sealed. However, nowhere is "hermetically" defined; so it is left to the discretion of the code enforcer. Strictly speaking, a hermetic seal is one which is perfectly airtight, so that no gas or spirit can enter, or escape.

In practice, a hermetic seal is a controlled leakage. The tightest seal is designated Grade A by Military Standard S-8484. A Grade A seal will not leak at a rate greater than one standard cubic centimeter of air per year, per inch of seal, at a differential pressure of one atmosphere. Seals are graded down to Grade D.

■ *Encapsulation.* Encapsulation is the embodiment of a component or assembly in a solid or semisolid medium, such as tar, wax, or epoxy. It is considered safe if the material effectively seals the ignition source from the atmosphere. If the voids (empty spaces) within the encapsulated assembly are small, the assembly may be safe because it is explosion-proof.

Class II Hazardous Locations

Class II hazardous locations are those caused by combustible dusts. Such areas can be controlled by good housekeeping procedures to avoid dust accumulation, dust- and ignition-resistant enclosures, sealing, adding inert dusts and purging, and by the use of intrinsically safe and non-incendive equipment.

Dust tends to settle and be contained in an area; but gases and vapor tend to disperse. Even heavier-than-air substances trapped in low areas are eventually diffused. In a hazardous location, should combustible gases be released more than once, the effects are generally independent of the first release. Dust, however, because it tends to accumulate, has a different effect. If a small cloud of dust ignites, the resulting puff may be harmless and cause no damage. But it may dislodge more accumulated dust. This larger cloud ignites, resulting in a larger blast and more dust. Such a series of explosions can be quite destructive.

Fortunately, most dust hazards are visible and obvious. For example, the Lower Explosive Limit (LEL) of wheat flour is equivalent to about one-half teaspoon per cubic foot. Before this limit is reached, visibility is reduced to a few feet, and breathing is difficult.

Enclosures that just contain explosions or flames may not be suitable to dust hazards. Gaps that are permissible in gas or vapor hazards may let dust infiltrate. Some enclosures approved for Class I hazards may also be rated for Class II situations, but such dual ratings should not be taken for granted.

Enclosures that are both ignition-resistant, and that keep out dust, are suitable for areas containing a dust hazard. Such enclosures must be constructed substantially and are usually metallic. Metal-to-metal joints are preferred, at least to $\frac{3}{16}''$ wide. Spacing between mating surfaces must not exceed 0.0015″. There is an additional allowance of 0.001″ per $\frac{1}{8}''$, to a maximum of 0.008″ clearance.

Both Underwriters' Laboratories and the Canadian Standards Association have publications dealing with dust-hazard enclosures.

Requirements for sealing-in equipment to be used in dust hazards are essentially the same as those for gases and vapors. Oil immersion per se would not seem to be a useful technique for dust hazards, unless the enclosure were also dust-tight.

Inerting practices in dust-hazardous areas are similar to those for gases and vapor, except that the inerting material can be either dust or gas. For example, mining operations add inert dusts (rock) to combustible dusts to decrease ignitability.

Such a technique is not practicable for foodstuffs, however. Gas inerting is applied to foodstuffs and other materials where solid inerting is impossible.

Purging uses positive pressure to exclude dust and prevent buildup of insulating dust layers on contacts.

Intrinsically safe and non-incendive equipment was described previously. Application of such equipment to dust hazards has been described as obvious, practical, and relatively easy, compared to most gas and vapor applications. This is because most dusts have minimum ignition energies of several millijoules (mJ).

Some factors that influence the degree of hazard of dusts include: the chemical composition and the size, shape, and concentration of the dust, and the chemical composition of the suspending medium.

■ *Dust composition.* A dust's chemical composition is the prime determinant of minimum ignition energy, minimum explosive concentration, pressure developed by the explosion, the rate of pressure rise, and ignition temperature.

The minimum ignition energy of even the most easily ignited industrial dusts is 20 times greater than that of typical Class I, Group D materials—a significant difference between Class I and Class II hazardous materials.

However, some dusts, such as zirconium and thorium ignite at energies below 10 millijoules. Under some conditions, they can ignite spontaneously at room temperatures.

Chemical composition provides some guides for subdividing broad classes of dusts. For example, among carbonaceous dusts, ignition hazard is closely related to volatile content. If it is below 8%, there may be a fire hazard, but essentially no explosion hazard. The volatile content of most carbons, charcoals, and cokes is below this figure. But lignite, pitch, and soft coals typically are 30 to 40% volatile. These present severe explosion hazards. Modified material—such as chromate-treated charcoals—is an exception to the rule.

■ *Size, shape, concentration.* Irregularly shaped particles are produced by milling and grinding operations. As far as size goes, the finer the dust, the more homogeneous the dust cloud and the greater the surface area for reaction.

The following statements illustrate the role of particle size in creating a danger of explosive atmospheres:

1. Ignition temperature is relatively independent of size.
2. Maximum pressure is slightly dependent on size.
3. A minimum concentration—below 0.003″ has little dependence; above that, concentration is inversely proportionate to the size of the particle.
4. Minimum ignition energy and rate of pressure rise are inversely proportionate to particle diameter.

With vapors and gases, the ignition energy changes markedly with concentration, and there is a well-defined, easily ignited concentration. This is not true of dusts.

The ignition energy of hazardous dusts drops from a relatively high value at the LEL, to a minimum value at several times the LEL. The energy requirement is not materially changed by further increase in concentration. Cloud ignition temperature also drops to a low value as concentration increases, and remains essentially constant.

Another difference between gas/vapor and dust hazards is that there is no well-defined Upper Explosive Limit (UEL) for dust.

■ *Chemical composition of suspending medium.* Reducing the oxygen content of a substance has an effect similar to that of adding moisture or dry inert material. The explosion characteristics change, slowly at first, and then very rapidly as the oxygen concentration approaches the limiting value for the particular dust.

A substance used to reduce oxygen content is effective in proportion to its molar heat capacity. For example, carbon dioxide is a more effective diluent than nitrogen. But water vapor—at high temperatures—is as effective as carbon dioxide.

OSHA

OSHA (OH-sha) is the abbreviation for the Occupational Safety and Health Act. This act covers all commercial and industrial employers except government employers. Government employers are the federal, state, county, city, and other governmental agencies. OSHA is, of course, subject to amendment; so changes are expected.

OSHA is concerned with all aspects of industrial safety and working conditions. In the electrical field it applies to such things as:

- Electrical connections.
- Guarding of live, arcing, or suddenly moving parts.
- Marking and identification of circuits and equipment.
- Overcurrent protection.
- Cord-and-plug connected equipment.
- Grounding.
- Appliances.
- Hazardous locations.

In brief, OSHA requires additional or corrective wiring so that existing installations are brought up to date or at least up to the 1971 National Electrical Code requirements. Installations made before 1971 have to meet the 1971 requirements. Any new installations made after that date will have to comply with the National Electrical Code book that is current at the time of installation. OSHA is a federal law. Therefore it applies to the entire nation.

QUESTIONS

1. Which kills—current or voltage?
2. What are the physiological effects of electrical shock?
3. Why is low voltage considered dangerous in some cases?
4. How can you work around electricity safely?
5. How much resistance does the human body have?
6. How can you determine the amount of current through your body when wet?
7. What are safe distances for operating around live circuits?
8. What are some typical electrical shock hazards?
9. Why are extension cords unsafe in some locations?
10. Why is it important to choose the right extension cord for a particular electrical tool?
11. What part does preventive maintenance play in electrical work safety?
12. What is the difference between grounding and bonding?
13. Why is improper grounding dangerous?
14. What are ten basic factors considered important in grounding?
15. What does the NEC require in grounding systems?
16. What is a bonding jumper?
17. What is a grid grounding system?
18. Where is bonding required?
19. How are conduit, EMT, and raceways bonded or grounded?
20. How is continuity of ground maintained at equipment connections?
21. Name some safety devices for electrical equipment.
22. How do GFIs protect you?
23. What is purging?
24. What is inerting?
25. What is the necessity for oil immersion of some electrical equipment?

KEY TERMS

bonding
bonding jumper
encapsulation
grid grounding system
grounded conductor
ground-fault circuit interrupter
grounding
grounding conductor
inerting
OSHA
purging

SECTION THREE CIRCUITS and ELECTRICAL PLANNING

CHAPTER 6

HOUSEHOLD CIRCUITS

Objectives

After studying this chapter, you will be able to:

- Identify the minimum service capacities for various house sizes.
- Know how to read a kilowatt-hour meter.
- Identify the conductor capacities for a room.
- Test a wiring circuit.
- Identify the two common causes of voltage spikes.
- Recognize the wiring connections for wall switches and convenience outlets.

In order to make an electrical system operational, it is necessary to design the system so that the proper amount of power is available to each plug or circuit whenever it is needed.

Most power is distributed locally within a neighborhood by overhead wires, located in most cases in the rear of the house lot. The wires carry the power to a transformer, where it is stepped down from more than 4000 volts to a usable 240 volts. Figs. 6-1 & 6-2.

Power is brought from the pole or transformer into the rear of the house (in some cases the lines are underground) by means of three wires—one black, one red, and one white (uninsulated). These wires may be three separate ones, or they may be twisted together to look like one cable. Fig. 6-3.

Once the cable is connected to the house, it is brought down to the meter by way of a sheathed cable with three wires: one red, one black, and one uninsulated (ground or neutral). Fig. 6-4. Wire size depends upon the load to be applied to the line. The square footage of the house determines the amount of minimum service capacity needed (125 amperes, 150 amperes, or 200 amperes). See Table 6-A.

The three-wire cable from the top of the house down to the meter bracket is insulated

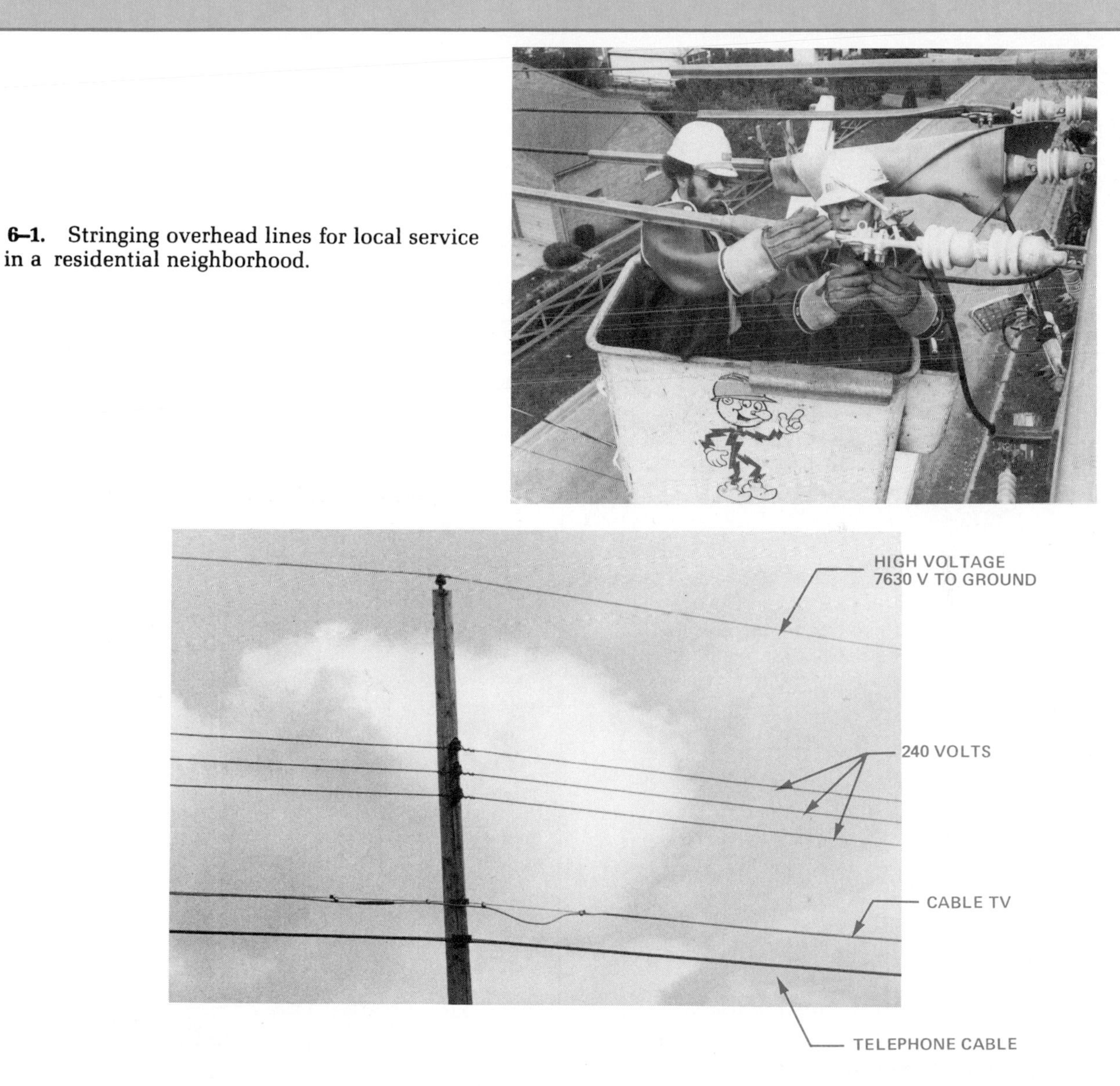

6–1. Stringing overhead lines for local service in a residential neighborhood.

6–2a. The top line that carries the power for home distribution has 7630 volts from the hot wire shown here to ground. A transformer is attached to this line and ground. From the transformer come 240 volts for connection to homes. Telephone lines and cable television lines are also attached to the pole. Thus, fewer poles are needed and services are less expensive to maintain.

6–2b. Transformer, on pole located at rear of house. Note high-voltage line on top and low-voltage (240-V) line on the bottom, with a take-off for the house. Bottom line is cable television coaxial cable.

6–3. Power from the transformer is fed through the wires and down the side of the house to the meter, then through the wall at a lower level to a distribution box in the basement.

and mounted against the wall to prevent movement and damage. The same size and type of wire is brought from the meter on the outside to the distribution box and main disconnect inside the house. Fig. 6-5 is the type that appears in older homes. Fig. 6-6 shows a newer circuit breaker type distribution panel.

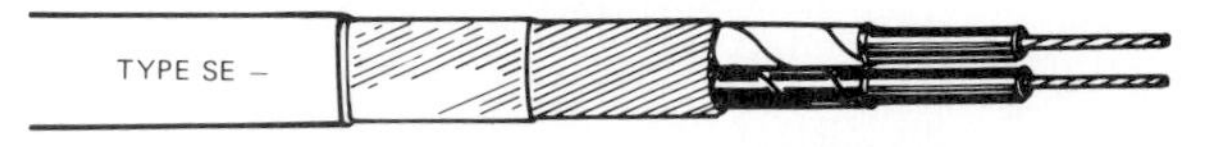

6–4. Type SE cable. This cable has three wires: one red, one black, and an uninsulated wire that forms a protection for the other two. The stranded, uninsulated wire is twisted at the end to make a connection.

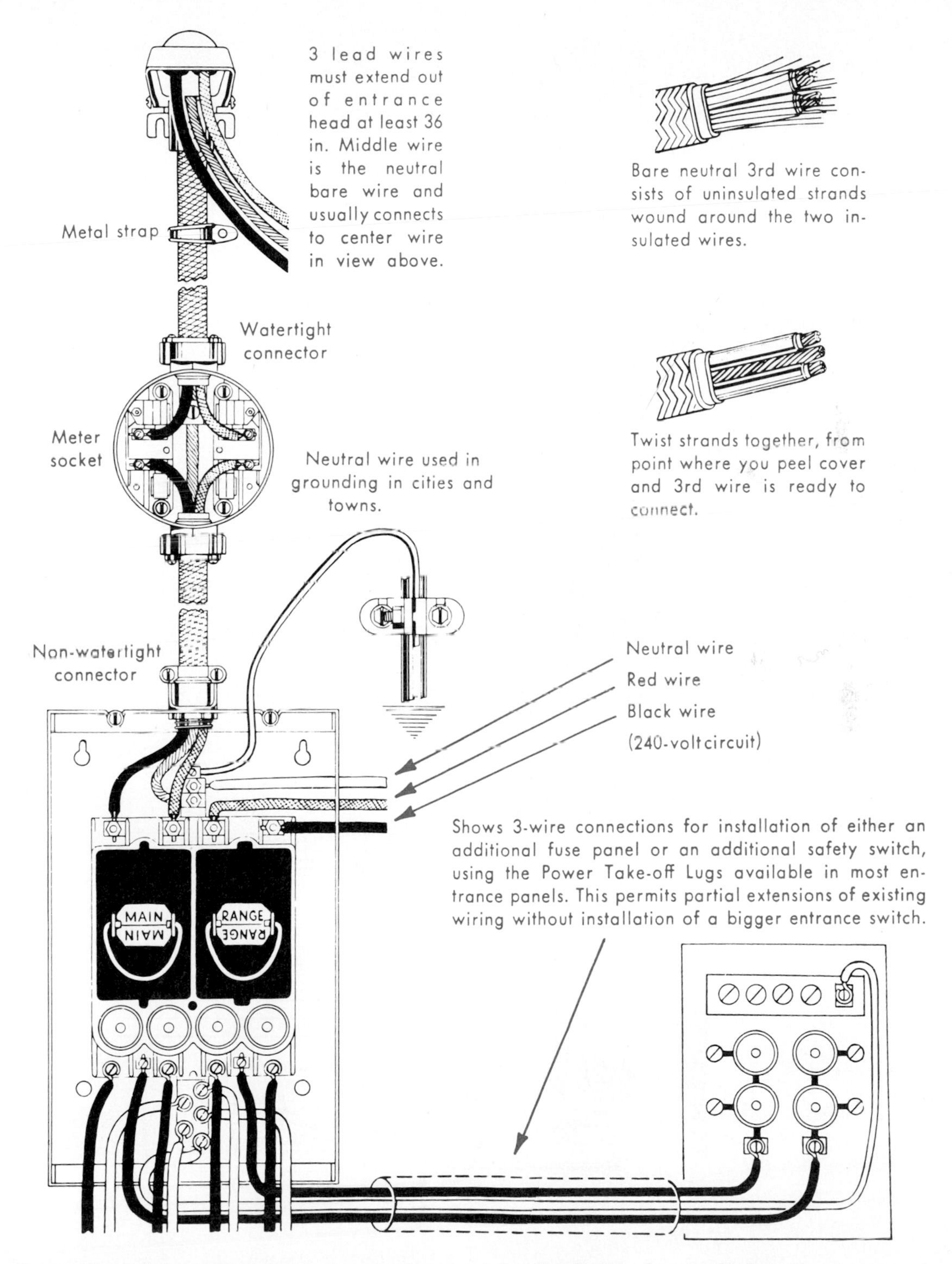

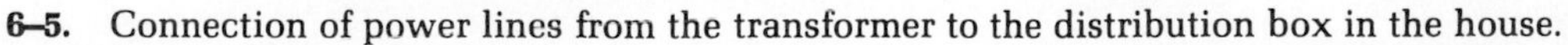
6–5. Connection of power lines from the transformer to the distribution box in the house.

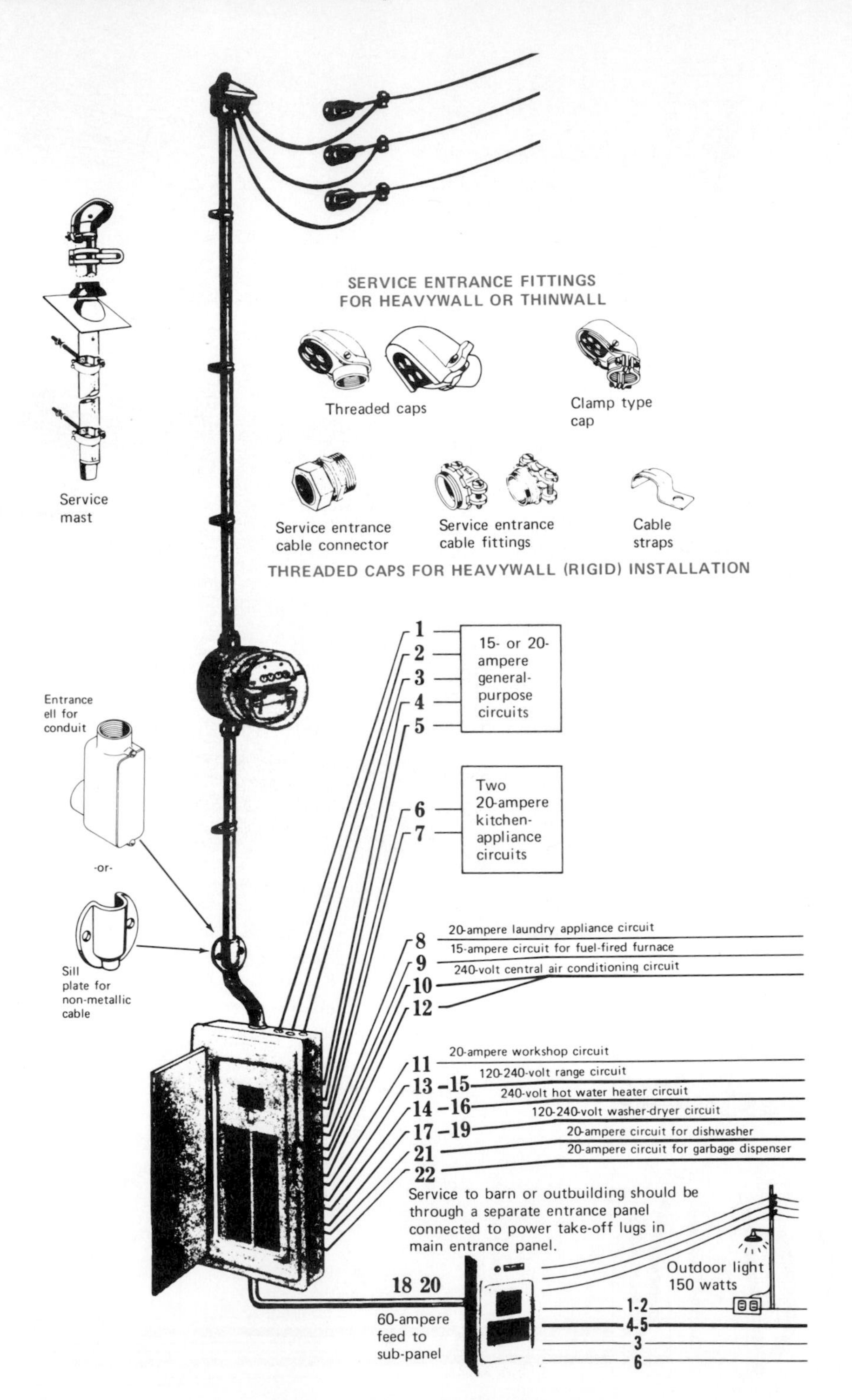

6–6. Service entrance.

SERVICE ENTRANCE

All services should be 3-wire. The capacity of service entrance conductors, and the rating of service equipment, should not be less than as shown in Table 6-A.

These capacities are sufficient to provide power for lighting, portable appliances, equipment for which individual appliance circuits are required (listed elsewhere in this text), and for electric space heating of the individual room type and air conditioning, or both. Fig. 6-7.

A larger service may be required for larger houses, or if electric space heating is of a central furnace or central hydronic-boiler type.

Because of the many major appliances in the kitchen requiring individual equipment circuits, it is recommended that, where practicable, electric service equipment be located near or on a kitchen wall to minimize installation and wiring costs. In addition, such a location will often be convenient to the laundry, thus minimizing circuit runs to laundry appliances as well.

KILOWATT-HOUR METERS

Meters measure power used by a household, business, or industry. There are several types of *kilowatt-hour meters.* Kilo means one thousand. Kilowatt-hour means 1000 watts of energy used for one hour. The unit of time for this type of meter is the hour. This means that whenever the meter reader records the meter reading—it may be a month from one reading to the other—the entire period in which power has been used is equated to one hour. The energy consumed is based on one hour. The number of watts would, in most cases, be thousands; so the kilowatt-hour is used for measuring the power consumed, no matter what the length of time.

Sometimes the meter reader makes the rounds every 28 days and sometimes every 31 days. This means the reading is higher or lower in terms of usage, but the time period for basing the consumption is still one hour. In a regular wattmeter, where there is no accumulative

6-A. Minimum Service Capacities for Various House Sizes

SQ FT FLOOR AREA	MINIMUM SERVICE CAPACITY
Up to 1000 sq ft	125 amperes
1001-2000 sq ft	150 amperes
2001-3000 sq ft	200 amperes

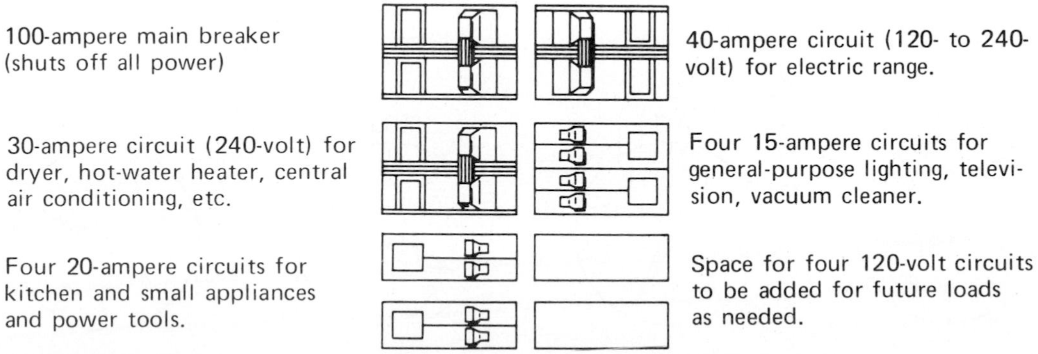

100-ampere main breaker (shuts off all power)

40-ampere circuit (120- to 240-volt) for electric range.

30-ampere circuit (240-volt) for dryer, hot-water heater, central air conditioning, etc.

Four 15-ampere circuits for general-purpose lighting, television, vacuum cleaner.

Four 20-ampere circuits for kitchen and small appliances and power tools.

Space for four 120-volt circuits to be added for future loads as needed.

6–7. An example of a fuseless service entrance panel capable of handling a 100-ampere service.

factor included, the reading and the watts consumed are in terms of the second, or what is being read at the moment the meter is used. This type of meter reading is usually done in the laboratory, or under test conditions where a definite consumption rate is necessary for a device.

In Fig. 6-8 the meter movement is composed of two coils, one across the line to check the voltage, and one in series with the load to check the current being drawn. This arrangement accounts for the voltage times the current, converting to watts, or power consumed. P = E × I, or power is equal to the product of the voltage and current.

Reading a Kilowatt-Hour Meter

Take a look at Fig. 6-9 and read the amount of power consumed on the meter. If the dial pointer has just passed a number, read that number and not the next higher one. For example, in Fig. 6-9, the reading is 7432 kilowatts. Read from left to right and write down the numbers.

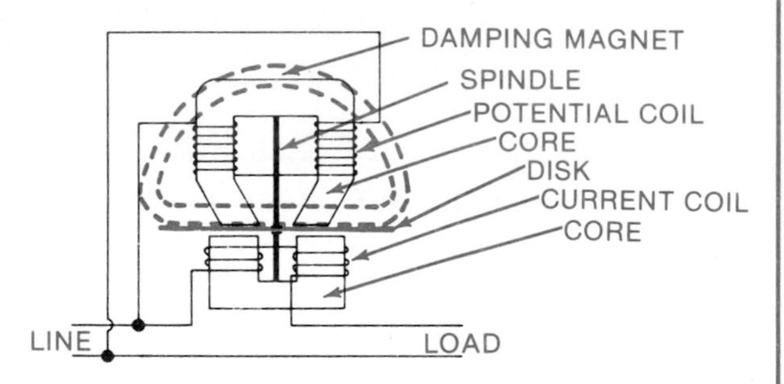

6–8. Parts of a kilowatt-hour meter and hookup.

Fig. 6-10 is a three-phase meter which is used with 240-volt, 3-wire, three-phase power. Fig. 6-11 is a 480-volt, three-phase meter with a mechanical demand register, for use with specialized industrial equipment. There is no need in home wiring for 480 volts, as there is in industry. Fig. 6-12 is a polyphase meter, also for 480 volts, with two stators and a 3-wire connection. It, too, is a mechanical demand meter.

Distribution Panels

There are many types of distribution panels made for the home. One type has fuses that screw in and must be replaced when they are blown, or open. See Fig. 6-5. Another type is the circuit-breaker box. Fig. 6-13b. A circuit

6–10. Three-phase kilowatt-hour meter, designed for 240 volts, 3 wires. This one has two stators.

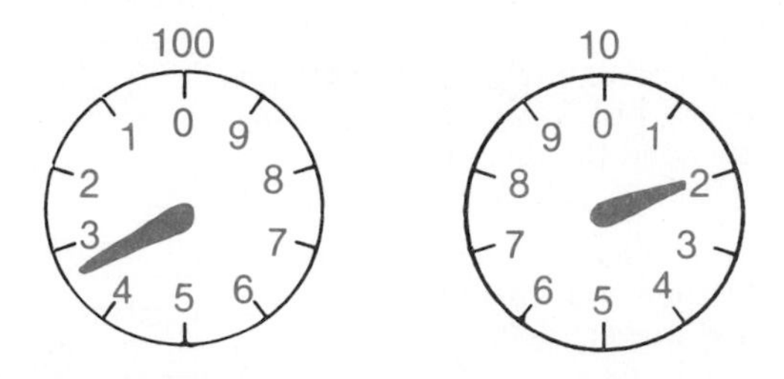

6–9. Read the dials to see if you get 7-4-3-2 kilowatt-hours.

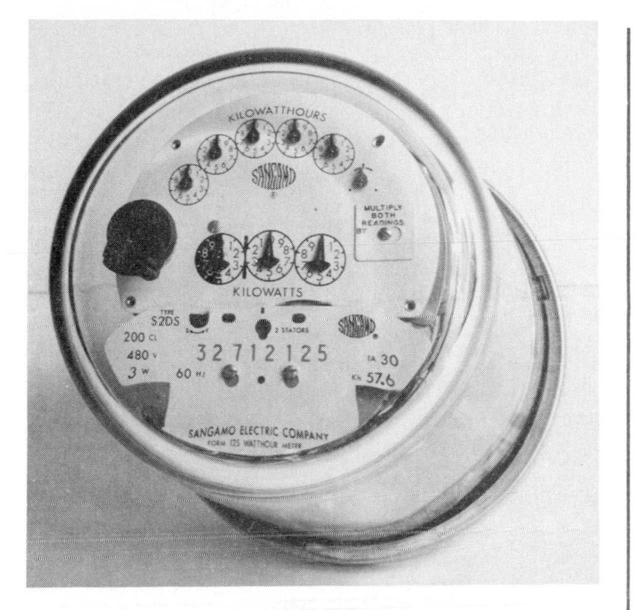

6–11. Another three-phase kilowatt-hour meter, designed for industrial applications where 480 volts is common.

6–12. Three-phase kilowatt-hour meter for 480 volts, 3 wires; mechanical demand meter.

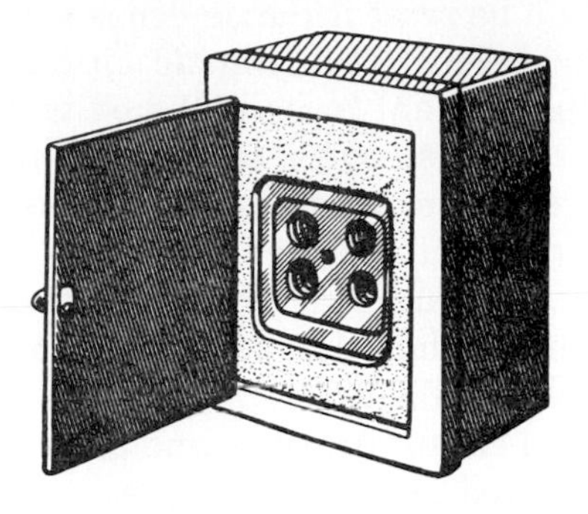

6–13a. Distribution box with screw-in fuses.

6–13b. Distribution box with circuit breakers.

breaker can be reset if the device is tripped by an overload. This type is made for home use. People do not want to be bothered with looking for a new fuse every time one blows. The circuit breaker may be reset by pushing it to the "off" position and then to the "on" setting. If it trips a second time, the circuit trouble is still present. It should be located and removed before resetting the breaker again.

Industrial circuit breakers and control boxes will be covered in another chapter.

Romex Cable

Romex cable is used to carry power from a distribution panel box to the individual outlets within the house. This nonmetallic sheathed cable has plastic insulation covering the wires to insulate it from all types of environments. Some types of cable may be buried underground. Fig. 6-14 shows the type most commonly used in house wiring.

Most new homes are wired with 12/2, or No. 12 wire with two conductors—one white and one black. Inside the cable is an uninsulated ground wire of the same size, No. 12. In the past, the smallest wire size used in homes was a No. 14, with 2 conductors (written on the cable as 14/2 with ground, or 14/2WG). Single conductors may be used in conduits, stranded or solid conductor. Fig. 6-15.

Wire Size. The larger the physical size of the wire, the smaller the number. Fig. 6-16. Number 14 has been used for many years as a 15-ampere circuit wire, and No. 12 for 20 amperes. Today, the Code calls for No. 12 for a 15-ampere circuit with aluminum or copper-clad aluminum wire. This is a safety factor.

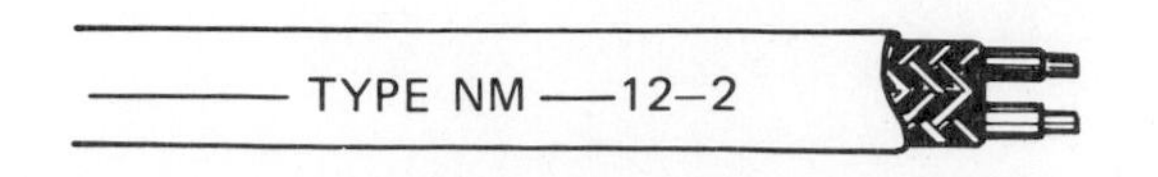

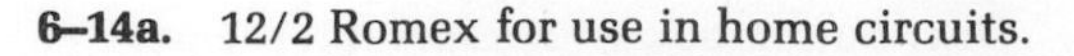
6–14a. 12/2 Romex for use in home circuits.

6–14b. Romex which can be used for underground equipment.

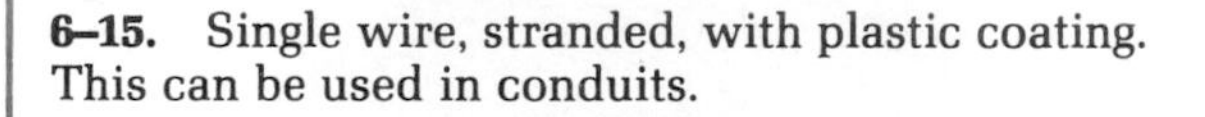
6–15. Single wire, stranded, with plastic coating. This can be used in conduits.

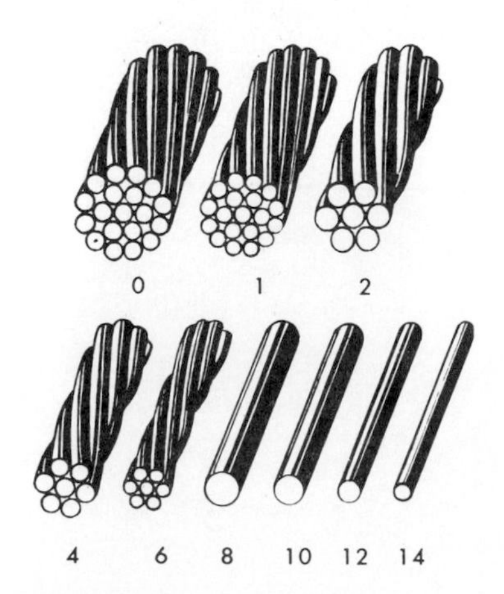

6–16. Sizes of wires. Note the relationship of size and number.

INDIVIDUAL EQUIPMENT BRANCH CIRCUITS

Circuits should be provided for the equipment listed in Table 6-B. Spare circuit equipment should be provided for at least two future 20-ampere, 2-wire, 115-volt circuits, in addition to those initially installed. If a branch circuit or distribution panel is installed in a finished wall, raceways should be extended from the panel cabinet to the nearest accessible unfinished space for future use.

Consideration should also be given to the provision of circuits for the commonly used household appliances and equipment on Table 6-C. (The chart does not list all the equipment available.)

In some instances, one of the circuits recommended may serve two devices not likely to be used at the same time, such as electric space heating and air conditioning. The majority of appliances for residential use are made for 110 to 120 volts. There is, however, a growing tendency to make fixed appliances for use on 220- to 240-volt circuits. The higher voltage is usually preferred in those cases where a choice exists.

If specified, circuits should be supplied for patios, or outdoor living areas, and for other exterior decorative or flood lighting. Fig. 6-17. Don't forget GFCIs.

Feeder Circuits

It is strongly recommended that consideration be given to the installation of branch-circuit protective equipment served by appropriate-sized feeders located throughout the house, rather than at a single location.

Small-Appliance Branch Circuits

There should be in home wiring service at least one 3-wire, 115/230-volt, 20-ampere small-appliance branch circuit which is equipped with split-wired receptacles for all convenience outlets in the kitchen, breakfast and dining room, and family room. Two 2-wire, 115-volt, 20-ampere branch circuits are equally acceptable.

6-B. Conductor Capacities for Household Equipment

ITEM	CONDUCTOR CAPACITY
Range (up to 21-kW rating) *or*	50 A-3 W-115/230 V
Built-in oven	30 A-3 W-115/230 V
Built-in surface units	30 A-3 W-115/230 V
Combination washer-dryer *or*	40 A-3 W-115/230 V
Electric clothes dryer	30 A-3 W-115/230 V
Fossil-fuel-fired heating equipment (if installed)	15 A or 20 A-2 W-115 V
Dishwasher and waste disposer (if necessary plumbing is installed)	20 A-3 W-115/230 V
Water heater (if installed)	Consult local utility

6-C. Conductor Capacities for Other Household Equipment

ITEM	CONDUCTOR CAPACITY
Room air conditioners *or*	20 A-2 W-230 V
Central air-conditioning unit *or*	40 A or 50 A-2 W-230 V
Attic fan	20 A-2 W-115 V (switched)
Food freezer	20 A-2 W-115 or 230 V
Water pump (where used)	20 A-2 W-115 or 230 V
Bathroom heater	20 A-2 W-115 or 230 V
Work shop or bench	20 A-3 W-115/230 V

6–17. Wires, run through the siding of a new house, will be attached to outlet boxes later.

The use of 3-wire circuits for supplying convenience outlets in the above locations is an economical way to divide the load. Such circuits provide greater capacity at individual outlet locations and lessen voltage drop in the circuit. They also provide more flexibility in the use of appliances. For maximum effectiveness, the upper half of all receptacles should be connected to the same side of the circuit. Fig. 6-18.

Laundry Receptacle Circuit

For a laundry receptacle outlet, one 2-wire, 115-volt, 20-ampere branch circuit is required.

BRANCH CIRCUITS

General-purpose circuits should supply all lighting and all convenience outlets throughout the house, except those convenience outlets in the kitchen, dining room (or dining areas of other rooms), breakfast room or nook, family room, and laundry areas. General-purpose circuits should be provided on the basis of one 20-ampere circuit for not more than every 500 square feet, or one 15-ampere circuit for not more than every 375 square feet of floor area. Outlets supplied by these circuits should be divided equally among the circuits. Fig. 6-19.

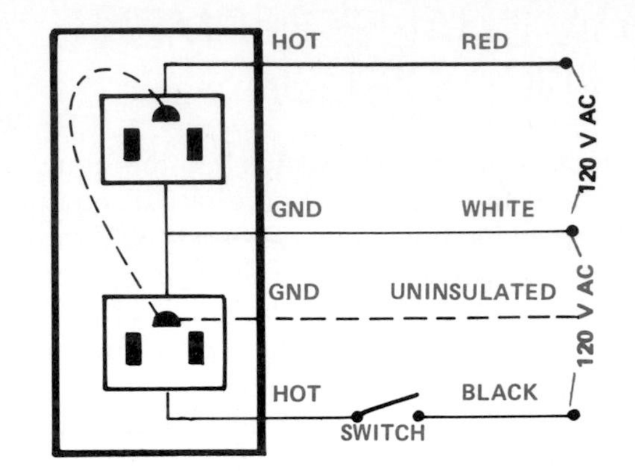

6–18a. Three wires feeding a single receptacle. Switched lower outlet—top outlet hot all the time.

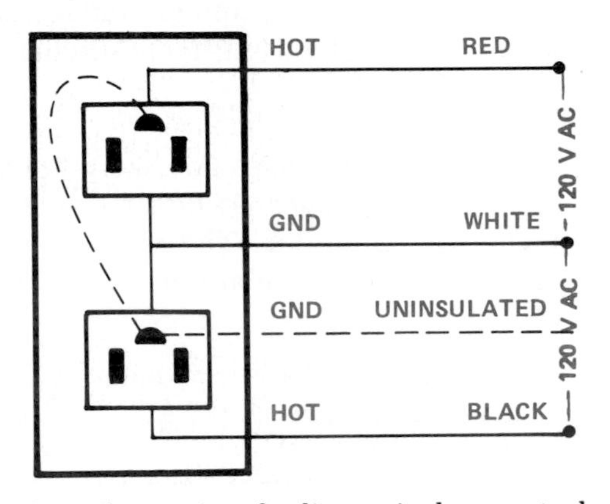

6–18b. Three wires feeding a single receptacle.

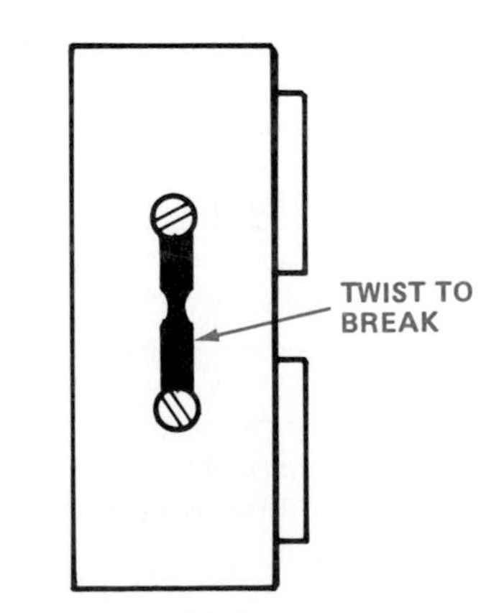

6–18c. Breakaway connection between the two outlets allows the red and black hot wires to be connected to the same duplex outlet.

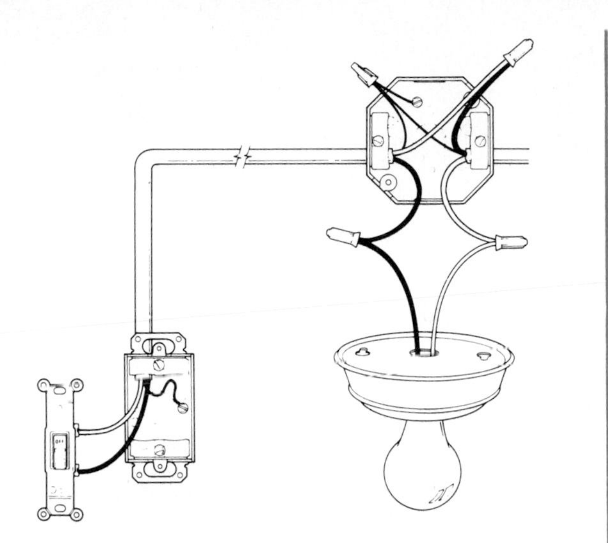

6–19a. To add a wall switch for control of a ceiling light at the end of the run.

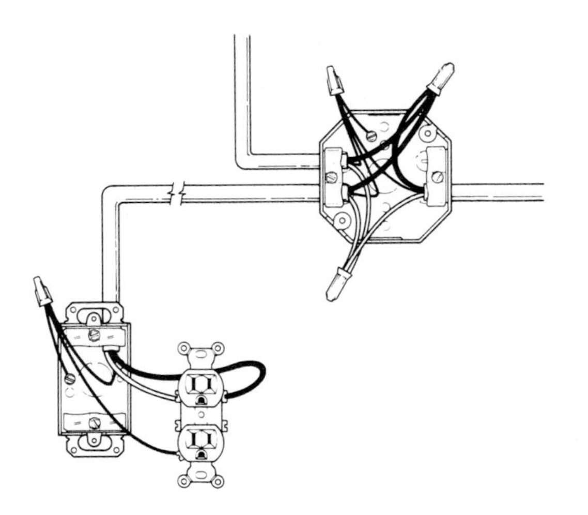

6–19b. To add a new convenience outlet from an existing junction box.

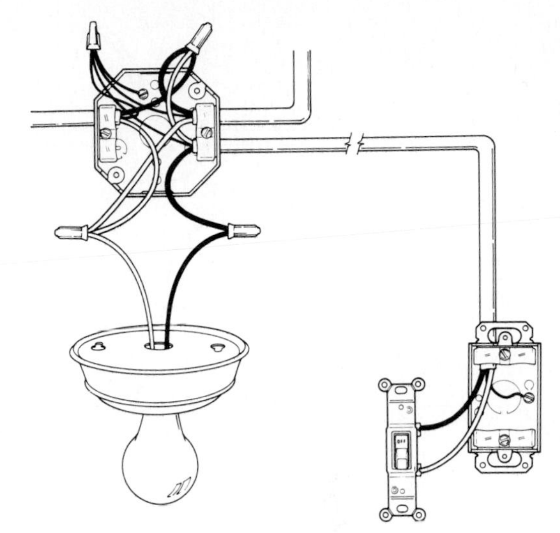

6–19c. To add a wall switch to control a ceiling light in the middle of a run.

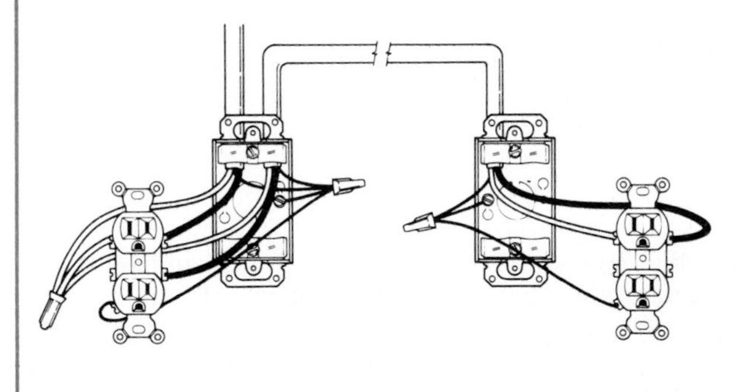

6–19d. To add new convenience outlets beyond old outlets.

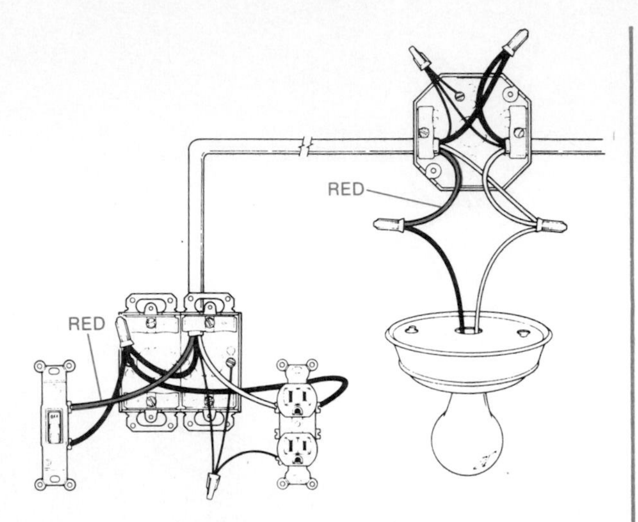

6–19e. To add a switch and convenience outlet in one box, beyond existing ceiling light.

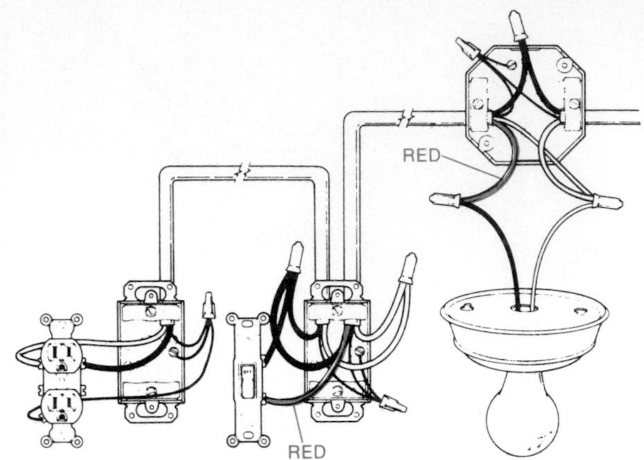

6–19g. To add a switch and convenience outlet beyond existing ceiling light.

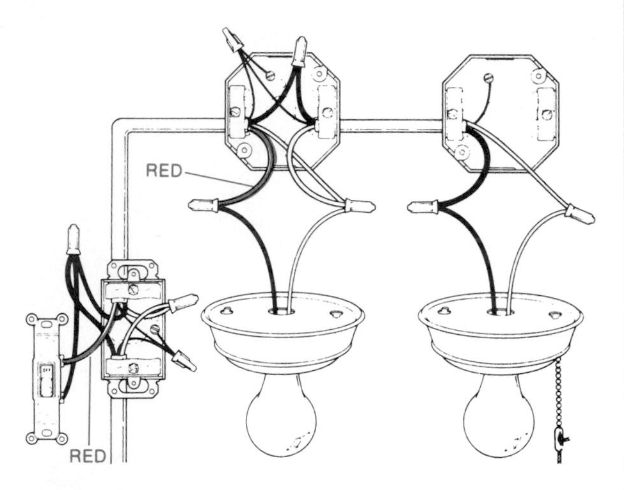

6–19f. To install one ceiling outlet and two new switch outlets from existing ceiling outlet.

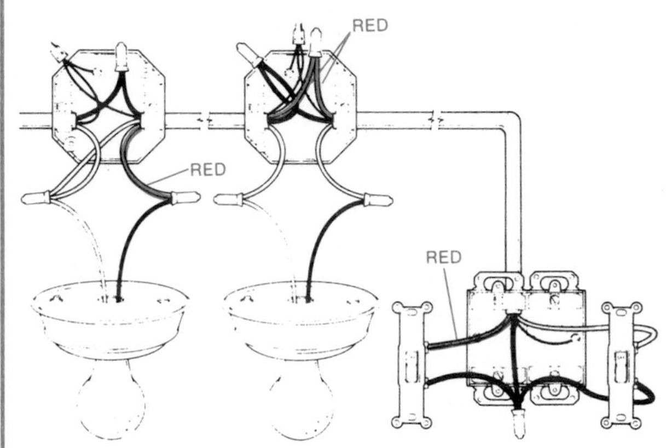

6–19h. To install one ceiling outlet and two new switch outlets from existing ceiling outlet.

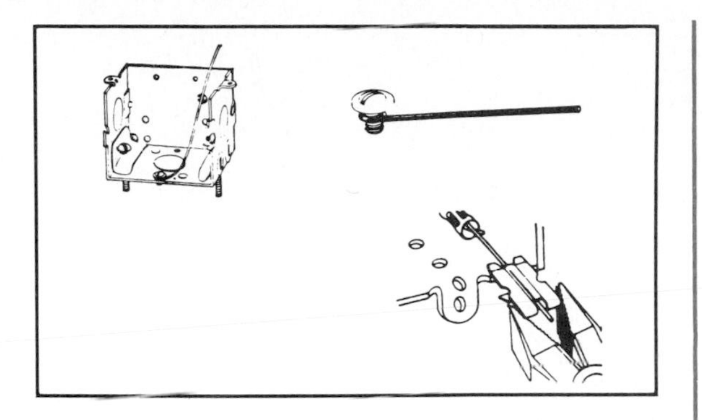

6–19i. Two methods of attaching ground wire to metal box.

These requirements for general-purpose branch circuits take into consideration the provision in the current edition of the National Electrical Code. Floor area designations are in keeping with present-day usage of such circuits. Figs. 6-20 & 6-21.

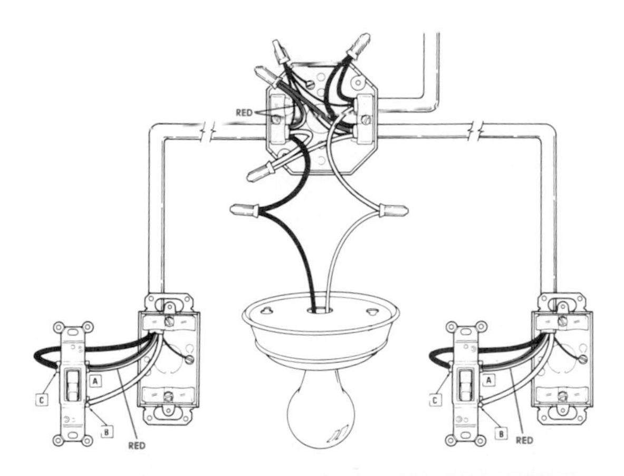

6–20a. Three-way switches control a single lamp. Feed is through the center lamp box. The terminals marked A and B are the light colored points to which red and white wires must be connected. Terminal C is the dark colored (brass screw) point to which the black wire must be connected. Note the extra wire coming out of the box. This is the uninsulated wire which has to be properly grounded to the box or switch terminal marked with green paint.

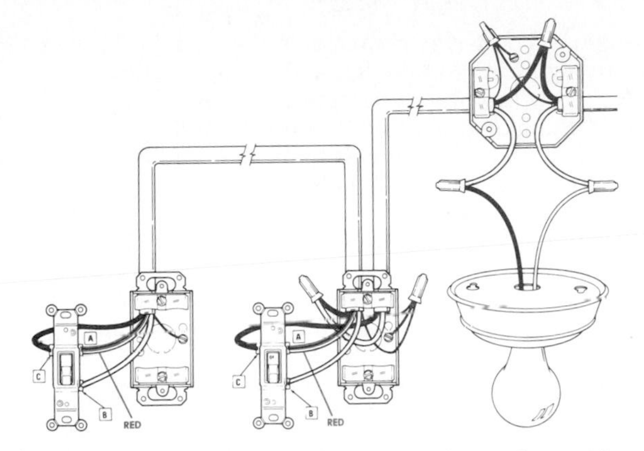

6–20b. Two 3-way switches are tied together with a 3-wire Romex cable. The lamp and switch box are fed with a 2-wire cable. The terminals marked A and B are the light colored screws to which red and white wires must be connected. Terminal C is the dark colored (brass colored screw) point to which the black wire must be connected. Note the extra wire that is coming out of the box. This wire is the uninsulated ground wire which must be properly grounded to the box or switch terminal marked with green paint.

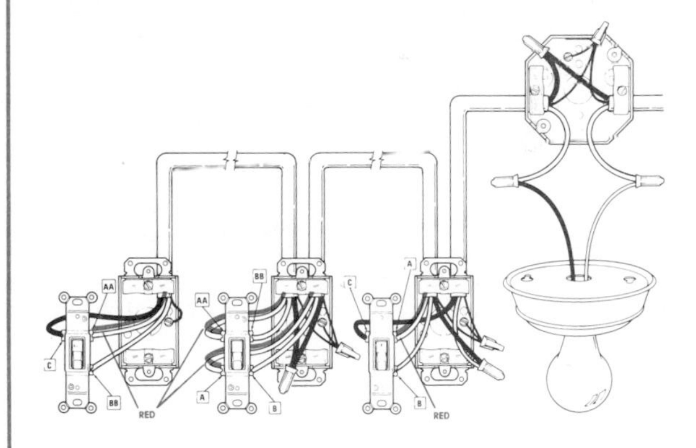

6–20c. Two 3-way switches are used to control the ceiling lamp and the outlet that is located beyond both switches. The receptacle is always hot. The terminals marked A are connected by a black wire, shown here by an interrupted red wire. That means the connections between the two switch boxes are made with two pieces of 14/2 Romex. The dotted *white wire* is part of this piece of Romex. It is shown in dotted fashion to distinguish it from the other piece of Romex, which has a normal black wire and a normal white wire. To connect the switch box on the right and the lamp, use a piece of 3-wire Romex (14/3 Romex). Connections between the lamp and the receptacle are made by a piece of 14/2 Romex.

6–20d. The 4-way switch is in the middle. There is a 3-way switch on the left and a 3-way switch on the right. Two 3-way switches and one 4-way switch make it possible to control a lamp from three locations. Terminals marked A and B are the light-colored screws to which red and white wires must be connected. Terminal C is the dark-colored screw to which the black wire must be connected. Terminals AA show where the two ends of the red wire are connected between the 4-way switch and the 3-way switch on the left. Terminals BB show where the two ends of the white wire are connected to the 4-way switch and the 3-way switch on the left.

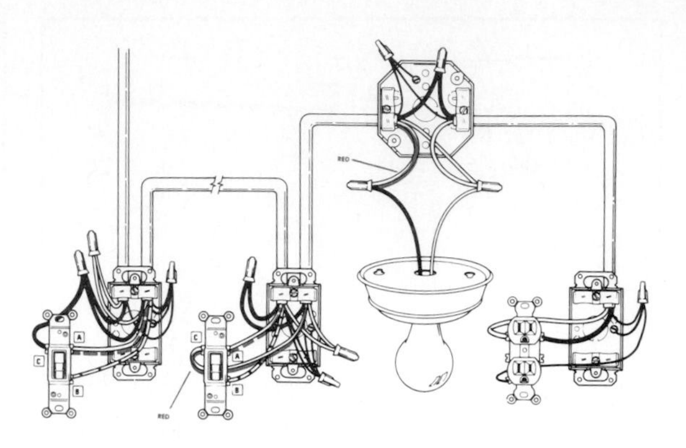

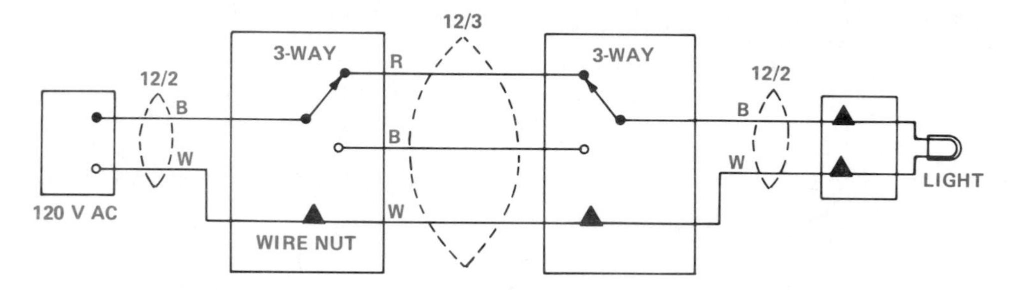

6–21a. Schematic drawing of two 3-way switches controlling a single lamp.

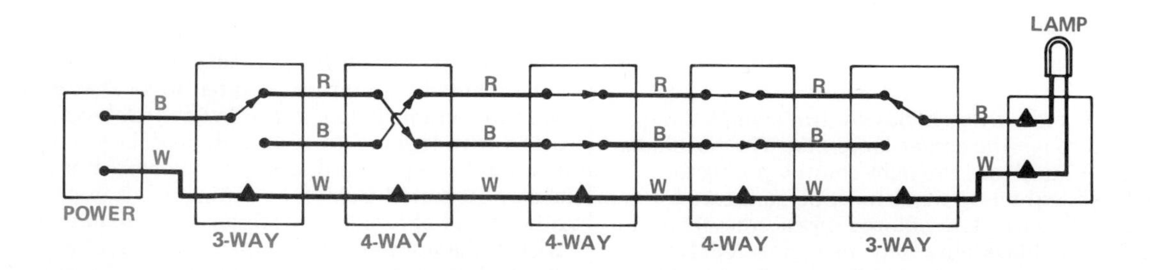

6–21b. Schematic diagram of five locations to control a single lamp. Note there are never more than *two* 3-way switches. Each time another location for control is needed, a 4-way switch is added.

It is recommended that separate branch circuits be provided for both lighting and convenience outlets in living rooms and bedrooms, and that the branch circuits servicing convenience outlets in these rooms be of the 3-wire type, equipped with split-wired receptacles.

For a close look at the internal wiring of a house, before it is covered with drywall or plaster, see Figs. 6-22 through 6-26. Wiring of switches and receptacles, and locations within the circuits, is shown in Figs. 6-27 through 6-36 (pages 177-179).

6–22. Location of an outlet where there is no insulation. This is the basement entrance and the switch box is located on the top right of the photo. It will serve to turn the basement light on and off, for light on the steps. Note how the cables have been stapled about 6″ to 8″ above the box and about 36″ above that. The cable is held approximately in the middle of the stud.

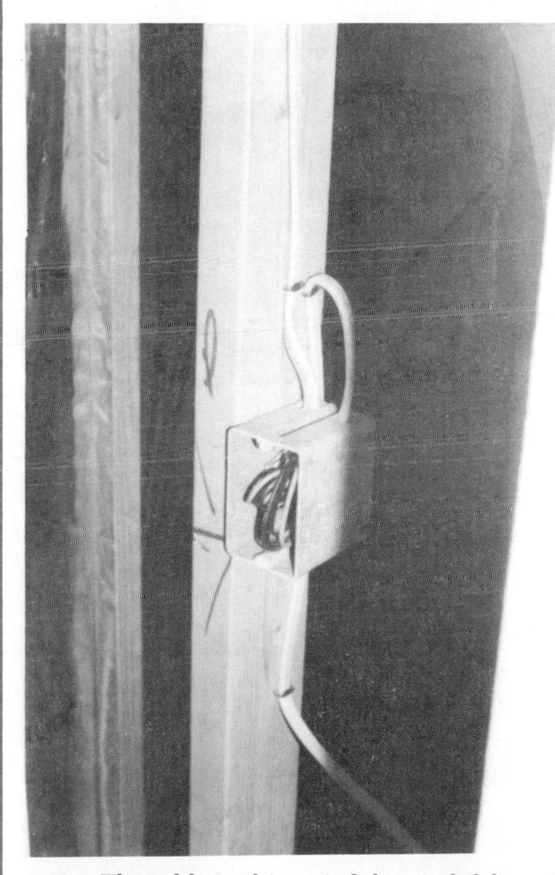

6–23. The cable in this switch box is fed from the top. It is then brought out from the top and stapled. Next it leads down and is stapled again below the box. Since this is an uninsulated box, it has no clamp or connector; the staple above the box serves as a strain relief.

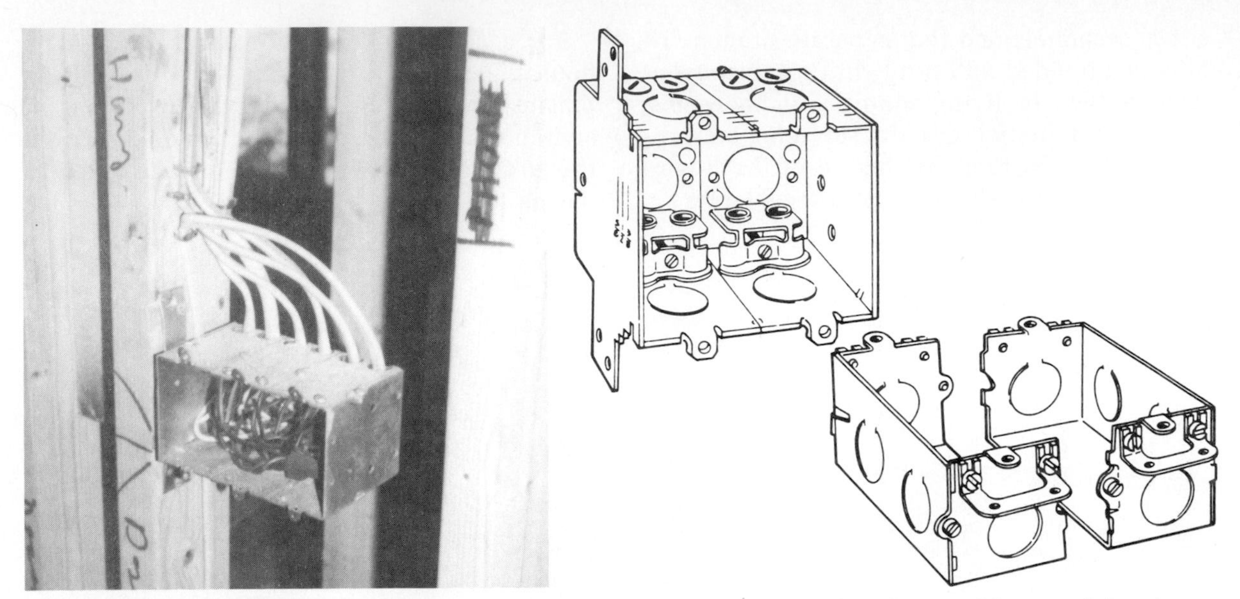

6–24. Three-ganged switch boxes. Note how the wires are stapled to anchor them and how each box has a cable entering and leaving. This will be a 3-switch control center.

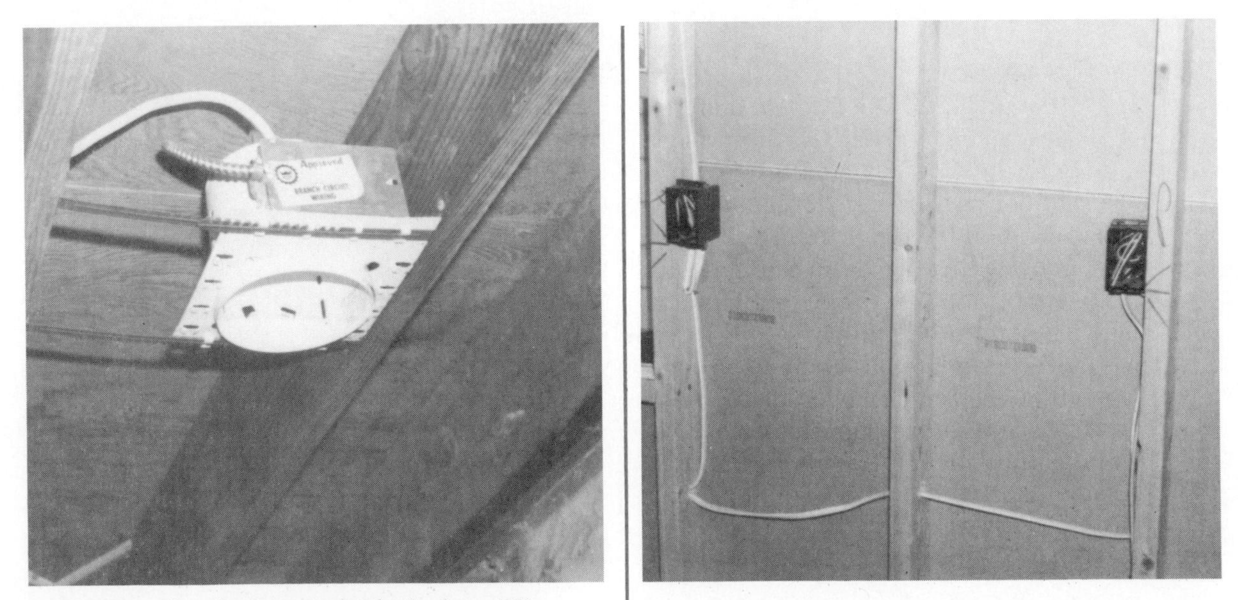

6–25. Note that the Romex feeds the box. BX (cable sheathed in metallic armor) is used to protect the asbestos-covered wires from the box to the lamp socket.

6–26. Note that the studs are drilled and the wires fed to switch boxes.

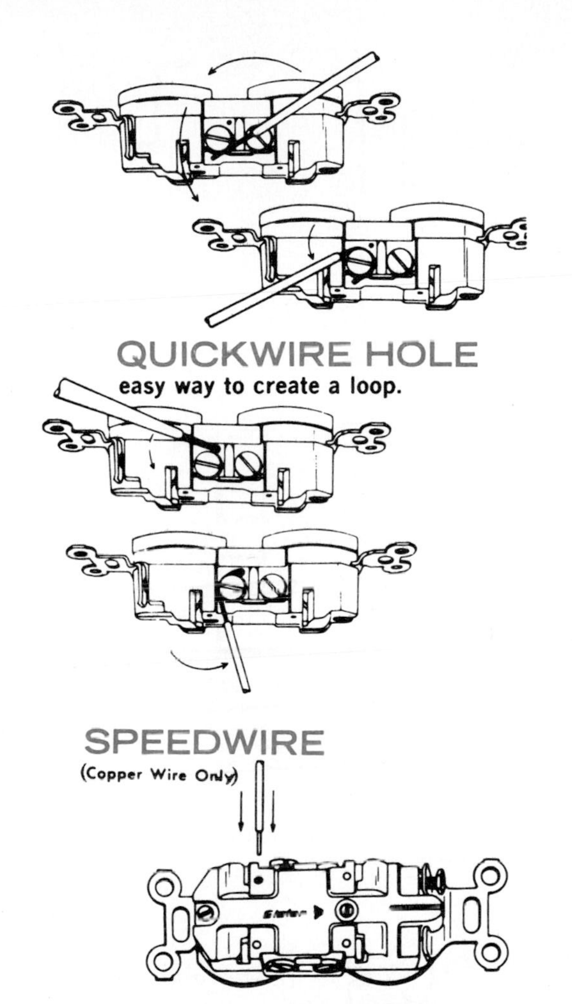

6–27. Duplex receptacles being wired. A speedwire connection means that the wire is pushed into the hole to make contact.

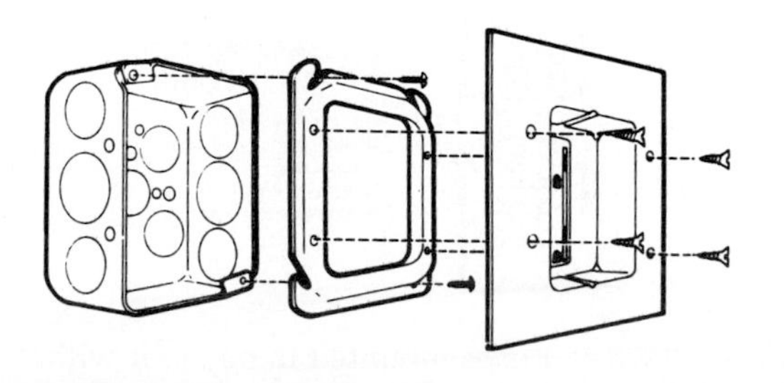

6–28. Recessed outlet mounted in a 4″ square box.

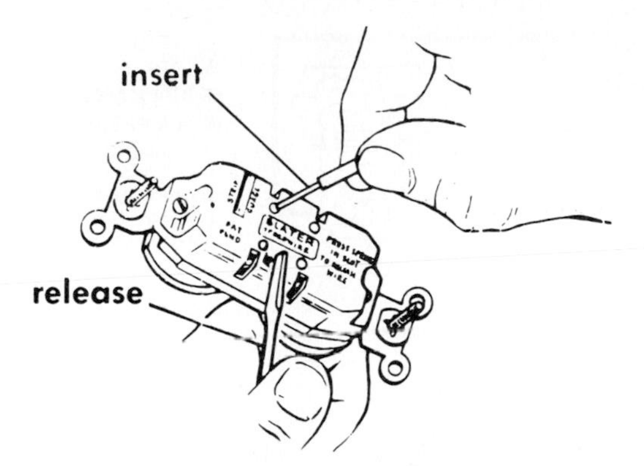

6–29. Note how the wire is released by pressing with a screwdriver blade.

COMMON FEED ONLY
When plug is inserted into receptacle, pilot light automatically lights up.

6–30. Receptacle with pilot light.

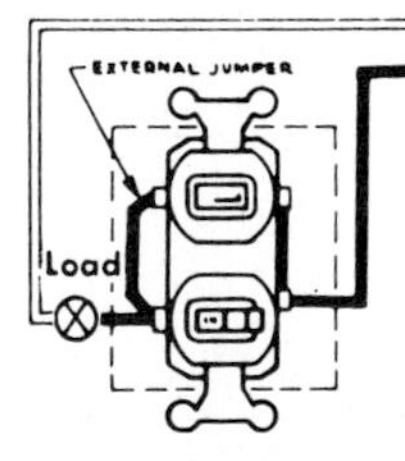

WIRED AS NIGHT LIGHT
When switch is OFF, indicator light glows.

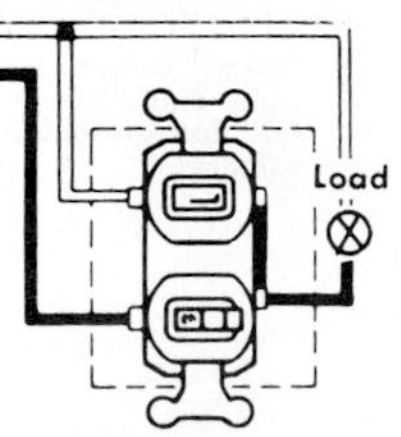

WIRED AS PILOT LIGHT
When switch is ON, pilot light glows.

6–31. Switch and pilot light.

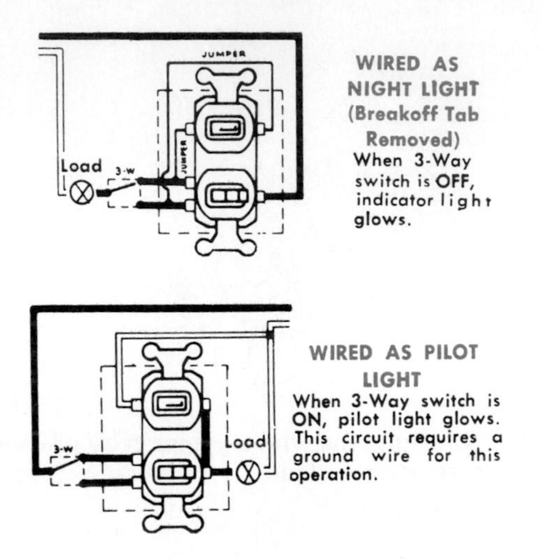

6–32. Switch and pilot light device wired with a 3-way switch.

Grounded power outlet controlled from 3-Way switch and 3-Way switch at another location.

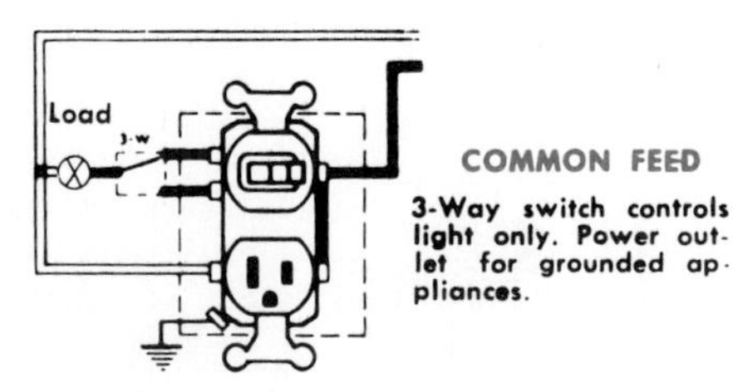

6–33. Power outlet and switch can be used in a number of configurations for circuit control.

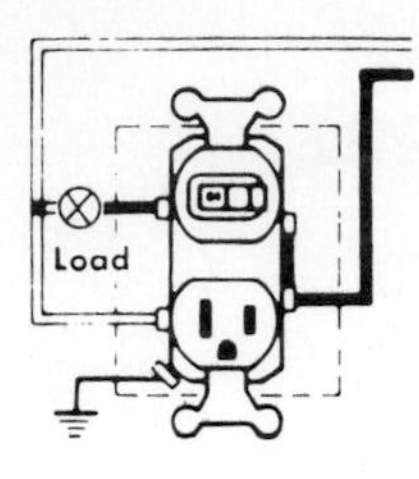

COMMON FEED
S. P. switch controls light only. Power outlet for grounded appliances.

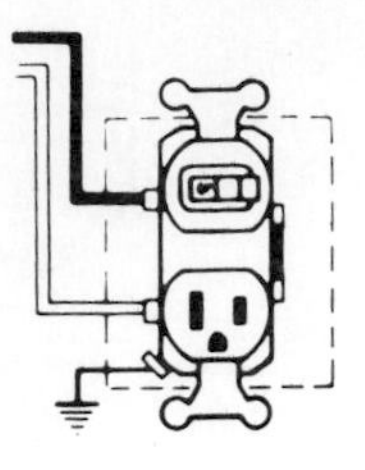

Grounded power outlet controlled by S.P. switch.

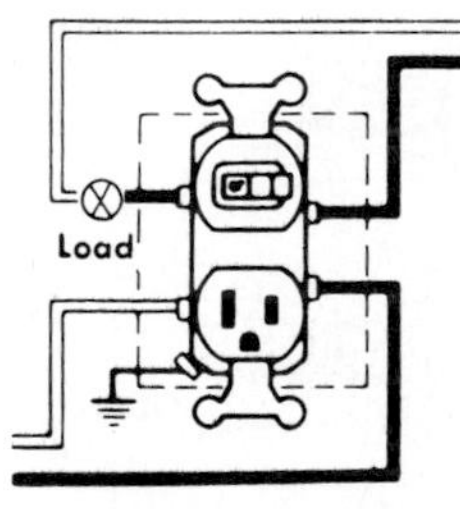

SEPARATED FEED (Breakoff Tab Removed)
S. P. switch and grounded power outlet on separate circuits.

6–34. Switch and outlet with a number of arrangements.

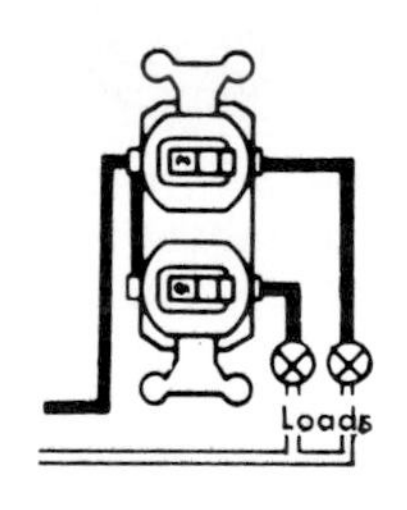

COMMON FEED
Two S.P. switches on same circuit. Each switch controls an independent light.

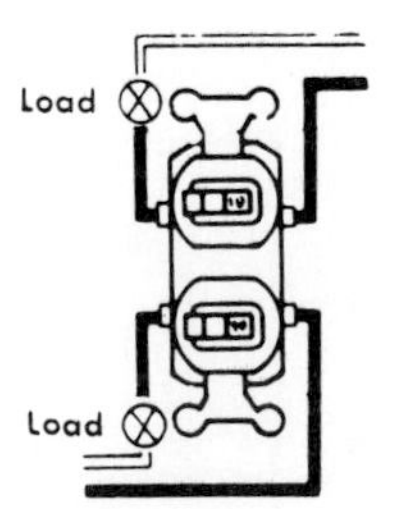

SEPARATED FEED (Breakoff Tab Removed)
Two S.P. switches on separate circuit. Each switch controls an independent light.

6–35. Two switches mounted in one unit with possibilities for controlling two loads in different ways.

6–36. Switch and night-light or pilot light with an addition of an *external* jumper.

Testing the Wiring

After the circuit has been energized, you will need to test the circuit. For this, a small *neon circuit tester* is available. This tester can be used to check the circuit to see if it is live. It can check for three-wire grounding. It can check for a ground with a two-wire receptacle. It can test three-wire grounding extension cords. Fig. 6-37.

Here are several common applications for the neon circuit tester:

Determine if circuit is live by inserting tester leads into hot (narrow slot) and neutral slots. Tester will glow if circuit is live.

For three-wire grounding receptacles, insert one lead into the hot (narrow slot) and the other lead into the U-shaped ground slot. Tester will glow if circuit is live and grounding slot is properly grounded and polarity is correct.

To check for ground with two wire receptacles insert one lead into hot (narrow slot) and touch the other lead to bare metal wallplate mounting screw. Tester will glow at full brightness if circuit is live and box is properly grounded and polarity is correct.

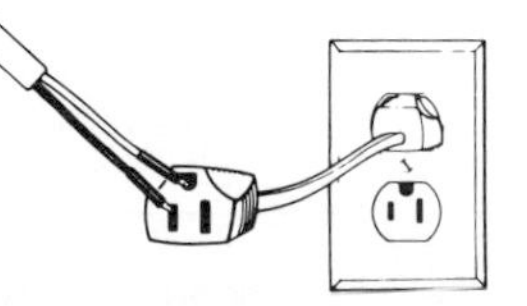

Test three-wire grounding extension cords in the same manner as grounding.

6–37. Here are several common applications for the neon circuit tester:

The neon tester is made from a NE-2 neon lamp with a resistor of at least a 100 000 ohms in series with one leg of the lamp. The neon lamp usually ionizes at 55 volts. If both electrodes glow, it is AC. If only one electrode glows, it is DC. The one that glows is the negative terminal of a DC source. This device can also be used to check the voltage. First, test the 120-volt known circuit and observe the brightness of the lamp. Then, plug it into a 240-volt circuit source and see how much brighter it is. This will serve as a reference for you when checking voltages of unknown value.

Space-Heating and Air-Conditioning Circuits

Wiring capacity for either electric space heating or for air conditioning, whichever is the larger load, is to be provided in accordance with the following.

Electric Space Heating. Capacity required for electric space heating should be determined from Table 6-D, which shows maximum winter heat loss, based on the total square feet of living space in the home.

If electric space heating is installed initially, wiring should be as follows:

- For a central furnace, boiler, or heat pump: a 3-wire, 115/230-volt feeder, sized to the installation.
- For individual room units: ceiling cable, or panels, sufficient 15-, 20-, or 30-ampere, 2-wire, 230-volt circuits to supply the heating units in groups, or individually. Figs. 6-38, 39, & 40.

6-D. Maximum Heat Loss Values*
(Based on Infiltration Rate of $\frac{3}{4}$ Air Change Per Hour)

DEGREE DAYS	Watts/sq ft	Btuh/sq ft
Over 8000	10.6	36
7001-8000	10.0	34
6001-7000	9.5	32
4501-6000	9.2	31
3001-4500	8.9	30
3000 and under	8.4	29

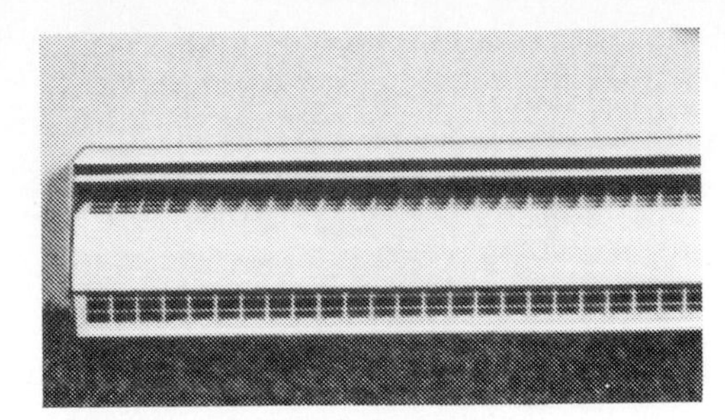

6–38. Baseboard electric heater.

6–39. Electric heater showing themostat.

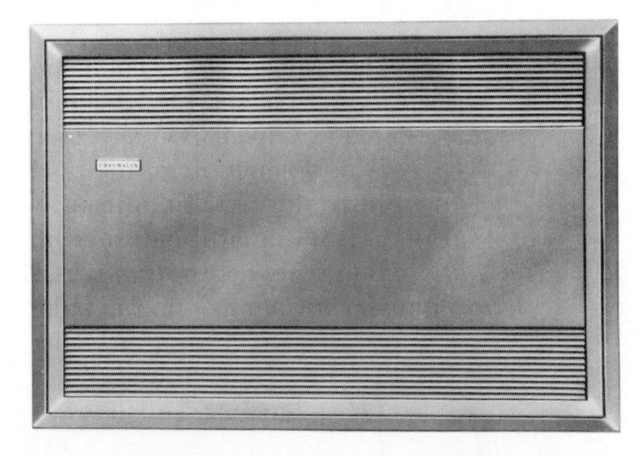

6–40. Smooth finish, flush installation of an electric heater.

Electric Air Conditioning. Capacity required for electric air conditioning is to be determined from the chart of maximum allowable summer heat gain, based on the total square feet of living space in the home (Table 6-E).

If electric air conditioning is installed initially, wiring should provide for:

- A heat pump, providing both winter heat and summer air conditioning; a 3-wire, 115/230-volt feeder, sized to the installation.
- A central air-conditioning unit, a 3-wire, 115/230-volt feeder sized to the installation.
- Individual room air-conditioning units, sufficient 15-, 20-, or 30-ampere, 3-wire, 230-volt circuits to supply all units, and a 20-ampere, 230-volt, 3-wire outlet in each room, on an outside wall and convenient to a window.

If neither electric space-heating nor electric air-conditioning equipment is installed initially, wiring capacity for the larger load should be provided as follows:

The service entrance conductors and service equipment should include capacity required, in accordance with the appropriate chart. Spare feeder or circuit equipment should be provided, or provision for these should be made in panelboard bus space and capacity.

The plan should allow bus space and capacity for a feeder position which can serve a central electric heating or air-conditioning plant directly, or can supply a separate panelboard to be installed later. The panelboard would control circuits to individual room heating or air-conditioning units. Space heating and air conditioning are considered dissimilar and non-coincident loads. Service and feeder capacity need be provided only for the larger load, not for both.

Space-Heating and Air-Conditioning Outlets. Since many different systems and types of equipment are available for both space heating and air conditioning, it is impractical to show outlet requirements for these uses in each room. Electrical plans and specifications for the house should indicate the type of system to be supplied and the location of each outlet.

Entrance Signals

Entrance push buttons should be installed at each commonly used entrance door and connected to the door chime, giving a distinctive signal for both front and rear entrances. Electrical supply for entrance signals should be obtained from an adequate bell-ringing or chime transformer.

A voice intercommunications system, which permits the resident and a caller to converse without the necessity of opening the door, is convenient and is an added protection. It may be designed for this purpose alone or it may be part of an overall intercommunications system installed for an entire house.

In a smaller home, the door chime is often installed in the kitchen, providing it will be heard throughout the home. If not, the chime should be installed at a more central location,

6-E. Maximum Allowable Heat Gain (Btuh)

AREA TO BE CONDITIONED, sq ft	DESIGN, DRY-BULB TEMPERATURE, °F			
	90	95	100	105
750	15 750	18 000	19 500	21 000
1000	20 500	23 000	24 750	26 500
1500	27 000	30 000	31 500	33 000
2000	36 000	40 000	42 000	44 000
2500	45 000	50 000	52 500	55 000
3000	54 000	60 000	63 000	66 000

usually the entrance hall. In a larger home, a second chime is often necessary to ensure its being heard throughout the living quarters.

Entrance-signal conductors should be no smaller than the equivalent of a No. 18 AWG copper wire. (AWG is American Wire Gage.)

Communications

An intercommunications system, with a station in each room of the living quarters, is recommended. It may be arranged for voice communication to the entire house, or selectively to an individual room only. Music from AM or FM radio, stereo, or recordings may be fed into the system. It may also be used to communicate with a caller without opening the door.

For the larger home with accommodations for a resident servant, a dining room-to-kitchen signal is also recommended. This can be operated by a push button attached to the underside of the dining table or by a floor tread placed under the table. In some cases a built-in annunciator in the kitchen, with push-button stations in various rooms of the living quarters, or a separate intercommunication telephone system, is desirable.

Intercommunication systems and telephones should be operated from a power unit recommended by the manufacturer of the system, and supplied with a minimum of 115 volts. Fig. 6-41.

For telephones, it is recommended that raceways and outlets be installed at the time of construction. Outlets, if desired, should be located in the kitchen, living room, bedrooms, and other rooms as requested. The telephone company installs concealed wiring in residential units during construction, on a selective basis.

It is suggested that the local telephone company be consulted for details of service prior to construction, particularly in regard to the installation of raceways and the location of protector equipment.

6-41. Intercoms must have a source of 120-V AC. However, low-voltage speaker wiring should *not* be terminated in the same box as 120-V AC.

Television

An outdoor television antenna and lead-in connections require a nonmetallic outlet box at convenient location for connection of the master antenna system to the various other locations in the house. Allow for a 115-volt outlet near the antenna rotor control. It is more convenient, and the appearance is more pleasing, if the antenna lead-in wire and the control circuits for the rotor are installed before the house has been closed in by drywall or by plaster. Follow the rules and regulations for installing lead-in wire:

- Staples are to be used to hold only the insulated type of lead in place. Do not use staples on twin-lead.
- No right-angle turns.
- Do not run the antenna lead-in cable parallel to 60-hertz house current lines.
- Use an uninsulated box for making the termination connections to the plug-in receptacle. Fig. 6-42.

6–42. TV outlets wired into boxes with receptacle for 120 volts to power the rotor.

Televison Line Problems

Voltage spikes can damage television sets and computer monitors. Voltage spikes are high-voltage surges that can occur in any electrical system. The two common causes in home circuits are lightning strikes near power lines and switching *off* or *on* heavy-duty appliances such as as air conditioners.

A *voltage spike protector* can eliminate much of the electrical interference that can damage a TV set. Fig. 6-43.

Extreme spike voltages can destroy unprotected solid-state components in television sets and other home entertainment electronic equipment. The voltage spike protector absorbs excess spike energy, allowing only a safe voltage level to enter the equipment it protects. Fig. 6-44 shows how its clamping action works.

Some TV receivers and other electronic equipment came from the factory with built-in protection. It is difficult to determine the presence of such safeguards. For electronic equipment owners who would rather be safe than sorry, a voltage spike protector is the answer.

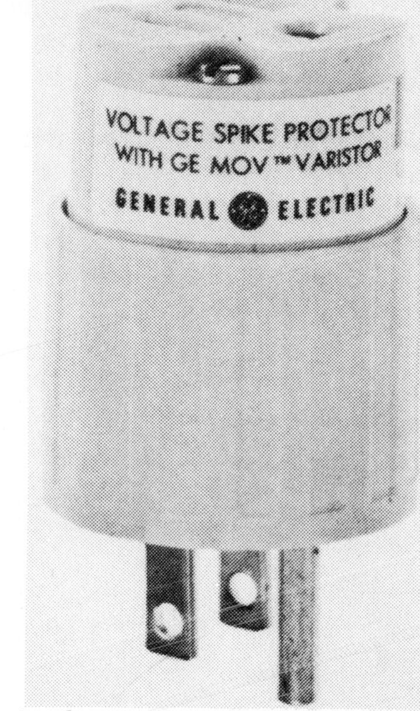

6–43. A voltage spike protector.

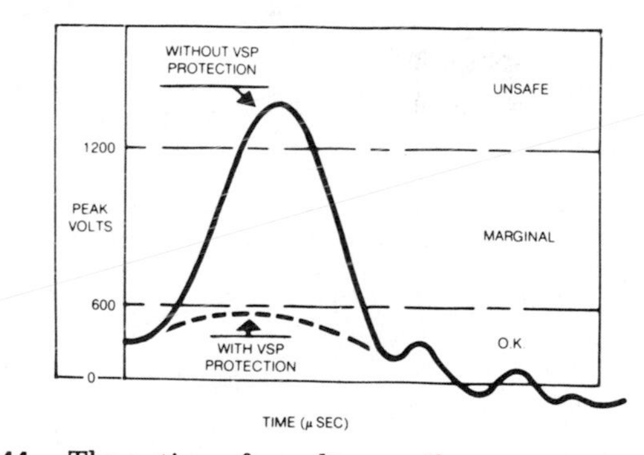

6–44. The action of a voltage spike protector.

The voltage spike protector shown is easy to install. It is plugged into the wall receptacle. The TV or electronic equipment is then plugged into the unit. The voltage spike protector will absorb the spikes and allow full current to flow in the circuit as needed. Remember that the voltage spike protector is not a lightning arrester. It will not afford protection in those rare instances when lightning actually strikes the house, the electrical service entrance, or the TV antenna.

A typical voltage spike or surge is characterized by a rapidly rising voltage of relatively short duration, usually 120 microseconds (0.000120) to one millisecond (0.001). In a 125-volt application, if the spike voltage stays below 600 volts, no harm is usually done. If it rises above this level, many solid-state devices begin to fail.

The solid curve in Fig. 6-44 shows a typical voltage "spike" to which an unprotected solid-state device might be subjected. The dotted curve illustrates the voltage clamping action of the voltage spike protector. The protector absorbs the energy in the spike and maintains the voltage to the solid-state device at a safe level.

Radio

In their modern form radio receivers used purely for the reception of broadcast communication seldom require antenna and ground connections. If FM reception is desired, it is recommended that provisions similar to those for television be included.

Household Warning Systems

An automatic fire warning system is recommended, including installation of both smoke and heat detectors. A smoke detector should be located in the immediate vicinity of, but outside, the bedrooms. Additional smoke detectors placed in strategic locations around the home, and in each bedroom, are recommended. Heat detectors should be provided in each room and each hall of the living quarters, as well as in the attic, furnace room, utility room, basement, attached garage, and each closet and partitioned-off storage area. Fig 6-45.

The fire alarm bell should be clearly audible in all bedrooms, even with doors closed. An outside alarm bell is also recommended. A test button should be installed in the master bedroom or other desirable location.

The primary power supply should be 115 volts, feeding a separate and independent transformer recommended by the manufacturer of the system. A "power-on" visible indicator should be provided. A battery-operated standby source of power is also desirable.

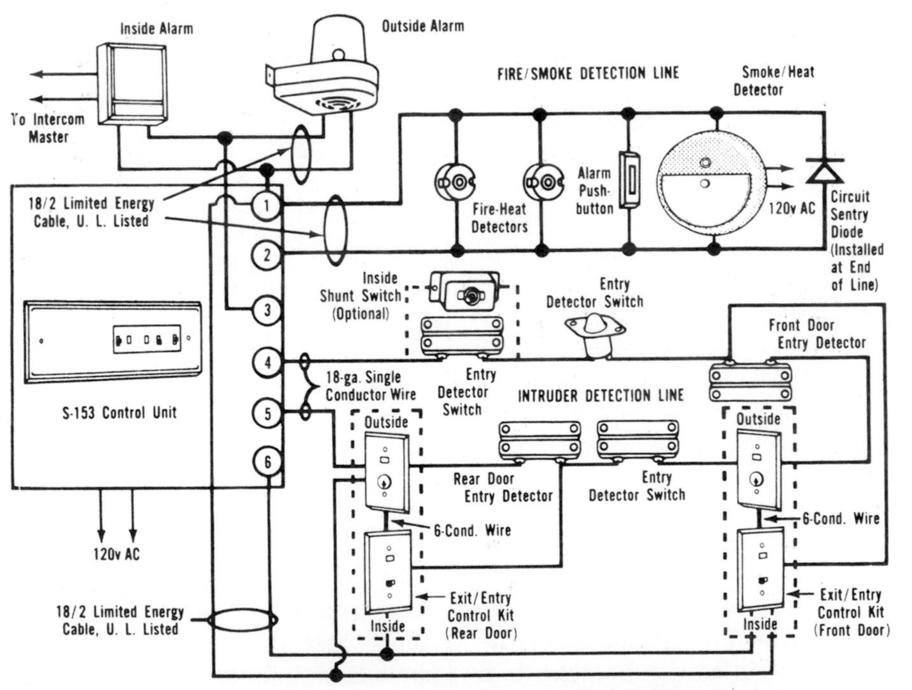

6–45. Wiring diagrams for home burglar alarm, smoke detector, and associated devices.

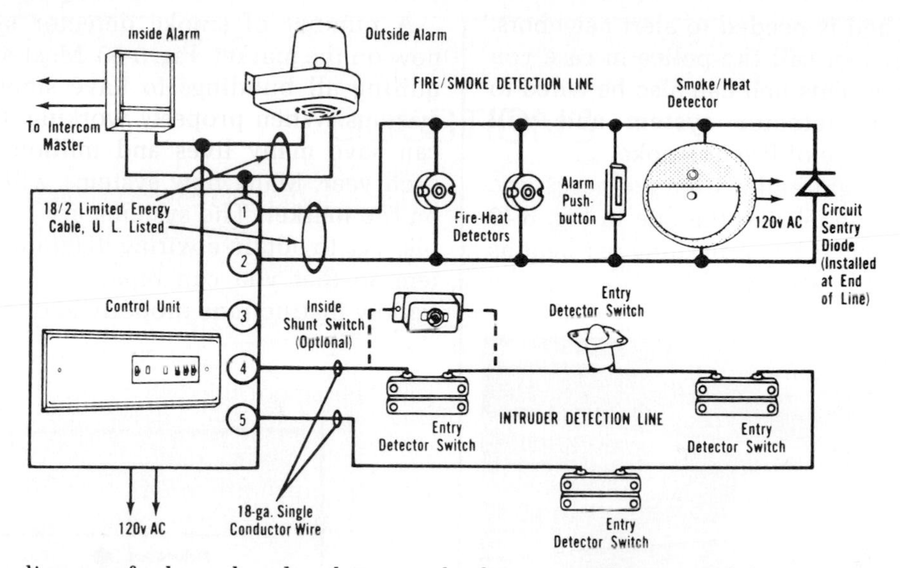

6-45. Wiring diagrams for home burglar alarm, smoke detector, and associated devices. (Cont.)

6-46. Fire-heat detector mounted on ceiling tile.

Fig. 6-46 pictures the newest type of fire-heat detector. It is mounted on the ceiling. It can detect heat and set off an alarm before a fire gets too large to stop. If the temperature rises above 135°F, the fire-heat detector automatically activates an alarm. It should be used in living areas, bedrooms, and closets. It is only $1\frac{3}{4}''$ in diameter and has a depth of $\frac{3}{4}''$. Another model, which looks exactly like the 135°F model, is available. It activates only at 200°F. This is designed for use in attics and furnace rooms, primarily.

A number of different ways are needed to trigger an alarm in case of an illegal entry, and these require different switches. Some are used to detect the opening of a window. Others are connected to the door. And there is a mat with a switch to detect entry through a door. Fig. 6-47 shows some of the different types of switches available to work in conjunction with a burglar alarm system. (Chapter 16 has additional information about security systems.)

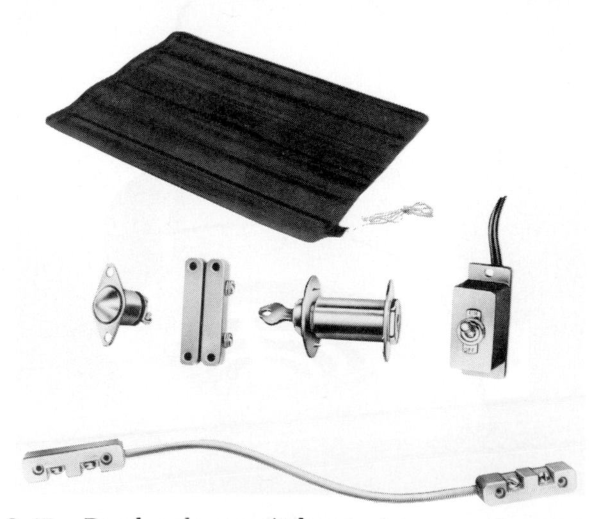

6-47. Burglar alarm switches.

CHAPTER

7

PLANNING AN INSTALLATION

Objectives

After studying this chapter, you will be able to:

- Figure the load of a circuit.
- Properly locate lighting outlets, convenience outlets, and wall switches.
- Identify the planning requirements for individual areas of a house.
- Explain the concept of demand metering.
- List the principal steps in estimating a job.

The correct wire size, from the transformer on the pole to the outlet on the wall, helps prevent such occurrences as blinking lights when the refrigerator turns on. Wire size is determined by the amount of current required by power-consuming devices. These devices present a load to the power source. The load is usually shown in watts on appliance nameplates. Power, which is measured in watts, is determined by multiplying the voltage times the current. If an overload exists, or if the wires are too small, a voltage drop caused by the small wires reduces power available inside the house.

The larger the physical size of the wire, the more current it can handle without line losses caused by an increase in the load. Wire size is stated in numbers. The larger the number, the smaller the wire.

FIGURING CIRCUIT LOAD

The load presented by a circuit may be figured by simply adding the wattage ratings of the devices to be attached to the line or plugged into the circuit. When planning a new house or rewiring an old one, it is best to figure the amount of wattage needed and then design the circuits to accommodate the load.

There are a number of reasons why you need adequate wiring in a house—new or old. If fuses blow, or a circuit breaker trips, probable overloads exist. If the lights dim, blink, or flicker whenever an additional device is turned on, then the circuits are overloaded. A television picture will shrink if it is on a line with an overload. If it is color television, the colors may change when something else turns on in the house. If an electric iron or electric range takes longer than usual to heat up, then chances are the circuits are overloaded. Air conditioners place quite a load on available circuitry. If the air conditioner is more than 5000 Btu's, it should be on a separate circuit.

Rewiring may be necessary to eliminate some of these problems. However, at the time a new house is being planned, it is best to make sure future electrical needs are taken into consideration. During the past 20 years, the use of electricity in the home has more than tripled. Probably, the future will hold more demands for electricity. Adequate planning is a must if blinking lights, shrinking TV pictures, and blown fuses are to be prevented.

General Requirements for Residential Wiring

A fundamental prerequisite for any adequate wiring installation is conformity to the safety regulations applicable to the dwelling. The approved American standard for electrical safety is the National Electrical Code. Usually, local safety regulations in municipal ordinances or state laws are based on the National Electrical Code.

Inspection service, to determine conformance with safety regulations, is available in most communities. Where available, a certificate of inspection should be obtained. In the absence of inspection service, a certificate should be obtained from the electrical contractor who installed the wiring. The certificate should say the wiring conforms with the applicable safety regulations.

Locating Lighting Outlets. Proper illumination is an essential element of modern living. The amount and type of illumination required should be adapted to the various seeing tasks in the home. There must also be planned lighting for recreation and entertainment. In many instances, best results are achieved by a blend of lighting from both fixed and portable light sources. Good home lighting requires thoughtful planning and careful selection of fixtures, portable lamps, and other lighting equipment.

Where lighting outlets are mentioned in plans, their types and locations should conform with the lighting fixtures or equipment to be used. Unless a specified location is stated, lighting outlets may be located anywhere within the area under consideration.

Locating Convenience Outlets. *Convenience outlets* should be located near the ends of wall space, rather than near the center. This reduces the likelihood of their being concealed behind large pieces of furniture. Unless otherwise specified, outlets should be located approximately 12 inches above the floorline.

Locating Wall Switches.

Wall switches should normally be located at the latch side of the doors, or at the traffic side of arches, and within the room or area to which the control is applicable. Some exceptions to this practice are: (1) control switches for exterior lights from indoors, (2) control of stairway lights from adjoining areas where stairs are closed off by doors at head or foot, and (3) the control of lights from the access space adjoining infrequently used areas, such as storage rooms. Wall switches are normally mounted approximately 48 inches above the floor.

Controlling from More Than One Switch. All spaces for which wall switch controls are required, and which have more than one entrance, should be equipped with a multiple-switch control at each principal entrance. If this requirement would result in the placing of switches controlling the same light within ten feet of each other, one of the switch locations may be eliminated.

For rooms lighted from more than one source, as when both general and supplementary illumination is provided from fixed sources, multiple switching is required for one set of controls only, usually to the general illumination circuit.

Principal entrances are those commonly used for entry to, or exit from, a room when going from a lighted to an unlighted condition, or the reverse. For instance, a door from a living room to a porch is a principal entrance to the porch. However, it would not necessarily be considered a principal entrance to the living room, unless the front entrance to the house is through a porch.

Number of Local Lighting or Convenience Outlets Necessary

If the standards require a certain number of outlets based on a linear or square-foot measure, the number of outlets shall be determined by dividing the total linear or square footage by the required distance or area, the number so determined to be increased by one if a major fraction remains:

Example:

Required: One outlet for each 150 square feet
Total square feet of area: 390
390 divided by 150 equals 2.6
Outlets required: 3

Dual-Purpose Rooms. Where a room is intended to serve more than one function, such as a combination living-dining room or a kitchen-laundry, convenience and special-purpose outlet provisions of these standards are separately applicable to those respective areas. Lighting outlet provisions may be combined in any manner that will assure general, overall illumination, as well as local illumination of work surfaces. In locating wall switches, a dual-purpose area is considered a single room.

Dual Functions of Outlets. In any instance where an outlet is located to satisfy two different provisions of the standard of adequacy, only one outlet need be installed at that location. In such instances, particular attention should be paid to required wall switch controls, as additional switching may be necessary.

For example, a lighting outlet in an upstairs hall may be located at the head of the stairway, thus satisfying both a hall-lighting outlet and a stairway-lighting outlet provision of Code standards with a single outlet. Stairway-lighting provisions will necessitate multiple-switch control of this outlet at both the head and the foot of the stairway. In addition, if the hallway is long, the multiple-switch control rule, previously stated, may require a third point of control in the hall.

INTERPRETING REQUIREMENTS

The following standards are necessarily general in nature. They apply to most situations encountered in normal house construction and are to be considered as *minimum* standards of adequacy for such construction. Situations will arise from time to time, because of unusual design or unusual construction methods, when it will be impossible to satisfy a particular provision of the standards. The following rules may serve as a guide for meeting such situations:

- Wiring installation should be fitted to the structure. If compliance with a particular provision would require alteration of doors, windows, or structural members, alternate wiring provisions should be made. For example, some types of building construction may make it desirable to alter the recommended outlet or switch heights presented here, or to resort to surface-type wiring construction.

- Each provision of the adequacy standard is intended to provide for one or more specific usages of electricity. If such usage is appropriate to the particular home under study, and cannot be provided in accordance with the standard, an alternate provision should be made. If a construction peculiarity eliminates the need for a particular electrical facility, no alternate provision is needed.

■ If certain functions are omitted from the plans, such as a workshop or a laundry, electrical wiring serving these functions naturally would be omitted.

■ If certain facilities are indicated as future additions to the plans, for example, a basement recreation room, the initial wiring should be so arranged that none of it need be replaced, or moved, when the ultimate plan is realized. Final wiring for the future addition may be left to a later date, if desired.

■ Standards given here may be applied to multifamily dwelling units. At the present time there are no nationally recognized standards for common-use spaces in such buildings, making good judgment on the part of the planner of electrical wiring a necessity. Particular attention should be paid to limiting voltage drop in feeders to individual apartments in order to assure proper operation of appliances and space-heating equipment.

Planning Requirements for Individual Areas

Exterior Entrance Wiring.

■ *Lighting provisions.* It is recommended that lighting outlets, controlled from wall switches, be installed at all entrances. Fig. 7-1. The principal lighting requirements for entrances are the illumination of steps leading to the entrance, and of faces of people at the door.

■ *Convenience outlets.* Outlets in addition to those at the door are often desirable for post lights to illuminate terraced or broken flights of steps, or long approach walks. These outlets should be wall-switch controlled from inside the house, near the entrance.

It is recommended that an exterior outlet near the front entrance be controlled by a wall switch inside the entrance, for convenient operation of outdoor lighting, lawn mowers, and hedge trimmers.

Living Room.

■ *Lighting provisions.* Provisions for general illumination of living rooms also apply to sun rooms, enclosed porches, family rooms, television rooms, libraries, dens, and similar living areas. Fig. 7-2. Installation of outlets for decorative lighting accents is recommended, such as picture illumination and bookcase lighting.

■ *Convenience outlets.* For switch-controlled outlets, it is recommended that split-receptacle outlets be used in order not to limit

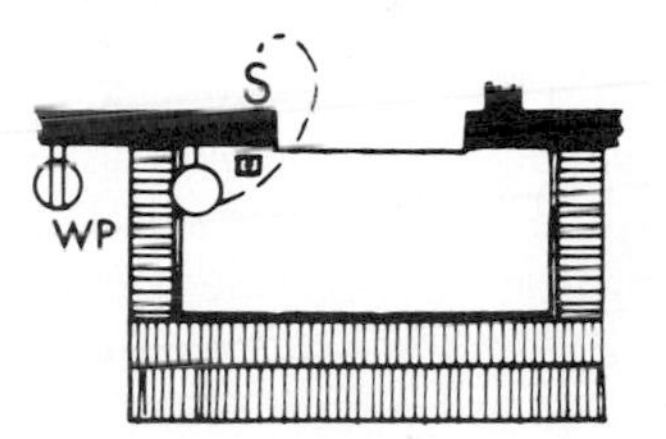

7–1. A weatherproof outlet, preferably near the front entrance, should be located at least 18″ above grade.

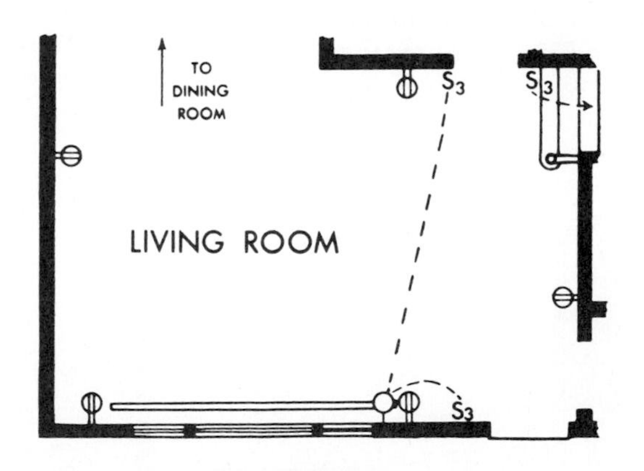

7–2. Some means of general illumination is essential. It may be provided by ceiling or wall fixtures, by lighting in coves, valances, or cornices, or by portable lamps. Provide lighting outlets, wall-switch controlled, in locations appropriate to the lighting method selected.

the location of radios, television sets, clocks, etc. Fig. 7-3. If construction plans permit, one convenience outlet should be installed flush in the mantel shelf.

In addition, a single convenience outlet might be installed in combination with the wall switch at one or more of the switch locations, for the use of a vacuum cleaner or other portable appliances. Outlets for the use of clocks, radios, decorative lighting, and so forth, in bookcases and other suitable locations are recommended.

Convenience outlets should be placed so that no point along the floorline in any usable wall space is more than six feet from an outlet in that space. Where windows extending to the floor prevent meeting this requirement by the use of ordinary convenience outlets, equivalent facilities should be installed using other appropriate means, such as floor outlets close to the window wall. If, in lieu of fixed lighting, general illumination is provided from portable lamps, two convenience outlets or one plug position in two or more split-receptacle convenience outlets shall be wall-switch controlled.

■ *Special-purpose outlets.* It is recommended that one outlet for a room air conditioner, and outlets for space heating, be installed if central systems are not planned.

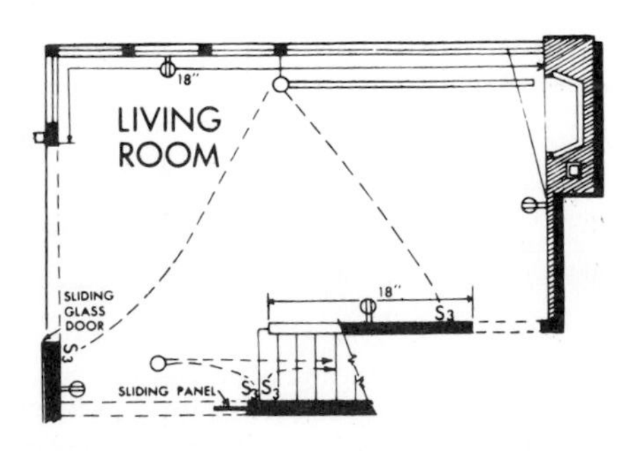

7–3. Location of convenience outlets in a living room.

Dining Areas.

■ *Lighting provisions.* Each dining room, or dining area combined with another room, or breakfast nook shall have at least one lighting outlet, wall-switch controlled. Fig. 7-4. When a dining or breakfast table is to be placed against a wall, one outlet should be placed at the table location, just above table height. Built-in counter space should have an outlet provided above counter height for portable appliances.

■ *Convenience outlets.* These should be of the split-receptacle type, for connection to small appliance circuits.

■ *Special-purpose outlets.* It is recommended that outlets for space heating, and one outlet for a room air conditioner be installed whenever central systems are not planned.

Bedrooms.

■ *Lighting provisions.* Light fixtures over full-length mirrors, or a light source at the ceiling located in the bedroom and directly in front of the clothes closets, may serve as general illumination. Fig. 7-5.

Master-switch control in the master bedroom, as well as at other strategic points in the

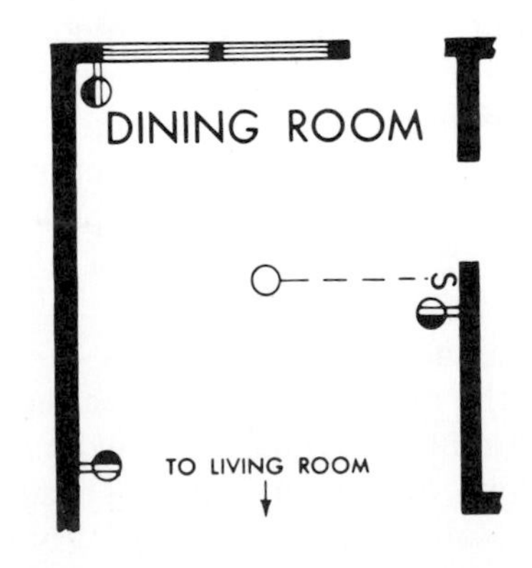

7–4. Lighting outlets for dining areas are normally located over the probable location of the dining or breakfast table to provide direct illumination.

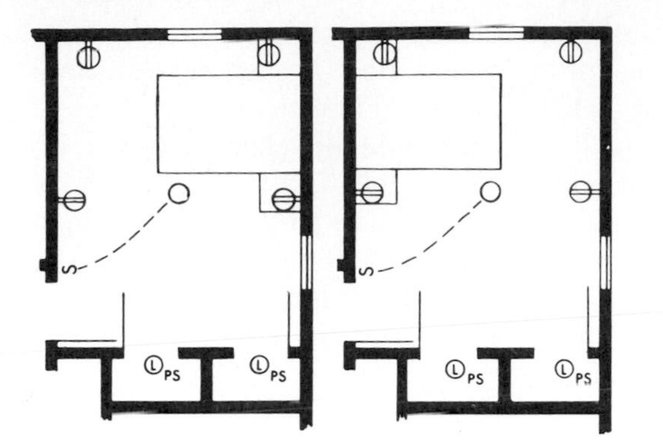

7–5. Good general illumination is particularly important in a bedroom. This can be provided from a ceiling fixture or from lighting in valances, coves, or cornices. Provide outlets, wall-switch controlled, in locations appropriate to the method selected.

home, is suggested for selected interior and exterior lights.

■ *Convenience outlets.* It is recommended that convenience outlets be placed only three to four feet from the centerline of the probable bed locations. The popularity of bedside radios and clocks, lamps, and electric bed covers makes increased plug-in positions at bed locations essential. Triplex or quadruplex convenience outlets are therefore recommended at these locations. Fig. 7-6.

Furthermore, at one of the switch locations, there should be a receptacle outlet for a vacuum cleaner, floor polisher, or other portable appliances. Figs. 7-7 & 7-8.

■ *Special-purpose outlets.* It is recommended that outlets for space heating and one outlet for a room air conditioner be installed in each bedroom if central systems are not provided.

Bathrooms and Lavatories.

■ *Lighting provisions.* A ceiling outlet located in line with the front edge of the washbasin will provide lighting for the mirror, general room lighting, and safety lighting for a combination shower and tub. Figs. 7-9 & 7-10.

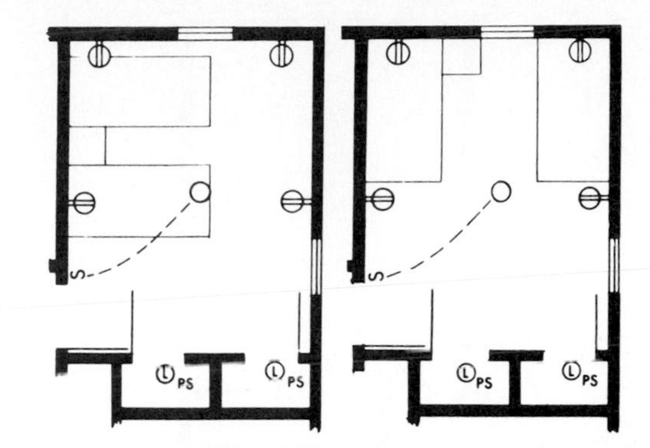

7–6. Start with an imaginary centerline through each probable bed location. Install outlets on each side of the centerline and within 6′ of it. No point along the floorline in any other usable wall space should be more than 6′ from an outlet on that wall. Add outlets as needed to achieve this.

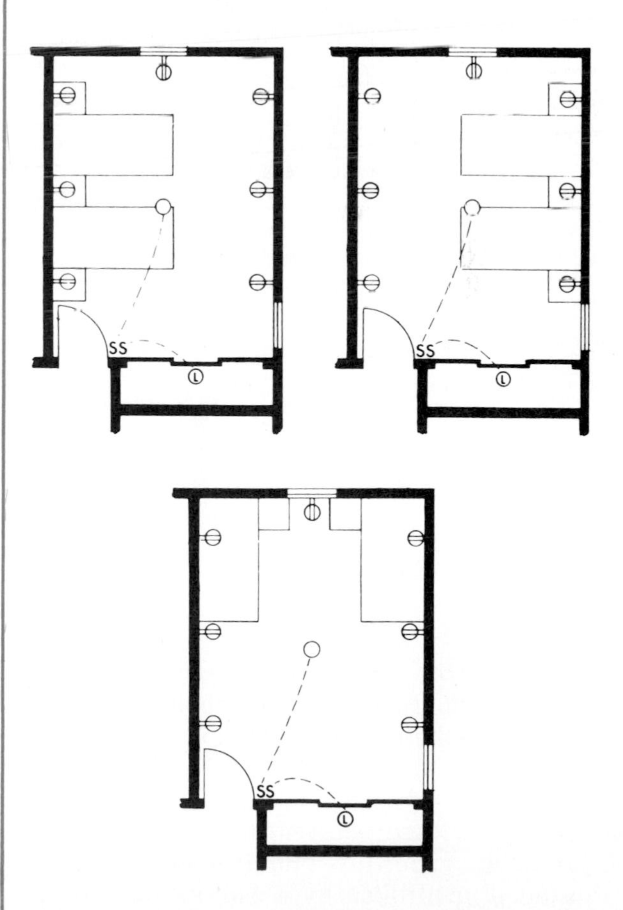

7–7. Wiring diagram for bedroom with twin beds.

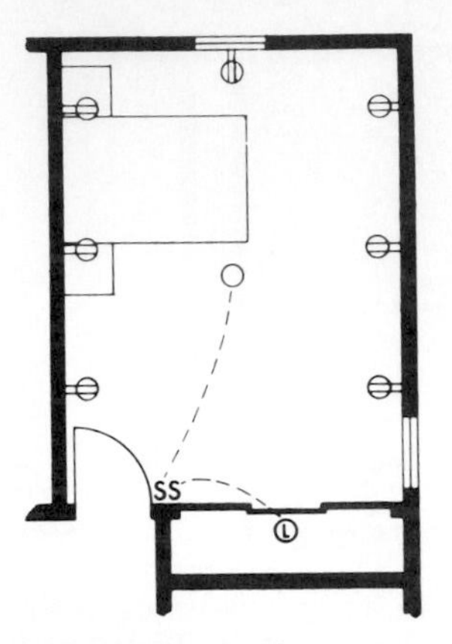

7–8. The same bedroom with wiring adaptation to double bed arrangement.

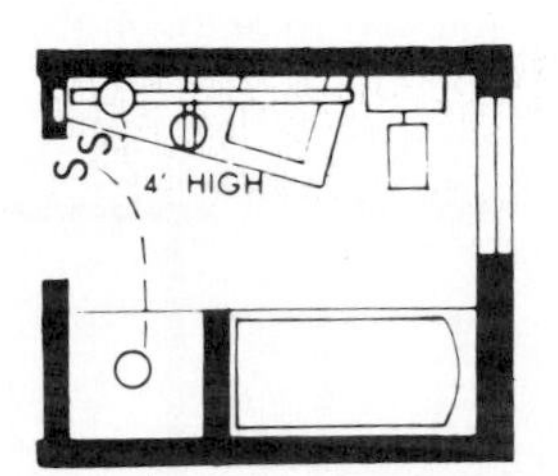

7–9. Illumination of both sides of the face when at a mirror is important. Keep in mind that a single concentrated light, either on the ceiling or side wall, is usually not acceptable. All lighting outlets should be wall-switch controlled.

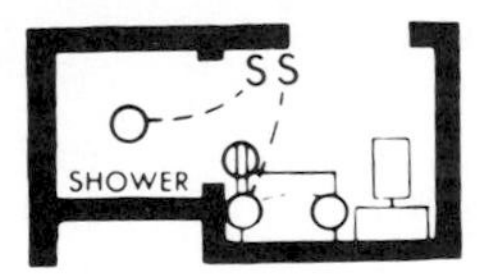

7–10. One outlet near the mirror, 3′ to 5′ above the floor.

When more than one mirror location is planned, equal consideration should of course be given to the lighting in each case. A switch-controlled night-light may be installed where desired.

If an enclosed shower stall is planned, an outlet for a vapor-proof luminaire is sometimes installed, controlled by a wall switch outside the stall.

■ *Convenience outlets.* It is suggested that an outlet be installed at each separate mirror or vanity space, and also at any space that might accommodate an electric towel dryer, electric razors, or electric toothbrushes. A receptacle which is part of a bathroom lighting fixture should not be considered as satisfying this requirement for small appliances, unless it is rated at 15 amperes and wired with at least 15-ampere rated wires.

Kitchen.

■ *Lighting provisions.* Provide outlets for general illumination and for lighting at the sink. These lighting outlets-should be wall-switch controlled. Lighting design should provide for illumination of work areas, sink, range, counters, and tables. Under-cabinet lighting fixtures, within easy reach, may have local-switch control. In some kitchens, consideration should also be given to outlets to provide inside lighting of cabinets.

One outlet for the refrigerator is needed; and one outlet for each four linear feet of work-surface frontage, with at least one outlet to serve each work surface. If a planning desk is to be installed, one outlet is needed for this area. (Work-surface outlets to be located about 44″ above the floorline.) Table space should have one outlet, preferably just above the table level. Fig. 7-11.

■ *Convenience outlets.* An outlet is recommended at any wall space that may be used for an iron or for an electric toaster. Convenience outlets in the kitchen should be of the split-receptacle type, for connection to small appliances.

■ *Special-purpose outlets.* One type of appliance which is helpful in the kitchen is a clock. It should be in a location easily visible from all parts of the kitchen. It requires a recessed receptacle, with a clock hanger.

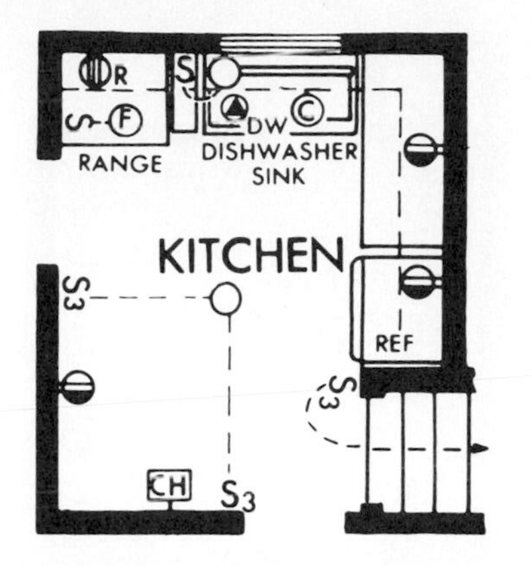

7–11. Wiring diagram for kitchen.

Fig. 7-12. An outlet for a food freezer, either in the kitchen or in some other convenient location, is often necessary.

Laundry and Laundry Areas.

■ *Lighting provisions.* It is recommended that all laundry outlets be wall-switch controlled. Fig. 7-13.

■ *Convenience outlets.* Convenience outlets in the laundry area should be connected to a 20-ampere branch circuit which serves no other areas.

■ *Special-purpose outlets.* Sometimes, plans require the installation of outlets for a ventilating fan and a clock.

Closets. One lighting outlet for each closet is suggested, except where shelving or other conditions make installation of lights within a closet ineffective. In this case, the lighting should be located in the adjoining space to provide light within the closet. Installation of wall switches near a closet door, or door-type switches, are recommended.

Halls. Lighting outlets, wall-switch controlled, should be installed for proper illumination of the entire hall. Particular attention should be paid to irregularly shaped areas.

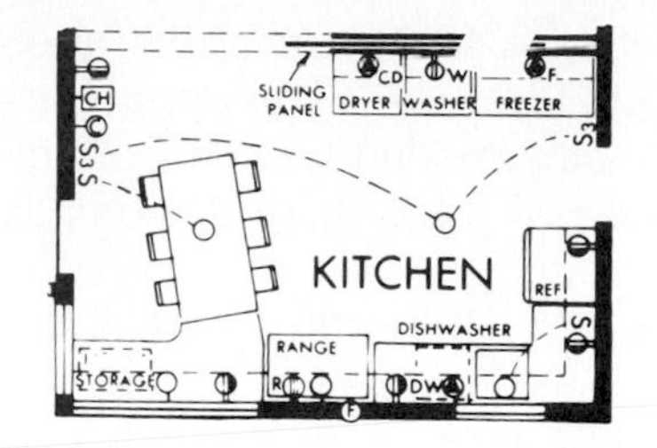

7–12. The following outlets are usually necessary: one for a free-standing range, or for built-in range surface units, and one for each built-in range oven (if each oven is a separate unit); one each for a ventilating fan, dishwasher, and food waste disposer (if necessary plumbing facilities are installed); and one for an electric clock.

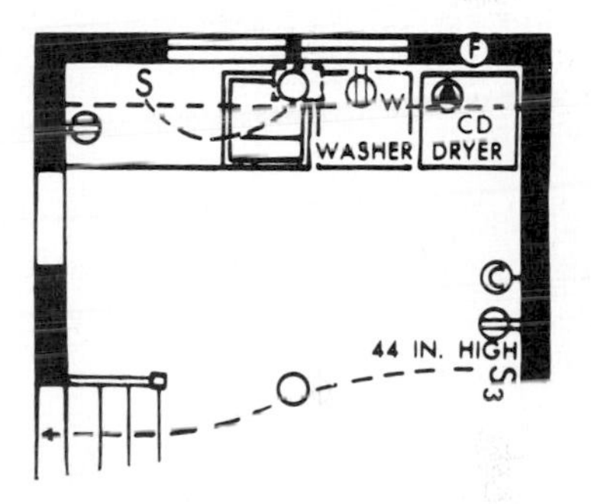

7–13. For a complete laundry, lighting outlets are installed over work areas such as laundry tubs, sorting table, washing, ironing, and drying centers. For laundry trays in an unfinished basement: one ceiling outlet, centered over the trays. One convenience outlet, or more if desired, shall be installed.

These provisions apply to passage halls, reception halls, vestibules, entries, foyers, and similar areas. It is sometimes desirable to install a switch-controlled night-light in a hall with access to bedrooms.

One convenience outlet for each 15 linear feet of hallway, measured along centerline is recommended. Each hall more than 25 square feet in floor area should have at least one outlet. In reception halls and foyers, convenience outlets shall be placed so that no point along the floorline in any usable wall space is more than ten feet from an outlet in that space.

It is further recommended that at one of the switch outlets in the hallway, a convenience receptacle be provided for connection of a vacuum cleaner, or other floor-cleaning device.

Stairways. Wall or ceiling outlets should provide adequate illumination of each stair flight. Outlets shall have multiple-switch control at the head and foot of the stairway, so arranged that full illumination may be turned on from either floor, but that lights in halls furnishing access to bedrooms may be extinguished without interfering with ground-floor usage.

These provisions are intended to apply to a stairway which connects finished rooms at either end.

Whenever possible, switches should be grouped together, and never located so close to steps that a fall might result while reaching for a switch.

At intermediate landings of a large stairway, (depending on planned usage) an outlet is often recommended for a decorative lamp, nightlight, or cleaning equipment.

Recreation, TV, or Family Rooms.

■ *Lighting provisions.* Selection of a lighting method for these rooms should take into account the types of activities for which the room is planned. Fig. 7-14. General illumination is essential in a family, recreation, play, or TV room. It may be provided by ceiling or wall fixtures, or by lighting in coves, valances, or cornices. Provide lighting outlets, wall-switch controlled, in appropriate locations. Convenience outlets should be placed so that no point along the floorline in any usable wall space is more than six feet from an outlet.

■ *Convenience outlets.* Convenience outlets in family rooms should be the split-receptacle type for connection to small appliance circuits.

If any of these rooms contains a fireplace, one convenience outlet should be installed flush in the mantel shelf. Outlets for the use of a clock, radio, television, ventilating fan, motion picture projector, and the like, should be located in relation to their intended use.

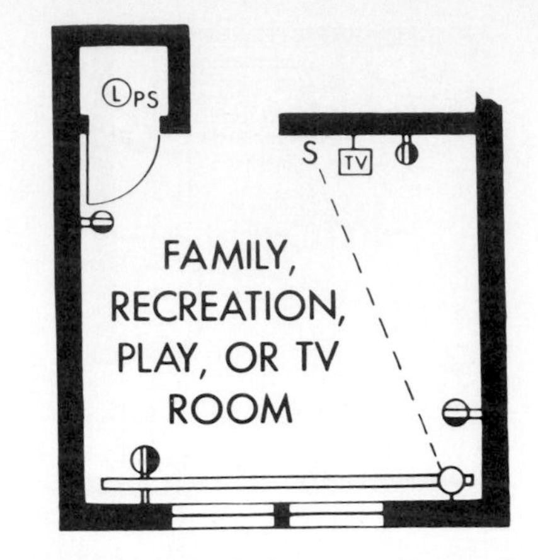

7–14. Lighting methods for recreation or family rooms should be selected according to the activities for which the room is planned.

■ *Special-purpose outlets.* Outlets for space heating, and one outlet for a room air conditioner, should be installed if central systems are not planned.

Utility Rooms or Space.

■ *Lighting provisions.* Lighting outlets in a utility room (or space) should be placed to illuminate the furnace area, and work bench, if planned. At least one lighting outlet is to be wall-switch controlled. Fig. 7-15. One convenience outlet, preferably near the furnace, or near any planned workshop, should be installed. Special-purpose outlets, such as one each for the water heater, boiler, furnace, or other equipment, should be located convenient to the appliance or equipment.

■ *Convenience outlets.* The washing machine outlet should be connected to a 20-ampere, laundry-area branch circuit.

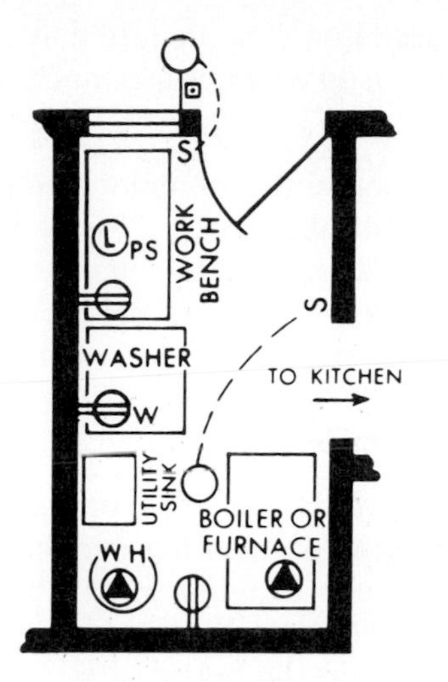

7–15. A wall-switched outlet is usually necessary to control the light in a utility room upon entering, because often this room has no windows.

Basements.

■ *Lighting provisions.* In basements with finished rooms, with garage space, or with other direct access to outdoors, the stairway-lighting provisions stated previously apply. For basements which will be infrequently visited, a pilot light should be installed in conjunction with the switch at the head of the stairs.

■ *Convenience outlets.* Basement lighting outlets should illuminate designated work areas or equipment locations, such as furnace, pump, and work bench. Fig. 7-16. Additional outlets should be installed near the foot of the stairway, in each closed space, and in open spaces so that each 150 square feet of open space is adequately served by a light in that area.

In unfinished basements, the light at the foot of the stairs should be wall-switched near the head of the stairs. Other lights may be chain-pulled for control. Convenience outlets in the basement should number at least two. If there is a work bench, one outlet should be placed at this location. In addition, one outlet for electrical equipment used in connection with furnace operation should be installed.

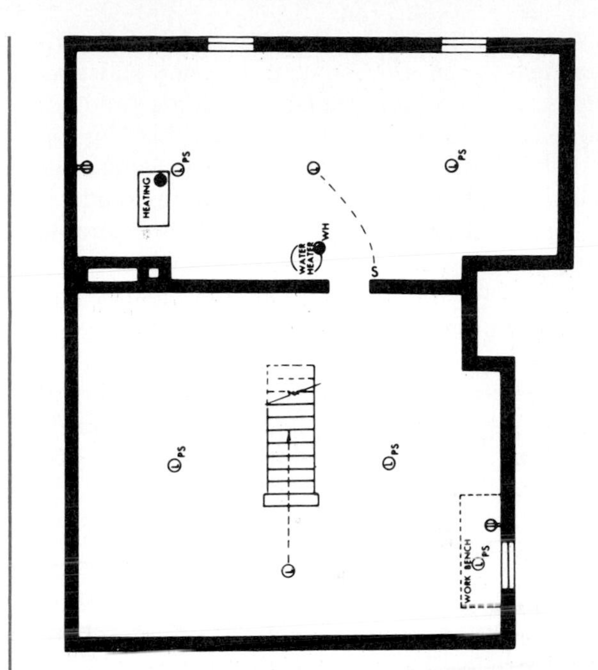

7–16. Basement convenience outlets are useful near the furnace, at the play area, for basement laundries, darkrooms, hobby areas, and for appliances such as dehumidifiers and portable space heaters.

■ *Special-purpose outlet.* If a food freezer is located here, an outlet for it must be installed. You may want to use the basement as a laundry or utility room. If so, check these headings for suggestions.

Accessible Attic.

■ *Lighting provisions.* One outlet for general illumination, wall-switch controlled from the foot of the stairs, is the minimum. When no permanent stairs are installed, this lighting outlet may be pull-chain controlled, if located over the access door. Where an unfinished attic is planned for later development into rooms, the attic-lighting outlet shall be switch controlled at the top and bottom of the stairs. One outlet for each enclosed space is recommended.

These provisions apply to unfinished attics. For attics with finished rooms or spaces, see the appropriate room classification for requirements.

■ *Convenience outlets.* One outlet for general use is the minimum. If an open stairway leads to future attic rooms, provide a junction box with direct connection to the distribution panel for future extension to convenience outlets and lights when the rooms are finished.

A convenience outlet in the attic is desirable for providing additional light in dark corners, and also for the use of a vacuum cleaner and cleaning accessories.

■ *Special-purpose outlet.* The installation of an outlet, switch controlled from a desirable point in the house, is recommended for use of a summer cooling fan.

Porches.

■ *Convenience outlets.* Each porch, breezeway, or other similar roofed area of more than 75 square feet in floor area should have a lighting outlet, wall-switch controlled. Large or irregularly shaped areas may require two or more lighting outlets. One convenience outlet (weatherproof, if exposed to moisture) for each 15 feet of wall that borders a porch or breezeway is recommended. Fig. 7-17. Multiple-switch control should be installed at entrances if the porch is used as a passage between house and garage.

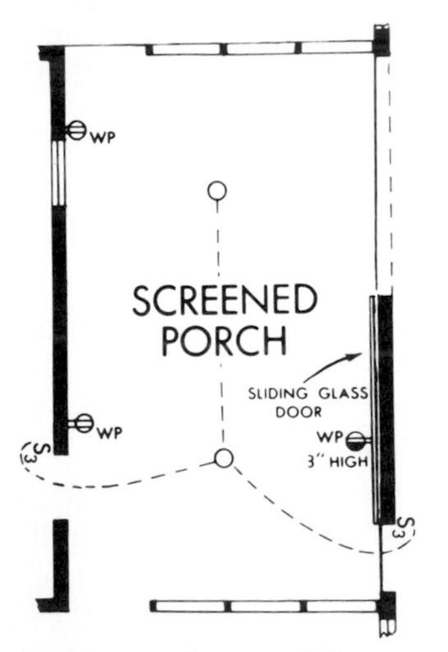

7–17. One outlet, weatherproof if exposed to moisture for each 15′ of wall bordering porch or breezeway, is recommended. One or more of such outlets should be controlled by a wall switch inside the door.

The split-receptacle convenience outlet is intended to be connected to a 3-wire appliance branch circuit. This area is considered an outdoor dining area. (See the relationship of a screened porch to the kitchen in Fig. 7-21.)

Terraces and Patios.

■ *Lighting provisions.* One or more outlets on the building wall, or other convenient location in the area to provide fixed illumination, are recommended. Wall-switch control shall be provided inside the house.

■ *Convenience outlets.* One weatherproof outlet located at least 18 inches above the grade line is needed for each 15 linear feet of house wall bordering terrace or patio. It is recommended that one or more of these outlets be wall-switch controlled from inside the house.

All outside outlets must have a GFI (ground-fault interrupter). At least one outside outlet is required for any newly constructed house.

Garages or Carports. If a garage is to be used for additional purposes, such as a work area, storage closets, laundry, attached porch or the like, the rules apply which are appropriate to those uses. Fig. 7-18.

An exterior outlet, wall-switch controlled, is recommended for all garages. Additional exterior outlets are often desirable, even if no specific additional use is planned for the garage. Long driveways warrant additional illumination, such as a post light, wall-switch controlled from inside the house.

If a food freezer or refrigerator, work bench, automatic door opener, or car heater is planned for the garage, outlets for these uses should be provided.

Exterior Grounds. In addition to terrace and patio lighting, outlets under the eaves on the exterior of a house and garage are sometimes

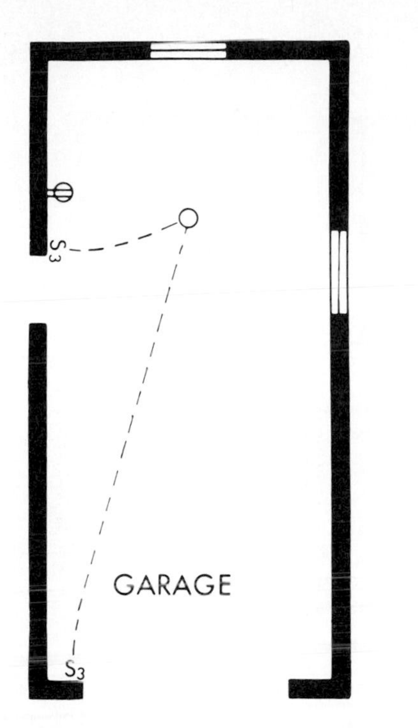

7–18. The garage or carport should contain at least one ceiling outlet, wall-switched, for a one- or two-car storage area. If the garage has no covered access from the house, provide one exterior outlet, multiple-switch controlled, from both garage and residence. One plug-in outlet for a one- or two-car storage area is necessary.

recommended for protective or decorative lighting. Switch control should be from within the house. Multiple and master-switch control is desirable also.

Weatherproof floodlights and spotlights are available to show the beauty of a garden by night. These are designed for use with either 115-volt lamps, or lamps at reduced voltage supplied through safety transformers. Fixtures may be permanently wired, and switched from inside the house, or they may be served from outdoor weatherproof convenience outlets. Lighting for sports (such as tennis, volleyball, or badminton) deserves consideration where the grounds are suitable for this purpose.

Underwater and area lighting for a swimming pool, and power requirements for a pool circulating pump and filter require careful design and installation by specialists in this field.

In general, it is recommended that all outdoor lighting outlets be switch controlled from within the house. In addition, certain ones, such as post lights and protective lighting, lend themselves also to photoelectric-cell or time switch control.

In climates where build-up of snow and ice on roofs is a problem, a "snow-melting" (heating) cable installation at the lower edge of the roof, and in the gutters and downspouts, provides excellent protection at low cost against water leakage into ceilings and outside walls of living quarters.

Motor-operated valves for buried lawn sprinkler systems are available. These can be arranged for manual or automatic operation. However, the water-operated ones are quieter.

Application of Outlet Requirements for Typical Homes

Figs. 7-19 through 7-23 (pages 200-202) show a method of drawing in outlets and switches for typical homes. Included here are the living room, basement, second floor, exterior, and a basement with unexcavated areas.

Other Requirements

Other considerations to be made in the wiring plan of a house include doorbells, chimes, communication systems (intercoms), household fire-warning systems, and television with special lead-ins and rotor wires, in addition to radio and the telephone. Table 7-A (page 202). For a discussion of these items, see Chapter 6.

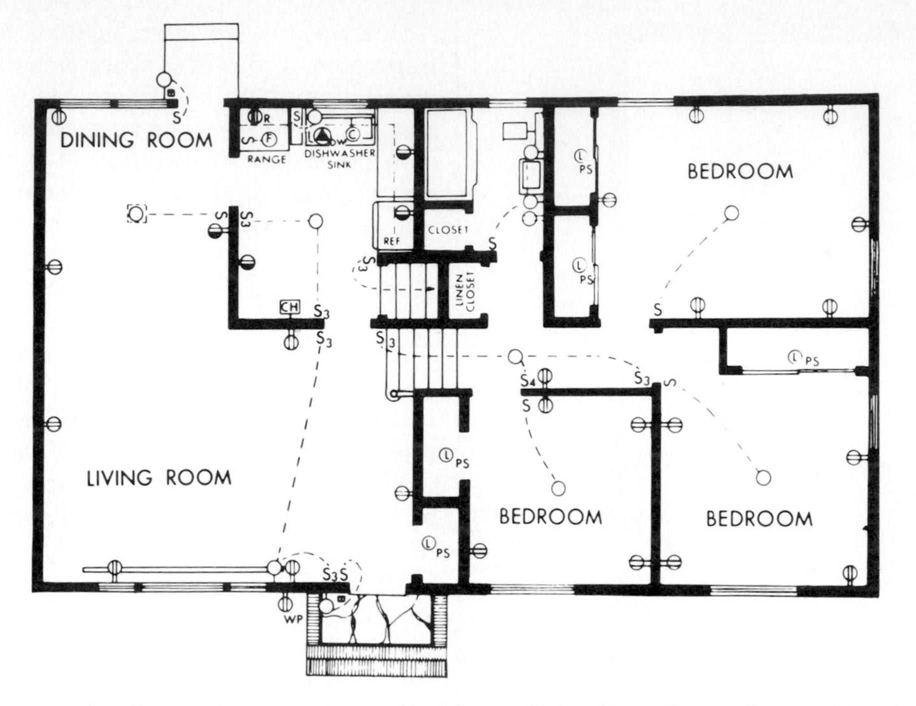

7–19. Application of outlet requirements in a typical home: living floor, first and second levels.

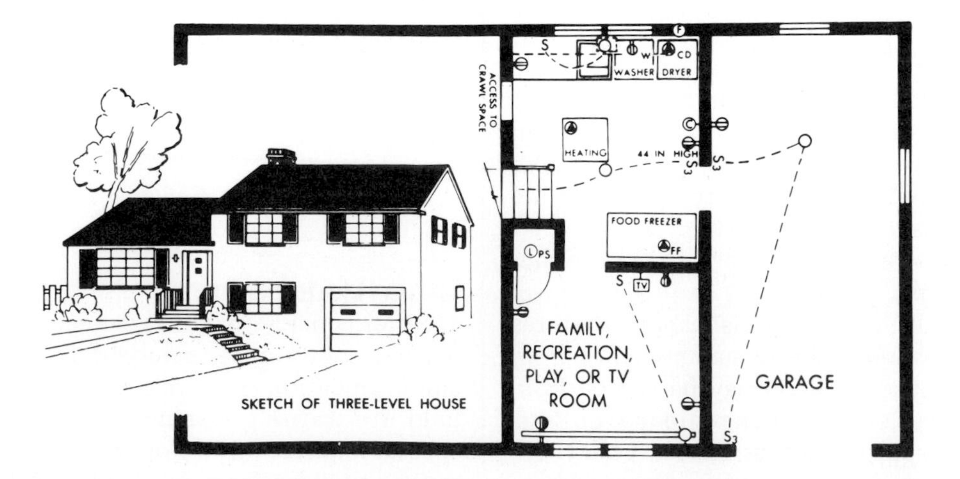

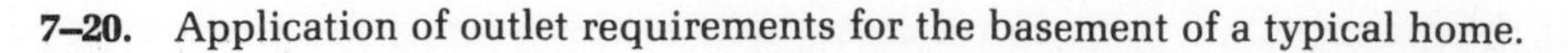

7–20. Application of outlet requirements for the basement of a typical home.

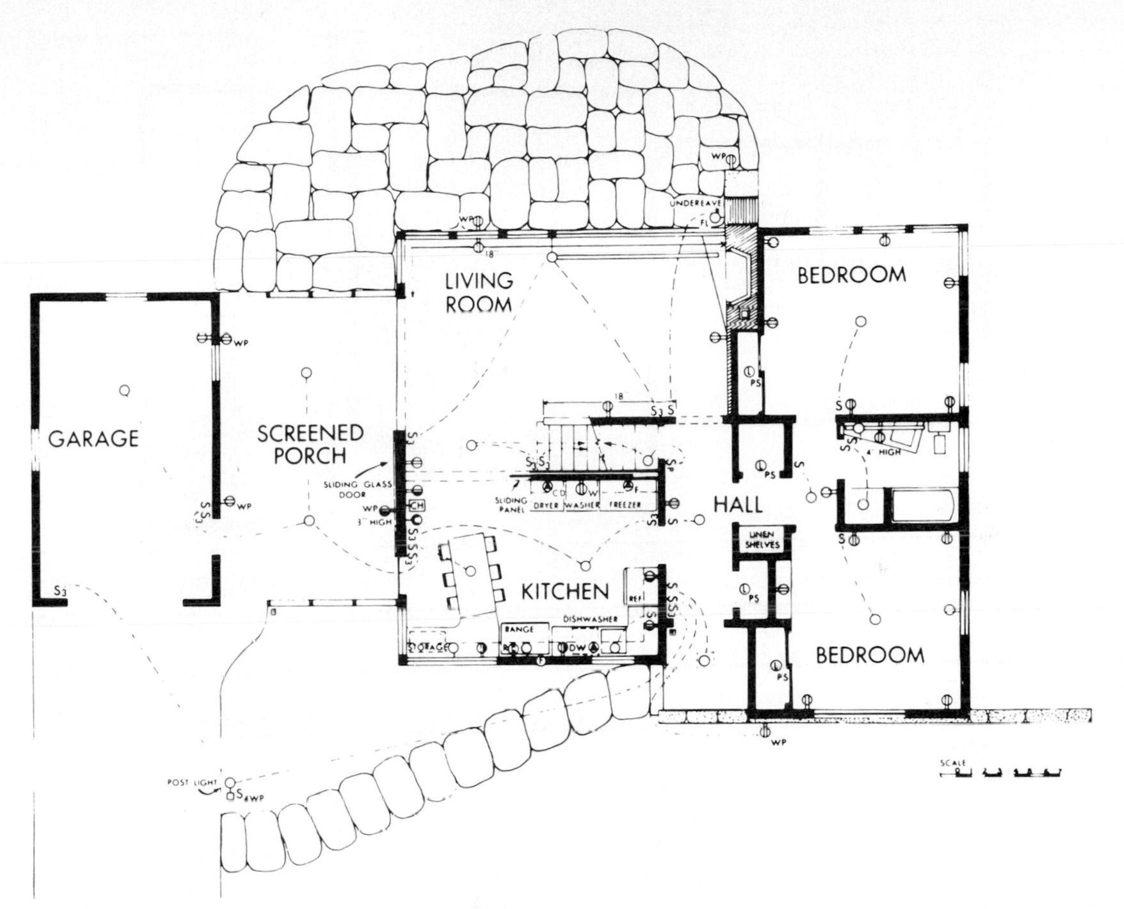

7–21. Application of outlet requirements for a home.

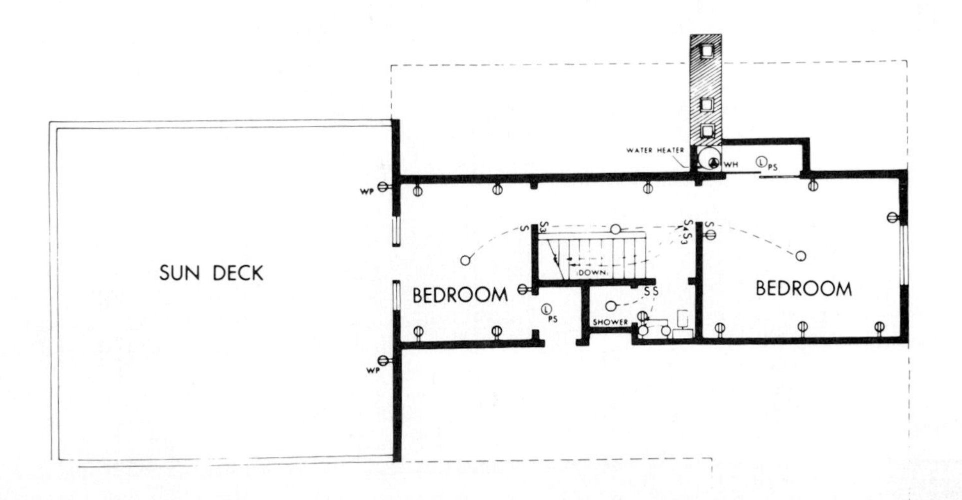

7–22. Application of outlet requirements on the second level of a home.

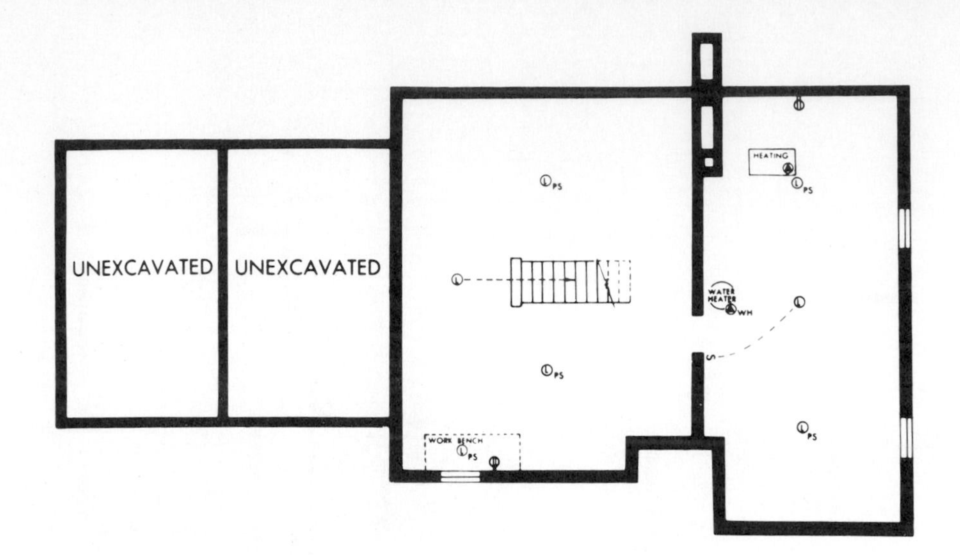

7–23. Application of convenience outlet requirements for the basement of a typical home.

7-A. Graphical Electrical Symbol for Residential Wiring Plans

General Outlets

○	Lighting Outlet
○ (in dashed outline)	Ceiling Lighting Outlet for recessed fixture (Outline shows shape of fixture.)
○ with rectangle	Continuous Wireway for Fluorescent Lighting on ceiling, in coves, cornices, etc. (Extend rectangle to show length of installation.)
Ⓛ	Lighting Outlet with Lamp Holder
$Ⓛ_{PS}$	Lighting Outlet with Lamp Holder and Pull Switch
Ⓕ	Fan Outlet
Ⓙ	Junction Box
Ⓓ	Drop-Cord Equipped Outlet
-Ⓒ	Clock Outlet

To indicate wall installation of above outlets, place circle near wall and connect with line as shown for clock outlet.

Switch Outlets

S	Single-Pole Switch
S_3	Three-Way Switch
S_4	Four-Way Switch
S_D	Automatic Door Switch
S_P	Switch and Pilot Light
S_{WP}	Weatherproof Switch
S_2	Double-Pole Switch

Low-Voltage and Remote-Control Switching Systems

S̲	Switch for Low-Voltage Relay Systems
M̲S̲	Master Switch for Low-Voltage Relay Systems
$○_R$	Relay-Equipped Lighting Outlet
- - · — · - -	Low-Voltage Relay System Wiring

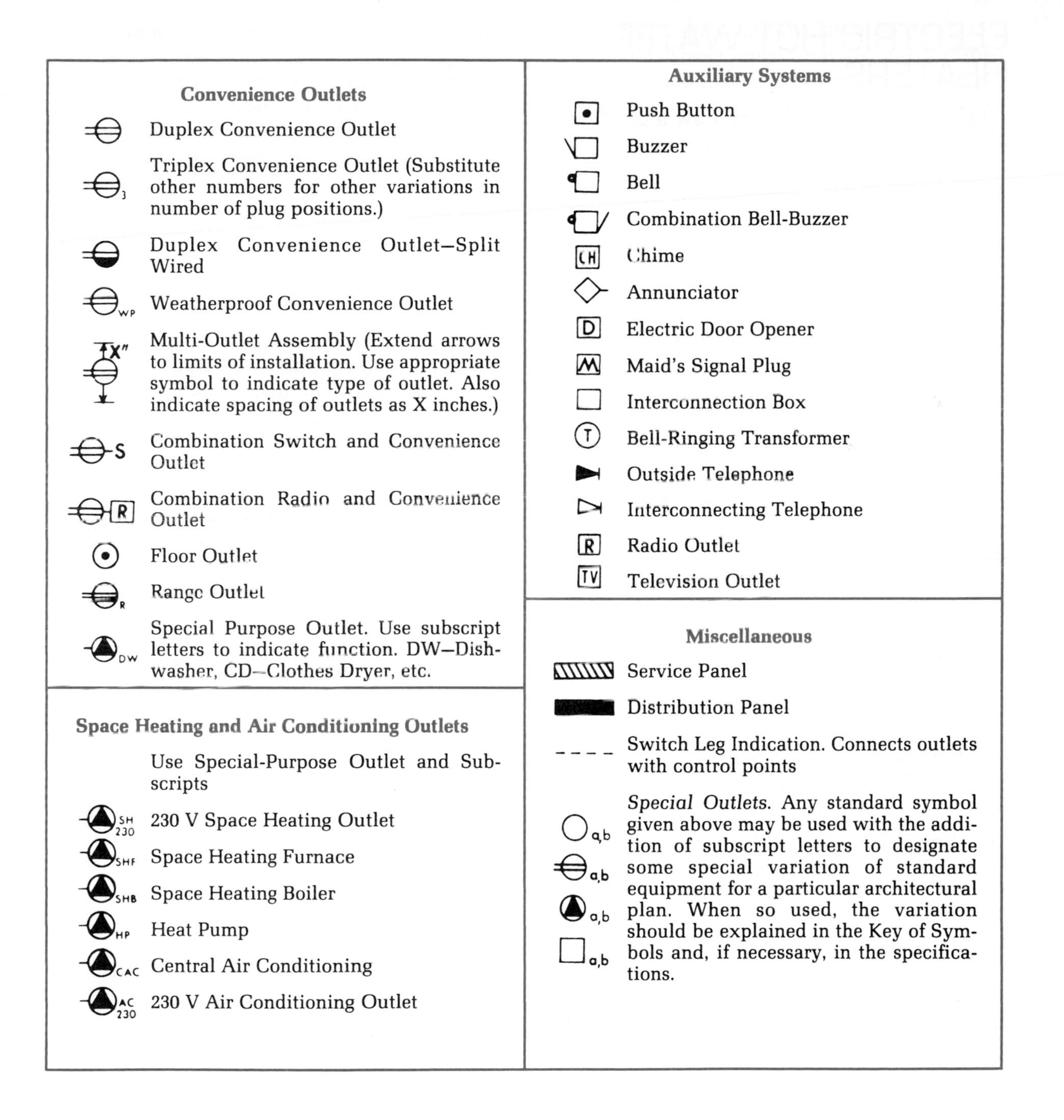
Convenience Outlets
Duplex Convenience Outlet
Triplex Convenience Outlet (Substitute other numbers for other variations in number of plug positions.)
Duplex Convenience Outlet–Split Wired
Weatherproof Convenience Outlet
Multi-Outlet Assembly (Extend arrows to limits of installation. Use appropriate symbol to indicate type of outlet. Also indicate spacing of outlets as X inches.)
Combination Switch and Convenience Outlet
Combination Radio and Convenience Outlet
Floor Outlet
Range Outlet
Special Purpose Outlet. Use subscript letters to indicate function. DW–Dishwasher, CD–Clothes Dryer, etc.
Space Heating and Air Conditioning Outlets
Use Special-Purpose Outlet and Subscripts
230 V Space Heating Outlet
Space Heating Furnace
Space Heating Boiler
Heat Pump
Central Air Conditioning
230 V Air Conditioning Outlet
Auxiliary Systems
Push Button
Buzzer
Bell
Combination Bell-Buzzer
Chime
Annunciator
Electric Door Opener
Maid's Signal Plug
Interconnection Box
Bell-Ringing Transformer
Outside Telephone
Interconnecting Telephone
Radio Outlet
Television Outlet
Miscellaneous
Service Panel
Distribution Panel
Switch Leg Indication. Connects outlets with control points
Special Outlets. Any standard symbol given above may be used with the addition of subscript letters to designate some special variation of standard equipment for a particular architectural plan. When so used, the variation should be explained in the Key of Symbols and, if necessary, in the specifications.

CIRCUIT	WIRE TYPE AND SIZE	LENGTH
1		
2		
3		
4		
5		
6		
7		
8		
9		
10		

QUESTIONS

1. How is circuit load figured?
2. How are convenience outlets located?
3. What determines the wire size of the service line from the pole to the house?
4. Where should wall switches be located?
5. Which lights should be controlled by a multiple-switch control?
6. List the basic steps in estimating the cost of an electrical job.

KEY TERMS

convenience outlet
principal entrance
special-purpose outlet
demand

CHAPTER 8

EQUIPMENT AND TOOLS

Objectives

After studying this chapter, you will be able to:

- Identify the basic tools needed for house wiring.
- Describe the use of bending equipment.
- Identify the principal ways of positioning a wire puller.
- Correctly identify the different types of drill bits.
- Install a PVC conduit system.

RESIDENTIAL WIRING

In order for an electrician to do the job and do it properly, it is necessary to have the right tools. Tool requirements vary with the job: some need only a minimum of tools, others require some rather complicated devices.

The more complicated tools are used by the electrician who works on commercial and industrial installations. Some of these tools, which are very expensive, are bought by the electrical contractor. It is necessary, however, for an electrician to have used such tools, or at least to have knowledge of their existence.

This chapter will endeavor to show both the simple and complicated tools which are used every day by an electrician. Some electricians have had to work with devices they improvised themselves, which have later become standard equipment for certain jobs. The ingenuity of electricians is well known to those people associated with them. Their ability to get a job done efficiently is also well known. Yet, most of this efficiency is due to the particular individual's skill in utilizing the equipment at hand.

Fig. 8-1 illustrates the minimum tools necessary for house wiring or for adding an outlet or switch to existing wiring.

■ *Hammer.* Although primarily for driving nails, staples, or for fastening hangers, a hammer may be used to anchor an object to concrete. Fig. 8-2. In Fig. 8-3 a hammer is being used to attach a box to a 2″ × 4″ stud. The hammer has many uses to an electrician. For example, it may be used when removing floor boxes from a concrete floor. Fig. 8-4.

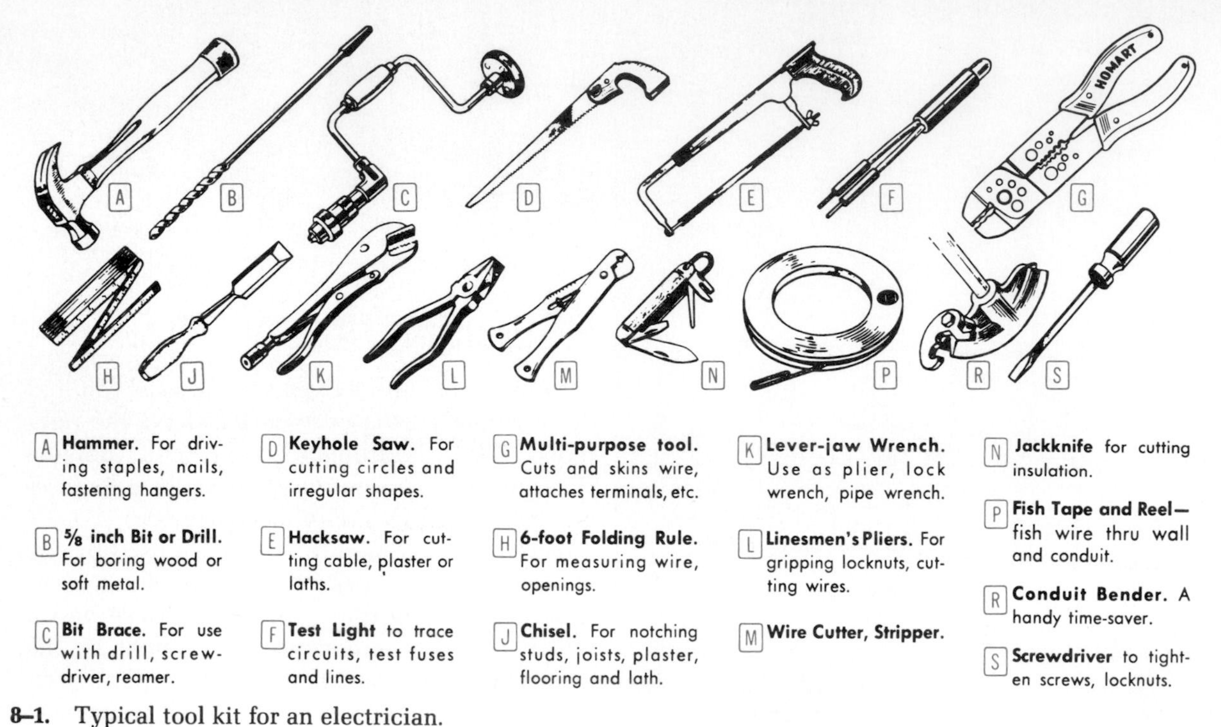

8–1. Typical tool kit for an electrician.

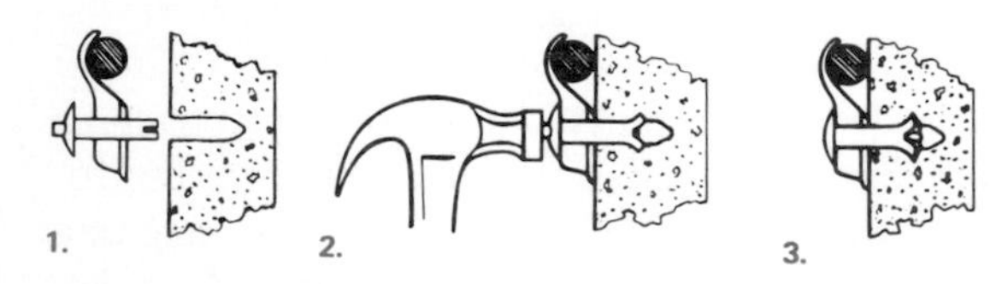

8–2. Using a hammer to set an anchor.

■ *Bit.* A drill bit may be used to drill holes in wood or soft metal. Some more extensive uses will be shown later in this chapter. The bit in Fig. 8-1 has been designed for use in a brace, for operation by hand.

■ *Bit brace.* This tool is very useful in drilling holes for the location of boxes and wires, and for fishing wiring out of hard-to-reach locations.

■ *Keyhole saw.* The keyhole saw cuts circles and irregular shapes.

8–3. Nailing a box to a stud.

8-4. Hammer used to remove extra cement from a floor box.

■ *Hacksaw.* Used for cutting cable (BX or Romex), the hacksaw may be used to cut plaster or laths, or any metallic conduit, thinwall or rigid.

■ *Test light.* A test light is used to trace circuits, test fuses, and test circuits to determine whether power has been applied. Fig. 8-5. It can also indicate the voltage and whether it is AC or DC.

■ *Multi-purpose tool.* Used to strip wires and crimp connectors, this tool will also cut screws without ruining the threads.

■ *Folding rule.* This standard, 6′ folding rule is in every electrician's tool kit. It is used to check the proper location of boxes, wiring, and fixtures.

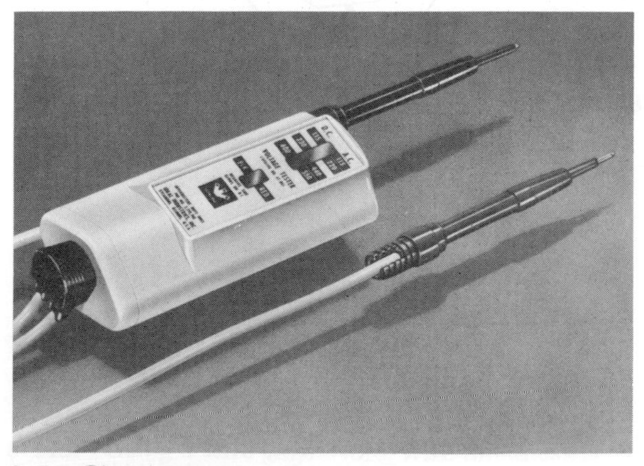

8-5. Circuit tester.

■ *Chisel.* A necessary tool for making the required small changes in the woodwork when mounting a fixture or box. A chisel may also be used to notch studs, joists, plaster, flooring, and lath.

■ *Lever-jaw wrench.* This vise grip can be used as a pliers, lock wrench, or pipe wrench.

■ *Linesmen's pliers.* This tool probably gets as much use as any other in the electrician's tool kit. It can be used for gripping locknuts, cutting wires, and breaking fins off dimmers. Fig. 8-6. It can be used to work on floor boxes. This tool also serves as a crimper of new-type connectors for Romex, Fig. 8-7, and for removing twist-outs and knockouts. Fig. 8-8.

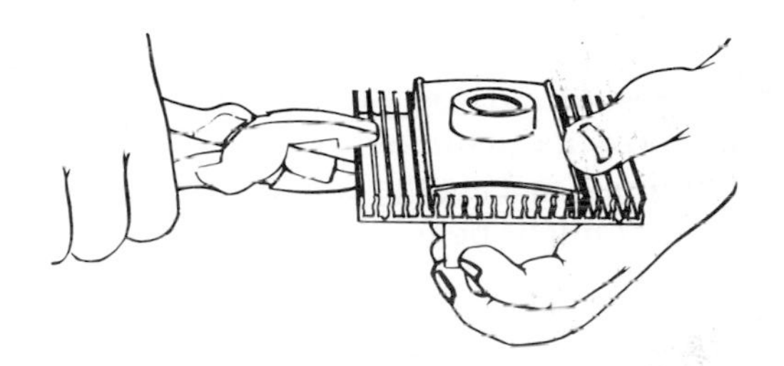

8-6. Pliers used to break off the fins on a light dimmer.

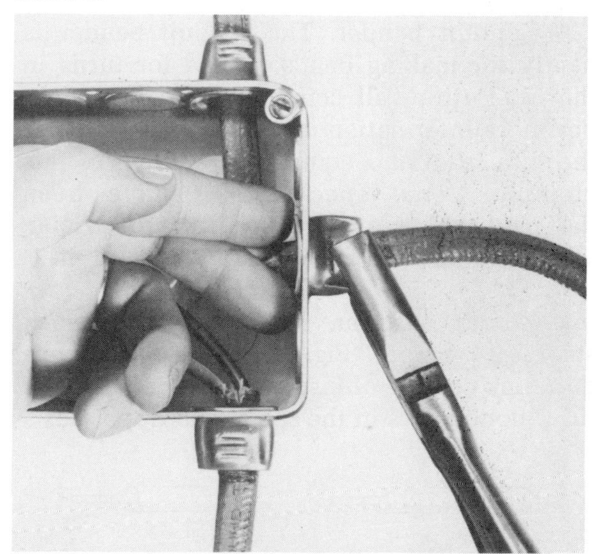

8-7. Pliers being used to twist new Romex connector.

■ *Hole saw.* A number of hole saws are available. The saw in Fig. 8-17a first drills a pilot hole, then cuts a hole the diameter of the curved saw blade. In Fig. 8-17b, you can see the different diameters available (only one saw blade is left in the tool when used).

■ *Spin-tite wrenches and screwdrivers.* Spin-tite wrenches are screwdrivers with sockets on the end. Fig. 8-18. Fig 8-19 shows the variety of screwdrivers probably included in an electrician's tool kit. Both the standard-type head and the Phillips head are included.

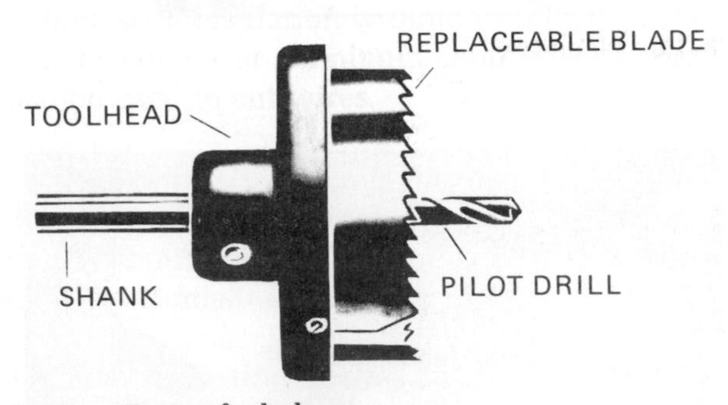

8–17a. Parts of a hole saw.

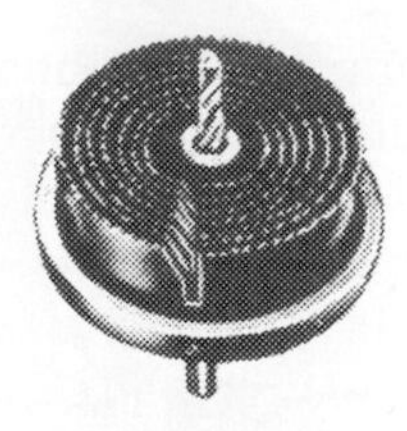

8–17b. Hole saw blades.

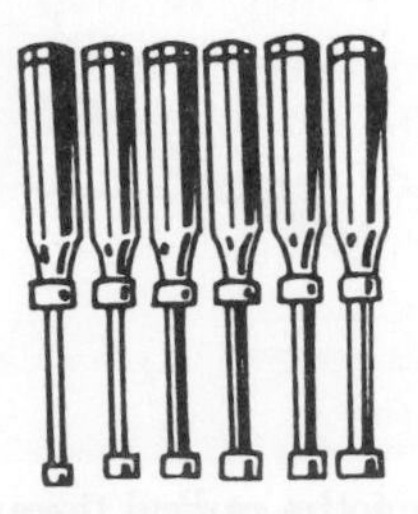

8–18. Spin-tite socket wrenches.

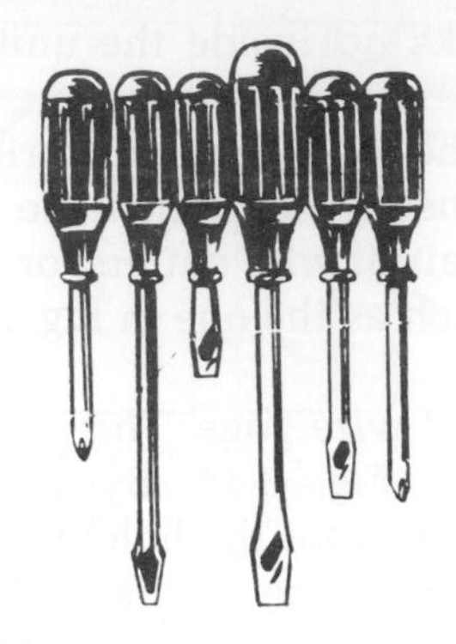

8–19. Screwdrivers.

■ *Chalk line.* This gadget comes in handy when a straight line is needed for a distance of more than a few feet. The string is covered with chalk as it is pulled from the case. When pulled tight and snapped, it produces a white line. Fig. 8-20.

■ *Fuse puller.* A plastic-handled fuse puller is a tool used to safely extract fuses from industrial or commercial distribution boxes. Fig. 8-21.

■ *Awl.* The awl has a variety of uses. It can be used to mark lines or make holes, and as a tool for prying staples loose. Fig. 8-22.

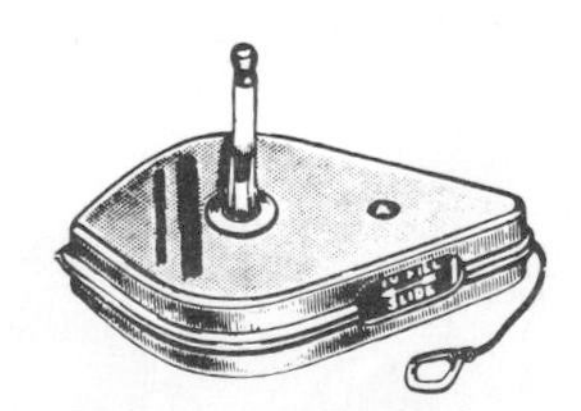

8–20. Chalk line.

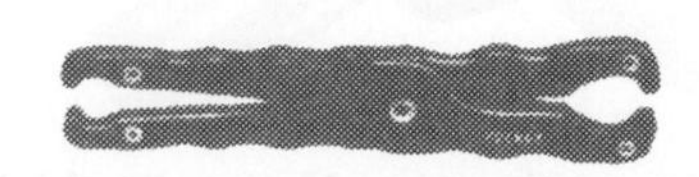

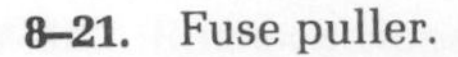

8–21. Fuse puller.

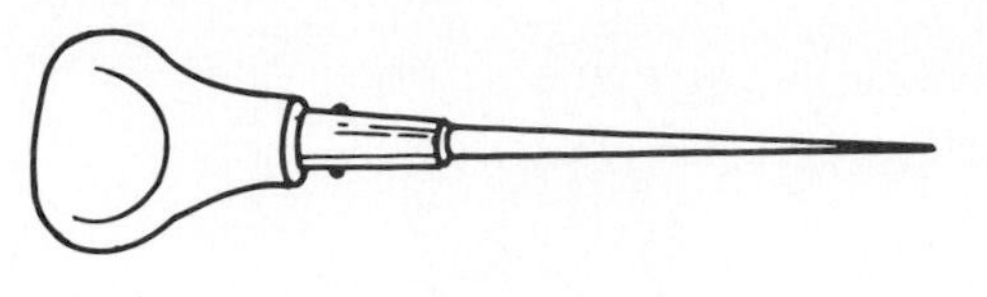

8–22. Awl.

■ *Soldering irons and torches.* Soldering irons and guns are common tools for electronics technicians, and electricians also use them on occasion. However, when there is no electrical power available at a construction site, it becomes necessary to find tools that operate on other power. A propane torch is efficient here. It may be used for soldering either small or large jobs. Fig. 8-23.

8–23. Propane torch.

There are other tools that the electrician can obtain for use on special jobs. Those listed here represent the necessities. A tool holder, as shown in Fig. 8-24, will also come in handy.

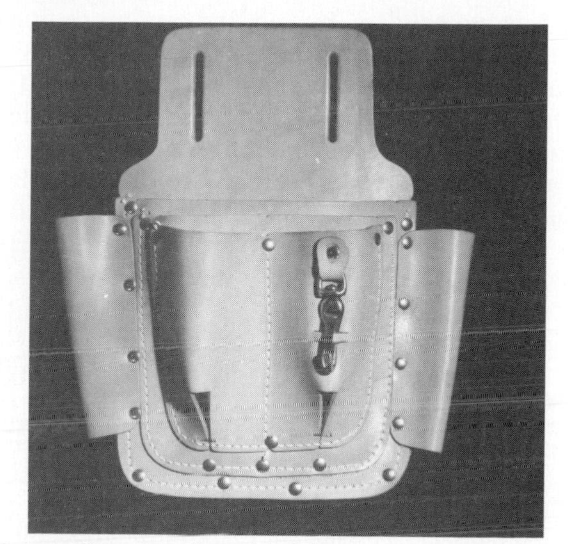

8–24. Electrician's tool holder.

INDUSTRIAL AND COMMERCIAL WIRING

Industrial and commercial wiring must, because of the large amounts of current required, be enclosed in large pipe to protect the wires from damage by equipment operating around them.

This large pipe, called *rigid conduit*, presents a number of problems, most of which arise whenever a bend has to be made. For conduit of small diameter, special hand benders or "hickeys" are used.

Large diameters, however, require hydraulic means of exerting pressure. In this situation the electrician must be sure of the machine's capabilities.

Dies are used to keep the pipe from collapsing as it is bent. The inside part of the pipe is compressed as the outside portion is stretched.

It is very easy to collapse the piece of pipe if proper care is not taken during the bending operation.

Bending Equipment

■ *Hand benders.* By bending conduit to fit a particular location, it is possible to eliminate the need for expensive elbows and other fittings. Fig. 8-25. Bent conduit is smoother inside and allows for easier pulling of the wire later. The conduit bender shown in Fig. 8-26 is designed to bend $\frac{1}{2}''$, $\frac{3}{4}''$, and 1″ rigid conduit. It is recommended for tight stub-ups or short bent ends needed in today's thinner slab floors. This eliminates the need to cut the conduit before mounting boxes or fixtures.

The hand bender in Fig. 8-27 is designed for EMT of $\frac{1}{2}''$ through $1\frac{1}{4}''$, but will accommodate $\frac{1}{2}''$ to 1″ rigid conduit. It can be used to correct overbends by simply reversing the direction of the applied force.

The offset needed to match up knockout holes in boxes to a piece of flat-mounted conduit can be accomplished with a special bender for $\frac{1}{2}''$ and $\frac{3}{4}''$ thinwall EMT.

8–25. Using a conduit bender to make a fit.

8–26a. Bending EMT with a hickey.

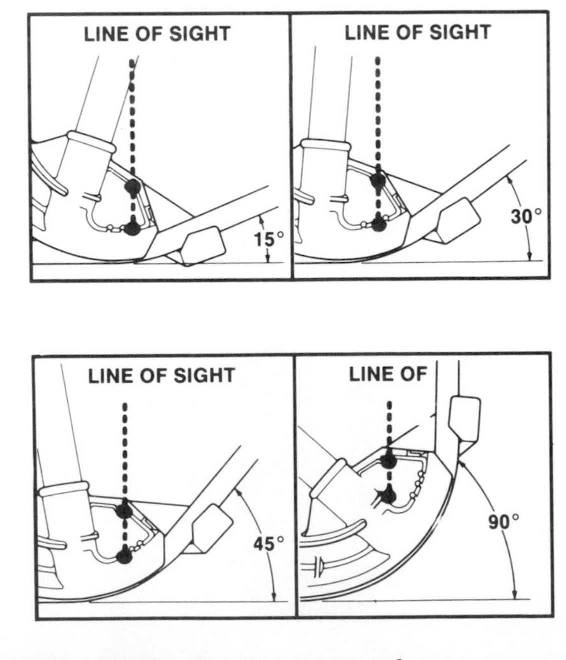

8–26b. Sighting for the correct angle.

To correct over- or under-bends, slip the top end of the handles over the stub and push. Bends straighten up. Conduit stays round. Handle is threaded to fit the pre-threaded handle sockets of Greenlee SITE-RITE® and hickey bender heads.

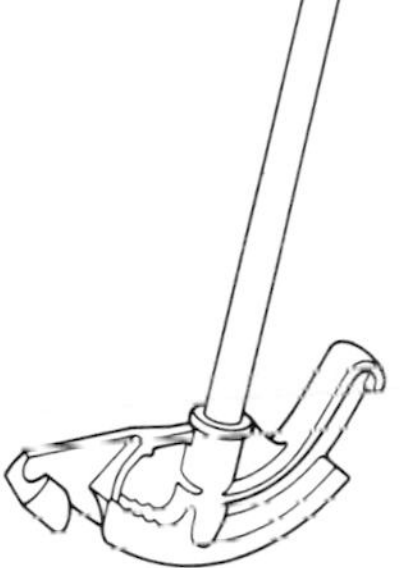

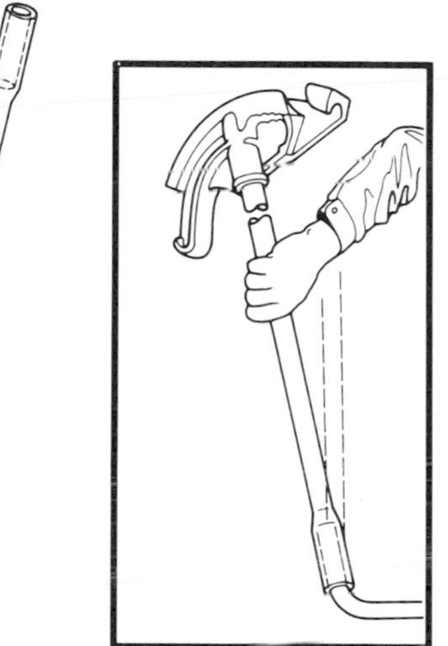

8–26c. Correcting over or under bends.

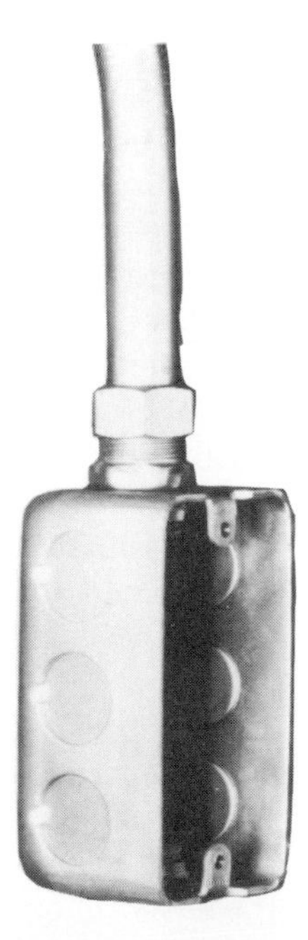

8–27a. Utility box surface-mounted with an offset bend.

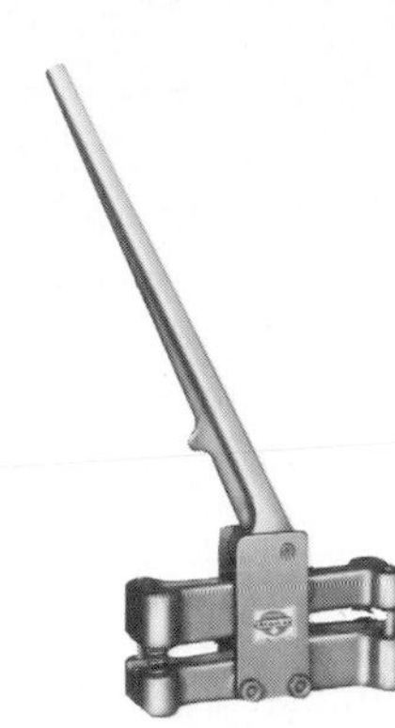

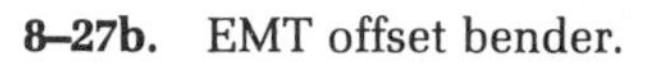

8–27b. EMT offset bender.

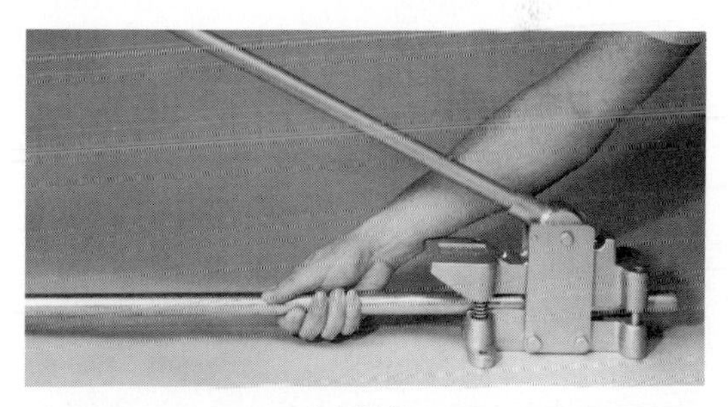

1. LOAD

2. BEND

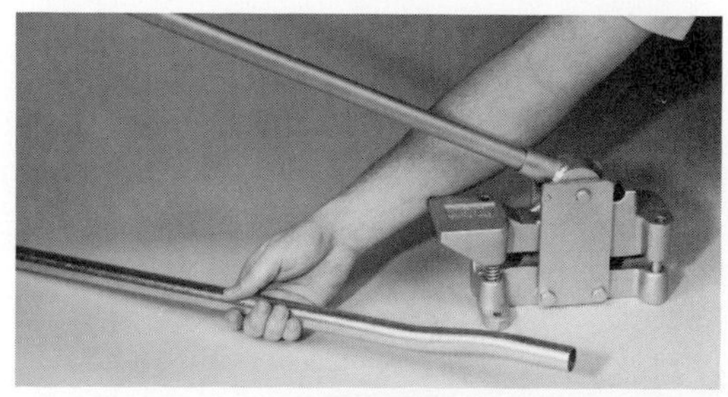

3. UNLOAD

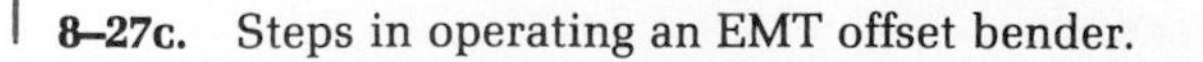

8–27c. Steps in operating an EMT offset bender.

A special offset bender has been designed so that the electrician can bend the $\frac{1}{2}''$ or $\frac{3}{4}''$ EMT without having to measure. Just load the EMT in the bender (Fig. 8-27) and bend. Unload the bender and you have a perfect offset for mounting the box on the wall.

This bend in the conduit eliminates the need for connectors whenever the EMT is used in exposed locations, thus preventing a lot of trial-and-error bending and cutting. Tools are available for all special jobs which warrant the investment, as you will see through our further examination of tools and equipment.

■ *Machine benders.* As mentioned, conduit too large to be bent by hand requires hydraulic pressure to do the job. Some examples of machine bending follow.

In Fig. 8-28 the EMT, or thinwall conduit, was offset to fit the externally mounted box. This type of bend may be made by hand for thinwall, but is not so easily done with rigid conduit. Fig. 8-29 shows a multipurpose hydraulic bender which can make an offset in rigid conduit. It can also make 90-degree bends. This arrangement can make both bends

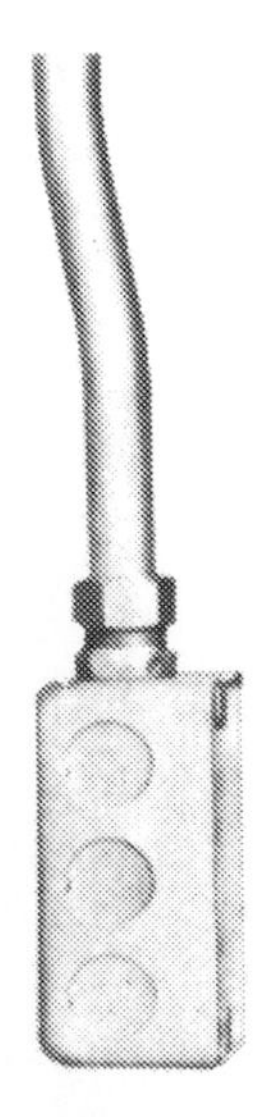

8–28. Utility box surface-mounted with an offset bend.

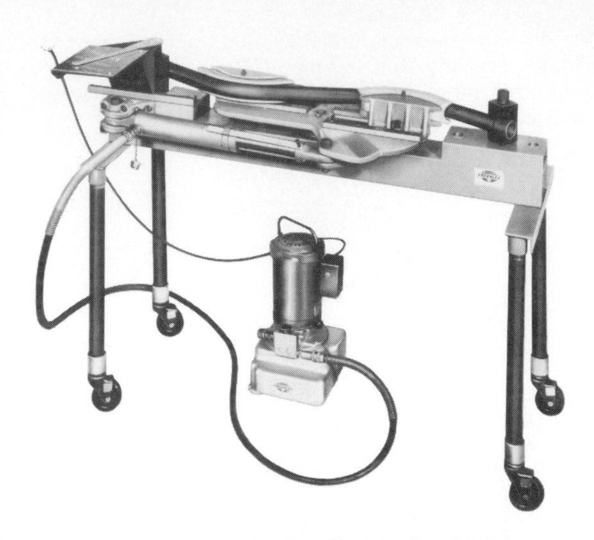

8–29. Multipurpose hydraulic bender. Makes offsets in seconds in one shot—also makes one-shot 90° bends. This unit makes two bends at once. Ram travel is used as an indicator of the degree of bend.

in an offset simultaneously with one stroke. The unit can handle $\frac{1}{2}''$ through 2″ rigid conduit. The hydraulic pump unit, which produces the power needed to make the bend, can deliver up to 10 000 psi (pounds per square inch) of pressure. It is powered either by an electric motor or a gasoline engine.

Some hydraulic units can also rely upon human power for operation. Fig. 8-30. The pump handle multiplies the force, using hydraulic principles to exert enough pressure to bend the large-diameter pipe or conduit. These benders may be mounted on tables and supplied with electric motors to furnish the energy to compress the hydraulic fluid. Figs. 8-31 & 8-32. A large disc, attached to the unit, indicates the conduit bend in degrees.

Wire Pulling. Once the conduit has been installed, it is necessary for the proper wires to be pulled through the pipe. Sometimes this is a difficult task, if some of the conduit runs are rather long, and if the wire has insulation that catches or hangs up on connectors or other rough spots in the conduit. Tools have been

8–30. Hydraulic hand-operated bender for large-diameter pipe or conduit.

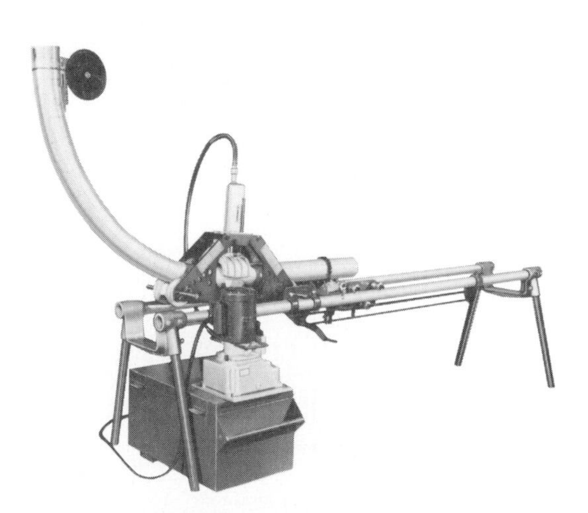

8–31. This bender is hydraulically operated. Note the table and the large black disc for indicating the number of degrees of bend in the large conduit (up to 5″).

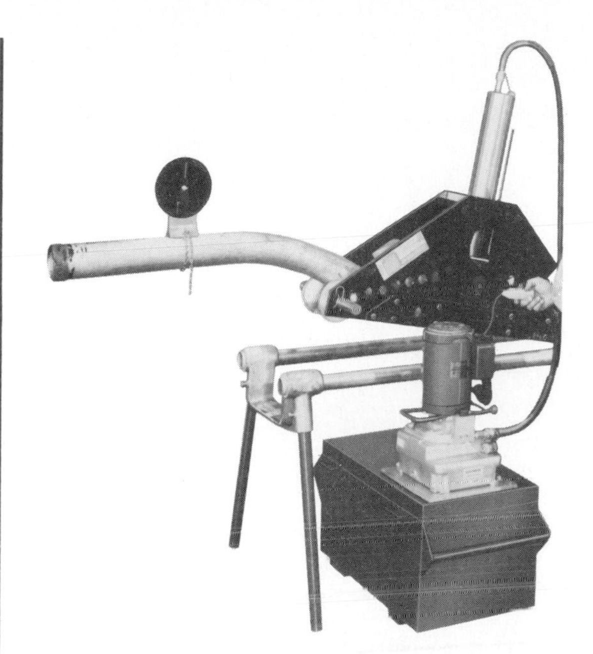

8–32. The bending-degree indicator (the large black disc) eliminates the need for leveling the table. The face of the protractor rotates to compensate for any surface that is not level. The dial is graduated in quadrants of 18, 20, 21, and 22 shots.

designed to make the job easier. Fig. 8-33 shows some of the equipment used in these jobs.

Before wire can be pulled through the conduit, a *fish wire* must be attached to it. This may be done by hand or by machine. Fig. 8-34 (page 224) illustrates how a vacuum, converted to a blower, can be used to blow a line through the conduit, and used as a vacuum to suck the fish line through the conduit as well as clean the conduit of dirt, dust, and debris.

Once the fish wire has been pulled through the conduit, it is necessary to attach the wires to it that are to be pulled through. Fig. 8-35 (page 224) shows how a rope is attached to the cone point (a device which clamps onto the wires to be pulled). A *lube spreader* is attached to clean the conduit of all water and debris ahead of the cable. Fig. 8-36 (page 224). It also spreads a uniform coating of pulling compound

8–33. Pulling wire with various types of setups.

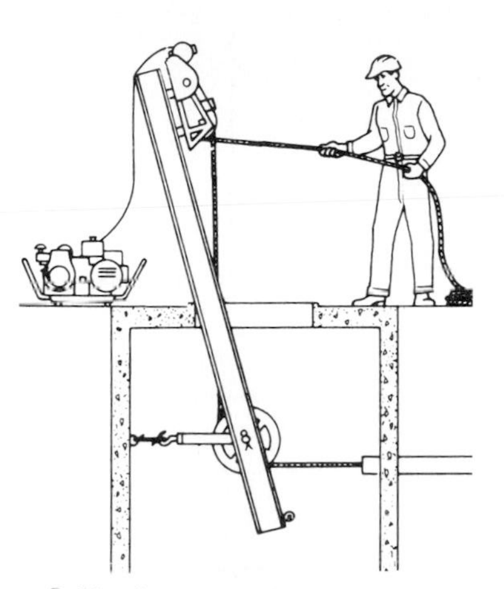

Pulling from manhole—with puller mounted on manhole sheave. Power is supplied by portable generator

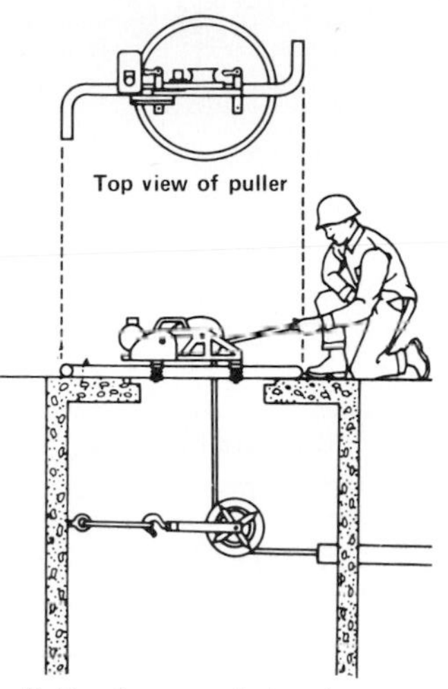

Pulling from manhole using hook sheave—Puller mounted on 4″ conduit with two 90 degree bends

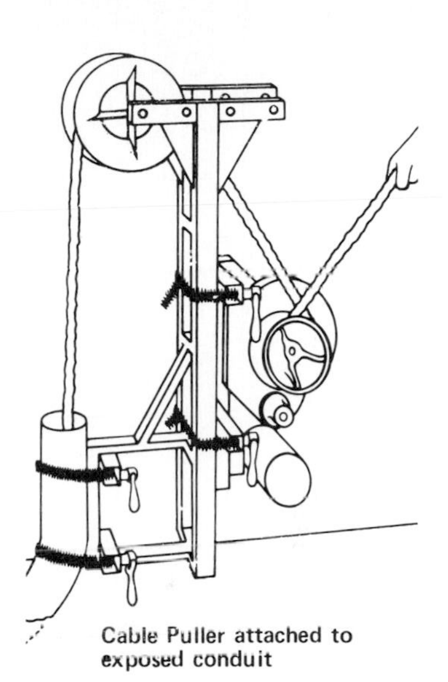

Cable Puller attached to exposed conduit

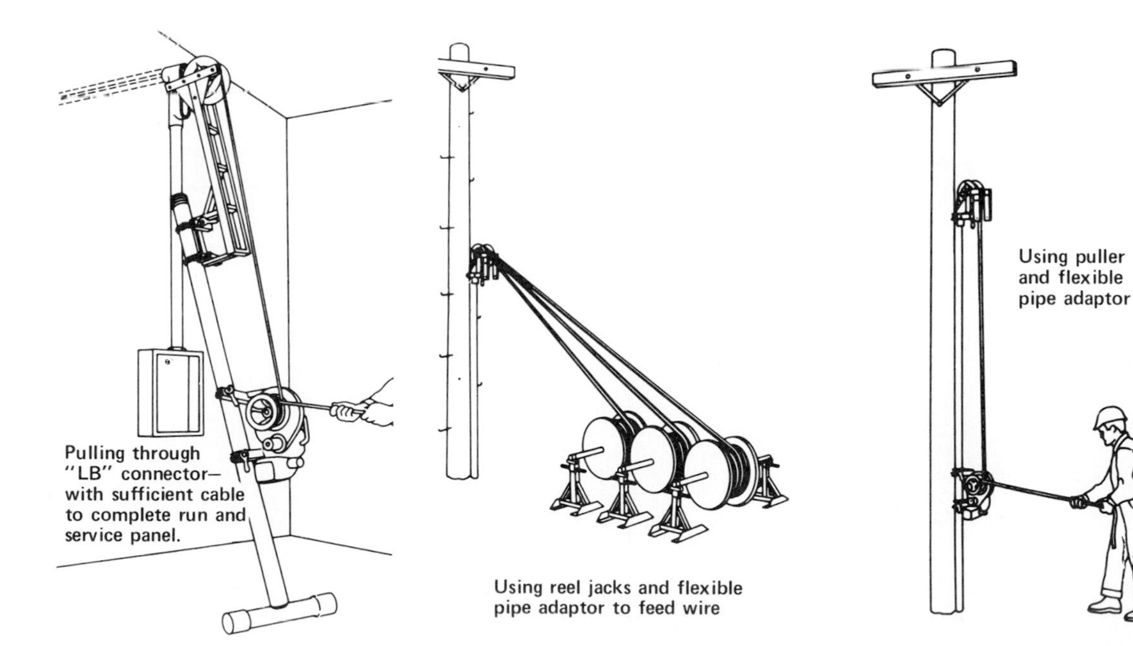

8–33. Pulling wire with various types of setups. (Cont.)

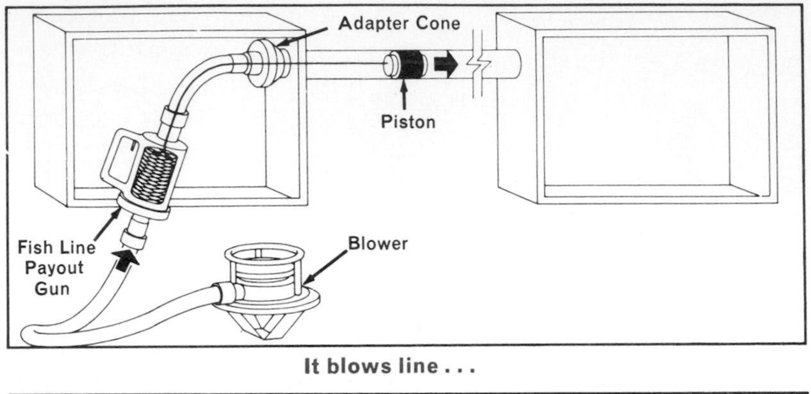

It blows line . . .

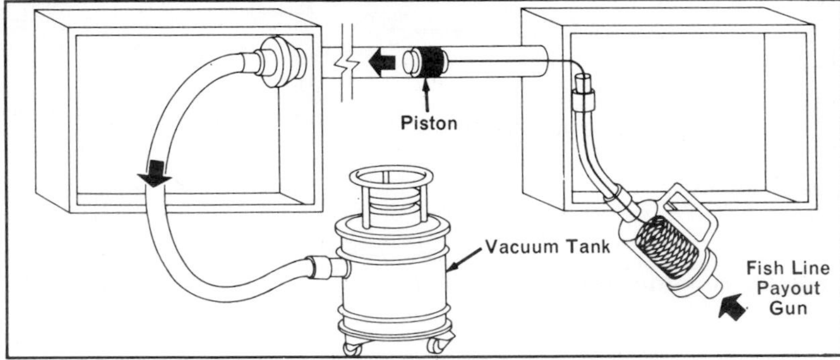

. . . or vacuums line

8–34. Using a vacuum to either blow or suck a fish line through the conduit.

8–35. Hookup for pulling cable through a conduit. The cutaway shows the rope clevis, wire grip, and cable inside a conduit.

8–36. Lube spreader lubricates the conduit and cleans it of water and debris.

inside the conduit to make the pulling process easier. The entire arrangement of wires being pulled through conduit is shown in Fig. 8-37.

Grips. *Grips* are flexible wire mesh holding devices used to pull electrical cable or conductor into place, to support cable or conductor after it has been installed, and to provide positive strain relief at the point of wiring connections. These easily assembled devices solve problems during the construction and installation of electrical systems. They make such systems significantly safer and more durable.

Pulling Grips. Pulling grips consist of a wire mesh that can be inserted over a cable end. Pressure applied to the end of the grip causes the grip to tighten around the cable and hold it firmly while the cable is pulled through a conduit or other obstacle. A wide assortment of grips is available individually or in kits to make wire pulling easier. Fig. 8-38. Note how the grip is removed from a conduit in Fig. 8-39. Once the cable has reached its intended destination, the pulling grip can be easily removed for use again.

8–37. Cutaway view showing the lube spreader, rope clevis, wire grip, and cable inside a conduit.

Grip Safety. The broad application of grips on a wide variety of objects requires that adequate safety factors be used to establish working loads. The approximate breaking strength of a grip represents an average calculation based on data established from actual direct tension testing done in the manufacturer's laboratories.

Each grip is designed to work on a specific range of cable diameters. Check the grip specifications to determine the style of grip best suited for your application.

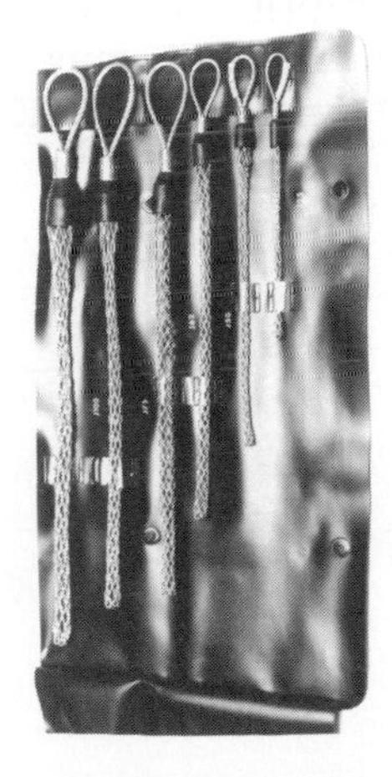

8–38. Junior pulling grips in a roll.

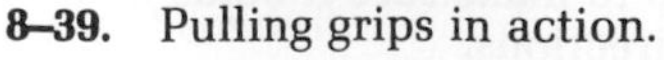

8–39. Pulling grips in action.

8–44. Pulling wire on the job. This is a *down-pull* with exposed conduit.

8–45. Portable conduit rack eliminates running back and forth to the stockpile for conduit. It holds up to 1000 lbs. of conduit, neatly arranged for easy selection.

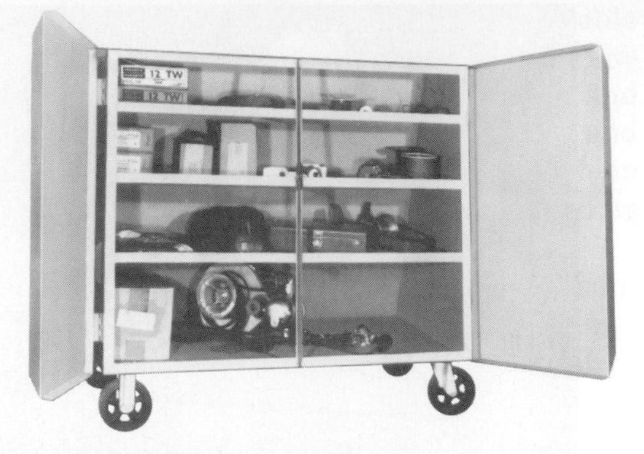

8–46. Utility cabinet provides storage on the job.

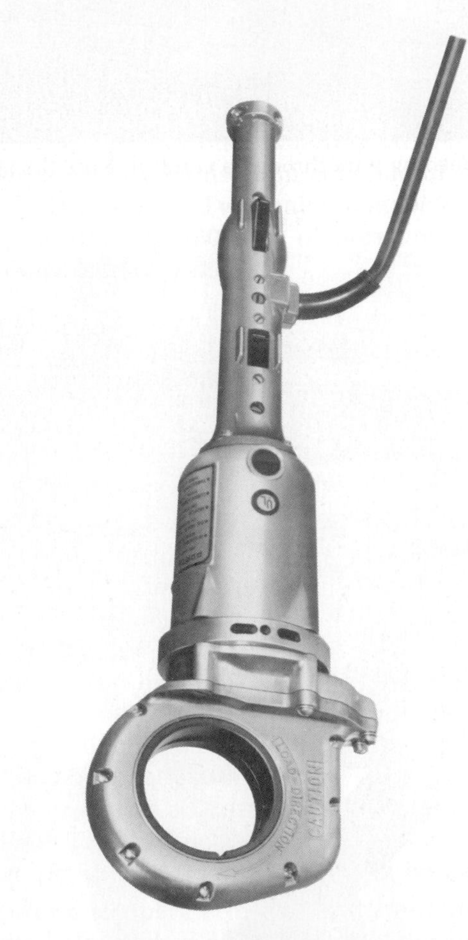

8–47. Portable power threader. This unit threads pipe or conduit in a variety of locations. It can be used to thread pipe or conduit already mounted. Just slip the end over the conduit and turn on the switch.

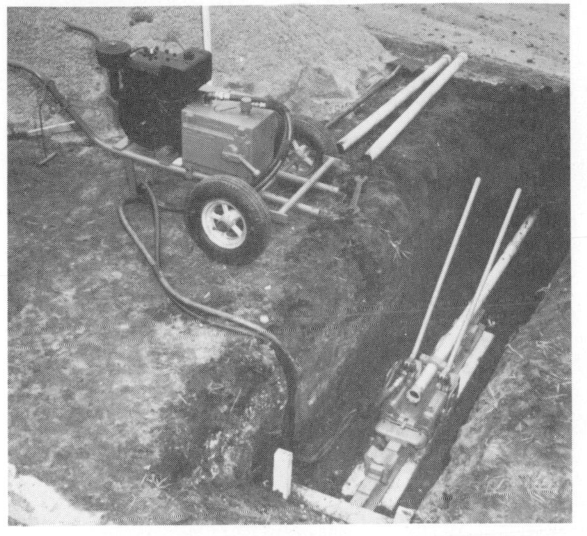

8–48. Hydraulic pipe pusher for installing pipe or conduit under streets, sidewalks, or railroad beds.

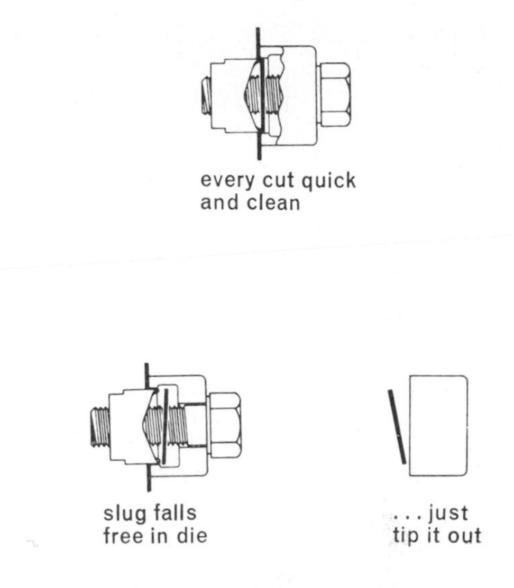

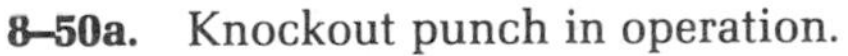

8–50a. Knockout punch in operation.

8–49. Pipe pusher. Each unit develops 75 tons of pressure. One can push pipe and duct up to 18″ in diameter; or two may be used, as shown here, to push concrete pipe.

8–50b. Using a punch in the field.

8–51. Close-up of the punch being used to cut a hole in a box.

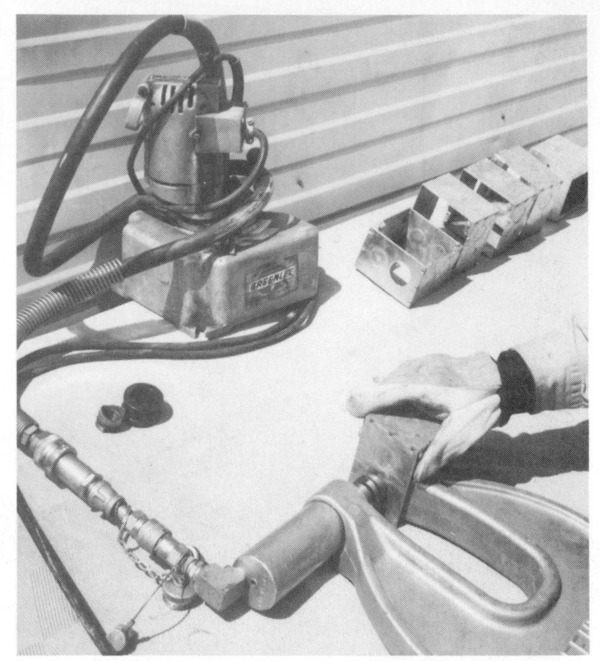

8–52. An electric motor drives a hydraulic unit which applies enough force to punch a hole in the heavy-gage boxes. Control button is in the worker's left hand.

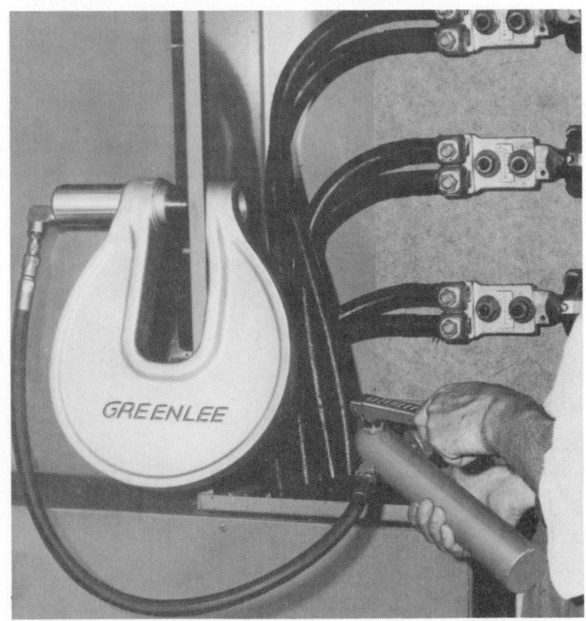

8–53. Knockout punch driver for $\frac{1}{2}''$, $\frac{3}{4}''$, and 1″ conduit. This unit eliminates predrilling or step-up punching. It can cut through 10-gage metal with a few strokes of the hydraulic pump.

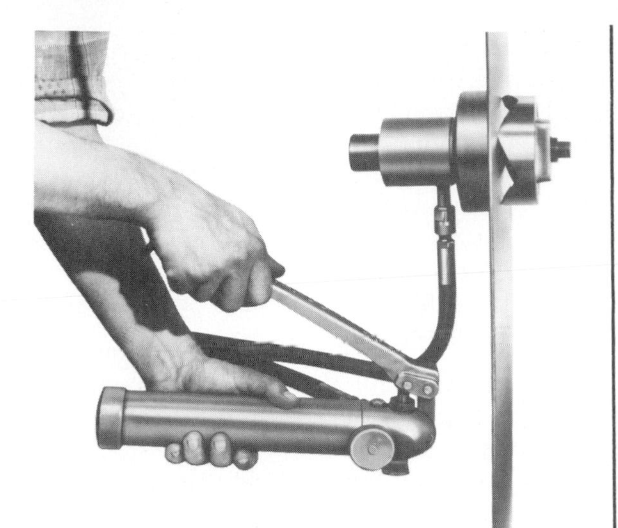

8–54. Close-up of a punch operated by a hand pump.

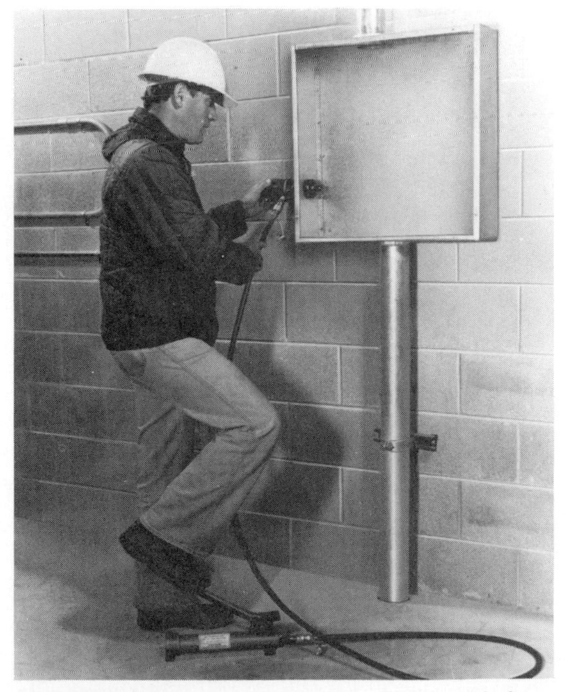

8–55. Hydraulic knockout punch driver with foot pump.

A hand-operated metal punch can also be used to punch holes for cable in metal studs. Fig. 8-56. Insulators are made to fit the punched hole. Note the different types of holes. In most cases, the rectangular shaped hole comes prepunched. The insulator protects the cable insulation from damage as it is pulled into position.

8–56a. Operating the metal stud punch.

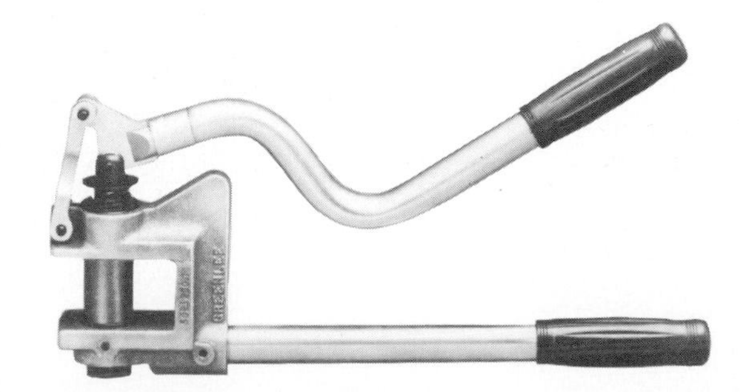

8–56b. The metal stud punch.

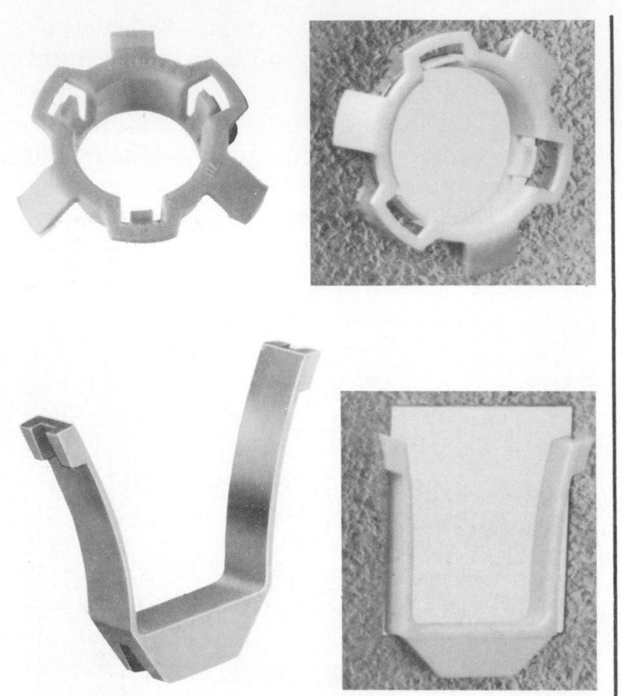

8–56c. Insulators made for the hole punched in the metal studs. Round one is for Romex. The other is for conduit.

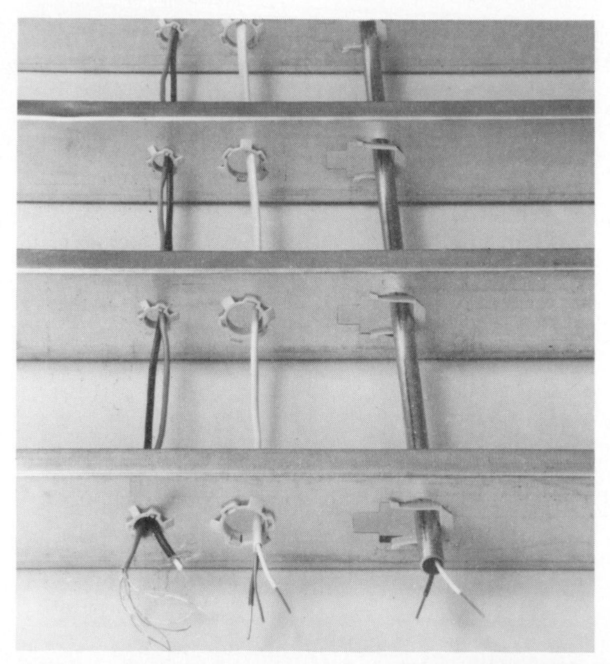

8–56d. Metal studs using various types of wiring.

■ *Other special purpose tools.* For a "T" splice, a mid-span strip is made. Fig. 8-57. The insulation is removed without scoring or damaging the conductor. Simply close the jaws on the cable and twist. A self-feeding feature assures positive progression down the cable to any position desired. Ends may also be stripped for insertion into lugs. This unit can strip larger cables $\frac{9}{16}$'' (14 mm) to $1\frac{7}{16}$'' (36.5 mm), outside diameter.

Larger cables call for a greater capacity in terms of cutters. The cable cutter shown in Fig. 8-58 has long handles to provide the leverage needed to cut large-diameter wire.

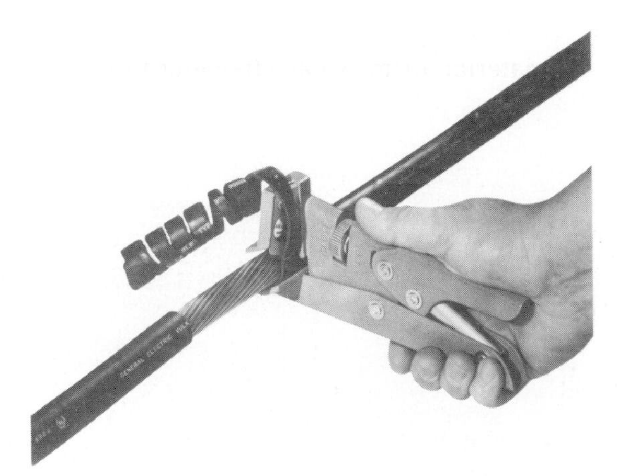

8–57. Cable stripper can remove the insulation without scoring the conductor.

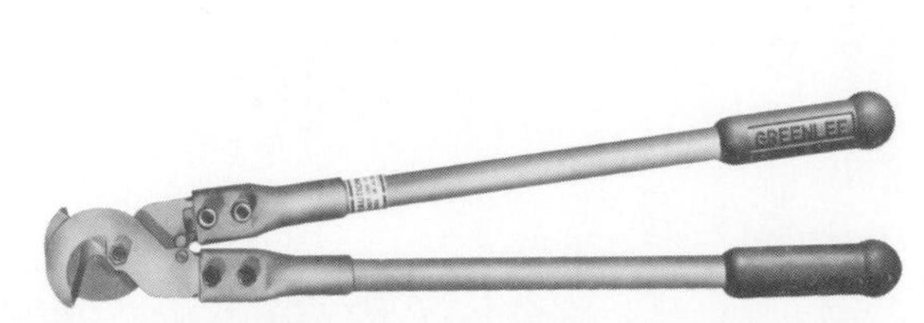

8–58. Cable cutter for soft copper and aluminum. Long fiberglass handles give better leverage. It can cut up to 750 MCM (750 000 circular mils) cable.

Tools for General Industrial Use

■ *Saws.* One of the most-used saws on the job is the hacksaw. Fig. 8-59. It comes in handy for cutting pipe, conduit, or any metal. A compass saw is used for cutting wood to make way for conduit, or a box, or cabinet. Fig. 8-60 shows a blade for a compass saw. The saw handle is similar to that of a hacksaw, and shaped to fit the hand.

The *flexsaw* is a compact saw for close-quarter jobs. Fig. 8-61. It uses a hacksaw blade, which flexes for a cut flush to the surface. It is used to cut metal.

Hacksaw blades suitable for the type of material to be cut should be chosen. They are available in 10″ and 12″ lengths. But the number of teeth per inch should depend on the shape and size of material being cut. The three teeth designations available are 18, 24, and 32 teeth per inch. The more teeth the better for thin material, or for small-diameter pipe.

■ *Chisels.* Chisels are used to remove wood from studs or framing to allow clearance for mounting conduit or metal boxes. There are a number of chisels available for electricians or carpenters. Here are some usually chosen by electricians or linemen.

The *utility chisel* in Fig. 8-62 is a special-purpose type used by utility and telephone line workers to cut daps in poles for crossarms. The blade is 2″ wide and heat treated for hard wear and use.

The *firmer chisel*, with beveled edges, is made of high-carbon tool steel. Fig. 8-63.

The *framing chisel* can be used for heavy framing, or as a wrecking tool. It can also be used like the firmer chisel to chip wood. Notice that it does not have beveled edges. (This is one way to distinguish it from the utility chisel.) Fig. 8-64.

The *gouge* is a woodworker's tool. The electrician uses it to remove excess wood for the installation of conduit. Fig. 8-65.

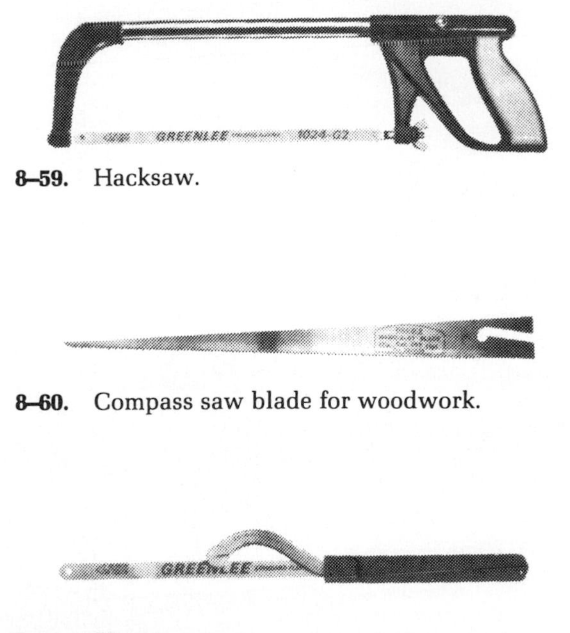

8–59. Hacksaw.

8–60. Compass saw blade for woodwork.

8–61. Flexsaw.

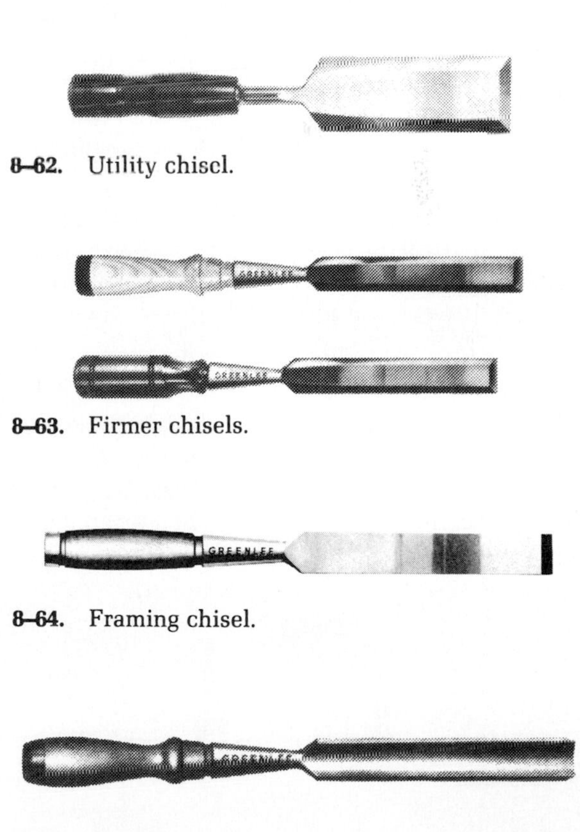

8–62. Utility chisel.

8–63. Firmer chisels.

8–64. Framing chisel.

8–65. Gouge.

Drill Bits. Woodworking tools are valuable to an electrician who must make holes in wood to pull wire or place conduit. A number of drills are available for boring holes in wood. A hand-operated drill, called a brace, takes a number of drill *bits* suitable for each job. Fig. 8-66.

The bit brace uses the principle of the wheel and axle in creating a driving force. The jaws of the chuck hold the bit in the brace during operation. Pressure is applied to the head of the brace (the rounded, wooden knob on the end) and a turning motion is applied by grasping the handle in the middle of the brace, and rotating it. Drill bits have a screw tip to catch the wood and pull the cutting edge forward. The twist of the bit causes smooth flow of wood chips to the outside of the hole.

Some braces have a ratchet wheel which allows for turning the bit in a clockwise direction only, but allows the bow (bent part of the brace) to be turned in the opposite direction while the bit is stationary in the wood. It is then possible to bore a hole in a piece of wood where the full swing of the bow is not possible. A *cam ring* can be turned to allow for ratchet action or for full operation.

An electrician should be able to identify the proper drill bit for a job. Some bits are more efficient than others due to their design and availability in a number of sizes.

■ *Machine spur bit.* The bit in Fig. 8-67 is properly referred to as a machine spur bit. It is a flat, wood-boring power bit designed for use in electric drills, to bore holes in any wood, at any angle, and at any speed. It is usually a little over 6″ in length and comes in cutter sizes ranging from $\frac{1}{4}$″ to $1\frac{1}{2}$″, in increments of $\frac{1}{16}$″.

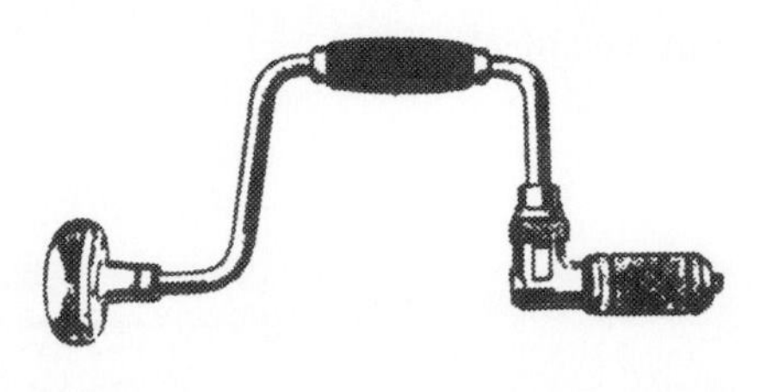

8–66. Brace.

■ The *multispur pipe bit* has three milled flats for use in portable electric or pneumatic drills, and in stationary boring machines. Fig. 8-68.

■ *Plain-type expansive bits* for use in a brace have a clamp and screw for locking the cutter in place. Fig. 8-69. This bit may be obtained in $\frac{1}{2}$″ to $1\frac{1}{2}$″, $\frac{7}{8}$″ to $1\frac{3}{4}$″, and $1\frac{3}{4}$″ to 3″ cutters.

■ *Auger bits.* A *bellhanger's drill* is used for telephone and other wiring installations. Sizes are $\frac{3}{16}$″ to $\frac{3}{4}$″, in $\frac{1}{16}$″ increments. It comes in 12″, 18″, and 24″ lengths. Fig. 8-70.

8–67. Machine spur bit.

8–68. Multi-spur pipe bit.

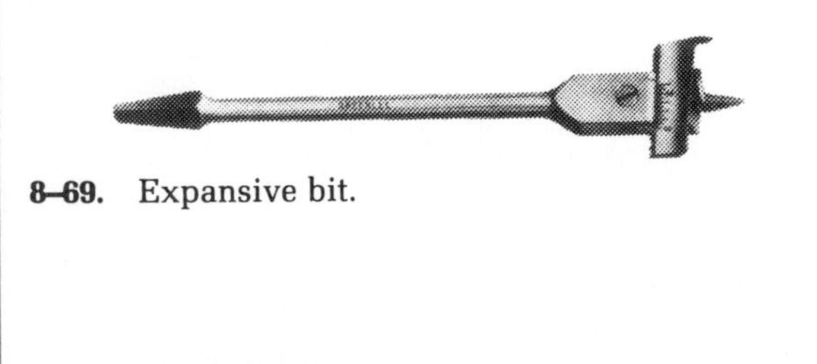

8–69. Expansive bit.

8–70. Bellhanger's drill bit.

■ The *uni-spur electrician's power bit* is a double-twist, smooth-boring bit used by electricians, plumbers, and pipe fitters. Fig. 8-71. It is especially designed for use in electric drills. The screw point feeds at 16 turns to the inch. Note the two set-screw holes for use in a power bit extension or brace adapter. The overall length is $8\frac{1}{2}''$. It comes in $\frac{5}{8}''$ to $1\frac{1}{4}''$ size with $\frac{1}{16}''$ and $\frac{1}{8}''$ intervals, depending upon the diameter needed.

■ The *single-twist power bit* is a smooth-boring bit designed for electric drills. Fig. 8-72. It has excellent chip clearance due to its single-twist design. It may be used to bore holes from $\frac{5}{8}''$ to $1''$ with $\frac{1}{16}''$ difference between various drill sizes. Overall length is from 5″ to 7″, depending upon the drill size. With extensions, it will drill through floor joists, studs, or through existing construction for installation of Romex.

■ A *solid-center power-car bit* has a solid center to add stiffness and strength. Fig. 8-73. It has a single twist for excellent chip removal from the cutting area. The screw point is of a medium pitch. It is designed for use in three-jaw chucks of $\frac{3}{8}''$ and larger. It can be used with a bit extension. Overall length is about 16″, available in sizes from $\frac{3}{8}''$ to 1″ in sixteenths.

■ The *power-pipe bit* is ideal for boring holes in wood to allow passage of pipe, conduit, and tubing, and for other uses requiring large holes. Fig. 8-74. It is designed for rough, rugged boring. Lengths run from 14″ to 17″ overall, with cutter sizes available from $1\frac{1}{4}''$ diameter for $\frac{3}{4}''$ pipe, $1\frac{1}{2}''$ diameter for 1″ pipe, $1\frac{3}{4}''$ diameter for $1\frac{1}{4}''$ pipe, 2″ diameter for $1\frac{1}{2}''$ pipe, and $2\frac{1}{2}''$ diameter for 2″ pipe.

■ An *impact-wrench bit* can be identified by the grooved portion near the end of the bit. Fig. 8-75. This single-spur (spur refers to the cutting edge near the screw tip) impact bit bores $\frac{3}{8}''$ to $1\frac{1}{16}''$ holes in telephone or power line poles quickly, easily, and safely, using an impact wrench. Two sizes fit all impact wrenches with the use of $\frac{7}{16}''$ or $\frac{5}{8}''$ hexagonal adapters. Overall length is 12″, 16″, or 24″, depending upon the twist length (8″, 12″, or 18″) needed. The diameter of the drill can be $\frac{3}{8}''$ to $1\frac{1}{16}''$, depending upon the bit chosen.

■ The *rafting auger* is sometimes referred to as a *boom auger*. This tool has a double twist with a flat-cut pattern head. Fig. 8-76. It is available in overall lengths of 14″ to 20″, with hole diameters ranging from $1\frac{1}{4}''$ to 4″.

8–71. Uni-spur power bit.

8–72. Single-twist power bit.

8–73. Solid-center power-car bit.

8–74. Power-pipe bit.

8–75. Single-spur impact-wrench bit.

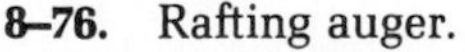

8–76. Rafting auger.

■ A *ship-auger bit* has a shank for use with $\frac{3}{8}''$ and larger electric drills. Electricians, plumbers, and other craftsmen employ it to drill deep holes or where extra reach is needed. Fig. 8-77. The ship auger has a medium-feed screw point and cutting head, with side lip to withstand rough usage. It can be used with a power bit extension to increase its length. Overall length is about 17″ normally, and it can be had in sizes from $\frac{3}{8}''$ to 1″.

The previous list is not by any means the extent of drill bits available. This list only illustrates some of the available types, and shows the difference between single-twist, double-twist, and solid-center bits. The spur side, or cutting edge, may make a difference in the job or the finish of the hole. The rate of feed and the ease with which a hole can be bored are also important with a hand-operated brace.

Extensions. Extensions for bits are basically of three types. They may vary by having either hand- or power-driven or round- or hexagonal-shaped shanks.

■ A *brace-bit extension,* Fig. 8-78, is used to extend the distance from the drill bit to the power source. It allows an electrician to get at places otherwise inaccessible because it allows for operation of the brace in a restricted area.

■ *Power-bit extensions* are needed to get at some locations where the power drill and normal drill bit cannot. The bit extension in Fig. 8-79 has one flat side with two set-screw holes permitting two or more extensions to be locked together for longer reach. They come in either 18″ or 24″ lengths.

8–77. Ship-auger power bit.

8–78. Brace-bit extension.

■ The *power-bit expansion* is an extension with a $\frac{1}{4}''$ hex end to take $\frac{1}{4}''$ *flat zip bits.* Fig. 8-80. The hex configuration makes for a secure bit fit in the extension. This one is available with an overall length of 12″.

Special Equipment.

■ *Nail puller.* The quick, easy removal of nails is important to an electrician. Wooden boxes, crates, or old construction frequently require nails removed so that the electrician can get the job done. Fig. 8-81.

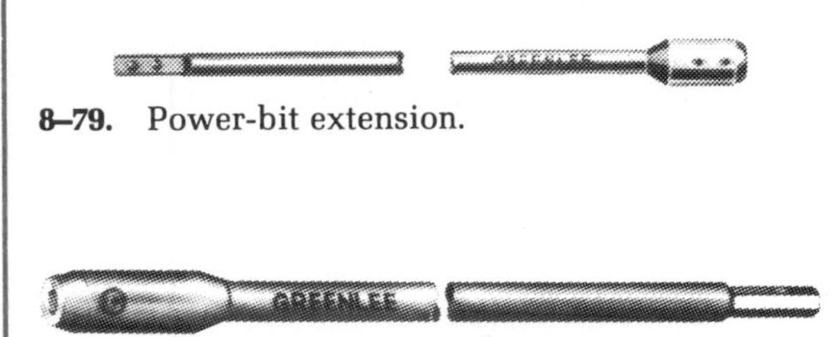

8–79. Power-bit extension.

8–80. Power-bit extension for use with hex shank bits.

8-A. Metric Measurements*

		Conversions
1 inch	=	25.4 mm
1 foot	=	0.3048 m
1 sq in	=	6.4516 cm^2
1 sq ft	=	0.09290304 m^2
1 cu in	=	16.387064 cm^3
1 cu ft	=	0.028316846592 m^3

*These units are given here so that the measurements in this chapter can be converted to metrics if the occasion arises.

It may be that in the future all tools will be metric. This will mean that a number of things will have to change first. The fasteners and the materials worked with by tools will have to be made metric.

No standards exist at the moment, but it may not be too long. Until metric standards are established, the boxes and other equipment will continue to contain such requirements as "only 6 # 14 conductors can be placed in a box with 12.5 cubic inches." The entire National Electrical Code handbook will have to be converted to metrics before the standards are found to be applicable to those working in the field.

The impact handle slides on an upright claw bar for quick hammer action, driving the claw under the nailhead. The claw hook is extended to form a leverage foot and to guide the handle for fast, safe positioning of claws, and to provide maximum pull on the nail.

PVC CONDUIT SYSTEMS

Nonmetallic conduit and raceway systems are covered by Article 347 of the National Electrical Code. The use of this plastic conduit in electrical wiring systems has decided advantages. Nonmetallic conduits weigh one-fourth to one-fifth as much as metallic systems. They can be installed in less than one-half the time and are easily fabricated on the job. Fig. 8-82.

PVC has a high impact resistance to protect wiring systems from physical damage. It is resistant to sunlight and approved for exposed or outdoor usage. The use of expansion fittings will allow the system to expand and contract with temperature variations. Fig. 8-83. PVC conduit will expand or contract approximately four to five times as much as steel and two and one-quarter times as much as aluminum. Installations where the expected temperature variation exceeds 14°C (25°F) should use expansion joints. The manufacturer furnishes the formulas for figuring out the expansion joint size and how often it is needed in any given installation. Since there are variations due to manufacturing differences, check the manufacturer's information before designing a complete system.

Any plastic conduit should always be installed away from steam lines and other sources of heat. Support straps should be tight-

8–81. Nail puller.

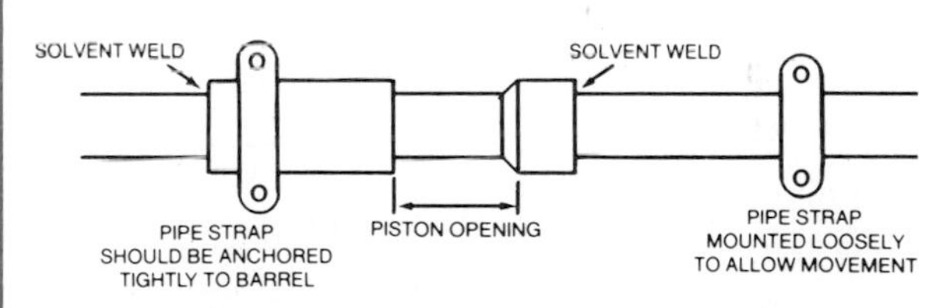

8–83. Allowing for expansion and contraction in a PVC conduit.

1. WRAP HEATING BLANKET AROUND PVC.

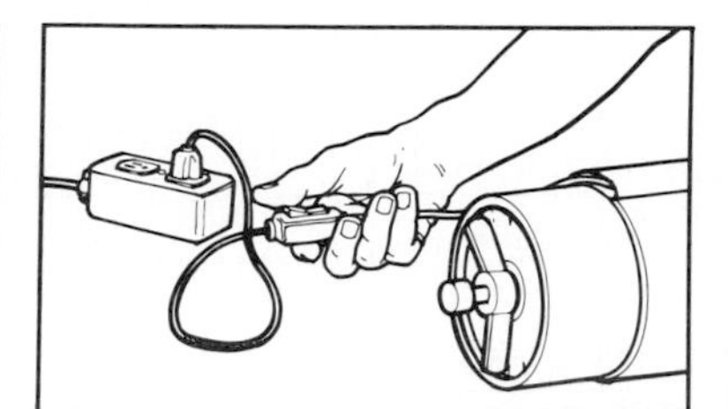

2. HEAT PVC.

3. REMOVE BLANKET AND BEND PVC.

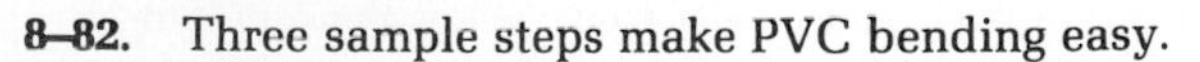

8–82. Three sample steps make PVC bending easy.

ened only enough to allow for lineal movement caused by expansion and contraction.

The most widely used rigid conduit in the United States is Plus 40. It is UL listed for use underground, encased in concrete, or direct buried, and for use exposed or concealed in most conduit applications above ground. See Table 8-B. It is rated for use with 90° conductors and is sunlight resistant. Check NEC Article 347 for approved locations. Plastic tends to degenerate when exposed to direct sunlight for long periods of time.

Plus 80 (see Table 8-C) is designed for above ground and under ground applications where PVC conduit with an extra-heavy wall is needed. It is frequently used where subject to severe physical abuse such as for pole risers, bridge crossings, and in heavy traffic areas. Typical applications are around loading docks, in high traffic areas, and where threaded connections are required.

PVC Bending

Conduit sections can be bent. First, though, they must be heated evenly over the entire length of the curve. Electric heaters are designed specifically for the purpose. These devices employ infrared heat energy, which is quickly absorbed by the conduit. Small sizes are ready to bend after a few seconds in the *hotbox*. Larger diameters require two or three minutes, or more, depending on conditions. *The use of torches and other flame-type devices is not recommended.* PVC conduit exposed to excessively high temperatures may take on a brownish color. *Sections showing evidence of such scorching should be discarded.*

8-B. Plus 40 Heavy-Wall Conduit

NOM. SIZE	CATALOG NO.	O.D.	I.D.	WALL	WT. PER 100′ PIN END	FT. PER BUNDLE
½	49005	.840	.622	.109	17	100
¾	49007	1.050	.824	.113	23	100
1	49008	1.315	1.049	.133	34	100
1¼	49009	1.660	1.380	.140	46	50
1½	49010	1.900	1,610	.145	55	50
2	49011	2.375	2.067	.154	73	50
2½	49012	2.875	2.469	.203	116	10
3	49013	3.500	3.066	.216	162	10
3½	49014	4.000	3.548	.226	195	10
4	49015	4.500	4.026	.237	231	10
5	49016	5.563	5.047	.258	313	10
6	49017	6.625	6.065	.280	406	10

(Industrex, Division of Carlon)

Rigid non-metallic conduit is normally supplied in standard 10′ lengths, with one belled end per length. For specific requirements, it may be produced in lengths shorter or longer than 10′, with or without belled ends.

8-C. Plus 80 Extra-Heavy Wall Conduit

NOM. SIZE	CATALOG NO.	O.D.	I.D.	WALL	WT. PER 100′ PIN END	FT. PER BUNDLE
½	49405	.840	.546	.147	22	100
¾	49407	1.050	.742	.154	29	100
1	49408	1.315	.957	.179	43	100
1¼	49409	1.660	1.278	.191	59	50
1½	49410	1.900	1.500	.200	72	50
2	49411	2.375	1.939	.218	99	10
2½	49412	2.875	2.323	.276	151	10
3	49413	3.500	2.900	.300	210	10
4	49415	4.500	3.826	.337	306	10
5	49416	5.563	4.813	.375	425	10

(Industrex, Division of Carlon)

Rigid non-metallic conduit is normally supplied in standard 10′ lengths, with one belled end per length. For specific requirements, it may be produced in lengths shorter or longer than 10′, with or without belled ends.

Forming the Bend

When properly heated the conduit is very flexible and can be shaped to almost any configuration. For the production of most bends, a variable jig is available. This adjusts for any radius from 5″ to 12″, for any segment of bend from 0° to 90°. It can also be adjusted for offsets.

After adjusting the job for the desired bend, place the heated conduit in the clamp sections to provide a smooth curve. Cool the conduit by sponging with water, and the bend is ready to install. See Fig. 8-82.

Conduits 2″ and larger in diameter and ducts require internal support to prevent crimping or deforming during the bending process. Bending plugs are available. These are inserted in each end of the conduit section before heating. The plugs are expanded to provide an airtight seal.

As the conduit is heated, the retained air expands. The increased internal pressure allows the conduit to be bent without deforming. The conduit must be cooled by sponging with cold water before the plugs are removed.

For special bends such as "blind" bends or compound turns in a conduit run, the heated conduit may be solvent cemented in place while still flexible. You must allow for expansion and contraction in a conduit. Fig. 8-83.

Conduit Terminations

There is more than one method available to terminate the PVC conduit. Terminations may be made in any electrical box or enclosure using standard size knock-outs or drilled holes.

■ *Method One.* Fig. 8-84. This is the permanent termination method. Apply solvent cement to the shoulder and shank of the box adapter and insert it through the knock-out from inside the enclosure. Push the coupling over the shank of the box adapter. Make sure it is tight against the enclosure wall. Rotate the coupling about one-half turn while installing. Hold it in position for a few seconds to permit the setting of the solvent cement. The coupling is now ready for the conduit to be installed. Only the shoulder of the box adapter extends outside the enclosure. Fig. 8-85.

■ *Method Two.* Fig. 8-86. If a watertight construction is required, place a flat washer over the threads of the terminal adapter. Secure it against the shoulder. Insert the adapter threads through the knock-out and secure it using either a standard locknut or threaded bushing. If watertight construction is not required, eliminate the flat washer.

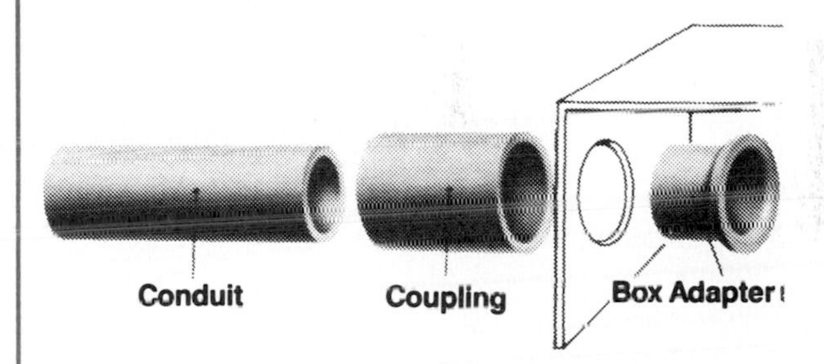

8-84. Permanent termination of conduit in a box.

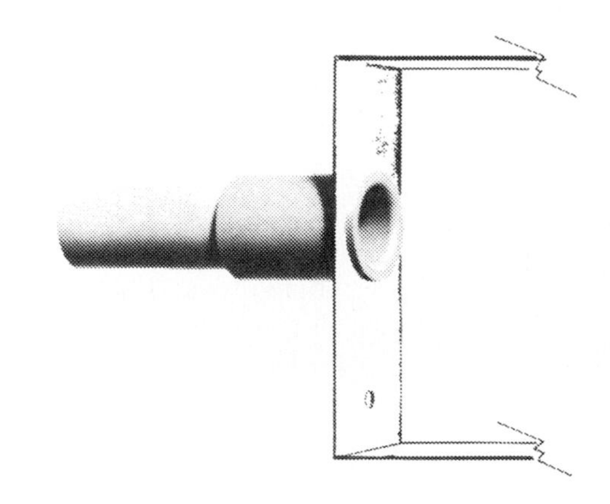

8-85. Properly terminated conduit with only the small amount of the end of the box adapter located outside the box.

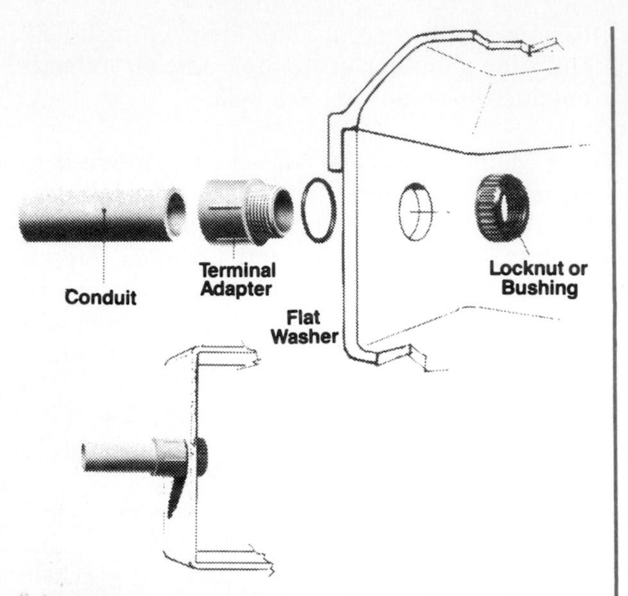

8–86. Making a waterproof termination of PVC conduit.

Cementing the Joints

Lengths of nonmetallic conduit ($\frac{1}{2}''$ to $1\frac{1}{2}''$) that are to be joined by solvent cement should be cut square using a fine-tooth handsaw. See Fig. 8-87. Deburr the ends. For sizes with a diameter of 2″ through 6″, a miter box or similar saw guide should be used to keep the material steady. After cutting and deburring, wipe the pipe ends clean of dust, dirt, and shavings.

Be sure the conduit ends are clean and dry. Apply a coat of solvent to the end of the conduit, the length of the socket to be attached. Use a natural bristle brush or the brush may disappear into the solvent. Push the conduit firmly into the fitting while rotating the conduit slightly about one-quarter turn to spread the cement evenly. Allow the joint to set approximately ten minutes.

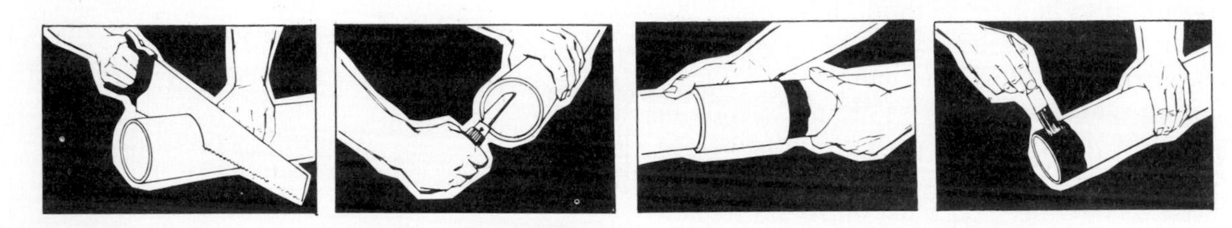

8–87. Using solvent to cement PVC conduit joints.

QUESTIONS

1. From the list of tools necessary for putting in a switch or outlet box, see if you can think of additional uses for each item.
2. Why are conduit benders needed?
3. What is a hickey?
4. Why are hydraulic benders sometimes needed to make conduit bends?
5. What is a fish wire? Why is one needed?
6. What are grips?
7. What is a lube spreader? How is it used?
8. What is a knockout punch?
9. Boxes come with a knockout insert. Why, then, is a knockout punch needed?
10. Why must one be careful when stripping insulation from a cable?
11. Tell where you may expect to find the following tools and why: hacksaw, flexsaw, chisel, gouge.
12. What is a bit brace?
13. How do you identify the impact-wrench bit?
14. Why are brace-bit extensions necessary in the electrician's tool box?
15. Describe two ways of terminating PVC conduit.

KEY TERMS

bit
fish tape
fish wire
fuse puller
grips
hand bender
knockout punch
linesmen's pliers
lube spreader
machine bender
pipe pusher
PVC conduit
Romex stripper
uni-spur electrician's power bit

CHAPTER 9

SPECIAL DEVICES

Objectives

After studying this chapter, you will be able to:

- List the main steps in installing self-drilling anchors.
- Identify the special tools needed to anchor an item to wood or concrete.
- Explain the function of a limit switch.
- Describe the three basic types of float switches.
- Describe four types of motor starters.
- Describe the operation of a circuit breaker.

This chapter deals with devices that an electrician might encounter in his work. Some of the devices will be used only by industrial electricians; others by those who specialize in house wiring. The aim of the chapter is to present for you a variety of the different types of devices available. This, it is hoped, will enable you to extend your skill-development opportunities to include a wide range of electrical devices and equipment with special applications.

ANCHORING SYSTEMS

Every electrician will probably have to anchor or attach an object to a wall, floor, or ceiling of concrete. A number of companies specialize in manufacturing anchoring devices especially suited to electrical installations.

■ *Self-drilling anchors.* This type of anchor is made to be inserted into a drill and become the device which will make its own hole in concrete. Fig. 9-1. It eliminates the need for expensive, easily damaged carbide drills. Fig. 9-2 illustrates some of the jobs that can be handled with self-drilling anchors. Other applications are shown in Figs. 9-3 & 9-4.

■ *Stud anchors.* Stud anchors may be installed without cleaning the hole. Their load capacity is due to the same expansion principle used in self-drilling anchors. Fig. 9-5 shows how to drill a hole, insert the anchor, and expand it with a hammer. Fig. 9-6 gives some possibilities for the stud anchor in electrical work.

■ *Lead anchor.* A wood screw anchor for quick installation of lightweight fixtures in masonry is the lead anchor. Fig. 9-7. Made of lead,

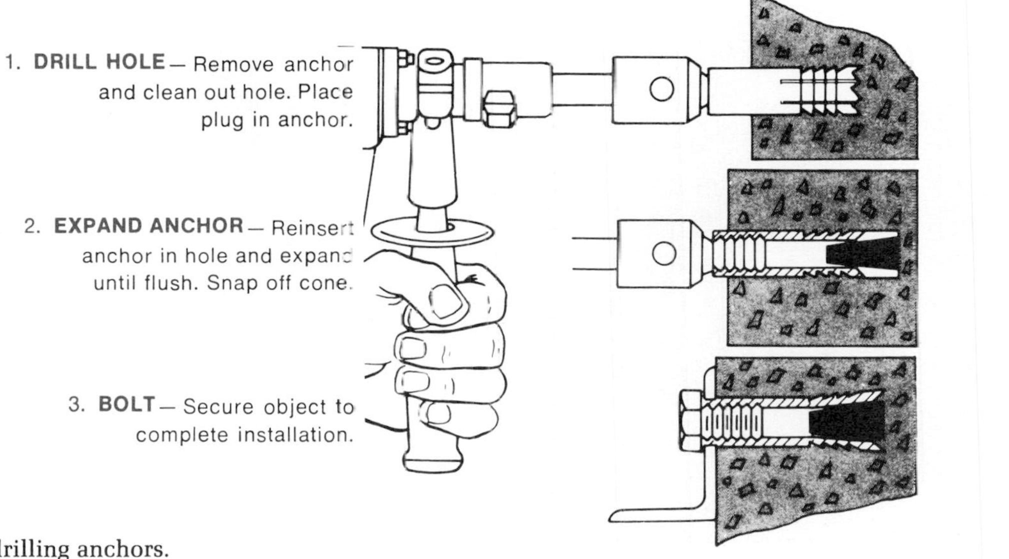

9–1. Installation of self-drilling anchors.

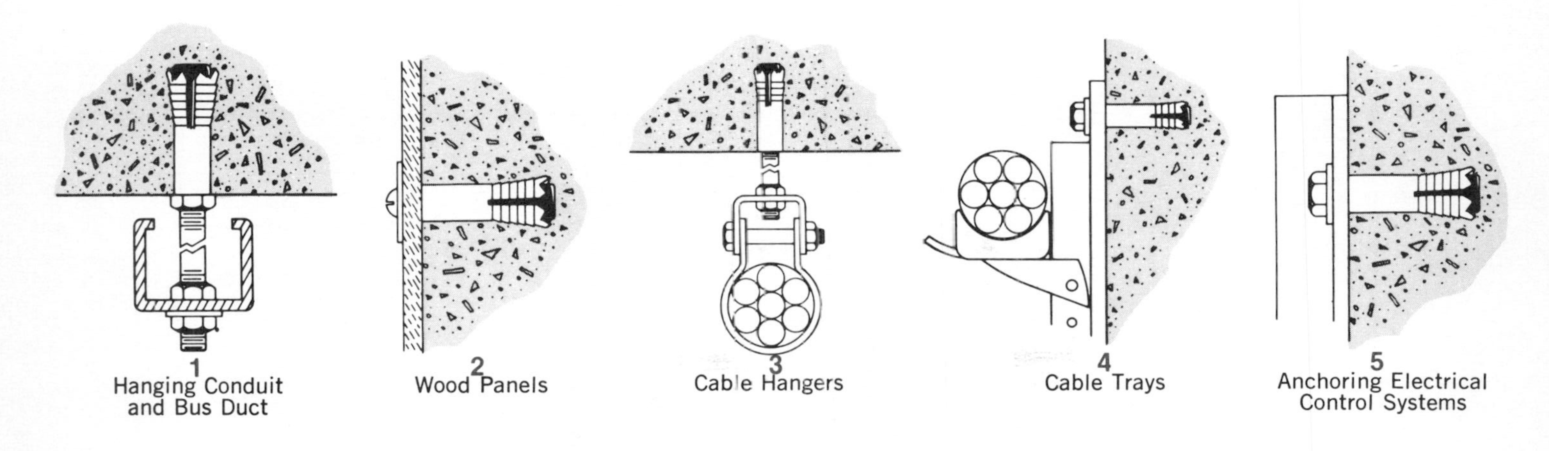

9–2. Uses for self-drilling anchors.

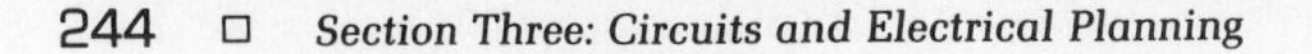

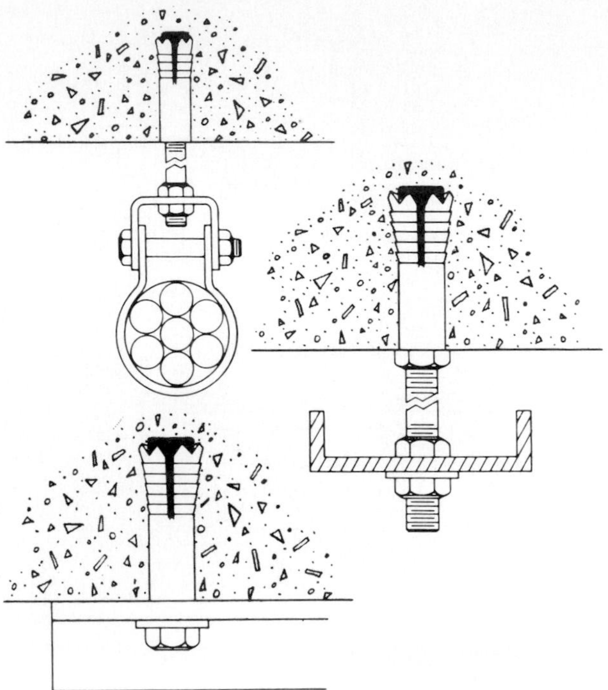

9–3. Examples of anchors used in the electrical trade.

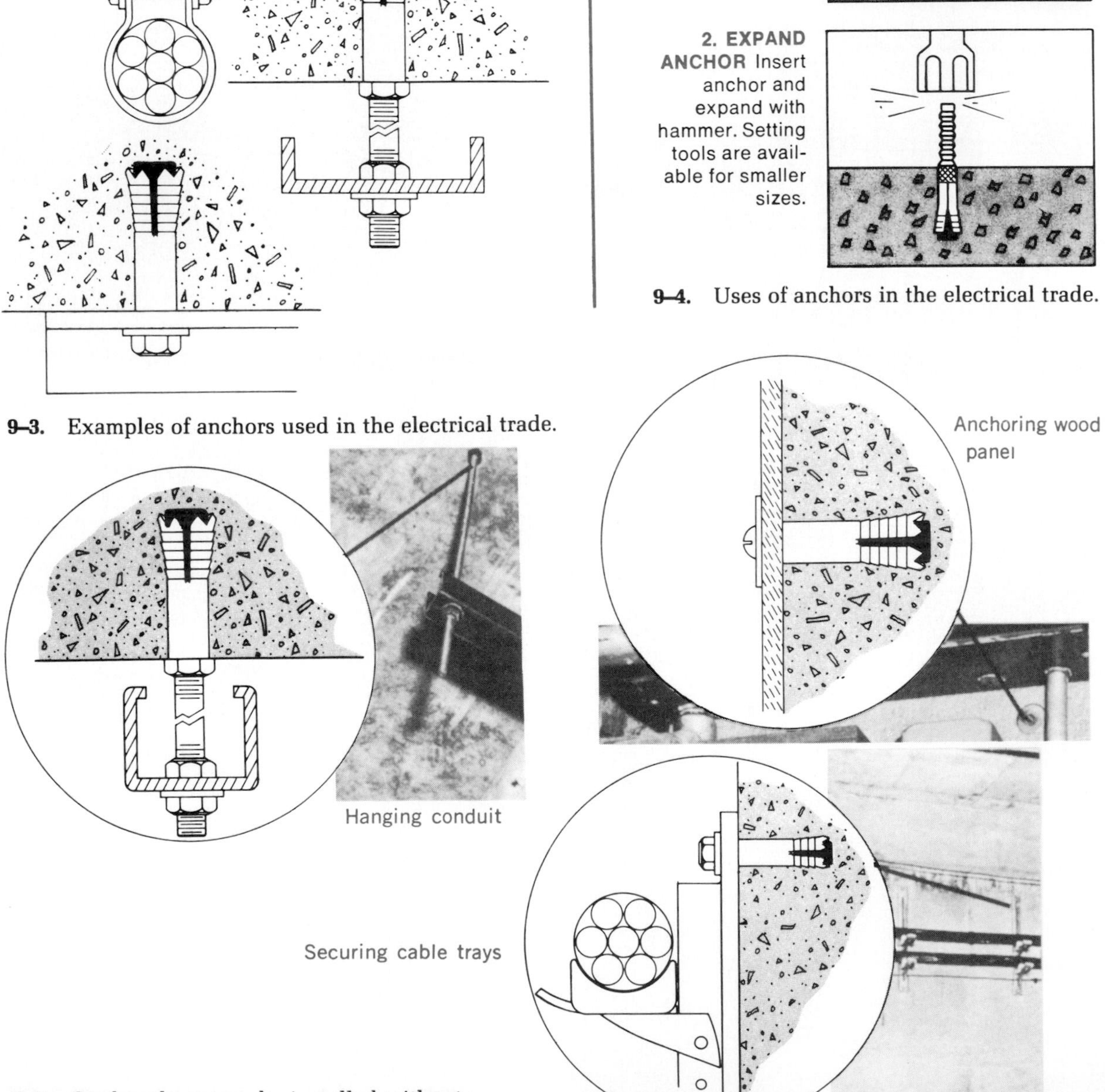

9–4. Uses of anchors in the electrical trade.

9–5. Stud anchors may be installed without cleaning the hole.

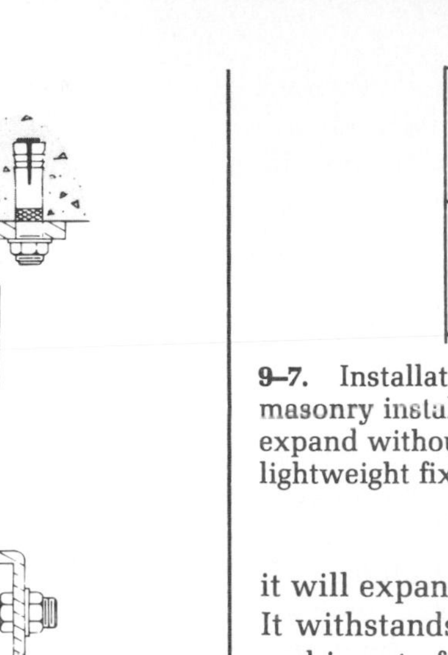

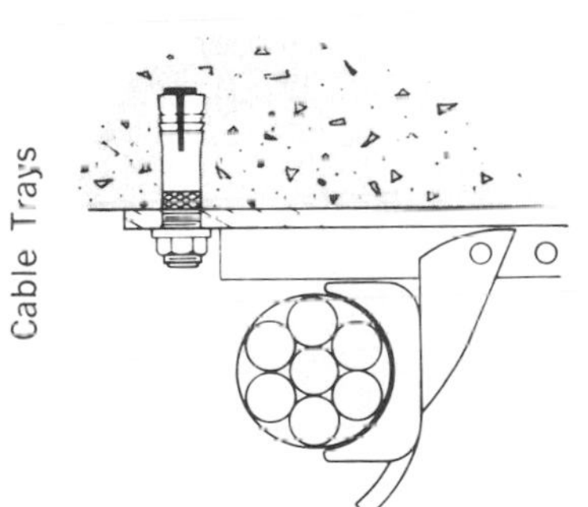

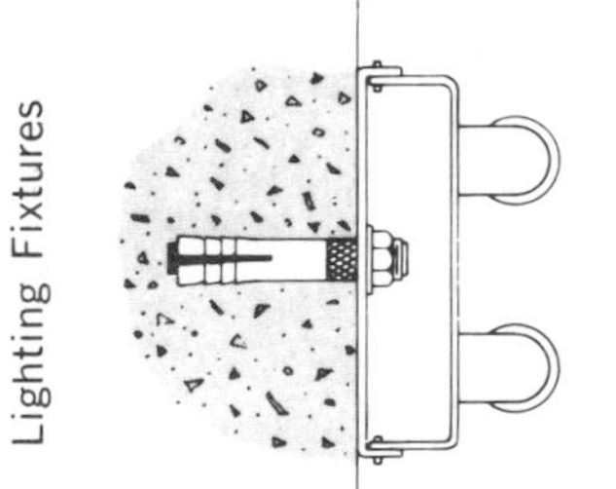

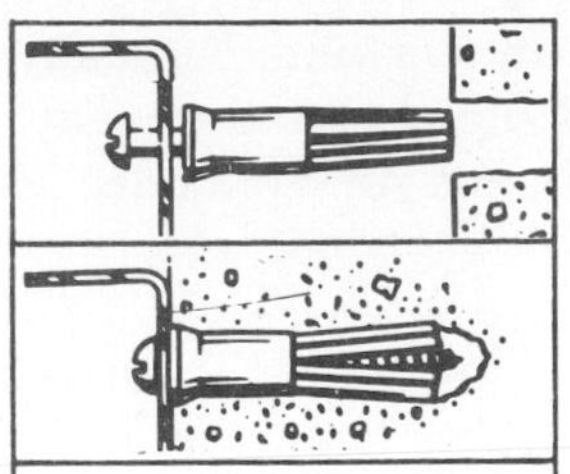

9–7. Installation of a wood screw anchor for quick masonry installation. It is made of lead and will expand without turning in soft material. Used for lightweight fixtures.

it will expand without turning in soft material. It withstands extreme changes in temperature and is not affected by moisture or chemicals.

■ *Plastic anchor.* These anchors are designed to be used with either wood or sheet metal screws. They are quickly installed in a drilled hole. Fig. 9-8.

■ *Pin-grip, hammer-driven blind rivets.* These rivets eliminate the need for explosives, special tools, or special skills. The aluminum alloy body is assembled with a stainless steel, knurled drive pin. When the pin is driven flush with the head, the slotted shank expands and grips firmly. Simultaneously, the knurled pin

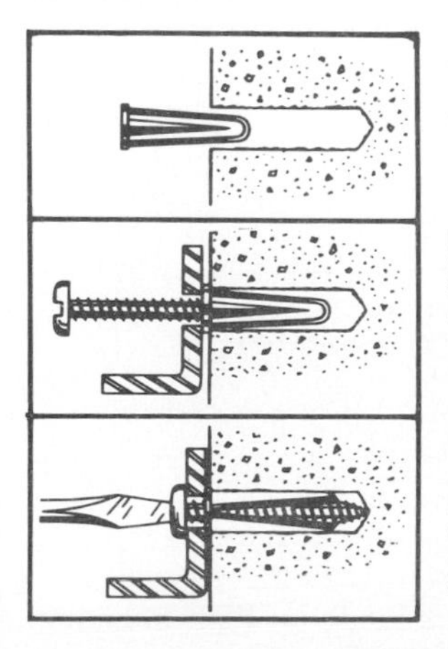

9–8. Installation of plastic anchors designed for use with either wood or sheet metal screws.

locks itself in position. Available in various lengths and diameters, they may be used for mounting boxes in concrete, as shown in Fig. 8-2.

Anchoring Tools and Devices

There are some heavy-duty applications for anchors that call for special tools to anchor the item to wood or concrete.

■ *Power tools.* The firing tool shown in Fig. 9-9 is adaptable for light to heavy duty. It can fasten wood (up to 2″ × 4″ size) or steel plates to concrete or steel. An adjustable guard allows firing from the center, corners, or sides. It uses .22 calibre shells to fire the pin in place. In some cases a .38 calibre cartridge is used. A number of pins are available for attaching electrical equipment to steel, concrete, or wood. Fig. 9-10.

■ *Hand-operated tools.* A hand-held pin and stud-setter holds fasteners true and straight and keeps them from buckling or bending as they are driven into wood, concrete, or other hard materials. Fig. 9-11.

Fixtures may be permanently, uniformly secured with a nail-type pin. Fig. 9-12. The stud with threads in Fig. 9-13 is just as tough, but leaves a threaded surface to attach equipment which may need to be removed or mounted at

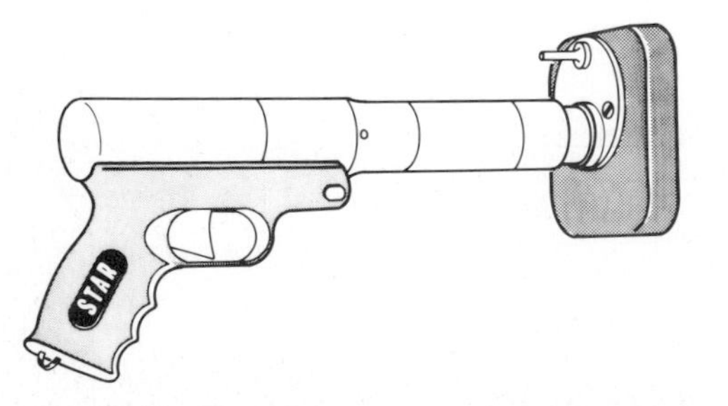

9–9. Universal power tool. Made for light to heavy duty, it can fasten wood up to 2″ × 4″ or steel plates to concrete or steel. Load with a .22-long cartridge and fire as you would a gun.

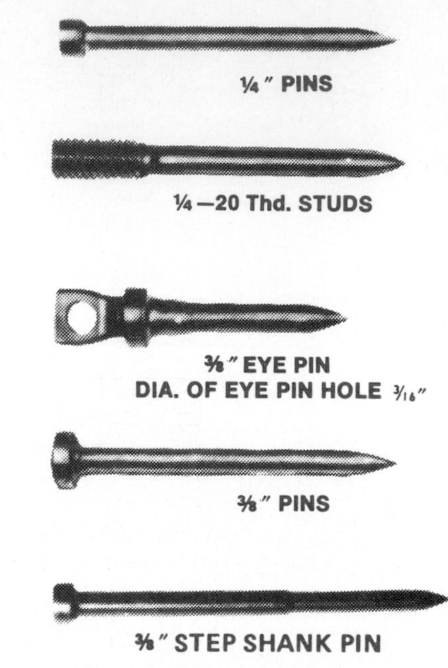

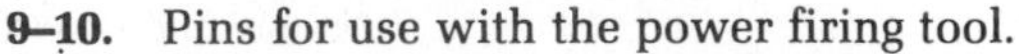

9–10. Pins for use with the power firing tool.

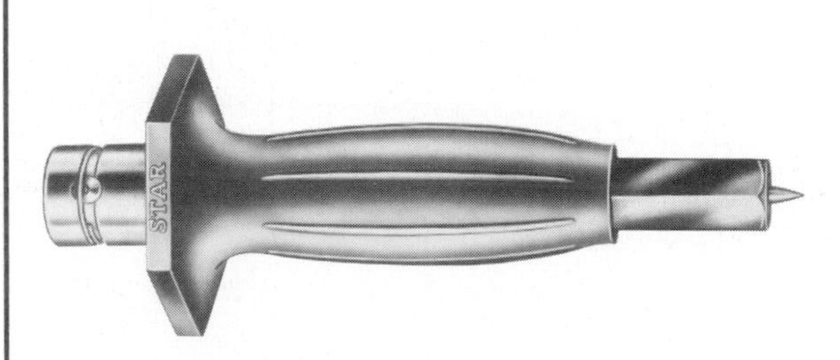

9–11. Hand-operated pin and stud-setter which holds fasteners true and straight and keeps them from buckling or bending as they are driven home. Insert the pin and strike the head with a hammer to fasten to concrete, masonry, or concrete blocks.

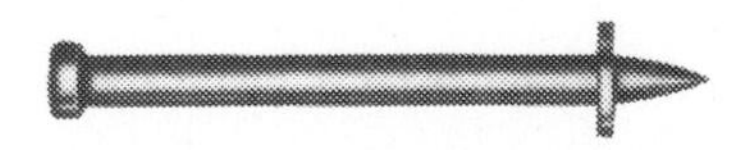

9–12. Pins for use with a hand-operated stud-setter.

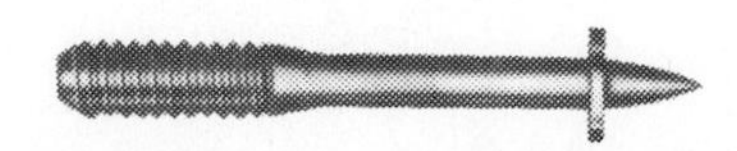

9–13. Studs for use with the hand-operated stud-setter.

a later date. Both these pins may be driven into concrete and through low-carbon steel, up to a thickness of $\frac{3}{16}$", without the aid of a drill. A hammer is used to hit the top of the setter to drive the pin into the material for mounting.

■ *Drill bits.* Carbide tipped drill bits are used to drill into concrete, masonry, or concrete blocks. Fig. 9-14. Notice the difference between this bit and the ones used for wood and metal. This one has a piece of carbide brazed into place at the tip of the bit to cut through the concrete. It is designed for use with all common types of rotary electric drills and hand braces.

■ *Wall screw anchors.* These are sometimes called Molly® bolts. They are used to secure almost anything to a hollow wall of plaster or drywall, wood or metal lath, wallboard, or cinder block. Just drill a hole and insert the anchor. Use a screwdriver to turn the screw until the anchor expands. Fig. 9-15.

■ *Toggle bolts.* The gravity-type toggle bolt has a toggle head which is pivoted off-center. It will fall (by gravity) into cross position when it is thrust through a hole, where the full length of the head then bears against the interior wall surface. Fig. 9-16.

The spring-wing toggle bolt is for use in hollow walls, hung ceilings, and horizontal surfaces where gravity-action toggles would be difficult to set. The winged head opens automatically by means of a spring within the head. All toggle threads are National Standard Thread, not stove-bolt threads. Fig. 9-17.

The spring-wing toggle bolt shown in Fig. 9-18 is also used similarly to that in Fig. 9-17. However, it has special tempered spring-steel wings which fold back against the bolt when passed through the hole. The spring then opens when the bolt is tightened. If you remove the screw after the unit has been inserted, however, it is possible to lose the spring-loaded wing, and another bolt (and entire unit) will have to be inserted.

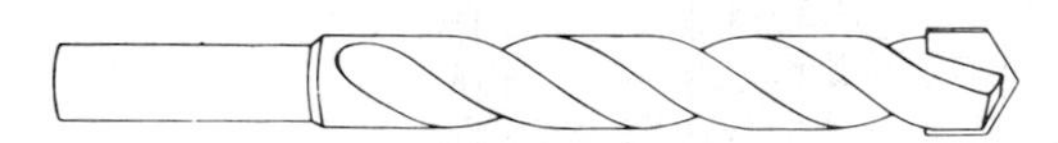

9-14. Rotary, carbide masonry drill.

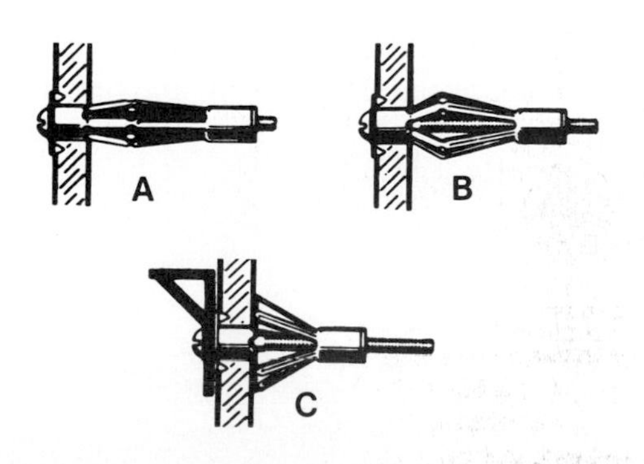

9-15. How to install hollow-wall screw anchors.

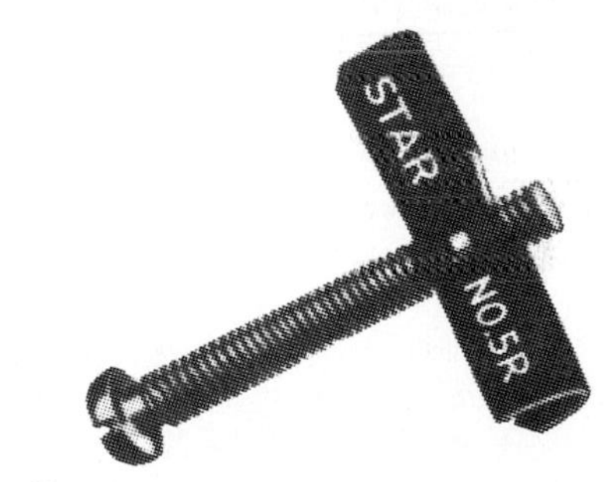

9-16. Gravity-type toggle bolt.

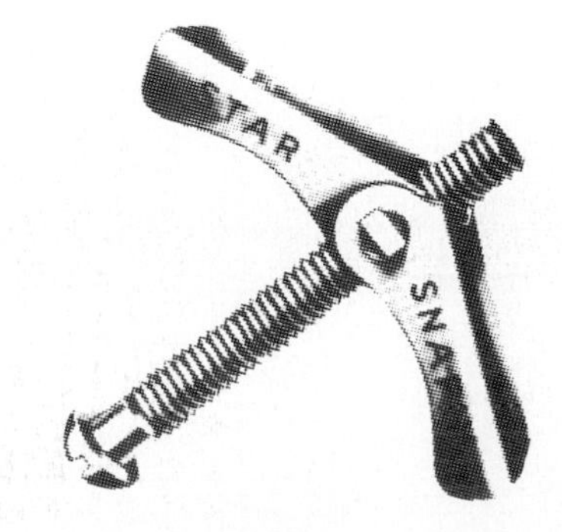

9-17. Spring-wing toggle bolt.

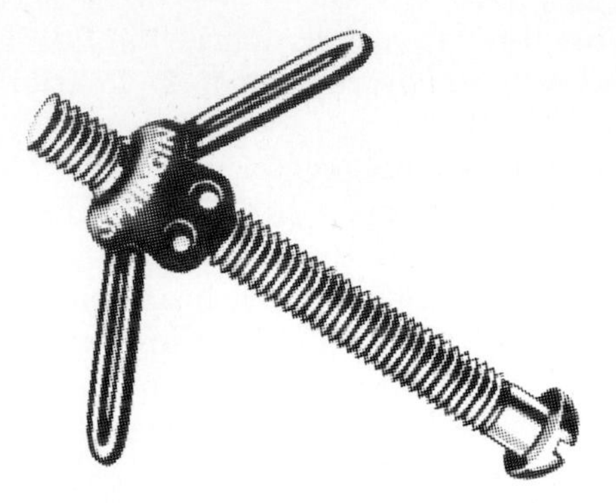

9–18. Another type of spring-wing toggle bolt.

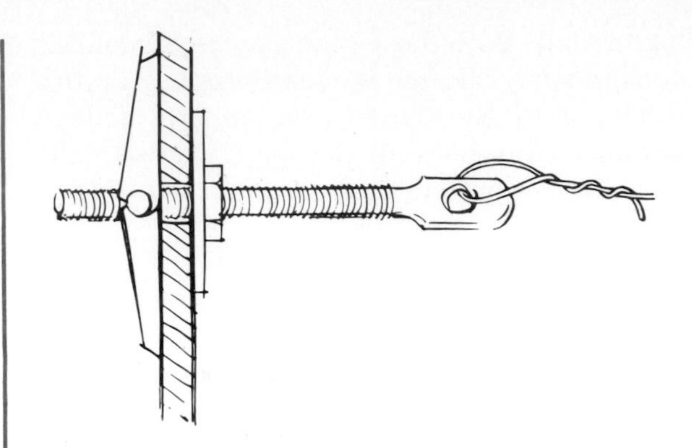

9–20. Tie-wire toggle bolt.

■ *Special toggles.* A time and money saver during remodeling is the giant-toggle fixture hanger. Fig. 9-19. It spreads the load over an 11″ length. It is suitable for fastening junction boxes or for hanging lighting fixtures. A $\frac{3}{8}$″ nipple fits through the knockout hole. A locknut then tightens the fixture in place. To remove it, just insert a screwdriver in the nipple and push the key up. Then pull down to collapse the wings and remove.

The tie-wire toggle bolt is another specially designed anchor for remodeling work. It comes complete with a spade end bolt. Several applications for this type of unit can be found in the installation of electrical fixtures, wiring, and equipment. Fig. 9-20.

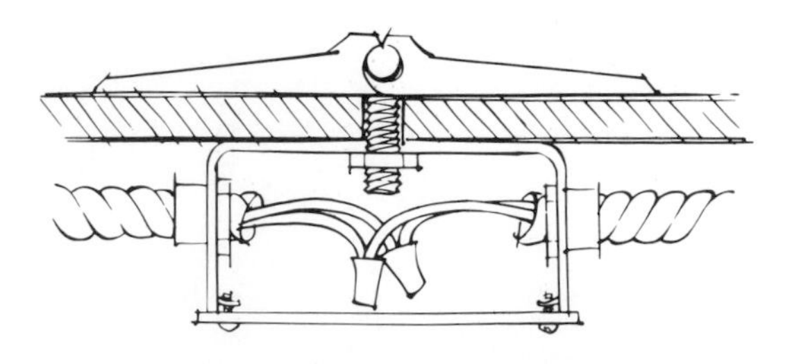

9–19. Giant-toggle fixture hanger. Note: ground wires are not shown here.

FLOOR BOXES

Conduit boxes designed for mounting in the floor are available in many sizes and configurations. Federal specifications ensure that boxes meet or exceed certain minimum requirements. Boxes are classified into two categories: shallow and normal. *Shallow* boxes are used where cement pour does not exceed 3″ in depth and where $\frac{1}{2}$″ and $\frac{3}{4}$″ conduit will be used. They are most commonly used in reinforced floors, or above grade where a two-pour system is used. *Normal* boxes are used in thicker cement pours (above 3″ thickness) and are usually found in "on grade" locations where $\frac{1}{2}$″ through 2″ conduit will be used.

■ *Type I conduit boxes.* These are specified as being adjustable, constructed so the cover plate may be adjusted vertically (not less than $\frac{3}{8}$″ from the plane of its base), and with an angular leveling adjustment of not less than 10 degrees for use after the box body is secured permanently. Fig. 9-21.

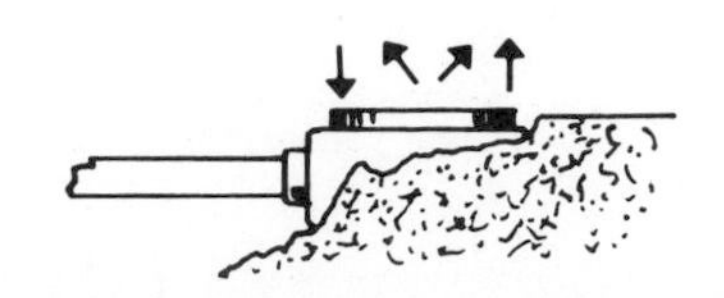

9–21. Type I conduit box.

■ *Type II conduit boxes.* These are semi-adjustable floor conduit boxes which are constructed with an adjustable cover plate (vertically, not less than $\frac{3}{8}''$ from the plane of its base). They should have an angular leveling adjustment of not less than 10 degrees for use at the time of installation. Fig. 9-22.

■ *Type III conduit boxes.* This type of conduit box is nonadjustable, with no means for angular or vertical adjustment of the cover plate. Fig. 9-23. These Type III boxes are further classified as Class 1, or *watertight*, and Class 2, or *concrete tight.*

Power and Communications Outlets

Power and communications outlets are designed to give the specifier flexibility in placing above-floor power or communication outlets. They are usually installed with a floor box, flush mounted, to handle modifications as needs change. Fig. 9-24 shows the type of outlet that may be used in floor boxes. It is, however, also possible to use duplex receptacles with floor boxes.

Fig. 9-25 shows the adjustment of a box to level it. This procedure eliminates the need to regrout boxes after their final leveling.

Fig. 9-26 shows the method of assembly for an outlet in a round cast box. Fig. 9-27 illustrates a duplex plate assembly in the box.

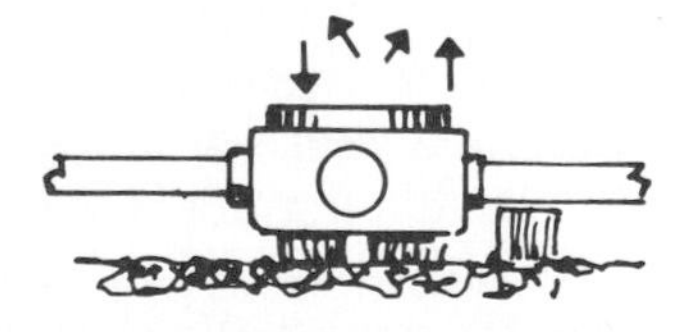

9–22. Type II conduit box.

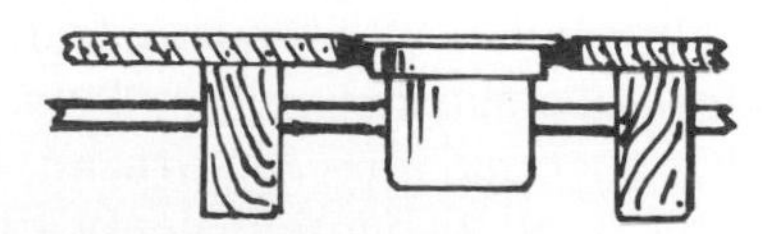

9–23. Type III conduit box.

9–24. Power outlet.

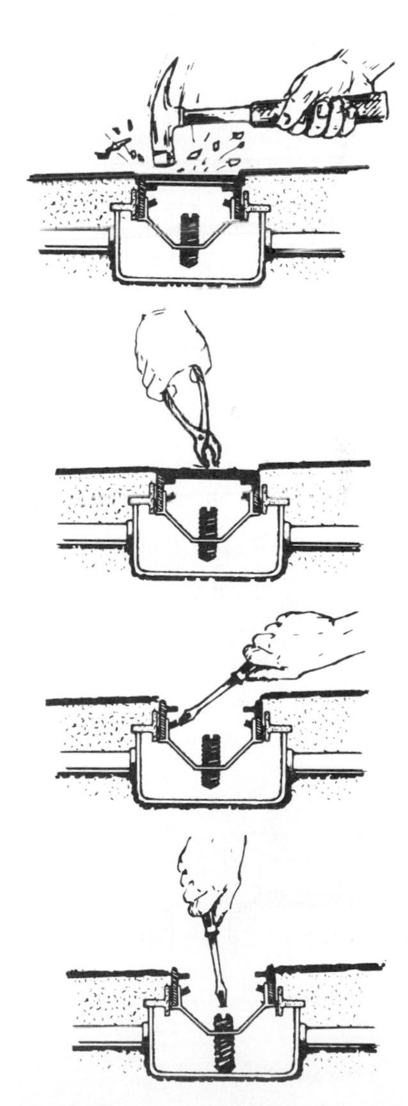

9–25. Internal adjustment method of making final leveling adjustments.

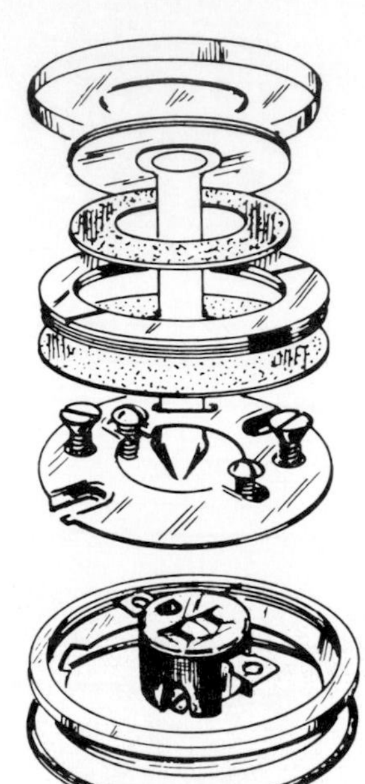

9–26. Floor plate assembly.

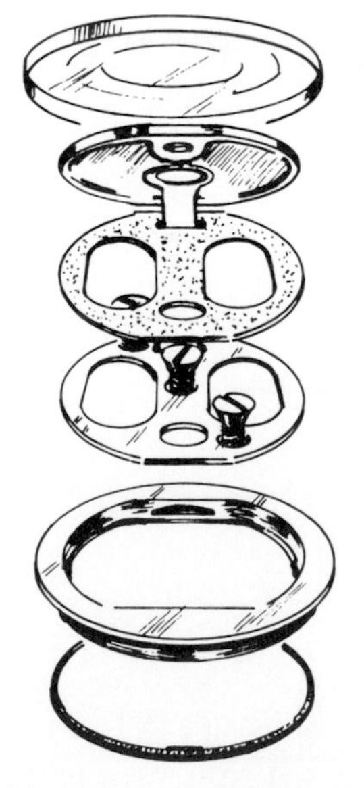

9–27. Floor plate assembly, duplex plate.

Three possibilities for mounting floor boxes to all types of underfloor systems are shown in Fig. 9-28.

These box installations are found in poured concrete floors, in laboratories, and in industrial plants. Some commercial jobs specify this type of equipment access to power on a set of blueprints, before the building is constructed.

Underfloor Raceways

Underfloor raceways may be installed as single- or multiple-run duct groupings with a wide range of available accessories. Connections to the ducts are made by means of headers extending across the cells. A header connects only to those cells that are to be used as raceways for conductors. Two or three separate headers, connecting to different sets of cells, may be used for different systems, such as light and power, signaling systems, and public telephones. Fig. 9-29.

Article 354 of the NEC covers underfloor raceways. They were developed to provide a practical means of bringing conductors for lighting, power, and signal systems to office desks and tables. They are also used in retail stores to make it easy to make connections for display-case lighting at almost any location. Fig. 9-30.

There are also systems for cellular metal decks. See NEC Article 356. Figs. 9-31 & 9-32 (page 254). This type of installation gives good mechanical protection for cables. Outlets are easily changed or rearranged. A clean, uncluttered look is obtained, and it provides office or store flexibility. By having the cables in metal trenches, it is possible to eliminate much of the electrical interference normally experienced in a large plant, store, or school.

POWER TOOLS

There are a number of power tools designed for use in special electrical applications. Industrial and commercial jobs require an electrician to know how to operate many different types of tools. For instance, the tool shown in

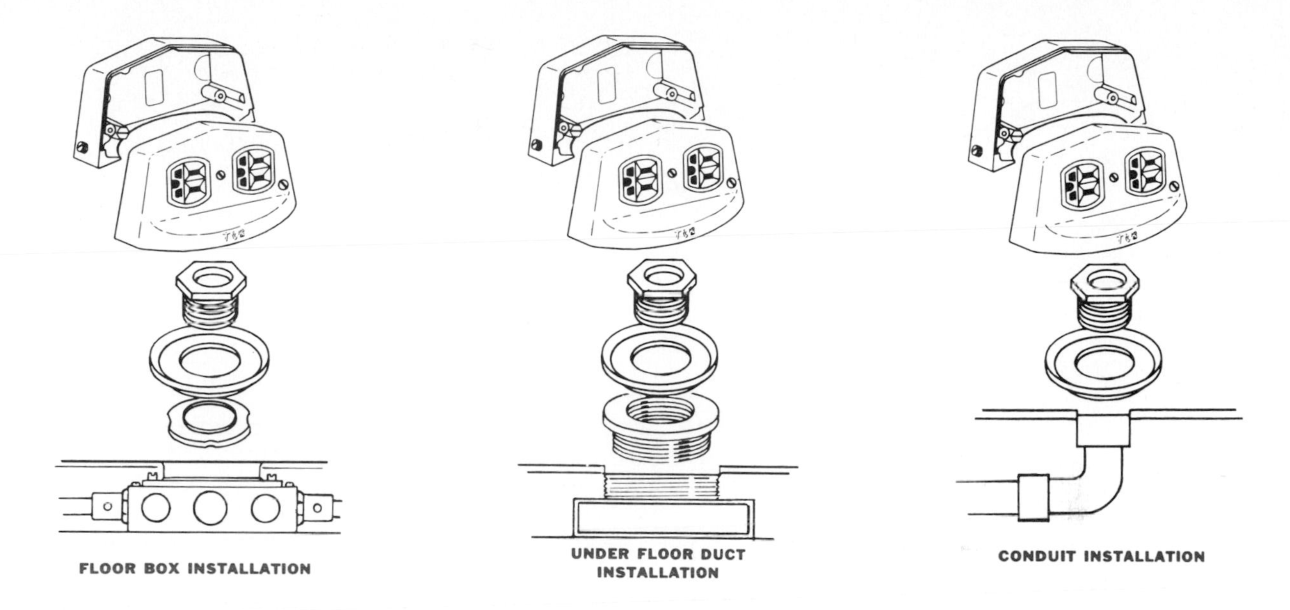

9–28. Mounting power box to floor system.

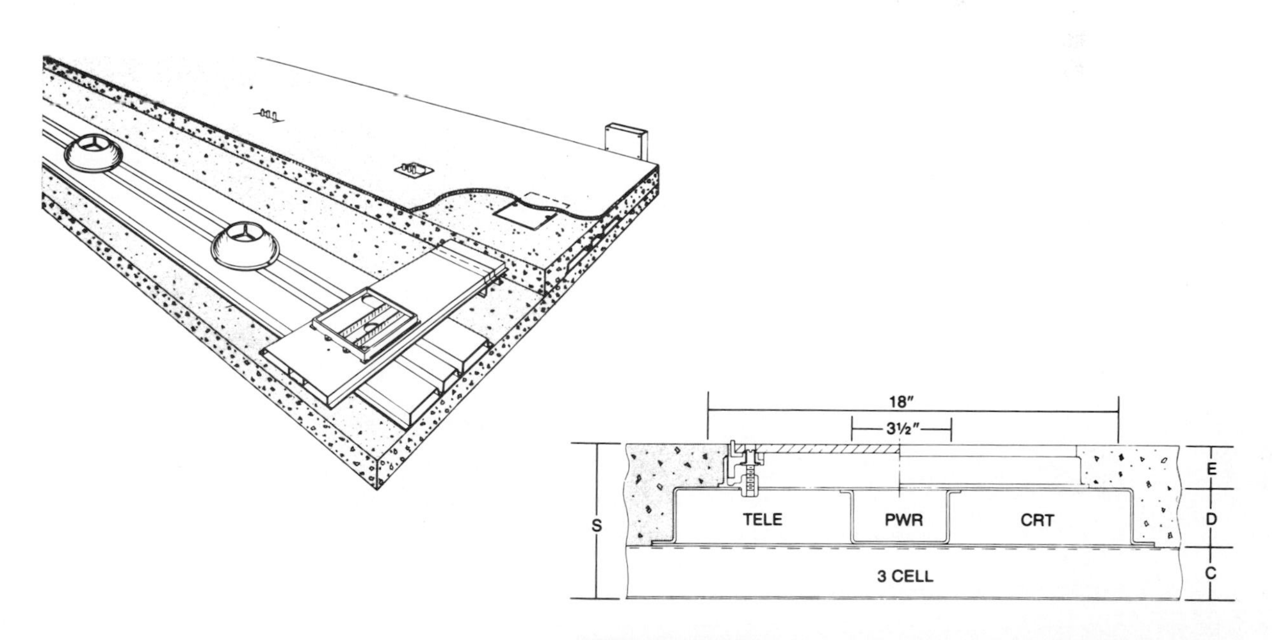

S SLAB	C CELL	D DUCT	E CONCRETE COVER	CAPACITY TELE	PWR	CRT
4½″	1½″	1⅝″	1⅜″	11.2 in.^{2}	5.4	11.2 in.^{2}
5″	2″	1⅜″	1⅜″	15.0	7.4 in.^{2}	15.0 in.^{2}

9–29a. Header with duct feed.

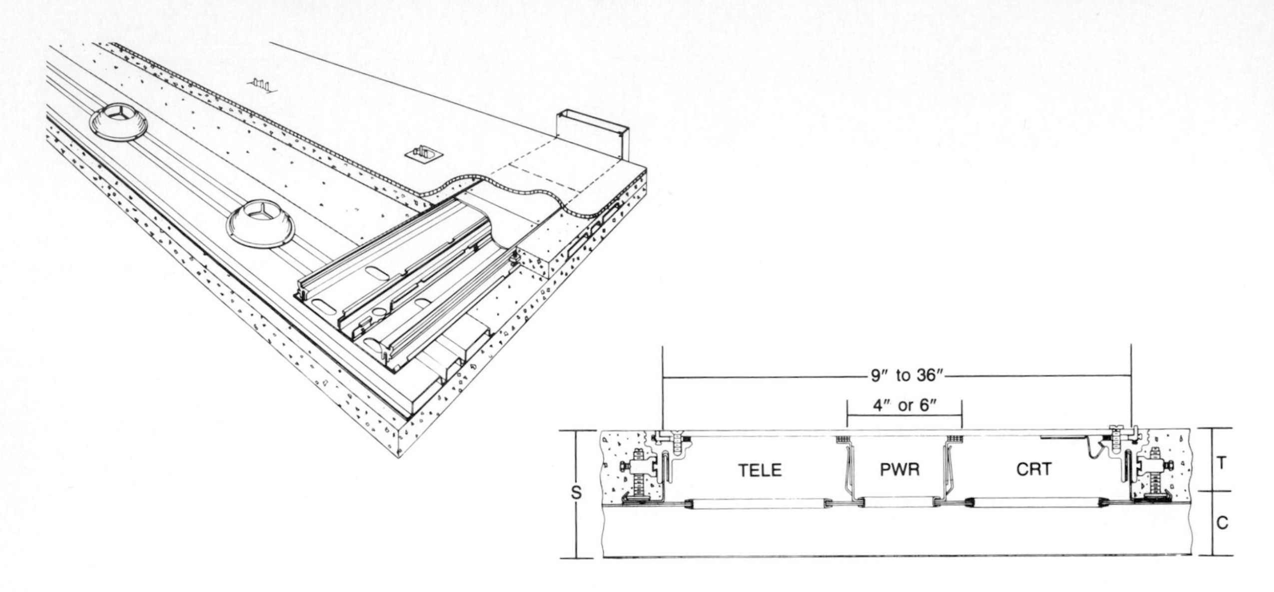

S SLAB	C CELL	T TRENCH	36″ WIDE TRENCH CAPACITY TELE	 PWR	 DATA
4″	2″	2″	27.2 in.^{2}	7.0 in.^{2}	27.2 in.^{2}
4½″	2″	2½″	34.9 in.^{2}	9.0 in.^{2}	34.9 in.^{2}

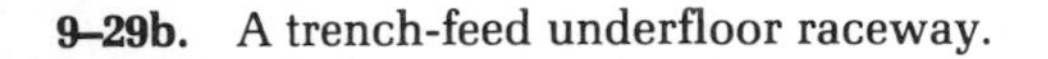

9–29b. A trench-feed underfloor raceway.

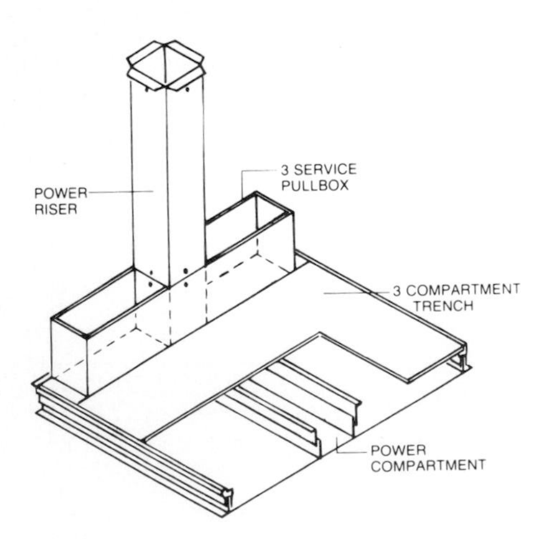

Preset Inserts

1. Available for 2″ to 3½″ concrete fills.
2. Compatable with all Conduflor electrical and telephone service fittings.

2″-3½″ Conc. Fill

1½″-2″ Cell

9–30. Trench wall ell is one of the accessories available to make it easier to wire the system.

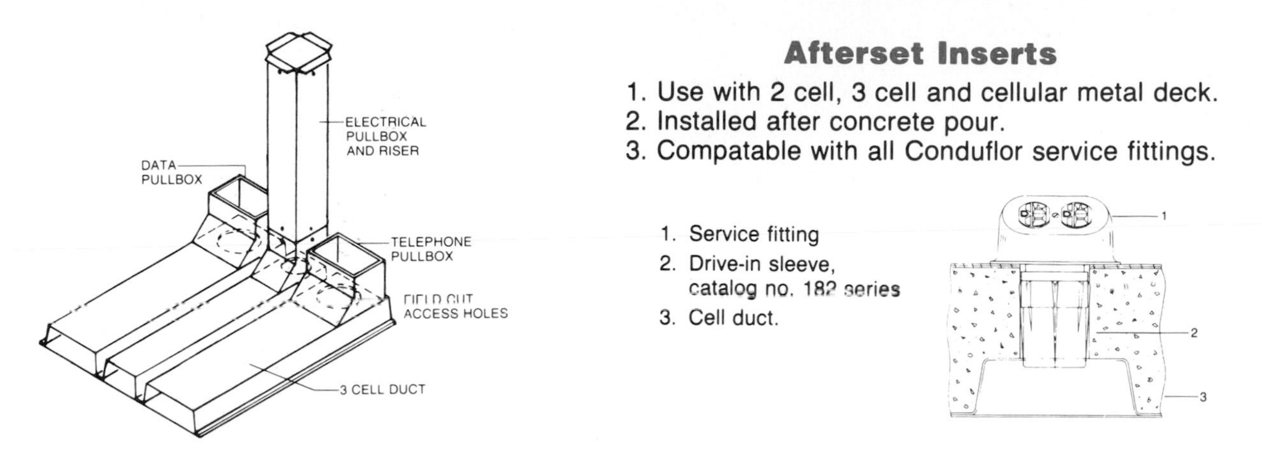

9–30. Trench wall ell is one of the accessories available to make it easier to wire the system. (Cont.)

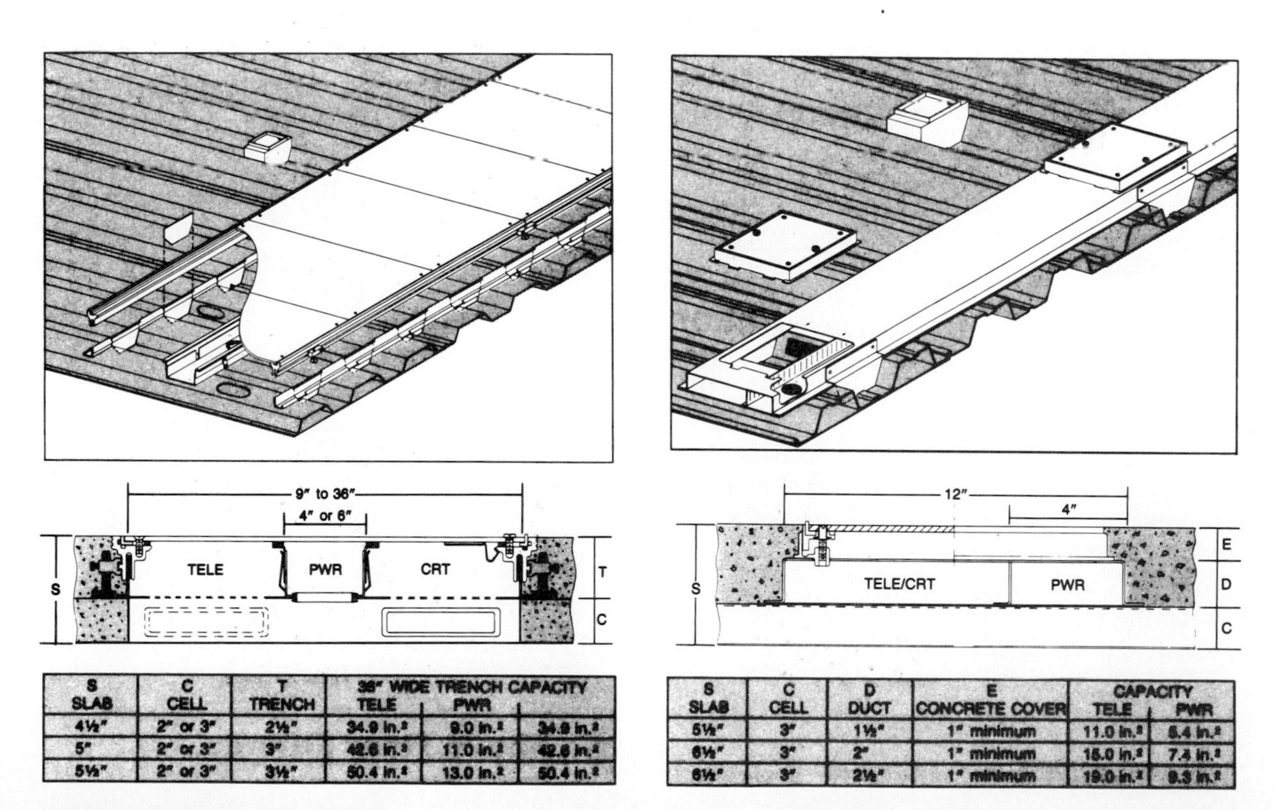

S SLAB	C CELL	T TRENCH	36″ WIDE TRENCH CAPACITY TELE	PWR	
4½″	2″ or 3″	2½″	34.9 in.²	9.0 in.²	34.9 in.²
5″	2″ or 3″	3″	42.6 in.²	11.0 in.²	42.6 in.²
5½″	2″ or 3″	3½″	50.4 in.²	13.0 in.²	50.4 in.²

9–31a. The bottomless trench.

S SLAB	C CELL	D DUCT	E CONCRETE COVER	CAPACITY TELE	PWR
5½″	3″	1½″	1″ minimum	11.0 in.²	5.4 in.²
6½″	3″	2″	1″ minimum	15.0 in.²	7.4 in.²
6½″	3″	2½″	1″ minimum	19.0 in.²	9.3 in.²

9–31b. The header duct.

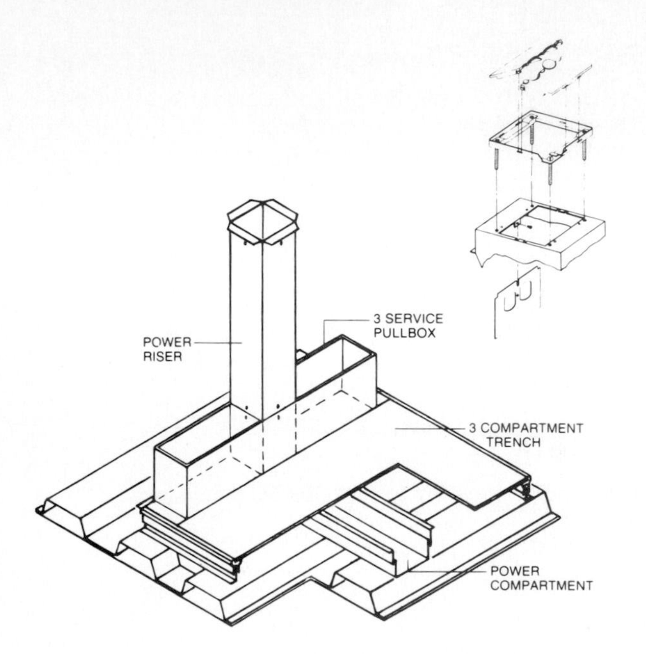

9-32. Trench wall ell for the cellular metal deck.

Fig. 9-33 is an open-yoke power installing tool which makes it possible to install an aluminum conductor in a conventional copper connector. Fig. 9-34.

Connectors

■ *Cutting-tooth connector*. The design features of cutting-tooth connectors are significant when connecting aluminum magnet wire to copper. When thermal expansion occurs, the aluminum expands into small pockets at the base of the pyramid-shaped teeth. During thermal stressing, the expanding conductor metal occupies these pockets instead of deforming the body of the connector or the wire itself. When the conductor cools and the wire contracts, the metal in the pockets will contract to normal size without disturbing the electrical contact between the wire and the teeth. The staggered pattern of the teeth and their pyramid-like shape reduces cold flow, or creep, by distributing the compression force within the conductor.

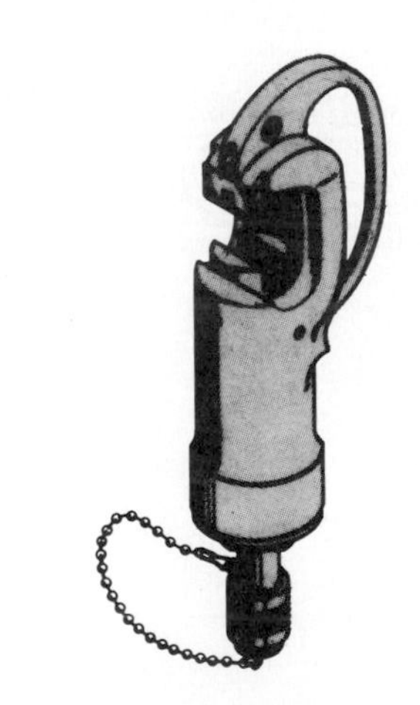

9-33. Open-yoke power installing tool for aluminum connectors.

Although high-temperature insulations for magnet wires have solved many problems for the electric motor or transformer manufacturer, they have created new problems in the splicing and terminating of these same wires. The durability of the magnet-wire insulation

Aluminum conductor installed in conventional copper connector

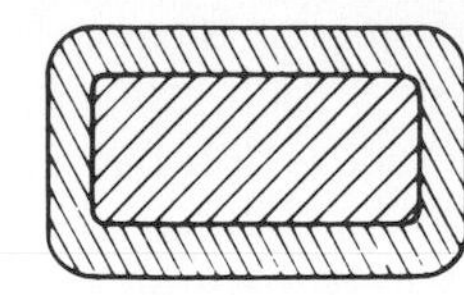

Before heating

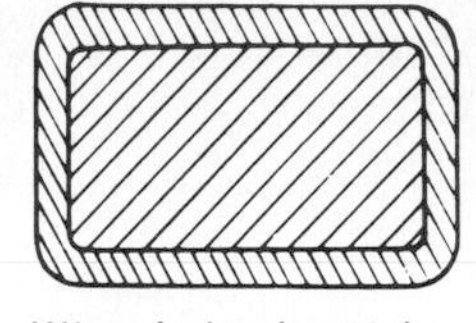

When being heated the copper connector is stressed higher and some of the conductor material creeps out.

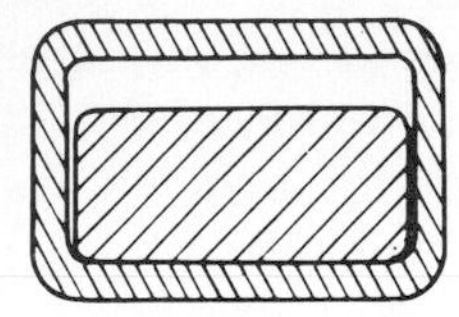

After heating, a space between the connector and the conductor is generated to destroy the contact.

Aluminum conductor installed in cutting tooth connector

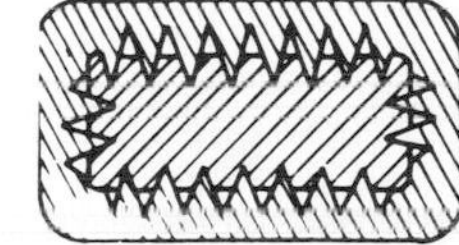

Before heating

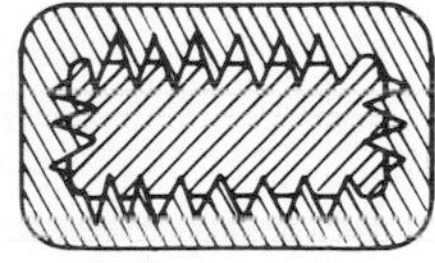

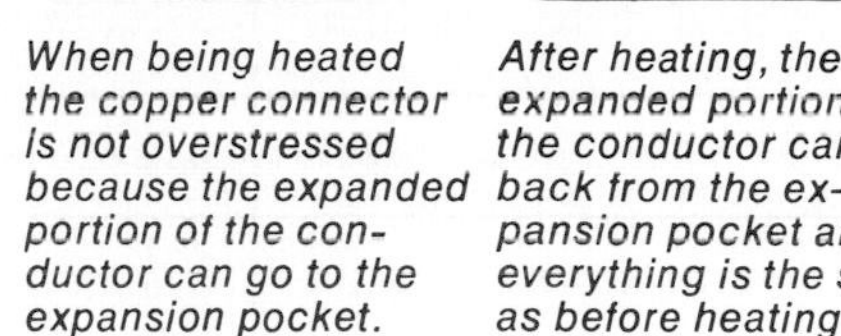

When being heated the copper connector is not overstressed because the expanded portion of the conductor can go to the expansion pocket.

After heating, the expanded portion of the conductor came back from the expansion pocket and everything is the same as before heating.

Cutting tooth connector transform the perpendicular compression force which would normally contribute to conductor creep into distributive forces that effectively resist cold flow.

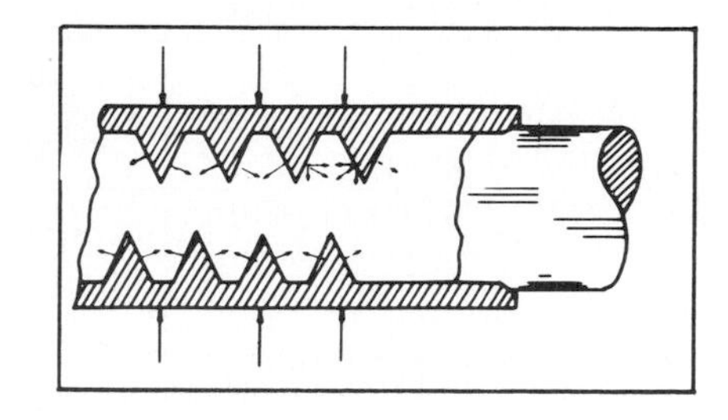

9–34. Aluminum conductor installed in a conventional copper connector, and in a cutting-tooth connector.

has made dip-soldering or brazing almost impossible without first stripping the insulation. Temperatures high enough to remove the insulation are almost equal to the melt point of the wire if the conductor is aluminum.

A further complication of the problem is the fact that much aluminum, as well as copper wire, is now used for magnet-wire applications. Aluminum magnet wire, which must be connected to copper, causes problems because of different coefficients of thermal expansion, galvanic corrosion, cold flow, and oxide film on the surface of the metals. For such problems, a cutting-tooth type of connector, applied to the wire with a power tool, is used. Fig. 9-34.

The cutting-tooth connector is a one-piece, copper alloy, tin-plated connector with a large number of pyramid-like teeth on the inner surface. When the connector is compressed onto an insulated magnet wire, the sharp, hardened edges of the teeth cut through the insulation and into the conductor metal.

■ *Self-riveting connectors.* A self-riveting connector is a two-piece, copper alloy, tin-plated connector. It is composed of a rectangular outer shell with parallel accordion-shaped side walls, and an insert with integral, arched cutting blades. When the connector is installed, the wire is simply inserted through the open end of the connector and compressed with the proper tool. Fig. 9-35.

The arch-shaped cutting blades pierce through the film insulation and form a permanent low-resistance electrical connection in the wire metal. The thrust force of compression causes the blades to cut through the insulation and literally rivet the connector to the magnet wire. Fig. 9-36.

Shape and flexibility of the cutting blades are responsible for overcoming thermal expansion and cold flow of aluminum wire. Tin plating counteracts galvanic corrosion, and the deep cut of the blades into the virgin conductor metal overcomes the threat of oxide-film buildup. Fig. 9-37 shows a hand tool needed to exert the correct force in the proper direction to ensure that the self-riveting connector is correctly applied.

Other special equipment may later become your standard tools as you become more specialized in the area of electrical work in which you are interested.

KILOWATT-HOUR METERS

The purpose of a watt-hour meter is to measure the electric energy transferred from one point to another by means of an electrical circuit. The electrical circuit may comprise two or more conductors, and the source may supply the energy in several different forms. Therefore the watt-hour meter is also required in several different forms.

The rate at which energy is transferred from the source to the load is called power. It depends upon the voltage across the line and the current flowing through it. In a direct current (DC) system the power, in terms of watts, is the product of the line potential, measured in volts, multiplied by the current, measured in amperes. The power, in watts, multiplied by the time, in hours, is the energy delivered to

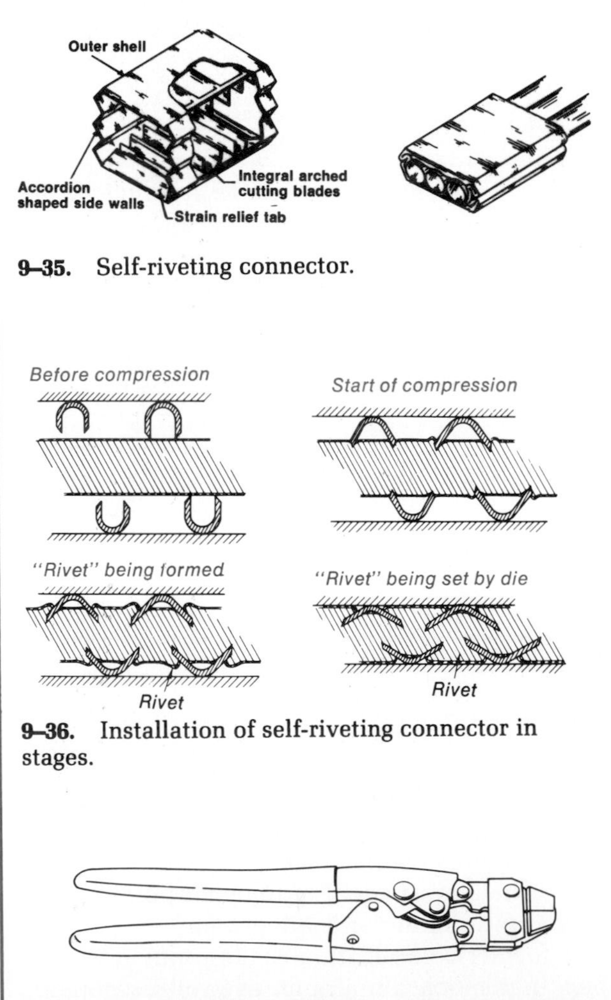

9-35. Self-riveting connector.

9-36. Installation of self-riveting connector in stages.

9-37. Hand tool for installation of the connectors.

the load during that time and is measured in watt-hours. A more common unit for expressing electric energy is the kilowatt-hour, which is 1000 watt-hours.

When an alternating voltage is impressed on a resistance load, the resulting current also alternates in polarity at 60 Hz and reaches its maximum value in each direction at exactly the same time as does the voltage. When this occurs, the current (I) and the voltage (E) are said to be in phase, and power (P) at any instant is the product of the current and voltage at that time.

However, if the load is not purely resistive, but includes some inductive load such as a motor, or a capacitive load, the current and voltage usually will not be in phase. For an inductive load, for example, the maximum value of current occurs somewhat after the corresponding maximum voltage Fig. 9-38.

Power, at any instant, is the product of the voltage and current at that instant. But because of the difference in phase between current and voltage, power will be negative during part of the cycle. The average power, therefore, will be less than in a resistive load circuit. Accordingly, the average power will not be simply the product of the voltage and current, but will be the product of these quantities multiplied by some other number, always equal to or less than 1, which is called the power factor (PF). In this case, the power is equal to the product of the voltage and current, times the power factor, or $P = E \times I \times PF$.

The time it takes the voltage or the current to go from one maximum value to its next maximum value in the same direction, $\frac{1}{60}$th of a second, is called one cycle, or hertz (Hz), or 360 electrical degrees. In Fig. 9-38a, the current lags 60 electrical degrees behind the voltage. If the current and voltage have the sine wave, or *sinusoidal*, shape as shown in any AC waveform for 60 hertz power, the power factor (PF) can be shown to be equal to the cosine of this angle by which current lags voltage. In any right triangle the ratio of one side to the length of the hypotenuse is called the cosine of the angle between them. Fig. 9-38b.

The value of the cosine of 60° (written cos 60°) is shown in a trigonometry table as 0.50; so the power factor for the example in Fig. 9-34

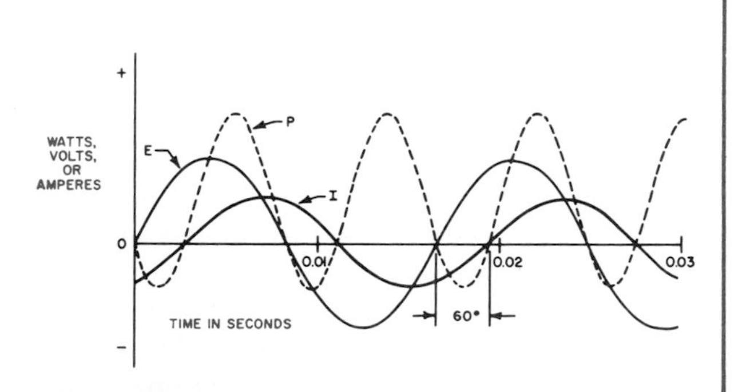

9–38a. Relationship among watts, volts, and amperes.

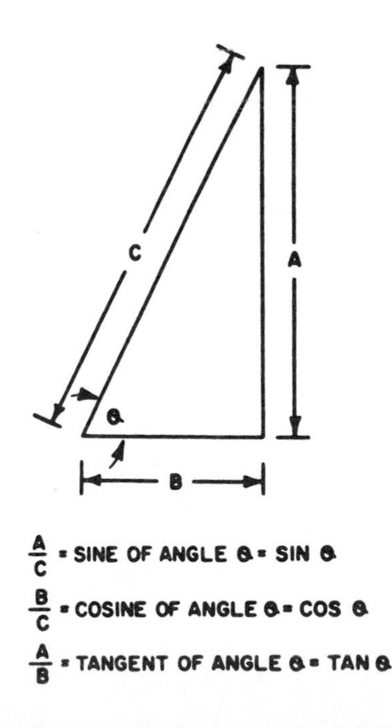

9–38b. Right triangle showing sides labeled, and sine, cosine, and tangent relationships.

would be 0.50, or, as it is frequently expressed, 50%. The general expression for AC power, then, is: $P = E \times I \times \cos\theta$. This relationship can be shown in *vectorial*, or *phasor*, form. Fig. 9-39.

In the process of measuring the transfer of electric energy, a watt-hour meter measures power. And, basing your knowledge on the previous discussion, you know that to do so it must somehow measure both the voltage and the current and multiply these quantities, taking into account the power factor.

A single-phase, 2-wire watt-hour meter—to take a simple case—has one coil connected across the line voltage (to measure the voltage) and another in series with the line to measure the current. These coils, mounted on a common magnetic structure, constitute a stator, which is like the stator of a motor. It drives an aluminum disc which is pivoted to rotate in an air gap.

In a properly designed and adjusted watt-hour meter, the *torque* produced to turn the disc is proportional to $E \times I \times \cos\theta$, or to the power measured. The disc turns through the air gap of a permanent magnet, which exerts a retarding, or damping, torque on the disc. This torque is proportionate to the speed of the disc. The disc therefore turns at a rate that is proportional to the power measured. The distance it rotates—the number of revolutions it turns—in any period of time is a measure of the energy transferred during that period.

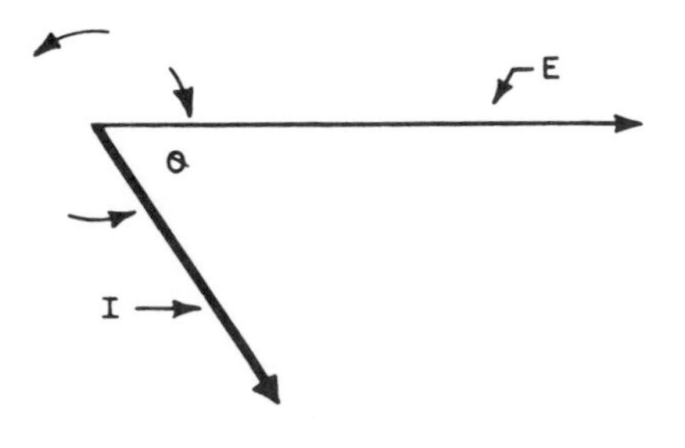

9–39. Vector.

Polyphase Meters

The use of more than two wires, for a circuit delivering electrical energy, offers advantages in economy and in the ways in which the energy can be utilized. As a result, several types of circuits using more than two conductors have been developed and are in common use. These may be either single-phase or polyphase.

In a single-phase system having more than two wires, the voltage between any pair of wires will be substantially in phase with the voltage between any other pair of wires. Voltages added in series may be added numerically (taking into account polarity). In the single-phase, 3-wire circuit shown in Fig. 9-40. For example, if the line-to-neutral voltages are each 120 volts, the line-to-line voltage is 240 volts.

In a polyphase system (*poly* means *many*), such voltages usually are not in phase. In a 3-phase system (by far the most common polyphase system), three voltages are generated which are equal in magnitude, but 120° apart in phase. Plotted against time, these voltages would look like (a) in Fig. 9-41. They would be represented simply, however, by phasors as shown in (b) of Fig. 9-41. The length of the phasor represents the magnitude of the voltage; and its direction, in respect to some reference, represents its phase. The phasors are considered as rotating counterclockwise, so that $E_{0\text{-}2}$ (for example) is equal in magnitude to $E_{0\text{-}1}$ but lags behind it in phase by 120°.

The use of double subscripts for the voltage is frequently helpful in keeping directions straight. Voltage $E_{0\text{-}2}$ means that the voltage is in such direction that current resulting from it will flow from point 0 to point 2 inside the source or generator. Similarly, at the load, where this same current would flow from line 2 to line 0, the voltage across the load would be called $E_{2\text{-}0}$.

Voltages (or currents) which are not in phase do not add algebraically but must be added vectorially. In Fig. 9-42(a) two voltages are shown 60° out of phase, plotted against time. If these voltages are combined in series, their

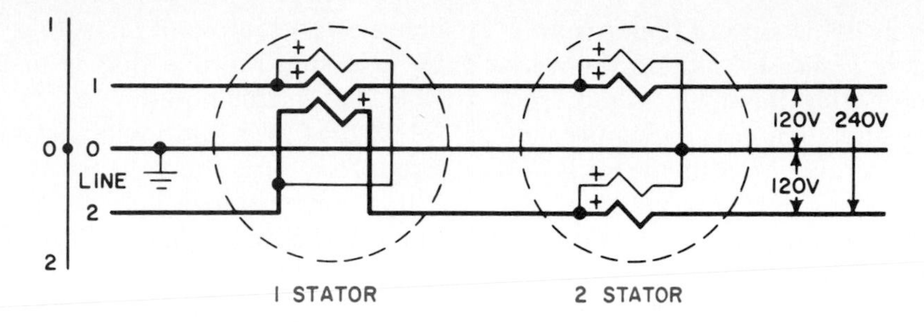

9–40. Single-phase, 3-wire, 240-volt meter hookup. (Jagged line represents a coil.)

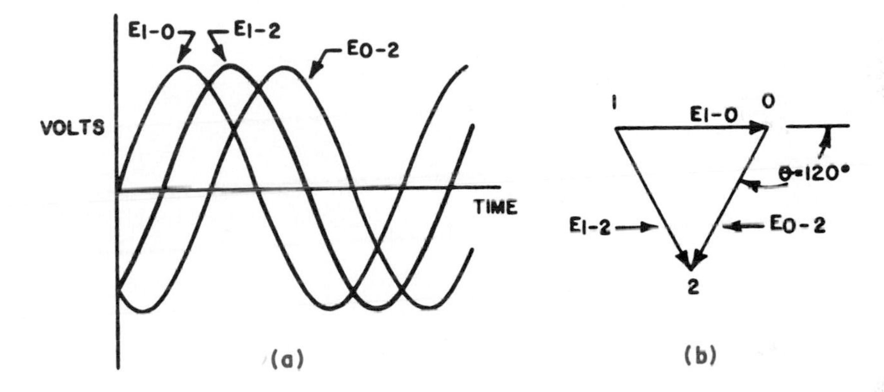

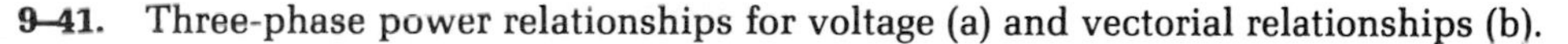

9–41. Three-phase power relationships for voltage (a) and vectorial relationships (b).

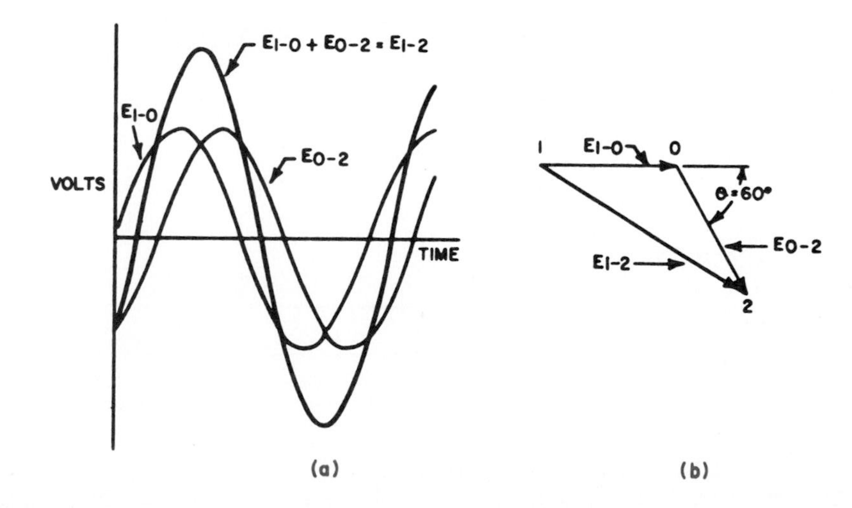

9–42. Three voltages, 60° apart (a), and vectorial relationships (b).

sum at any instant is the sum of their instantaneous voltages, and may be plotted point-by-point. It is much easier, however, to represent them as phasors and then add by simply connecting them together as shown in (b) of Fig. 9-42.

The sum of the two voltages, E_{1-2}, is seen to be for this particular example, $\sqrt{3}$ times as great as the individual voltages and to lag 30° behind E_{2-1}.

Fig. 9-43 shows the addition of two Voltages 120° apart; their sum is a voltage of the same magnitude (but halfway between them in phase). Subtraction of phasors is accomplished by reversing the one that is to be subtracted. Note that in this notation, $E_{1-2} = E_{2-1}$.

Fig. 9-44 is a three-wire network being metered. Note how the voltages are distributed.

In a conventional 3-phase, 3-wire system, all line voltages are equal in magnitude, usually at 240 volts or 480 volts; or, for primary metering, at higher distribution voltages. A 2-stator meter is required, which is accurate regardless of voltage unbalance. Such a system, usually employed for large commercial or industrial loads, may or may not be grounded. If one line

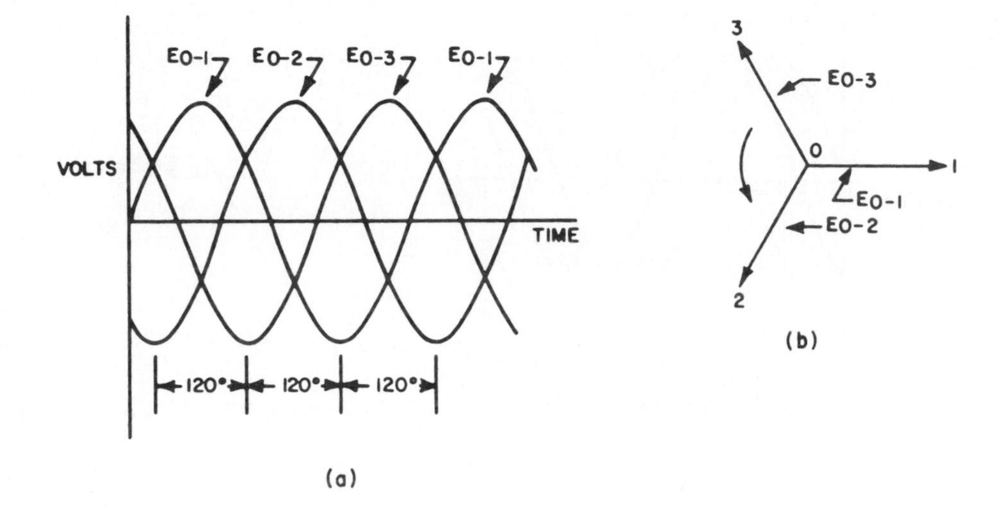

9–43. Addition of two 120° voltages (a), and vectorial representation (b).

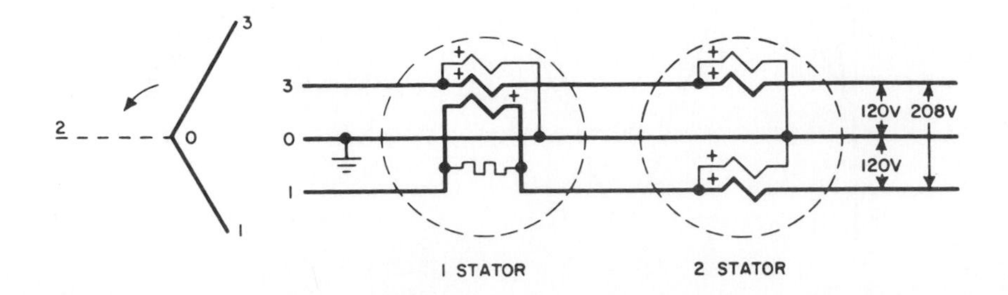

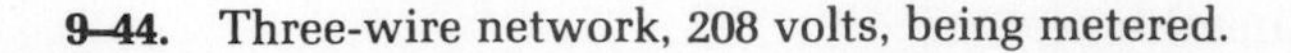

9–44. Three-wire network, 208 volts, being metered.

is grounded, the current coils must, of course, be in series with the ungrounded lines. Fig. 9-45.

If a center tap is brought out from one leg of a 3-phase delta, the system becomes a 3-phase, 4-wire delta, such as that frequently used to supply polyphase power at 240 volts for motors, and 120 volts, single-phase, for lighting. The 3-stator meter (Fig. 9-46) is rarely used, partly because of the complicated testing procedure. Instead, the 2-stator, 4-wire delta meter shown in Fig. 9-45 is almost universally used in this service.

Although a 4-wire delta system is normally used only for 240/120-volt service, the 3-phase (3Ø), 4-wire Y system covers a wide range of voltages from 208/120-volt service, up to the highest transmission voltages. Fig. 9-47. The neutral is usually grounded. It is measured with either a 3-stator or 2-stator meter, depending upon the importance of the load and the degree of accuracy required. Note the distribution of voltages between any two of the conductors.

For details of meter reading, with pictures of the different types of meters mentioned here, see Chapter 6.

Demand Meters

The cost of supplying electric service depends upon the amount of energy provided and upon the rate at which it is consumed. This factor is ultimately reflected in the amount of generating and distributing capacity which must be kept available. Therefore, in addition to a charge based on the amount of energy consumed, a charge is sometimes made on the basis of the maximum sustained rate at which energy is used. Such a charge, known as a "*demand charge*" because it is based on the

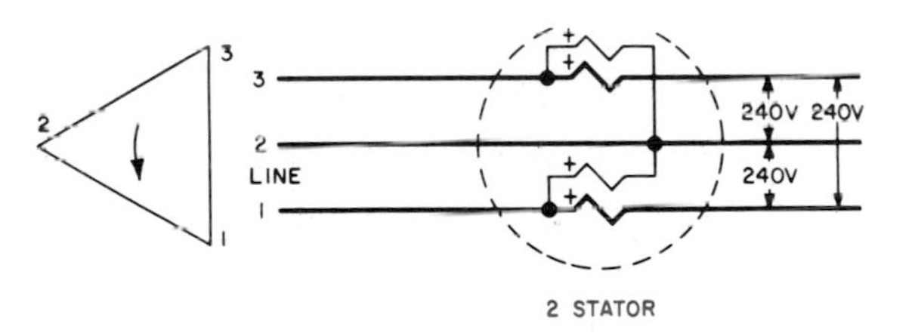

9–45. Three-phase, 3-wire, meter hookup, delta.

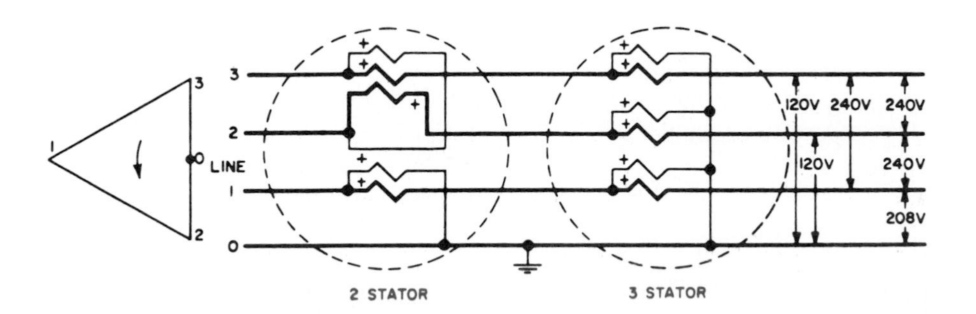

9–46. Three-stator meter; rarely used.

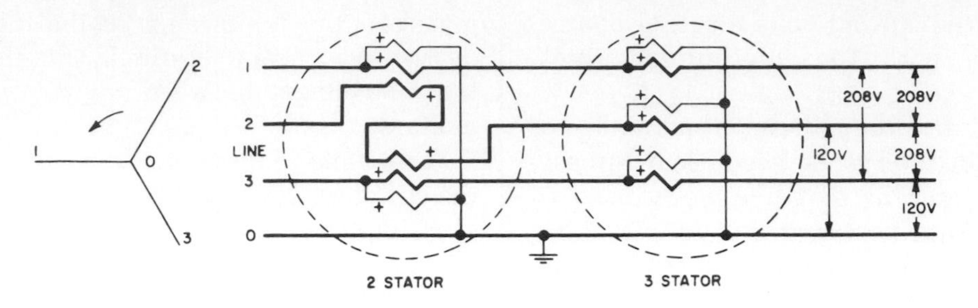

9-47. Four-wire wye system being metered.

demands made on the electric system by the customer, is determined by the maximum power, or rate at which energy is used, averaged over some predetermined period of time.

Like the conventional kilowatt-hour register which it replaces, the mechanical demand register provides a series of dials and pointers, driven through suitable gearing from the watt-hour meter, to record kilowatt-hours. In addition, from the same worm-wheel take-off staff, other gearing is provided to drive a demand pusher arm proportionally to the kilowatt-hour measured. This latter gearing, however, is arranged so that it may be momentarily disengaged (under the control of a timing motor and a reset mechanism) at the end of each demand interval, allowing the pusher arm to return to zero through the action of gravity.

During each demand interval, then, the pusher arm is advanced through an angle proportional to the kilowatt-hours measured during that interval. At the end of the interval the arm is reset to zero to start over again for the next demand interval. Its position just before reset is a measure of the kilowatt-hours measured during the preceding interval. Also, through a suitable combination of gearing and readout dial or dials, its reading provides a measure of the average power, or demand, for that interval.

If, for example, 12 kilowatt-hours were recorded by the watt-hour meter during a fifteen-minute interval, the position of the pusher arm should indicate a demand of 48 kilowatts (12 kilowatt-hours divided by $\frac{1}{4}$ hour) for the interval.

In establishing a new maximum demand, the pusher arm pushes ahead of it a maximum-demand indicator, which it leaves at the point of maximum advance. At each reading, the indicator is manually reset to zero by the meter reader after the maximum demand is read.

The relationship between the instantaneous power of the load, the pusher-arm position, and the maximum-demand reading is illustrated in Fig. 9-48. The solid line indicates the value of a hypothetical load plotted against time, and the area under this curve is proportional to energy. The position of the pusher arm for a 15-minute register is indicated by the dashed line, and the position of the maximum demand indicator by the dotted line. Wherever the dashed and dotted lines coincide, the pusher arm is in contact with and is pushing the maximum-demand indicator. Thus in the first interval, when the maximum-demand indicator is starting from zero, the pusher arm advances the maximum-demand reading to 5 kW, the average load for that interval. At the end of that interval the pusher arm returns to zero, but the maximum-demand reading remains at 5 kW. During the second and third intervals, because the demand does not exceed 5 kW, there is no further advance of the maximum-demand indication. In the fourth interval, however, due to an increase in load, the indicator is advanced to 15 kW. It remains there, in spite of a subsequent reduction in load, until late in the thirteenth interval when a new maximum demand of 16.7 kW is recorded.

Figs. 9-49 through 9-51 illustrate the types of kilowatt-hour meters currently in use. Check

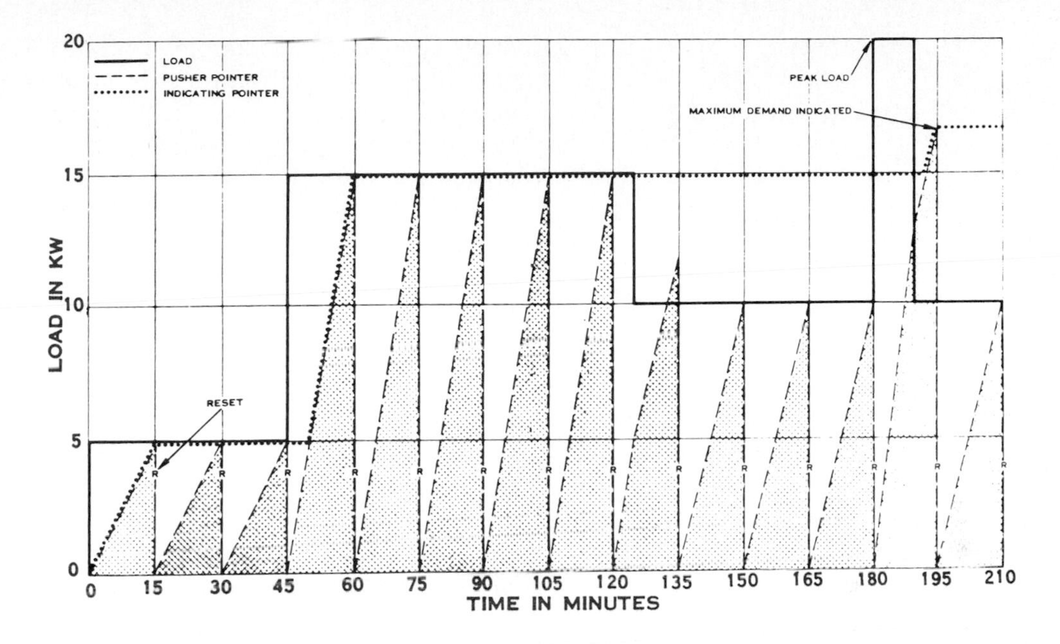

9–48. Relationships of a demand meter to load, pusher pointer, and indicating pointer.

9–49. Kilowatt-hour meter, 240 volts, 1-phase, 3-wire.

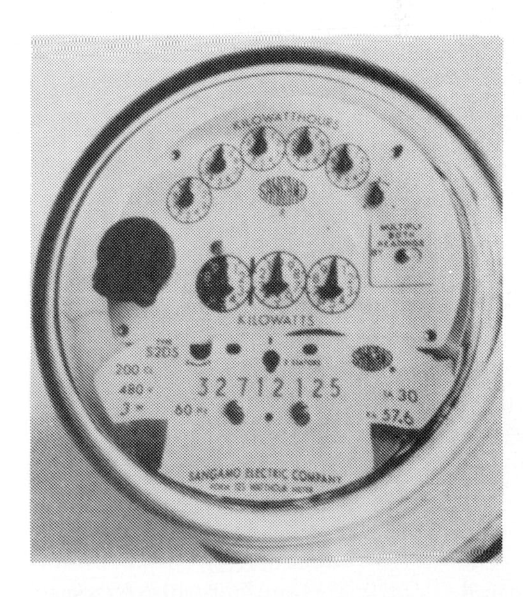

9–50. Kilowatt-hour meter, 480 volts. Note the two sets of dials.

9-51. Demand meter with pointer, 3-wire, 480 volts.

the captions under the pictures for identification and differences among meters. See if you can identify them with descriptions given in this discussion.

INDUSTRIAL CONTROLS

Industrial controls consist of nothing more than relays and switches. They are relays and switches designed for a specific job. Switches may come in the form of thermostats or pressure-sensing devices. Relays may have one set of contacts, or any number, even 40.

The electrician should become familiar with various manufacturers and what each has done to adapt a device to a special job. In this part of the chapter we shall introduce you to some of the devices and equipment associated with industrial control functions in a number of different fields.

Switches

To ensure the proper switch for each job, the National Electrical Manufacturers Association (NEMA) has designed different switches and relays for starting motors and for controlling other devices. For example, a NEMA Type 1, general-purpose station for turning a motor on and off is shown in Fig. 9-52. Note its construction and the type of box that encloses the switch for protection. A cast-metal box, die-cast of aluminum, will be specified as NEMA Type 4, and Type 7-9. The cover of the station in Fig. 9-52 contains the entire contact mechanism. Wiring terminals are located in the base.

Spring-type, silver-plated contacts connect the two assemblies. The operating buttons may be rotated 90° to make them suitable for horizontal mounting.

Limit Switches. *Limit switches* are used to automatically control production machine tools, and other industrial machines where the available mounting space is small, and where motion to operate the limit switch is measured in thousandths of an inch. Limit switches are especially valuable in working to very close tolerances. Precision-limit switches may be

9-52. Standard duty push-button switch.

used as self-contained units, or as several mounted in an assembly, to control the operating cycle of a machine tool. Fig. 9-53.

■ *Precision-limit switches.* These are of the single-pole, double-throw type with a common connection. They are designed to prevent "dead centering" and provide positive switching action, no matter how slowly the actuating force is applied.

9–53. Precision-limit switches.

There are two basic operating means for limit switches–the *roller lever* type and the *push* type. Contacts are either spring-return or maintained type. The contact in the spring-return type is restored to its initial position when the operating force is removed. Maintained type contacts will remain in the operated position after the operating force is removed. Pushing the external reset button restores the contact to its initial position.

■ *Sealed-contact, oil-tight limit switches.* These have a rugged contact hermetically sealed in a glass envelope and are, therefore, an ideal switch for environments containing dust, gas, or similar contaminants. Fig. 9-54.

9–54. Sealed-contact, oil-tight limit switch.

These limit switches may be equipped with any of six different operating levers: roller lever, adjustable roller lever, micrometer adjustment roller lever, rod lever, one-way rod, and fork lever. These may be used interchangeably on all lever-type switches except the low-operating force switch.

There are as many as one hundred different types of microswitches and limit switches for use in industrial control situations. A good way to learn some of their limitations and adaptations is to obtain a manufacturer's catalog and use it as a source of information.

Automatic Float Switches. *Automatic float switches* provide automatic control for motors operating tank or sump pumps. They are built in several styles and can be supplied with several types of accessories that provide rod or chain operation, either wall or floor mounting. Two pumps control the switch, which is a mechanical alternator, designed to automatically alternate the use of two pumps in a duplex system as peak loads demand. Fig. 9-55. Such a method of operation equalizes wear on both pumps.

Various applications of switches in use with automatic float controls are shown in Fig. 9-56.

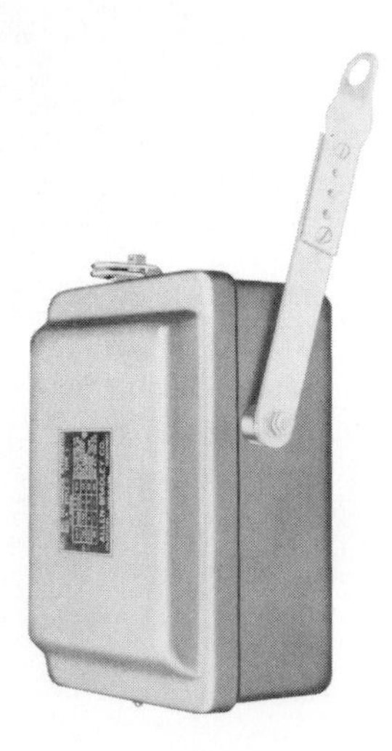

9-55. Automatic float switch.

AC Automatic Transfer Switch. *Automatic transfer switches* are used in industrial and commercial installations where it is essential to maintain continuous power to protect against process stoppages, panic conditions, or property damage. (See Chapter 14.)

The load is connected to either the normal or the emergency power supply by means of mechanically and electrically interlocked contactors. Undervoltage relays (one for single-phase, three for three-phase) respond to failure of the normal supply by automatically opening the normal supply contactor and closing the emergency supply contactor. The action is then reversed when normal power is restored. Fig. 9-57.

Full phase protection provides automatic transfer to the emergency supply when voltage fails on any of the phases or drops below 70% nominal, and the return to the normal supply will occur only when all phases are restored to 90% normal voltage. Fig. 9-58.

Relays

Relays are used for any situation that requires remote control. Keep in mind the voltage and current necessary for causing the solenoid to energize, the number and arrangement of the relay contacts, and the amount of current which can be safely handled with the contacts.

There are DC and AC relays. They are available to operate on DC at 6 volts, 12 volts, 24 volts, 32, 48, 64, 115-125, and 230-250 volts. Contact ratings are from 1 ampere at the higher end of the voltage range to 10 amperes for 64 volts and less. AC relays operate on the same voltages: 6, 12, 24, 32, 48, 64, 115-125, and 230-250 volts AC.

Fig. 9-59 is a NEMA Type G, or *general-purpose* relay. It has nonconvertible, fixed contacts. These relays are ideally suited for applications where space limitations require a smaller relay. They are available in 2-pole construction only.

The relay in Fig. 9-60 is a Type E and has fixed contacts like Type G, but in a wide vari-

Type	Wall Mounting	Floor Mounting
Rod Operated These accessories can be used with rods up to 9 feet long on the Style A and D float switches and up to 18 feet long on the Style B, C and E float switches. The float is fixed to one end of the rod. Adjustable stop collars at the top of the rod operate the float.	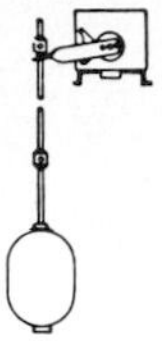 **WALL MOUNTING** Includes two 3 foot lengths of 3/8'' brass tubing with couplings, stop collars and copper float.	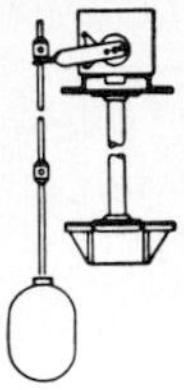 **FLOOR MOUNTING** Includes floor mounting bracket, 20'' length of 1'' pipe, mounting accessories, two 3 foot lengths of 3/8'' brass tubing with couplings, stop collars and copper float.
Long Rod — Free Float These accessories can be used with rods up to 33′ long. The double arm lever carries a counterweight to offset the weight of the rod. The float moves up and down between stops on the rod so that even though the liquid level varies greatly, the rod moves only a short distance. Top of rod is fixed to the switch lever.	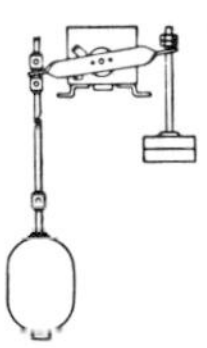 **WALL MOUNTING** Includes double arm lever for float switch, counterweight, two 3′ lengths of 3/8'' brass tubing with couplings, stop collars and copper float.	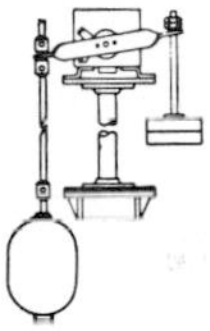 **FLOOR MOUNTING** Includes double operating arm for float switch, floor mounting bracket, 20'' length of 1'' pipe, mounting accessories, two 3′ lengths of 3/8'' brass tubing with couplings, stop collars and copper float.
Long Rod — Parallel Motion These accessories are used with rods up to 33′ long. The parallel lever arrangement keeps the rod vertical since a long rod might otherwise tend to move sideways. The float moves up and down between stops on the rod so that even though the liquid level varies greatly, the rod moves only a short distance. Top of rod is fixed to the switch lever.	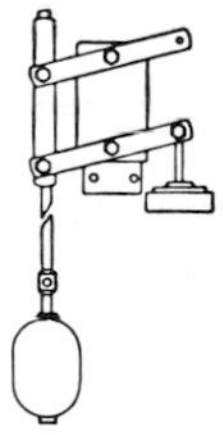 **WALL MOUNTING** Includes two double arms for float switch, counterweight, two 3′ lengths of 3/8'' brass tubing with couplings, stop collars and copper float.	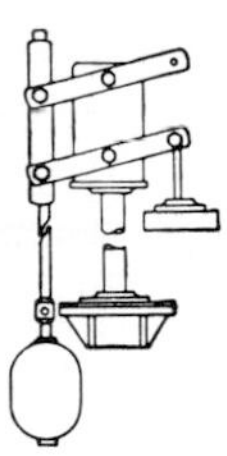 **FLOOR MOUNTING** Includes two double arm levers for float switch, counterweight floor mounting bracket, 20'' length of 1'' pipe, mounting accessories, two 3′ lengths of 3/8'' brass tubing, stop collars and copper float.

9–56. Types of automatic float switches.

9–57. Automatic transfer switch.

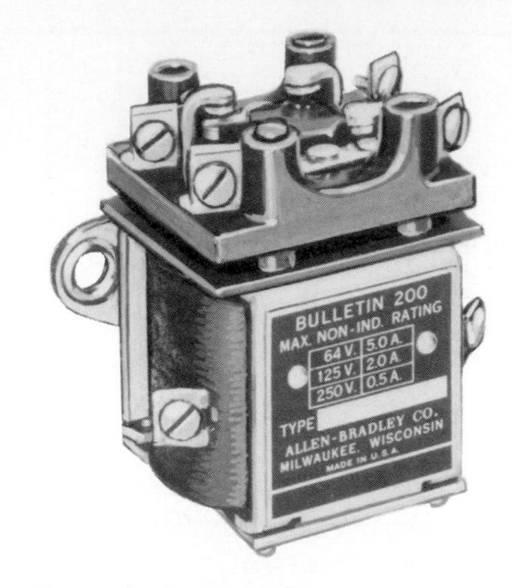

9–59. General-purpose relay.

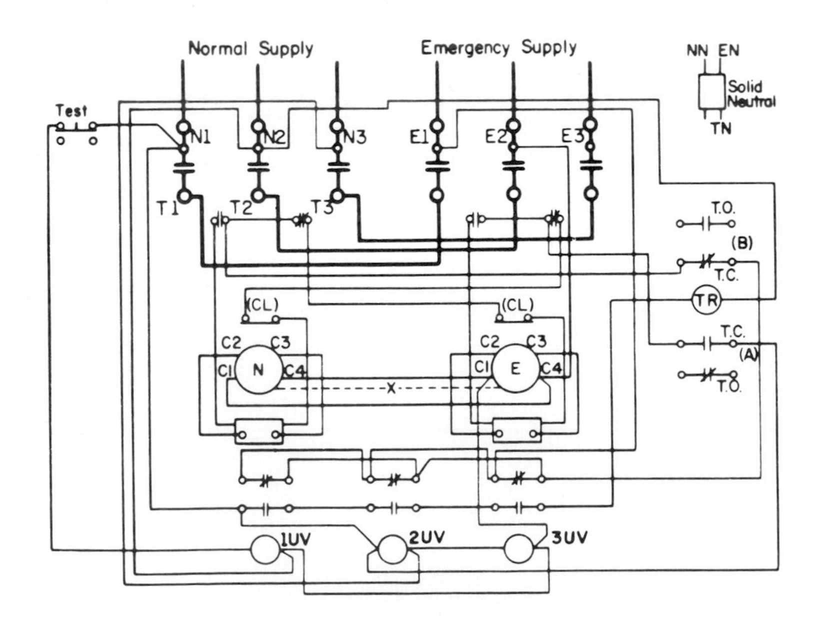

9–58. Three-phase, four-wire, solid neutral, automatic transfer switch, with time delay to emergency and normal supply transfer.

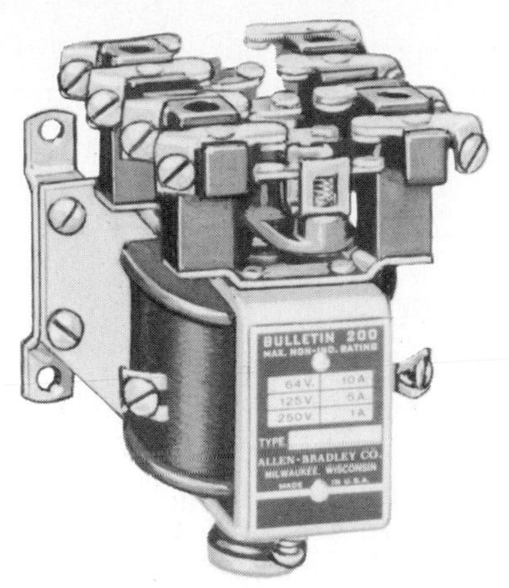

9–60. Relay.

ety of combinations. Type E has a voltage and current rating as follows (for contacts):

64 volts or less: 5 amperes
115-125 volts: 2 amperes
230-250 volts: 1 ampere

Note how the current rating for the contacts decreases as the voltage increases.

Type EX relays can have universal poles, with a maximum of 4. Fig. 9-61. Each universal pole may be used as a NO (normally open) or NC (normally closed) contact. Choice of contact action is simply a matter of making the proper wiring connections. Fig. 9-62.

A *thermostat relay* is shown in Fig. 9-63. Note the resistor across the top of the unit. The number of poles for this type is 2, 3, 4, or 6. It may be obtained in Type 1 (without enclosure), Type 4 (watertight enclosure), or Type 7-9 (hazardous locations enclosure).

MOTOR STARTERS

There are numerous types of *motor starters* for industrial and commercial installations. Here we shall look at the autotransformer-type of reduced-voltage starter, the part-winding starters, reduced voltage starters, and a starter for a synchronous motor.

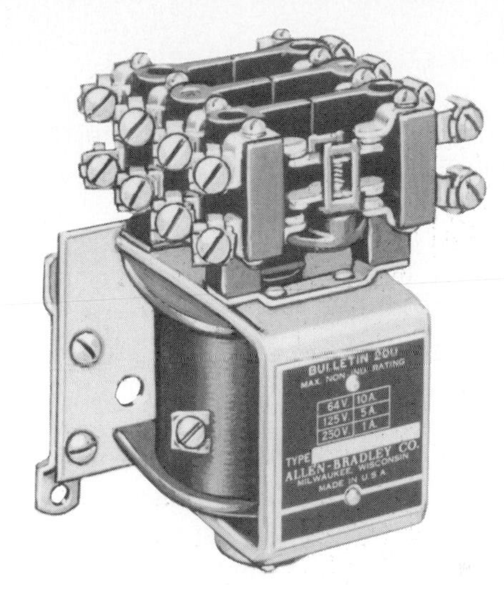

9–61. Relay.

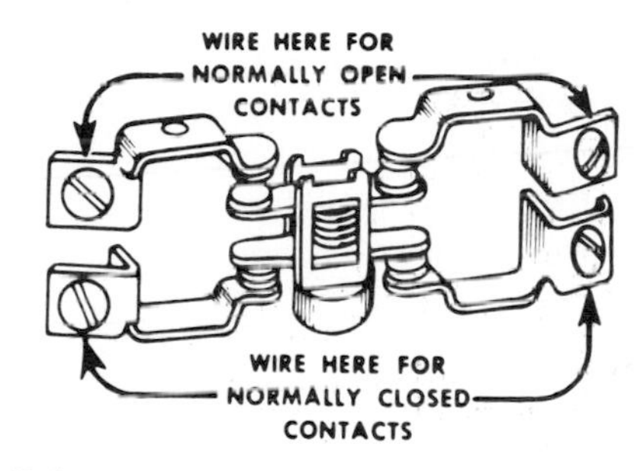

9–62. Relay contacts.

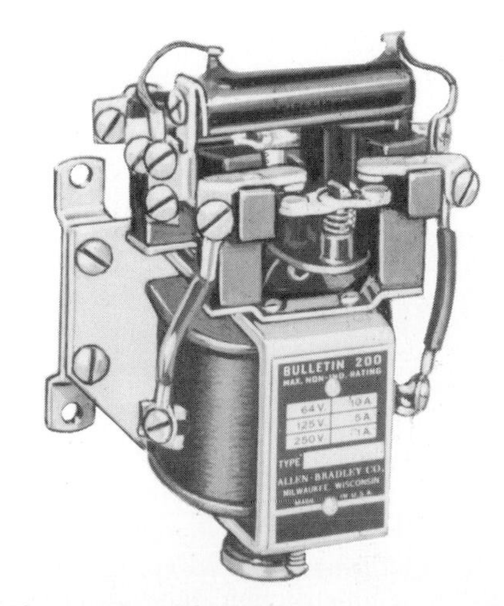

9–63. Thermostat relay.

Autotransformer Starter

A manual autotransformer-type reduced-voltage starter is used to start squirrel-cage, polyphase motors in cases where the characteristics of the driven load, or power company limitations, require starting at reduced voltage. Fig. 9-64.

The autotransformer on these starters is designed in accordance with medium-duty standards of the NEMA, which permit one start every four minutes for a total of four starts, followed by a rest period of two hours. Each starting period is not to exceed 15 seconds. Fig. 9-65 is a general-purpose enclosure with air-break construction. The transformer is located on the top shelf.

Where oil-immersion contacts are considered desirable, manual autotransformer starters are available in the oil-immersion construction. Fig. 9-66. The method of operation is identical to that of the air-break construction. Heavy-duty copper contacts are used to provide long life under oil. The tank, which is removable through the bottom of the enclosure, is fastened securely into place with bolts. These also serve as suspension supports for the tank while it is being filled with oil, or during inspection.

Part-Winding Starter

This type is used with squirrel-cage motors having two separate, parallel stator windings. Fig. 9-67. It provides a simple and economical method of accelerating loads for fans, blowers, or other loads involving low starting torque. These starters are generally used with motors having wye-connected windings, although motors having delta-connected windings are applicable, provided that neither winding is open during starting. Fig. 9-68.

Starters are available in 2-point and 3-point constructions. A 2-point starter includes two 3-pole contactors, a pneumatic timer, and six manual-reset, motor-overload relays. The 3-point starters have an additional contactor and timer, and a set of graphite compression-starting resistors.

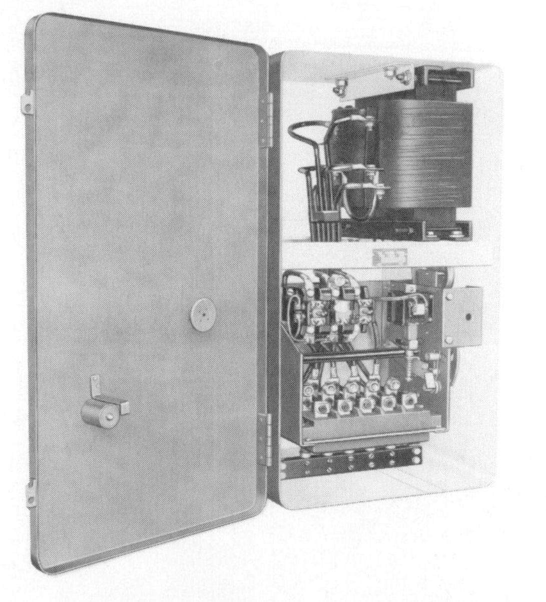

9-64. Autotransformer reduced-voltage motor starter, air-break construction, general-purpose enclosure.

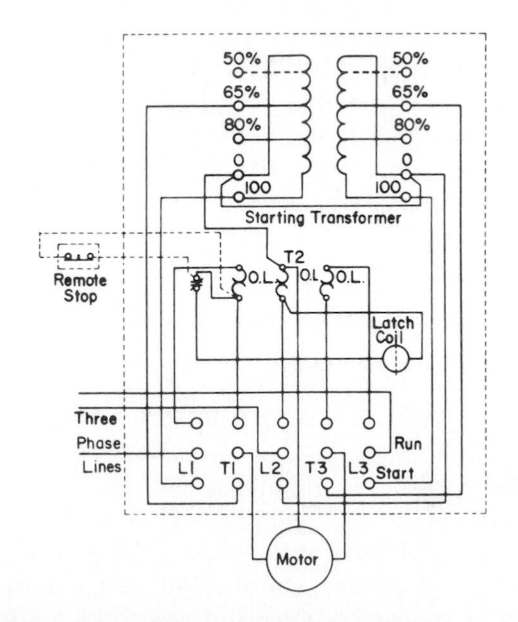

9-65. Wiring diagram for a reduced-voltage, autotransformer-type motor starter.

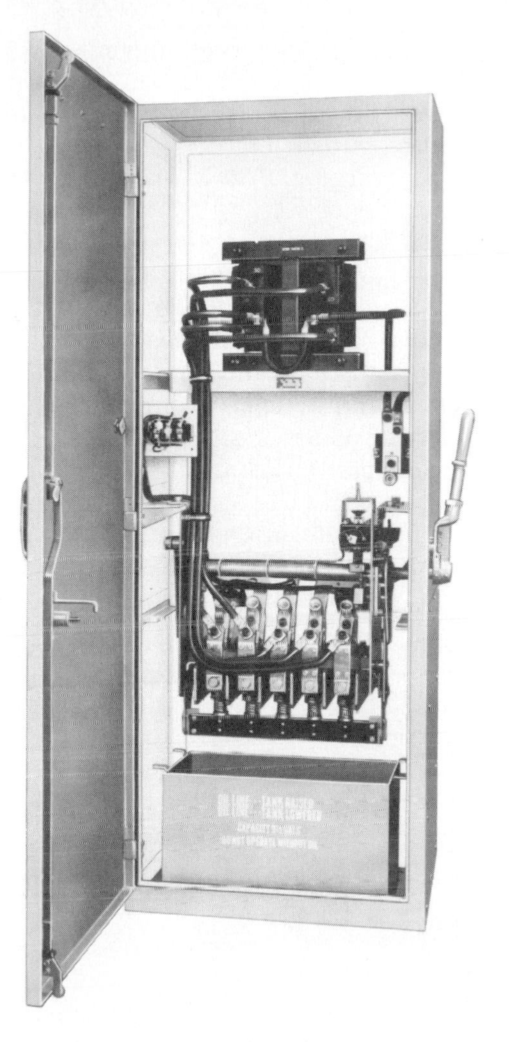

9–66. Oil-immersion construction autotransformer motor starter.

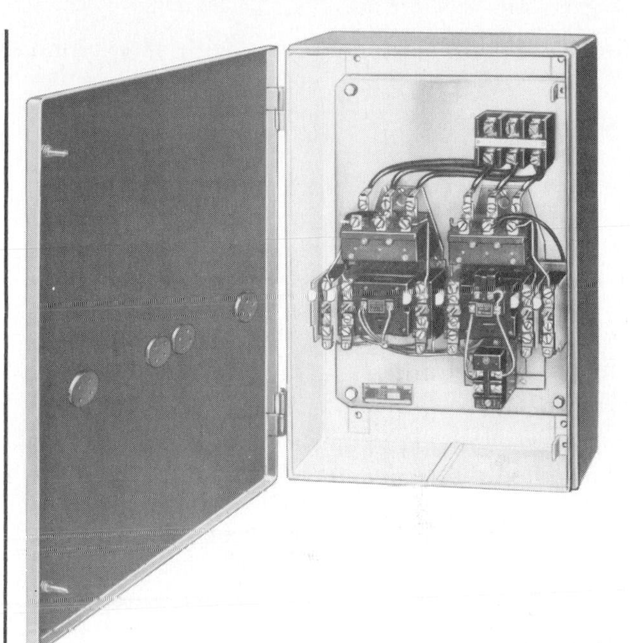

9–67. Part-winding starter, 2-point.

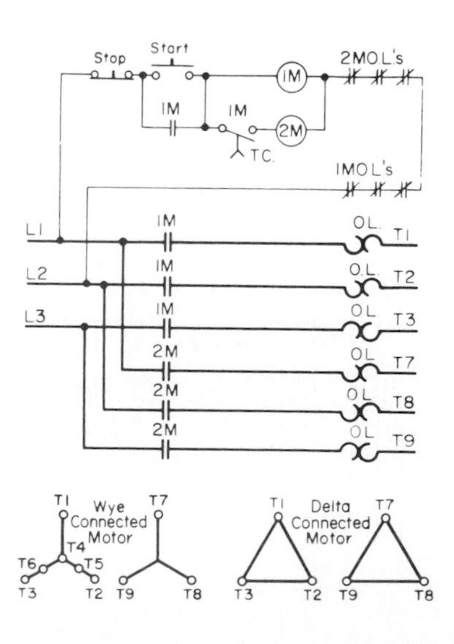

9–68. Wiring diagram for a part-winding starter.

Two-point starters operate with an external start button or pilot device, which closes the first contactor and connects one of the motor windings to the line. Assuming that the motor draws 65% of normal, locked rotor current, it will then develop approximately 45% of normal, locked-rotor torque. After a preset one-second interval, the second contactor closes, connecting the remaining winding to the line, in parallel with the first. The motor then develops normal torque.

Three-point starters have reduced voltage applied to the first winding through the series-connected resistance, resulting in less initial current and torque than that of the 2-point starter. After a preset one-second interval, the resistance is shorted out; after a second preset one-second interval, the second winding is connected in parallel with the first. Fig. 9-69.

Reduced-Voltage Starters

The *reduced-voltage starter* with graphite resistors is designed for automatic reduced-voltage starting of squirrel-cage motors. It has graphite compression resistors and provides unusually smooth acceleration. It is used where necessary to control starting torque, to limit starting inrush current

The starter in Fig. 9-70 is a 2-point type. It includes a 3-pole starting contactor, a set of graphite resistors with a thermal switch for protection against overheating, an adjustable pneumatic timer, a 3-pole running contactor, and a 3-pole, block-type overload relay. Three-point starters include an extra contactor, a set of resistors, and a timer to provide an additional starting point.

A 2-point starter motor is brought up to speed in two steps, one at reduced voltage, another at full voltage. Operating an external start button or pilot device closes the starting contactor, which connects the graphite resistors in series with the motor. After a preset in-

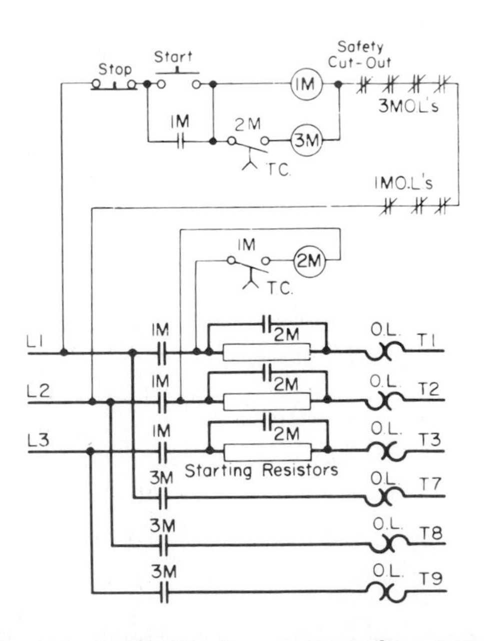

9-69. Wiring diagram for a part-winding starter, 3-point.

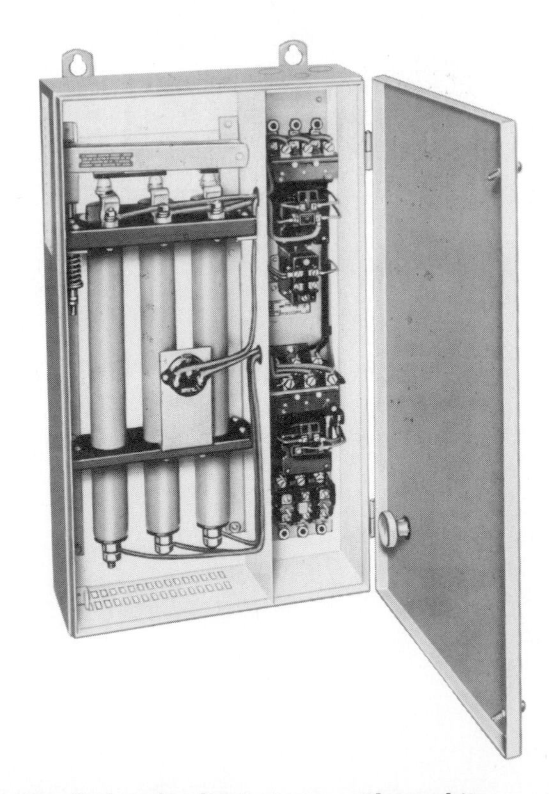

9-70. Reduced-voltage starter with graphite resistors.

terval, the timer contacts close the running contactor, shorting out the resistors and applying full voltage to the motor.

Three-point starters use an additional step at reduced voltage. Initially, all the resistance is connected in series with the motor. After a preset interval, part of the resistance is shorted out, and after a second preset interval, full voltage is applied. Fig. 9-71.

2-POINT STARTER

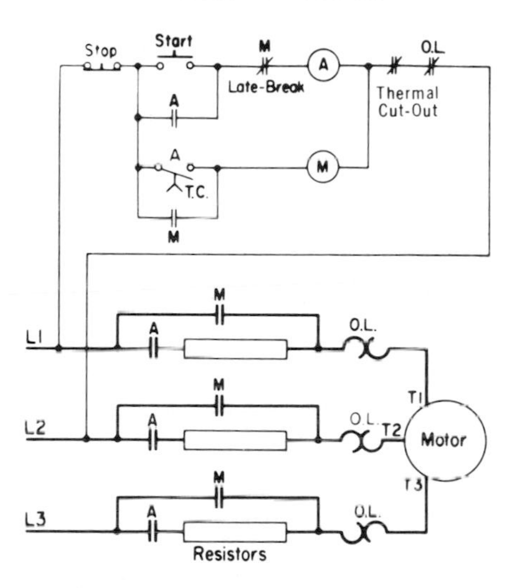

3-POINT STARTER

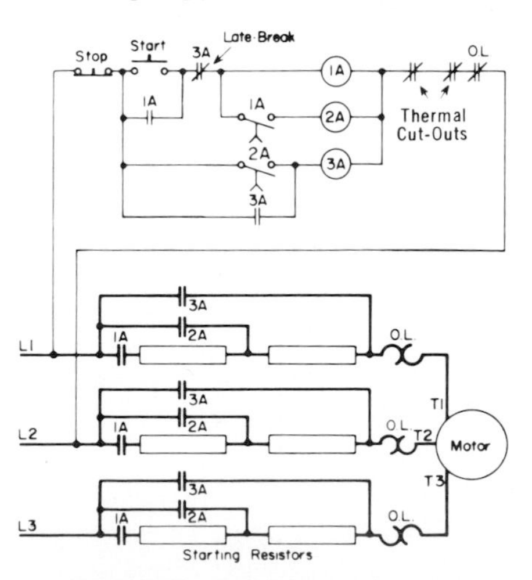

9-71. Wiring diagrams for a 2-point and 3-point reduced-voltage starter with graphite resistors.

Multipoint, resistance-type reduced-voltage starters are designed to limit the inrush current of the squirrel-cage motor in compliance with power company regulations for network distribution systems. With these automatic starters, starting current is built up in predetermined increments by connecting resistance in series with the motor, then short-circuiting successive sections of the resistance in timed steps.

Since the motor torque as well as current is increased from a low initial value, these starters are also suitable for applications where reduced starting torque is necessary to accelerate the load gradually. Fig. 9-72.

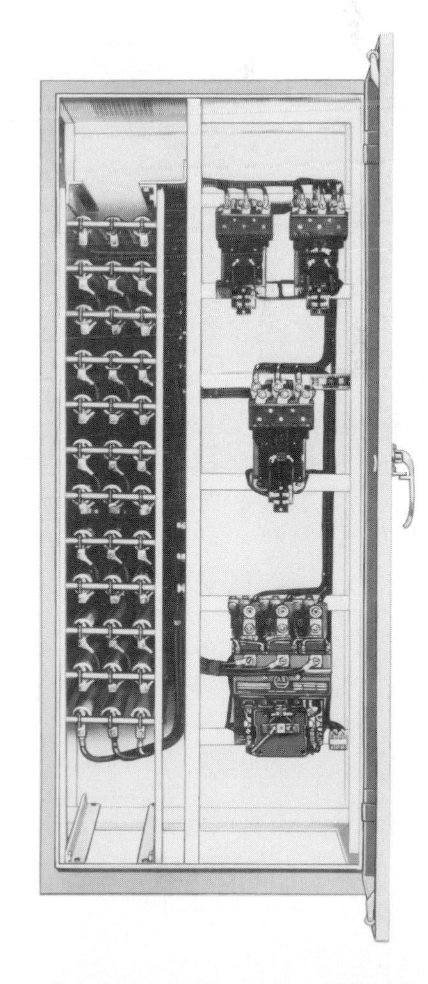

9-72. Multipoint resistance-type, reduced-voltage motor starter, 4-point type.

With 2-point starters, operation of an external start button or pilot device closes the accelerating contactor, applying reduced voltage to the motor through the series resistance. After a preset interval, the timer contacts close the running contactor, shorting out the resistance and applying full voltage to the motor.

With the 3-, 4-, and 5-point starters, closing the first accelerating contactor puts the resistance in series with the motor. The remaining accelerating contactors then short-circuit successive sections of the resistance in timed steps, with the final accelerating step applying full voltages to the motor.

The duty cycle of the starter is one 5-second start every 80 seconds. Fig. 9-73 shows a typical wiring diagram of the multipoint reduced-voltage starter.

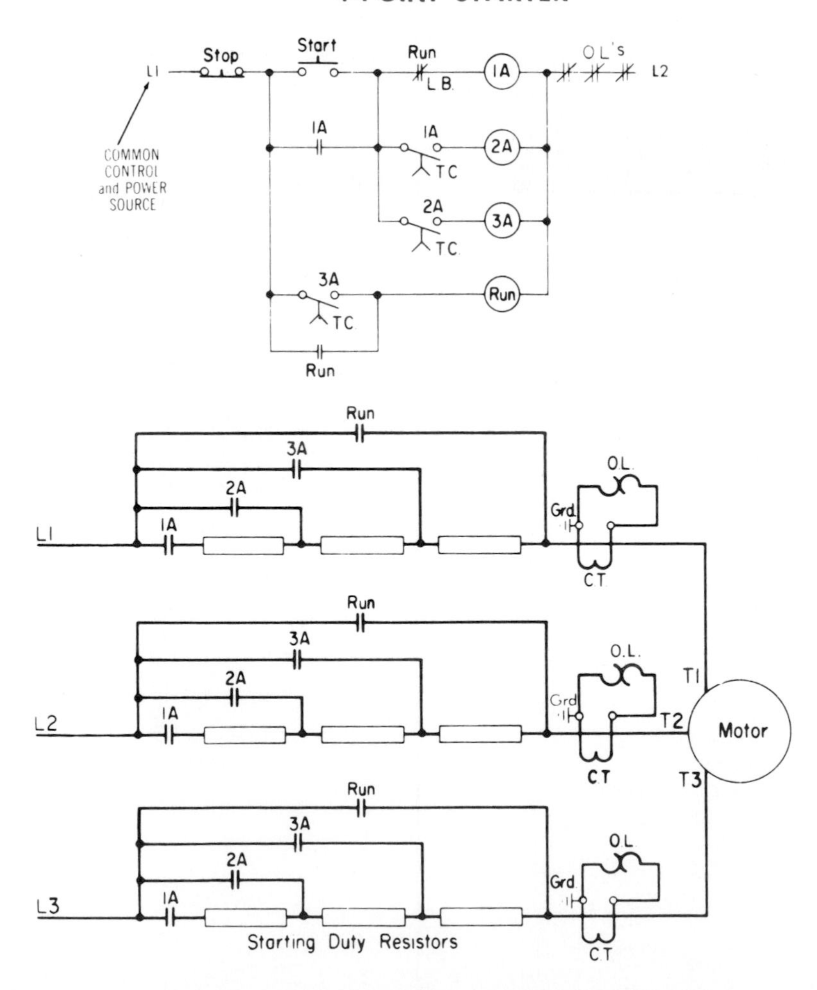

9–73. Wiring diagram for a multipoint, resistance-type, reduced-voltage motor starter.

Synchronous Motor Starters

Synchronous motors provide high efficiency, high power factor or power-factor correction, constant speed, and low operating speeds. They are used extensively in rubber mills, cement mills, flour mills, and ice plants to drive heavy machinery. Fig. 9-74.

A simple relay and reactor combination applies field excitation automatically. This is done at the correct motor speed and at the correct rotor angle to provide smooth synchronization of the motor with maximum synchronizing torque, and minimum stator current.

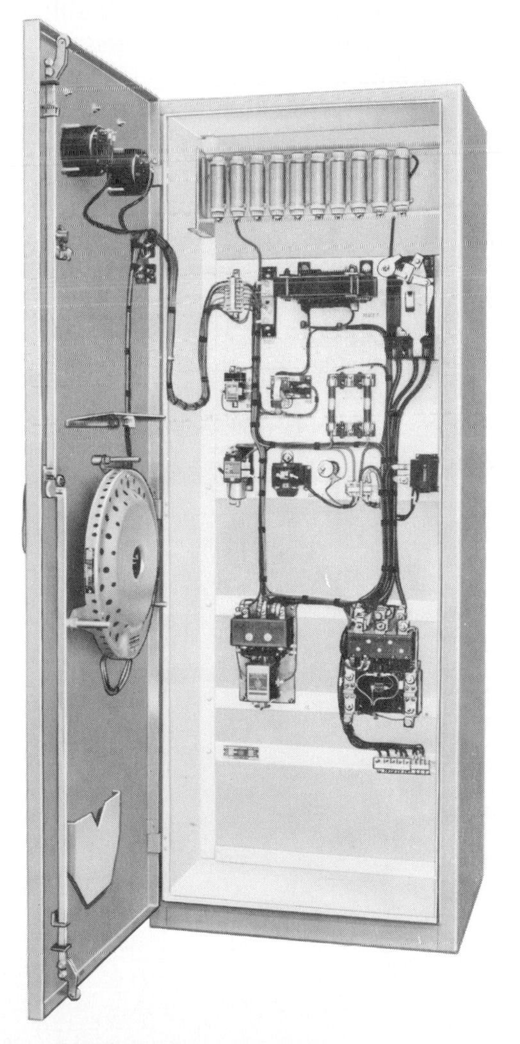

9–74. Synchronous motor starter.

The polarized-field frequency relay will automatically remove the field excitation if the motor pulls out of step. It will then resynchronize, provided the motor has sufficient pull-in torque upon restoration of the normal line voltage, or load conditions.

The squirrel-cage winding on the synchronous motor is designed for starting only. It is susceptible to damage through overheating if the motor operates a sub-synchronous speed for more than the safe, allowable starting time. To afford protection to the squirrel-cage winding, an out-of-step relay is provided on all the starters. This relay is essentially a fluid dashpot-type overload relay.

A DC ammeter and an AC ammeter are mounted in the door of the enclosure. Meters are of the rectangular type, flush mounted, and include shunts or current transformers where required.

CIRCUIT PROTECTIVE DEVICES

The circuit breaker and the fuse are the circuit protective devices most frequently involved in the job of making sure a circuit operates within its designed limits.

Fuse

The *fuse* is a device which destroys itself when it operates. That is, a fusible link is destroyed when it performs its designed function. Fig. 9-75. It must be replaced to make the fuse

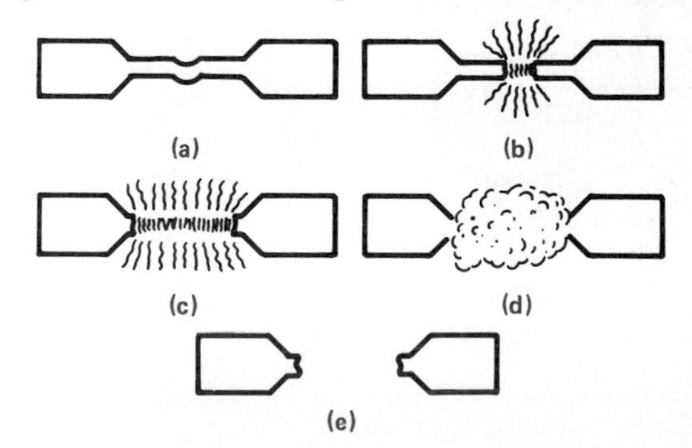

9–75. Progressive destruction of a fuse link.

unit operational again. An overcurrent will cause a piece of metal in the fuse to heat up and finally melt, thereby interrupting the circuit. This occurs because the fuse is inserted in series with the device it is to control.

An industrial-type fuse is shown in Fig. 9-76. The link may be replaced if the end is unscrewed and the fuse pulled apart to allow for the insertion of the fuse link. In some fuses of this type, it is not possible to replace the link—the whole unit must be thrown away.

Another type of fuse is shown in Fig. 9-77. It is a spring-loaded one which can disrupt the circuit if a unit overheats and is removed from its contact, as in (d), or if it burns or melts, as in (c). Part (a) of Fig. 9-77 shows the unit in its complete form, with a screw-in type base for insertion in a fuse box. A cutaway view with the construction of this type of fuse is shown in (b). These are normally found in outmoded wiring in old homes and businesses.

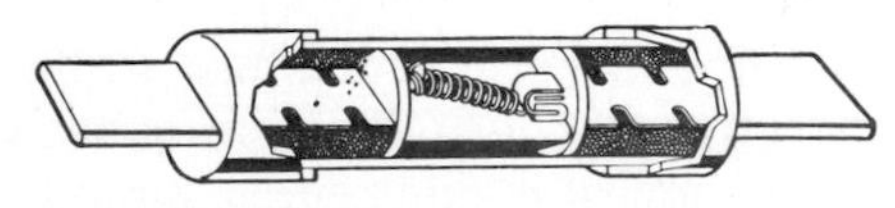

9–76. Industrial-type fuse.

Circuit Breakers

A *circuit breaker* is a device which, after breaking a circuit, can then be reset by turning the handle to off and then to on. Fig. 9-78.

The circuit breaker is mounted in a distribution box by snapping it into place. The hot wire (black or red) is attached by inserting it under the screw and tightening. A knockout blank in the distribution box must be removed, to allow for the handle and top of the circuit breaker to be exposed. Fig. 9-79 is a circuit breaker for 20 amperes and a single connection.

Another type is the 100-ampere double circuit breaker, with the two breakers tied together, or a common trip for a 2-pole arrangement. Fig. 9-80. It requires two spaces in a distribution box. Fig. 9-81 is a 3-pole, 20-ampere, common trip breaker which can control three lines. It requires three spaces in the box. A common trip, 3-pole 240-volt, AC breaker rated at 200 amperes is shown in Fig. 9-82. It requires six spaces in the distribution box.

Circuit breakers are snapped into place in the panelboard. Fig. 9-83 shows such a circuit breaker. Fig. 9-84 is a feed-through box. Note that the connections from the large cable to the box terminals are fed with a smaller wire. Cir-

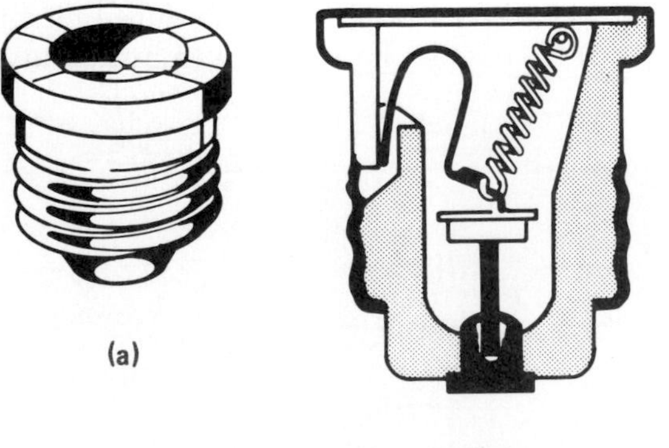

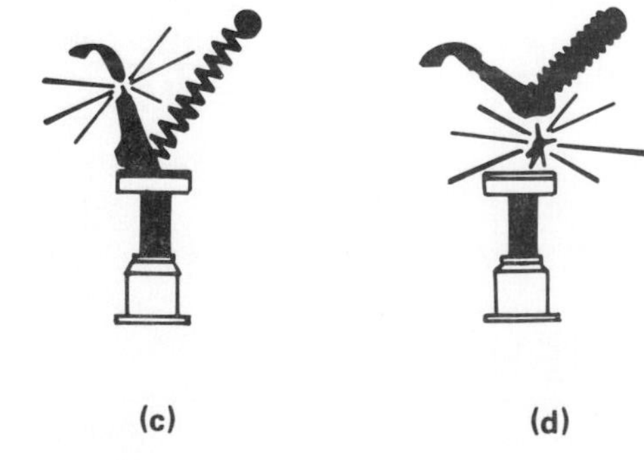

9–77. Screw-in fuse and its operation.

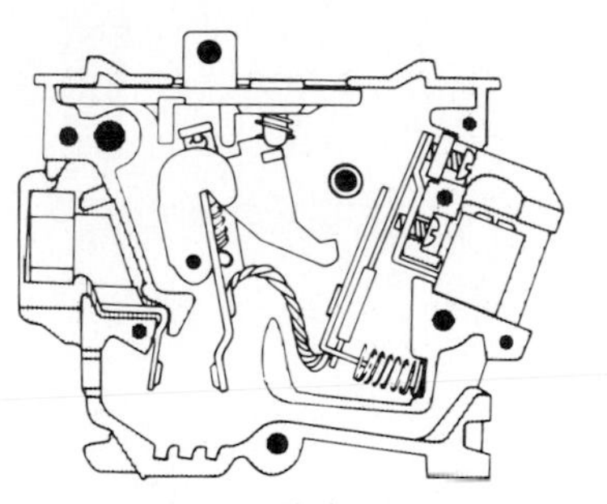

Tripped Position
- Contacts Open—no current flow
- Handle Stationary When Tripped

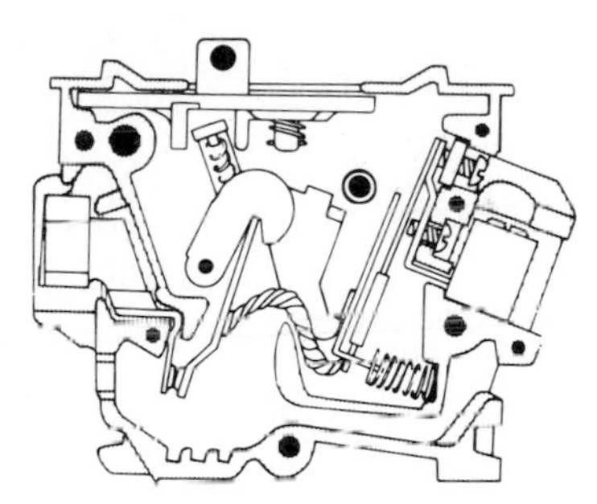

"ON" Position
- Contacts Closed—current on
- Handle in "On" position (shows "On")

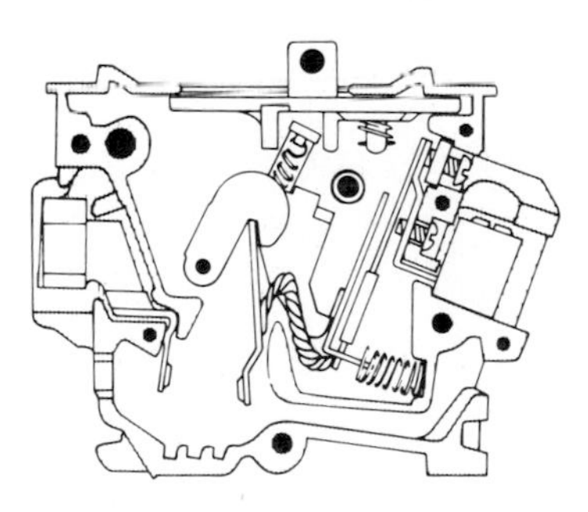

"OFF" Position
- Contacts Open—no current flow
- Handle in "Off" position (shows "Off")

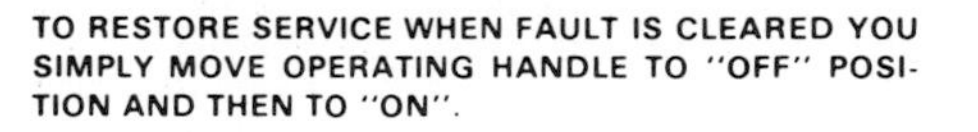

TO RESTORE SERVICE WHEN FAULT IS CLEARED YOU SIMPLY MOVE OPERATING HANDLE TO "OFF" POSITION AND THEN TO "ON".

9–78. Cutaway view of a circuit breaker in the tripped position, the on position, and the off position.

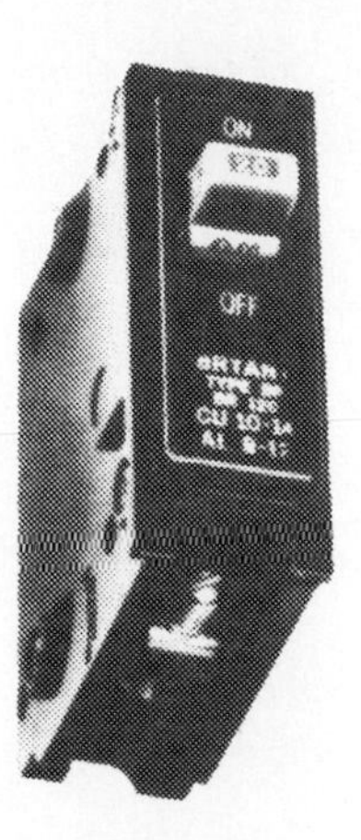

9–79. Single 20-A circuit breaker.

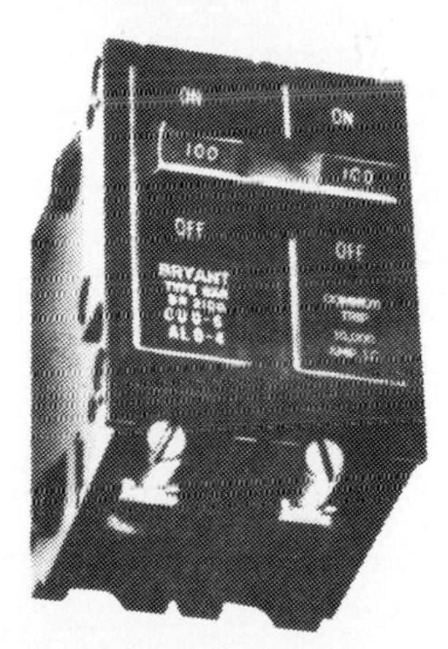

9–80. Double circuit breaker, 100-A, with common trip.

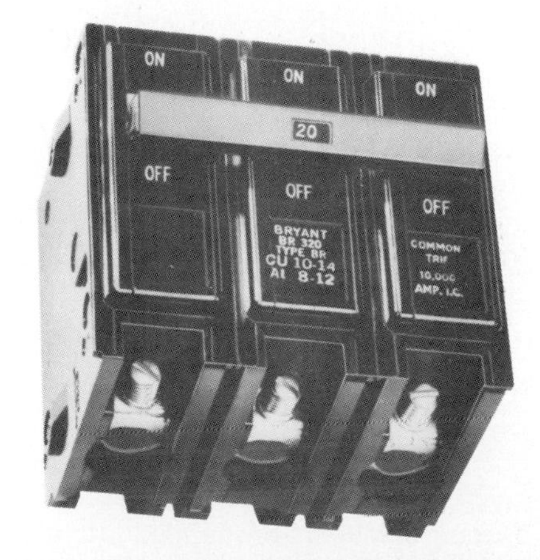

9–81. Common trip for three circuit breakers.

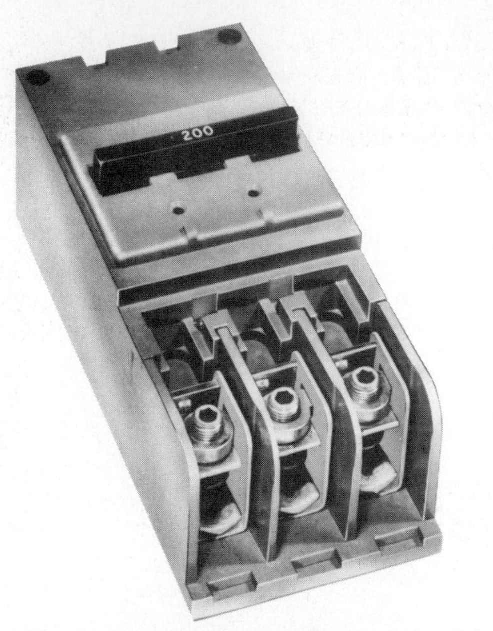

9–82. Three circuit breakers with a common trip, rated at 200 amperes.

9–84. Feed-through box.

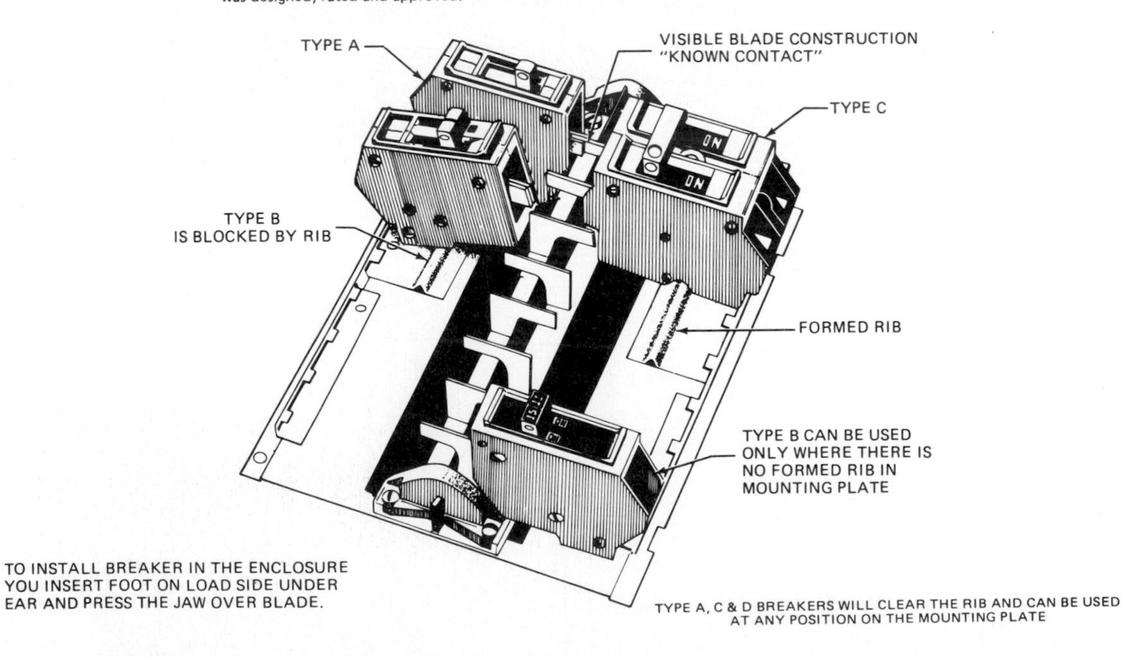

9–83. Circuit breakers are snapped into place in the panelboard. The wires are connected before the circuit breakers are snapped into place. Make sure the circuit breaker is in the off position while installing.

cuit breakers may be mounted in the box and feed a number of circuits. The large cable may then be used to feed another distribution panel or box.

Meter Centers

Meter housing is shown in Figs. 9-85 & 9-86. This is a two-meter case. The meter is merely pushed into the clips. Sealing rings are then mounted around the meter, to prevent its removal.

Some locations, such as apartments, need a number of meters so that each tenant will be responsible for the electricity used. A four-meter box, which will safely handle four meters of single-phase or three-phase power, is shown in Fig. 9-87. Each stack is provided with 800-ampere, 4-wire cross bus, with room for

9–85. Two-meter box.

9–86. Two-meter box with clips exposed.

9–87. Four-meter box.

circuit breakers to control the main power into an apartment, or any unit needing individually metered power.

In Fig. 9-88 are a number of meter-socket panel schematics, riser-panel schematics, and typical meter-center schematics.

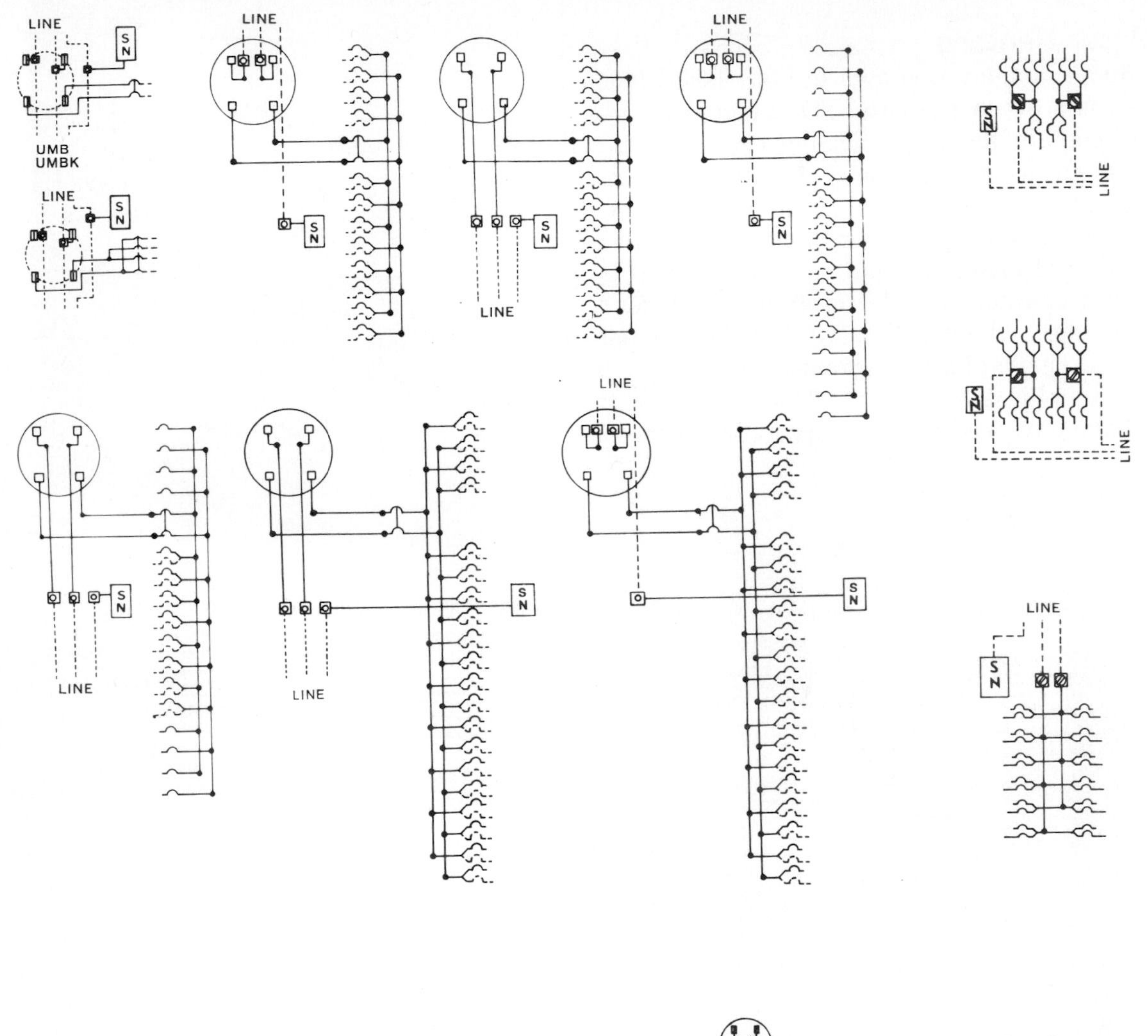

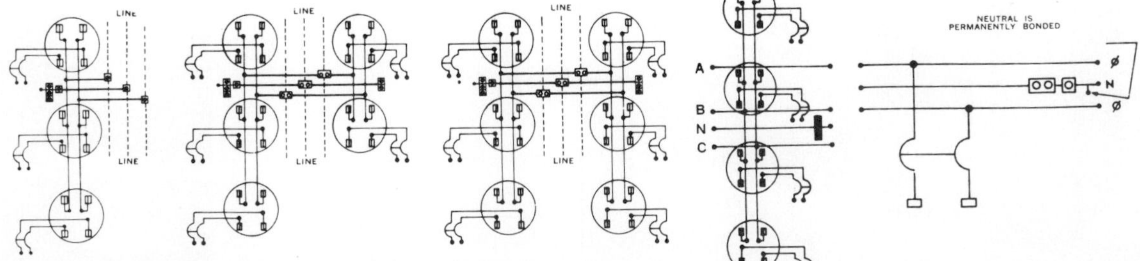

9–88. Meter schematics.

RAINTIGHT ENCLOSURES WITH INTERCHANGEABLE HUBS

Most Wadsworth raintight enclosures have interchangeable hubs. The series is designated SH or LH. This designation is stamped on the hub and on the end wall of each enclosure. There are four hubs that fit the SH mounting studs and four that fit the LH mounting studs. These hubs are positioned on the mounting plate as shown in Fig. 9-89. Adjustment space is allowed so the conduit can be aligned. All hubs have a built-in conduit end stop. No conduit bushings are required. Outside meter housings can use this type of connection for rainproofing.

Removing Knockouts and Twistouts

Knockouts and twistouts for meter boxes and circuit breaker boxes are sheared alternately in and out. The center knockout is always knocked in. Follow the instructions as shown in Fig. 9-90.

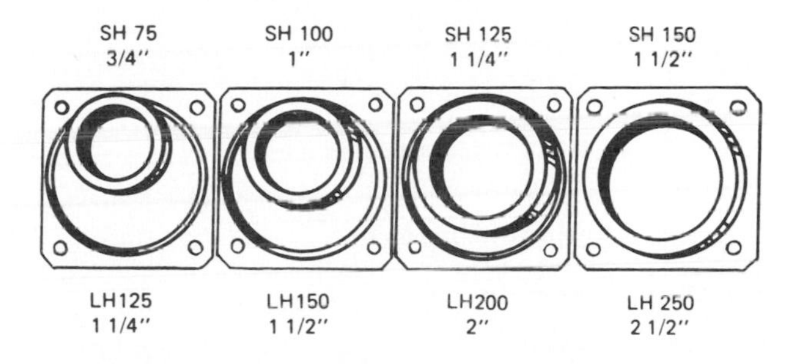

9–89. Raintight enclosures (boxes) have interchangeable hubs like those shown above. There are four hubs that fit the remaining studs. The hubs are positioned on the mounting plate to allow for adjustment in aligning with the conduit. All hubs have built-in conduit end stops. No conduit bushings are required.

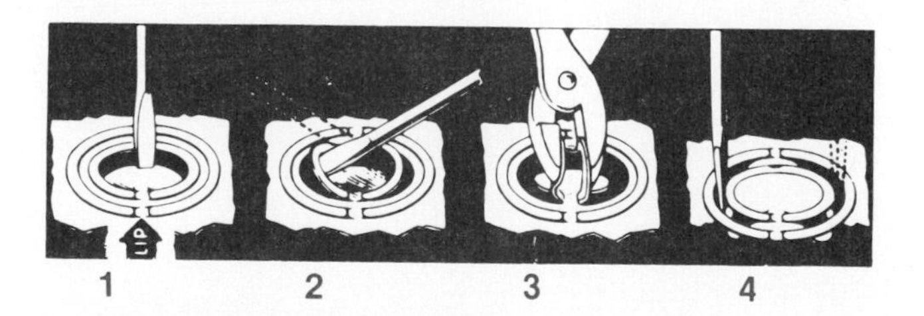

9–90. Knockouts and twistouts are sheared alternately IN and OUT. The center knockout is always IN. Knockouts and twistouts should be removed in the same sequence.

(1) First remove center knockout by placing screwdriver opposite lip, drive knockout *inward* and twist it in an open-and-closed direction to remove from single lip. To remove each succeeding twistout, screwdriver should be placed between lips as shown in 2 or 4 to ensure removing them one at a time.

(2) Remove all OUT twistouts by driving or bending each half of ring *outward* and then twisting it in an up-and-down direction to remove it from lips. (See 3.)

(3) Method of removing all twistouts from lips.

(4) IN twistout is removed by driving each half of ring *inward* and then twisting it in an up-and-down direction to remove it from lips. (See 3.) All single knockouts are driven *inward* and removed from lip by twisting it in an open-and-closed direction.

QUESTIONS

1. Describe at least five types of concrete anchors an electrician may use.
2. In what situations are floor boxes used? What are the three types of floor boxes?
3. In underfloor raceways, how are connections made to the ducts?
4. What is a self-riveting connector?
5. What is a demand kilowatt-hour meter?
6. What are limit switches? Where are they needed?
7. What is an automatic transfer switch? Where is it used?
8. Describe two different types of motor starters.
9. Why do electrical systems need fuses?
10. What is a circuit breaker? Who uses it?

KEY TERMS

anchors
automatic float switch
automatic transfer switch
circuit breaker
conduit boxes
demand charge
fuse
limit switch
motor starter
power
relay
underfloor raceways

CHAPTER 10

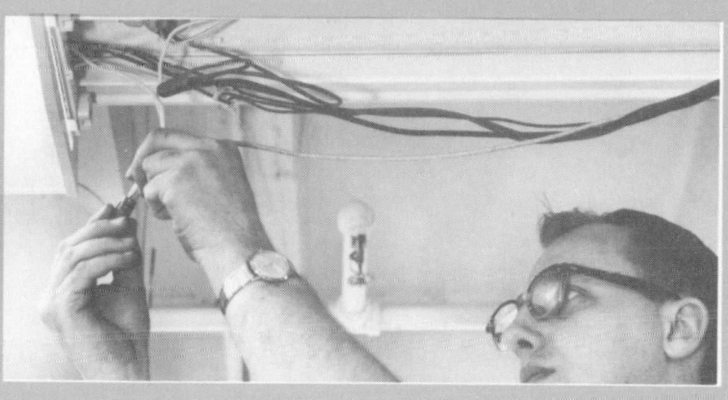

INSTALLING DEVICES

Objectives

After studying this chapter, you will be able to:

- Identify the boxes and other equipment commonly used with Romex.
- Replace a damaged lampcord.
- Splice wires.
- Terminate wires.
- Identify the two main types of aluminum connectors.

ROMEX

Romex cable of the No. 14 or No. 12 size is usually employed in the wiring of homes, light industry, and businesses. In order to safely utilize the nonmetallic sheathed cable in a manner consistent with wiring practices employed by electricians and others, it is necessary to use the switch and outlet box for terminations and connections. Splices are not allowed along the run of the cable unless the splice is housed in an appropriate box and totally enclosed with a cover plate.

When insulating material is used to make boxes and other equipment for wiring a house, it is not necessary to use a clamp or connector to hold the wire in place. Such a box is permitted when the wire or cable is supported within 8″ of the box. Cable enters the box by way of a knockout. All knockouts should be closed if not used for cable.

Box Volume

The volume of the box is the determining element in the number of conductors that can be allowed in the box. For instance, a No. 14 wire needs 2 cubic inches for each conductor. Therefore, if a two-wire Romex cable of No. 14 conductors is specified, it means that a ground wire (uninsulated) is also included which will count as a conductor, and the three wires, or conductors, will require 6 cubic inches of space. A box used for this installation should have at least this amount. A 3″ × 2″ × $1\frac{1}{2}$″ device box has only 9.0 cubic inches and will be allowed to contain three of either No. 14, No. 12, or No. 10 conductors, but only two No. 8 conductors. Boxes manufactured recently will have the cubic-inch capacity stamped on them. You should check the National Electrical Code handbook for the number of wires allowed until you are familiar with the standards for boxes.

In Figs. 10-1 through 10-14 you will find a representative sampling of devices made of insulating material that are acceptable in house wiring. Read the captions under each figure and identify the characteristics for future use in wiring buildings.

10–1. Insulated metallic grounding bushing.

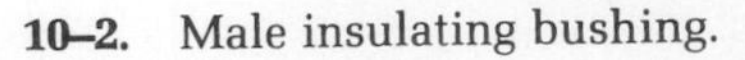

10–2. Male insulating bushing.

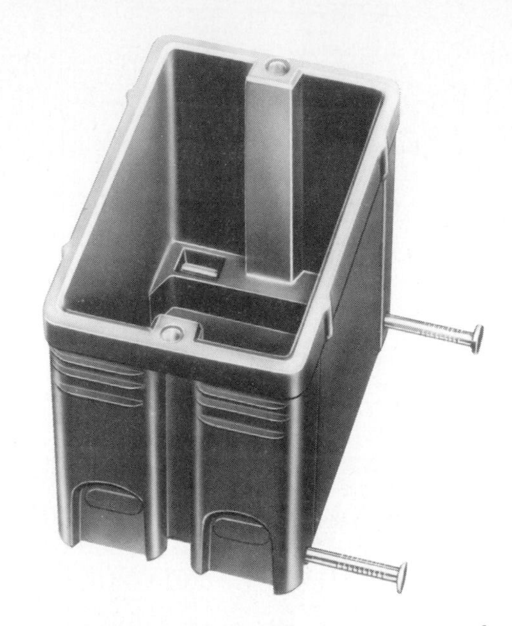

10–3. Nonmetallic switch box for houses and apartments.

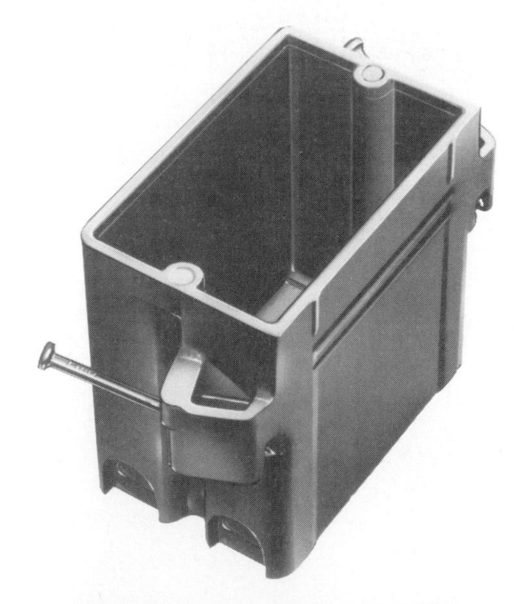

10–4. Switch box, nonmetallic, will hold 9 conductors of No. 14 wire, 8 conductors of No. 12 wire, or 7 conductors of No. 10 wire.

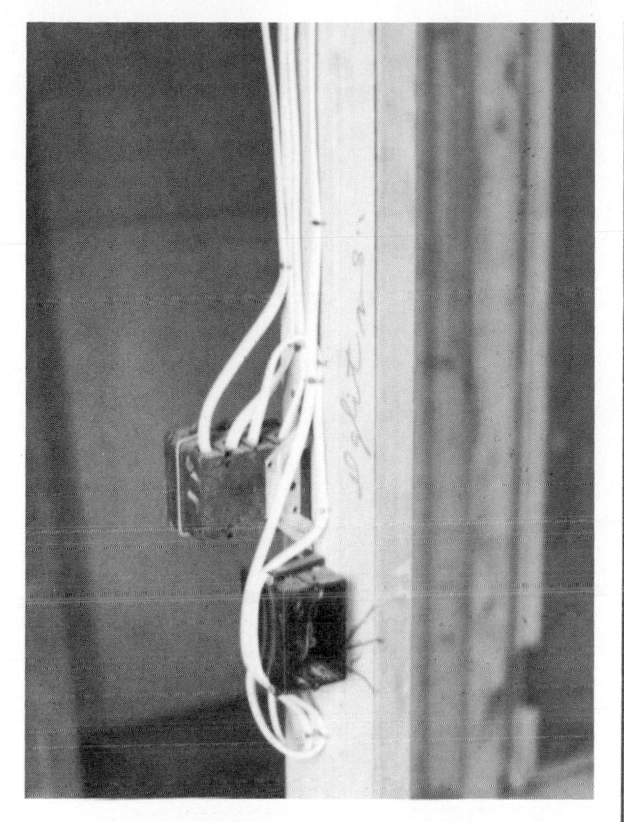

10–5. Nonmetallic switch box, with nails used to mount it. Will hold 7 conductors of No. 14; 6 of No. 12 or No. 10 conductors.

10–6. Nonmetallic switch box, designed with a bracket to mount to wood or steel studs with SST tool. The box will hold 15 No. 14 conductors, 13 No. 12, or 12 No. 10 conductors.

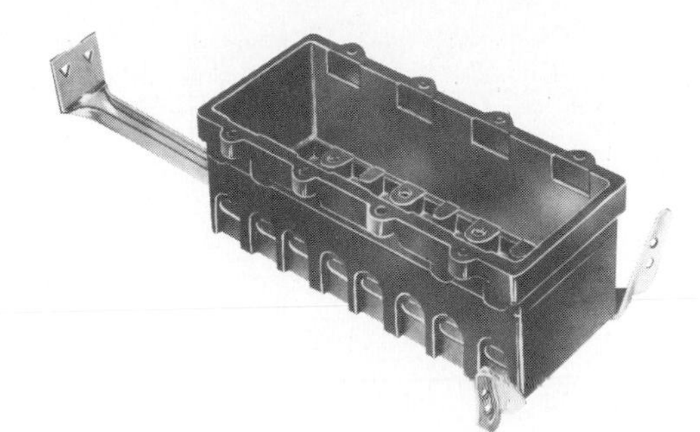

10–7. Nonmetallic switch box. Will hold four devices. The bracket will fit between two studs located on 16″ centers. The box will hold 20 No. 14 conductors, 17 No. 12, or 15 No. 10 conductors. Remember—the ground wire counts as a conductor here.

10–8. Cover for outlet or handy box for a duplex receptacle, nonmetallic.

10–9. Nonmetallic cover for a toggle switch.

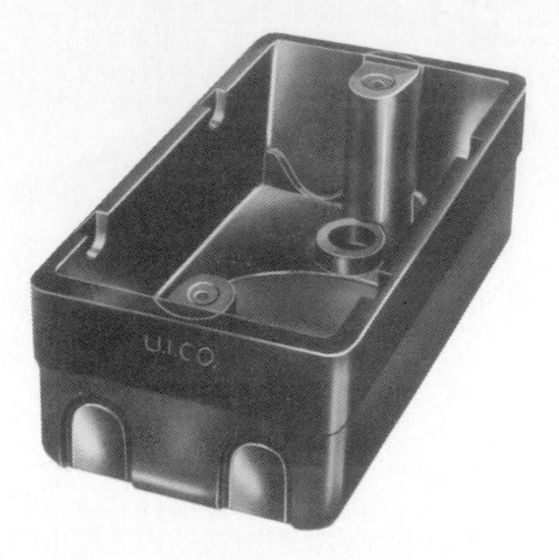

10–10. Surface box, nonmetallic, will hold only 3 No. 14 conductors, and 2 No. 12 or No. 10 conductors.

10–11. Nonmetallic conduit box, round, with four $\frac{1}{2}''$ threaded knockouts. Suitable for fixture mounting, it has 12.5 cubic inches of space and a threaded knockout. It is also available with $\frac{3}{4}''$ threaded knockouts.

10–12. Duplex receptacle (REC) cover, with gasket and stainless steel screws, may also be obtained with a single receptacle hole. Note the REC on the outside to identify the outlet.

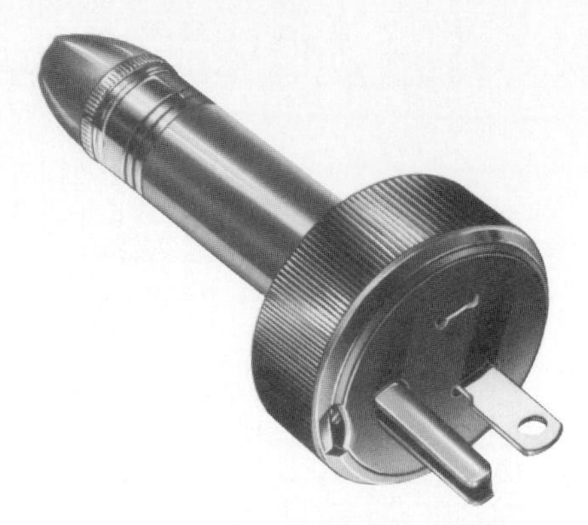

10–13. Ground continuity tester for use on nonenergized circuits only. It is equipped with ground and neutral blades only. To use, plug it into a grounding-type receptacle. If the light comes on, the ground continuity is complete. If the light does not come on, there is a fault in the circuit. It uses a penlite cell for power.

10–14. Box finder and cable tracer used to locate boxes or cable that are hidden from view. To use, plug it into one of the receptacles on the covered box circuit, or attach it to one of the circuit wires. The transmitter will produce a signal. Tune in a small transistor radio to pick up the transmitter signal. Follow the cable by checking sound build-up or fade. As the box is approached, sound builds up due to more wire being folded back in the box. The box will be located at maximum signal strength. The unit is more efficient if the circuit ground wire is disconnected. It uses a standard 9-volt transistor battery.

CONDUIT AND BX

■ *Connectors.* Conduit (both thinwall and rigid) and BX require connectors that make a clamping action which holds the protective coating rigidly in place. They also make a good electrical connection for grounding purposes. The design of these connectors will vary with different manufacturers. However, the group of connectors shown in Figs. 10-15 to 10-20 will represent those made by a particular manufacturer, and will serve here as examples of connectors available for switch boxes, utility boxes, and both square and octagonal outlet boxes.

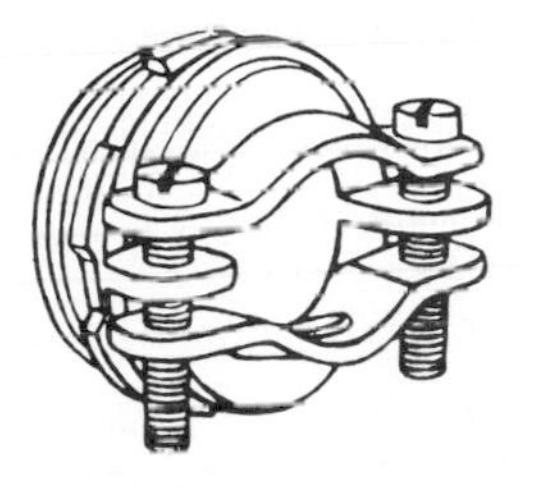

10–15. Box connector for entrance service, for nonmetallic sheath cable or nonmetallic flexible tubing. The connector does *not* make a weatherproof connection to the box.

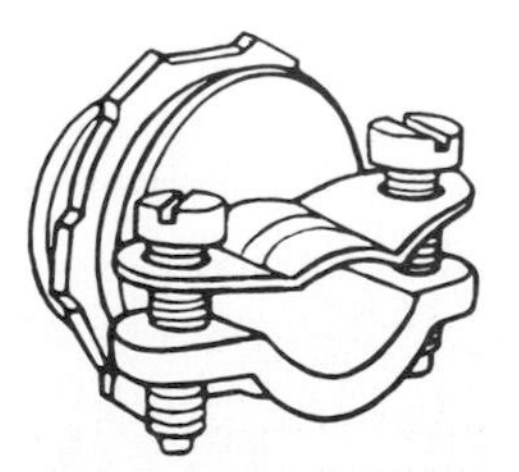

10–16. Romex box connector designed for nonmetallic sheathed cable (Romex) or nonmetallic flexible tubing. It will accommodate 14/2, 14/3, 12/2, 12/3, or 10/2 cable. (Romex for home use is usually 14/2, or 2 wires of #14. Ground wire is assumed to be there, but designation will usually be 14/2 WG.)

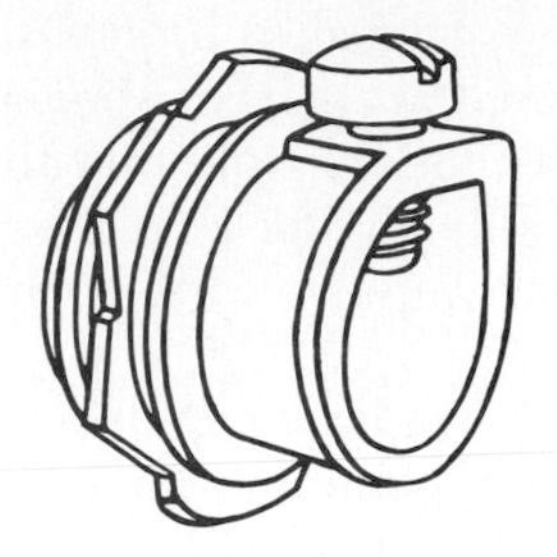

10–17. BX box connector. Armored cable, or $\frac{3}{8}$″ flexible steel conduit, may be accommodated with this connector. It can handle 14/3, 14/4, 12/2, 12/4, or 10/2 cable.

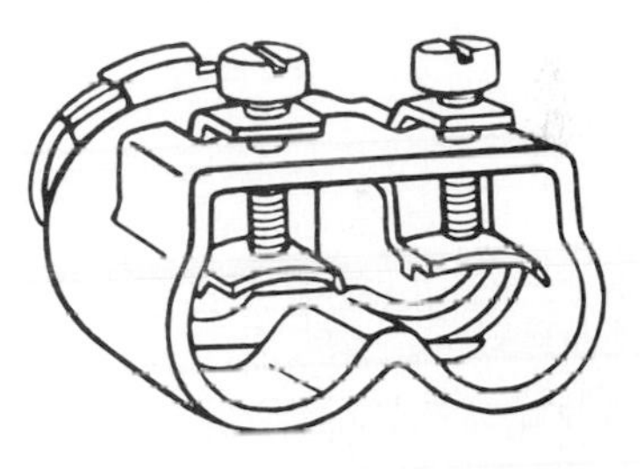

10–18. BX box connector. A duplex connector for armored cable, and for $\frac{3}{8}$″ flexible tubing.

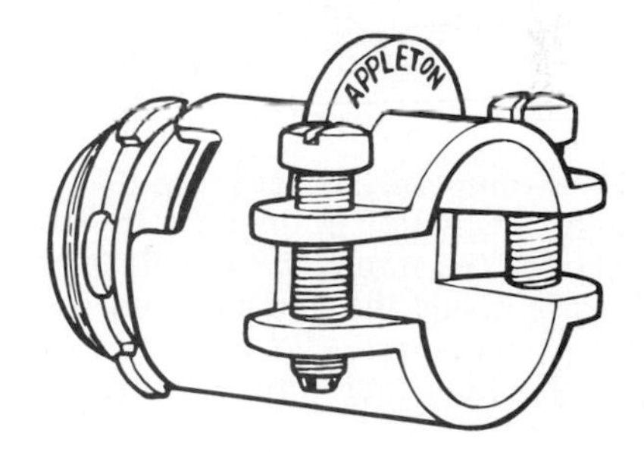

10–19. BX box connector, with two screws for $2\frac{1}{2}$″ flexible steel conduit.

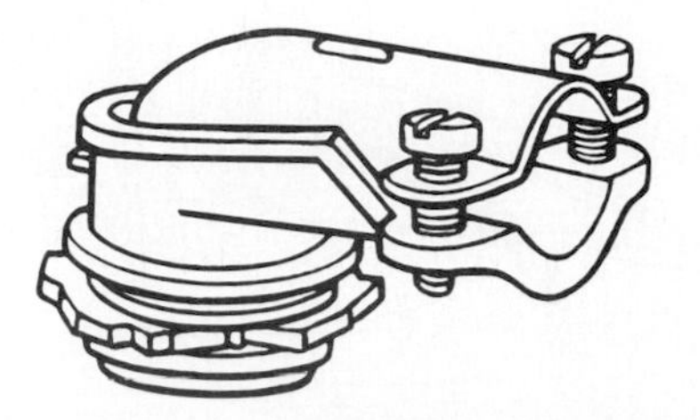

10–20. A 90° BX box connector, used also for $\frac{1}{2}$″ flexible steel conduit. It will handle 8/2, 8/3, 8/2 WG cable.

A *bushing* is needed to protect the wires inside the armored cable from abrasion and from cuts in the insulation. Note that the connector is mounted in the box for a positive holding action. Fig. 10-21 shows a fiber bushing inserted in a BX.

Fig. 10-22 shows how a connector is used to hold the Romex cable in place. *Note how the grounding clip is attached to the metal box.* Note also that part of the insulation of the Romex is extended through the connector into the box.

A number of sizes and configurations are available to the electrician for use as receptacles. A sampling of the different configurations available for installation in switch boxes and outlet boxes, both metallic and nonmetallic, is shown in Fig. 10-23.

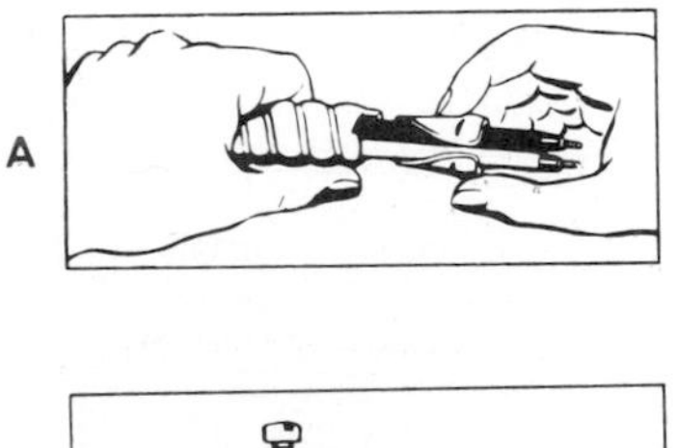

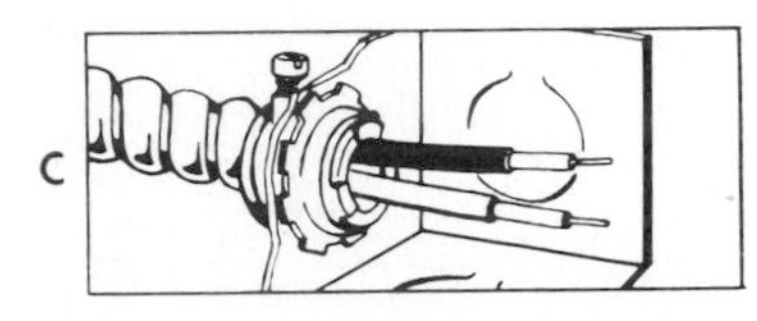

10–21. (A) Bushing inserted into the BX to prevent damage to the insulation of the wire. (B) BX connector is slipped into place over the bushing and metal armor, and the screw tightened to hold the connector in place. (C) BX connector is held in place through a knockout in the box, by use of a locknut.

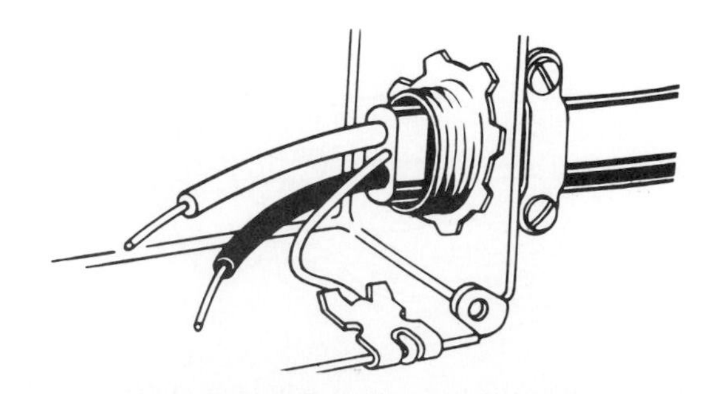

10–22. Romex connector attached to a box. Note the ground clip for the uninsulated ground wire.

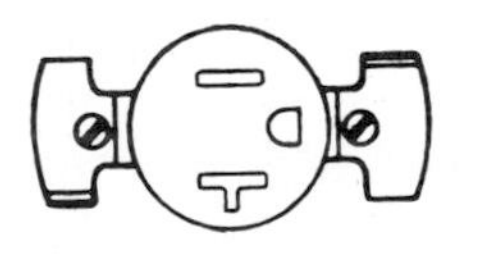

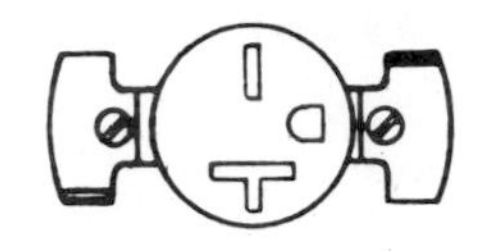

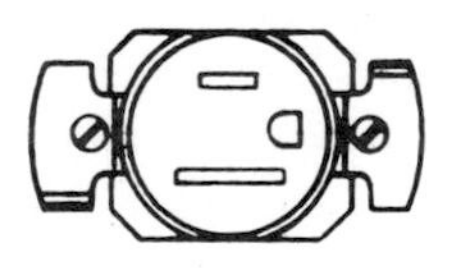

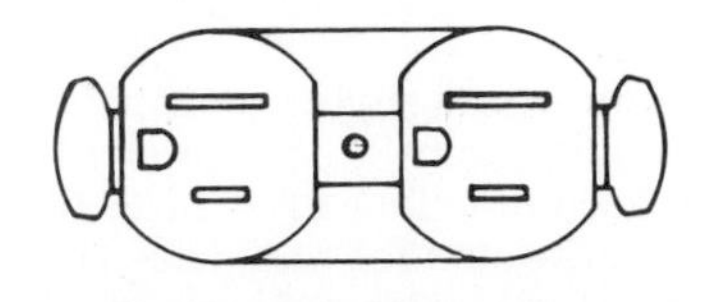

10–23. Single and duplex receptacles for various current ratings and voltages.

In Figs. 10-24 through 10-32 you will be able to see the different types of boxes available for use as switch boxes. Some installation procedures are also depicted.

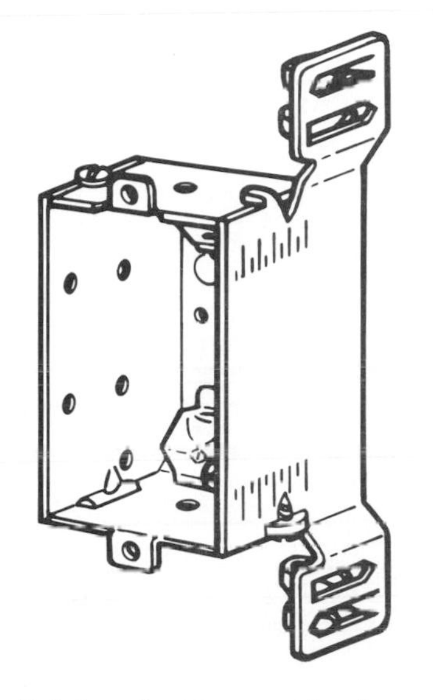

10–24. Switch box for Romex. Can be mounted without nails. The bracket on the side has an "eagle claw" so that the tabs may be hit with a hammer and driven into the stud without using nails. Note the beveled corner and clamps for the Romex.

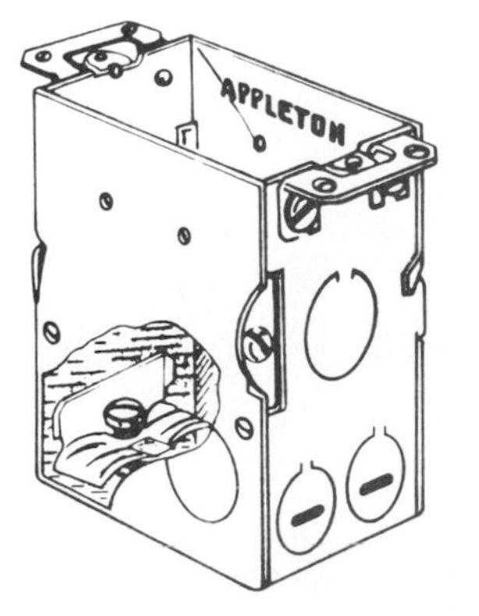

10–25. Square-corner Romex switch box, with clamps. This one is deeper than the standard size.

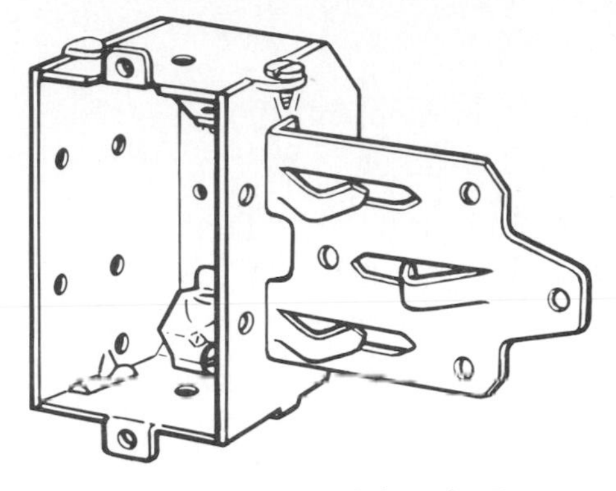

10–26. Beveled-corner switch box for Romex may be mounted without nails.

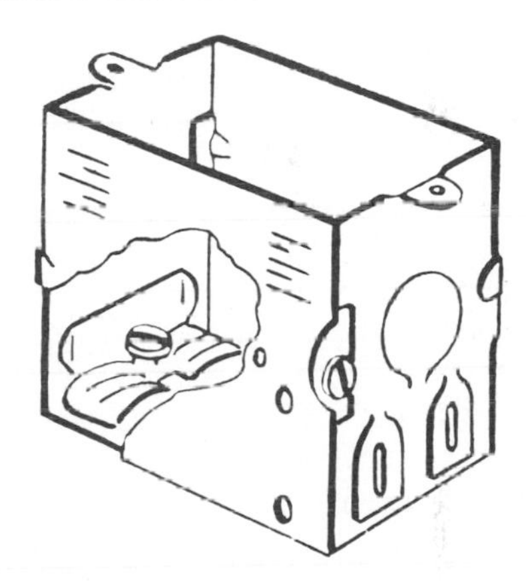

10–27. Square-corner switch box with Romex clamps inside. This one does not have plaster ears, but has side leveling ridges and tapped grounding holes.

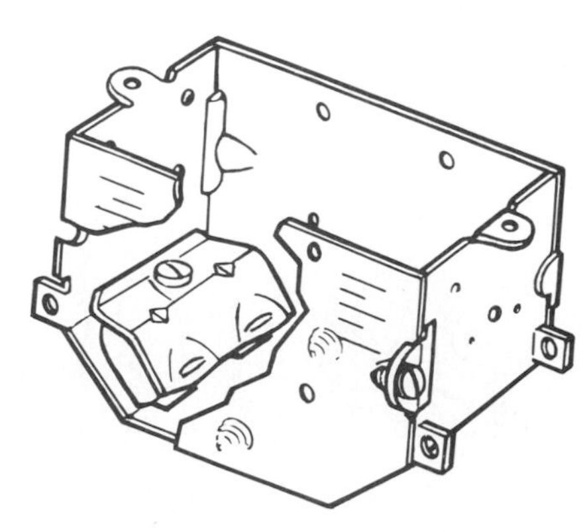

10–28. Bevel-corner Romex switch box, with tapped grounding holes and side leveling ridges. Nail holes are located on the brackets outside the box. A Romex clamp is included.

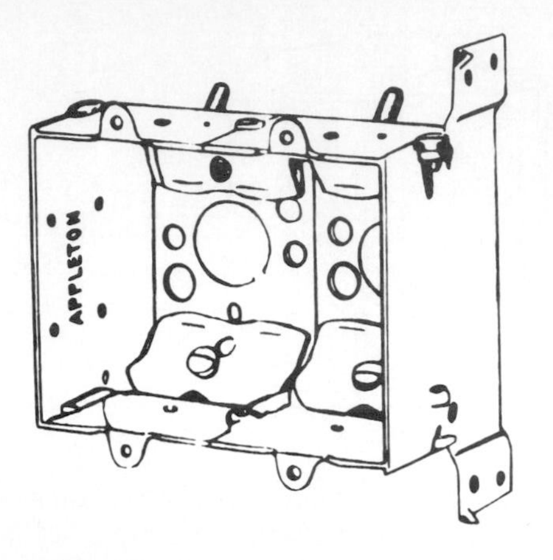

10–29. Two switch boxes, *ganged*. They have Romex clamps and claws, and nail holes. The claws may be hit with a hammer to hold the box in place until nails can be inserted.

10–30. Two switch boxes, ganged and inserted in a concrete block wall. Note that no ground wire is showing here.

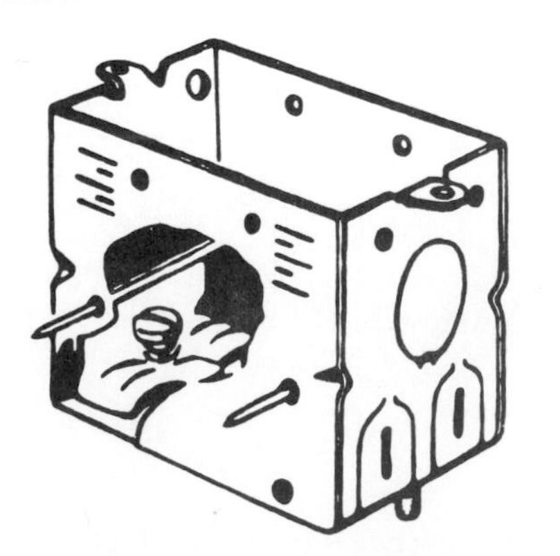

10–31. Square-corner Romex switch box without plaster ears, but with side leveling ridges and the side nail capability.

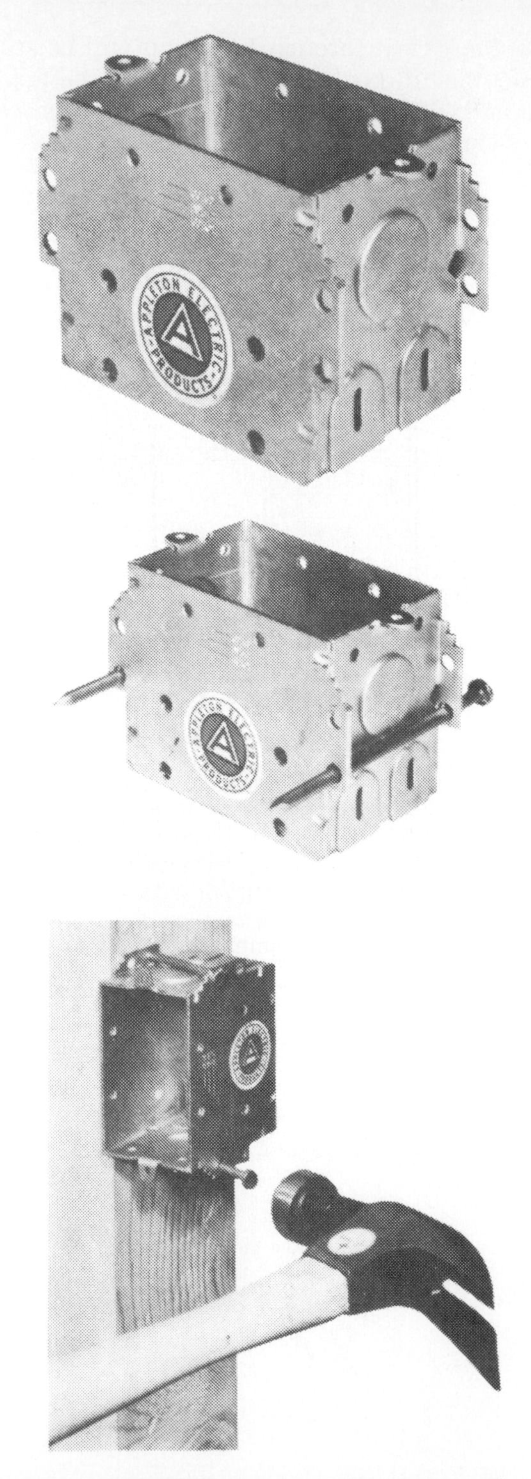

10–32a,b,c. Switch box with nailing holes and hammer driving the nails into the stud.

10–32d. Beveled corner switch box mounted with two nails.

10–32e. Front view of mounted beveled-corner switch box.

10–32f. Insulated wire grounded to the switch box by using the tapped grounding hole provided. Insert shows the wire attached to the box by a screw.

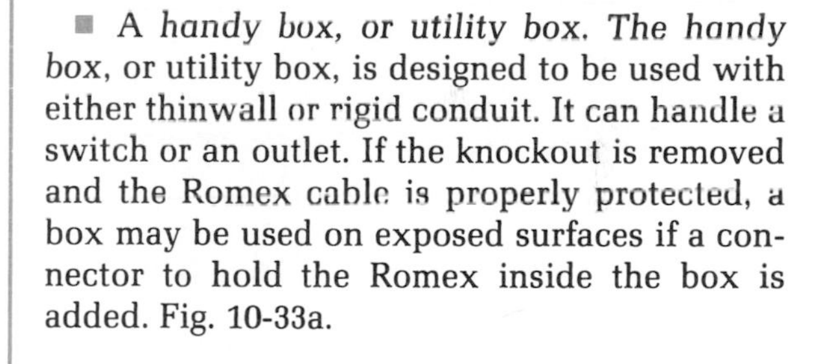

■ A *handy box, or utility box. The handy box*, or utility box, is designed to be used with either thinwall or rigid conduit. It can handle a switch or an outlet. If the knockout is removed and the Romex cable is properly protected, a box may be used on exposed surfaces if a connector to hold the Romex inside the box is added. Fig. 10-33a.

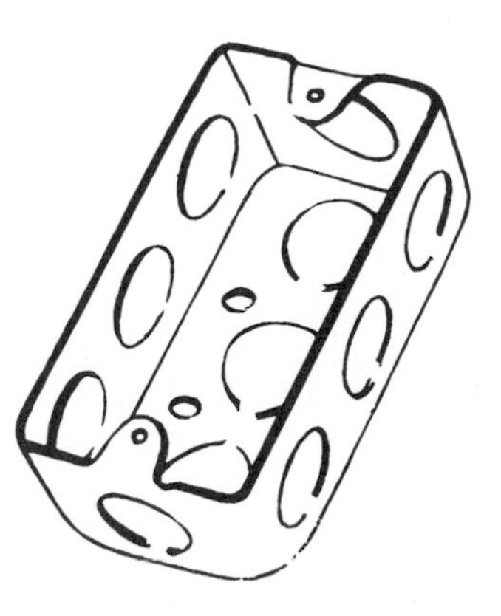

10–33a. Handy box (utility box) with or without nailing brackets for thinwall or rigid conduit. Galvanized.

In Fig. 10-33b the following procedures are shown: "A" shows how to connect the existing third wire (green insulated or bare copper) to the green grounding screw on a receptacle; "B" shows how the ground wire from the green grounding screw should be connected to a water pipe if no ground is available in the existing system; and "C" shows that the outlet is

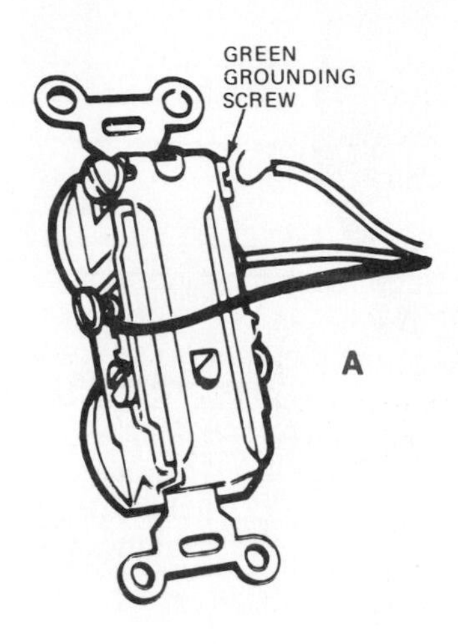

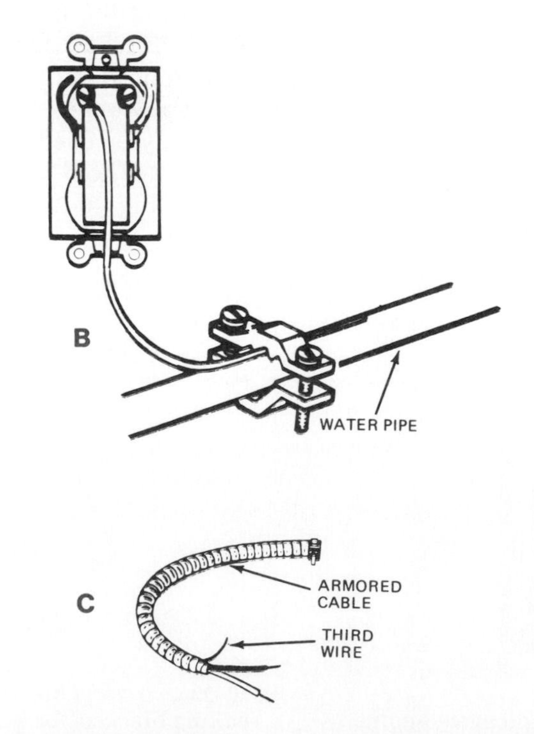

10–33b. Grounding of receptacle.

automatically grounded if properly grounded armored cable (BX or EMT) is used throughout the system. The grounded outlet provides protection against shock hazards, such as defective internal wiring on a hand drill.

Romex is protected in an exposed position against a wall in a basement. Fig. 10-34.

■ A *handy-box extension.* Sometimes this is added if the proper volume is not available for the necessary number of conductors making their entrance and exit from the utility box. An extension simply slips onto the screws of the utility box and is tightened into place. A cover plate or switch is then placed in the box, or over the extension. Fig. 10-35.

■ *Cover plates.* Because utility boxes must be totally enclosed, several plates are available to fit individual situations. Figs. 10-36 through 10-39a show the types available.

Fig. 10-39b gives directions for installing a switch plate on a box. Make sure the power is turned off if you are installing a plate on existing wiring. Note the different types of plates: double switch plate, combination switch plate and double-outlet plate, double-outlet plate, and switch plate and single-outlet plate.

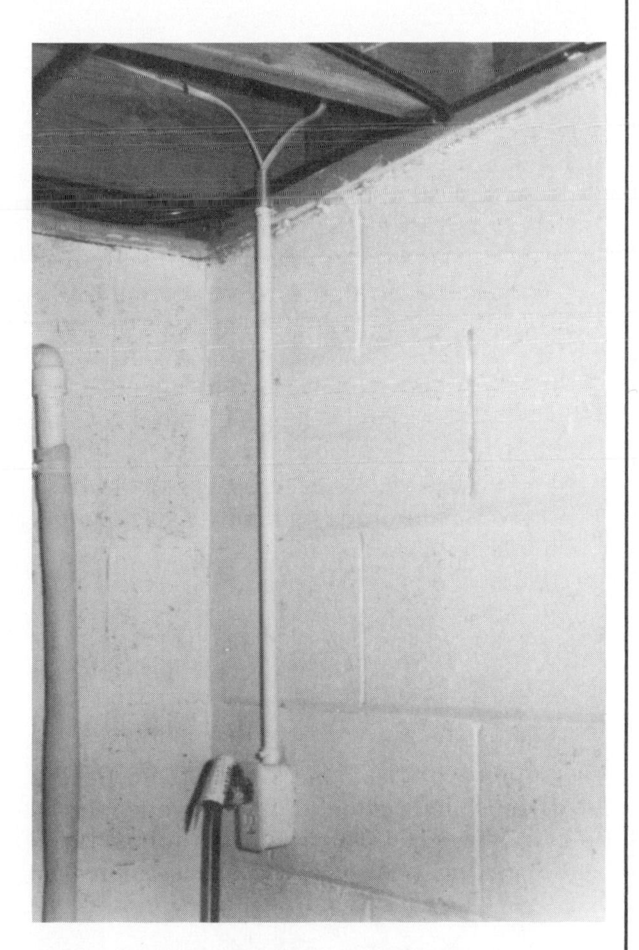

10–34. Handy box being used with a ½″ thinwall to protect the Romex from damage. This type of installation is usually found in a home in the basement where there is the possibility of damage to the unprotected Romex by the movement of heavy objects.

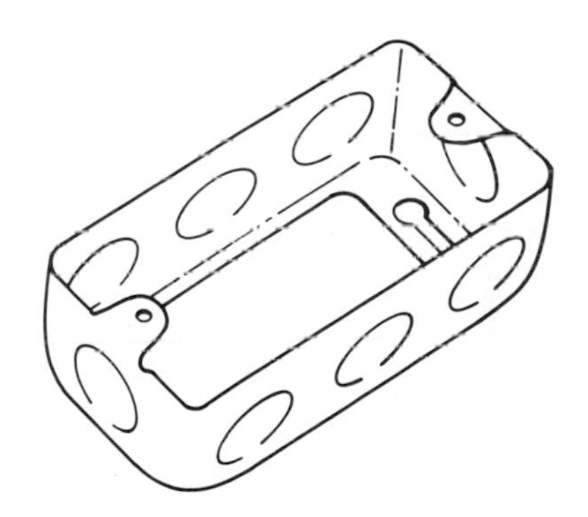

10–35. Handy box extension, needed where the box capacity is not sufficient to allow the number of wires or conductors required. This extension slips onto the two screws on the handy (utility) box. It is an exact duplicate of the handy box, but without a bottom.

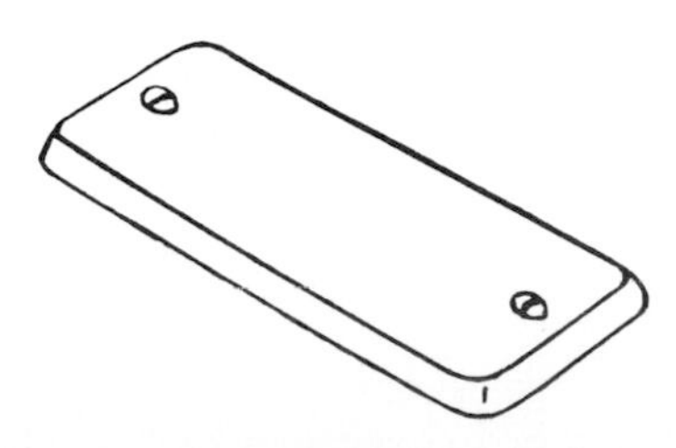

10–36. Blank cover for the handy box.

Octagonal boxes of 3″ and 4″ sizes are available for use with Romex, BX, thinwall, or rigid conduit. They come with or without clamps and have knockouts of various sizes. All of them have two screws for attaching a cover plate. Extensions are available for increasing the space to hold a number of conductors. Fig. 10-40 through 10-45.

Sometimes it is necessary to place a utility box or outlet box between two studs or joists. For this situation, an expandable bar hanger comes in handy. It expands to fit between

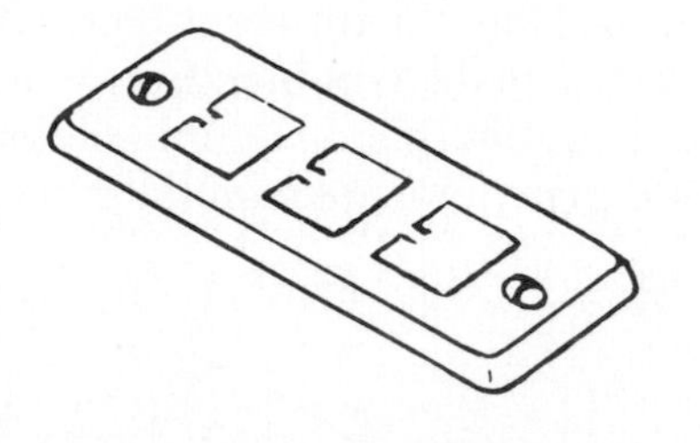

10–37. Handy box cover with 3 knockouts. Captive screws and a galvanized finish make up the total cover. "Captive screws" means they will not fall out when unscrewed from a handy box. They must be intentionally unscrewed to remove them from the cover.

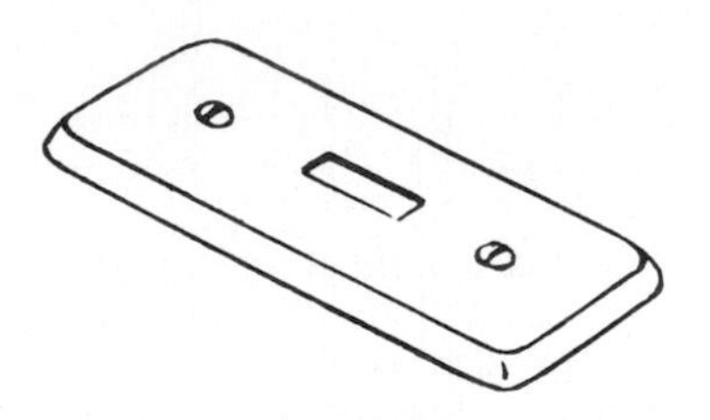

10–38. Handy box switch box cover for tumbler or toggle switch with square handles.

10–39a. Handy box cover for a 20-A, single-receptacle outlet.

10–39b. How to install a switch plate.

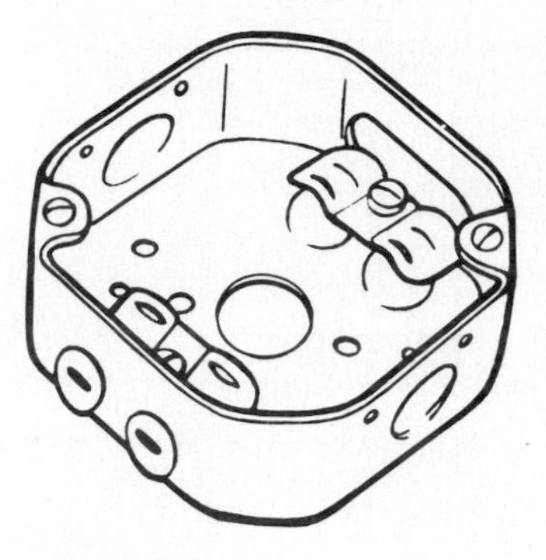

10–40. 4″ octagonal outlet box for Romex or BX with tapped grounding screw holes. Clamps are included.

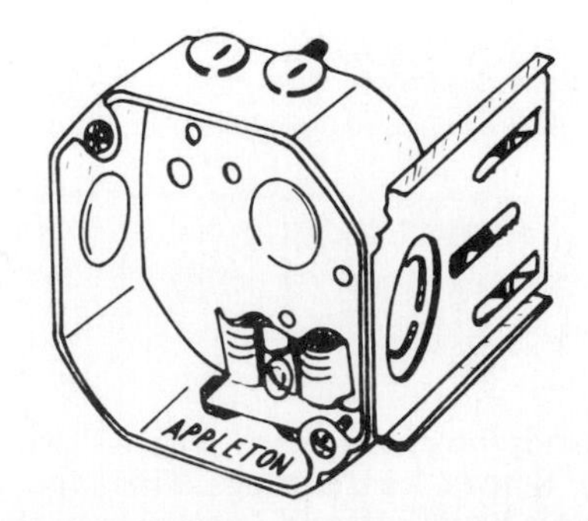

10–41. 4″ octagonal outlet box for Romex and BX with bracket, no nails needed. This one includes the clamps and has a tapped grounding screw hole.

studs or joists centered 14″, 16″, 18″, or 20″. The center clip is adjustable to any location between the maximum and minimum setting. Fig. 10-46. The bar hanger in Fig. 10-47 has nail holes for mounting between studs or joists. The sliding clip facilitates easy location of the box.

10–42. Outlet box used as a junction for Romex and flexible metal conduit. Note cover plate is blank, making this a junction box.

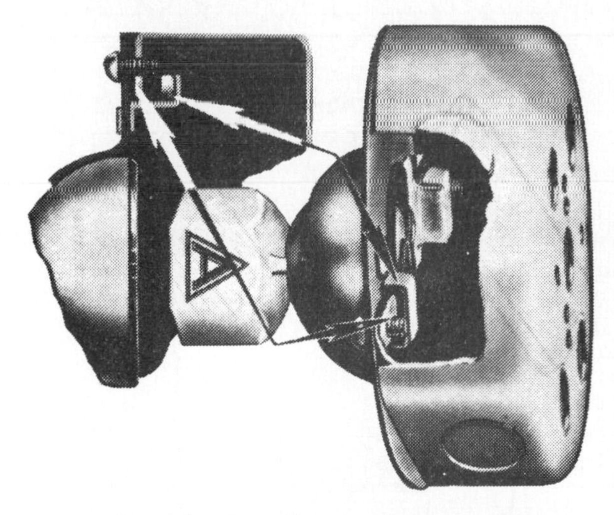

10–43. Special safety clips are shown on this swivel hanger cover, which is designed for use on a 4″ octagonal outlet box. The clips are permanently fastened to the cover. If the outlet box screws corrode through or vibrate out, the clips will support the weight of the fixture. The fixture hanger provides free swing of 20° from plumb in all directions. Two pear-shaped holes in the cover permit installation without removing outlet box screws. Just loosen the screws, slip on the swivel cover plate, turn 10° clockwise to position clips, then tighten.

Fig. 10-48 shows a Romex bar set with an outlet box. It will expand to fit joists on 14″, 16″, 18″, or 20″ centers.

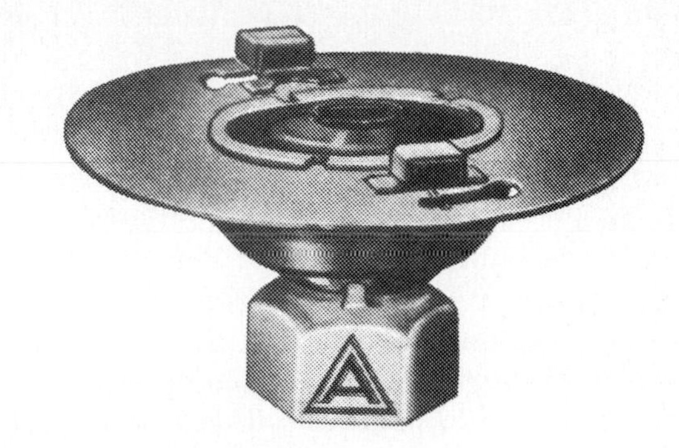

10–44. Fixture hanger cover with safety clips.

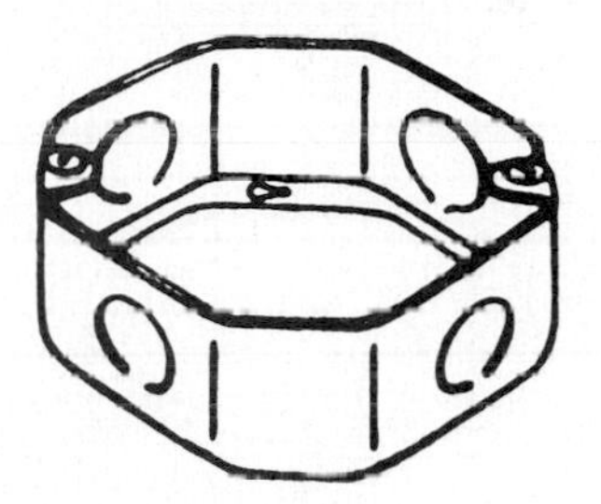

10–45. 3″ octagonal box extension ring.

10–46. Expandable bar hanger. Its range is 11½″ minimum to 18½″ maximum to be used between joists or studs on 14″, 16″, 18″, or 20″ centers. Galvanized.

10–47. Bar hanger, 18″ long with sliding clip.

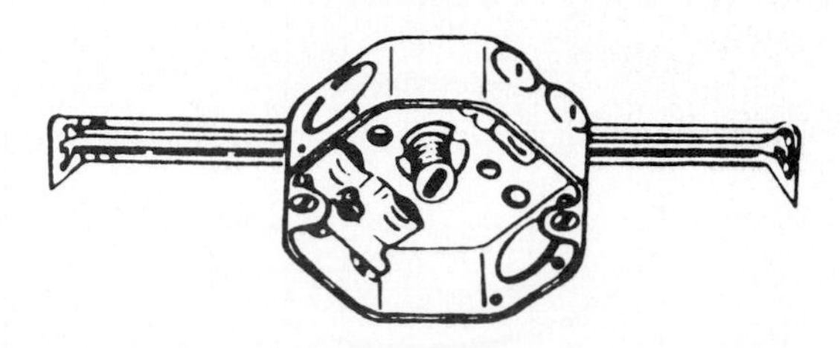

10–48. Outlet box mounted on an expandable bar hanger.

Occasionally, a cover is needed which will extend a box to fit flush with the wall. A raised cover ($\frac{5}{8}$″) for a 4″ octagonal outlet box is shown in Fig. 10-49. It is also available in other sizes.

Sometimes an outlet with a steel cover plate is required for areas subject to possible mechanical damage of the plate. A steel cover plate for a 4″ octagonal box will accommodate a duplex receptacle. Fig. 10-50. Fig. 10-51 is a blank cover plate for a 4″ octagonal box. Because a blank box cover is not always the answer, especially if you want to mount an object from the box, a blank cover with a 1″ knockout might be utilized, so that a fixture may be mounted from the 4″ octagonal box it covers. Fig. 10-52.

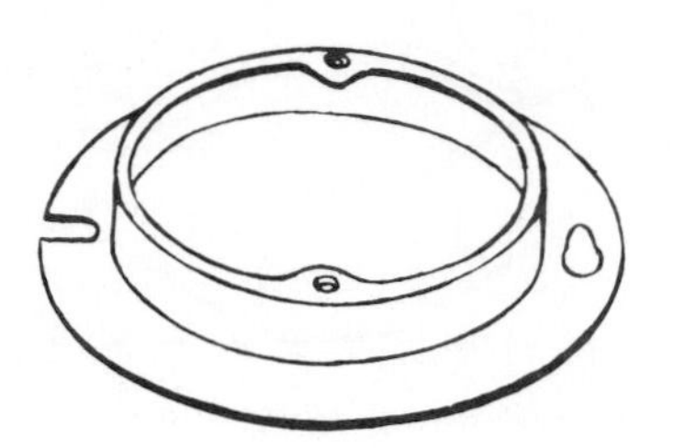

10–49. 4″ octagonal outlet box with a $\frac{5}{8}$″ raised cover.

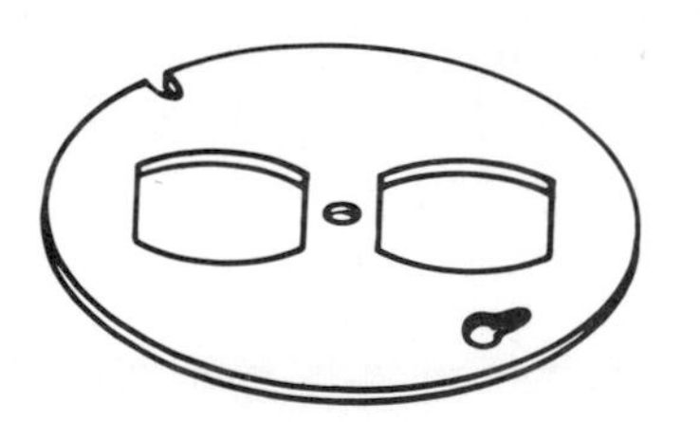

10–50. 4″ octagonal outlet box cover with flush mounting of a duplex receptacle.

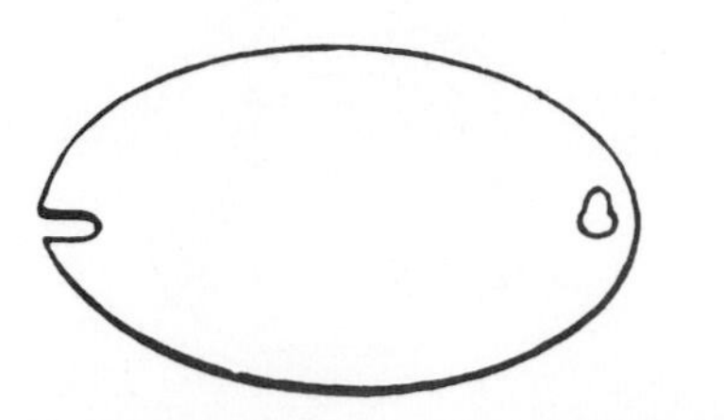

10–51. 4″ octagonal outlet box cover, blank.

In Figs. 10-53 through 10-57 you will find an assortment of boxes 4″ square, some with tapped, grounding-screw holes and some with clamps. A portion of these boxes have no brackets for mounting. Others have no-nail

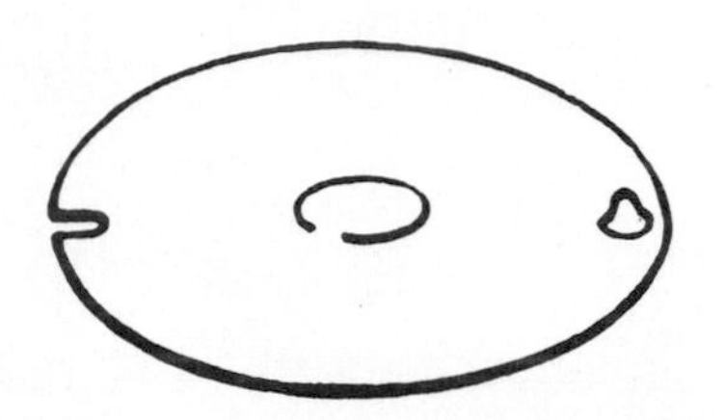

10–52. 4″ octagonal outlet box cover with $\frac{1}{2}$″ knockout.

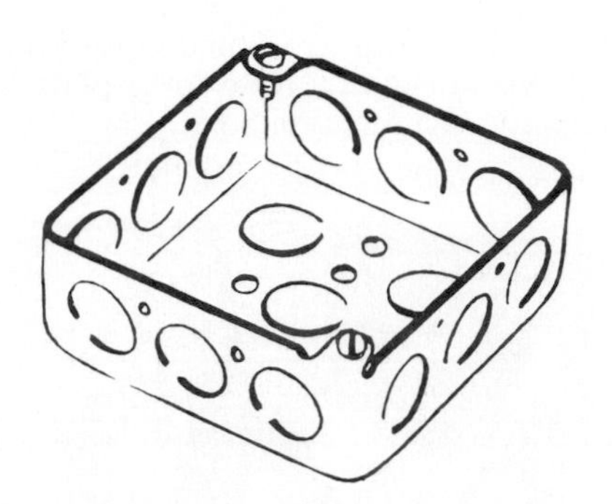

10–53. 4″ square drawn box with tapped grounding-screw holes and two screws for cover mounting. This one is made for use on rigid or thinwall conduit since it has both $\frac{3}{4}$″ and $\frac{1}{2}$″ knockouts.

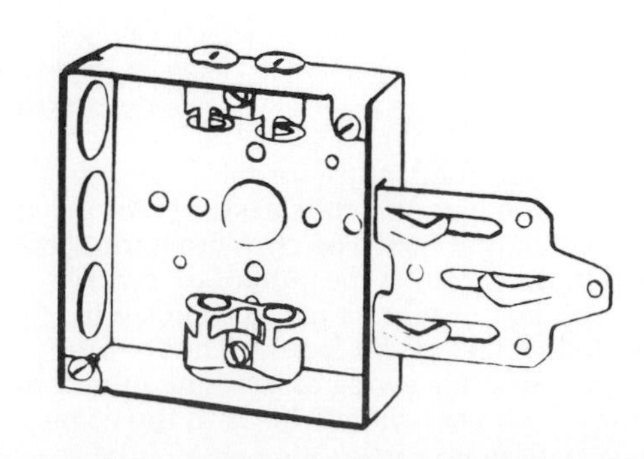

10–54. This square box with brackets does not need nails for mounting on studs. It comes with clamps for BX or flexible steel conduit.

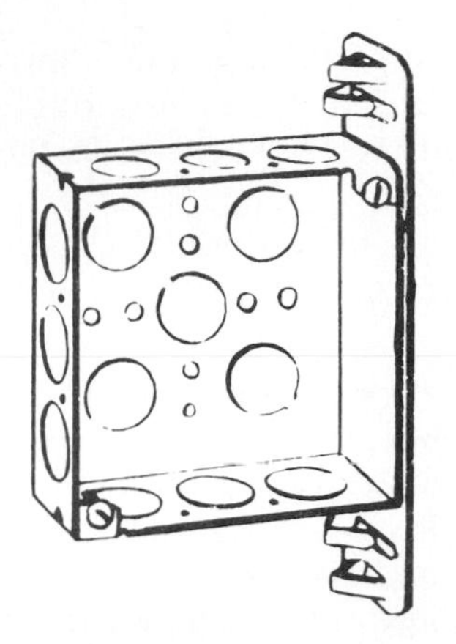

10–55. This 4″ square box has a bracket for mounting without nails and is designed for use with thinwall or conduit of the rigid design. No clamps are included; so if it is needed for use by Romex or BX wiring, it can be done by using the appropriate connector and knockout.

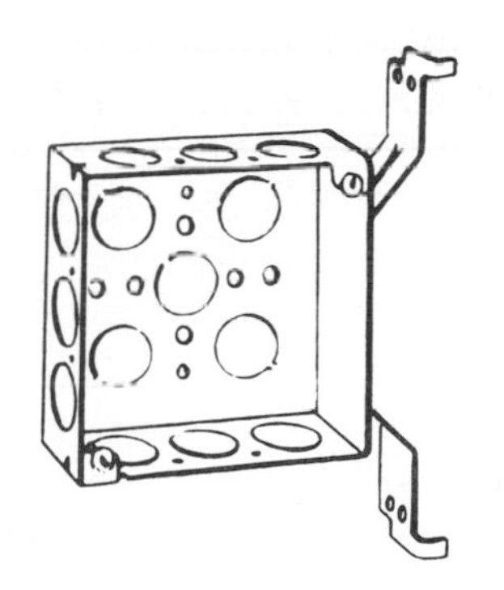

10–56. 4″ square box for thinwall or rigid conduit. This has a bracket which allows it to be aligned against the leading edge of the stud before nailing.

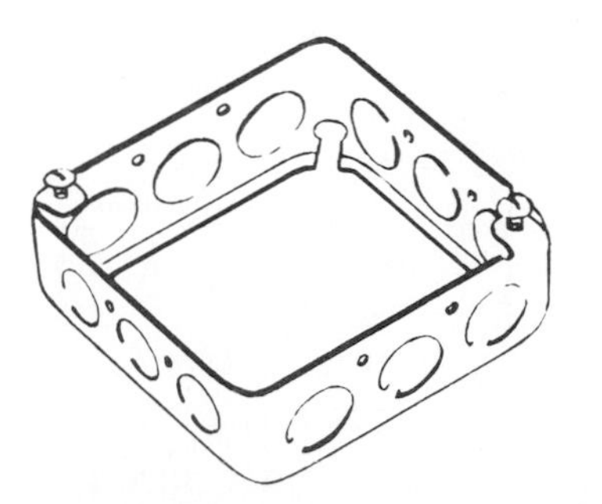

10–57. Four-inch-square box extension. It has two #8-32 tapped holes in the bottom in addition to the slide-on screw slots.

brackets and self-alignment brackets. Some use nails for mounting, while others will be used on conduit or thinwall by removing the knockout and inserting the proper connector. There is also an extension for these boxes. A box that is $4\frac{11}{64}''$ square has a greater cubic-inch volume for additional conductors. It is difficult at first to recognize the difference that only $\frac{11}{64}''$ makes. Once you are working with both the 4″ and the $4\frac{11}{64}''$ sizes, however, you will become very accustomed to the difference.

Covers for a 4″ square box are many and varied. Figs. 10-58 through 10-62 show various types and configurations. Note figure captions to obtain more details of size, function, and shape.

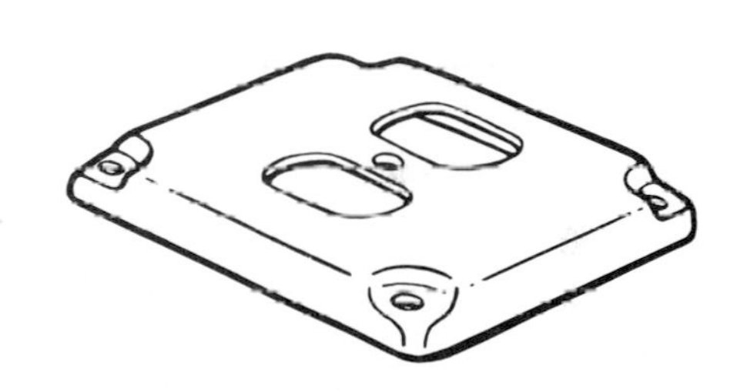

10–58. A $\frac{1}{2}''$ deep surface cover for a square box. This one will accommodate one duplex flush receptacle.

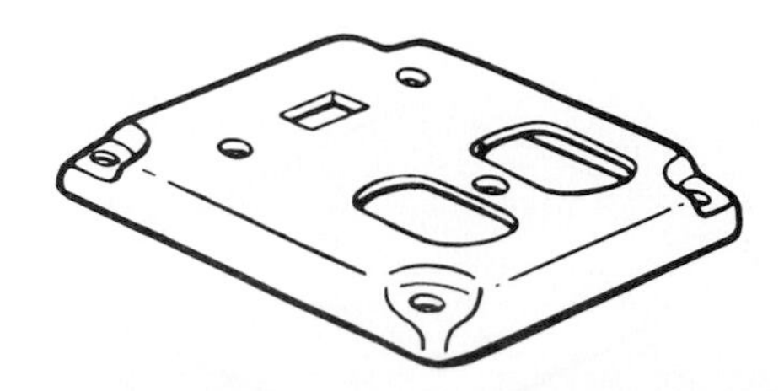

10–59. Surface cover for a 4″ square box. It is $\frac{1}{2}''$ drawn and is designed for a toggle switch and a duplex receptacle.

Thinwall conduit is shown with clamps, connectors, and a 90° bend with wires pulled through in Fig. 10-63. Empty conduit should be mounted in place and connected to boxes before insulated wires are inserted. Conduit must be used with steel boxes only. Nonmetallic boxes are shown in Fig. 10-64. Conduit comes in 10-foot lengths which are joined together with couplings. Cut shorter lengths with a hacksaw or tubing cutter. Ream the cut ends inside and taper with a file. Use a conduit bender to make all bends. Conduit should be supported with a strap every 6 feet on exposed runs.

When connecting conduit to boxes, fit the threadless end of the conductor over the conduit, and insert the connector through the box knockout. Then tighten the locknut.

After conduit and boxes are installed, pull the wires through the conduit into the boxes. Allow 8″ of insulated wire at each end of the box for connections. Use a white wire for neutral, and a black or red wire for the "hot" side.

In exposed work, conduit may be mounted on studs or rafters without additional protection. In concealed work, conduit must be supported.

Lighting Fixtures

Installing lighting fixtures may be accomplished in several ways. There are hanger supports that thread onto a threaded stud,

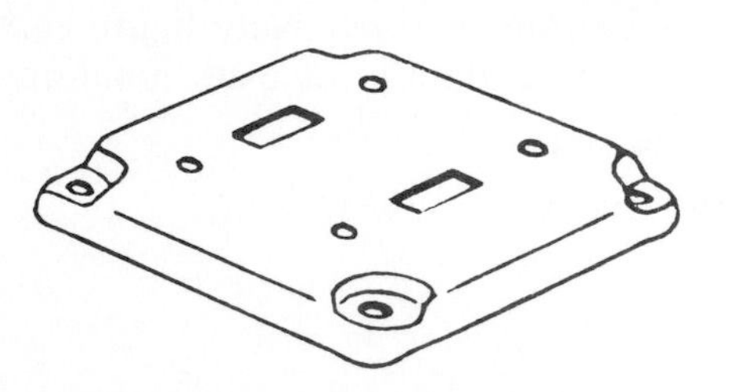

10–60. Switch cover for 4″ square box. This one is ½″ deep and will accommodate two toggle switches.

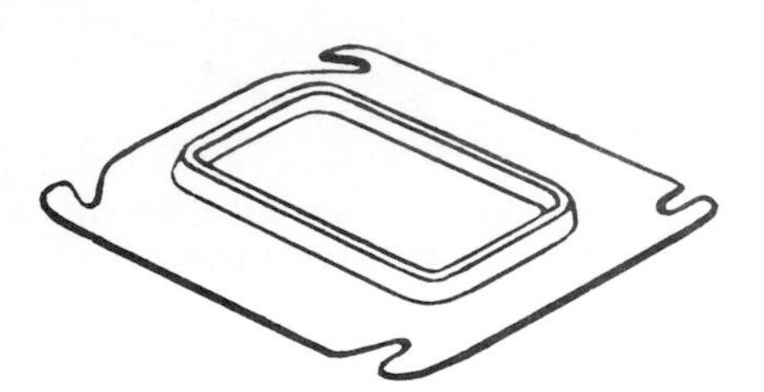

10–61. A ¼″ raised single-device cover for 4″ square box.

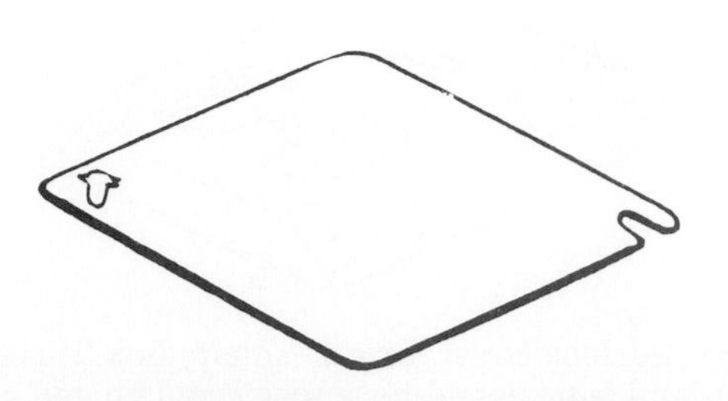

10–62. Flat cover, blank for 4″ square box.

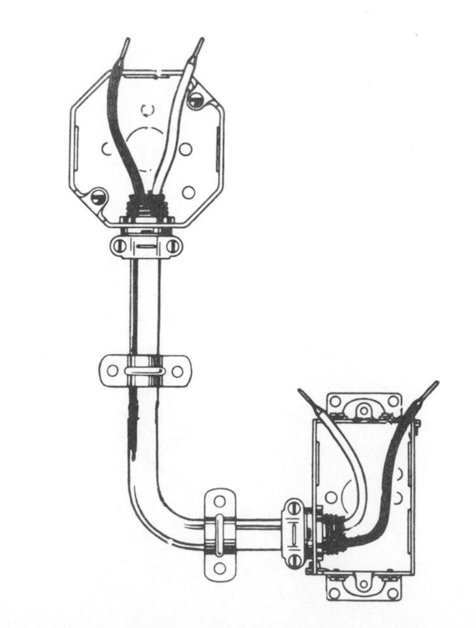

10–63. Thinwall conduit between switch box and octagonal outlet box. Note the straps and connectors.

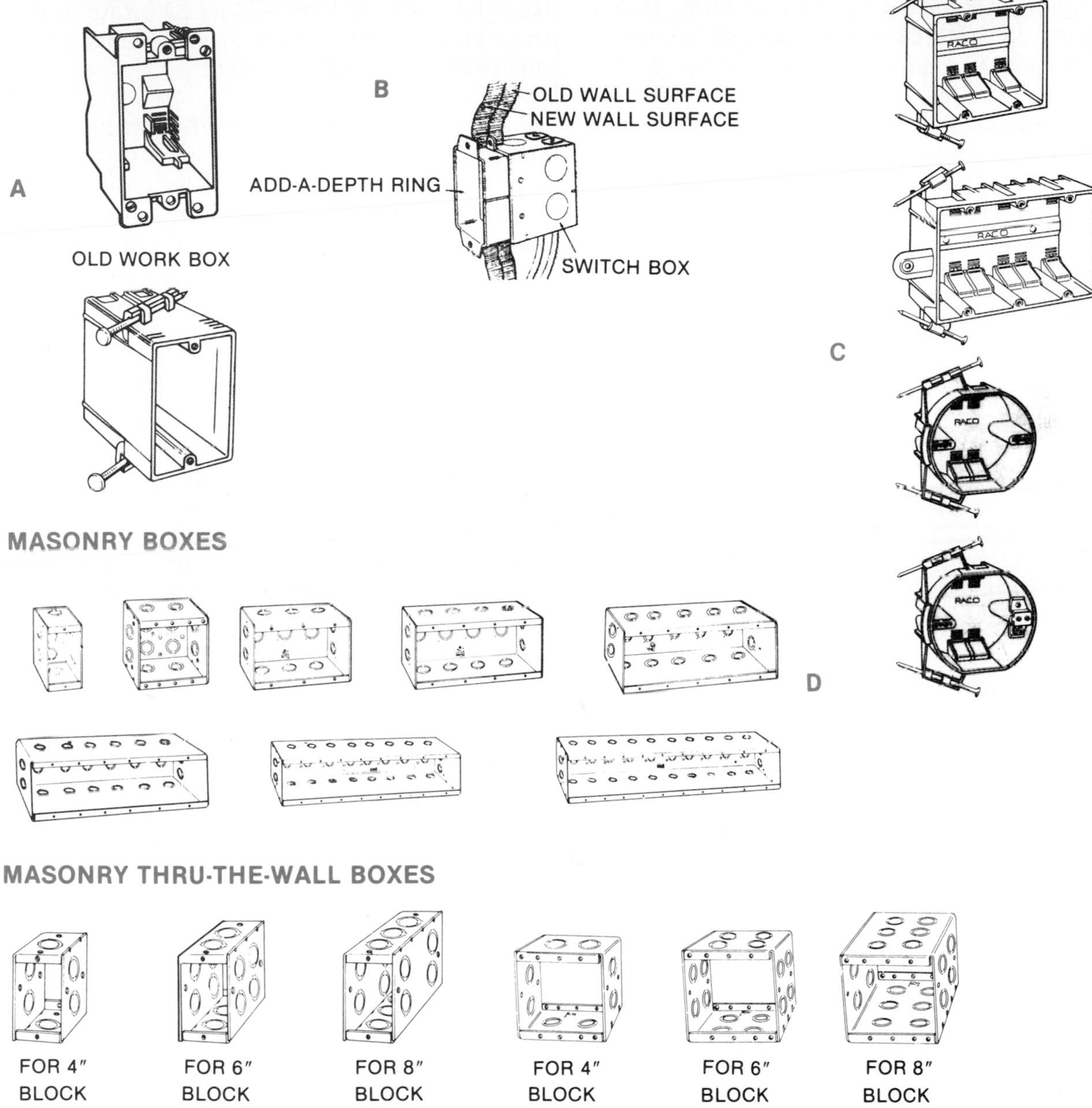

10–64. Nonmetallic boxes are made for old work and for new. See A. In B, the box has been adjusted for the addition of a new surface. The add-a-depth ring aids in making the extension fit the new surface. In C, the newer nonmetallic ganged boxes have nails for ease in attaching. Note the newer clips in the boxes so the Romex can be pulled through. The clamp holds it without any screws. In D, the masonry boxes come in a variety of sizes for all applications. In E, the masonry thru-the-wall boxes are also available in a number of sizes for 4″, 6″, and 8″ block.

mounted in the box, as in Fig. 10-65. Straps and machine screws are used to mount the fixture in Fig. 10-66. If there is no stud, the metal strap may be used to hold the canopy of the light fixture in place, as in Fig. 10-67.

Glass-enclosed ceiling fixtures are easily attached to the ceiling box by using a threaded stud. Fig. 10-68. Wall lights may be installed using the same method. Make sure the outlet found in the wall fixture is wired for full-time service—not controlled by the switch that turns the light on and off. Fig. 10-69.

Replacing a Damaged Lampcord

Plugs and cords sometimes need repair. Replacement plugs with a heavy-duty capability should be used. Fig. 10-70 shows how the *underwriter's knot* is used to relieve strain put on the connection when the cord is pulled. The two plugs are replacement types that do not have a third wire for grounding.

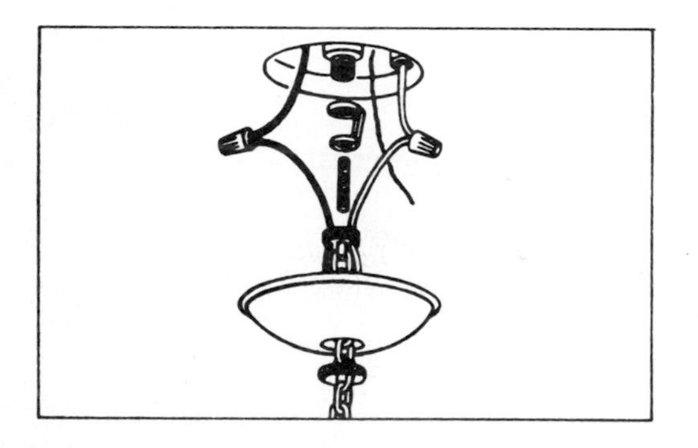

10–65. Mount large drop fixtures by simply using a screw hanger support onto the threaded stud in the outlet box. Use solderless connectors (wire nuts) to connect the electrical wires and a grounding clip for the extra uninsulated ground wire. Raise the canopy and anchor in position by means of a locknut.

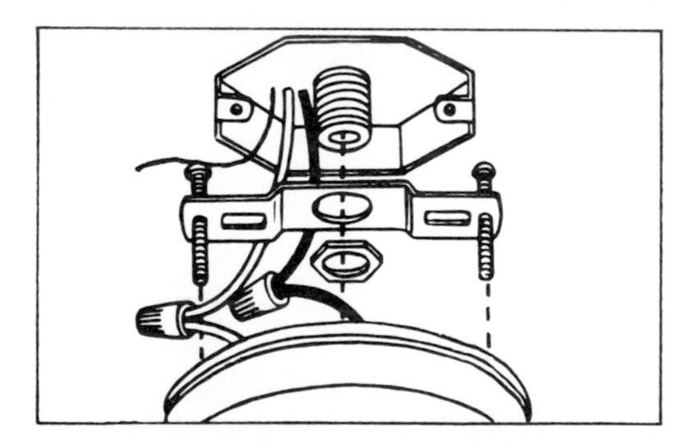

10–66. Outlet box has a stud in this case. Insert the machine screws in threaded holes of the metal strap shown. Slip the center hole of the strap over the stud in the outlet box. Hold the strap in position by a locknut. Connect wires with wirenuts and slip the canopy over the machine screws; fit flush and secure the fixture with two cap nuts. Don't forget to anchor the uninsulated ground wire to the box with a ground clip or to the box's threaded grounding screw.

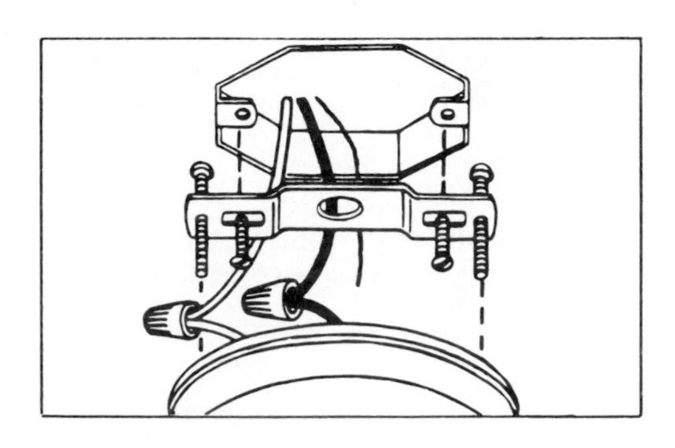

10–67. If there is no stud, insert the machine screws as shown here. Fasten the ears of the outlet box and the strap with screws. Then align the canopy onto the two screws pointing down and cap off with cap screws.

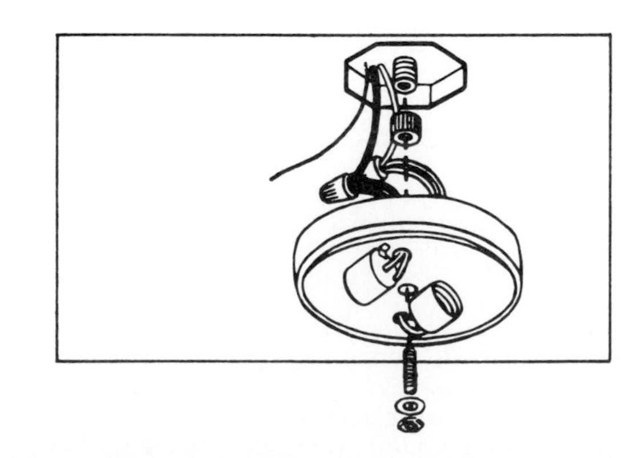

10–68. Glass-enclosed fixtures can be installed by the method shown here.

Some lamps use an in-line switch for control. It may be replaced easily, or installed, if needed, by following the illustrated steps in Fig. 10-71.

An adapter is available for three-pronged plugs to make them usable on older, two-wire systems. Fig. 10-72. The lug is inserted under the screw that holds the cover plate onto the outlet.

Splicing Wires

Manufacturers of a specific type of fastener or clip have specialized methods that they recommend to splice wire. Most of these will be covered in Figs. 10-73 through 10-108 (pages 302–310). Read the captions under the figures to learn the proper procedure for making a splice and insulating it.

The proper disposition of the grounding wires inside a nonmetallic or a metal box can be a problem. NEC Article 250-74 covers connecting a receptacle grounding terminal to the box. It is best to know how this is done properly. Take a look at Figs. 10-109 & 10-110 (page 311) to see how it is done with a wire connector that allows one of the wires to continue through and be terminated properly on the receptacle. In the metal box the ground wire is terminated at the receptacle and on the box. Fig. 10-111 (page 312) shows how the connector operates to allow the system grounding wire and through wire to be connected properly.

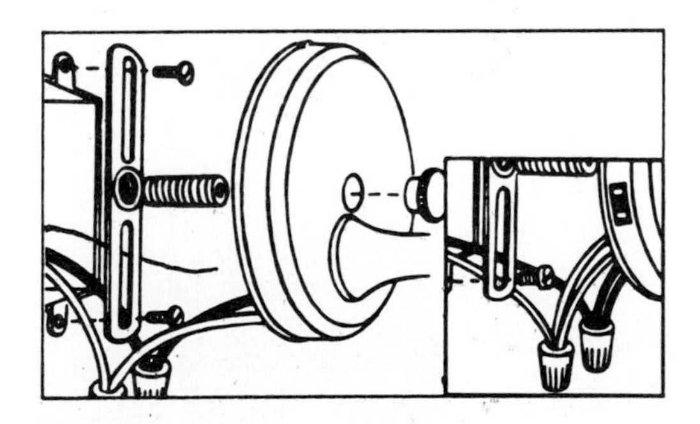

10–69. Wall brackets or lights are installed by strapping to the ears of the box, then using a nipple and cap to complete. Don't forget the ground wire and the ground clip. If there is an outlet which is "always on," wire according to the insert.

To make proper use of the connector, follow the steps below:

1. Make sure the power is off.
2. Strip short grounding wires 1″, if insulated. Strip "through" wire (the one that is to go through the grounding connector) 1″, if insulated, plus the amount needed to terminate the device.
3. Bunch the wires. Hold the short wire ends even at the desired distance from the end of the "through" wire. Insert all wires into the connector allowing the "through" wire to extend through the connector. Pre-twisting is not necessary.
4. Push the wires firmly into the connector when starting. Then screw on the connector until it is tight.

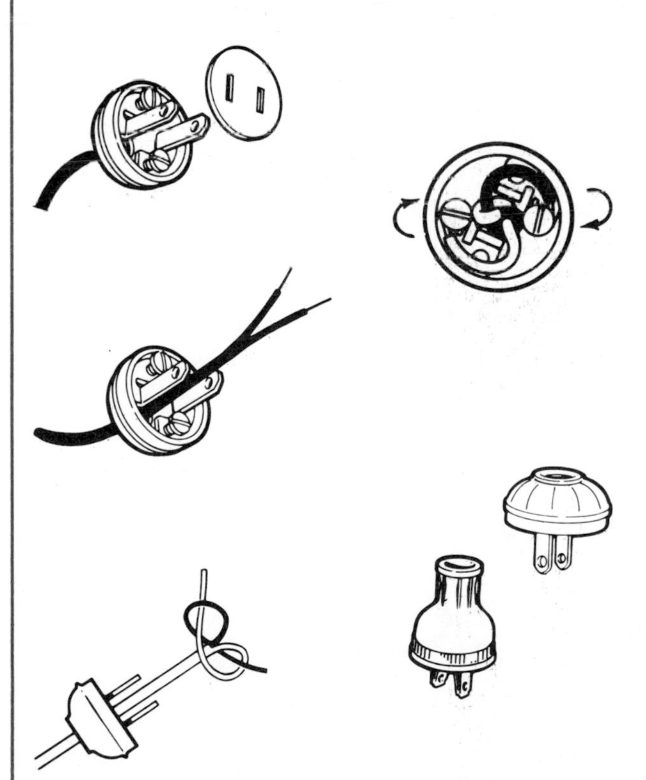

10–70. Replacing a damaged lampcord plug with an attachment plug. Note the Underwriters' knot and how it is looped to prevent strain on the connection when the cord is pulled.

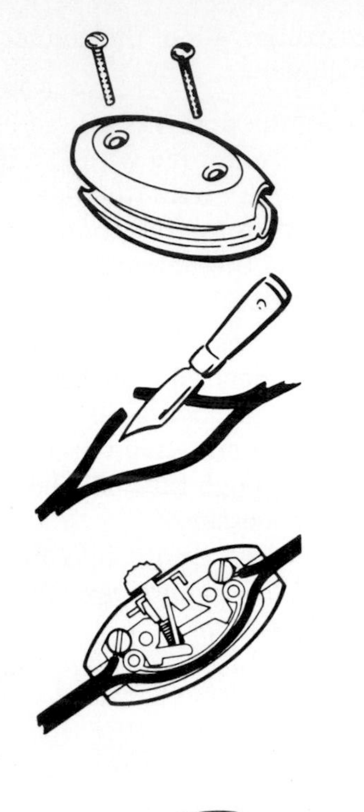

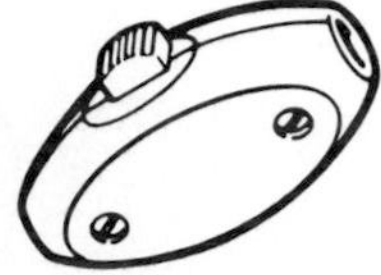

10–71. Some lamps have an in-line switch for controlling the light bulb. A switch can be put in the line easily as shown. Make sure the power is off or the plug has been pulled before attempting to work on the line.

10–72. This 3-wire grounding adapter makes it possible to use three-pronged safety grounded plugs with a conventional two-wire outlet. Make sure the green insulated grounding wire attaches to the outlet box by being placed under the screw which holds the cover plate in place.

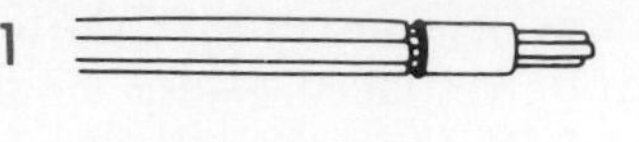

10–73a. To splice wires: 1. Insert untwisted stripped wires through the splice cap. 2. Twist wires. 3. Cut wires flush with cap. 4. Insert tool—squeeze to crimp. 5. Snap on a nylon insulator.

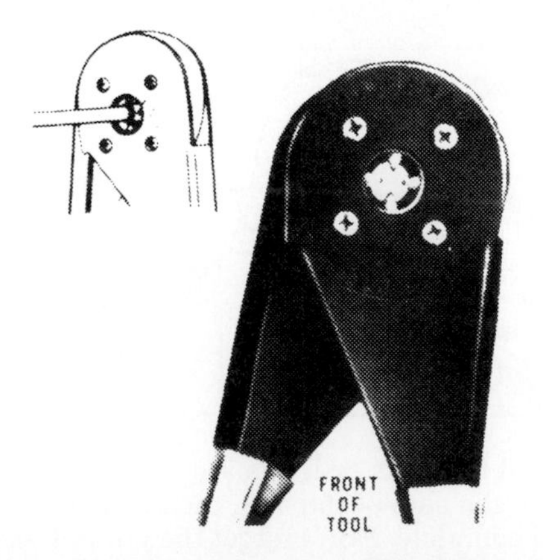

10–73b. Tool for splice caps. The wire shape which results can be seen in Fig. 10–101.

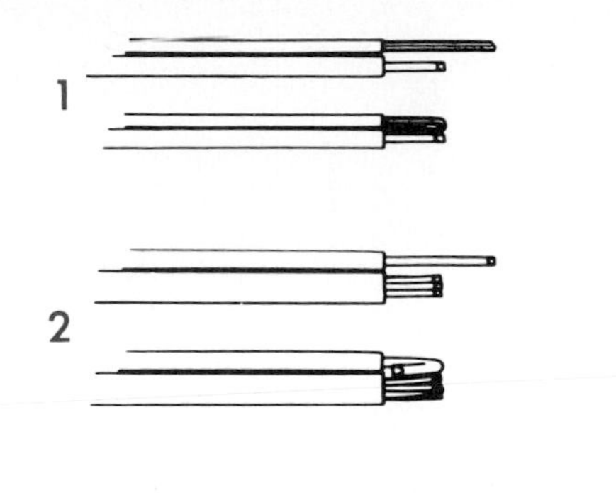

10–74. Splicing solid and stranded wires: 1. Loop stranded wire, if No. 16 or smaller. 2. Loop solid wire, if smaller than stranded wire. 3. When joining 2 or more solid wires to a larger stranded wire, twist the solid wires together.

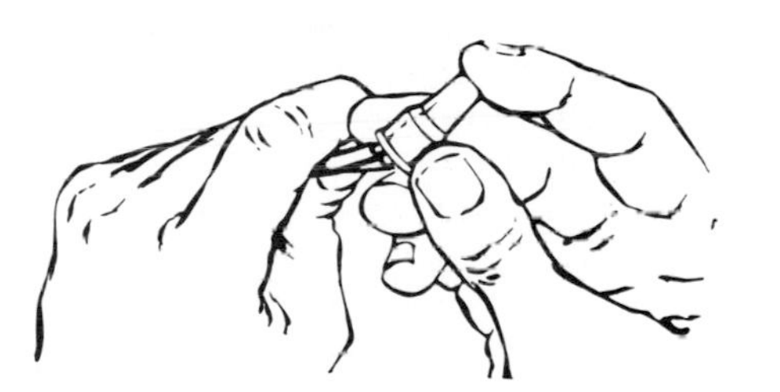

10–75. To insulate splices, just snap on the nylon insulator over the installed splice cap.

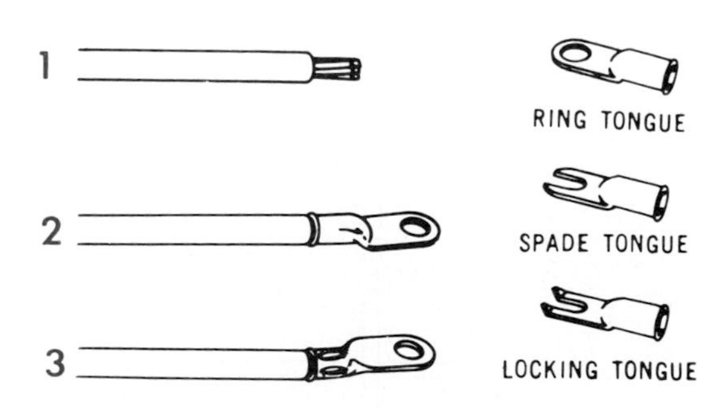

10–76. To terminate wires with lugs: 1. Strip wire(s) approximately $\frac{5}{16}''$. 2. Slip lug on wire with tool in position. 3. Insert in tool with flat side of the lug in the up position so that the tongue enters the slot in the latch. 4. Squeeze the tool to crimp.

10–77. To remove splicing caps or lugs without damaging the wires, snip splice cap at both ends of one crimp, counteracting cutting pressure with the index finger. Then peel off the cap. To remove the lugs, snip off tongue and proceed as for the splice caps.

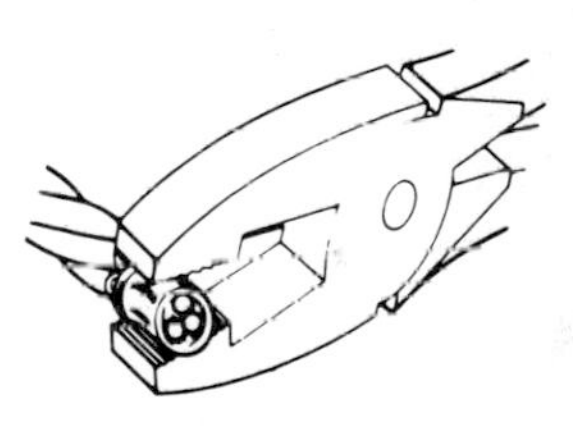

10–78. To remove "cold working" cap, apply pressure alternately at two points 90° apart and between crimps—then pull off the cap. This is most effective when the cap is not full of wire.

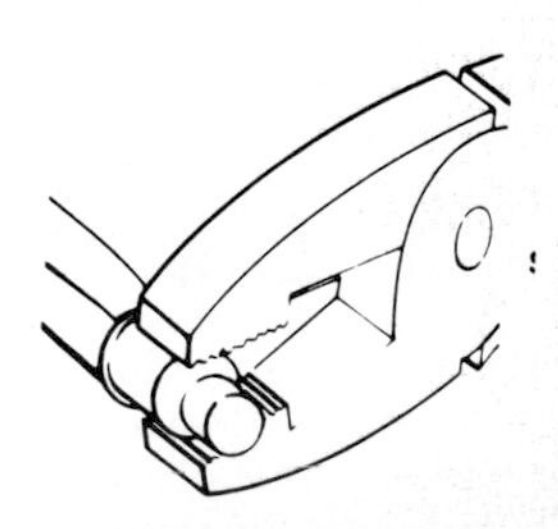

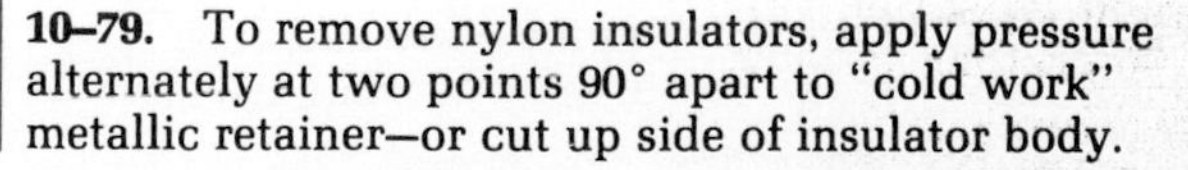

10–79. To remove nylon insulators, apply pressure alternately at two points 90° apart to "cold work" metallic retainer—or cut up side of insulator body.

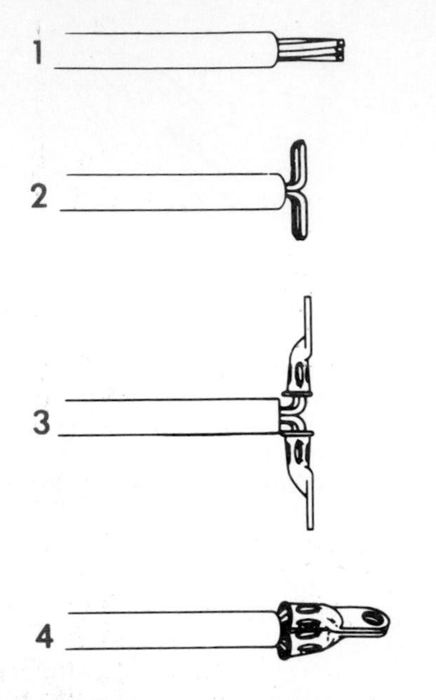

10–80. To terminate #6 wire using two lugs: 1. Strip wire approximately $\frac{3}{8}''$. 2. Untwist wire lay and separate strands into two approximately equal groups. 3. Crimp a lug on each group. 4. Bring flat sides of lugs together.

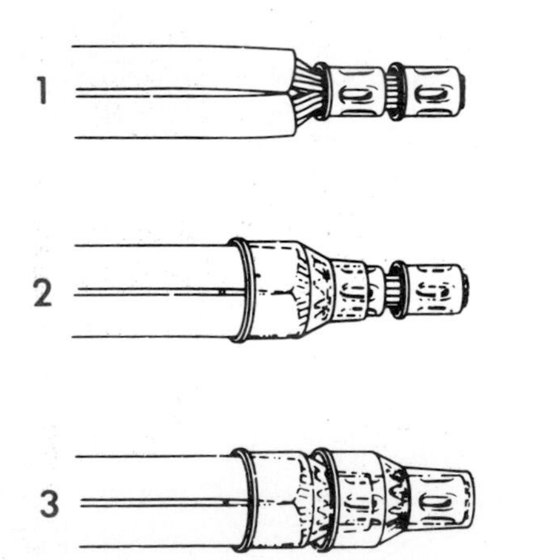

10–81. How to make a strain-relief splice for service entrance: 1. Strip both wires approximately $1\frac{1}{2}''$ and untwist wire lay; install two splice caps leaving about $\frac{1}{16}''$ between the two caps. 2. Cut off the top half of the tip of an insulator and push remainder over the cap. 3. Snap another insulator over the end cap.

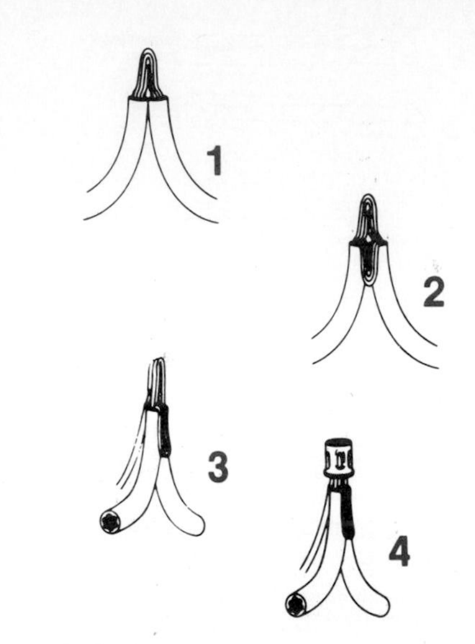

10–82. Reducer tap off larger wire (for parallel street lighting and similar installations): 1. Strip off insulation approximately 2″ and squeeze to make the wire loop as shown. 2. Bend enough strands down so the remainder will fit a splice cap. 3. Lay tap wire against the loop, put on a splice cap and crimp. 4. Insulate with tape or insulator if wire insulation permits.

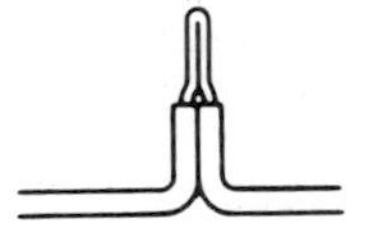

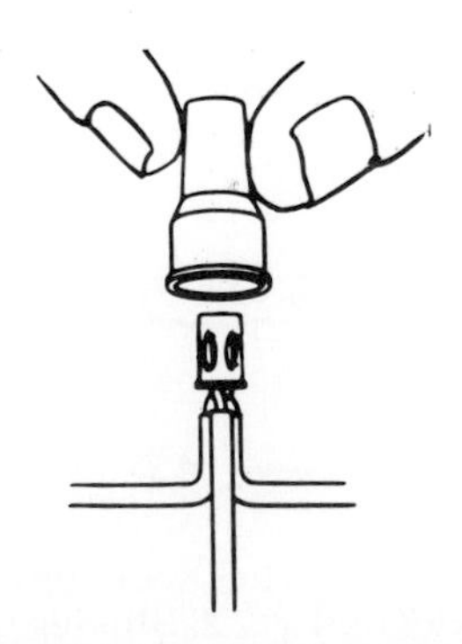

10–83. Tee-tap (where slack permits): 1. Strip and loop wire. 2. Add stripped tap wire, splice, and snap on the insulator.

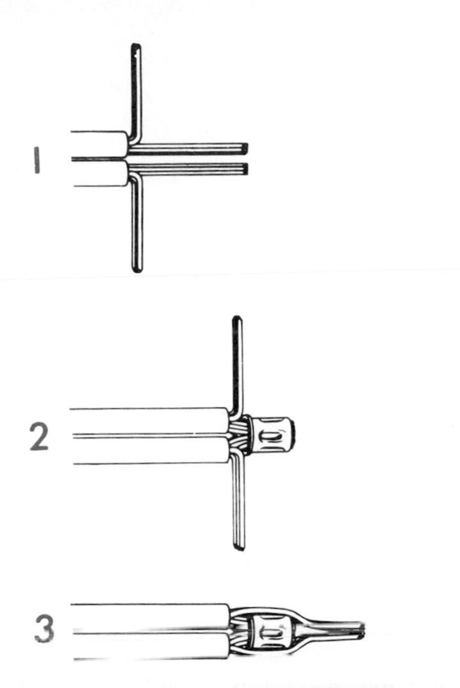

10–84. To splice two No. 4s (using two splice caps): 1. Strip both wires $1\frac{1}{2}''$, untwist wire lay and bend approximately $\frac{1}{2}$ of each wire at right angles. 2. Install splice cap as shown and trim off excess wire. 3. Shape remaining strands around the splice cap.

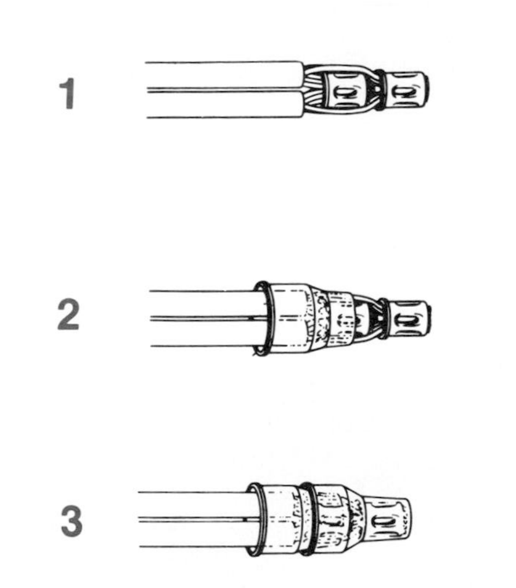

10–85. To complete the splice, install another splice cap: 1. Cut off the top half of the tip of an insulator and remove the retainer ring—force it over the lower cap. 2. Snap another insulator over the end cap.

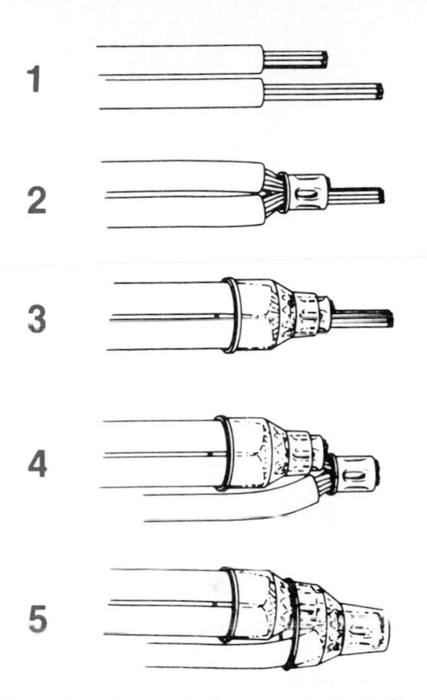

10–86. To splice beyond the range of a splice cap (3 or more No. 6s or 4 or more No. 8s): 1. Strip one wire $1\frac{1}{2}''$, another wire $\frac{1}{2}''$; untwist wire lay. 2. Install splice cap. 3. Cut off the top half of an insulator and push the remainder over the cap. 4. Strip the third wire $\frac{1}{2}''$ and splice as shown. 5. Snap on another insulator.

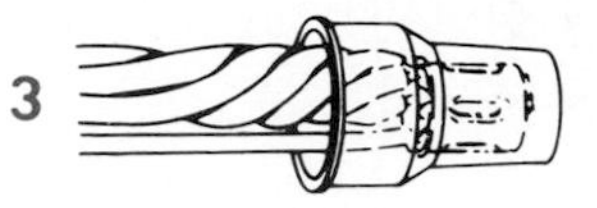

10–87. Adding wire (No. 12 or smaller): 1. Loop wire to be added and press into the crimped indentation of the cap. 2. Force the splice cap over the joint, then crimp. 3. Insulate with a snap-on insulator.

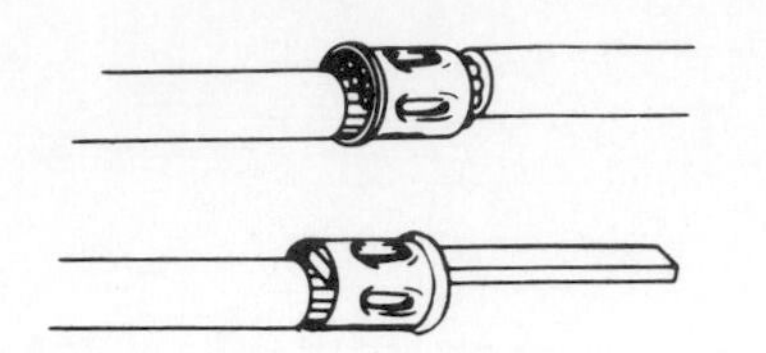

10–88. Parallel splice for attaching one short free length of wire, or for splicing short free length of flexible lead to solid conductor of various shapes as in coil windings, etc.: 1. Place tool in position to allow passage of wire through the tool. 2. Insert wires into opposite ends of the splice cap and crimp; insulate with tape.

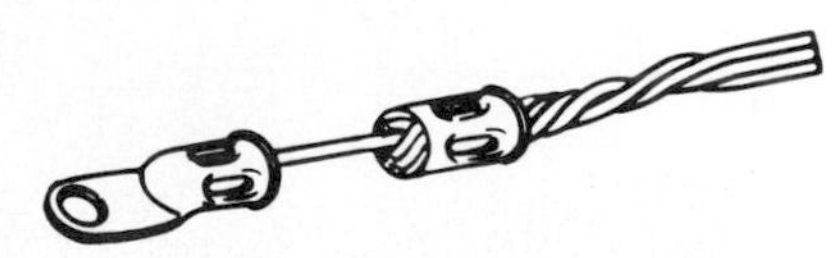

10–89. Grounding connection. Splice wires with a splice cap, leaving one wire extending through the cap to permit attachment of a lug.

10–90. Terminating two or more wires with a lug. This is a technique used if the barrel capacity of the lug is too small to hold the wires. Splice wires, leaving one or more wires extending through the cap to permit attachment of the lug as shown.

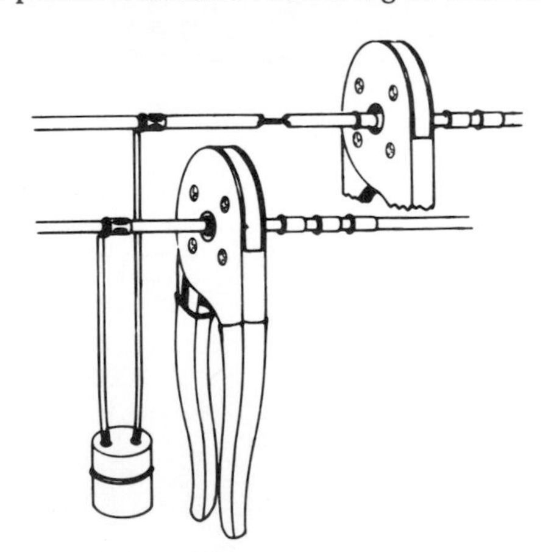

10–91. To splice pigtail sockets to streamers, strip streamer wires at required number of splice caps. Put tool latch in position and slide over the streamer. Crimp each cap in succession as shown; insulate with tape.

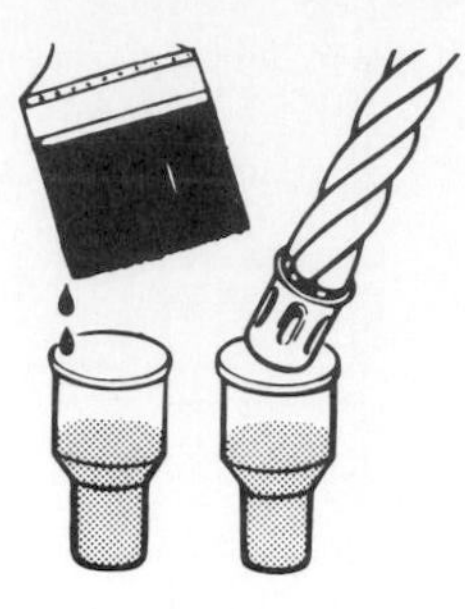

10–92. To hermetically seal a splice, fill the insulator to about $\frac{1}{3}$ of its capacity with insulating compound and push over the installed splice cap for a perfect hermetic seal.

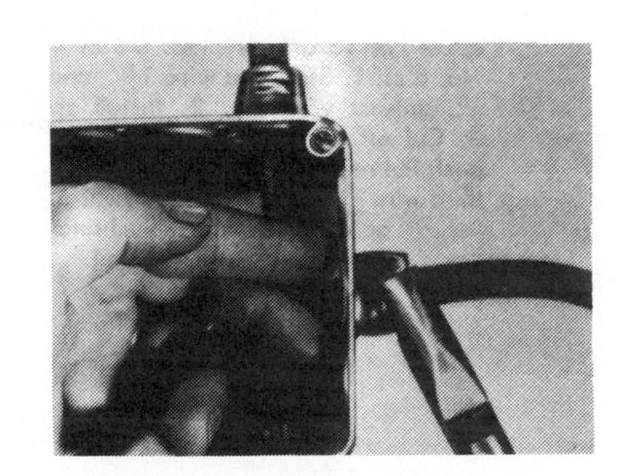

10–93. Romex connectors—no screws, no locknut. Just use pressure to hold the Romex in place.

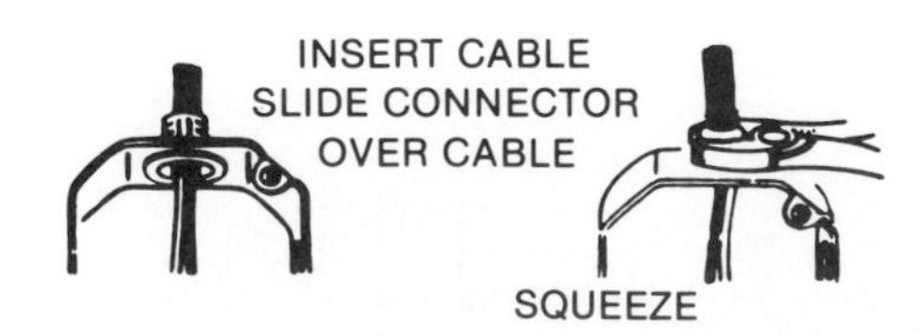

10–94. Insert the Romex cable; slide the connector over the cable from inside the box. Then squeeze the connector on the outside of the box to make a tight fit with no movement of the wire or connector.

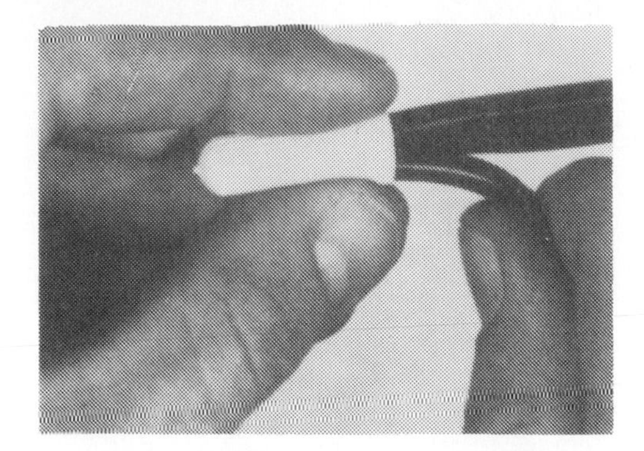

10–95. Wire nut insulator used for a splice of wires inside a switch box or outlet box.

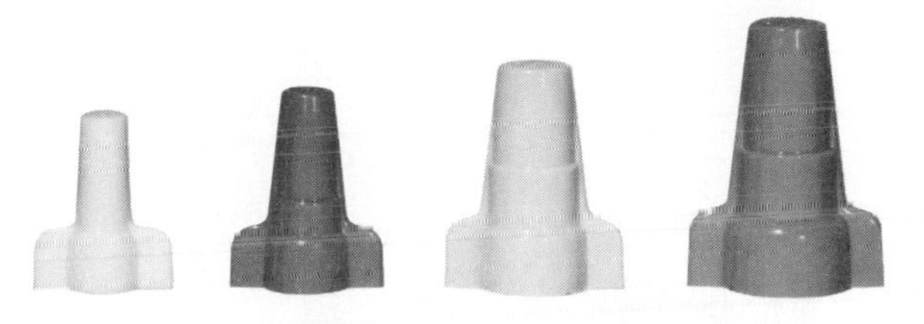

10–96. Wire nut insulators used in various sizes for different size wires. They are also available in color for color coding of wires.

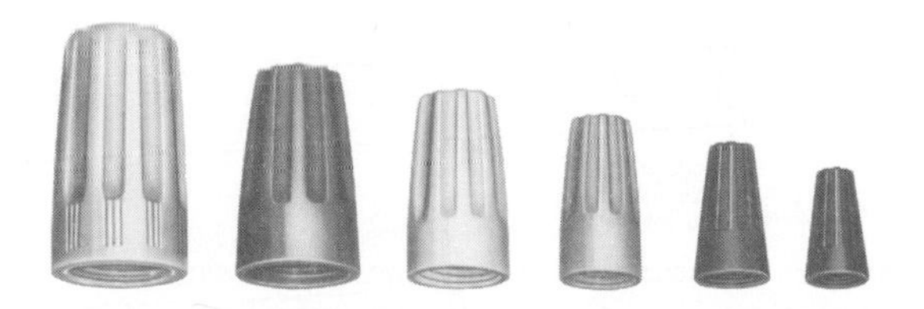

10–97. Wire nuts with a Bakelite case and copper inside.

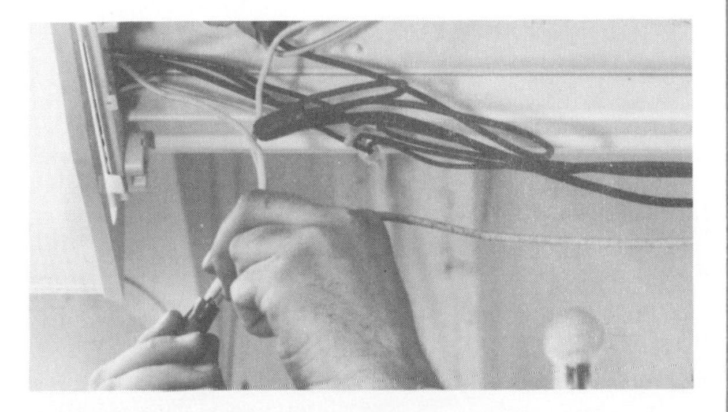

10–98. Using wire nuts for the splicing needed in the installation of a fluorescent fixture.

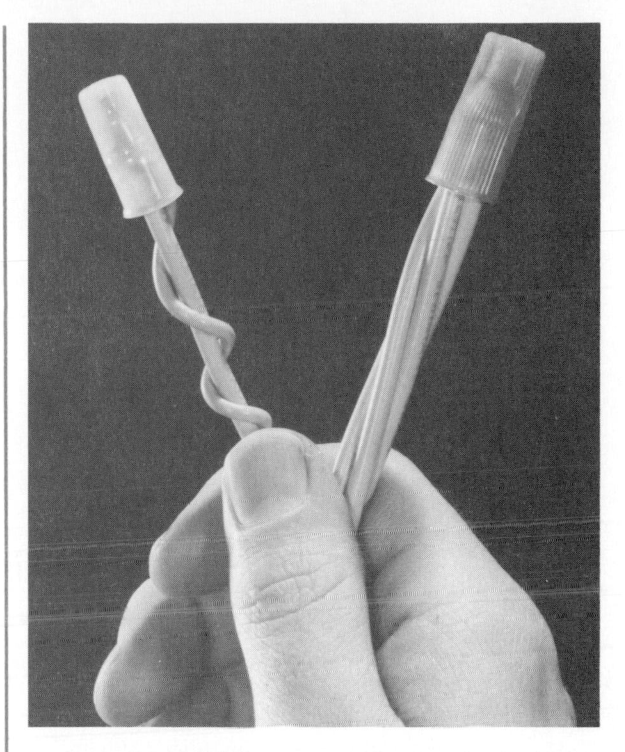

10–99. Two- and three-wire splices with plastic "see-through" insulators.

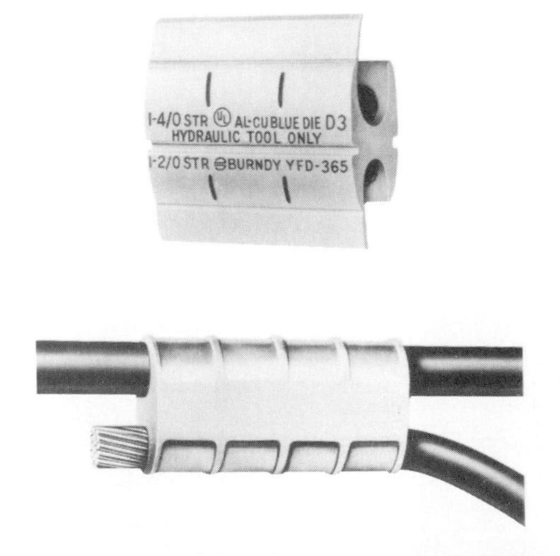

10–100. A splice using a hydraulic tool to make sure the tapped line and the other wire are firmly attached with maximum contact between the wire and the connector.

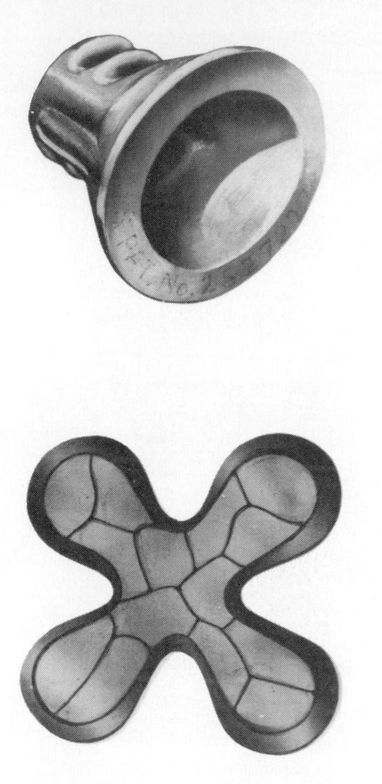

10–101. End cap and the resulting wire configuration after the splice has been made with a tool capable of exerting tremendous pressure. This way the surfaces of the wire are not able to contact the air and corrode.

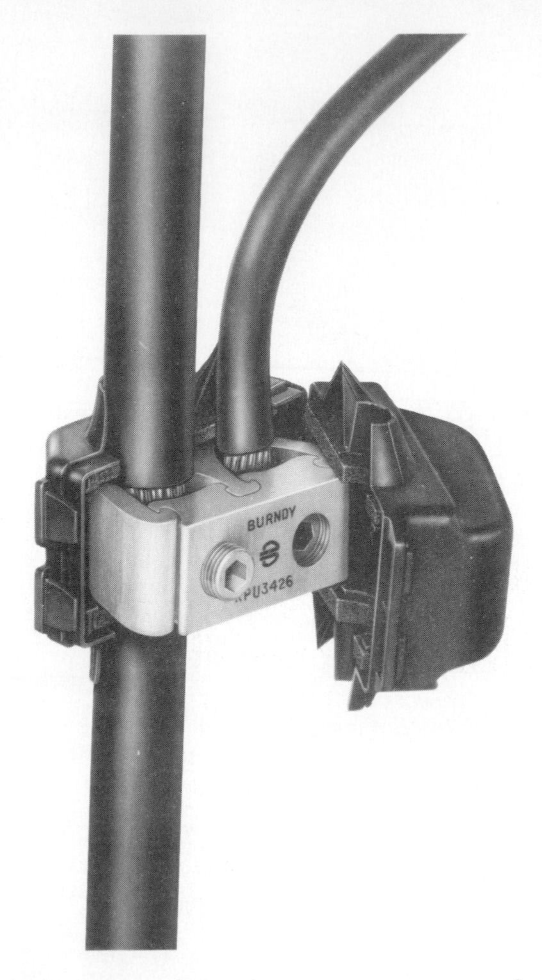

10–103. A tap splice using a screw-pressure and lock design. Note the plastic cap which snaps in place, thus insulating the splice.

10–102. Various sizes and shapes of end caps for splicing wires.

10–104. One-, two-, and three-wire terminations with screws to hold the wire in place against the walls of the connector. These are usually used in industrial applications where large-diameter wires are common.

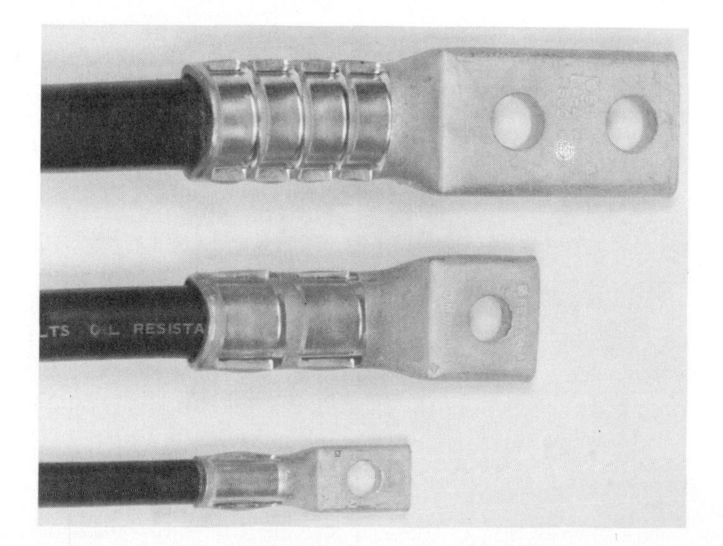

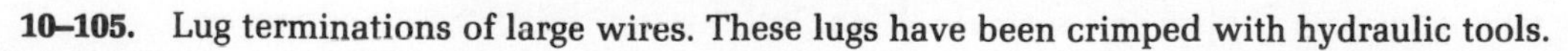

10–105. Lug terminations of large wires. These lugs have been crimped with hydraulic tools.

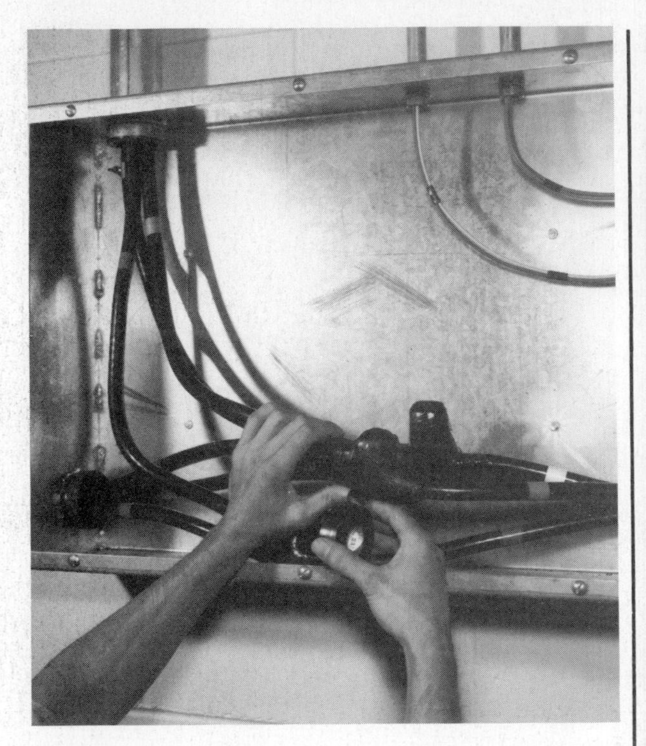

10–106. Using tape to insulate a splice inside a junction box.

10–107. Using tape to insulate a splice on a motor.

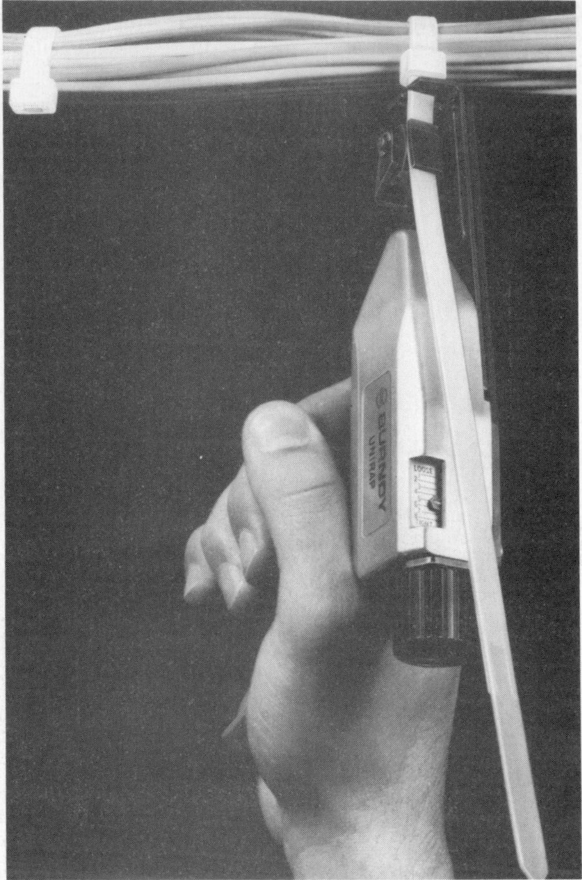

10–108. Using a tool to tie loose wires together.

Mounting Boxes in Existing Walls

A switch box may be mounted by using, the existing wall as part of the support device. Figs. 10-112 & 10-113. A special clip is made onto the box in Fig. 10-112 which will exert pressure from inside the wall when the screws are tightened. In Fig. 10-113 a piece of sheet metal is inserted to fold over inside the box and fit behind the dry wall to support the box. Note that the plaster ears on the box are important in adding to the support when a switch is turned on and off, or when a plug is pulled from an outlet mounted in the box. These are two of the many methods that are used to add an outlet to an existing wiring system.

ALUMINUM CONNECTIONS

Aluminum is a more active metal than copper. In the presence of moisture, the aluminum will erode. This susceptibility to galvanic corrosion, or electrolysis, is a problem because it will weaken an aluminum connection. Fig. 10-114.

The electrolytic action between aluminum and copper can be controlled by plating the aluminum with a neutral metal (usually tin). Such plating not only prevents electrolysis from taking place but also helps keep the joint tight. As an additional precaution, however, a joint sealing compound should be used. The compound contains fine zinc particles which break through the oxide film that forms on an aluminum connector. It also seals out air and moisture.

Upon exposure to air, aluminum immediately becomes coated with a film of oxide. This oxide film has the properties of a ceramic,

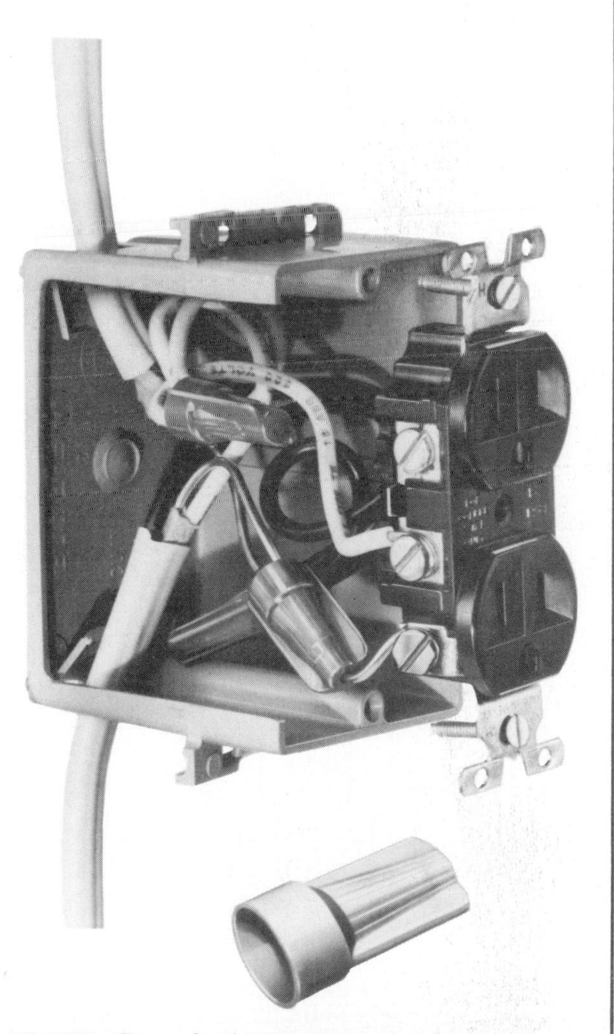

10–109. Ground wire connectors used to terminate the uninsulated wire in the nonmetallic box.

10–110. Metallic box with a ground wire connector. Note how three ground wires are twisted together and one emerges to attach to the ground screw on the receptacle.

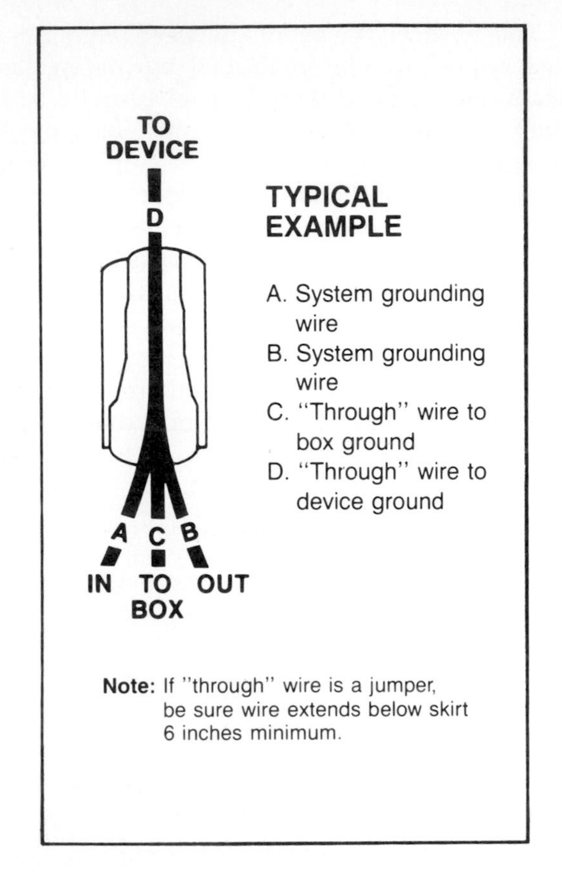

10–111. Grounding connector showing the "feed through" wire.

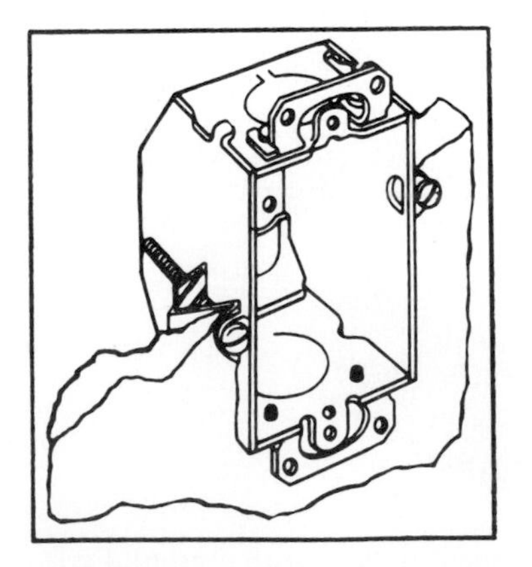

10–112. Using a box with "grip-tight" brackets to hold a box in place in the addition of a switch off outlet to existing wiring.

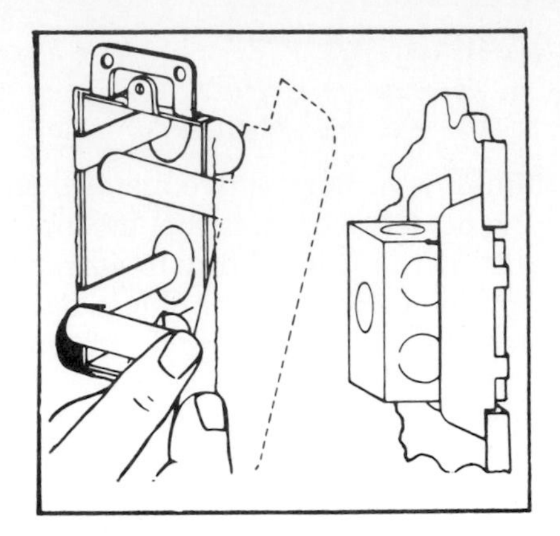

10–113. Metal box supports can assure a stronger job. Insert supports on each side of the box. Work supports up and down until they fit firmly against inside surface of the wall. Bend the projecting ears so that they fit around the box.

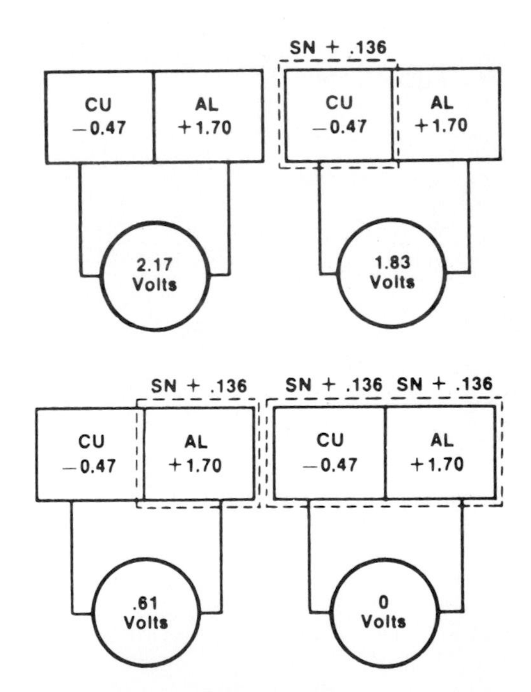

10–114. Electrode potential. (Values only show relative activity of metals. Activity decreases in going from copper to aluminum, tin-plated copper to aluminum, copper to tin-plated aluminum, tin-plated copper to tin-plated aluminum.) Cu = copper, Al = aluminum, Sn = tin.

therefore insulating the aluminum and increasing joint resistance. The film must be removed, or penetrated, before a reliable aluminum joint can be made.

Aluminum connectors are designed to bite through the film of oxide on the aluminum as the connector is applied to the conductors. It is further recommended that the conductor be wire-brushed and preferably coated with a joint compound to guarantee a reliable joint.

The design of some connectors counteracts most of the problems that can arise in terminating, splicing, or tapping aluminum conductors. For best results on No. 8 and larger conductors, a tool-applied, aluminum-bodied, compression-type connector should be used.

After compression, the cross-sectional area of the connector body and cable strands is substantially solid, with each strand permanently reshaped and compressed into intimate contact with its neighbor. The connector body itself is also permanently reshaped and compressed. Fig. 10-115.

Types of Aluminum Connectors

There are two types of aluminum connectors: compression and mechanical.

■ *Compression connector.* This type of connector is used for terminating aluminum conductors by cold swaging aluminum-alloyed connectors over the end of the conductor. Hand or hydraulic tools with precision dies apply pressure uniformly to the barrel of the connector.

10–115. Before compression, a typical cross section of cable and connector consists of about 75% metal and 25% air. After compression, the cross section looks like this—if a hydraulic compressor tool were used—100% metal with virtually no air spaces.

This type of connector has long been used in overhead and underground power transmission lines, secondary power distribution systems in aircraft, and in electrical locomotive power feeder cables. Aluminum connectors are also well known and accepted for their high reliability for building wiring, when they are properly selected and installed.

■ *Mechanical connector.* This type, like the compression connector, is also used for terminating stranded wire. The difference is that the wire is placed in the barrel of the connector and a screw or saddle is forced down on the conductor. The high-pressure wedge effect of the screw on the wire provides a positive interstrand contact. A standard hex wrench is the only tool required for installation. Fig. 10-116.

Aluminum Wire for Home Wiring

Aluminum wiring came into use when there was a shortage of copper wire. Involved in this wiring were houses built between 1965 and 1971. The aluminum wire was usually No. 12 instead of the copper No. 14.

Use of aluminum wire can be a safety hazard. The greatest problems have occurred where there are extreme temperature differences. Aluminum expands and contracts at a greater rate than copper.

Most of the potential problems can be eliminated by tightening all connections at outlets and switches. There are materials available now to make sure there will not be trouble where copper and aluminum are connected with wire nuts or by other means.

There are two types of aluminum wire in general use today: all-aluminum and copper-clad aluminum. Bonded to the surface of the copper-clad aluminum is a coating of copper. This eliminates some of the problems that occurred when copper and aluminum were placed together in a wire nut or some other connector.

When aluminum wire was first introduced, it was used just as copper wire. This created some problems since the aluminum wire was

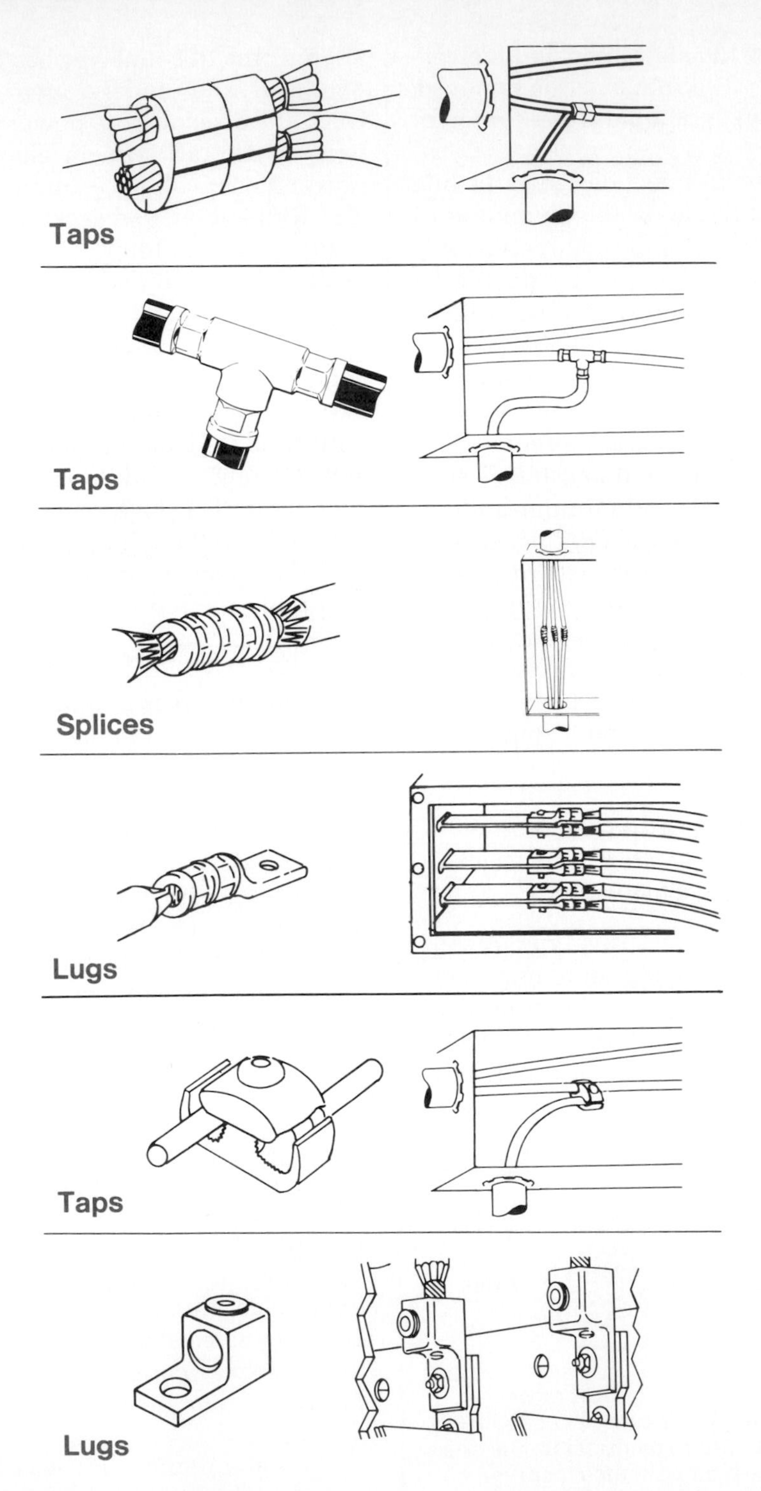

10–116. Types of industrial aluminum connectors.

clamped in a brass or copper terminal screw or connector. The heat from current flowing in the wire causes the aluminum to cold-flow out of the connector. When the connection cools, it becomes a little loose. As the current is turned on and off, the heating and cooling cause still more aluminum to become extruded. Eventually the wire can heat up and cause a fire, even when nothing is plugged in or when the switch is off.

In damp areas this can cause an electrolytic action to take place. Copper or brass terminal screws react with the aluminum.

All switches and receptacles which were intended for 15- or 20-ampere service are not marked. However, those intended for use with copper (CU) and aluminum (AL) are marked CU-AL. The UL changed their rating of 15- and 20-ampere switches and receptacles and introduced the CU-ALR designation for aluminum in the 15- and 20-ampere rating. The "R" stands for revised. The composition of the aluminum wire has been changed also. There are special chemical and physical properties specified by the UL before the wire gets approval.

Devices marked CU-AL should be used with copper wire only.

Devices marked with CU-ALR (or CO-ALR) may be installed with aluminum, copper-clad aluminum, or copper wire. Devices with a push-in connection must not be used with all-aluminum. They can be used on copper-clad aluminum or copper wire.

Devices over 30 amperes or more are not marked with CU-ALR or CO-ALR. Those not marked, and rated at 30 amperes or over, can be used with copper-clad aluminum or copper wire.

QUESTIONS

1. Where can you find the quantity allowed of a given size conductor in a switch or outlet box?
2. What is a BX bushing? Why is it used?
3. What is a connector?
4. What is a handy, or utility, box?
5. Why are handy box extensions sometimes used?
6. Why are cover plates necessary?
7. What is a bar hanger?
8. How are lighting fixtures installed?
9. What is an in-line switch?
10. What equipment is available to aid in the mounting of switch boxes in existing walls?
11. How can the electrolytic action between aluminum and copper wire be prevented?
12. What are the two types of aluminum connectors?

KEY TERMS

compression connector
handy box
handy-box extension
mechanical connector
underwriter's knot

CHAPTER 11

OTHER WIRING METHODS

Objectives

After studying this chapter, you will be able to:

- Identify the components of low-voltage systems.
- Know the correct procedure for checking a wiring installation.
- Install wire in raceways.
- Differentiate cable trays from raceways.

Other chapters have discussed various "standard" methods of wiring a house or building. Here we will discuss low-voltage remote control for home and industrial uses. Also included here will be the raceway method of wiring. This is usually known as Wiremold®, a name well known to all electricians who have worked in business or industry. However, Wiremold® is not the only manufacturer of raceways. There are also National and Square D, to name but a couple of others.

LOW-VOLTAGE REMOTE CONTROL

Low voltage is used in some applications where industrial plants, homes, or offices require remote control of a piece of equipment or a light. Usually a 24-volt source is used to control relays, which in turn, control the 120-volt or 240-volt circuits.

A *relay* is a device that has a coil of wire and contacts. The coil of wire has a loosely fitted core. This core can move when the coil is energized. The coil has a sucking effect which tends to pull the core into the coil. If a pair of contacts is attached to the core, the contacts can control the other circuit. These contacts can be controlled by energizing or deenergizing the coil. Low-voltage coils are used with small wires and switches. With low voltage it is possible to use small wires and a different type of wiring.

Control Flexibility

With the advent of the 3-way switch, the concept of multi-point switching gradually broadened into the many conveniences and economies of modern automated lighting controls. A single switch at more than one point can control a single lighting fixture, or several switches at one point can control many lighting fixtures at various locations. These may be either on the same or on different branch circuits.

Where desirable, a master switch can start a motorized control unit. It turns any number of lights *on* or *off* as required. The addition of a program timer allows several motor-master control units to be set into motion at one time. It will turn on selected lights, then turn off all lights left on. At the close of a working day, photoelectric controls can override these program timers. This occurs when light intensities fall below preset levels. The flexibility provided by low-voltage, remote-control switching is practically limitless.

Use in Modern Construction

Because of the low-voltage, low-current requirements of relay-switching circuits, small, flexible wires may be snaked in thin wall or steel partitions without the protection of metal raceways (except where local codes prohibit such installations). Wires can be dropped down through hung ceilings into movable partitions as easily as rewiring telephones. In fact, wiring can be run under rugs to switches located on desks, and easily changed when required. The switches are modern in appearance and very compact.

The low-voltage systems in this chapter are those designed by General Electric Company. However, there are other manufacturers, and the systems are basically the same. The products here have been chosen to give a relatively typical approach to the general idea of using low voltage for remote-control circuits.

System Components. There are a number of component parts familiar to the electrician. These low-voltage devices have been redesigned to accommodate particular situations that can easily utilize the convenience and reliability of such systems.

■ *Power supply.* All the components of the GE system operate on 24 volts. To obtain this voltage it is necessary to step down the usual 120 volts available in homes, businesses, and industry. For this purpose, special transformers with energy-limiting characteristics are available for either 120 VAC (volts AC) or 277 VAC operation. Fig. 11-1.

The transformer has two pigtail leads which can be spliced to a 120-volt line with wire nuts. Note that the cover-mounted transformer has to be mounted on the outside of the utility box. This means that the low voltage will be outside the box, and the high voltage (120 volts) will be inside, according to Code. Low- and high-voltage lines should not be included in the same box or conduit.

■ *Relays.* Low-voltage switching depends upon relays to do the work. The low-voltage switching system eliminates line-voltage wiring to all switches and replaces it with inexpensive, low-voltage wire. The heavy work of switching is done by dependable relays which can be controlled from various remote points. Master switches control many circuits from one, or from several convenient locations. Fig. 11-2.

11–1. Transformer. Available in 115 VAC or 277 VAC. Steps down to 24 volts.

11–2. The remote-control wiring system eliminates line-voltage wiring to all switches and replaces it with inexpensive, low-voltage wire.

Because relays used here are of the mechanical latching type, the switching circuits require only a momentary impulse—as short as a half-cycle or $\frac{1}{120}$th of a second. These relays have a coil design that resists burnout due to equipment or operational failure. This usually results in energizing the relay for extended periods at ambient temperatures not exceeding 140°F (60°C). They can control 20 amperes of tungsten, fluorescent, or inductive loads at 125 VAC and 277 VAC. They can also be used for $1\frac{1}{2}$ HP 240 VAC and $\frac{1}{2}$ HP 125 VAC. Some relays are available with a pilot light switch to indicate the on position in remote locations on a master panel.

Fig. 11-3 shows the internal wiring of a relay. The enclosed relay, shown in Fig. 11-4, may be mounted in a metal box by removing a 1″ knockout and sliding the relay through the hole. It will snap in place for permanent mounting.

■ *Switches.* One basic type of switch, the single pole, double-throw (SPDT) momentary-contact, normally open circuit, does the work of single-pole, double-pole, 3-way, or 4-way switches used in conventional wiring. It also uses much less wiring and of a smaller size. The switches are in the circuit for a split-second, just enough to energize the relay and have it move the contacts to the *on* or *off* position. Many switches may be wired in parallel to control either a motor or a light from a number of locations. Switches are available with or without a pilot light.

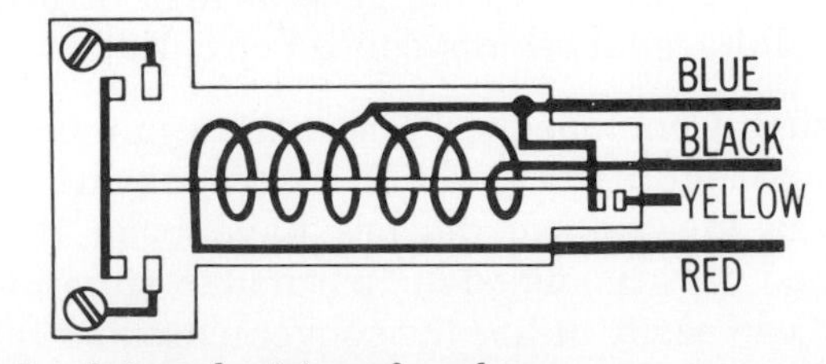

11–3. Internal wiring of a relay.

11–4. A relay for low-voltage control circuits.

11–5. Single-pole, double-throw, momentary-contact, normally open circuit switch.

11–6. This remote-control switch matches other kinds of interchangeable switches and fits standard or narrow mounting brackets and wall plates; normally open, single-pole, double-throw, momentary-contact.

Fig. 11-5 is a pilot light switch with push-button control action. Fig. 11-6 is an interchangeable line switch. It fits a standard or a narrow interchangeable mounting bracket. The center position does not make contact. Up is the *on* position; down, the *off* position. A heavy-duty switch with the possibility of fitting into a standard wall-box mounting is shown in Fig. 11-7. Here, again, the center position is not making contact. When pushed momentarily into the up position, it is *on*, down is *off*. The contacts can handle up to 30 amperes.

■ *Circuits.* Fig. 11-8 shows the simple utilization of one switch, one transformer, and one relay. The circuit is capable of switching a 120- or 277-volt circuit without an operator being near. This means that high-voltage switching (277) may be done outside the area, not in the control center where the operator contacts the switch for controlling action. The safety of such an arrangement is evident.

11–7. Heavy-duty switch for standard wall-box mounting. Available with up to 30-A rating.

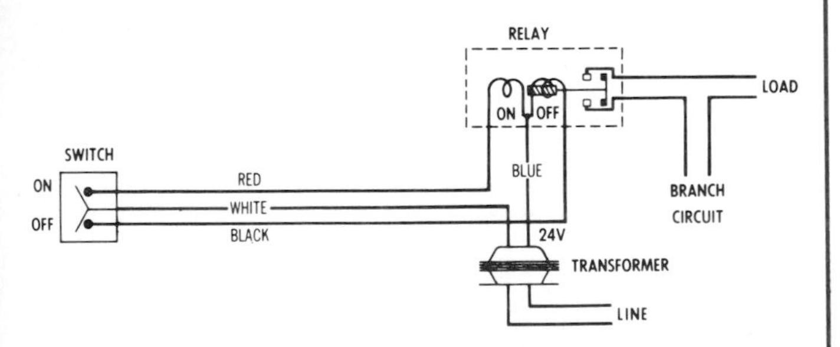

11–8. Simple remote-control circuit.

The red, white, and black wire from the switch to the relay and transformer is usually small—about No. 20 AWG (American Wire Gage). This means it can be run longer distances for less cost than conventional wiring.

Two relays sometimes control two different circuits. Fig. 11-9. Note that a rectifier is used to change the 24 volts AC to DC, for less noise in the operation of the relays. A rectifier is a device which changes AC to DC.

■ *Rectifiers.* A semiconductor diode is used as the rectifier to change the 24-volt, AC transformer output to DC. Note that it is connected to a terminal strip, for ease in mounting as well as making connections. This rectifier is capable of 7.5 amperes continuous duty, and 20 amperes for short periods. Fig. 11-10.

■ *Remote-control wire.* Fig. 11-11 is a 3-conductor insulated wire for low-voltage, remote-control wiring. The seven strands of copper wire make up a No. 20 AWG conductor having $\frac{1}{64}''$ of thermoplastic insulation. The 3-conductor wire has a red stripe on one outer conductor and a black stripe on the other outer conductor.

In old construction, wire can be snaked into walls, or stapled to wallboard surfaces. In plaster walls, it can be laid in a shallow groove and plastered over.

■ *Twisted indoor* wire may have 2-, 3-, or 4-conductor insulated wire for remote-control wiring. It, too, comes in 7-strand, No. 20 AWG. The solid colors used for coding coincide with color codings of remote-control devices. The 2-conductor wire, for instance, comes in blue and white, or in red and black. The 3-conductor wire has red, white, and black conductors. Four-conductor wire has red, white, black, and yellow conductors. This wire may be installed in the same manner as flat wire. Fig. 11-12.

■ *Flat, outdoor wire* in a 3-conductor, neoprene-insulated jacket for low-voltage, remote-control applications is available. It comes in No. 18 AWG and may be used overhead or on underground applications. Fig. 11-13.

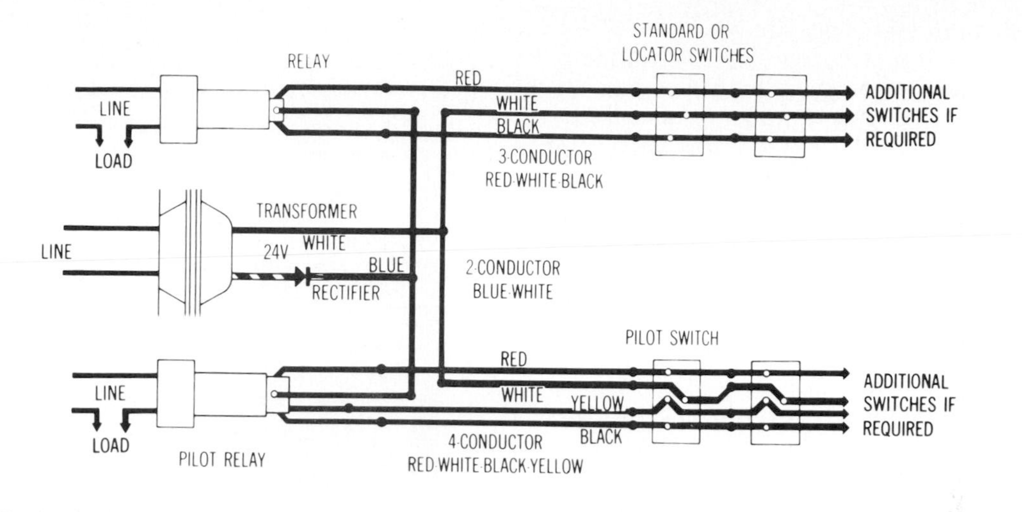

11–9. Basic circuitry for remote-control system.

11–10. Silicon rectifier.

11–11. Flat indoor wire, No. 20 AWG.

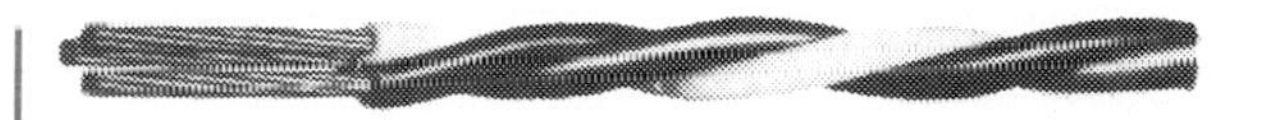

11–12. Twisted indoor wire, No. 20 AWG.

11–13. Flat outdoor wire, No. 18 AWG.

■ *Multiconductor* wire in No. 20 AWG is available in 19-conductor or 26-conductor cable. It is used for making runs from master-selector switches and motor-master control units. The 19-conductor cable is made up of nine pairs of red and black wires, with one white common wire. The 26-conductor wire has twelve pairs of red and black wires, plus one white common conductor and one blue wire for a locator lamp, or for a motor lead. Fig. 11-14.

■ *Push-button master switch.* Where many circuits must be controlled from one convenient location, these master-selector switches perform the necessary selection of only those circuits wanted. Or, by sweeping the *off* and *on* sides of these switches, a manual master control of all circuits may be used for one or more, or all eight of the switches, and may even be extended to other individual switches when desired. Fig. 11-15.

Also available is a "dial-type" master-selector switch for 12 circuits, where each dial permits individual control; or by pressing and then sweeping all circuits, master control is accomplished. Both push-button and dial types are available with or without pilot lights. Both have a directory for locating individual circuits. Fig. 11-16.

11–14. Multi-conductor cable, No. 20 AWG.

11–15. Push-button master switch.

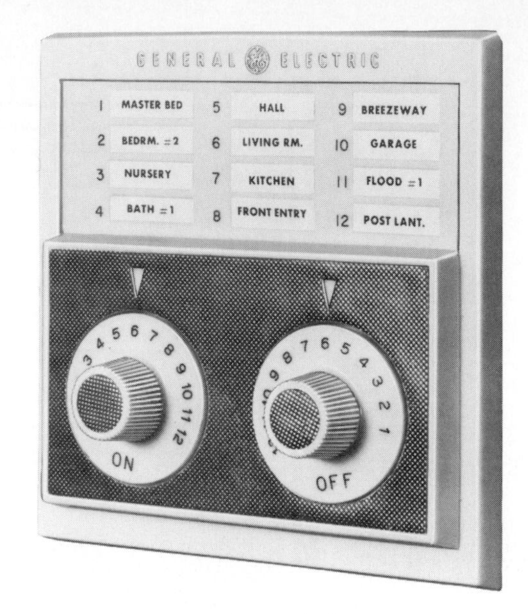

11–16. Dial-type master switch.

■ *Motor-master control units.* Motor-driven, master control units are used to turn *on* or *off* up to 25 individual circuits. When cascaded, the pressing of a single master switch can control any number of circuits, depending upon the number of motor-master units that are ganged together. Ganging is accomplished by having the last position of the first unit wired to start the second unit, the last position of the second unit wired to start the third, and so on. For automated lighting control, these units can be activated by program timers or photocell relays. Fig. 11-17.

Illustrated in Fig. 11-18 is the motor-master control unit starting switch, which is connected to lead (A) and white lead (B). When the starting switch is activated, the contact arm (C) starts to rotate in the direction indicated. The arm connects the white lead to contact points 1 through 25. It also picks up the slip ring connected to the white lead and keeps it completed. Complete revolution requires 17 seconds.

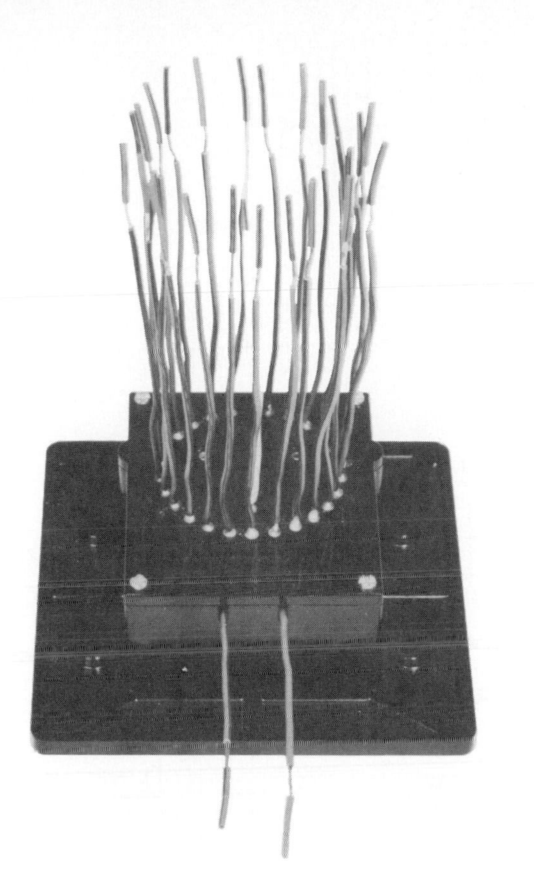

11–17. Motor-master control.

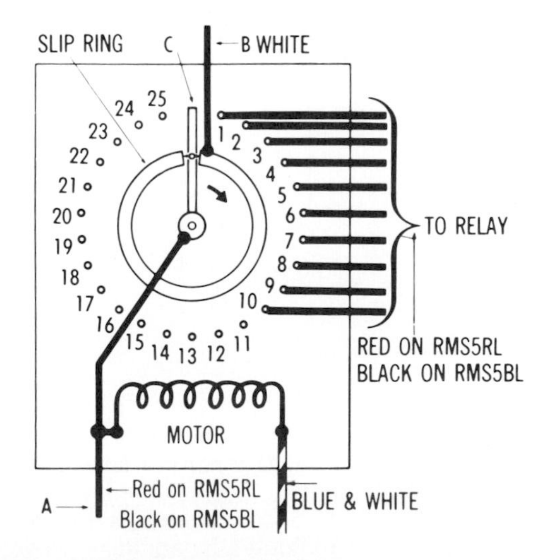

11–18. Internal wiring of motor-master control.

Fig. 11-19 shows a circuit that provides maximum flexibility of control. The circuit treats each floor as a unit for individual switch control. It uses master-selector control and motor master for turning off all floor relays. A central control, located in the superintendent's office, also has individual control of each floor motor master through a master-selector switch. This switch has an eighth position control. This eighth position is a motor master that starts all the floor motor-master units. This circuit can be extended. Simply add a program clock and/or a photoelectric cell relay.

■ *Components cabinet.* Cabinets for housing relays, motor masters, and power supplies make wiring easy. There are several sizes of cabinets available. Fig. 11-20 will hold up to 24 relays, a 24-volt transformer and rectifier assembly, two motor-master control units, and all other bus bars and terminal strips necessary for a neat-appearing, easily wired control center for the switching system. A steel separator is used to isolate the low-voltage section from the high-voltage section.

■ *Pilot light assembly.* An assembly of 12 pilot lights mounted on a bracket fits behind the master-selector switch. Each light illuminates the director tab corresponding to the circuit number indicated on the bracket. A plastic shield permits only a narrow slit of light to illuminate the printed tab. There are bracket mounts on the upper section of the two-gang plaster ring used for the master-selector switch. Fig. 11-21.

■ *Wall plates.* Wall plates are available to fit over the switches and give an attractive appearance to the wiring system. They are made of stainless steel or plastic. They range from a 1-gang standard size with one vertical opening for standard remote-control switches to a 6-gang standard size with six vertical openings.

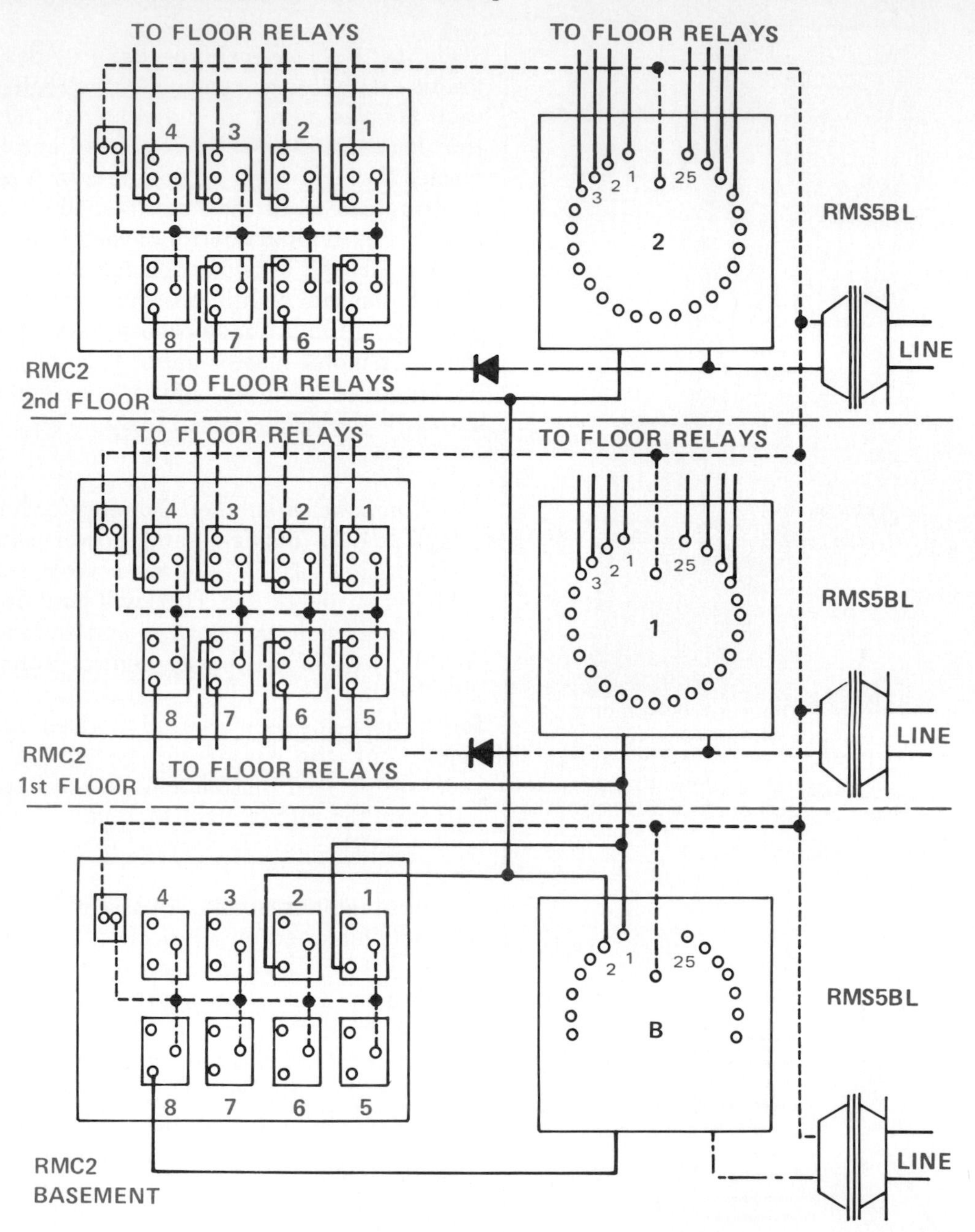

11–19. Multistory motor-master control. This circuit provides maximum flexibility of control, treating each floor as a unit for individual switch control, master-selector control, and motor-master control for turning off all relays. In addition there is a central control, located in the superintendent's office, for the individual control of each floor motor master through a master-selector switch. The eighth position controls a motor master that in turn starts all the floor motor-master units. This circuit can be extended by adding a program clock and/or a photoelectric cell relay.

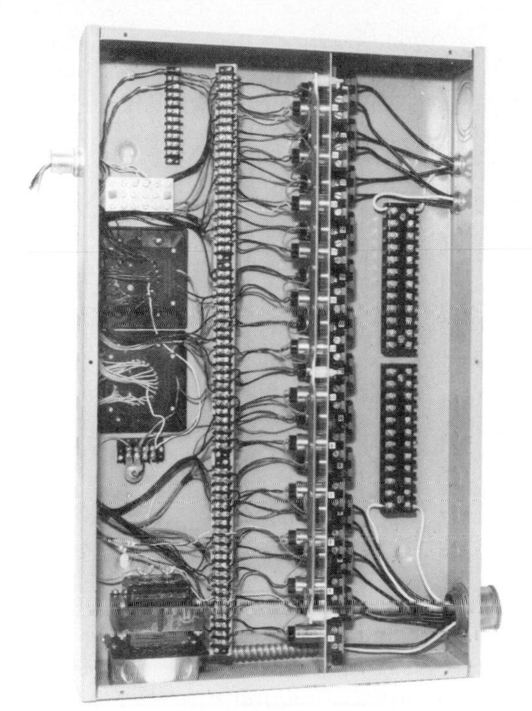

11–20. Component cabinet houses relays, motor masters, and power supply for ease of wiring.

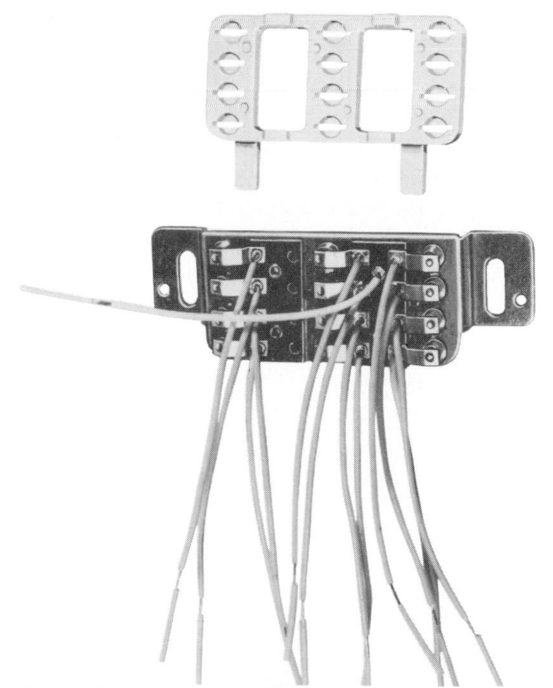

11–21. Pilot light assembly for master-selector switch.

Fig. 11-22. Fig. 11-23 shows narrow-design wall plates for metal partitions where a minimum of space is available.

■ *Extension switch.* This is a switch arrangement with a plug-in capability. It is designed to be used on top of a desk where you may want to control a number of remote operations without getting out of your chair. The cable, 8′ in length, will give accessibility to anyone within that distance of a receptacle. Fig. 11-24.

■ *Pilot lamp.* Fig. 11-25 is a pilot-lamp bulb. It has a voltage rating of 24 to 36 volts. It is designed to last indefinitely unless it is damaged physically.

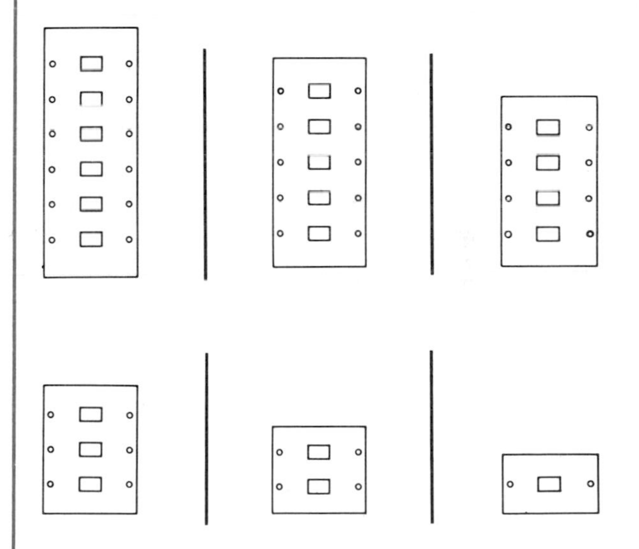

11–22. Stainless steel wall plates for standard grade, remote-control switches.

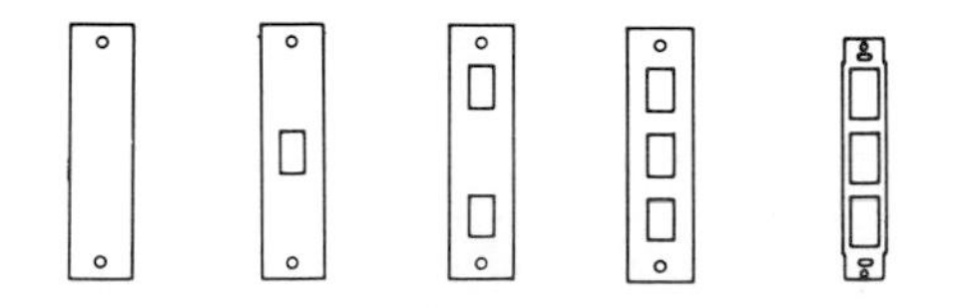

11–23. Narrow-design switch plates for narrow partitions.

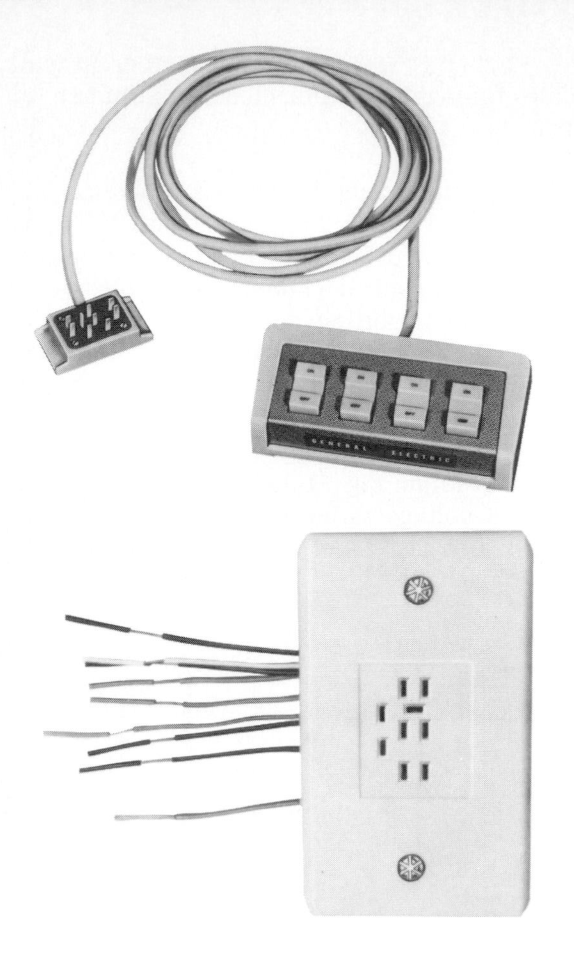

11–24. Extension switch and matching receptacle.

11–25. Replacement lamp for switches.

Basic Circuits. Some of the possibilities for remote-control applications of low-voltage relays and control switching are given in the next few illustrations to get you to start thinking. Figs. 11-26 through 11-33. Your imagination is perhaps the only limitation to the application of low-voltage, remote-control switching.

Residential Installation Pointers

Install frames in relay boxes. Frames can only be inserted one way. Frames (brackets) for standard relays are installed in a similar manner except that there are no bus bars for automatic connection to the relays, and wiring to each relay must be made through pigtails attached.

■ *120-volt wiring.* This is installed in the usual manner, running the branch circuits to the 120-volt side of the relay boxes. Plug in the stripped conductors (strip $\frac{1}{2}$ inch) into the pressure lock terminals, making sure that the white and black wires are plugged into the terminals as marked. Only one set of terminals on each bus section is used for branch circuit supply. The other set may be used to extend, or couple, the two sections together. Next, make runs from relay boxes to the receptacles. Plug in the $\frac{1}{2}''$ stripped conductors into the designated relay for that receptacle, making sure that the white wire is pushed into the "white" pressure-lock terminal on the relay, and that the black conductor goes into the other terminal of the relay.

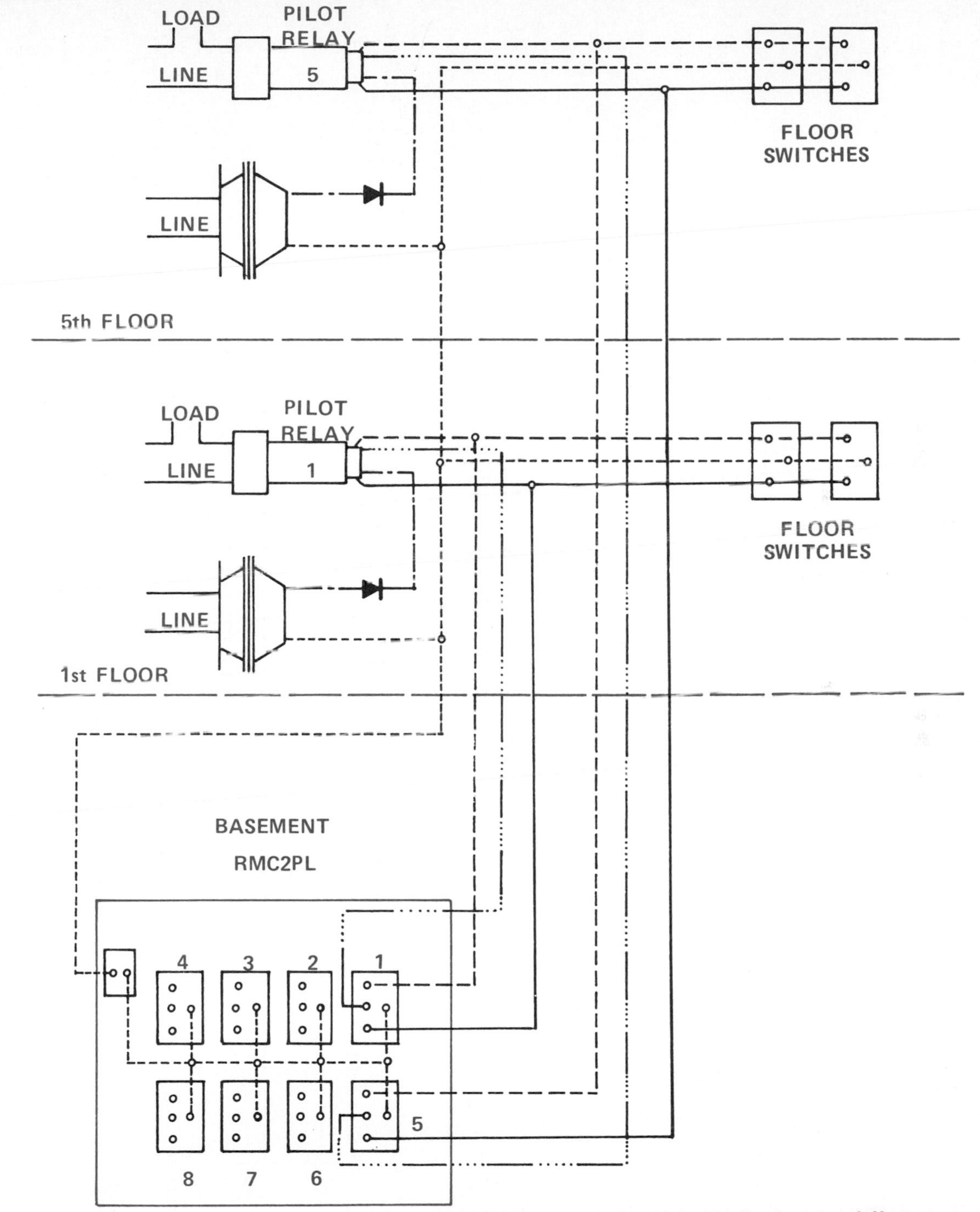

11–26. Multistory, separate transformers. For master-selector control of individual relays on different floors, where separate floor transformers and floor switches are also used, this circuit diagram explains the necessary wiring requirements. This circuit is especially useful for lights controlled from a security guard's station, or for control of corridor lights from a superintendent's office.

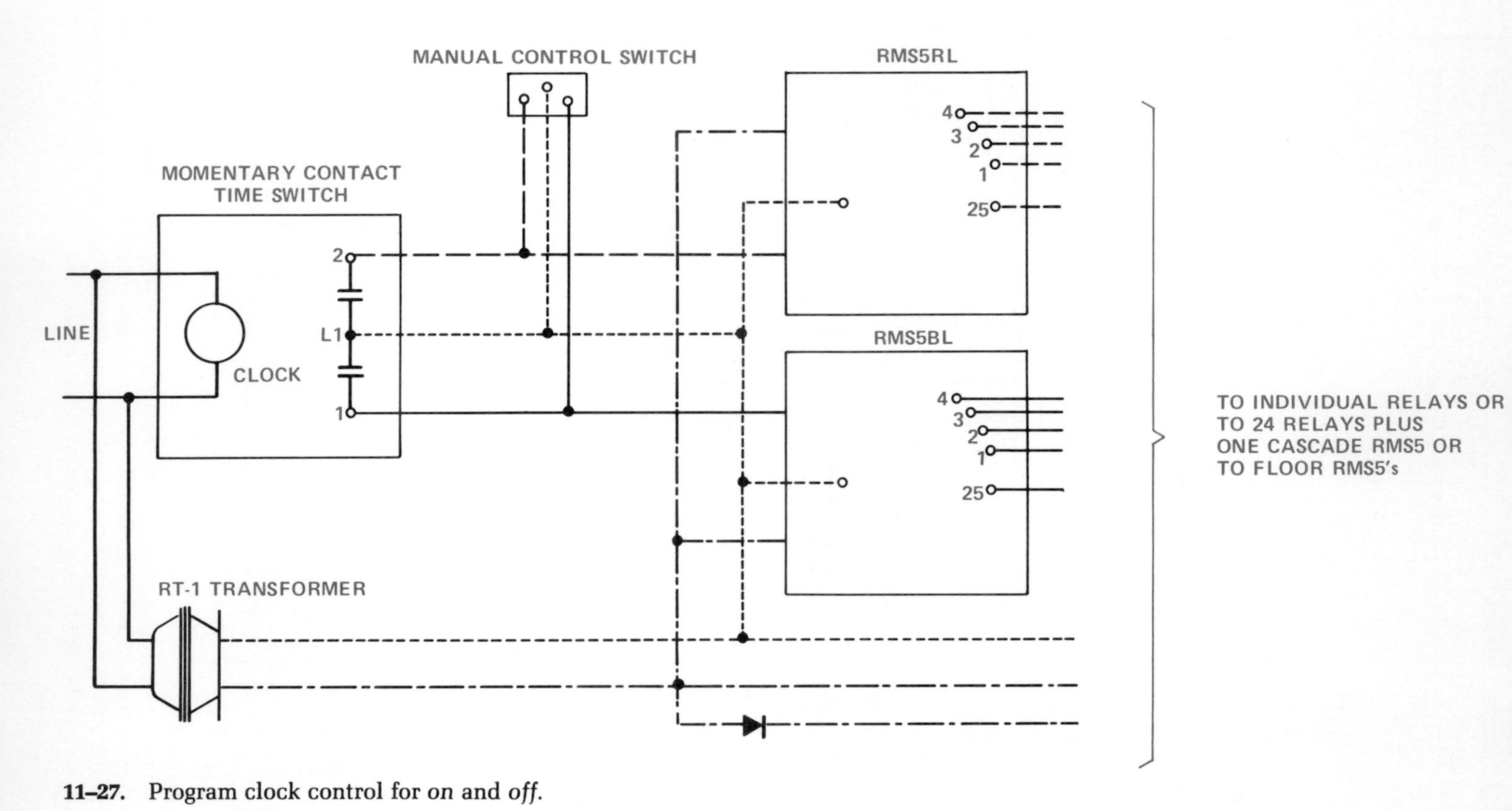

11–27. Program clock control for *on* and *off*.

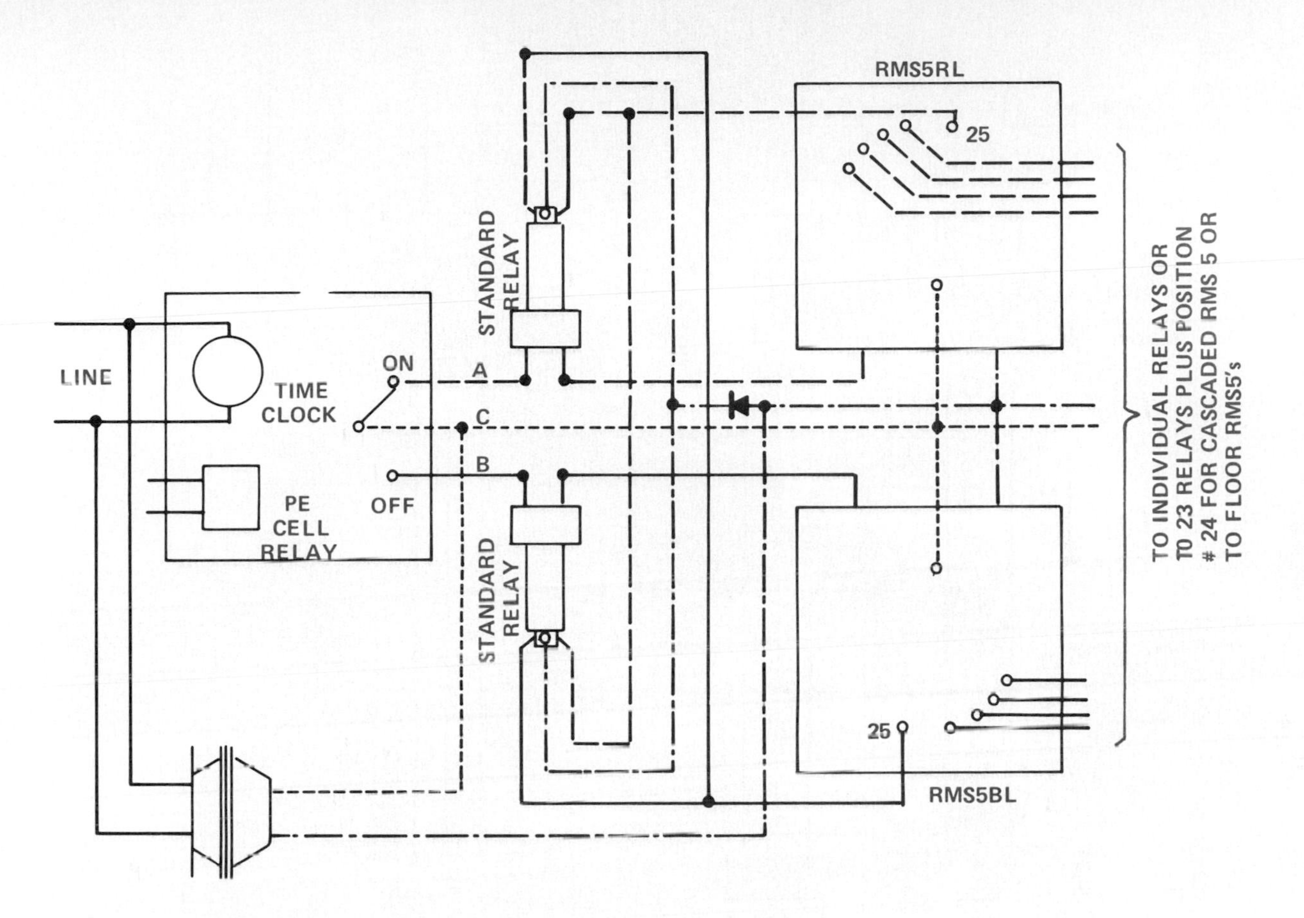

11–28. Time switch, or photoelectric cell control, for *on* and *off*.

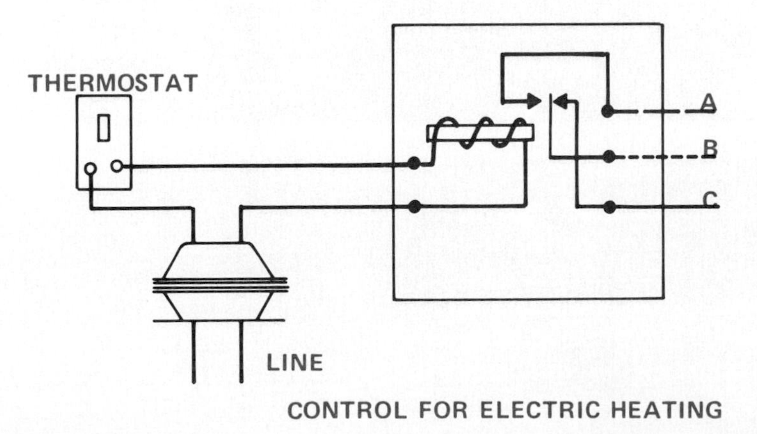

11–29. Thermostat-controlled relay for *on* and *off*.

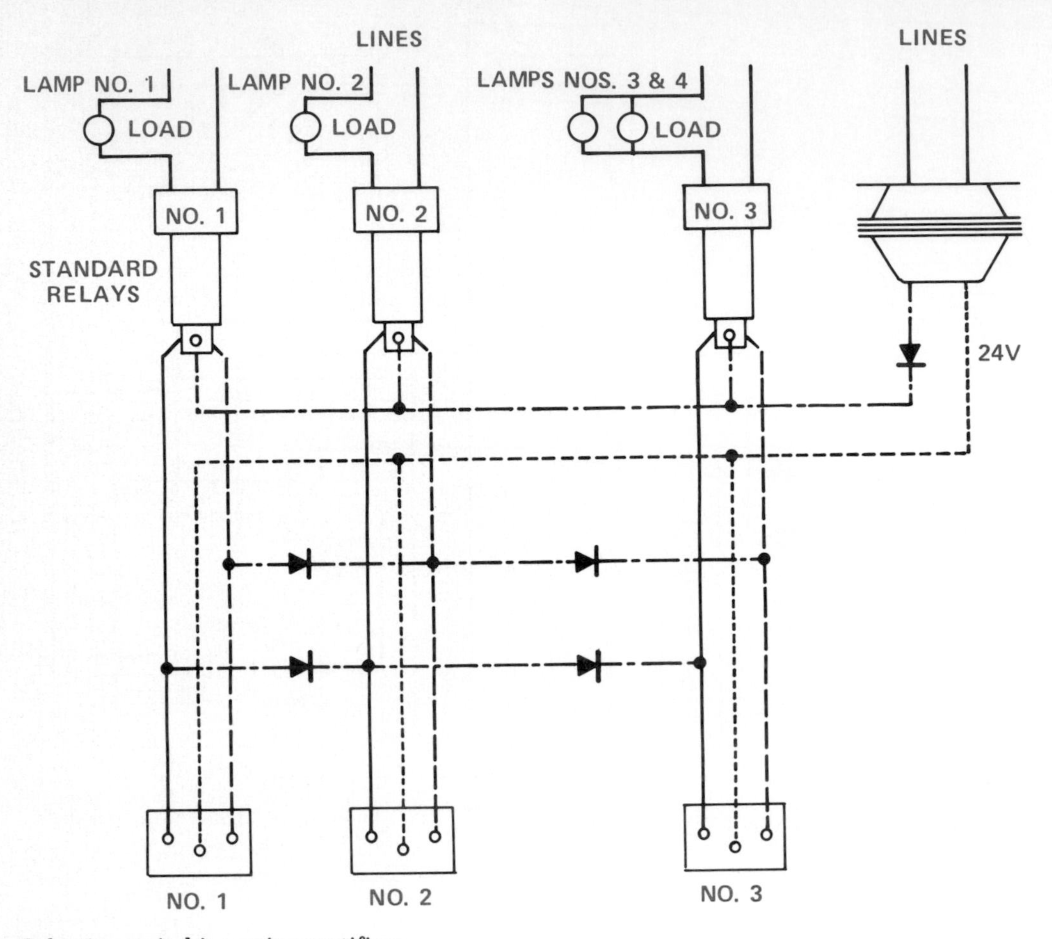

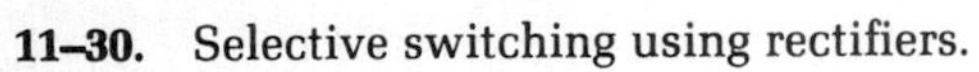

11–30. Selective switching using rectifiers.

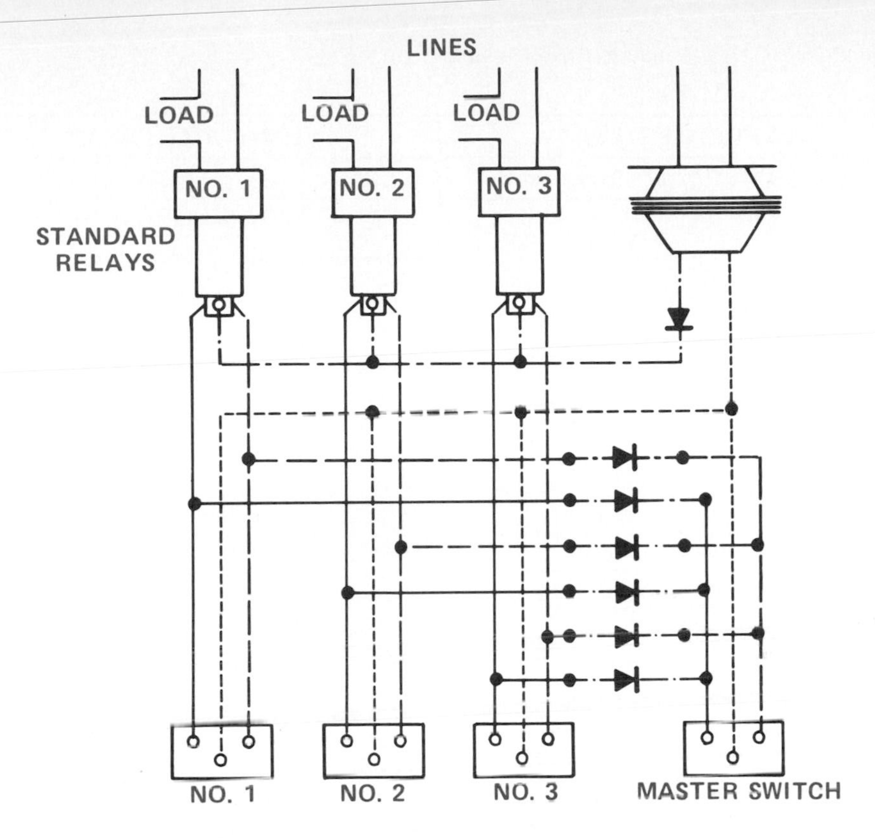

11–31. Master control using rectifiers.

11–32. Typical wiring diagram and layout for residential construction.

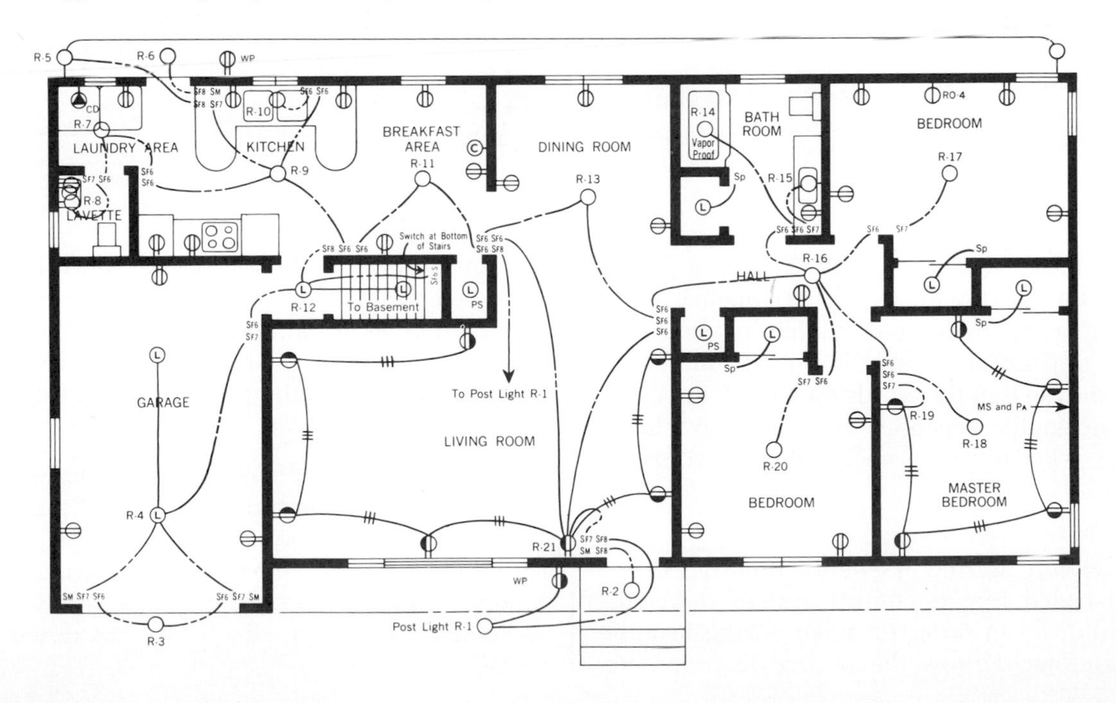

LINE-VOLTAGE SYMBOLS	
Symbol	**Description**
	2-Cond. 120V Wire or Cable
	3-Cond. 120V Wire or Cable
	Ceiling Receptacle
	Floodlight
	Valance Light
C	Clock Receptacle
L	Keyless Lampholder
L PS	Pull Chain Lampholder
	Double Receptacle, Split Wired
	Grounding Receptacle
WP	Weatherproof Grounding Receptacle
R	Range Receptacle
CD	Clothes Dryer Receptacle
S_L	Lighted-Handle Mercury Switch
S_P	Push-Button, Pilot Switch
S_D	Closet Door Switch

LOW-VOLTAGE SYMBOLS	
— - - —	Remote-Control Low-Voltage Wire
T	Low-Voltage Transformer
	Rectifier for Remote Control
B_R	Box for Relays and Motor Master Controls
R	Remote-Control Relay
R_P	Remote-Control Pilot-Light Relay
P_{11}	Separate Pilot Light, R.C. Plate
P_{10}	Separate Pilot Light, Inter. Plate
MS	Master-Selector Switch
MM_R	Motor Master Control for ON
MM_B	Motor Master Control for OFF
S_M	Switch for Motor Master
S_{F6}	R.C. Flush Switch
S_{F7}	R.C. Locator-Light Switch
S_{F8}	R.C. Pilot-Light Switch
S_{K6}	R.C. Key Switch
S_{K7}	R.C. Locator Light Key Switch
S_{K8}	R.C. Pilot-Light Key Switch
S_{T6}	R.C. Trigger Switch
S_{T7}	R.C. Locator Light Trigger Switch
S_{T8}	R.C. Pilot-Light Trigger Switch
S_{T4}	Interchangeable Trigger Switch, Brown
S_{T5}	Interchangeable Trigger Switch, Ivory
RO	Remote-Control Receptacle for Extension Switch

11–33. Suggested symbols.

■ *Mount switch supports.* Standard plaster rings for the thickness of the finished walls should be mounted at all switch locations. For best appearance, and for positioning the switches so that the words *on* and *off* appear in the normally accepted manner, the plaster rings should be mounted horizontally. Fig. 11-34.

■ *24-volt wiring.* General Electric has a color-coded remote-control system that must be followed in order to make wiring installations easier. Follow the wiring diagrams. Remember that the motor-master control unit must be connected to the transformer ahead of the rectifier as it cannot run on rectified current (DC). It is an AC-operated unit. Use the color coding to eliminate possible errors in wiring.

The order in which the components are wired depends upon the preference of the contractor and in no way affects the speed of installation. However, the use of multi-conductor remote-control cables, with numbered as well as color-coded leads, effectively cuts down on installation time.

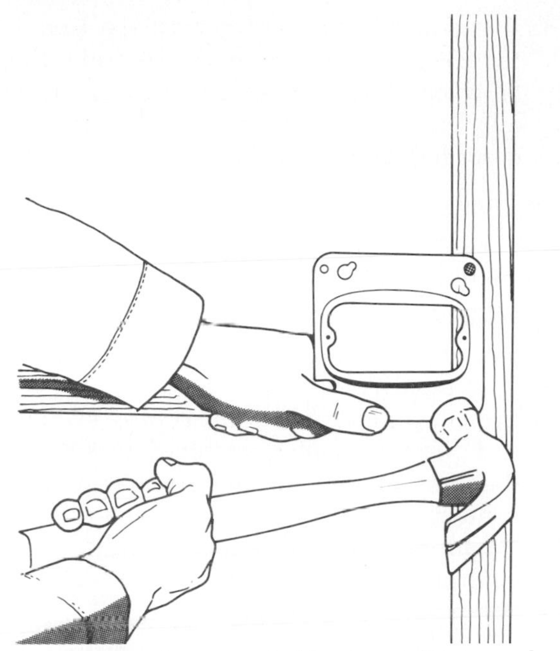

11–34. Plaster ring, mounted horizontally, is used to support one, two, or three flush switches.

Wires should be stapled, using tackers made for installing electric-heating wires and cables (not flat, sharp-edged staples that cut into the insulation). Stapling guns, such as the Arrow Type T25 tacker, with $\frac{3}{8}''$ staples, or the Bostitch T5-8W tacker, are recommended. It is important to allow at least 12″ of wire at the end of each run, placing a staple close to the plaster ring for the switch.

Adhesive tape, wrapped around the wires, can be numbered to correspond with the relay to which they are connected. Multi-conductor cables are already numbered and need no further identification, except perhaps where more than one master-selector switch is to be installed in the same location.

Important Safety Precaution. No electrician would ever think of bunching up bare line-voltage conductors in an outlet box during the plastering operations without first removing the fuses. Similarly, the 24-volt connections from a remote-control transformer must be disconnected before bunching up the low-voltage wires at the switch locations. This must be done to prevent workers shorting the wires in an attempt to get power or light. Standard grade relays are not designed to operate on extended energizing periods that may cause relay burnouts.

Checking the Installation. Every new wiring installation, including remote-control wiring, should be checked as soon as the power is turned on.

To check the system, each circuit should be tried from each switch point, including the motor-master control switches, to ensure that everything is operating correctly. If care was taken in following the installation procedures, as outlined, the system should work perfectly.

The following suggestions are offered as a guide to the proper procedure for locating and correcting installation errors.

■ *Entire system dead when power is applied.*

1. Check for power at transformer primary.
2. Check voltage of secondary.
3. Check to see if transformer is dead. If there is power at the primary leads (120-volt leads), short out secondary terminals with screwdriver. A spark indicates transformer is "alive."

■

This procedure does not harm the transformer if done only momentarily to produce a spark. Look for a short or open circuit in the transformer loop wiring as follows:

■

a. If there are separate zones for the location of relay-center boxes, or if there are several transformer wires to groups of relays mounted in outlet boxes, disconnect all but one.
b. Check to see if switches will work in the zone or relay center left connected. Connect each zone branch to the transformer and check to find the one branch that is shorted.

4. When the zone is found in which the relays fail to operate, look for a short or open circuit in the loop circuit. This may be done by disconnecting the transformer connections at the relay and checking with a "ring out" set (or continuity tester of any type) from location to location until the short or open circuit is found.

■ *Individual circuit stays on or off—relay hums and the relay barrel becomes warm.* This fault is caused by a continuously energized relay coil.

1. Check to see whether an individual switch, or the switch on the 9-position master selector, is binding with the wall plate, preventing it from returning to the neutral position. (If the switch is not free to move, it may be held in either the on or off position and thus heat up the relay.)
2. Check for a shorted control switch by disconnecting the wires and shorting first the red wire to the white wire, then the black wire to the white wire. If this operates the relay, then the control switch should be checked and replaced if necessary. It may be necessary to allow the relay time to cool before making this test if the coil has been energized for an extended period of time.
3. Inspect the relay for possible short circuits between conductors at low-voltage soldered connections at the end of the cap, near the point where the wires are connected.
4. If trouble persists, check relay.
5. If this check does not correct the trouble, there is a short circuit in the control wiring between relay and control switches which may be found in much the same manner as described before for the transformer loops.

■ *Individual circuit stays on or off—relay does not hum or become warm.* This fault is caused by an open circuit in the control wiring.

1. Inspect the connections at the low-voltage end of the relay to be certain there are no open circuits.
2. Inspect connections at the control switches and master-selector switches to see that they are not broken and are making good contact.
3. If fault persists, check for continuity in the control wiring.

■ *Circuit turns on when control switch is turned off.*

Reverse connection to the red (on) and the black (off) terminals of the relay or switch. This error will not occur if color-coded wire is used and colors are matched up in making the connections.

LOW-VOLTAGE WIRING

Built-in Home Entertainment, Intercoms, Door Chimes and Openers, Climate Control, and Burglar Alarms

An electrician is called upon to install a variety of electrical devices. Most home entertainment systems require a low-voltage wiring installation in addition to a 120-volt circuit. Some multipurpose systems can operate doorbells, intercoms, a source of music, or a burglar alarm, as well as smoke and fire detectors. Much of the wiring for such a system should be done before plaster or wallboard is applied. (See also Chapter 6.)

■ *Entertainment and intercoms.* Fig. 11-35 is an example of a communications center for the home. It requires 6-conductor wire, 14/2 Romex, and 2-conductor No. 18 or 16 wire.

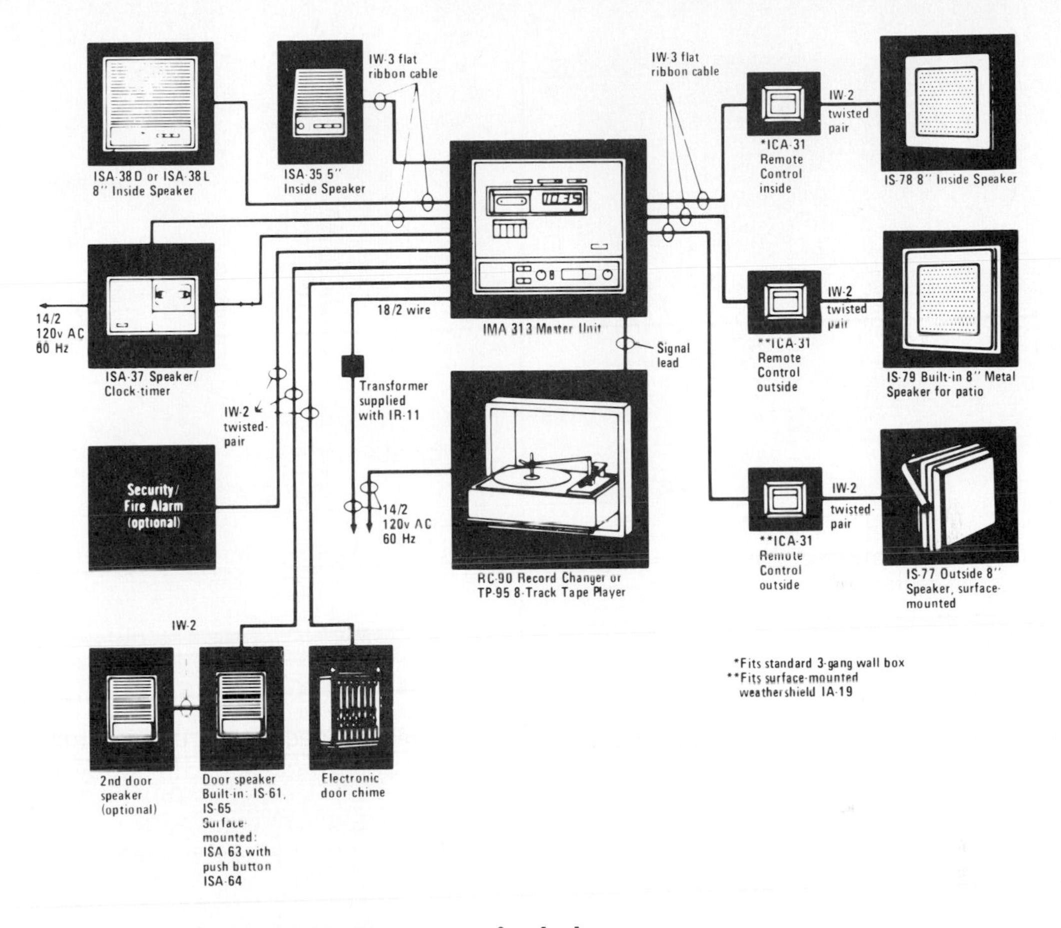

11–35. Block diagram of a communications center for the home.

Fig. 11-36 shows a "communi-center" radio intercom with digital clock and tape player. A record player may be added. This unit can have a complete series of door, patio, and inside speakers.

■ *Door chimes and openers.* Doorbells are the most common of the low-voltage circuits encountered by an electrician. Fig. 11-37. In this circuit the transformer supplies 10 or 16 volts, depending on the design of the bell (or buzzer, or chime). Fig. 11-38. The push button energizes the buzzer or bell. Figs. 11-39 & 11-40. The transformer may be mounted on or near the house distribution-panel box.

In some instances a low-voltage door opener is necessary to allow access to a room only upon a button being pressed from a remote location. This is usually an installation for a store, bank, or industrial plant where controlled access to a specific area is needed. Fig. 11-41.

■ *Climate control.* Most thermostats require low-voltage controls, usually 24 volts. The wire from the thermostat to the furnace is 2-wire No. 18. Fig. 11-42. This wire may also be used for bell wiring or other low-voltage applications. Do not run low-voltage and 120-volt wiring in the same conduit, however. Thermostat wire is also available in 3-conductor, 4-conductor, and 5-conductor cables. It usually has solid copper conductors, is thermoplastic insulated

11–36. Communi-center radio intercom. It has a family message center, a cassette tape player/recorder, and a digital clock. A light indicates a message has been recorded and is awaiting playback. This is a 3-wire centralized system. Auxiliary jacks are provided to add record changer and an eight-track tape player.

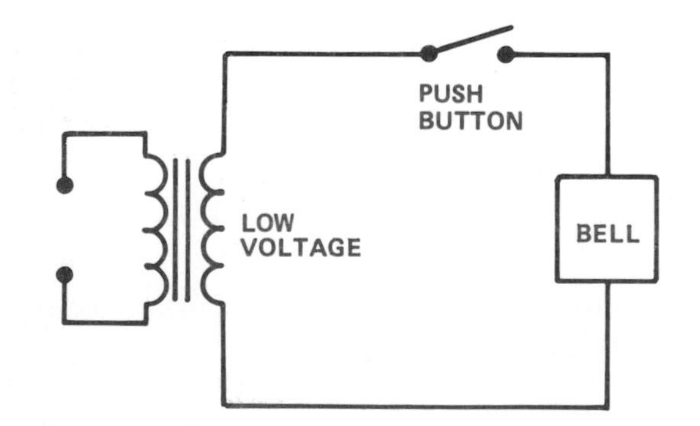

11–37. Doorbell circuit.

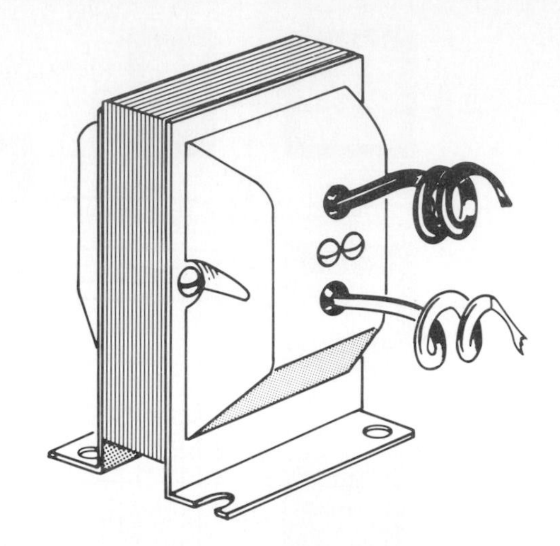

11–38. Doorbell or chime transformer.

11–39. Chime. One sound or note for rear entrance and two notes for the front entrance.

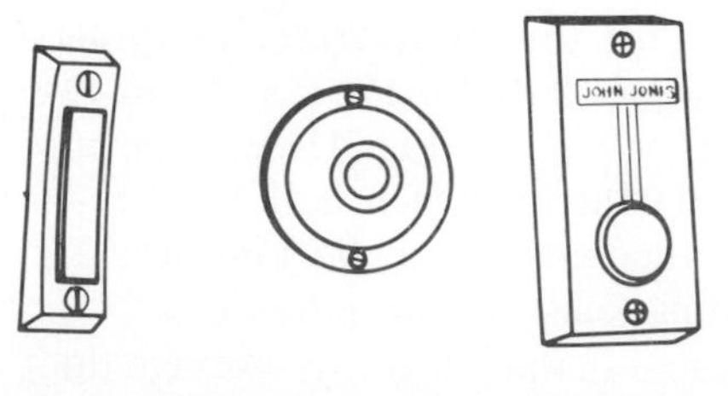

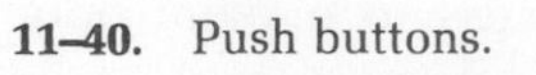

11–40. Push buttons.

11–41. Electric door release. This flush-mounted mortise type installs in place of a lock strike plate on wood, or on a light-gage metal doorjamb. The entrance door remains locked and secure unless it is opened with a lock key, activated by a lock-release control on an apartment speaker, or opened by a special postal lock provided within the main-entrance panel. It connects to a 16 VAC, 1-ampere source.

11–42. Thermostat wire. Used for thermostat controls in heating and air-conditioning installations. Also used for bell and annunciator systems. Designed for use on low-voltage systems. Solid copper conductor, thermoplastic insulation, color coded; conductor twisted, overall jacket of plastic.

and color coded, and is twisted with an overall plastic jacket. Use insulated staples, being careful not to break the insulation on the cable when pounding the staple in place.

Air-conditioning units require more conductors than a simple furnace control. If the furnace and air-conditioning units are combined for a central air-conditioning and heating capability, it is probable that the cable should be 5-conductor. This specification varies with the manufacturer; so be sure to check the unit for specific directions before installing the wire. Fig. 11-43. Remember also that this low-voltage wiring is only for the control circuits. The 120- or 240-volt wiring is separate.

■ *Burglar alarms, smoke detectors, and fire alarms.* Because burglar alarms vary in design, you should know exactly which unit is being installed before beginning to wire the switches and detectors. Fig. 11-44 is a typical wiring diagram for a device to detect forced entry, as well as smoke and fire. Note the different voltages

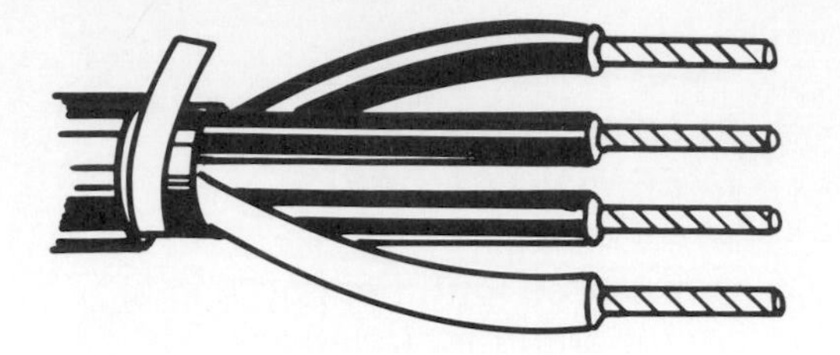

11–43. Although this is a 4-conductor cable, there are available to the electrician almost any number of conductors. They come color-coded or in pairs, with or without shielding.

and wire sizes necessary to get the job done properly within the local and national codes. You will find more on this subject in Chapter 16.

SURFACE-MOUNTED WIRING

Raceways

Metal raceways (channels for wiring) are used in dry locations for external mounting. They are easily installed and give protection to

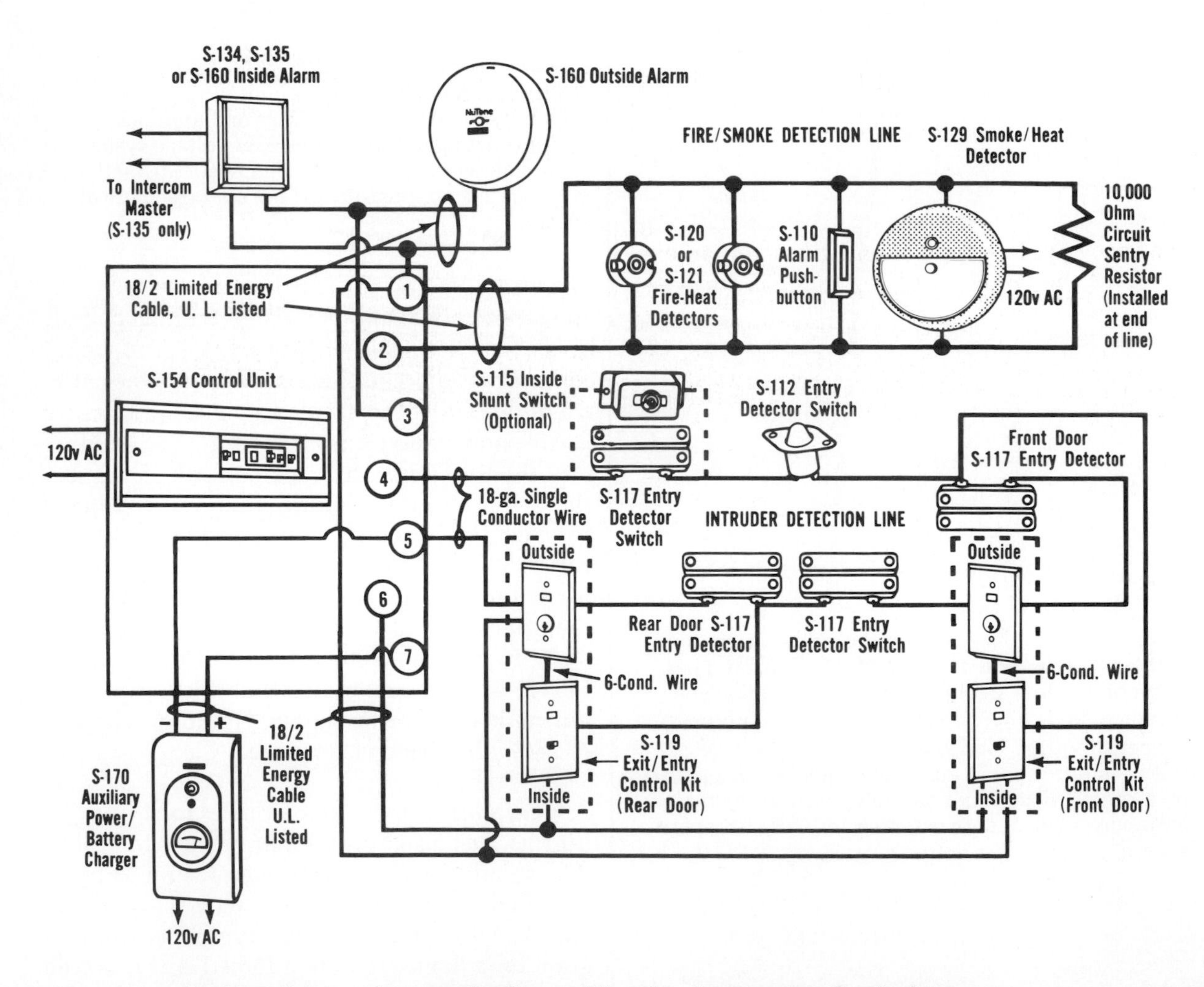

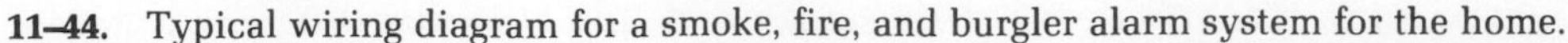

11–44. Typical wiring diagram for a smoke, fire, and burgler alarm system for the home.

wires located inside the metal enclosures. Specific installation methods are used to make sure that electrical and mechanical continuity are maintained throughout the system.

An entire group of fittings and couplings is available for making the system a complete unit. Metal raceways are used in old and new construction, and for additions where it would be too expensive to pull wires through solid walls or other obstructions.

Code. The National Electrical Code has a section dealing with this type of installation. The Code also regulates the size and number of conductors that can be used in a particular size of raceway.

Wiremold® Method. Here we will be concerned with the Wiremold® method of wiring in a raceway. It is typical and gives you an example of the completeness of the system with its various fittings and outlets.

Fig. 11-45 shows five sizes of raceway used for external mounting. There are, of course, other sizes and shapes for raceways. These show the variety available for use in such construction.

A *pancake overfloor raceway* carries power and communications wiring in separate runs to busy office areas. Fig. 11-46.

Tele-Power® poles deliver electric power to a desired location in a classroom. This model of the Tele-Power® pole is simple to install and simple to move to new locations. It has two channels to separate the electrical and low-potential services. Fig. 11-47.

Such systems are furnished with the components shown in Fig. 11-48. A clamp assembly secures the pole to a dropped ceiling T-bar or concealed runner and allows structural bracing. The foot assembly is for positive anchoring to carpeted or hard-surfaced floors. The trim plate is used to conceal any irregularities in openings cut in ceiling panels.

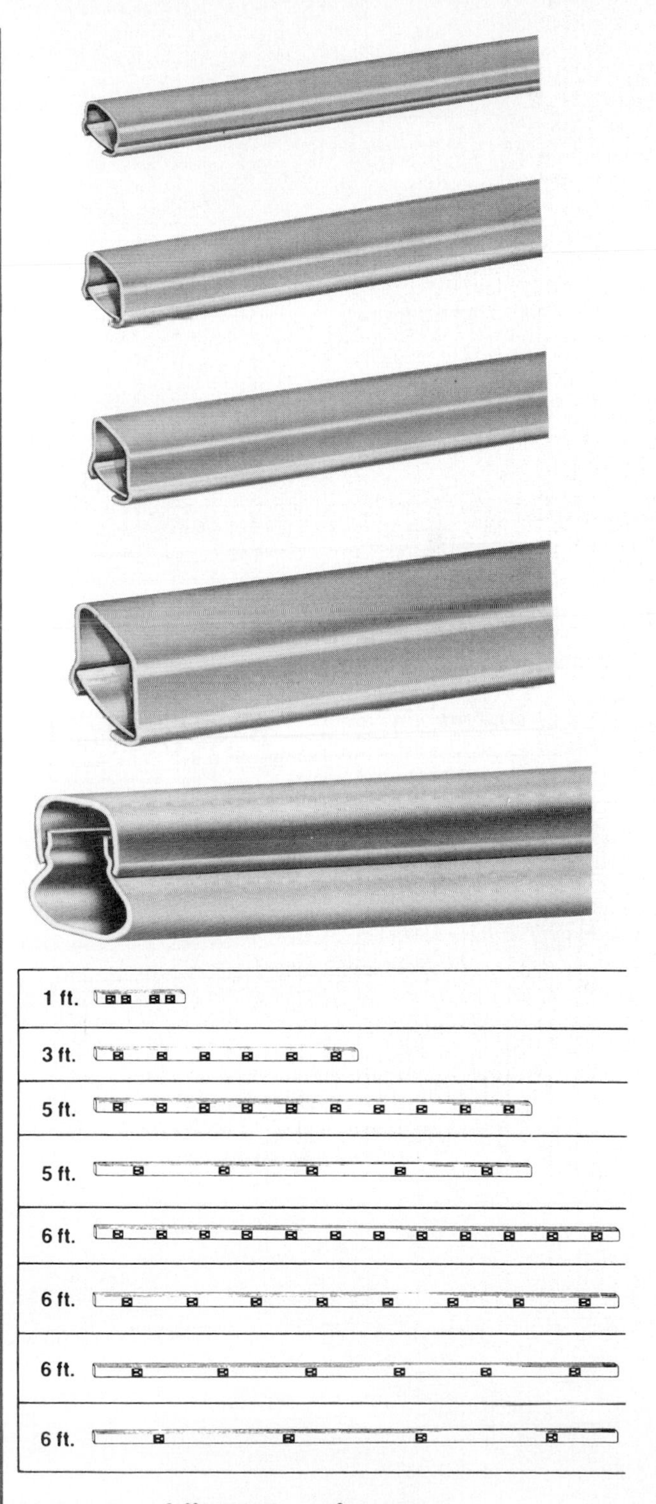

11–45. Five different sizes of raceway.

11–46. Pancake raceway being used to place power at the point of greatest convenience.

Tools.

■ *A raceway bender* is used whenever a small offset or an angle of 90° is required.

■ *Wire pulleys*, one at each internal corner, are illustrated in Fig. 11-49.

■ *Fish-tape leaders* are used to make sure there is a minimum of bulk. They make sure conductors lie parallel to one another, as they are fished into the raceway more smoothly and easily. No crossovers or kinks are produced. A fish-tape leader is 6″ long. It is used for pulling conductors through raceway. One leader is furnished with each wire pulley. To use, stagger-strip the ends of the conductors and attach each conductor to one of the eight holes at one end of the leader. Fish tape through the far end of the raceway and attach it to the leader through the hole provided. The leader and con-

11–47. The Tele-Power® pole makes a convenient source of power and blends in with office partitions.

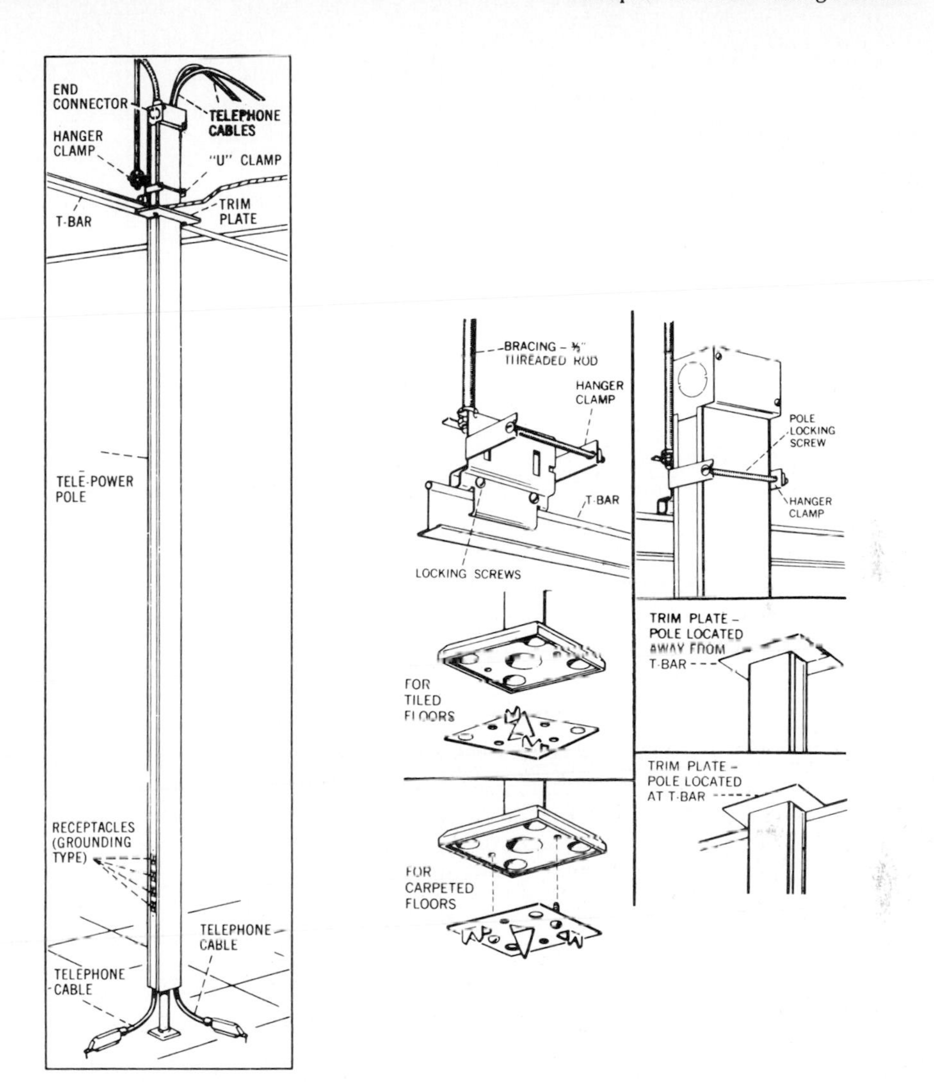

11–48. Tele-Power® pole installation details.

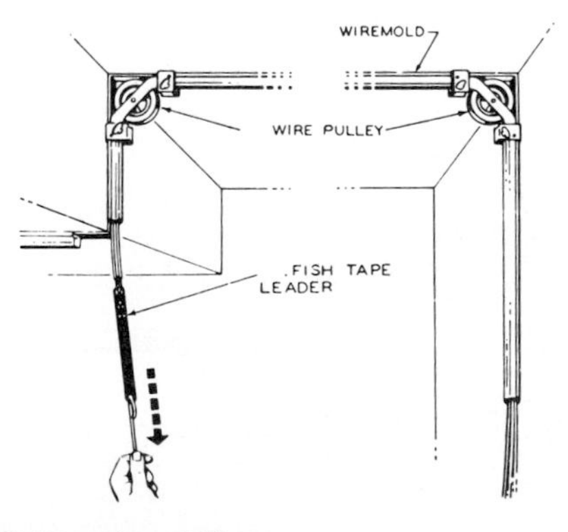

11–49. Using wire pulleys.

ductors are then pulled through the raceway in the same fashion as with conduit. Wiring is done after the entire raceway system has been installed, in accordance with the recommendations of the National Electrical Code. Fig. 11-50.

■ *Shears* may be used instead of a hacksaw to cut the raceway. By attaching the shears to a block of wood it is possible to get better leverage. Fig. 11-51. Or it may be mounted on a bench for benchwork. To operate, merely insert the cover or base of the raceway in the proper place and with the handle as a lever,

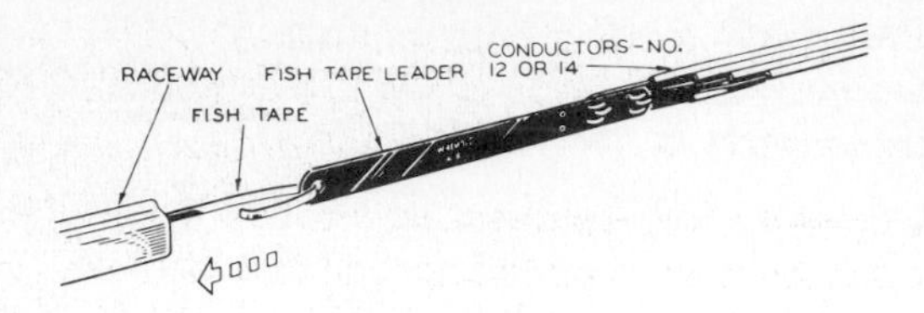

11–50a. Fish-tape leader for pulling wires through a raceway.

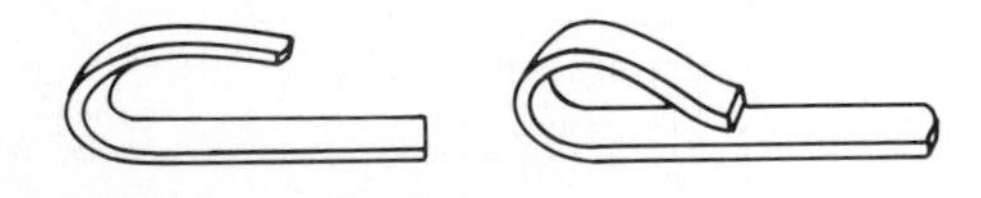

11–50b. Fish tape. All fish tapes (or snakes) must have a hook for fishing through conduit, ceilings, or partitions. This hook is made by heating end of fish tape on open flame until wire becomes red hot. The end is then bent with pliers to the correct size hook.

11–51. Using the shears to cut raceway.

make the desired cut. A quick motion assures a clean, burr-free, square cut, without any distortion of the metal.

■ A *cover-removal tool* is designed to speed up removal of cover from raceway without twisting, kinking, or bending the cover section. Fig. 11-52. Each end of the tool is made to fit one of the raceways and is stamped according to the type of raceway it fits. Simply slide the end of the tool over the cover and push upward to remove the cover.

■ A *canopy cutter* is used for cutting fluorescent fixture canopies to fit raceway. With this tool, it takes only a moment to cut accurate, standard openings in any fluorescent fixture canopy. It eliminates tedious and often haphazard cutting of canopies with snips or hacksaws and assures a perfectly fitted job. Just place it on the canopy where the hole is desired, and punch. It cuts a neat, accurate, raceway opening. Fig. 11-53.

■ *Power saws* may be used to cut the raceway and covers when fitted with special blades. Remember that this is steel being cut—the saw blade should not be one for wood.

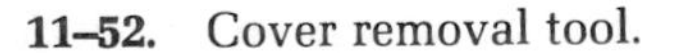

11–52. Cover removal tool.

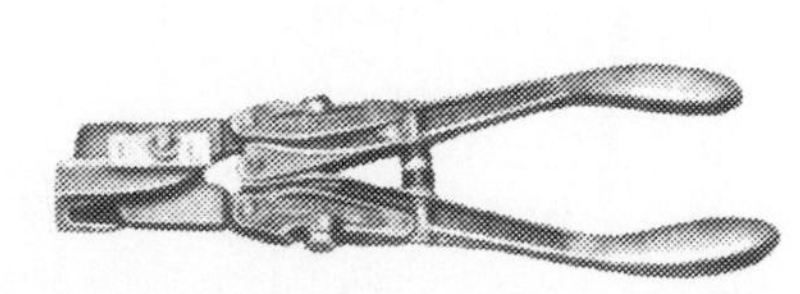

11–53. Canopy cutter.

Fig. 11-54 shows a radial-arm saw used for a job requiring several cuts of wider raceway materials.

Fittings. In order to make the Wiremold® system complete, there is a need for fittings to handle some special applications.

■ A *flexible section*, for instance, is available on special order in lengths greater than 18″, or in coils for getting around offsets, side bends, twisted turns, or semicircular or curved surfaces. Fig. 11-55.

■ *Bushings* are used to protect wires from abrasion. Fig. 11-56. The bushing simply slips into the open end of the raceway, where it enters a terminal fitting.

■ A *supporting clip* is used to keep the raceway close to the wall. Just mount the clip against the wall; then push the raceway into the clip. Fig. 11-57.

■ A *strap* is used to hold the raceway in place. Fig. 11-58.

■ A *connection cover*, Fig. 11-59, is used to cover openings where two lengths of raceway, not squarely cut, come together. Just snap the cover over the joint.

■ A *flat elbow* with a 90° angle is used for right-angle turns on the same surface. Fig. 11-60. The elbow may be installed very quickly. Fig. 11-61. Fig. 11-61A: slide the raceway onto the base of the elbow; Fig. 11-61B: pull wires in; Fig. 11-61C: snap the elbow cover on.

■ An *external elbow* is used for surfaces at right angles. Fig. 11-62. A single-pole switch and box are used in Fig. 11-63. The single-pole switch is a standard one. (Any other standard size switch, including 3-way, may be used.) The cover has twist-outs on each end and each side for the raceway to enter.

11–54. Using a radial-arm saw to cut raceway. This saw uses a special metal-cutting blade.

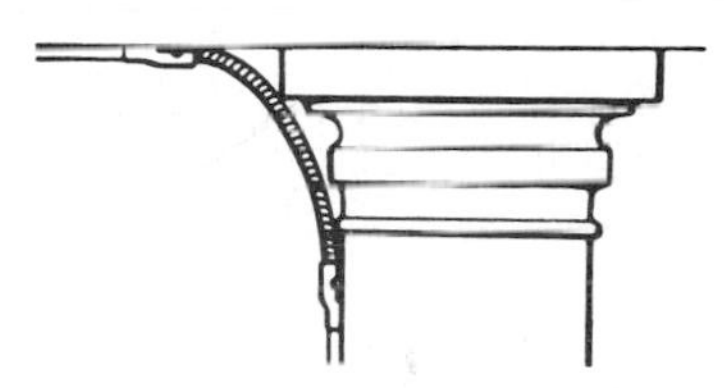

11–55. A flexible section to get around objects not easily adaptable to raceway.

11–56. A bushing.

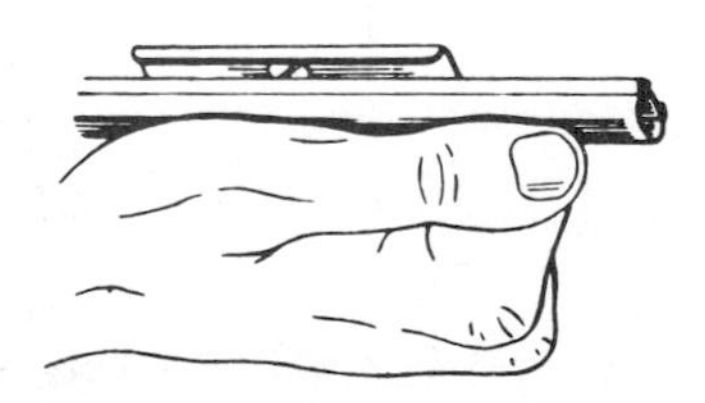

11–57. Supporting clip.

11–58. Strap.

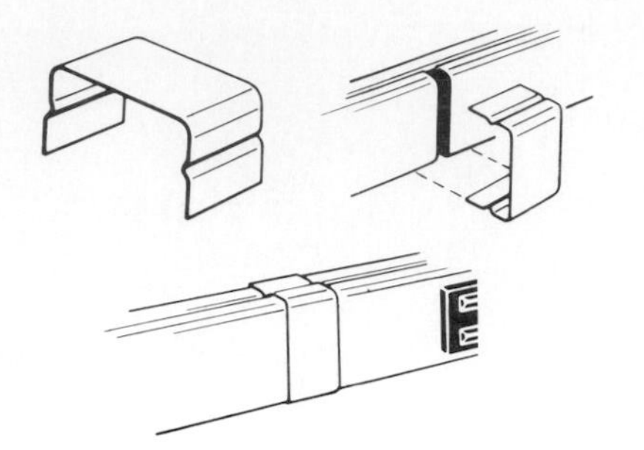

11–59. Connection cover.

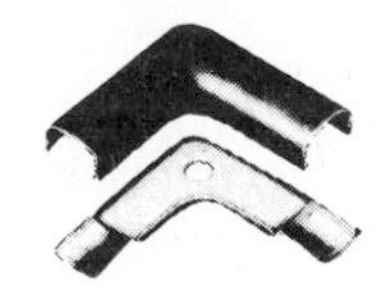

11–60. Flat elbow.

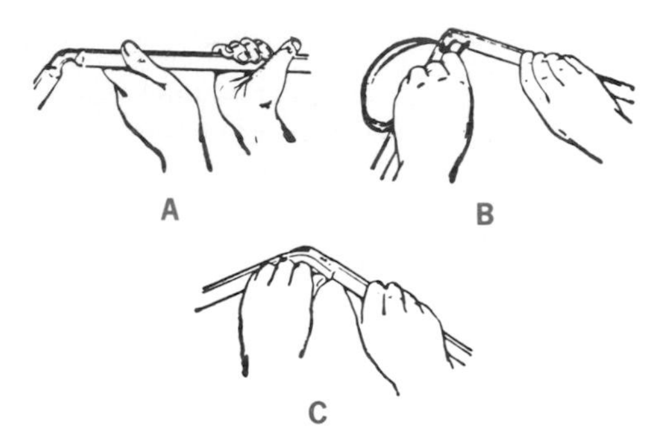

11–61. Installing an elbow.

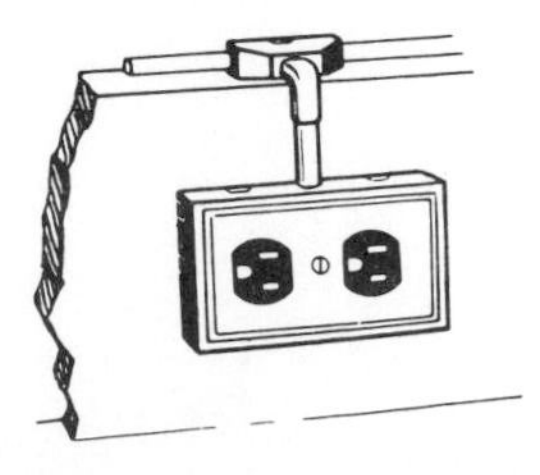

11–62. External elbow.

11–63. Single-pole switch and box.

■ A *utility box* is available for use as a tee, cross, pull box, junction box, blank box, or for drop cords. Fig. 11-64. The position of the twist-outs on the ends permits running the raceway close to interior trim.

■ A *duplex grounding receptacle and box,* Fig. 11-65, are used for standard, duplex grounding receptacles. A standard, single-gang duplex receptacle may be substituted. The cover has twist-outs on each end and each side. The location of the twist-outs permits the raceway to be placed close to interior trim.

11–64. Utility box.

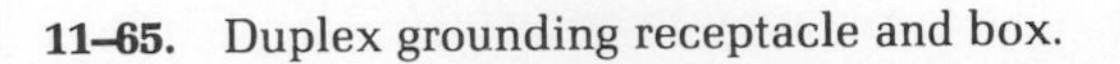

11–65. Duplex grounding receptacle and box.

Fittings used for special purposes in the installation of raceways are pictured in Fig. 11-66 through 11-78. The caption under each illustration gives details for that particular device.

The selection of raceway should be done according to the location of the work and the number and size of wires to be used in the installation.

Once the choice has been made, the actual work of placing the raceway where it belongs is a matter of measuring for length, and the selection of the proper device to make the corner or connection without losing the ground, or cutting the insulation on the wires.

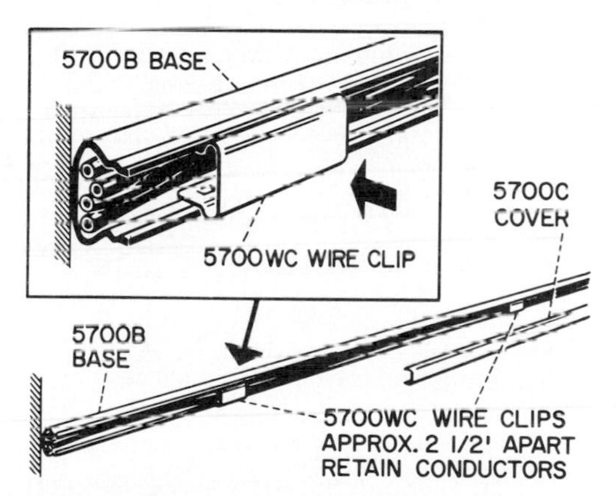

11–66. Installing a 1″ long wire clip to hold conductors in place before the cover is installed.

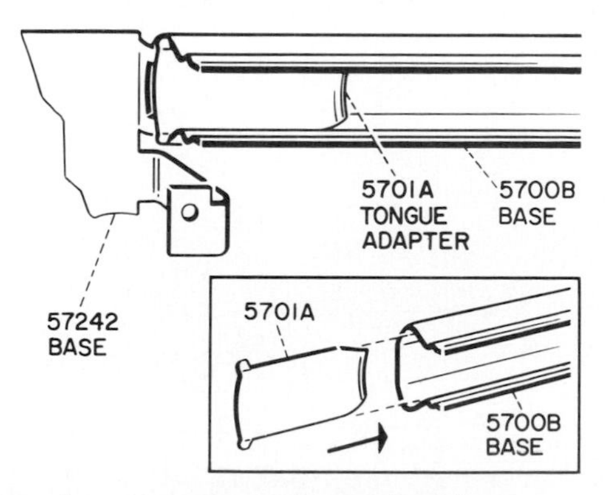

11–67. A tongue adapter is required for connections of different types of fittings and raceways.

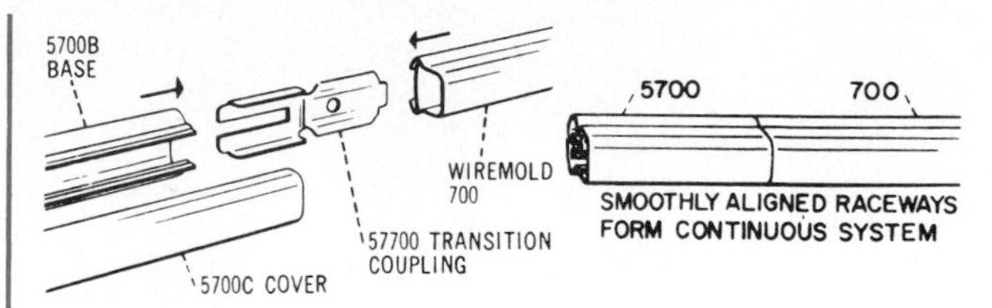

11–68. The transition coupling is used for making direct connections between different types of raceways.

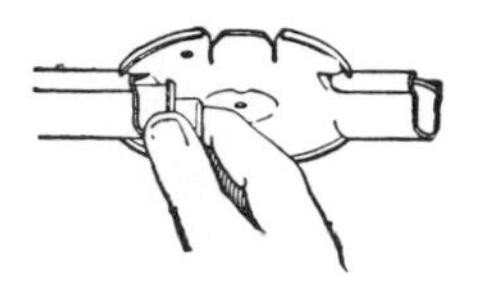

11–69. A bushing is inserted to protect wires from abrasion. The bushing slips into the open end of the raceway, where it enters the terminal fitting.

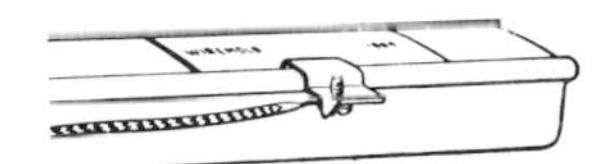

11–70. A ground clamp is inserted into the raceway before installing. Fasten the wire with a screw as required by the NEC. This ground clip is required when the system is not otherwise grounded.

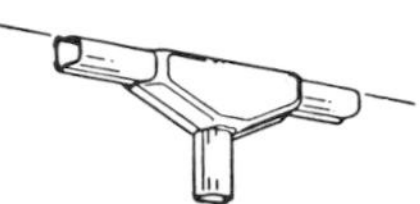

11–71. Installing a tee for branches at right angles.

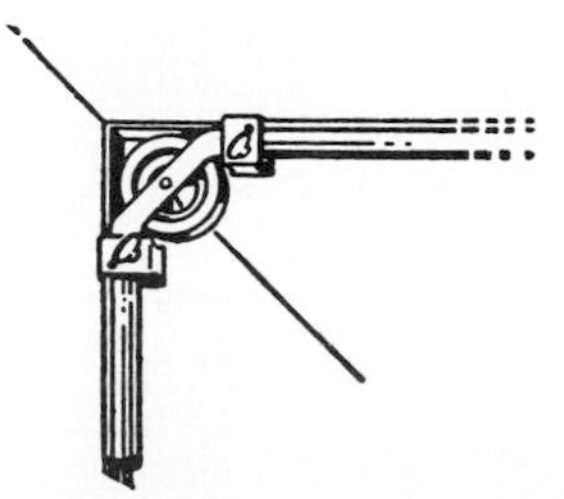

11–72. Using a wire pulley with internal elbows.

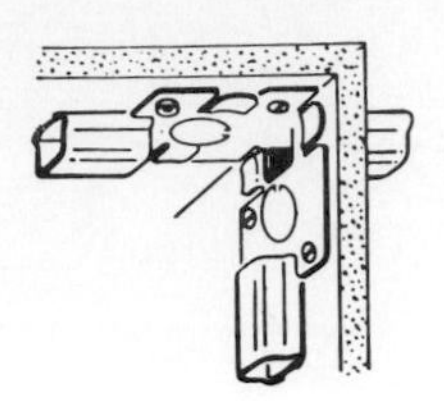

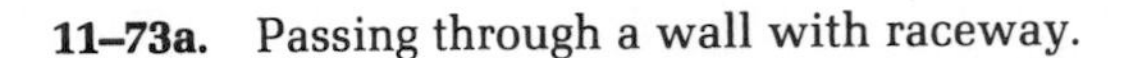

11–73a. Passing through a wall with raceway.

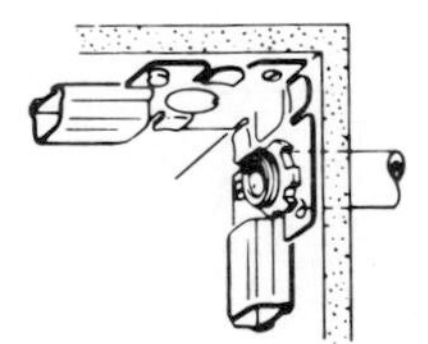

11–73b. Passing through a wall with conduit.

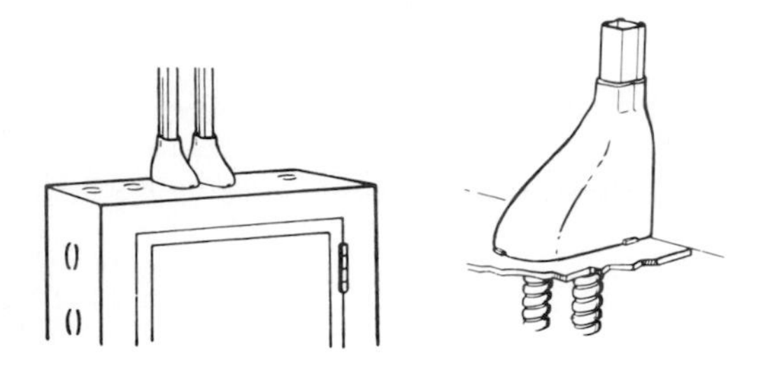

11–74. Adjustable offset connector. This connector eliminates offsetting raceway in connecting with surface-type panel cabinets.

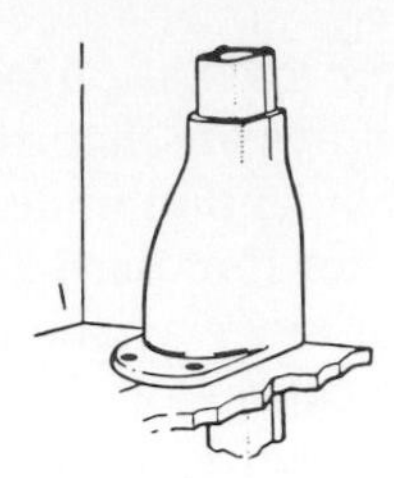

11–75. Kick plate for protection of raceway passing through the floor.

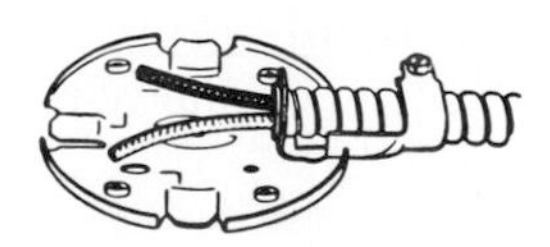

11–76. Armored-cable connector for connecting 14/2 up to 12/3 armored cable to boxes.

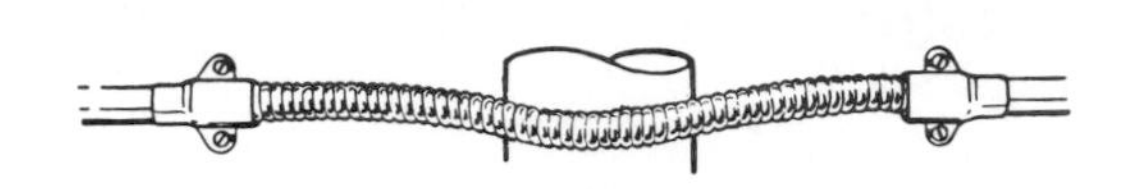

11–77. Flexible section, available in 18″ lengths. It is used for semicircular or curved surfaces.

Figs. 11-79 through 11-101 (pages 348-352) illustrate special instructions that should be followed for installations to ensure that the finished job will pass inspection and function properly.

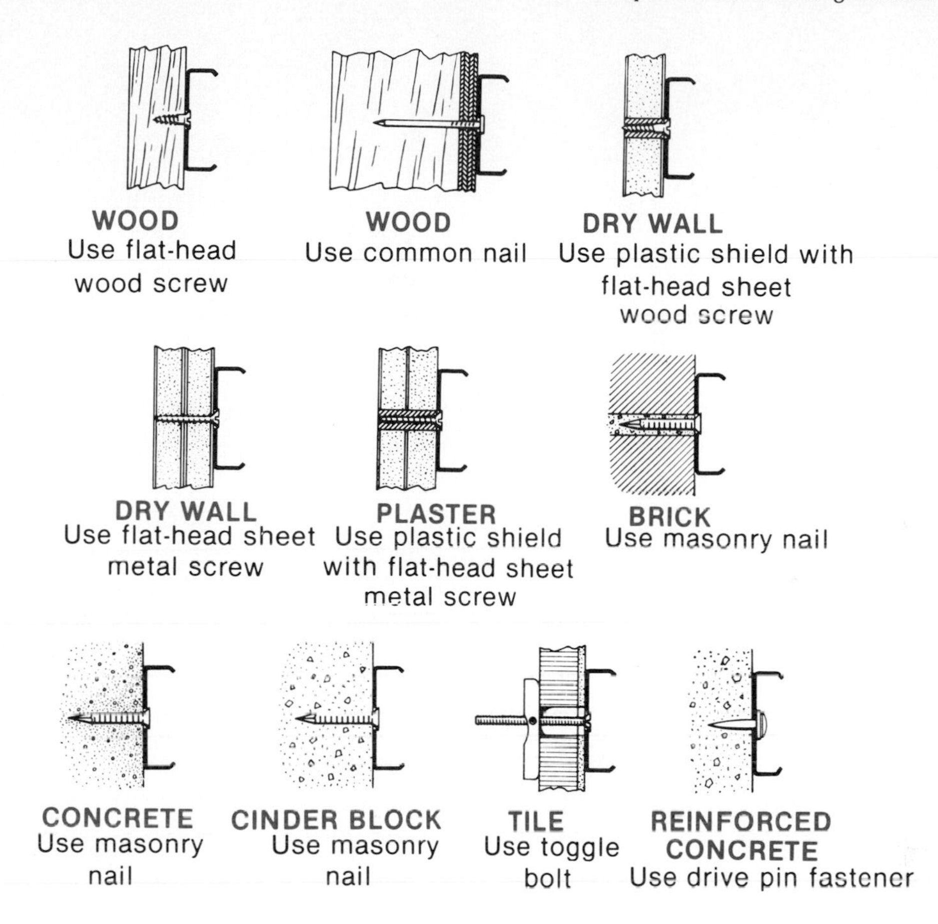

11–78. Mounting raceways. Raceways can be mounted to any type of surface through the use of the fasteners shown. *Caution:* Make sure the correct size screw or nail is used. Straps for the one-piece raceways are the only items which take *round*-head screws; all other mountings call for *flat* heads.

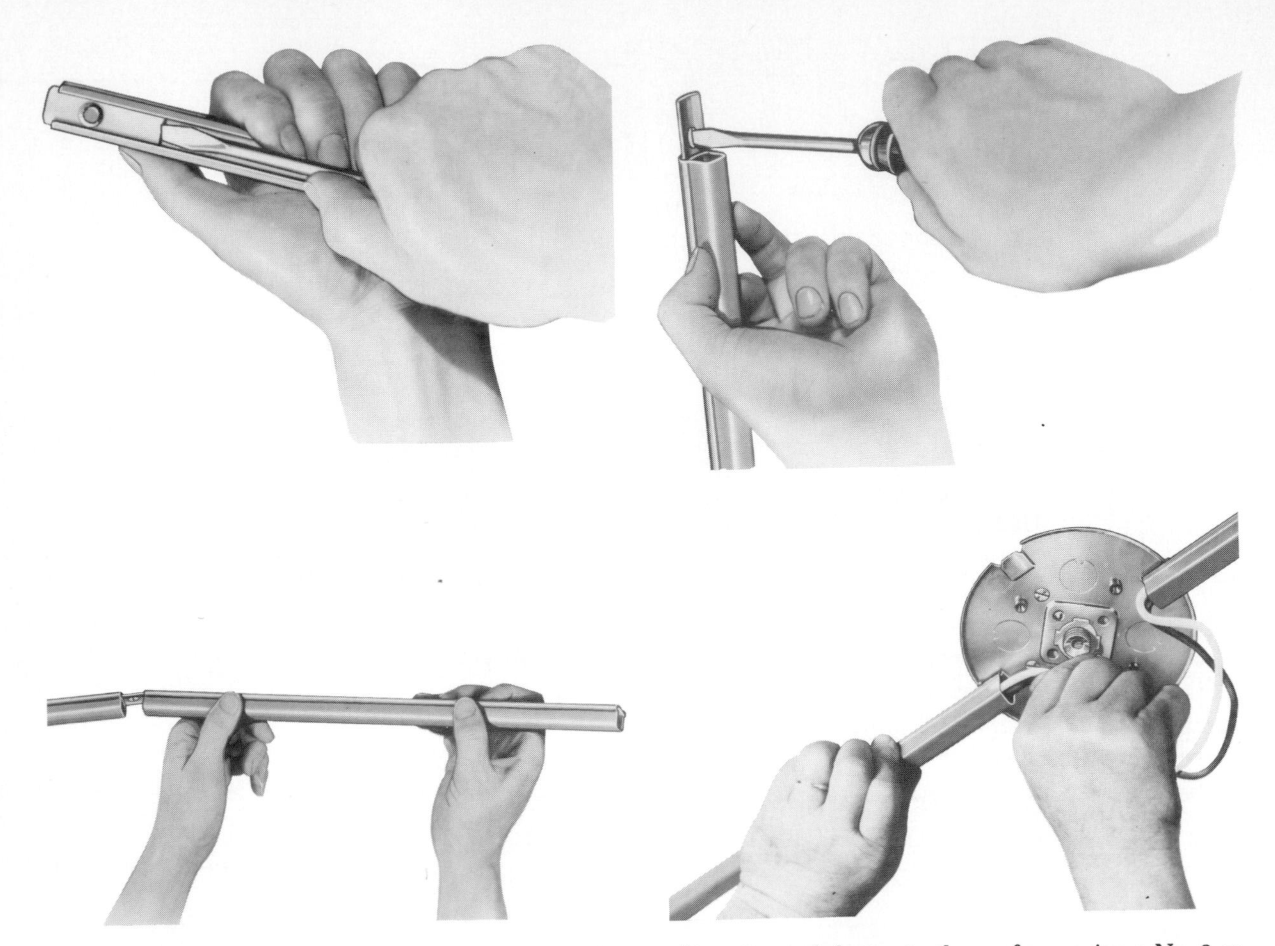

11–79. How to install raceways. *First*, push out the coupling. *Second*, fasten to the surface, using a No. 6 or No. 8 flat-head wood screw. *Third*, couple the lengths together. *Fourth*, pull wires in.

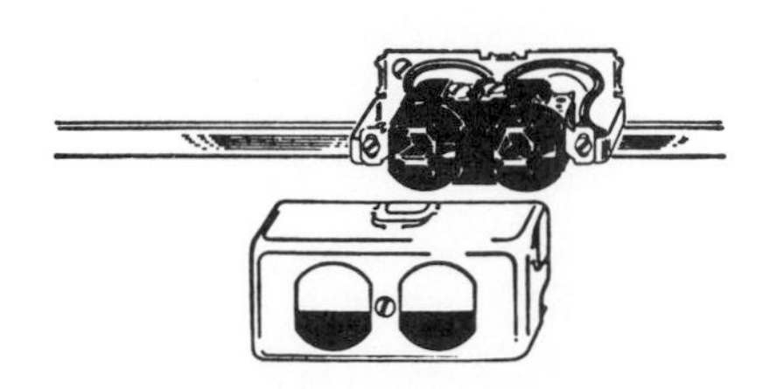

11–80. Making a duplex receptacle connection and final assembly.

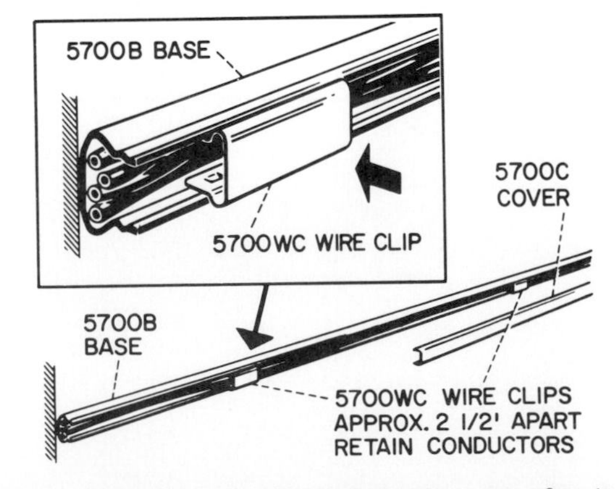

11–81. Insert a wire clip to hold the wire after it has been placed into the base.

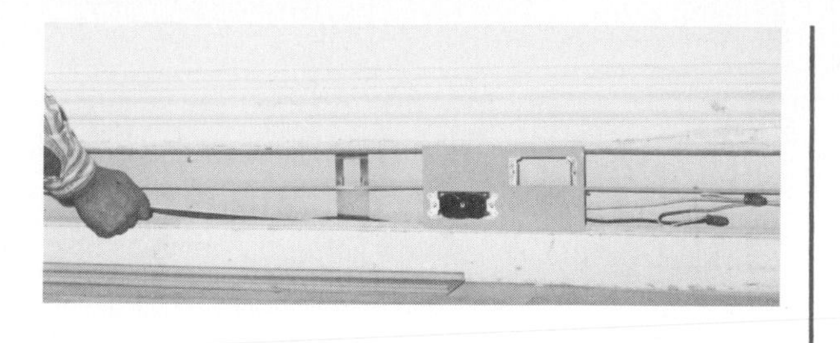

11–82. Electrician wires the power compartment of the divided raceway.

11–83. Single cover snaps in place after completion of the wiring in both sections of the raceway.

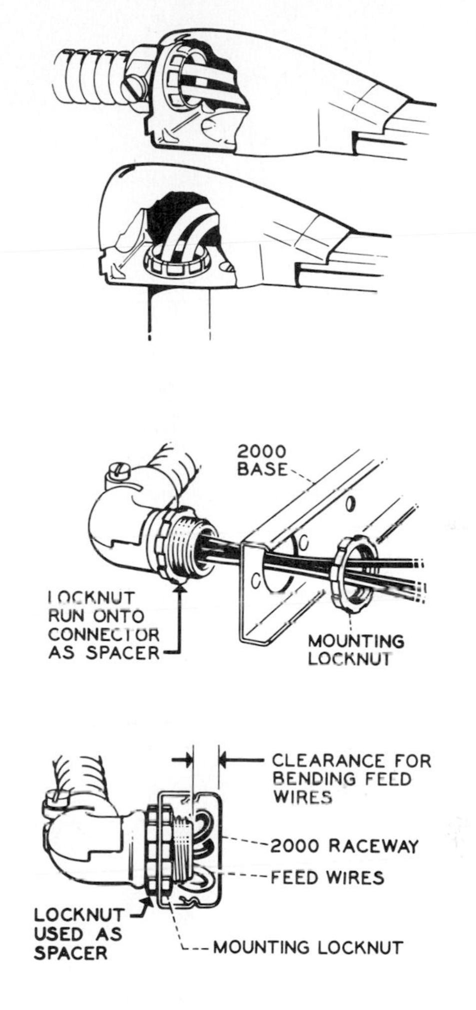

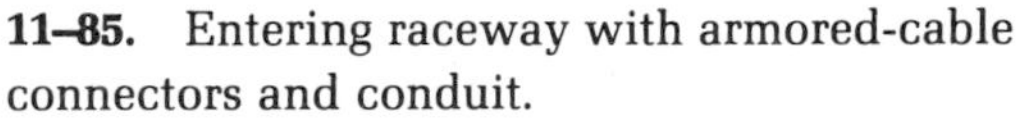

11–85. Entering raceway with armored-cable connectors and conduit.

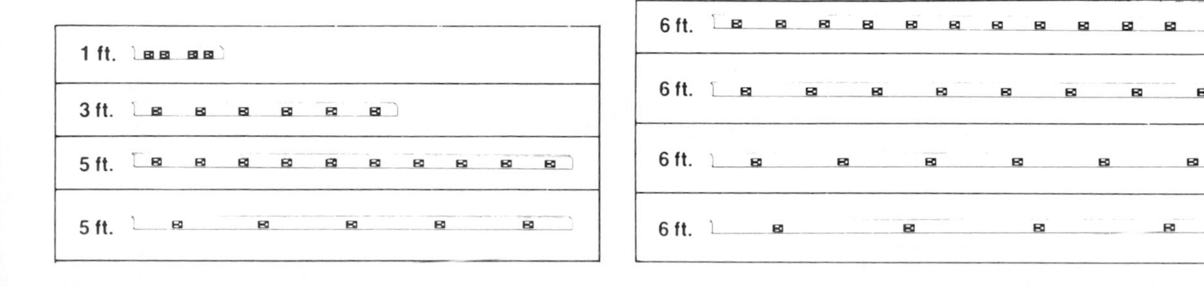

11–84. These prewired sections are excellent for adding to an existing electrical system. They come in several lengths and spacings.

11–86. A student working in a woodshop has power for his lathe furnished through a raceway large enough to handle the proper size wires for the lathe, and for the other machines in the room.

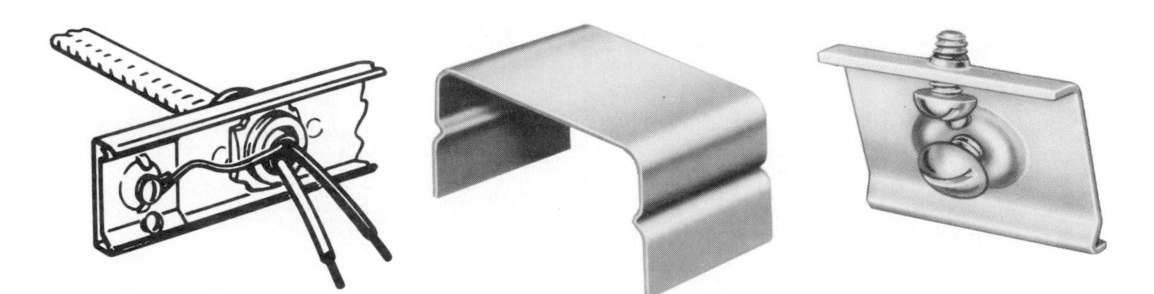

11–87. A cover clip and ground clamp are used to cover a joint where two lengths of cover or base join and the Romex needs to be grounded to the raceway.

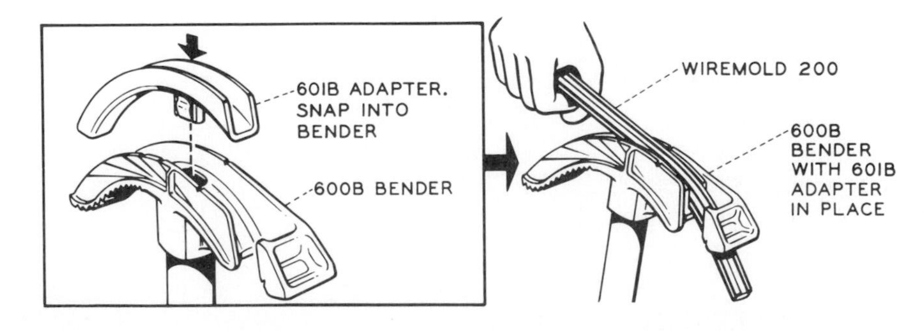

11–88. Raceway can be bent by using a bender, or a bender with an adapter to fit the size of raceway needed.

11–89. This room is lighted with keyless sockets, 12″ on center. (That is, centers of sockets are spaced 12″ apart.) The lighting arrangement is in a private club. Electrical service is delivered in raceways on the bottom of exposed wooden ceiling beams, all of which come from one corner of the room.

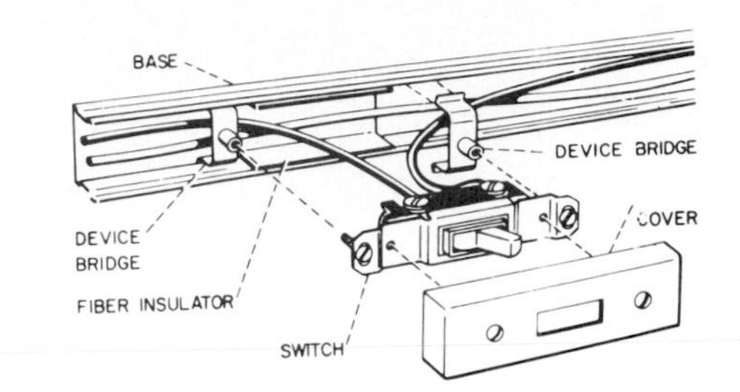

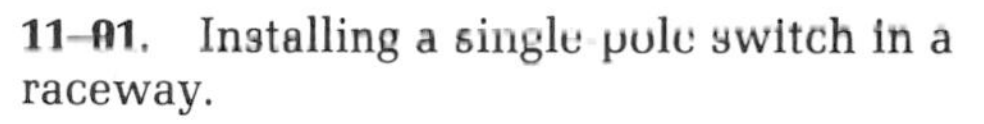

11–91. Installing a single pole switch in a raceway.

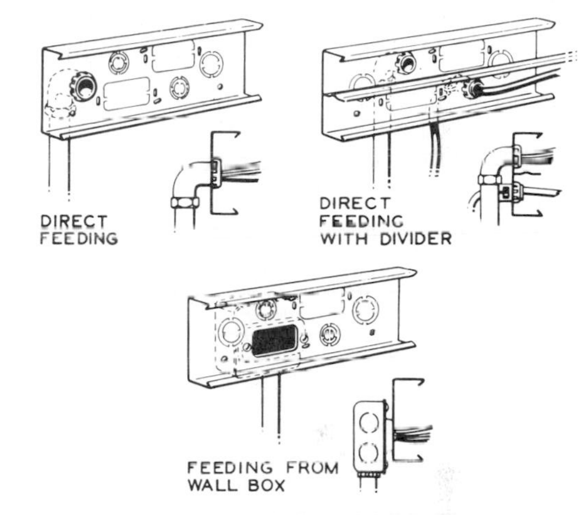

11–92. Methods of feeding raceway.

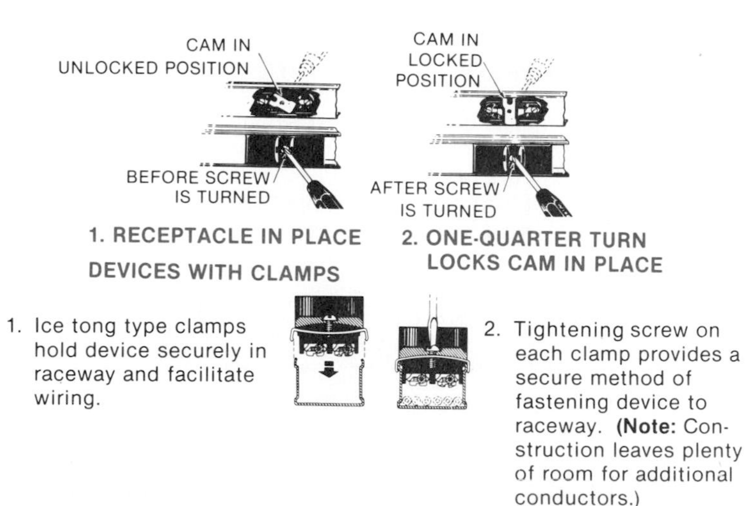

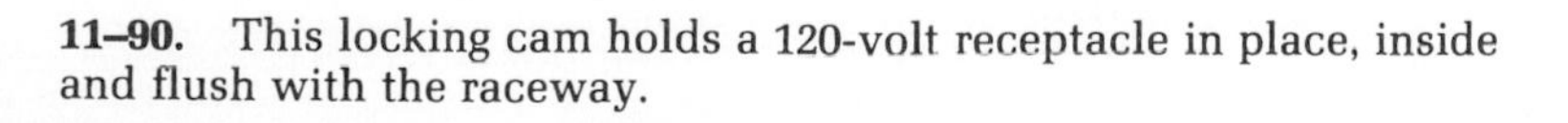

11–90. This locking cam holds a 120-volt receptacle in place, inside and flush with the raceway.

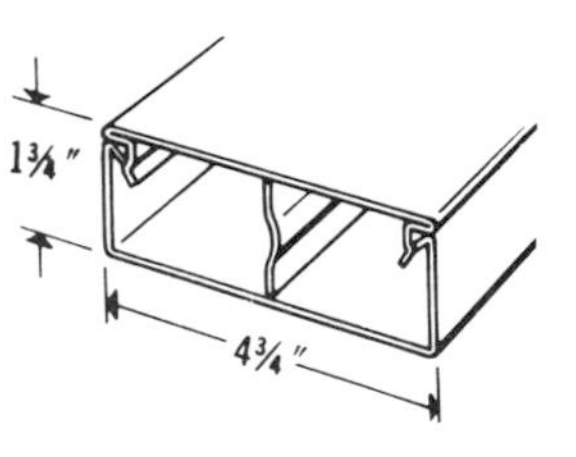

11–93. Larger size 2-channel raceway.

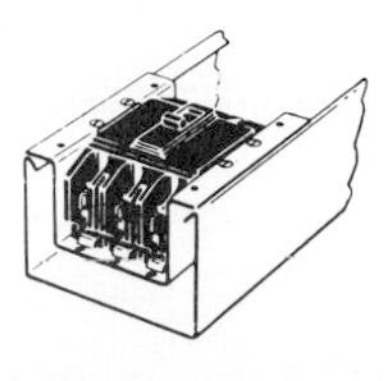

11–94. Circuit-breaker housing; 3-pole circuit breaker shown.

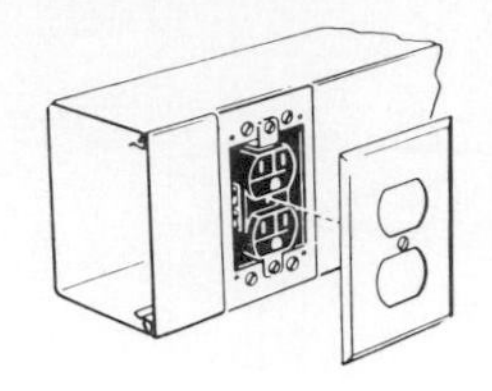

11–95. Mounting a duplex receptacle into a raceway.

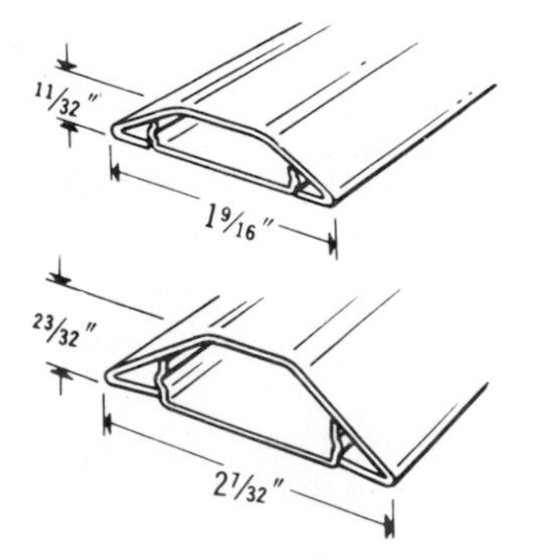

11–96. Two sizes of pancake raceway. These can be mounted on the floor where traffic rolls over them without damage to the wires inside.

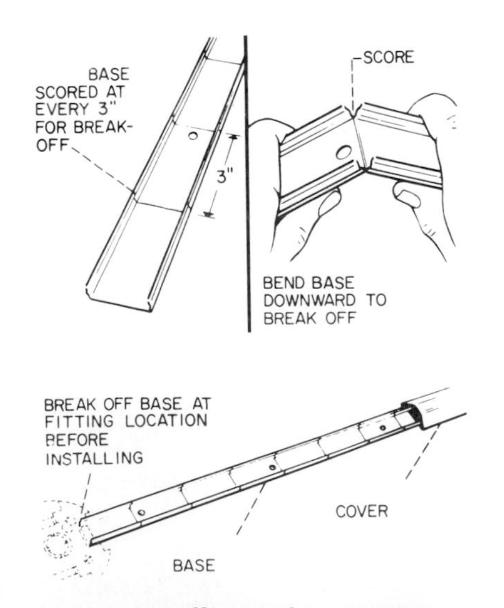

11–97. How to install pancake raceway.

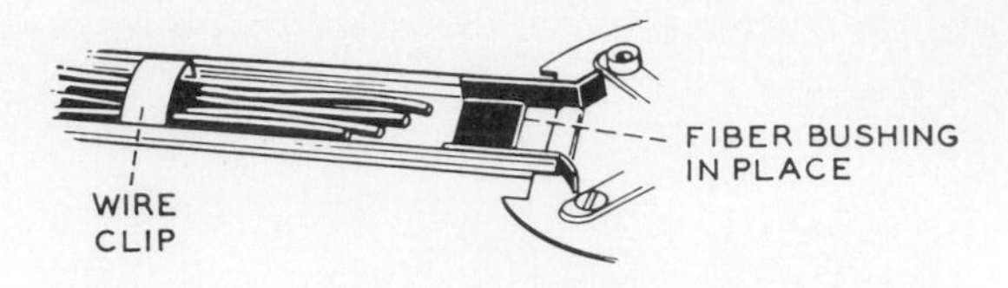

11–98. Lay conductors flat in base of the raceway. Use wire clips to hold conductors in place when run exceeds 5′. To protect wires from abrasion when entering a fitting, use bushing, or bend edges of raceway base outward.

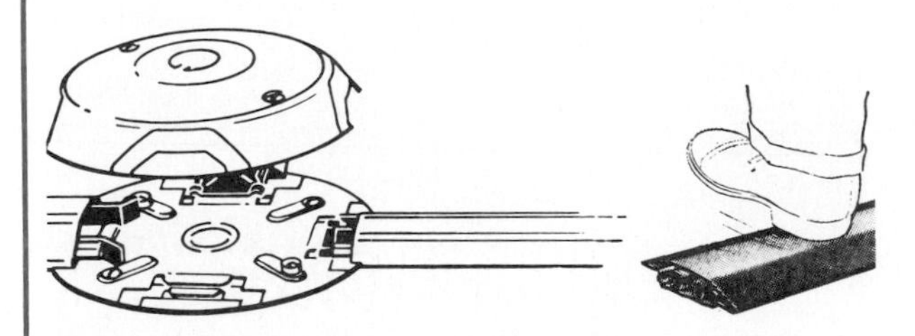

11–99. Snap the cover over the base of the raceway. Hook one side under the bead of base and apply pressure by stepping on the cover with a glancing blow of the heel. Fasten the cover of the fitting.

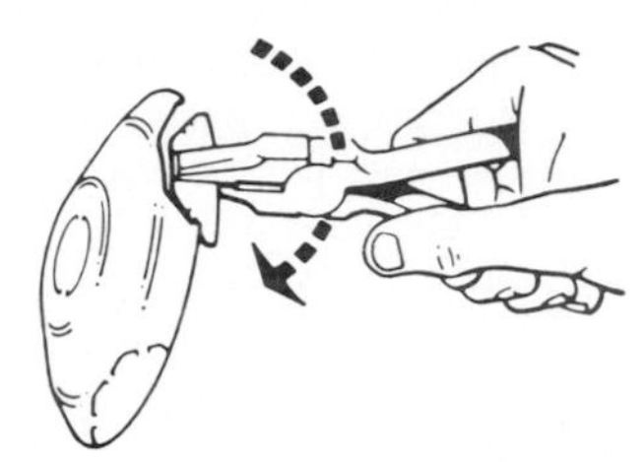

11–100. To break out the twist-outs in the cover, first make certain that the screw holes in the cover line up in position to correspond with the brass inserts in the base. Then grip twist-out at one end and work pliers up and down until first edge is free. Third, repeat same operation on the other edge and break out the twist-out.

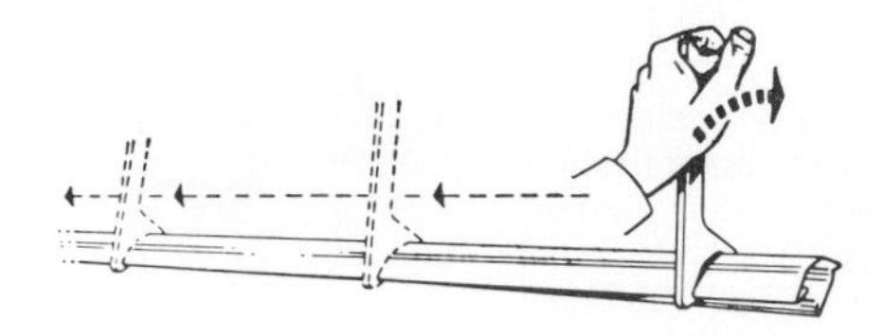

11–101. To remove the cover of a raceway, use a cover-removal tool designed for the purpose. It has two ends, each designed for a different type of raceway. Choose the correct end for the raceway you're working with.

COMMUNICATIONS AND DATA PROCESSING WIRING

The computer and its need for interfacing have increased the demand for wiring. The electrician is expected to be up on the latest in communications wiring. A number of systems are available for installation to provide both power and telephone circuits for the computer and data processing equipment.

The following example shows how a system is installed or added to for the installation of communications, data processing, and computer links.

Overhead Distribution System

With an overhead distribution system the wire is out of sight and out of the way. In most instances, this system results in putting the raceway in the ceiling. There it is not seen. The power and communications channels are brought down to where they are needed by poles.

A *header raceway* supplies power from a panel into an area. Fig. 11-102. Take-off connectors from the header direct the power into smaller raceways called *laterals*. Fig. 11-103. Receptacles within these laterals accept *whips*. These, in turn, wire into overhead fluorescent fixtures or bring the power sources to points of use below, either down through a wall or via Tele-Power® poles. Fig. 11-104.

The overhead distribution system offers some design advantages. One advantage is the modular plug-in system. Raceways, fittings, and connectors are all "contractor designed" for easy installation. Even the pre-wired harnesses just snap into place within the laterals. Handwiring, too, is kept to a minimum. Only a few tools are required to install the entire job smoothly and quickly. This system allows you to rearrange offices or work stations during regular working hours with little disruption of routine.

Phone service, CRT, intercom, and power can all be wired into the system. Communications wiring is directed through its own header, which runs parallel to the power header. From a lateral, also hung parallel to its power counterpart, all communications wiring can easily be brought down to work level through a wall or via poles. Isolated ground power is also available to protect CRTs from transient current. Fig. 11-105 shows a telephone connector housing for exit of telephone and data cable from the raceway. It permits use of less expensive standard communications cable within the raceway. This reduces the need for the more expensive, flame-resistant, low smoke-producing cable required in open air plenums.

Tele-Power® Poles

Power and communications are dropped from the overhead system to the user level by using poles. Fig. 11-106 (page 356). Most Wiremold Tele-Power® poles are dual channeled. A few have three channels. One side of the pole carries electrical power circuits, plus receptacles. The other carries telephone and data communications circuits. The power side is prewired with 6″ leads at the top. The communications side is furnished unwired—you add the cables.

Installation. Though pole details and capacities may be different, the installation technique is basically the same with all models. Stand it up, secure it, and connnect it overhead. A notch in the ceiling tile lets you bring those cables down where you need them. No floor drilling or carpet cutting is needed. A variety of knockouts and twistouts at the bottom of the poles permits easy access for updating and wiring—oftentimes without moving the pole at all.

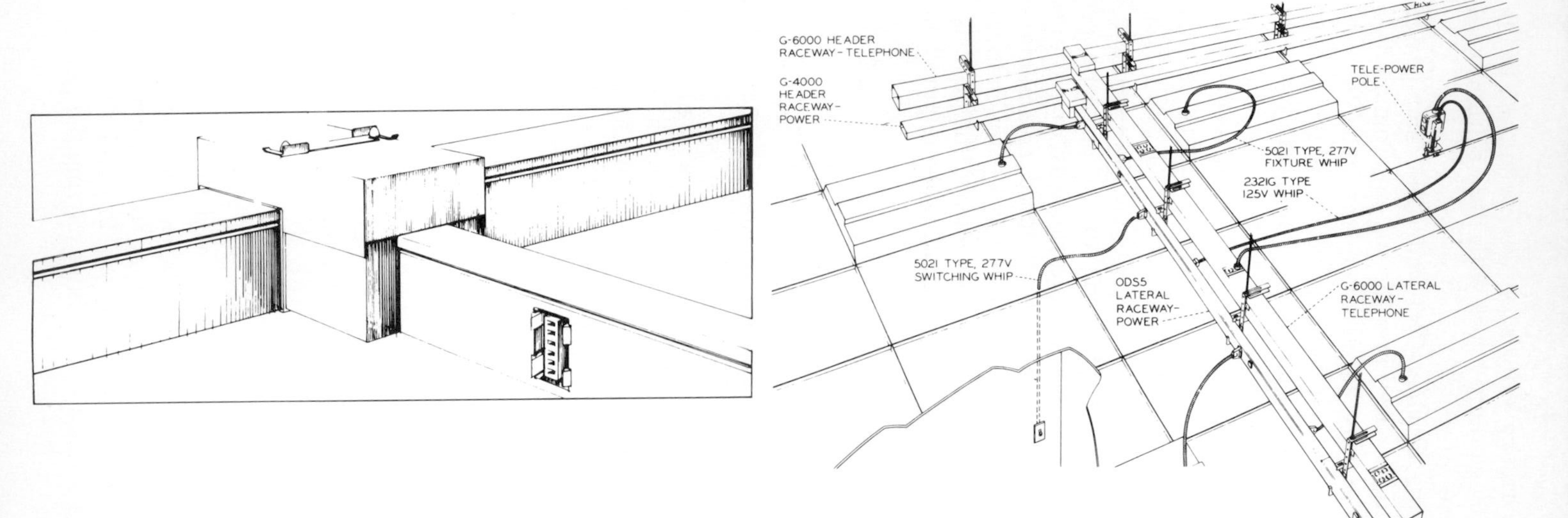

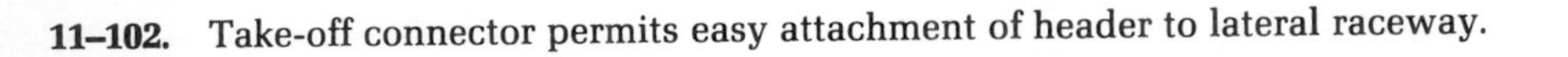
11–102. Take-off connector permits easy attachment of header to lateral raceway.

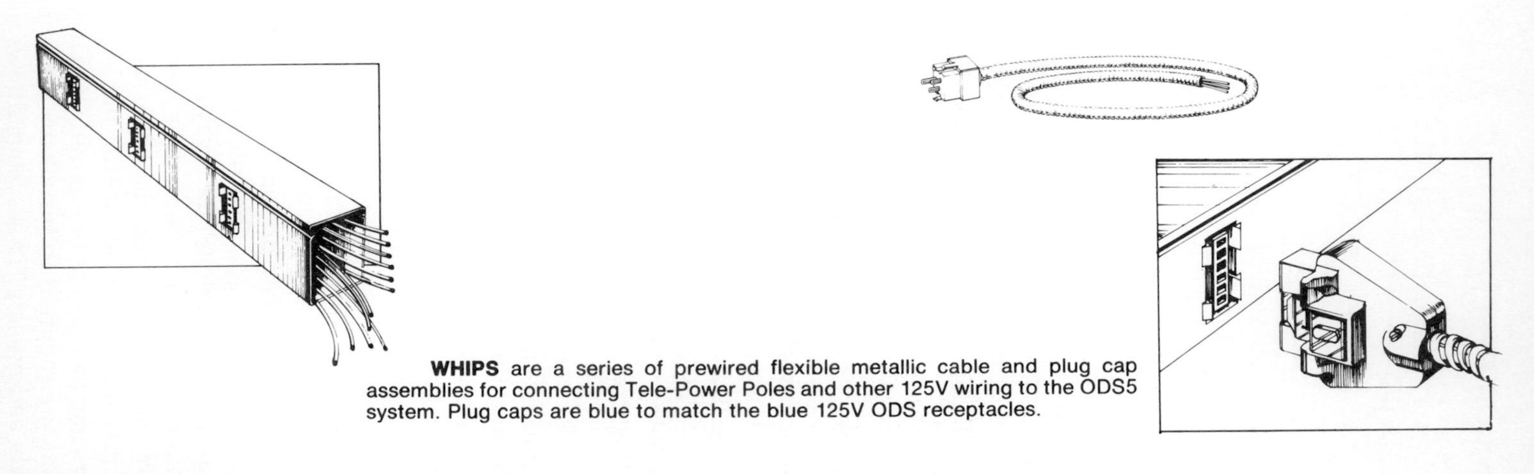

WHIPS are a series of prewired flexible metallic cable and plug cap assemblies for connecting Tele-Power Poles and other 125V wiring to the ODS5 system. Plug caps are blue to match the blue 125V ODS receptacles.

11–103. Receptacles on the overhead distribution system are located on alternate sides of the laterals. The 277-V receptacles are spaced 24″ apart; the 125-V receptacles are 36″ apart. They are properly color coded. Blue is used for 125 volts and orange for 277-V.

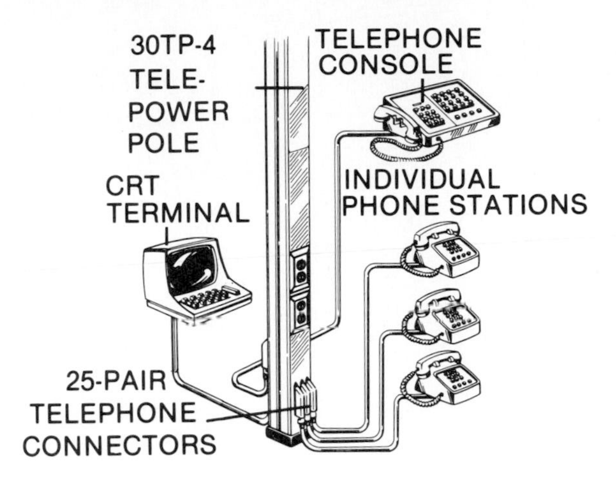

11–104. Poles provide electrical power and capacity to accept from four to one hundred pairs of telephone cables as well as coaxial cables for data communication.

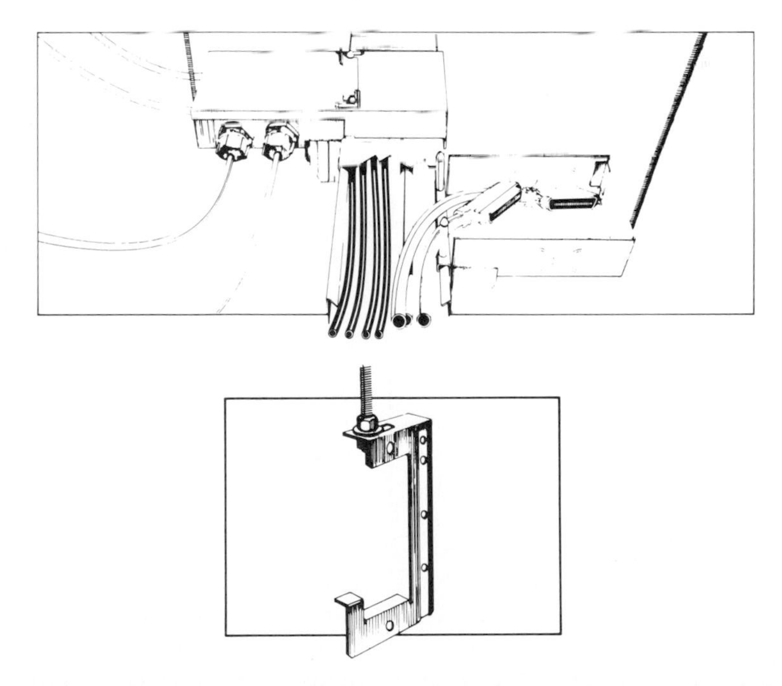

11–105. Telephone connector housing for exit of telephone and data cable from the raceway permits use of less expensive standard communication cable within the raceway, reducing the need for more expensive flame-resistant, low smoke-producing cable required in open-air plenums. C-hangers suspended from ceiling rods hold the laterals and/or header raceway.

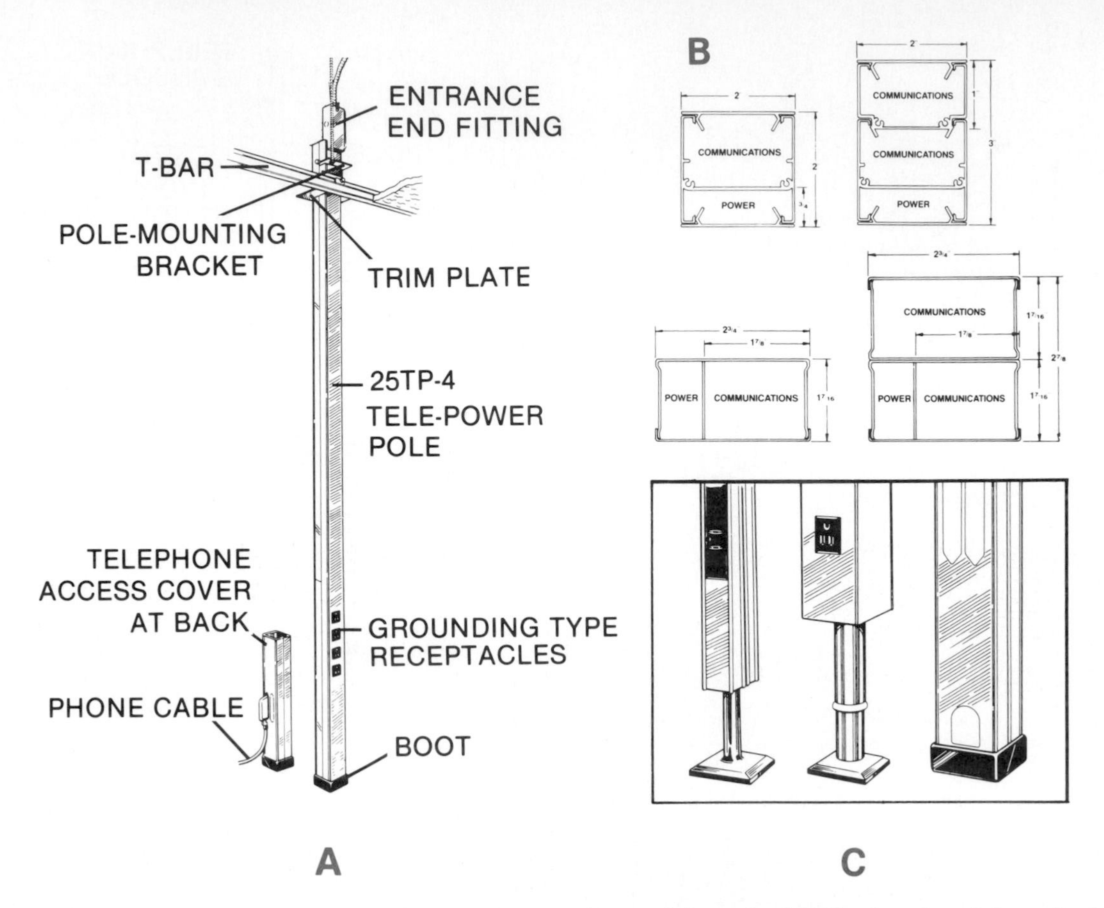

11–106. Installing the Tele-Power® pole. (A) The main parts of the pole. (B) The interior of the pole. (C) Left to right: fixed stanchion (or pedestal) base, adjustable stanchion foot, and fixed boot.

Add-on Compartment. An add-on compartment has been made available for Tele-Power® poles. It allows you to double the estimated communications capacity of an existing pole. At the same time, it allows the separation of telephone and data communications lines if you desire. Fig. 11-107. The advantage of this type of add-on is its ability to reduce cross-talk.

Electronically sensitive equipment can be protected from RF (radio frequencies) and EMI (electomagnetic interference) when plugged into an isolated ground receptacle. The poles can be ordered with one isolated ground receptacle and three branch circuit receptacles. Refer to Fig. 11-107b, which shows how the isolated ground receptacle works. These receptacles, and subsequently the instruments plugged into them, have no connections to the building ground circuit. Thus, electrical noise will not be fed back to the equipment. This is important when dealing with medical instruments, digital testing equipment, computers, and other sensitive scientific instruments and equipment.

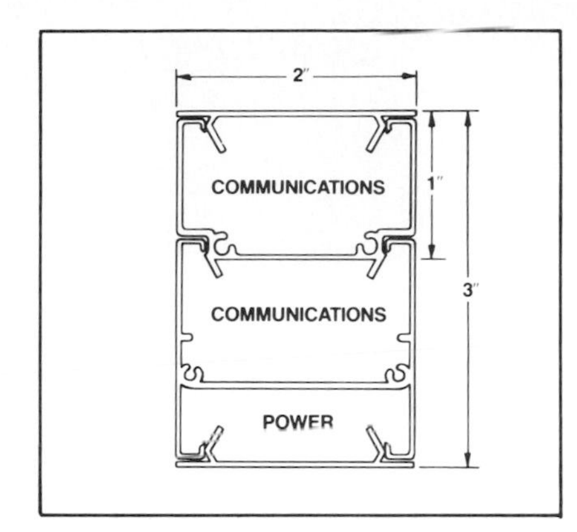

11–107a. Another channel for the Tele-Power® pole produces two communication compartments and one power compartment isolated by metal raceway.

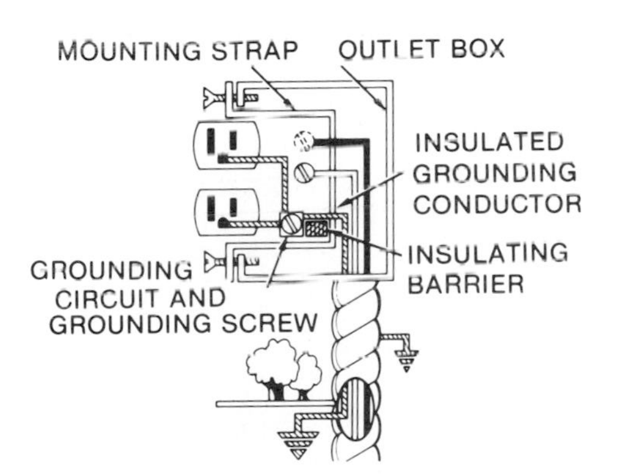

11–107b. The isolated ground receptacle makes it easier to use electronics equipment with sensitive circuitry.

Plastic Raceway

Wiremold has introduced a textured ivory plastic raceway system particularly suited to computer and telecommunication wiring needs. The raceway protects the wiring of electronic instrumentation and security systems, and organizes unsightly loose CATV cables. It is ideal for intercom, public address, and audio-visual cabling in schools and hospitals. Fig. 11-108.

The raceway comes in lightweight 5′ lengths. It is made of impact and corrosion resistant plastic.

The raceway is easy to cut with a hacksaw or utility knife. It is approximately $1\frac{5}{16}''$ wide by $\frac{3}{8}''$ deep. All you have to do is place the base of the raceway where you want it, put the cables or wiring into place, and then snap on the cover. Once installed, the cables or wires are readily accessible for repairs or system changes. Fig. 11-109.

The plastic raceway comes in two versions. One has an adhesive backing and one does not. The adhesive-backed raceway may be used for

11–108. Installation of plastic raceway when used with power electrical receptacles.

11–109. Plastic raceway carries telephone cables in this installation.

low voltage installations (under 50 volts). Acrylic-based-backing strips allow for quick installation. To install, lay out the route, peel off the backing strip, and press the base of the raceway onto the surface.

For power applications (up to 300 volts) the plastic raceway is UL listed and meets Article 352, Section B, of the National Electrical Code for nonmetallic surface raceways. The raceway also meets UL test standards 94V-0 to qualify as a self-extinguishing plastic that does not support combustion.

When using the plastic raceway to carry electrical power, the Code requires that the base be mechanically fastened to the surface. So, for power, you must mechanically attach the raceway to the wall and use the nonadhesive raceway.

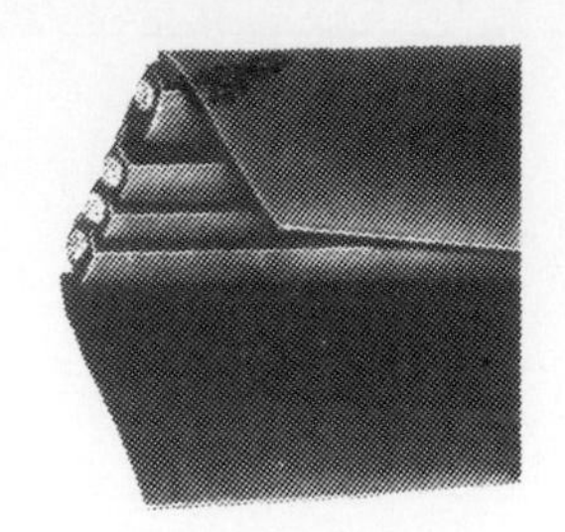

11–110. The cable has three conductors closely spaced with a ridge separating the ground wire.

Flat Cable Wiring

One of the faster methods for wiring fixtures in warehouses and other large areas is *flat cable wiring*. One system utilizes cable with a 600-volt rating, 4 conductor No. 10 AWG. The cable has three phase conductors, plus a neutral. Fig. 11-110. Three lighting circuits may be served by using the three color-coded taps to balance the load between the three phases. Fig. 11-111. Each phase may be connected to its own breaker to control a group of lights.

Article 363 of the Code covers flat cable. Each tap for hanging lights or a box is equipped with two pairs of pins. These puncture the cable insulation to contact one phase and neutral. The tap serves both as power connection and as a hanger for lighting fixtures. Fig. 11-112 shows how the tap and fixture hangers are installed.

System components are shown in Fig. 11-113. The cable is pulled through the strut by using a cable pull-in guide and cable leader. Fig. 11-114. The raceway comes in 20′ lengths and is supported by rods and hangers at intervals not exceeding 10′. The strut can support a

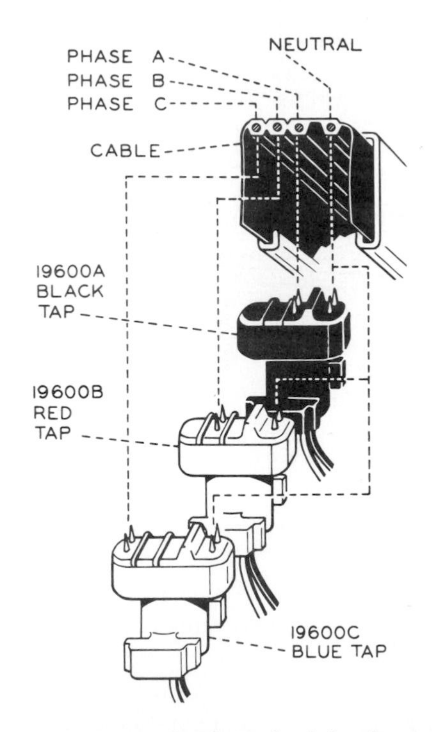

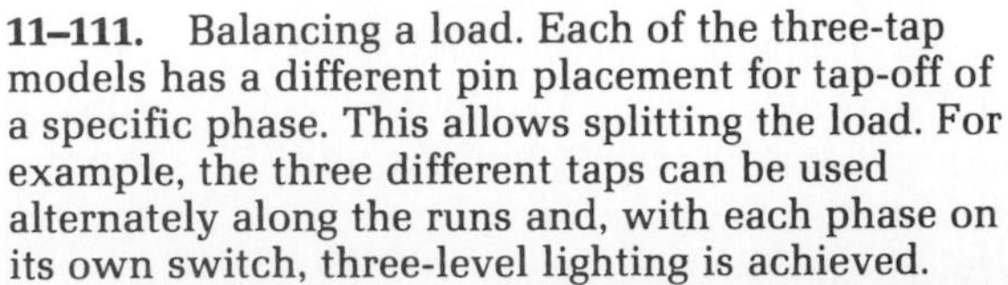

11–111. Balancing a load. Each of the three-tap models has a different pin placement for tap-off of a specific phase. This allows splitting the load. For example, the three different taps can be used alternately along the runs and, with each phase on its own switch, three-level lighting is achieved.

11–112. How tap and fixture hangers are installed.

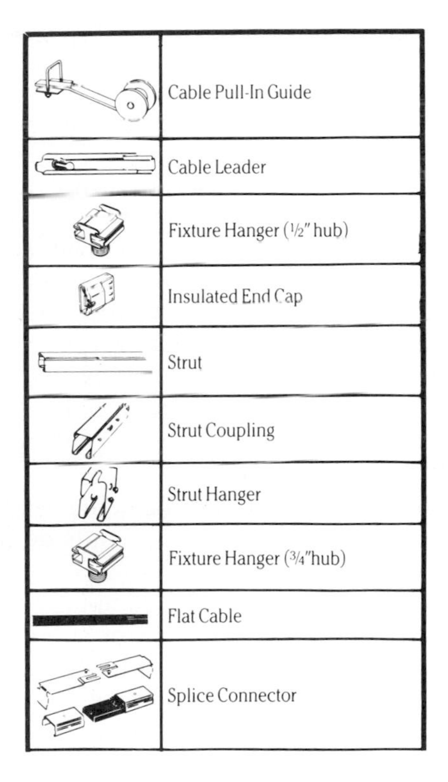

	Cable Pull-In Guide
	Cable Leader
	Fixture Hanger (½" hub)
	Insulated End Cap
	Strut
	Strut Coupling
	Strut Hanger
	Fixture Hanger (¾"hub)
	Flat Cable
	Splice Connector

	Outlet Box
	Dual Feed Outlet Box
	Phase A (Black) Tap (½" hub) Phase A (Black) Tap (¾" hub) Phase B (Red) Tap (½" hub) Phase B (Red) Tap (¾" hub) Phase C (Blue) Tap (½" hub) Phase C (Blue) Tap (¾" hub)
	Phases A/B (Black/Red) Tap (½" Phases A/B (Black/Red) Tap (¾" Phases B/C (Red/Blue) Tap (½") Phases B/C (Red/Blue) Tap (¾") Phases C/A (Blue/Black) Tap (½" Phases C/A (Blue/Black) Tap (¾"
	Phase A Tap and Box Phase B Tap and Box Phase C Tap and Box

11–113. Chan-L components.

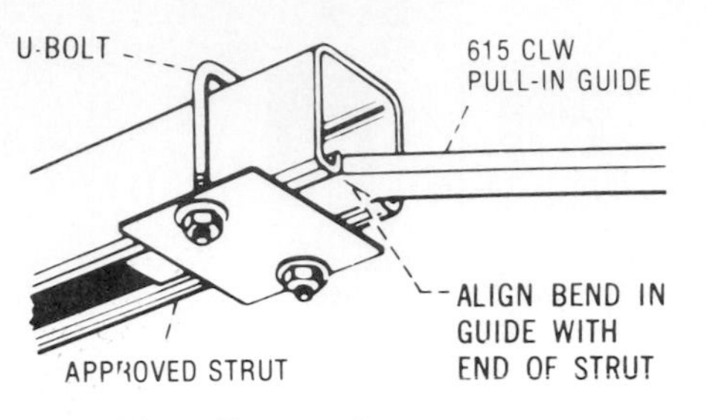

11–114. Cable pull-in guide in use.

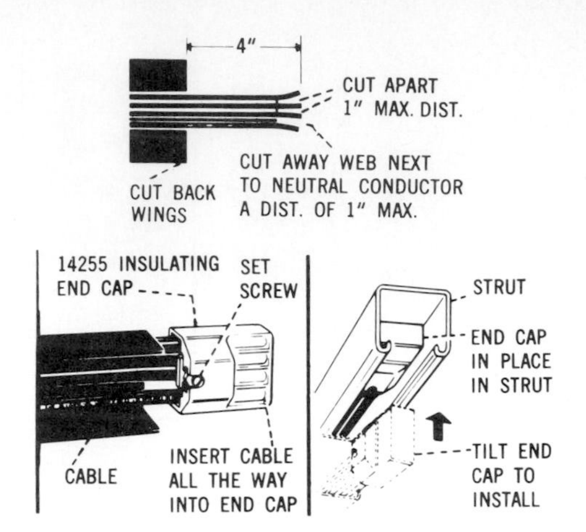

11–115. Installing the insulated end cap on the flat wire system.

load of up to 180 pounds between hangers if spaced at 10′ intervals. An end cap is used to insulate the end of the cable. Fig. 11-115.

Track lighting is very popular in warehouses and industrial plants. The light can be moved easily whenever the operations within the building require it. All types of lighting fixtures can be attached to the system. The flat wire makes it easier for the electrician to run the wires and make connections, thereby shortening the installation time and reducing labor costs.

Cable Trays

Cable trays are not raceways. They are covered by Article 318 of the National Electrical Code. *Cable trays* are open raceway-like assemblies made of metal, or a suitable nonmetallic material. They are used in buildings to route cables and support them out of the way of normal building activities. A strong, sturdy support for cables is provided with troughs and ladders to route the cables to their destination or termination.

Fig. 11-116 provides examples of how the trays can be routed and used to support heavy cables. The trays are made in straight sections, with matching fittings to accommodate all changes of direction or quantity of cables. They are made of aluminum or zinc-coated steel.

Trough-type trays protect cables from damage and give good support and ample ventilation. Solid-bottom fittings generally create no ventilation problems since they are a small part of the system. Cables are adequately ventilated through straight sections. Ladder trays provide maximum ventilation to power cables and other heat-producing cables. However, cables are vulnerable to damage.

Covers are available. Various parts are needed to support the trays and covers. Fig. 11-117 shows the necessary accessories for installation of a system.

Cables suitable for use in cable trays are marked CT (cable tray) on the outside of the jacket.

The cable system must be complete. It must be used as a complete system of straight sections, angle sections, offsets, saddles, and other associated parts to form a cable support system that is continuous and grounded as required by the Code in Section 318-6(a). The system must be grounded as any raceway system must also be grounded. The Code treats the cable tray as a raceway and a wiring method. Limitations are placed on the number, size, and placement of conductors inside the tray. These limitations can be obtained by checking the Code.

The use of cable trays needs to be studied more closely before attempting to install a system that meets the Code requirements.

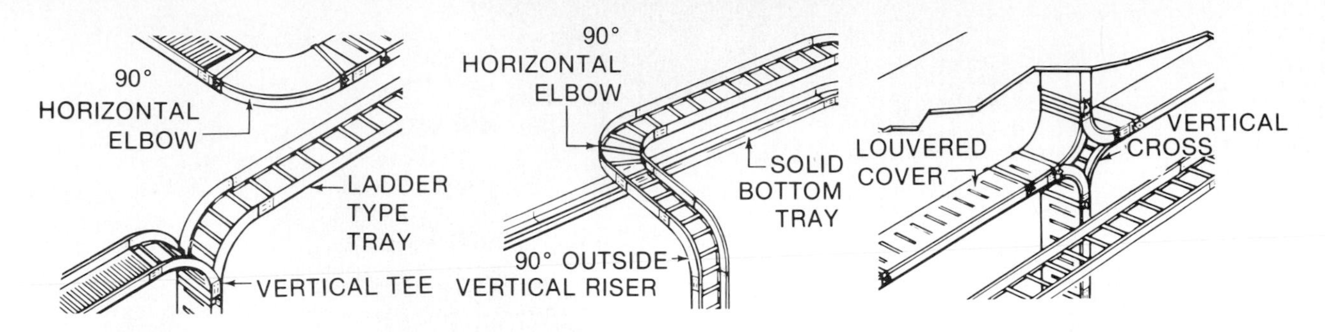

11–116. Various cable tray configurations.

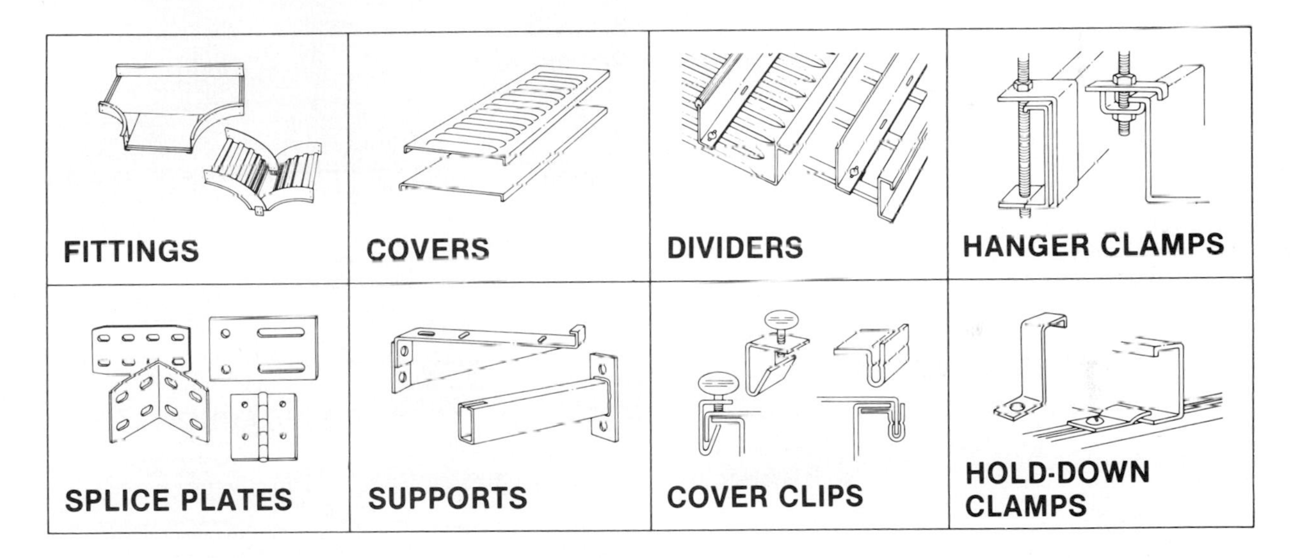

11–117. Accessories needed to complete a cable tray installation.

Marina Pedestals

Dockside marina services can be provided in the latest service center. Corrosion-resistant and durable pedestals are made of polycarbonate. These pedestals can be bought with metered electrical power, light, water, and telephone and cable TV connections. Units are factory wired and assembled, leaving only the pedestal installation and service connections for the jobsite. Fig. 11-118.

Power inlets and molded cable sets have been designed for use with the *marina pedestal*. Fig. 11-119. Various types of ship-to-shore plugs, cords, and cables have been designed for use with boats to make it safe to operate electrical equipment and power the boat safely. Fig. 11-120.

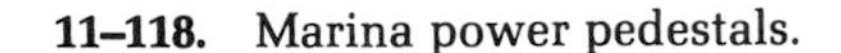
11–118. Marina power pedestals.

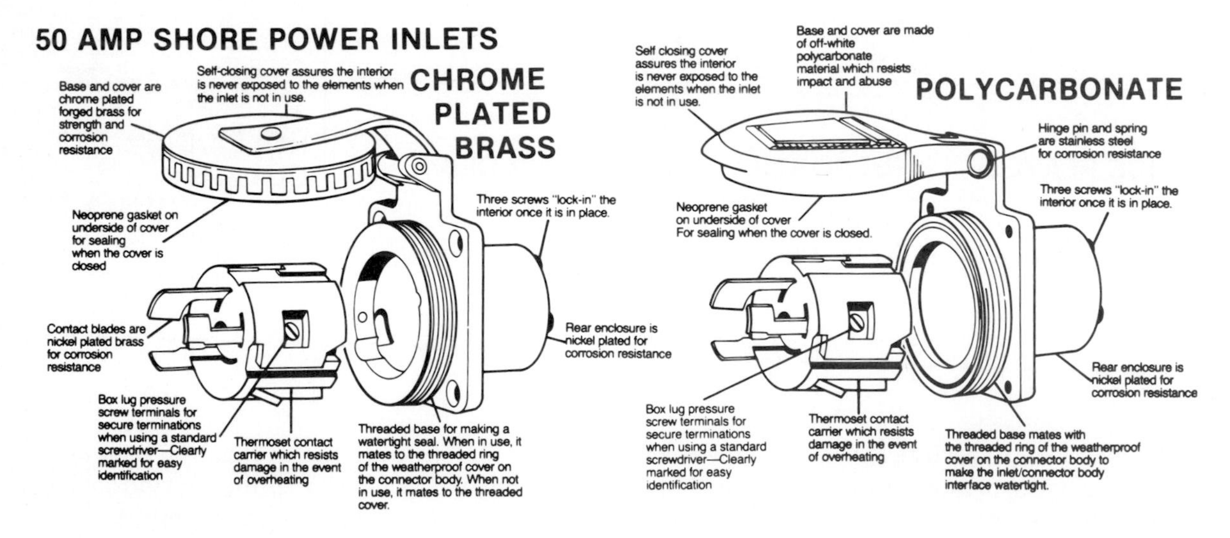

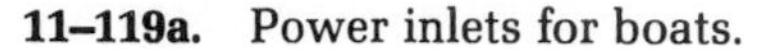
11–119a. Power inlets for boats.

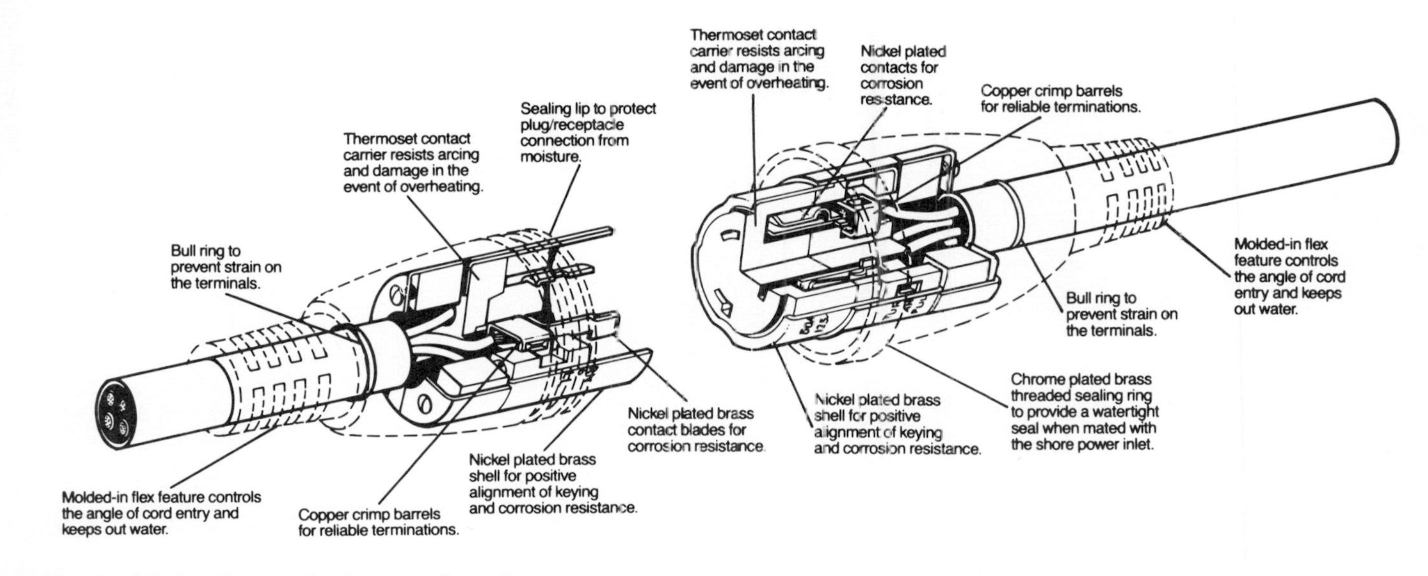

11–119b. Molded cable sets for boats and marinas.

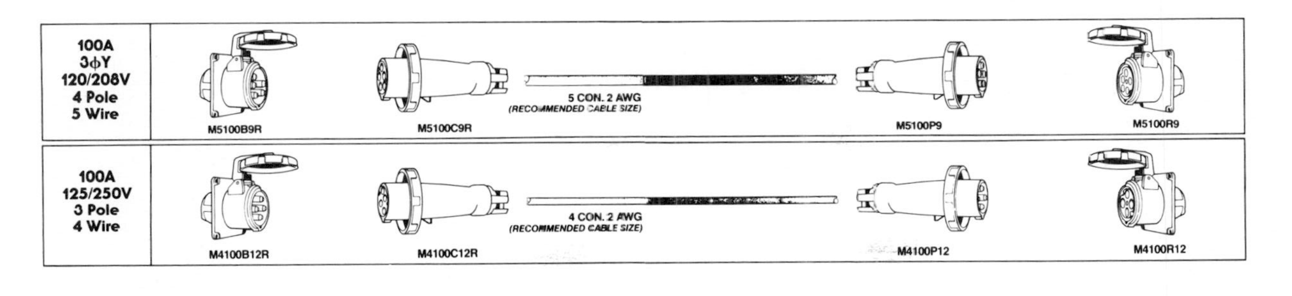

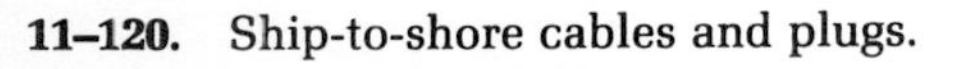
11–120. Ship-to-shore cables and plugs.

QUESTIONS

1. Describe a relay and its operation.
2. What are the advantages of a low-voltage wiring system?
3. How is a low-voltage system checked before power is applied?
4. What is the most common low-voltage system encountered by an electrician?
5. What is a metal raceway? Where is it used?
6. How does the NEC affect raceway installation?
7. List the special tools needed for Wiremold® installation.
8. What are cable trays?
9. What is a marina pedestal?

KEY TERMS

cable trays
canopy cutter
fish-tape leaders
flat cable wiring
lateral
marina pedestal
raceway
raceway bender
rectifier
relay
whips
wire pulley

CHAPTER 12

LAMPS: LIGHTING AND CONTROLS

Objectives

After studying this chapter, you will be able to:

- Identify the eight types of incandescent lamps.
- Identify the basic ballast circuits.
- Troubleshoot fluorescent lamps.
- Identify the various types of mercury vapor lamps.
- Identify the basic controls for incandescent lamps.
- Identify the various types of lighting fixtures needed on a farm.
- Identify the principal types of portable lighting.

TYPES OF LIGHTING AND LAMPS

Incandescent Lamps

Lamps that glow white-hot are *incandescent.* By heating an element until it glows white-hot, it is possible to get the filament, or element, to give off light as well as heat. The idea is to cause more light than heat.

Thomas A. Edison introduced the first practical carbon-filament lamp in 1879. The carbon filament was used because of its ability to glow with such intensity when white-hot. It wasn't until 1893 that the world got its first glance at practical electric lighting. The newly developed alternating-current dynamo, or generator, was put to work at this time to produce electricity for 20 000 "stopper" electric lamps which were ordered to cover the fairgrounds at the Chicago World's Fair.

The vacuum-type light bulb, modified to a gas-filled bulb in 1913, gave rise to extensive home use of electric lights. Today the envelopes of bulbs of more than 40 watts are gas filled, an improvement which effectively increases the filament life of the bulb. Tungsten filaments, Fig. 12-1, were substituted for carbon ones in 1912.

Incandescent lamps are most popular in homes and in industrial lighting of small areas. An incandescent lamp needs a filament to operate. At a temperature of 4750°F (2621°C), it is necessary to prevent the filament from becoming oxidized. Therefore the air is removed in bulbs of less than 40 watts, and in those of more than 40 watts the air is replaced by an inert gas, usually argon.

Types of Incandescent Lamps. There are eight types of incandescent lamps, each with its own characteristics and application. Different filaments are designed to handle various applications. Fig. 12-2.

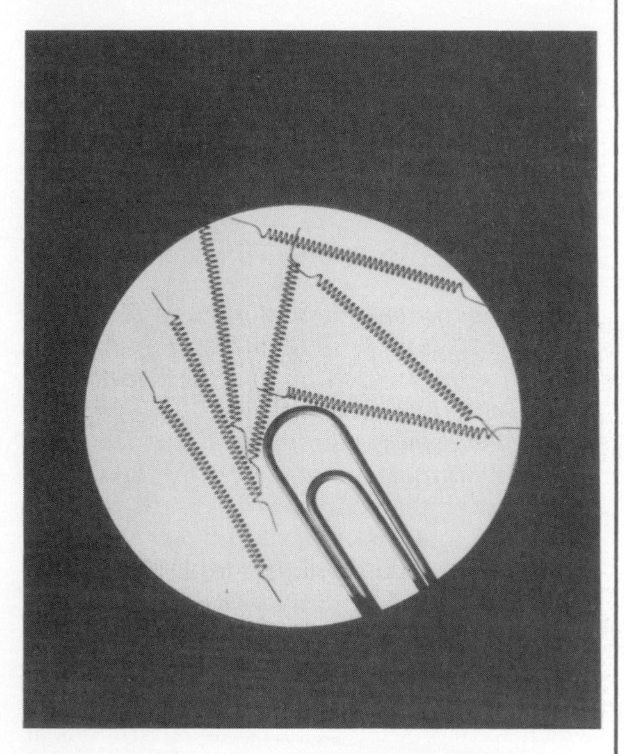

12–1. Filaments for lamps are precision wound. Compare the size of the paperclip with the precision-wound filaments for light bulbs.

■ *General-service lamps.* These are usually used in the home and are available in sizes ranging from 7 to 1500 watts. Although a 200-watt light bulb is usually the largest used in the home, some 300-watt bulbs are used for special service areas.

■ *Three-light lamps.* Some lamps have two filaments. Fig. 12-3. The two filaments are wired so that each can be used separately or both can be used together. The light switch makes the selection. They may be obtained in watt sizes of 50-100-150, 100-200-300, and other

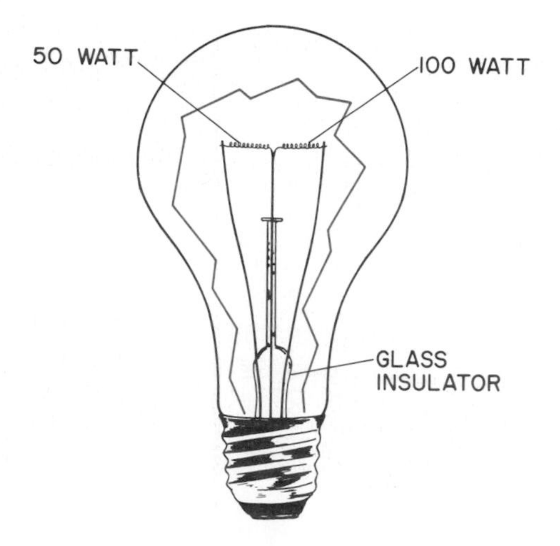

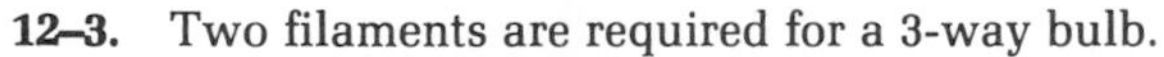

12–3. Two filaments are required for a 3-way bulb.

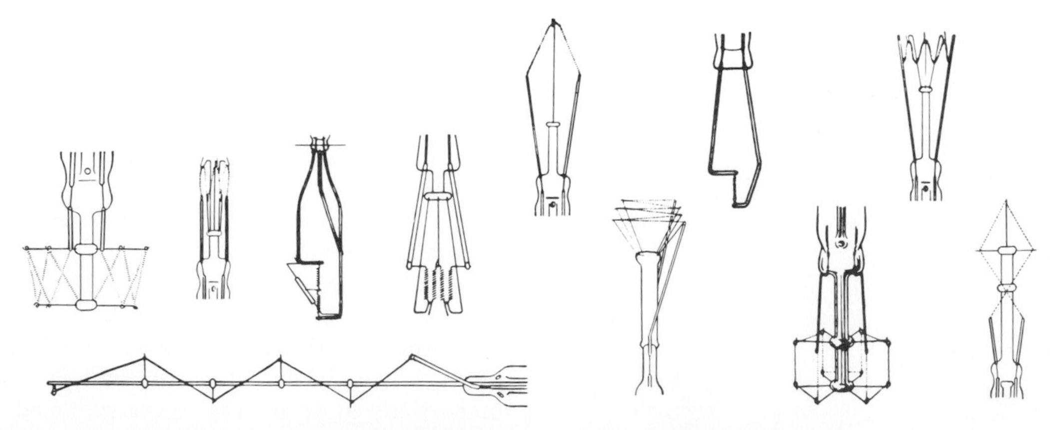

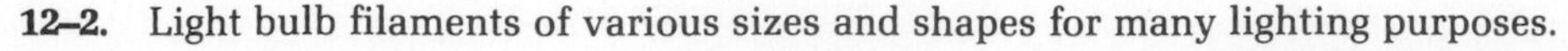

12–2. Light bulb filaments of various sizes and shapes for many lighting purposes.

combinations. They are referred to as 3-way bulbs. A special socket is required for 3-way lamps.

■ *White indirect-light bulbs.* Table and floor lamps without diffusing bowls use this type of light bulb. They are available in several sizes. The bowl has a heavy white coating on the sides, which permits only 20% of the light to be cast down, directing 80% of it upward for illumination of a room.

■ *Reflector bulbs.* Reflector bulbs are shaped like the white indirect-light bulb, but instead of a white coating, the sides are coated with an opaque silver finish. All of the light is directed out of the flat bowl-end of the bulb. These may be interior floodlights, or spotlights, depending upon the frosting. This type of bulb is available in 75-, 150-, and 300-watt sizes. Reflector bulbs are designed for indoor use only.

■ *Projector bulbs.* These are the outdoor equivalent of the reflector bulb. Made of hard glass, they are available in 75- and 150-watt sizes for use as exterior flood- or spot-lights. They resemble the reflector bulbs since they too are silver coated from the base up to the top or lens.

■ *Tubular bulbs.* These are long, tube-like bulbs available in two different base shapes. All sizes come in either clear, inside frosted, or in various colors. A coil filament extends the full length of the tube. Small tubular bulbs are known as showcase lights and are available in 25- and 40-watt sizes.

■ *Daylight bulbs.* These are standard bulbs with a blue-green glass which absorbs some of the red and yellow rays to produce a whiter light.

■ *Reflector-type bulbs.* These have a silver or aluminum reflector built into the lamp, either inside or outside the lamp. They are available in over 150 different lamps.

Fig. 12-4 illustrates a variety of incandescent lamp shapes. The letter under the bulb indicates the code printed on the package for easy identification of shape and base. Methods used for diffusing light from incandescent lamps are shown in Fig. 12-5.

Fluorescent Lamps

Since the first fluorescent lamp was introduced at the New York World's Fair in 1939, the bulb has undergone many changes in size and shape. Among these changes has been the starting method. Fig. 12-6.

A *fluorescent lamp* is an arc-discharge device which has no inherent resistance. Therefore unless controlled, current flow will rapidly increase until the lamp is burned out. This problem is solved by a ballast which is connected between the lamp and the power supply—limiting current to the correct value for proper lamp operation. The name "*ballast*" stems from its stabilizing or ballasting function. In addition, most ballasts perform the function of transforming supply voltage to the proper value required for starting the lamp.

Limiting Current to the Lamp. A simple but impractical way to limit lamp current is to put a resistive element in series with the lamp. This element could be a piece of iron wire, carbon, or even a small low-voltage incandescent lamp. There is a disadvantage in this arrangement, however. While current would be effectively limited, the power dissipated by the resistor in the form of heat would be approximately equal to the power consumed by the lamp. The overall efficiency, therefore, would be low.

Another way of limiting current to a lamp is by use, in the proper combination, of such reactive components as inductance and reactance. From the standpoints of economy and performance, these ballasting combinations have proved to be the most practical way to limit current to the lamp in an AC application. Fig. 12-7.

Simple Inductive Ballast. Some simple inductive ballasts are basically coils of copper wire wound around iron cores. This type of ballast uses alternating current which passes through the turns of copper wire, creating a strong magnetic field. The magnetic field reverses its polarity 120 times per second when operating on 60 Hz power. The resulting inductance opposes a change in current flow and limits the current to the lamp. Only a few commercial ballasts are as simple as this, but the basic principle of operation is the same for all ballasts.

Here is the sequence of operation in a simple, switch-start ballast circuit (such as might be found on a desk-type, push-button fluorescent fixture).

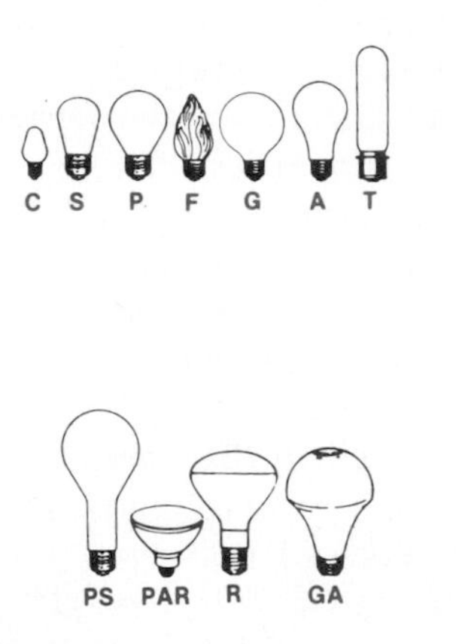

12–4. Various shapes and sizes of incandescent light bulbs are available. Note the letter under each bulb. This code refers to a particular type bulb.

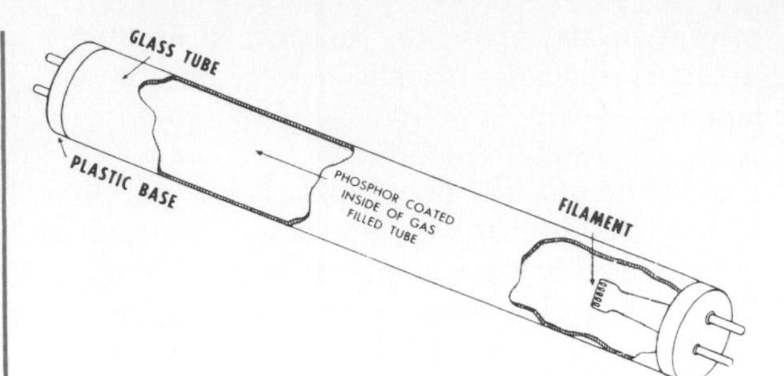

12–6. Construction of a typical fluorescent lamp.

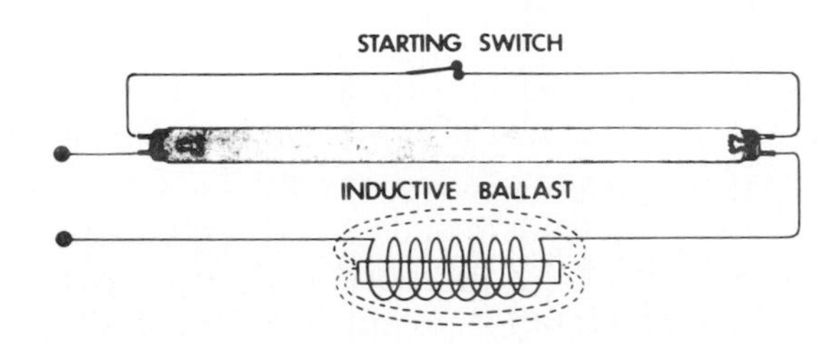

12–7a. The first step in the ballast-starting sequence of a fluorescent lamp is to preheat the cathodes. The current heats cathodes until they "boil off" clouds of electrons.

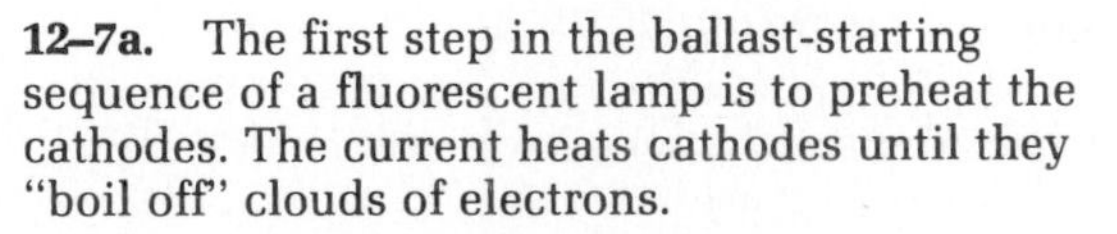
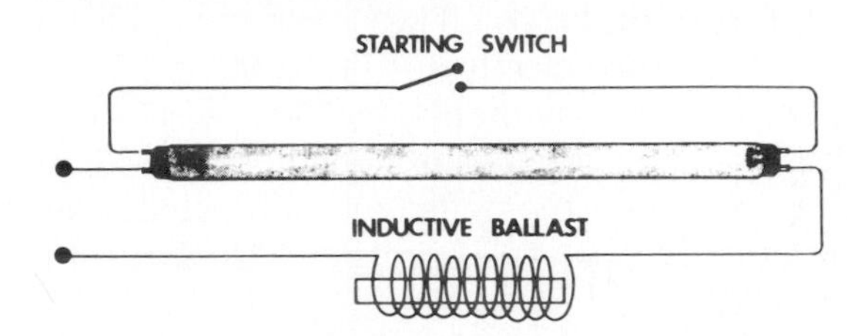

12–7b. Starting the arc. When the starter switch opens, the magnetic field of the ballast "collapses." This action generates an inductive voltage "kick" which strikes an arc across the lamp, causing it to light.

12–5. Various methods of making a diffuser within the light bulb evolved from the pointed bulb to the rounded one with a frosted interior to reduce glare.

■ *Preheating the cathodes (switch-start system).* Pushing the starting button causes the starting switch to close, allowing current to pass through the ballast, the starting switch, and the cathodes at each end of the lamp. The current heats the cathodes until electrons "boil off" and form clouds around each cathode. When the starting switch is opened (by releasing the push button), a higher-voltage inductive "kick," caused by the sudden collapse of the ballast's magnetic field, strikes the arc and lights the lamp.

■ *Ballast stabilizes lamp.* Once the lamp starts, it tends to draw more and more current. This flow of current reestablishes the magnetic field, and the ballast performs its chief function of stabilizing the operation of the lamp at the correct current level.

In the circuit described, no mention is made of a power capacitor, which is included in many ballast circuits. The power capacitor is an energy-storage device used to improve, or correct, the ratio of working power (watts) to the total input power (volt-amperes). This ratio is known as the power factor. A ballast with corrected power factor is commonly referred to as a high power-factor ballast.

Some Basic Ballast Circuits.

■ *Preheat circuit.* The preheat, or switch-start, circuit is the oldest one in use today. It is used with preheat or general-line lamps requiring starters, and is particularly well suited for low-wattage, low-cost applications. Both high and low power-factor designs are available.

The preheat circuit operates similarly to the basic, inductive-ballast circuit described earlier. One difference in the two circuits is that line voltage in the preheat circuit is frequently not high enough to start a 30-, 40-, or 90-watt fluorescent lamp. To boost the starting voltage, an autotransformer is added within the ballast. Fig. 12-8.

■ *Instant-start circuit.* The instant-start circuit, developed in the late 1940s, is used with slimline and other instant-start lamps. This system requires no starter. Starting is accomplished by applying a high starting voltage between the lamp cathodes. The lamp is thus started by "brute force" without preheating the cathodes. There are two basic types of 2-lamp, instant-start circuits: the series-sequence circuit and lead-lag circuit. The series circuit has a lower initial cost, but does not offer independent lamp operation, as does the lead-lag design. The latter type also provides reliable, low-temperature starting. Fig. 12-9.

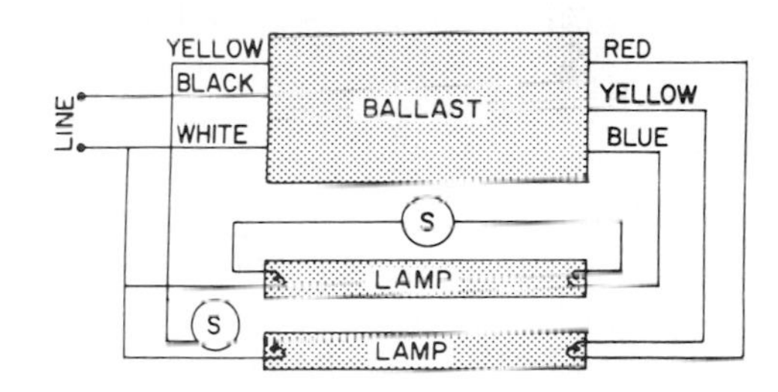

12–8. Basic preheat (switch-start) circuit.

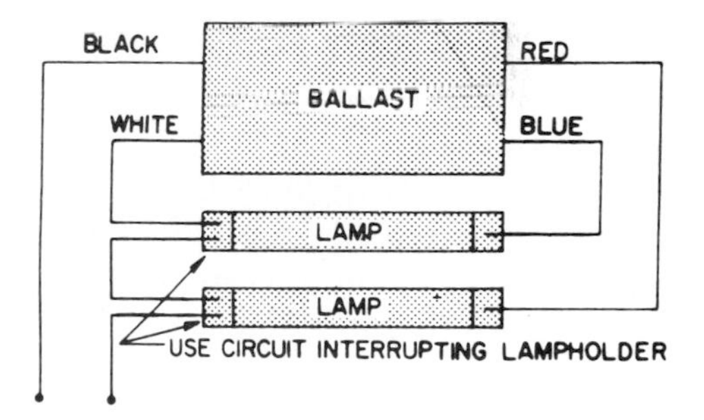

12–9a. Lead-lag circuit.

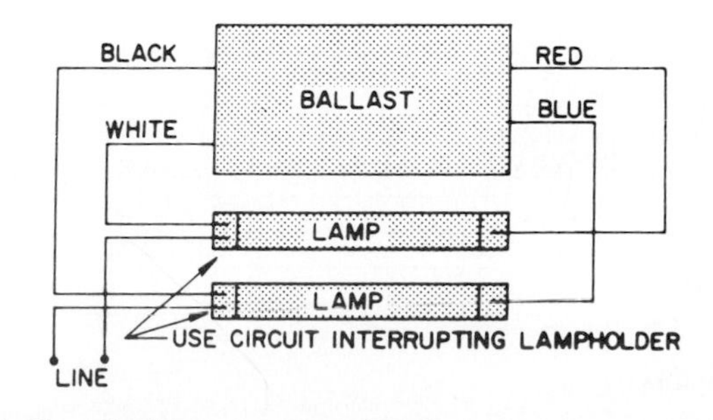

12–9b. Series-sequence circuit.

■ *Rapid-start circuit.* The rapid-start ballast circuit is the most popular in use today. Fig. 12-10. It is used with rapid-start (430 mA), high-output (800 mA), and extra-high output (1500 mA) lamps that employ cathode heating continuously during lamp operation.

In the rapid-start system, the lamp is lighted, without the use of starters, in two phases. First, relying on the metal fixture as a starting aid, the lamp is "conditioned" for starting by preheating the cathodes and pre-ionizing the lamp gases. Such conditioning, of one or two seconds' duration, permits the second step, the actual striking of the arc. This is performed with a much lower open-circuit voltage than is required for a comparable instant-start lamp. With the lower open-circuit voltage, a resultant decrease in ballast volt-ampere requirements makes possible a smaller core and coil, and a generally quieter and more compact ballast. The rapid-start ballast is particularly suitable for shallow fixtures because of its small size. Some specially designed rapid-start ballasts permit flashing or dimming of lamps, due to the continuous cathode-heating feature of the rapid-start system. Certain rapid-start systems are designed for low-temperature starting.

■ *Trigger start.* A trigger start is similar to the rapid-start circuit in that no starter is required, and the cathodes are continuously heated. The triggerstart ballast, however, is designed for use with preheat-type lamps rated 20 watts and under. Both low and high powerfactor ballasts are available. Trigger-start systems are frequently installed in kitchens and bathrooms.

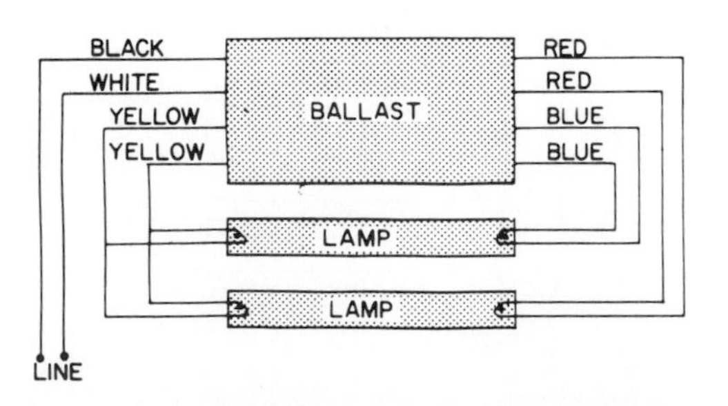

12–10. Rapid-start (and trigger-start) circuit.

The Effect of Heat on Ballast Life. The best way to assure full ballast life is to properly apply the ballast/fixture combination so that ballast case temperatures do not exceed 90°C.

Ballasts, like other electromagnetic devices such as motors and transformers, generate heat during normal operation. Ballast heat is transferred from the internal components to the ballast case through the medium of the filling compound. Once the heat reaches the ballast case, it is dissipated by conduction, convection, and radiation to the surrounding air or mounting surfaces. If these normal means of heat dissipation are not adequate, the ballast may overheat. If this occurs, coil insulation may break down, resulting in internal short circuits and in ballast or capacitor failure.

Because such overheating is the greatest problem of long ballast operation, the obvious satisfactory answer is both a fixture design and an installation that permit the ballast to operate without overheating.

To ensure long ballast life, follow these suggestions:

1. Ventilate the ballast compartment of the fluorescent lamp fixture.
2. If lamps are located directly under the ballast compartment, some means should be provided to ventilate the lamp compartment so that minimum heat from the lamp reaches the ballast compartment.
3. A ballast should be mounted directly against the metal surface of the fixture. If possible, more than one surface of the ballast should be mounted against the fixture. This will permit additional heat dissipation.
4. The use of heat-conducting "radiators," which firmly contact the ballast case, will help lower temperatures.

5. Maximize air circulation around the fixture by eliminating insulation around or above the fixture and, when possible, by lowering the fixture from a flush mounting on the ceiling.

Other Factors Affecting Ballast Life.

■ *Applied voltage.* This is an important determinant of ballast life length. In most ballast circuits, a 1% increase over the nominal design voltage will cause a rise in the ballast's operating temperature of 1°C.

■ *Duty cycle.* The median life expectancy of fluorescent ballasts is approximately 12 years. This means that 50% of the units installed are expected to have failed after a period of 12 years.

This 12-year figure should be qualified, however. For example, a typical duty cycle might be 16 hours per day, 6 days per week, 50 weeks per year. This "typical" duty cycle, totaling approximately 60 000 hours of operation, would cause half the ballasts to wear out in 12 years. However, since it is assumed that about 4 hours of each day's cycle are required for bringing the ballast up to temperature, such a duty cycle actually totals only 12 hours a day, or about 45 000 hours of full, 90°C case temperature operation during the 12 years.

Outdoor applications, on the other hand, often have a low ballast-case temperature during winter months, which usually results in reduced operating temperatures and therefore extended ballast life.

■ *Lamp maintenance.* Failed lamps in most instant-start and preheat (switch-operated) systems should not be allowed to remain in the fixture for extended periods of time. This practice will cause ballast overheating and result in a significant reduction in ballast life. Ballasts are designed to withstand abnormal conditions resulting from a failed lamp for periods of up to four weeks. Rapid-start ballasts, however, are not affected by failed lamps in the fixtures.

Ballast Sound. The problem of ballast "hum" is one which has been subjected to considerable investigation. By installing ballasts in accordance with a sound-control program sponsored by the manufacturer, it is possible to predict the likelihood of audible ballast hum in advance. This can result in reducing or eliminating potential sound problems.

As long as fluorescent-lamp ballasts are electromagnetic devices, they will have magnetic hum. However, their 60 Hz hum can be controlled by selecting and installing the proper number and type of sound-rated ballasts.

Fig. 12-11 shows a ballast protector. Fig. 12-12 shows a protector in the capacitor. The automatic-resetting thermostat was intro-

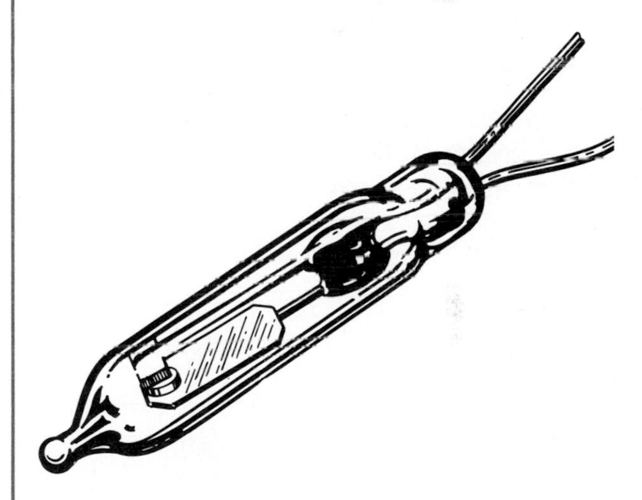

12–11. Automatic resetting thermostat.

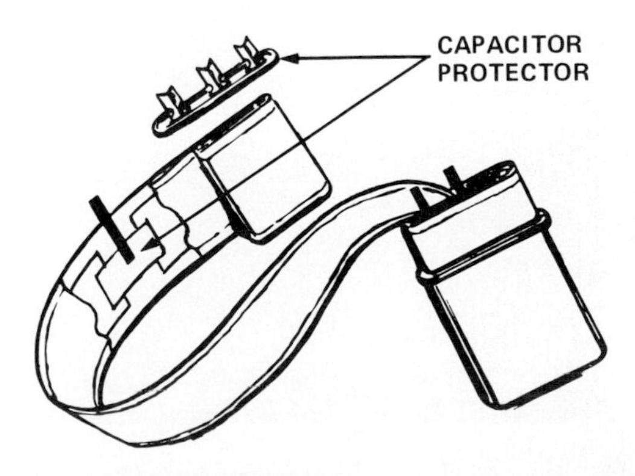

12–12. Capacitor protector.

duced in 1957. It functions to protect the ballast against adverse application conditions, taking the ballast out of the circuit when coil temperature is excessive. It will also remove most failing ballasts from the circuit. It is important that supply voltages to the ballast circuit be maintained within the limits shown in Table 12-A.

Fig. 12-13 is a special type of fluorescent lamp. It has more surface area as a result of the indentations. Thus it will give more light relative to its length than other similar designs.

Outdoor Applications. Most standard indoor-type ballasts are designed to operate with reliable starting in ambient temperatures of 50°F (10°C) or above. For lamp starting in outdoor, weather-protected fixtures, or in plastic signs, special low-temperature ballasts are available. Many of these will reliably start lamps in ambients as low as −20°F (−29°C).

For locations where they will be exposed directly to outdoor weather conditions, special weatherproof ballasts are available.

Rapid-start lamps may be flashed, without reduction in lamp life, by using ballasts especially designed for flashing. These ballasts provide different electrical characteristics from those designed for continuous burning, to assure a satisfactory lamp life. A special flashing device is required.

12-A. Ballast Temperature Chart

COMPONENT	MAXIMUM TEMPERATURE
Ballast winding	105°C
Capacitor case	70°C
Ballast case	90°C

Troubleshooting. The troubleshooting chart (Table 12-B) will identify some of the problems to look for when a symptom develops. Study the chart to locate the possible cause of the trouble. It could save you many hours of work.

Flourescent Lamp Colors.

- *Daylight*—very blue-white; seldom used in homes because it grays complexions as well as the reds and pinks of room decor.
- *Standard cool*—blue-white light; dulls red and pink surfaces.

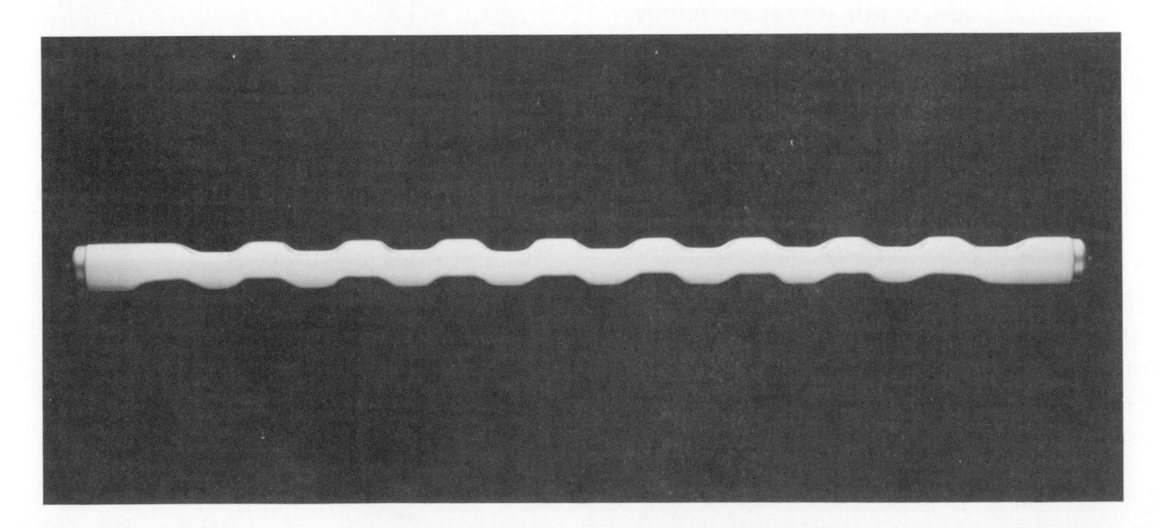

12–13. Special-design fluorescent tube. The indentations make it capable of giving off more light since it has more surface area.

12-B. Troubleshooting Chart

CONDITION	POSSIBLE CAUSES TO INVESTIGATE
I **Lamps won't start**	1. Lamp failure 2. Poor lamp-to-lampholder contact 3. Incorrect wiring 4. Low voltage supply 5. Dirty lamps or lamp pins 6. Defective starters* 7. Low or high lamp bulb-wall temperature 8. High humidity 9. Fixture not grounded 10. Improper ballast application 11. Ballast failure
II **Short lamp life**	1. Improper voltage 2. Improper wiring 3. Poor lamp-to-lampholder contact 4. Extremely short duty cycles (greater than average number of lamp starts per day. Check with lamp mfgr.) 5. Defective starters* 6. Defective lamps 7. Improper ballast application 8. Defective ballast
III **Lamp flicker (spiraling or swirling effect)**	1. New lamps (should be operated 100 hours for proper seasoning) 2. Defective starters* 3. Drafts on lamp bulb from air-conditioning system (lamp too cold) 4. Defective lamps 5. Improper voltage 6. Improper ballast application 7. Defective ballast
IV **Audible ballast "hum"**	1. Loose fixture louvers, panels, or other parts 2. Insecure ballast mounting 3. Improper ballast selection 4. Defective ballast
V **Very slow starting**	1. Improper voltage—too low 2. Inadequate lamp-starting-aid strip† (Refer to fixture mfgr.) 3. Poor lamp-to-lampholder contact 4. Defective starter* 5. Defective lamp 6. Improper circuit wiring 7. Improper ballast applicaton 8. High humidity 9. Bulb-wall temperature too low or too high

12-B. **Troubleshooting Chart (Continued)**

CONDITION	POSSIBLE CAUSES TO INVESTIGATE
VI **Excessive ballast heating (over 90°C ballast case temperature)**	1. Improper fixture design or ballast application (Refer to fixture mfgr.) 2. High voltage 3. Improper wiring or installation 4. Defective ballast 5. Poor lamp maintenance (instant-start and preheat systems 6. Wrong type lamps 7. Wrong number of lamps
VII **Blinking**	1. Improper fixture design or ballast application (Refer to fixture mfgr.) 2. High voltage 3. Improper wiring or installation 4. Defective ballast 5. Poor lamp maintenance (instant-start and preheat systems) 6. Wrong type lamps 7. Wrong number of lamps 8. High ambient temperature

*Applies only to preheat (switch-start) circuits.
†Applies only to rapid-start and trigger-start circuits.

- *Standard warm*—orange-white light; somewhat yellows red and pink surfaces being lighted.
- *White*—compromise between standard cool and standard warm.
- *Deluxe cool*—blue-white, but does not dull reds and pinks; has tendency not to distort colors.
- *Deluxe warm*—orange-white light that accents reds and pinks; blends well with incandescent lighting; recommended for home use.
- *Soft white*—pinkish-white light which emphasizes reds and pinks.

Fluorescent tubes in red, pink, blue, green, and gold are available for special lighting effects.

Fig. 12-14 pictures some of the standard bases for fluorescent lamps; Fig. 12-15a shows starter socket mountings.

Fig. 12-15b illustrates the internal elements of a fluorescent lamp starter. Types of cathodes are shown in Fig. 12-16.

Mercury-Vapor Lamps

Mercury-vapor lamps are just that—mercury vapor. They usually require a ballast to serve as a current-limiting device, once the arc is struck and the unit is emitting light. There are various types of mercury-vapor lamps, of which we shall give a cross section here, as well as information helpful for selection of the proper type for the job to be done.

Usually, mercury lamps need time to warm up to full brilliance. This is obvious at sundown in shopping centers, in parking lots, and along highways where they are found in abundance. Fig. 12-17 shows a screw-in type mercury lamp, with parts identified.

Some mercury lamps are *self-ballasted*. Fig. 12-18 shows such a lamp. It is a 250-watt, with a ballast contained within the lamp (there is no separate ballast). Changing the bulb is the only maintenance. The reflector in the package is plastic, and the fixture arm is of lightweight aluminum. A photoelectric eye automatically turns the fixture on at dusk and off at dawn.

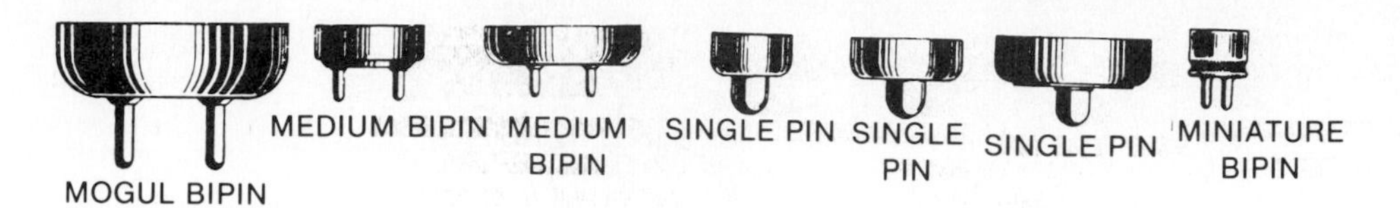

12–14. Bases for fluorescent tubes.

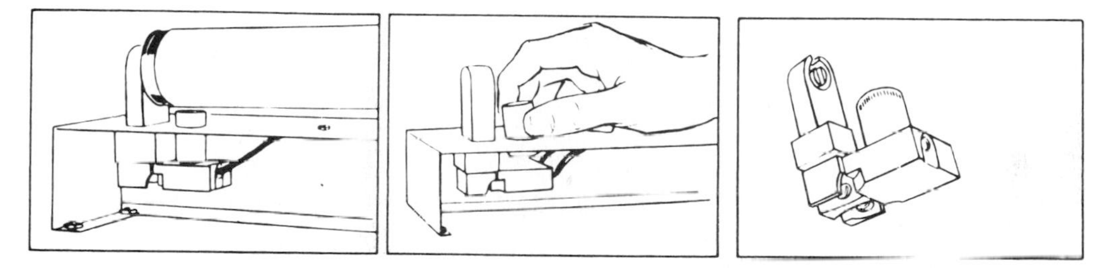

12–15a. Placing a starter in a preheat-type fluorescent.

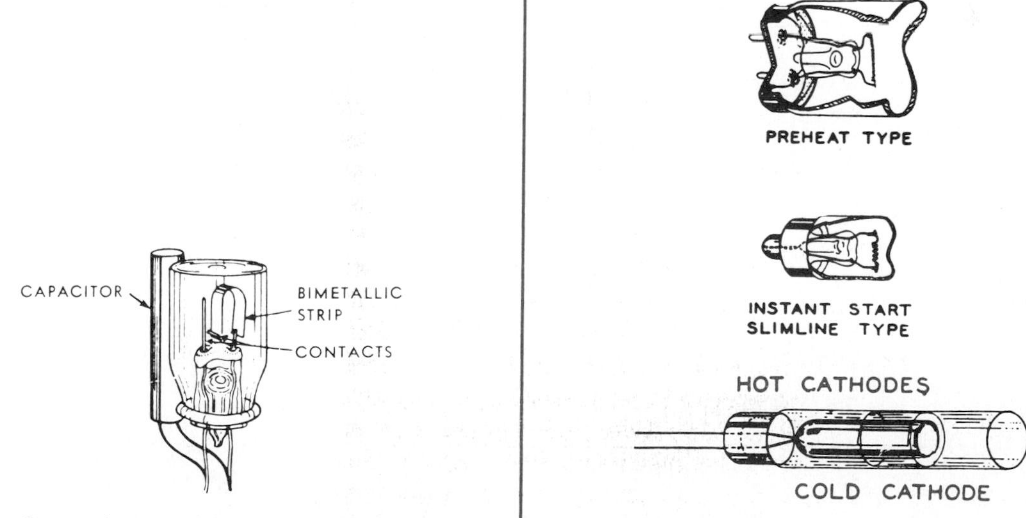

12–15b. Internal parts of fluorescent tube starter.

12–16. Types of cathodes for fluorescent tubes.

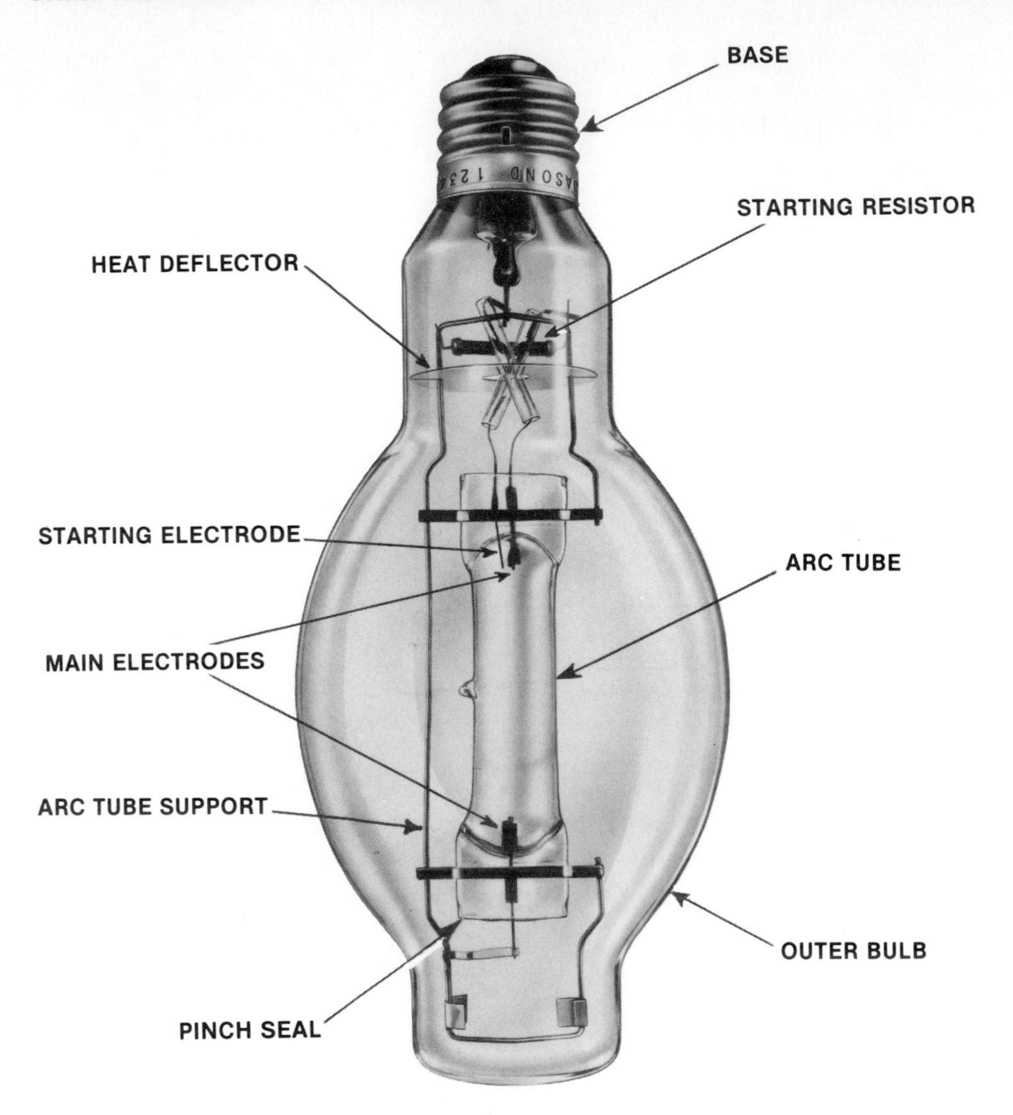

12–17. A typical screw-in type of mercury-vapor tube.

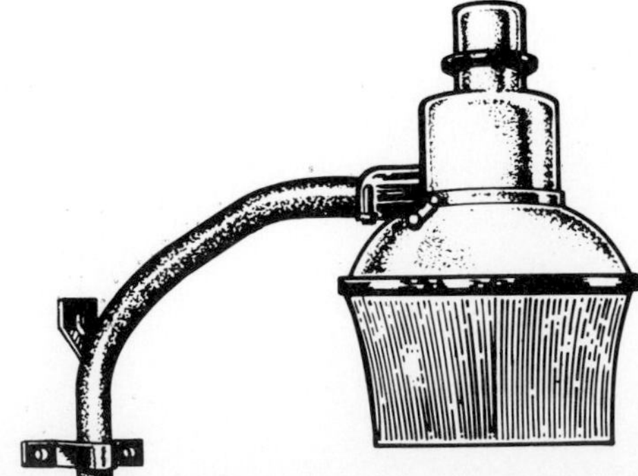

12–18. Self-ballasted mercury-vapor lamp for mounting on poles.

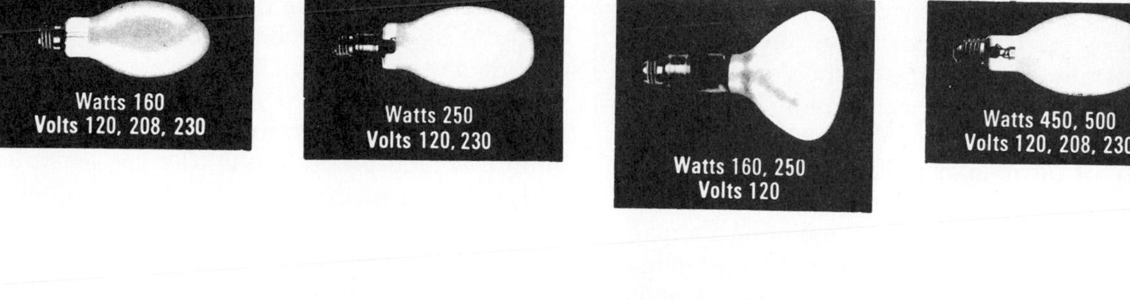

Watts 750
Volts 120

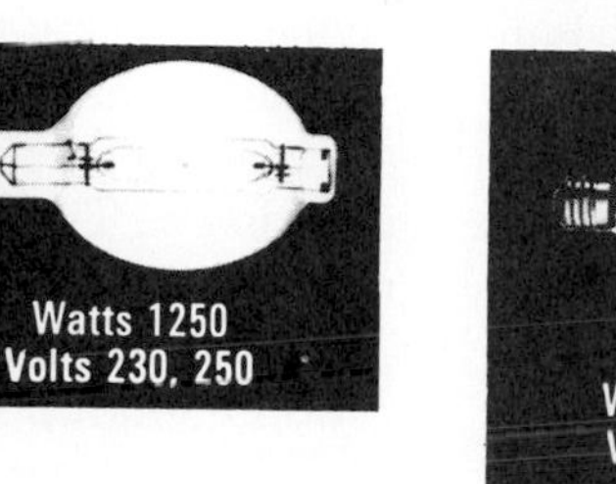

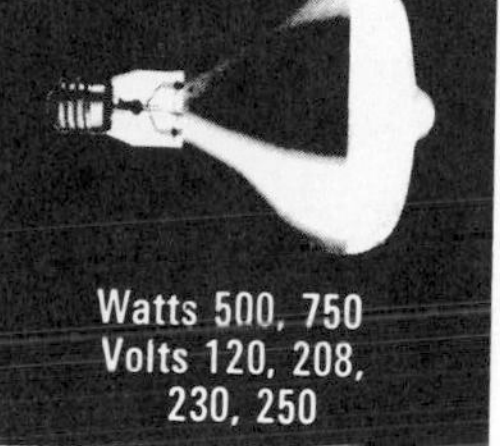

12–20. Various voltages, shapes, and wattages of mercury-vapor lamps.

Compact-Source Iodide Lamp. There are a number of features which merit consideration in a bulb of this type. It uses an AC source and is ideal for professional color slide projectors and for microscope lighting. It has a useful life of about 400 hours. The lamp produces about 60 lumens per watt. Iodide lamps have very simple controls and produce a highly uniform screen illumination. Fig. 12-21.

Iodide lamps must be protected from temperatures in excess of those specified in their design. Therefore proper electrical contact must be ensured and forced-air cooling may be required. The air should not be blown directly onto the bulb because it may cause change in the spectral distribution due to condensation on one or more of the metal additives. A ballast is required. Ignition usually takes 10 seconds. The bulb should be operated in a vertical position. Fig. 12-22 shows the circuit diagram for an iodide lamp.

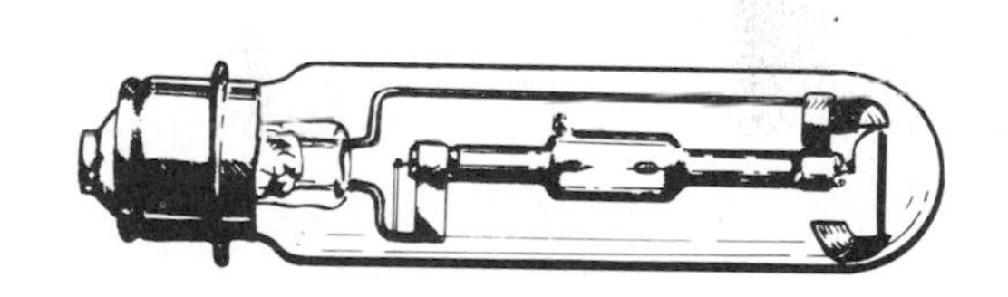

12–21. Compact-source iodide lamp.

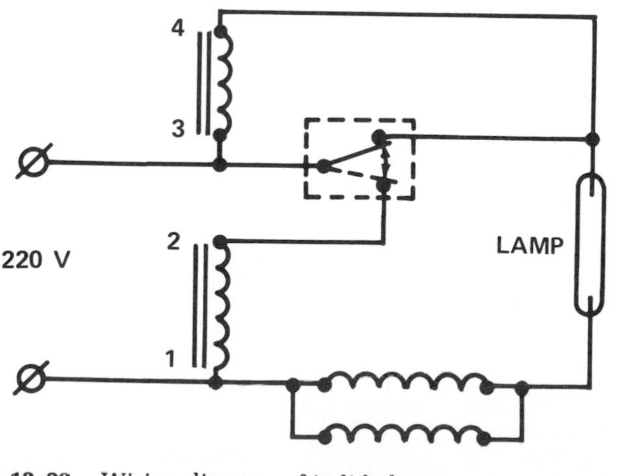

12–22. Wiring diagram of iodide lamp.

off at dawn. Keep in mind that the type of bulb for this lamp is self-ballasted. Do not screw the ballasted type (shown in Fig. 12-17) into the outlet.

Each manufacturer of mercury-vapor lamps has its own designation and trade name. One of these, the Merco-Matic®, has certain distinct features. Its operation requires 120 volts. Instantly upon energizing, the lamp emits light, as opposed to the required warm-up time for normal-discharge lamps. It operates in the following manner:

1. Current flows through a bi-metal switch (which is in a closed position), the starting coil, and the ballasting coil. The effect is to produce not only instant usable light, but primarily to ionize the ignition gas inside the burner. The heat emitted from the coils opens the bi-metal switch. This creates a state in which the entire line voltage is impressed across the main electrodes. This phenomenon occurs in a microsecond of time. At this point the current no longer flows through the bi-metal switch or the inner coil within the burner. Fig. 12-19.
2. The current is now flowing through the ballasting coil, setting up a resistance for the operating voltage of the burner, which, in the case of 120-volt operation, is 60 volts, compared to a starting voltage requirement of 100-120 volts.
3. While the arc stabilizes, an incandescent ballasting filament continues to emit a large number of lumens. As the arc becomes stable, the incandescent ballasting filament "steps down," continuing to build resistance. It eventually settles to a glow state, which essentially contributes more color balance to the light source than lumens. For this reason, the life of the filament ballasting coil can be expected to exceed the economic life of the burner, which is up to 16 000 hours. The total time required to achieve normal operation from ignition to stabilization varies. The time is from 1 to 3 minutes, depending on primary line voltage, burning position, and ambient temperature.

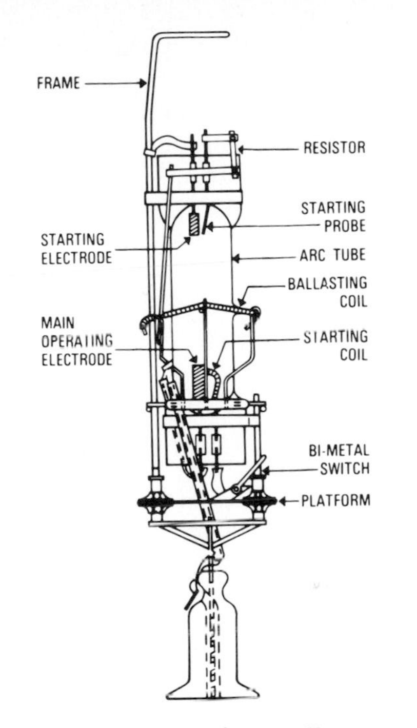

12–19. The frame support is actually a part of the stem, creating a one-piece indestructible unit which is firmly affixed to the inverted dimple at the top of the lamp.

Voltage or cycle interruptions may cause the lamp to go out. As in other discharge sources, a time factor of 4 to 11 minutes, depending on conditions, is required for re-ignition. Self-ballasted lamps may flicker momentarily upon hot re-ignition until the vapor pressure reduces and the bi-metal switch closes, thereby allowing normal ignition.

Such lamps are recommended for operation on AC current, but can be operated on DC current with an anticipated reduction in efficiency and life of approximately 25%. For best results on DC, polarity should be reversed, as lamps are cycled by the use of a reversing-polarity switch.

There are high-voltage types of this particular lamp. They require voltage from 208 to 250 to ignite, but operate at about 135 volts, as do standard-discharge mercury lamps. The ballasting coil functions as a ballast by setting up resistance as the arc stabilizes. When the arc is fully stabilized, the ballast coil remains in a glow state, emitting low lumens per watt while contributing a high percentage of red, a color missing from the visible light emitted by the basic mercury lamps.

High-voltage lamps will operate on DC as well as AC because of the elimination of the bimetal switch. There is only a modest reduction (approximately 15%) in life and light output. Table 12-C compares various lamp systems to sunlight in light output. Fig. 12-20 indicates some of the Merco-Matic® bulbs available, with their voltage and wattage.

This type of mercury-vapor lamp is unique in its output. It is also unique in its ability to be substituted for an incandescent lamp. The lamp simply screws into a lamp socket. The socket may have been originally planned for an incandescent high-wattage lamp. You can convert explosion-proof fixtures to mercury vapor by merely changing lamps. Make sure, however, that it is a self-ballasted lamp. (The word lamp and the word bulb as used here have the same meaning.)

Ballasted Mercury-Vapor Lamp. Most mercury lamps do require a ballast, or some type of current-limiting device. Depending on the lamp type, this can be an inductive coil, an autotransformer, or a resistance. In addition to the current-limiting device, a source of high voltage is needed to initiate the discharge.

Some bulbs may be operated on DC as well as AC. The characteristics of the rectifier to produce the DC should be capable of meeting the lamp requirements. Compact-source mercury lamps require a supply voltage that is well above the arc voltage. If the smoothing circuit consists of a capacitor and an inductive coil, a capacitor of 500 microfarads must be connected directly across the rectifier terminals. To prevent difficulties arising from pulse discharging of this capacitor, a resistance of 15 ohms should be connected between the lamp and capacitor, and in series with the lamp.

The ripple percentage is caused during rectification to produce the DC supply. It has a significant influence on lamp life and stability. A high ripple percentage is liable to cause unstable operation and premature discharge-tube blackening due to electrode sputter. High ripple means a large amount of AC is present on top of the DC. The effect is progressive. This means the distance between the electrodes increases with the consequent rise in arc voltage and lamp wattage.

12-C. Operation of Light Sources—Colors Present and Amount of Light Available

ENERGY	VIOLET	BLUE	GREEN	YELLOW	RED
SUNLIGHT					
MERCO-MATIC®					
BALLASTED MERCURY					
FLUORESCENT LAMPS					
INCANDESCENT LAMPS					
ANGSTROMS	4000	4500	5000	5500	6000 6500

High-Pressure Mercury-Vapor Lamp. High-pressure mercury-vapor lamps have a small quartz tube in which the discharge takes place. It contains a bipolar connection to the electrodes. The lamps are characterized by a high degree of ultraviolet radiation and are mainly applied for photochemical process work. Persons exposed to ultraviolet radiation are subject to erythema and conjunctivitis (inflammations of the skin and eyes). Therefore, adequate skin and eye protection should be worn when working with the lamps. Fig. 12-23a. Fig. 12-23b is the schematic for a high-pressure mercury-vapor lamp with ballast. The ballast is located in series with the lamp, and with a capacitor across the line (in some cases).

Repro-Lamps. Fig. 12-24 illustrates a mercury lamp designed especially for use as a floodlight. (When used with a separate Wood's glass filter it is a "black light" source.) The reflector directs the beam of light. It produces a bluish-white light with strong actinic radiation. This makes it suitable for black-and-white reproduction and for copying processes.

12–23a. Mercury lamps, high pressure.

12–23b. Wiring diagram for a 125-watt bulb, such as the one in Fig. 12–23a.

Black Light. A mercury-vapor black-light lamp is a high-pressure one (the gas inside is under high pressure). It consists of a quartz discharge tube in an outer envelope of black Wood's glass. Fig. 12-25. It constitutes a source of invisible radiation for producing luminescence. The lamp will burn in any position. However, there is a slight risk when operating the lamp in a vertical position—that is, with the cap down. The mercury vapor will condense and a drop of mercury may settle between the main and auxiliary electrodes. Under this condition the lamp would not start.

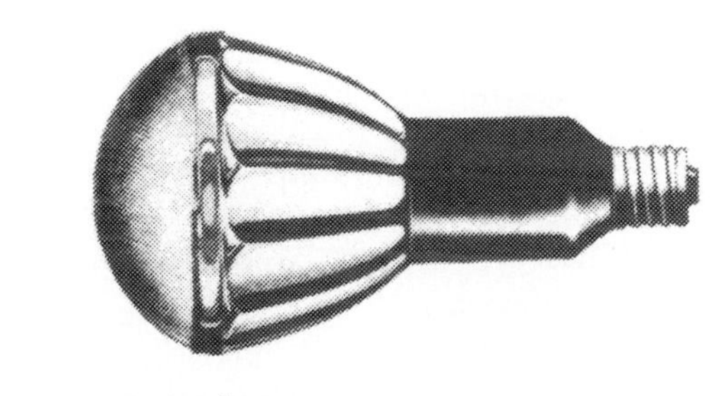

12–24. Repro-lamp.

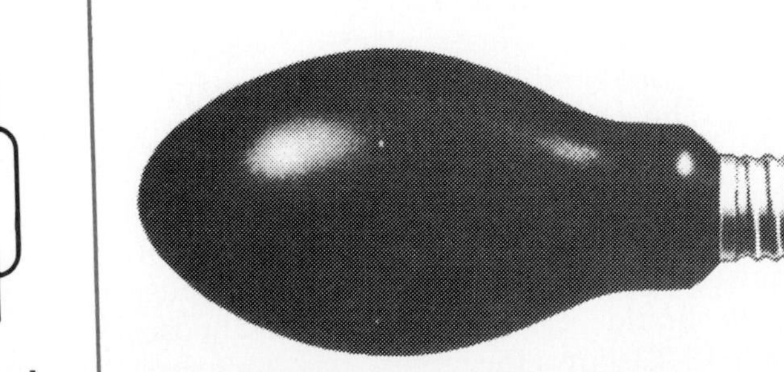

12–25. Mercury lamp, black-light type.

Because of their ease in mounting and simplicity of operation, black-light lamps are used in such diverse fields as:

- The chemical industry.
- Food production.
- Mineralogy.
- Philately (stamp collecting).
- Criminology.
- Banking.
- The textile industry.
- In laundries and dye works.
- In medicine.
- For illumination.

The longwave ultraviolet radiation from this lamp is harmless to the eyes. (Longwave and ultraviolet radiation refers to the position of this light's wavelength on the light spectrum charts.)

Fig. 12-26 shows a commercially mounted ultraviolet lamp. The diagram shows a resistor in series with it. The resistor makes sure the lamp consumes exactly 250 watts.

Water-Cooler, Super High-Pressure Mercury Lamp. This type of lamp, Fig. 12-27, is cooled with water. It is a highly concentrated discharge type. The discharge takes place in a quartz capillary tube around which cooling water flows. Light from the lamp is whiter than that produced by ordinary mercury lamps. This is due to internal pressure that is very high when the lamp is in use. Full light output is reached at once. Another feature is immediate re-ignition.

The water-cooled mercury lamp, moreover, is distinguished by a high level of luminance and by high efficiency. In addition to a visible light, the discharge produces a considerable quantity of ultraviolet radiation. Most of the UV is absorbed by glass parts in the lamp jacket. If desired, these glass parts may be replaced with corresponding parts made of quartz, which transmit the ultraviolet radiation almost in its entirety.

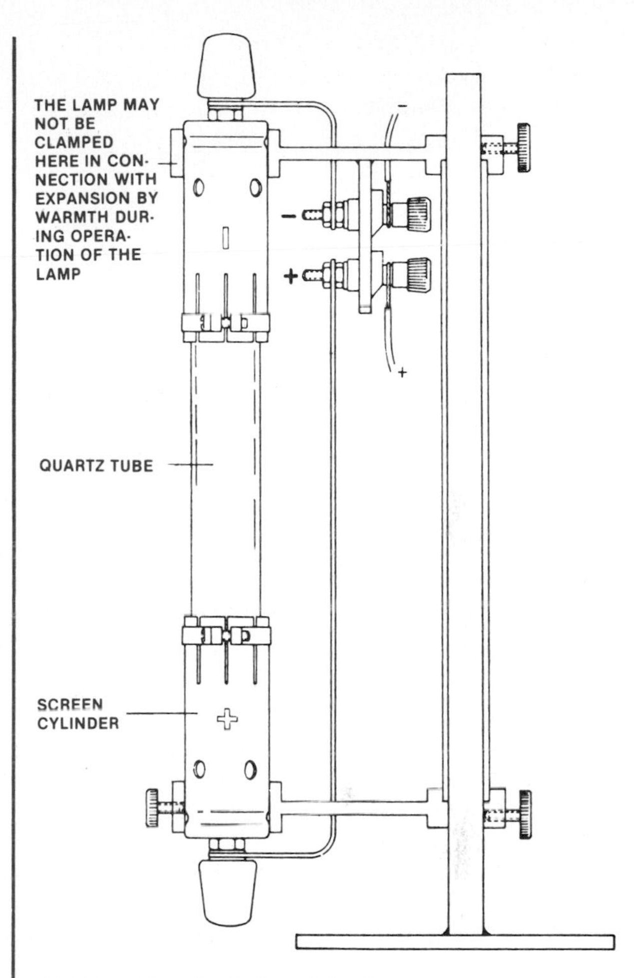

12–26a. Standard ultraviolet lamp.

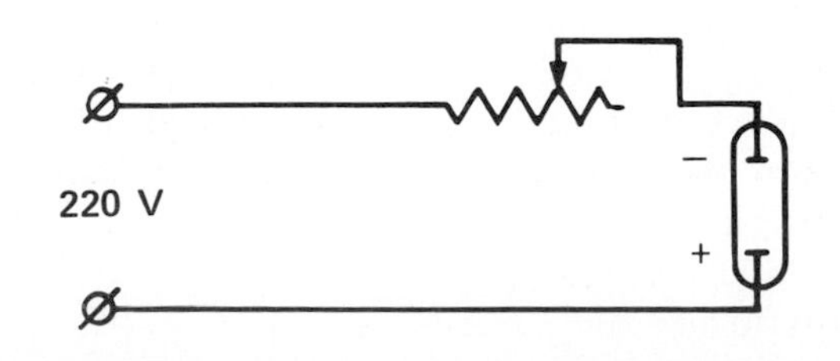

12–26b. Wiring diagram. The resistor must be adjusted in such a way that the discharge consumes 250 watts exactly.

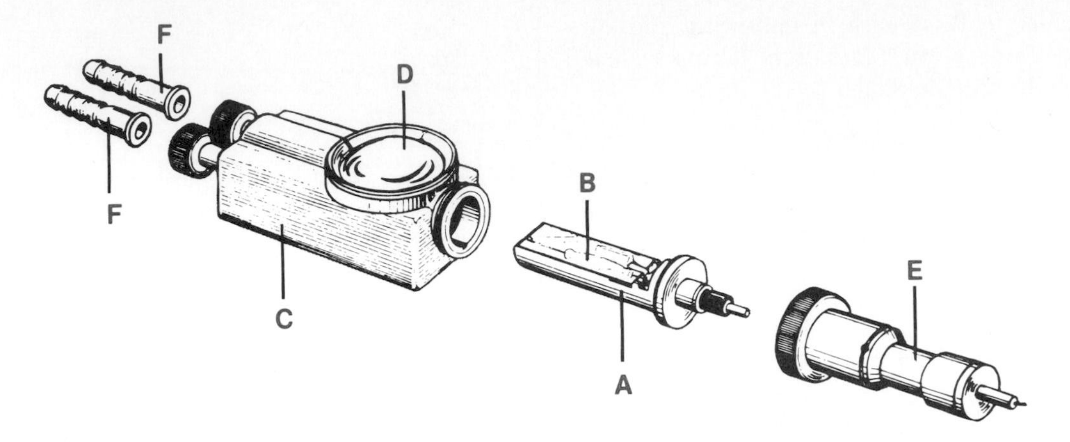

12–27. Water-cooled, super high-pressure mercury lamp.

Part "A" in Fig. 12-27 is the reflector. It is fitted with a glass or quartz cover. Part "B" directs the cooling water past the discharge tube. Part "C" is a metal lampholder which also serves as a cooling jacket. It has either a glass or a quartz convex window, "D." Both the reflector cover and window are removable to allow for the use of either glass or quartz parts, as required. Quartz windows have a small triangle etched on their rim for easy recognition.

The electrical connection is made to a contact pin. The pin forms part of the lamp. The body of the lampholder must be grounded. The contact pin fits into a single-pole female plug, "E." The lampholder cooling jacket has a flat mounting base with screws. It is provided with coupling nuts for connecting the cooling water pipes. Separate nipples, "F," can be supplied for either flexible rubber or for suitable plastic tubes.

The water-cooled, super-high-pressure mercury lamp is used for copying processes in the graphic arts industries. It is also used for projection of microfilm and regular film and in photography.

A 500-watt lamp is operated with a leak transformer (a British term for autotransformer). The primary circuit of the transformer has several terminals for connection to AC lines of voltage between 105 and 380. One of the two secondary terminals must always be grounded (*earthed* in British terminology). If desired, the power factor can be improved by using a capacitor of 35 microfarads at 250 volts. Fig. 12-28. The diagram in Fig. 12-29 shows connections to the lamp from a transformer.

Air-Cooler, Super High-Pressure Mercury Lamp. This lamp has a highly concentrated discharge which takes place in a quartz capillary tube. Forced-air cooling is necessary to keep the tube in the proper operating condition. Light from these lamps is whiter than that produced by ordinary mercury lamps. The whiter light is due to the high internal pres-

12–28. Ballast for water-cooled, super high-pressure mercury lamps. This one is designed for 500 watts.

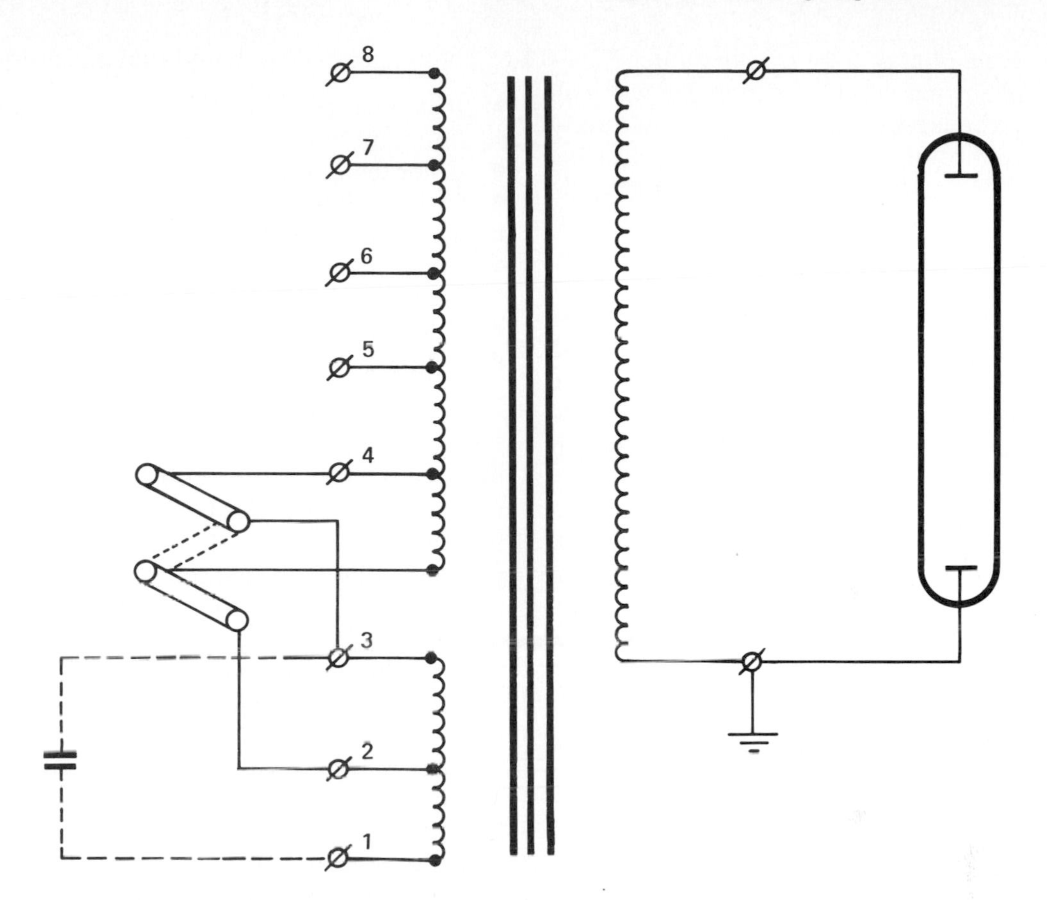

12–29. Connection diagram for the water-cooled mercury lamp.

sure. Full light output is reached at once. It will also re-ignite immediately. It is highly efficient. The discharge produces a high level of ultraviolet radiation and visible light. Fig. 12-30. The transformer needed for operation, and the connection diagram of the transformer to the lamp, are shown in Figs. 12-31 & 12-32.

Light-Printing Lamp. The quartz tube is transparent to ultraviolet (UV) radiation. The tube has an admissable, high-watt load per unit of length. This leads to high efficiency for photoprinting techniques. Where very high printing speeds are required, this lamp is very practical. In many applications other than light printing, this lamp is a very economical item.

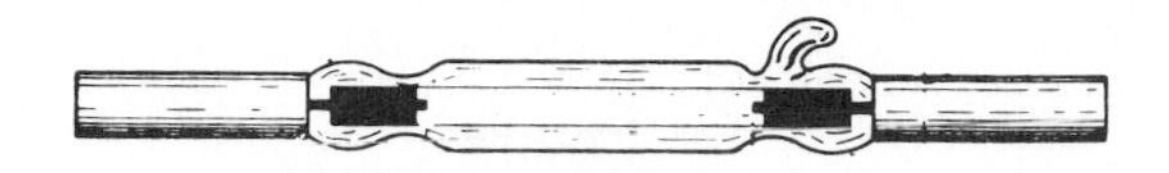

12–30. Air-cooled, super high-pressure mercury lamp.

A light-printing lamp also lends itself extremely well to sterilization processes in the medical field. It also produces ozone. It should be noted, however, that ozone may be harmful, depending on the degree of concentration. Users of this lamp should also be protected from ultraviolet radiation.

The lamp may be operated in any position. Insulation or ventilation must be applied frequently. This is necessary to achieve an equilibrium between the heat developed by the lamp and the dissipation of that heat (see engineering design data). If there is too much cooling in the starting period, the mercury will not vaporize. The lamp then will not operate properly. Therefore in such cases cooling must be restricted during the starting period. It is gradually increased to the necessary capacity. The tube requires a ballast. Fig. 12-33.

Pulsed-Xenon Lamp. This lamp is a low-pressure, pulsed, xenon-discharge type developed for the graphic arts industry. One of its most important features is that the spectrum approximates average daylight. This makes the lamp perfect for copyboard lighting, particularly with respect to color exposures.

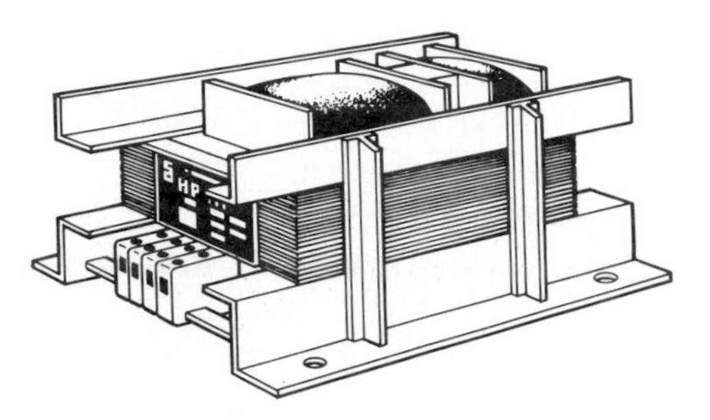

12–31. Ballast for the high-pressure, air-cooled lamp.

These lamps have the following features:

- Instant start and restart (no warming up).
- Full light output immediately after starting.
- Color temperature remains constant.
- Light output remains constant.
- Very clean operation.
- Ideal for reflector design, due to the very small diameter.
- Uniform light output during exposure.
- Long life—hence, low operating cost.
- High efficiency.

Small, horizontal copyboards, as well as large vertical ones, can be lighted very evenly with either two or four linear lamps. Due to their spectral energy distribution, pulsed-xenon lamps are suitable for color reproduction and for black-and-white reproduction. They are superior to almost any other light source. Fig. 12-34 shows a looped-design type; Fig. 12-35, the linear type. Fig. 12-36 shows the connection diagram.

This type of lamp requires forced-air cooling, mainly to keep the temperature of the surrounding area under a certain limit, but also to cool the quartz tube. The temperature of the surrounding area should not exceed 400°C. The tube temperature should remain below 750°C. Normal operation cycle is 3 minutes on, and 1 minute off.

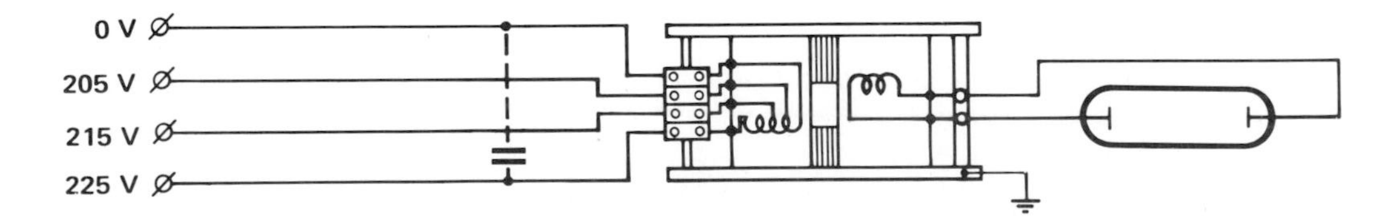

12–32. Connection diagram for the air-cooled lamp shown in Fig. 12–30.

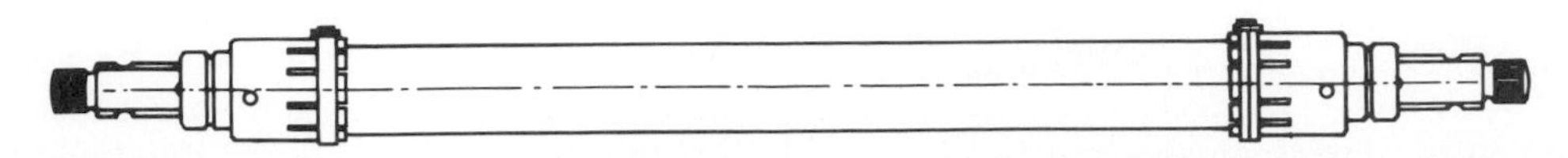

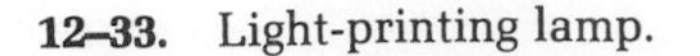

12–33. Light-printing lamp.

The discharge capacitor (C in Fig. 12-36) is charged by a transformer, T via an induction coil, L_1 (or a resistor) to a voltage of about 500. This voltage, however, is well below the breakdown voltage of the lamp. In order to discharge the capacitor initially across the lamp, a high-frequency pulse is introduced into the lamp lead. This makes the lamp conductive.

The transformer, T, recharges the capacitor and the lamp flashes again. The charging inductance, L_1, must be of such a value that the capacitor is charged again within a half cycle. This means that on a 60-hertz line the lamp flashes 120 times per second.

12–34. Pulsed-xenon lamp, loop design.

The induction coil, L_2, is connected in the circuit for limiting the current. These lamps do emit ultraviolet radiation. They must be screened by a suitable UV filter to protect the user's eyes and skin.

Super Actinic Lamp. The *super actinic* lamp emits longwave UV radiation. Because the energy distribution accords well with the spectral sensitivity of diazo papers used in photoprinting machines, this is where most super actinic lamps are used. Forced cooling is necessary to keep the tube wall at 43°C or slightly above. Fig. 12-37. A ballast is required for proper operation.

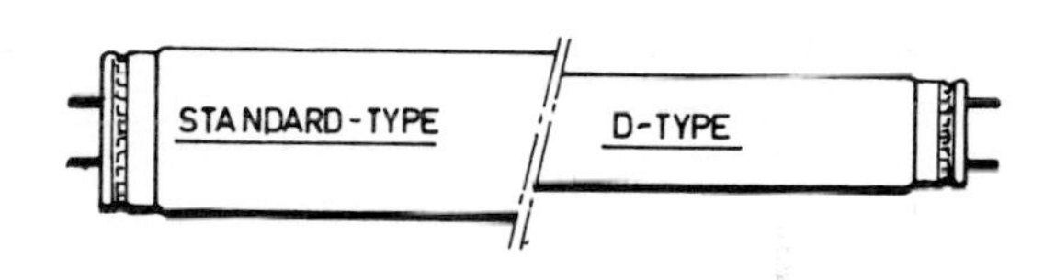

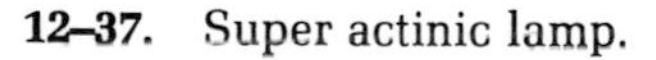

12–37. Super actinic lamp.

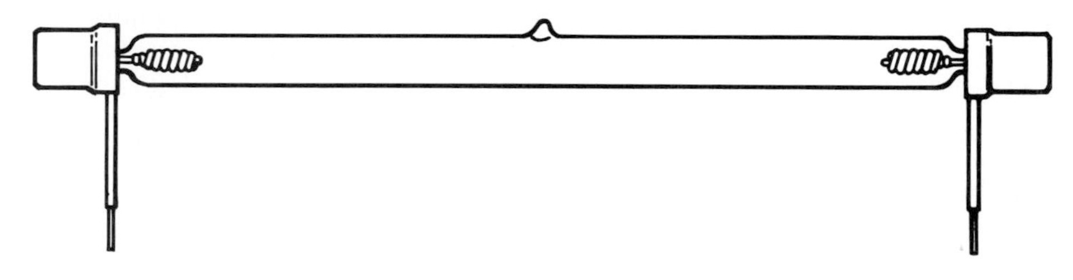

12–35. Pulsed-xenon lamp, linear.

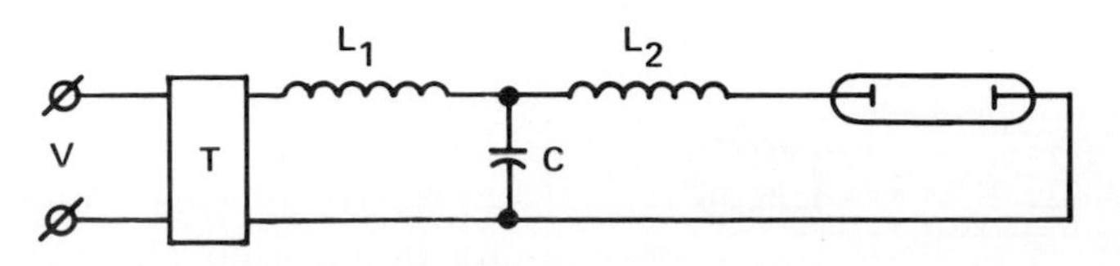

12–36. Circuit diagram of a pulsed-xenon lamp.

Spectral Lamp. Strong monochromatic sources (sources which emit a number of monochromatic lines in a small range of known wavelengths) are important in physical and chemical research where visible or ultraviolet radiation plays a part.

To provide such a source, the *spectral lamp* has been developed. It consists of a small discharge tube surrounded by a cylindrical outer bulb. The discharge tube contains either a gas, a metallic vapor, or a mixture of both in a very pure state. The electrodes permit a very high current density. In this way a light source is obtained which is capable of emitting considerable energy in a single spectral line, or in a few lines.

All lamps have identical outer dimensions as well as an identical light-center length. This ensures complete interchangeability. Fig. 12-38. For those applications where ultraviolet radiation is to be considered, lamps are available with a quartz discharge tube mounted in a quartz outer bulb. This lamp emits ultraviolet radiation extending to the short UV range.

The lamps are filled, either with one of the rare gases—helium, neon, argon, krypton, or xenon—or with argon to which a metal has been added. (The metal is vaporized by the heat of the discharge.) For this purpose, alkali metals are suitable—sodium, potassium, rubidium, and cesium. Other metals such as zinc, cadmium, mercury, indium, gallium, or thallium, are also used.

Mercury lamps are made for low as well as high pressures. In the high-pressure lamps the amount of mercury is such that the metal is entirely vaporized at the operating temperature. The spectrum of a high-pressure mercury lamp shows, in addition to the lines, a relatively weak continuum. The continuum covers the UV and visible region of the spectrum. Conversely, the low-pressure mercury lamp shows no addition of a continuum.

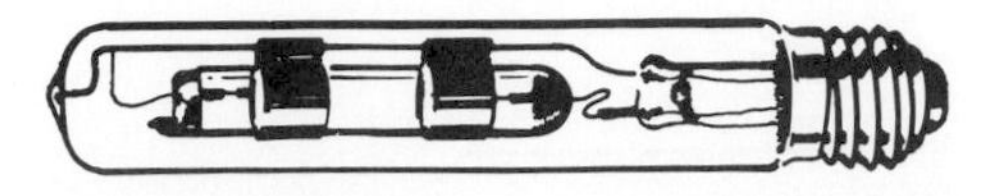

12-38. Spectral lamp.

These lamps are used in the sciences of biology and chemistry, and in the fields of interferometry, polarimetry, refractometry, and spectroscopy.

If it is desired to separate a part of the spectrum, filters can be used. In favorable cases these can be so arranged that only light of one wavelength is emitted. If conditions are such that this isolation cannot be achieved with filters, a monochromator will have to be placed in front of the lamps.

Fig. 12-39 shows the transformer needed for operating spectral lamps, with an accompanying diagram.

The *deuterium spectral lamp* does not belong to the range of spectral lamps previously described. Fig. 12-40. This lamp produces a continuous spectrum from about 200 nanometers up to the infrared. The filament current must be kept constant in order to obtain constant intensity. A deuterium spectral lamp is to be used on DC, according to the circuit diagram given in Fig. 12-41.

The very low working pressure in this lamp results in an absence of continuous radiation. Thus, the energy is dissipated in the mercury lines only. This factor makes the lamp very useful for calibration purposes—that is, for spectrophotometric and spectroscopic equipment.

Neon Lamps

Neon lamps are used for many applications, from night-lights to dial indicators and pilot lights. They draw very little current and last almost indefinitely. Various sizes and shapes of neon glow bulbs are shown in Fig. 12-42. Some are so small they are almost invisible. The small ones draw only 1/40 watt of energy from a line source. Neon signs have been used for years. Today, small lamps using neon or argon are being used more frequently for low current-drain, pilot-light indicators.

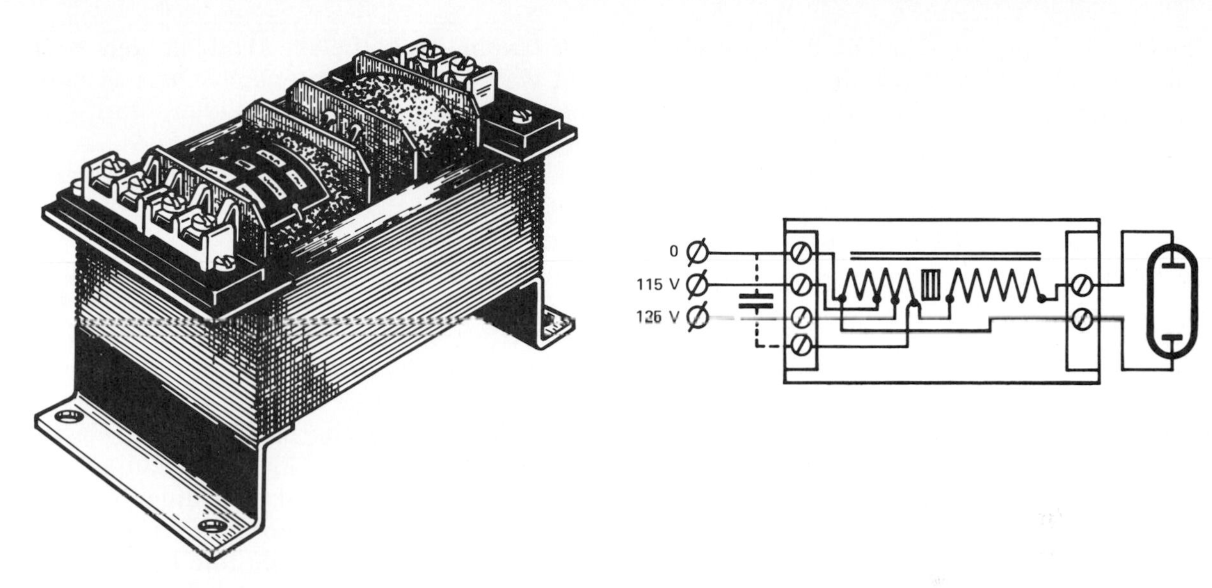

12–39. Auto-leak transformer for operating spectral lamps. The resistors in the schematic are actually transformer coils. Industrial electronics often uses the heavy-lined resistor symbol to indicate a coil or transformer.

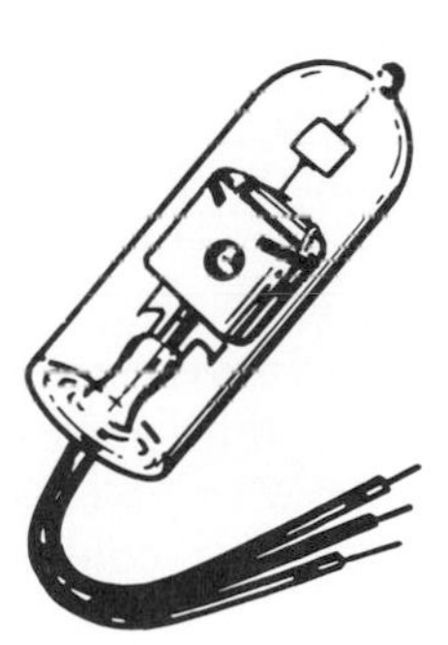

12–40. Deuterium lamp.

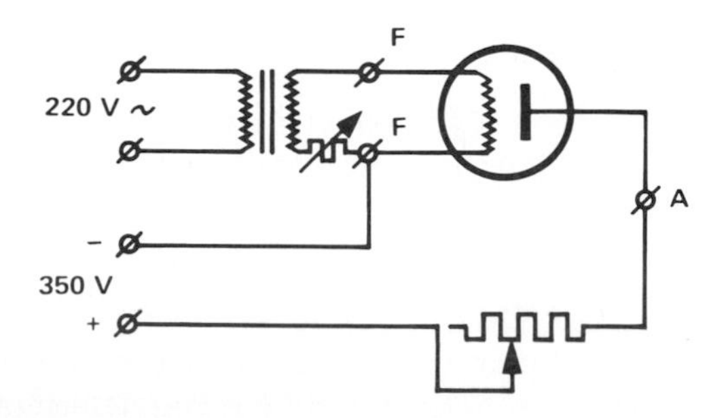

12–41. Circuit diagram for the deuterium lamp.

12–42. Neon, or glow, lamps.

The NE-2, the smallest lamp shown in Fig. 12-42, will ionize or glow when approximately 55 volts are applied to its electrodes. This means that it will need a resistor in series in order to operate on 120-volt lines. If one electrode glows, the current is DC; if both electrodes glow, the current is AC. The one that glows in DC operation is the negative electrode. This means the lamp can be used to tell whether a circuit has DC or AC, as well as the polarity of the DC.

The neon glow lamp is sensitive to temperature. It should not be operated in ambient temperatures below −60°F (−50°C) or above 165°F (74°C).

Home Lighting

There are at least two functions of home lighting—to provide light for reading and seeing, and to serve as decoration. Use of light as decoration, either inside or outdoors, is usually referred to as "architectural" lighting.

Wall Lighting. Some ways to light walls to make a room attractive are:

■ The *valance method* is used with a window to provide "nighttime" sunshine. Light is directed both upward toward the ceiling and downward on the walls and drapery.

■ A *wall bracket of fluorescent lighting* is similar to the valance. It is used mainly on inside walls away from windows. The wall bracket can be used in any room. For example, there are high wall brackets. These are often used to balance the illumination from a window valance on the opposite wall. Low wall brackets are used for local lighting and as a "working light" where specific seeing tasks are performed close to a wall. They are also used to highlight fireplaces and pictures, as well as above desks and sofas.

■ *Cornice lighting* is positioned at the junction between the ceiling and the wall, pointing downward and lighting the surface below the light fixture. It emphasizes wall coverings and textures and will light pictures and other wall hangings. Because the cornice creates the impression of a higher ceiling, it is used frequently in basement recreation areas.

■ *Luminous wall elements.* Luminous walls and wall panels are an effective way to make rooms appear brighter, cheerier, and even larger. Fig. 12-43.

The electroluminescent light, with the entire unit mounted on a ceiling or wall, may eliminate many fixtures as used today. Since electroluminescent wall panels are still a future product, however, the fluorescent lamp is often used with a diffuser to create the effect of a full wall of light. Fig. 12-44.

Light Control. In the early days of electric lighting, the relatively low light output was commonly used without shielding the eyes from the direct rays. In most cases it was used without redirecting the light into useful zones. Improvements in electric lamps brought about the development of more powerful light sources with higher output.

These lamps were immediately put into use in an attempt to satisfy the demands for more light. At the same time, however, it became evident that lamps in the field of view must be shielded in order to reduce their brightness and to minimize glare. Reflectors and other light control methods were very useful in this respect. In addition, they gave better distribution of light.

Common methods for controlling light are by:

- Reflection.
- Diffusion.
- Transmission.
- Absorption.
- Refraction.
- Polarization.

All of these methods involve physical phenomena which may be observed every day. Fig. 12-45.

Check Your Own Lighting Efficiency. Light output of all light bulbs is measured in lumens. This basic unit of light will help you determine the lighting in any room.

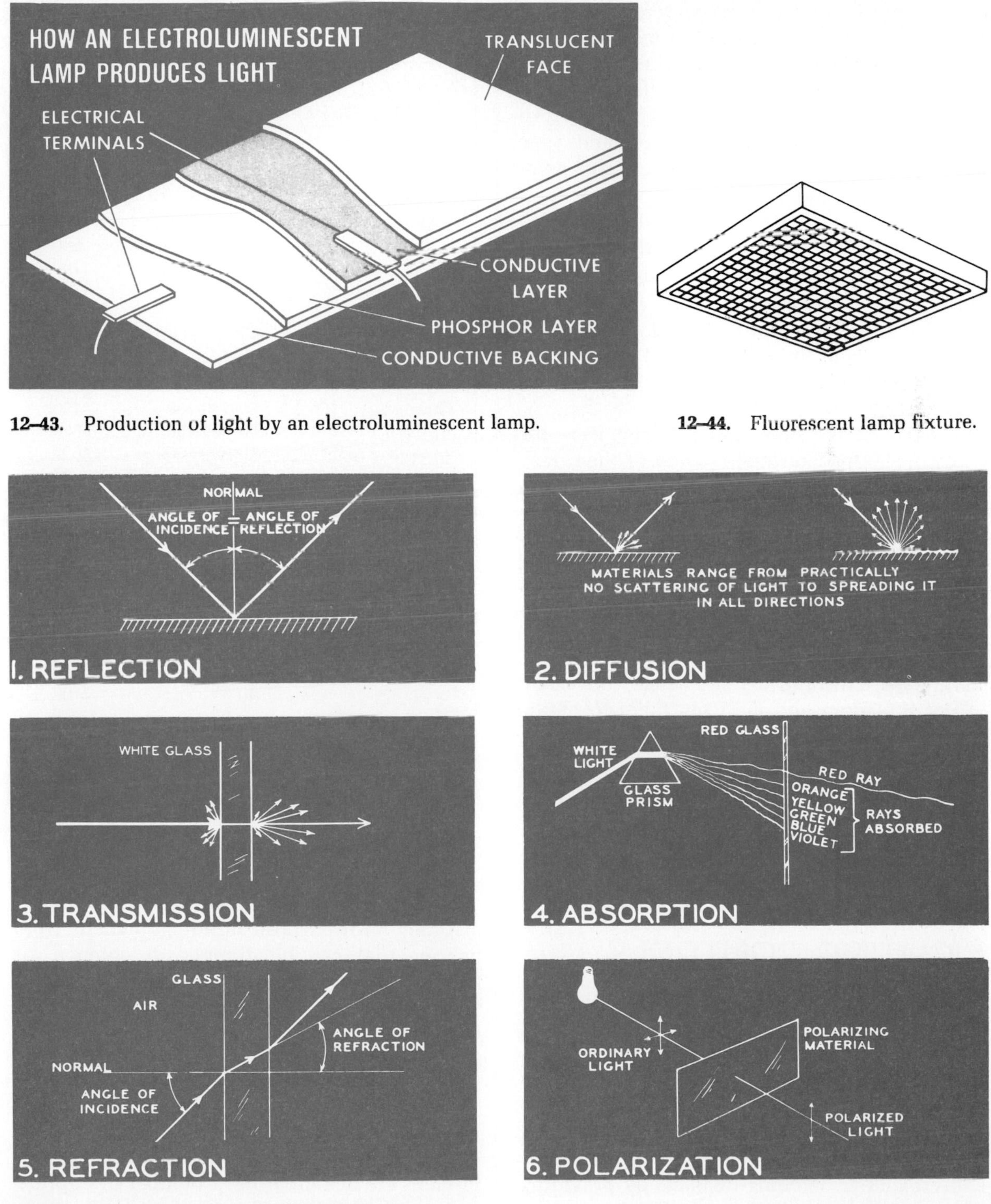

12–43. Production of light by an electroluminescent lamp.

12–44. Fluorescent lamp fixture.

12–45. Six methods of controlling light rays.

A *lumen* is a unit of measurement of luminous flux. It is approximately equal to the light output of one candle. It is defined more precisely as a flux on a unit surface (for example, one square foot or square centimeter). All points on the surface are at unit distance (one foot or one centimeter) from a uniform point source of one candela intensity.

Candela is the basic unit of light intensity. It is now defined as the intensity of light given off by five square millimeters of platinum at its solidification temperature of 1773.5°C.

Light bulbs are now required to have imprinted on them their lumens and average life. Check Table 12-D to see how your room lighting measures up.

If some of the light bulbs do not have the lumens stamped on them (some of them may have been manufactured prior to the stamping requirement), check the chart for watts to obtain the bulb size, then look at the opposite column to find the approximate lumens. To find the lumen count for a room:

1. Add the lumens output of all the bulbs in the room.
2. Measure the approximate length and width of the room and multiply them to obtain the room area.
3. Find the number in the first column of Table 12-E that is closest to the area of your room.

12-D. Approximate Lumen Values to Use in Self-Check of Home Lighting

BULB TYPE	WATTS	APPROX. LUMENS
General Household (Soft White or Inside Frost)	15	140
	25	250
	40	450
	50	650
	60	840
	75	1150
	100	1700
	150	2700
	200	3900
(Highest Setting) Three-Way, Soft White, or Inside Frost	25-35-60	600
	30-70-100	1250
	50-100-150	2150
	50-200-250	3850
	100-200-300	4600
Decorative (Round & Flame Shaped) GA Bulb Indoor Spot and Flood Lamps	25	250
	40	450
	50	600
	75	750
	150	1900
Flourescent Tubes (Deluxe Warm White)	14	450
	15	550
	20	800
	30	1500
	40	2100
Circline Flourescent (Deluxe Warm White)	22 ($8\frac{1}{4}''$)	750
	32 (12″)	1250
	40 (16″)	1750

LIGHTING CONTROLS

Control of lighting can make the difference between a successful store and an unsuccessful one. In auditoriums, restaurants, and theaters—wherever people congregate—the control of lighting affects the mood and atmosphere. Offices and other places of business control light to stimulate interest, dramatize appointments, and create eye-catching displays by providing the right intensity of light to feature items most effectively. School auditoriums, meeting halls, or other multipurpose buildings benefit from lighting that can be dimmed, brightened, or blended to meet the needs of various activities. Controlled light is appropriate, wherever people get together, to provide suitable lighting for the occasion.

Controls for Incandescent Lamps

Transformers. Light controls using transformers are available for a number of applications and wattages. *Transformers (autotrans-*

formers to be exact) are designed to completely control light intensity by controlling the voltage applied to lamps. Rotating the knob moves a brush contact over a commutator, producing any desired intensity of light, ranging from complete darkness to full brightness. Because the brush is always in contact with the commutator, operation is smooth, silent, and flickerless. Only the amount of power required to produce the desired illumination is used. Operation is always cool, safe, and economical, unlike resistance-type controls that regulate by dissipating power in the form of heat.

Transformers are made for a number of loads and voltages. Fig. 12-46. They can be installed in a wall box and protected according to the NEC requirements. Fig. 12-47 shows a transformer for either home or commercial use. Note the thermal overload protector. Wall plates are used to give a finished appearance.

When replacing an existing wall switch with a light control transformer, it may be found that a "switch loop" has been used between the lighting load and the switch, as shown in Fig. 12-48a. This usually does not present a problem because the dimmer can be incorporated by changing the circuit according to Fig. 12-48b.

If on/off control from a number of locations is desired, a 3-way wall switch may be in-

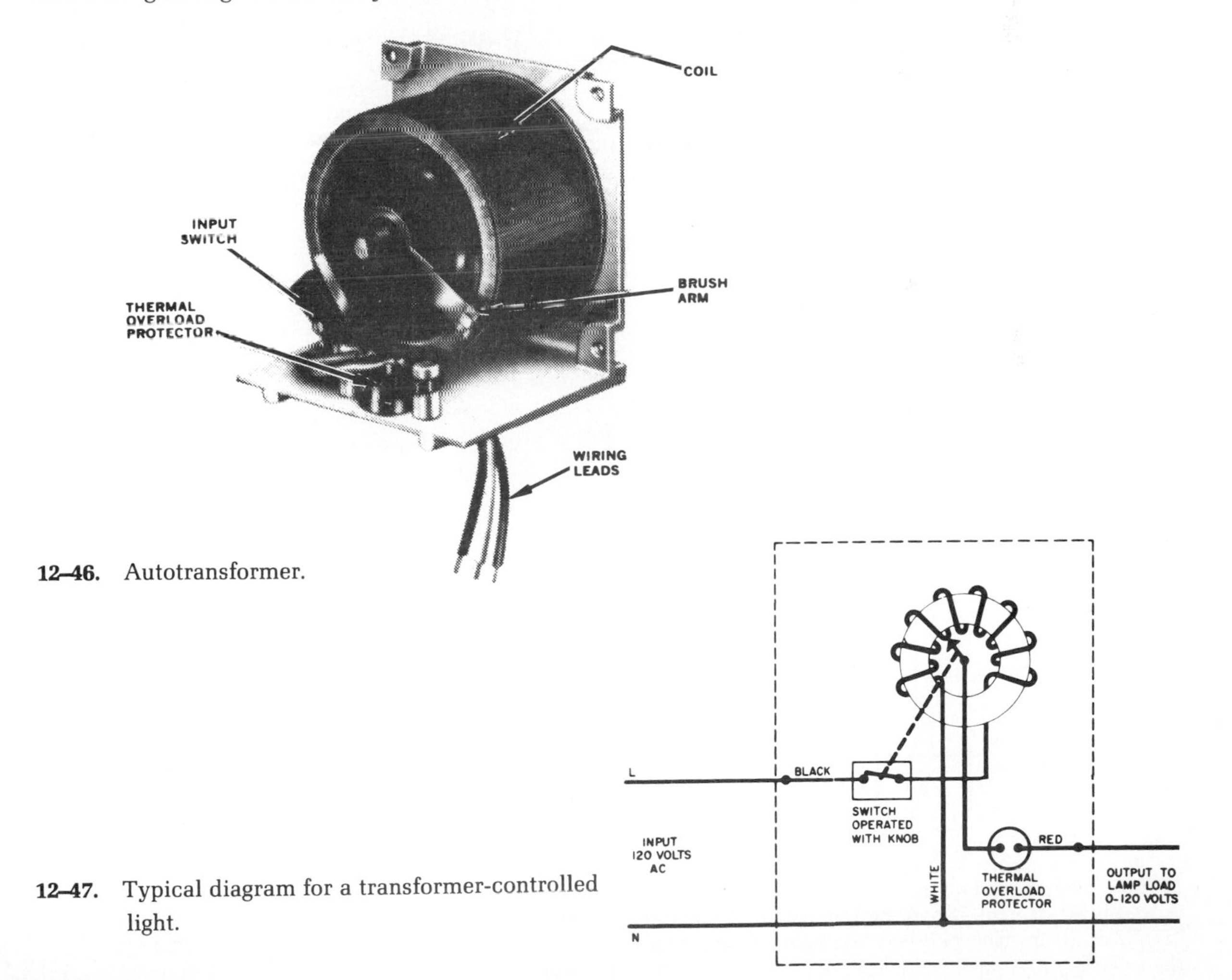

12-46. Autotransformer.

12-47. Typical diagram for a transformer-controlled light.

12-E. Determining the Lumen Count* for a Room

IF YOUR ROOM AREA (in sq. ft.) IS	AND THE TOTAL LUMENS IN YOUR ROOM IS							
40	400	800	1200	1600	2000	2400	2800	3200
60	600	1200	1800	2400	3000	3600	4200	4800
80	800	1600	2400	3200	4000	4800	5600	6400
100	1000	2000	3000	4000	5000	6000	7000	8000
120	1200	2400	3600	4800	6000	7200	8400	9600
140	1400	2800	4200	5600	7000	8400	9800	11 200
160	1600	3200	4800	6400	8000	9600	11 200	12 800
180	1800	3600	5400	7200	9000	10 800	12 600	14 400
200	2000	4000	6000	8000	10 000	12 200	14 000	16 000
220	2200	4400	6600	8800	11 000	13 200	15 400	17 600
240	2400	4800	7200	9600	12 000	14 400	16 800	19 200
260	2600	5200	7800	10 400	13 000	15 600	18 200	20 800
280	2800	5600	8400	11 200	14 000	16 800	19 600	22 400
300	3000	6000	9000	12 000	15 000	18 000	21 000	24 000
320	3200	6400	9600	12 800	16 000	19 600	22 400	25 600
340	3400	6800	102 000	13 600	17 000	20 400	23 800	27 200
360	3600	7200	108 000	14 400	18 000	21 600	25 200	28 800
	THE LUMEN COUNT INDEX FOR YOUR ROOM IS							
	10	20	30	40	50	60	70	80

*INDEX–

Recommended Lumen Count: Living Room, 80; Dining Room, 40; Kitchen, 80; Bathroom, 60; Bedroom, 70.

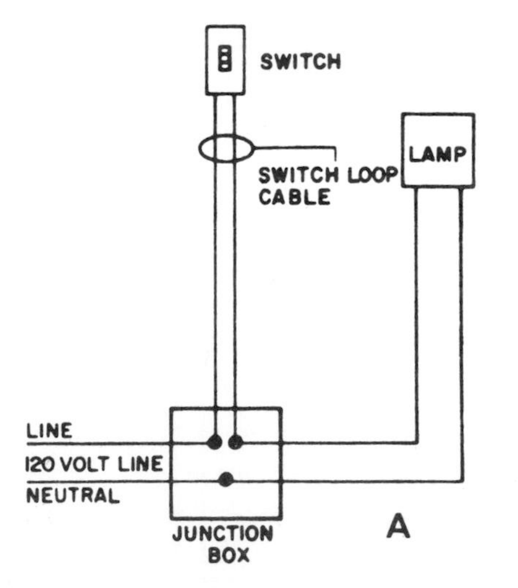

12–48a. When replacing an existing wall switch with a light control, it may be found that a "switch loop" has been used between the lighting load and the switch.

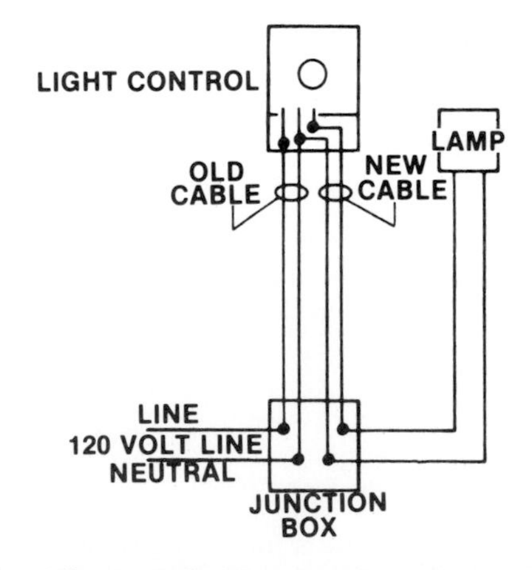

12–48b. This usually does not present a problem because the dimmer can be incorporated by changing the circuit.

stalled in the input line to the dimmer to turn the lights on or off from more than one location. Fig. 12-49. When turned on, the lights will be at the brightness level previously set on the dimmer knob. Other variations are possible that will meet the requirements of the installation.

Electronic Dimmers. In Figs. 12-50 & 12-51, *electronic dimmers* are used to control incandescent lamps. These dimmers are small enough to fit into a standard wall box. The wall plate can be any standard type used for a toggle switch. Fig. 12-52. These units have an SCR (silicon-controlled rectifier) which controls the voltage by the use of a small variable resistor. This potentiometer in turn controls the resistance the SCR puts into the circuit. These are also available in a 3-way configuration.

The switch in Fig. 12-53 has a "hi-lo" type of control with "center-off." ("Hi" is up, "center" is off, and "lo" is down position of toggle switch handle.) A diode is usually placed into the circuit for the "lo" position, with full voltage on the lamps for the "hi" position. They may be mounted in the vertical position with the "hi" at the top or in a horizontal position. However, the plate reads correctly only in the vertically mounted direction.

Some rotary dimmers have break-away fins that allow for fitting into the space available. The fins serve to dissipate the heat generated by the electronic circuitry. Fig. 12-54.

Photoelectric-Control Switches. *Photoelectric switches* are available for use in automatic circuits, for the person who does not want to be bothered with remembering to turn a certain light on at dusk and off at sunrise. Fig. 12-55.

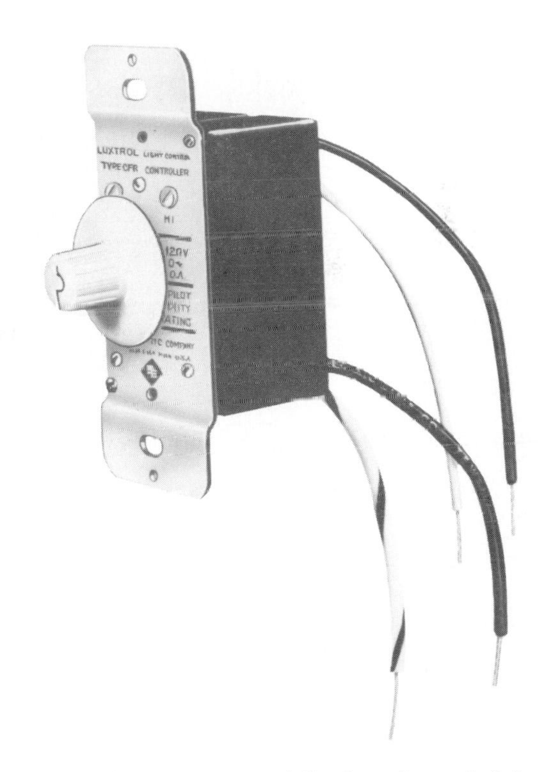

12–50a. Electronic control fits into the switch box.

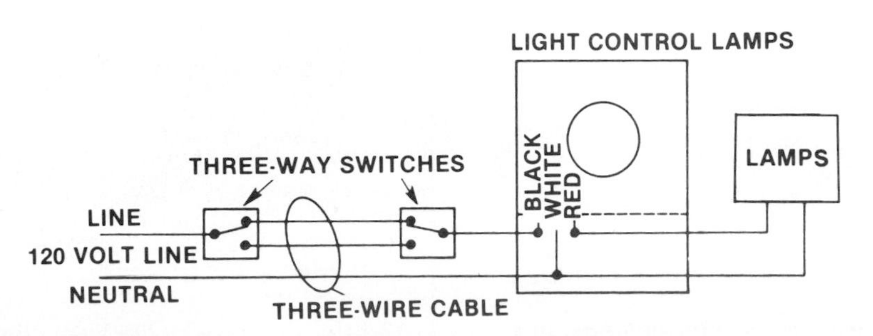

12–49. Control from multiple locations.

12–50b. Wall plate and knob mounted on the electronic contol make a neat appearance.

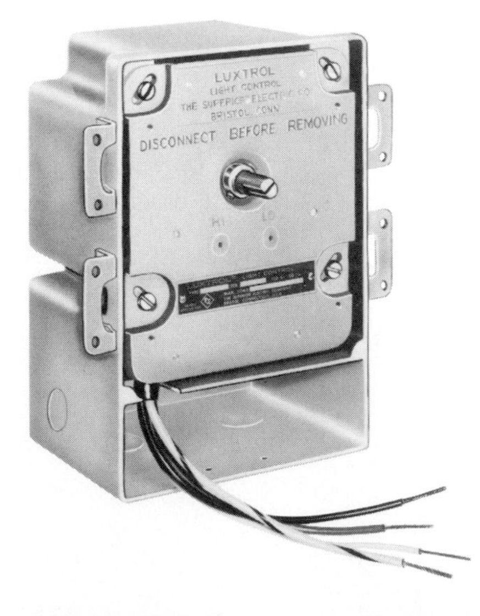

12–51a. Light control, electronic, mounted in a box.

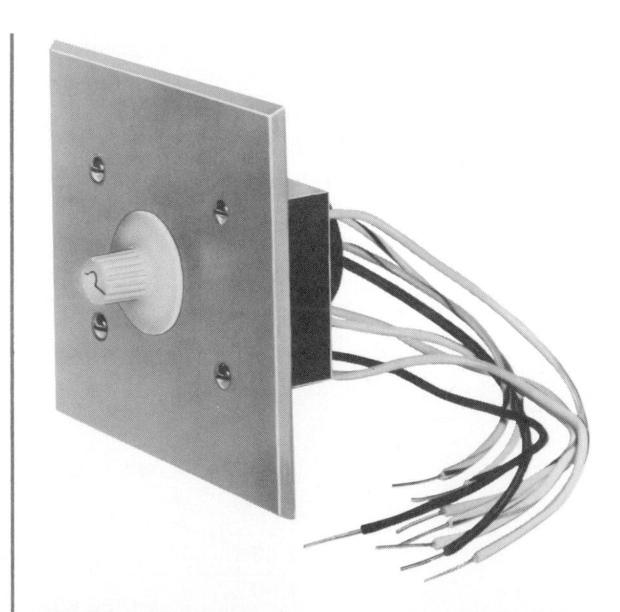

12–51b. Three-way electronic light control with wall plate and knob.

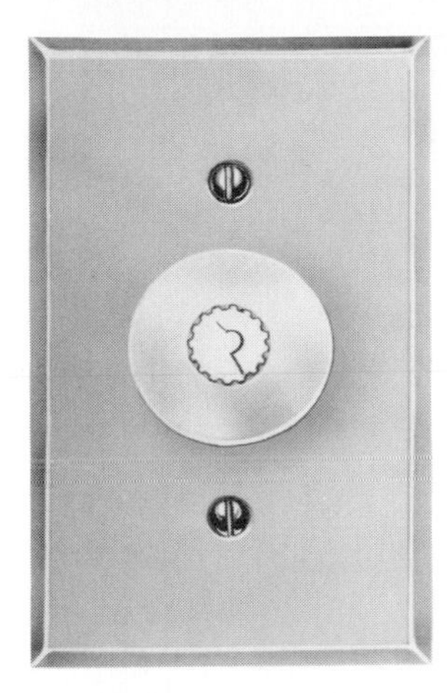

12–52. Note that the wall plate and knob fit in place the same way that a regular wall switch goes into a box.

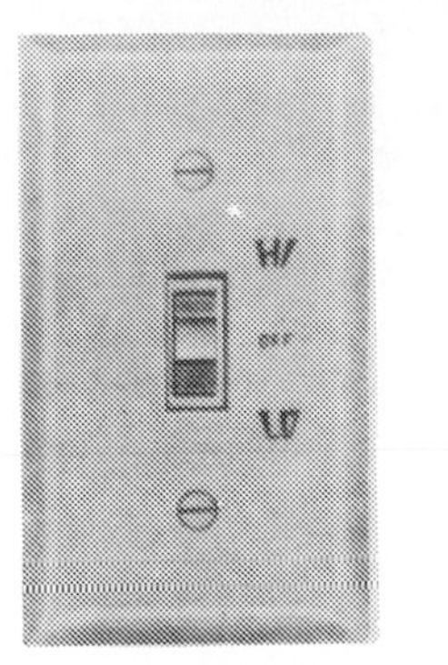

12–53. Hi-lo switch.

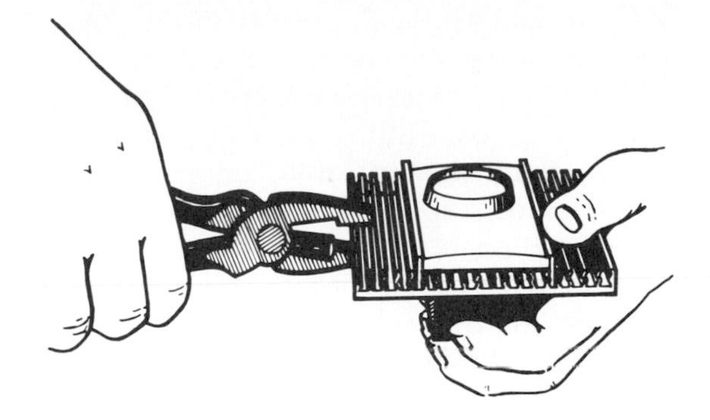

12–54. Snap-off fins on electronic dimmers make it easy to fit into almost any combination of wall plates.

"Nite-on" photoelectric switches operate by turning lights on at dusk and off at dawn, automatically. They have a built-in delay in each unit to prevent flickering or flashing of lights. Flickering might be caused by sudden extraneous light sources such as headlights or flashlights.

A wiring diagram for a photoelectric switch is shown in Fig. 12-56. The photocell is a sealed unit made of cadmium sulfide. It will operate at −40°F (−40°C) to 160°F (71°C). Most will operate 1000 watts of tungsten lamps or 450 watts of mercury-vapor lights at 120 volts AC. Photocells have a rubber gasket that weatherproofs them when they are mounted in a standard metal or insulated box.

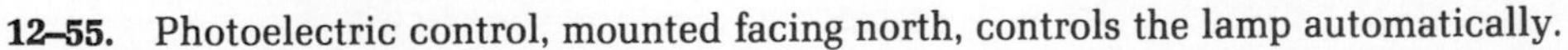

12–55. Photoelectric control, mounted facing north, controls the lamp automatically.

Manual- and Motor-Control Devices. Manual and motor-driven controls are continuously adjustable transformers which control lighting intensity by controlling the voltage applied to lamps. Rotating the knob, or actuating the drive motor, moves a brush-contact over a bared portion of the winding, producing any desired range of light intensity from complete darkness to full brightness.

Light controls are available in knob-operated units for direct manual control. They are also available in motor-driven models for remote control. Positioner-drive systems are for applications where it is not possible to observe the lighting when adjusting intensity, or where presetting is required. Fig. 12-57.

Wirings for incandescent lamp control and for a typical circuit are shown in Fig. 12-58.

Momentary-contact switches, which mount in a standard box, are often used to remotely control motor-driven devices. Figs. 12-59, 12-60, & 12-61. Figs. 12-62 & 12-63 are pictures of remote-control boxes. Fig. 12-64 shows a diagram of a simple motor-driven control circuit.

Special instruments, designed to check light level, are sometimes used in large buildings. Fig. 12-65. Theater lighting calls for high-wattage controls. Fig. 12-66. Note the size and shape of the unit. The wiring diagram, Fig. 12-67, shows a master unit used to control stage lighting up to 1200 watts. The controls are either ganged or individually connected. The master switch can control all the lights at once, or the individual controls can take care of a separate string, one at a time. Many combinations and shapes are available.

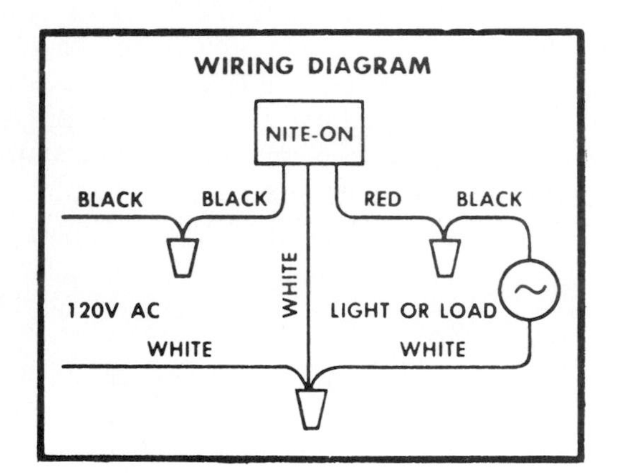

12–56. Wiring diagram for an automatic, outdoor weatherproof switch, controlled by light.

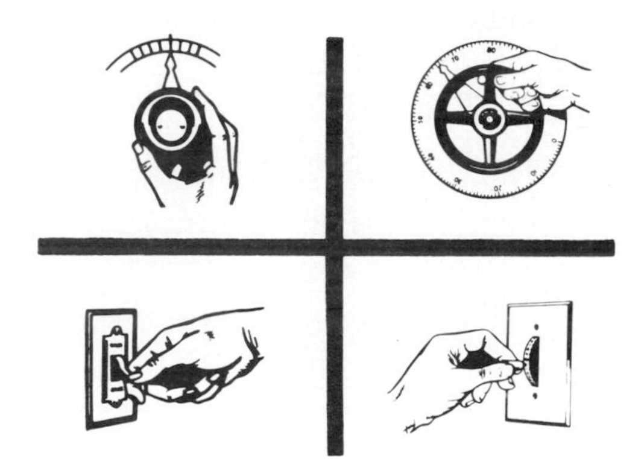

12–57. Types of manually controlled lights and their knobs.

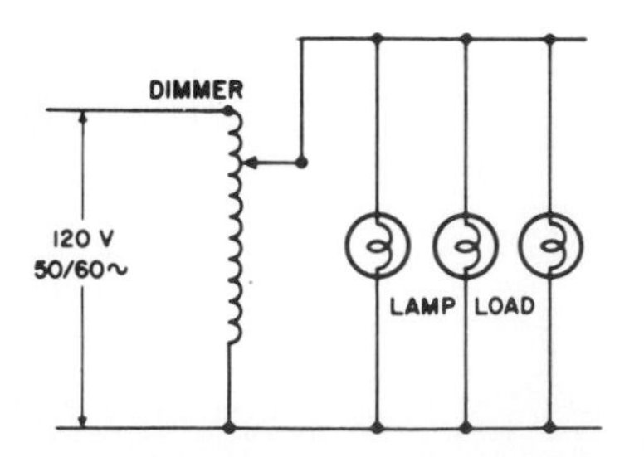

12–58a. Incandescent lamp control circuit.

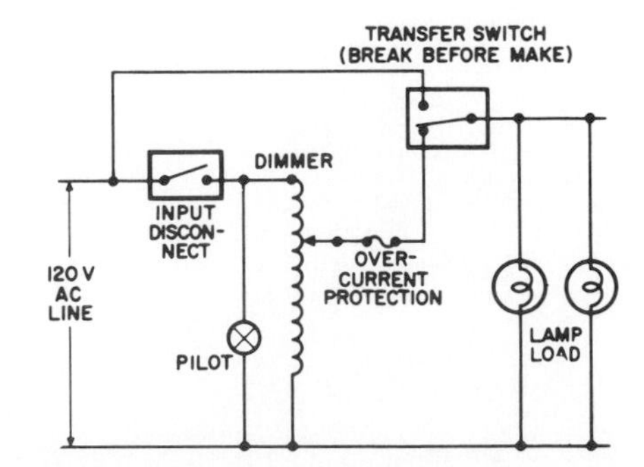

12–58b. Typical transfer circuit.

12–59. Motor-driven light control.

12–60. Motor-driven light control (reverse side of Fig. 12–59).

12–61. Motor-driven light control.

12–62. Remote-control box.

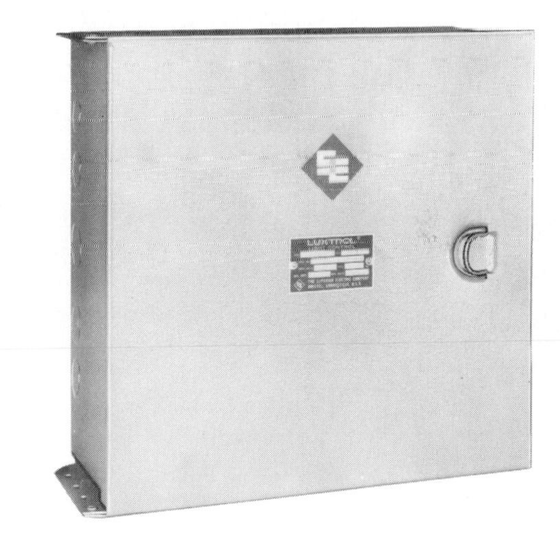

12–63. Remote-control box.

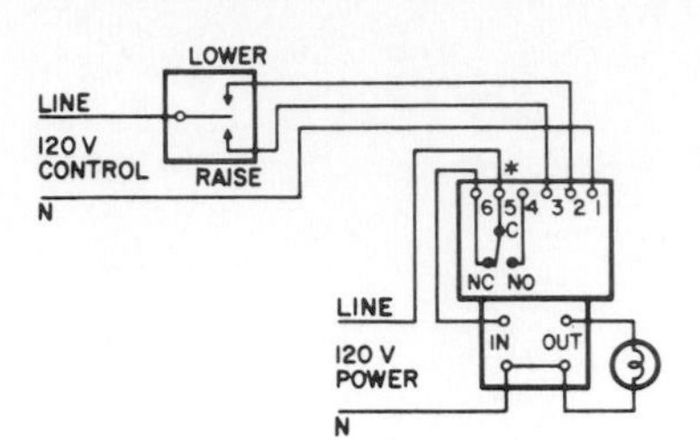

*Switch in NC position from full bright to near off.

12–64. Motor-driven control diagram.

12–65. Checking light levels in a bowling alley.

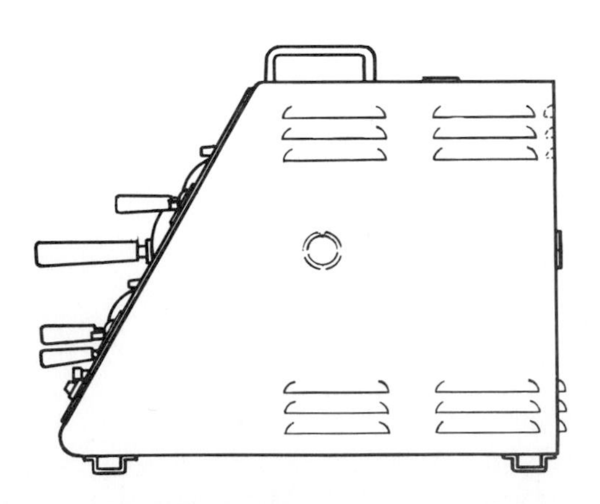

12–66. Theater-type control box.

Solid-State Light Dimming

Solid-state dimming systems can be used to control lighting intensity. Rheostats and autotransformers can be used to control the voltage available to incandescent lamps. They can thereby control the intensity of the light. The controlled voltage also restricts the amount of current available for lighting.

Solid-state dimmers use a method that pulses both voltage and current. The duration of each pulse is shortened or lengthened to decrease or increase the average current flow. For full lighting intensity current flows during the full time of each half cycle. Fig. 12-68(1). For one-half light, the electronic dimmer control is turned up to approximately half its maximum setting. Solid-state switching components delay the startup of current in each half cycle so that current starts halfway between A and B and again halfway between B and C as seen in Fig. 12-68(2). The average current is the sum of the shaded areas and is equal to 50% of the normal average current. For lower light levels, current flows for a proportionately shorter part of each half cycle as shown in Fig. 12-68(3). Therefore the average current flow or sum of shaded areas is proportionately less. Since current flows only for the duration of the pulse, the cost of electric power is proportionately less as lighting is dimmed.

The electronic device used to control lights is the silicon-controlled rectifier (SCR). It can be controlled easily with a simple circuit. The SCR produces the pulsating DC current from the line AC. However, it also produces electrical interference for radios, PA systems, and other noise and hum sensitive devices.

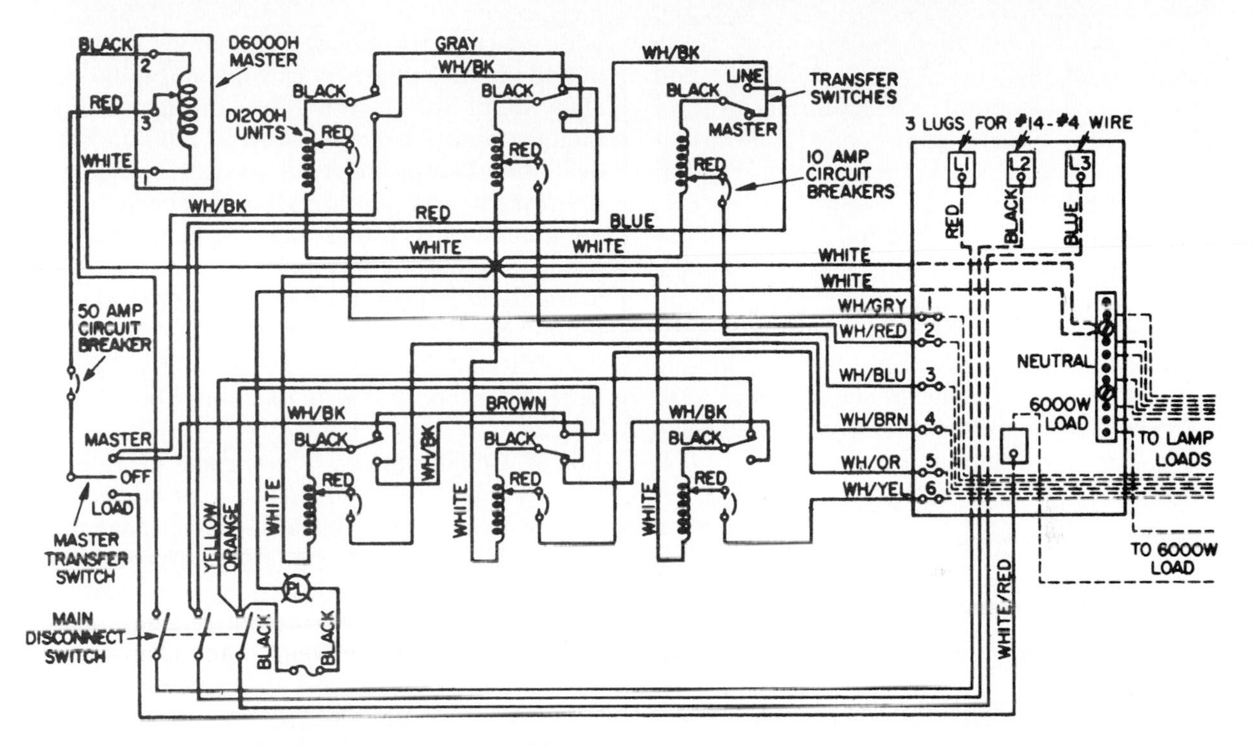

12–67. Circuit diagram for the theater control box.

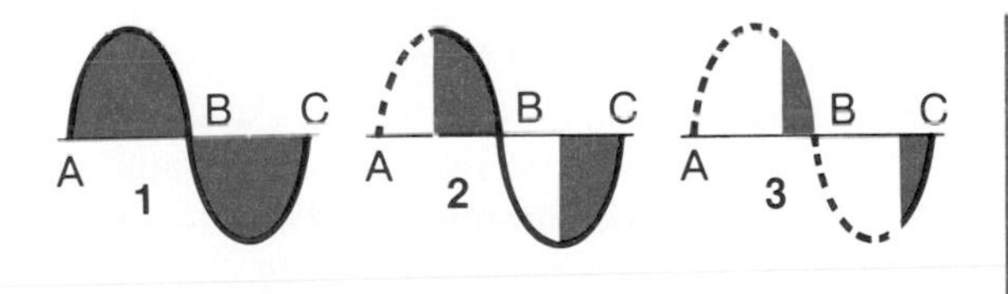

12–68. Electronic dimming.

An advantage of this type of dimming is that it can be motorized and the motor unit controlled by low-voltage switches. Any number can be connected to the same motorized unit for multi-point control. These switches are designed to be used independently or in conjunction with a remote control low-voltage switching system to provide additional control flexibility to the complete lighting system in any given installation.

To prevent lamp hum, the lighting fixtures should be non-parabolic to prevent focusing of lamp noise. Install a debuzzing coil in the load circuit. Place the coil in a location where the slight hum from the filter coil will not be objectionable.

Install wall dimmers and modular dimmers at least 12′ from intercoms, public address systems, and radios. Wire the intercom, PA systems, or radios on separate branch circuits in individual metal EMT or conduit. If possible, wire dimmers and communications systems on different phases of a three-phase system. For best results, wire dimmers to a different distribution cabinet from that used for communications systems or radios, keeping the two cabinets at least 12′ apart.

Keep intercom leads at least 12′ away from AC lines feeding dimmers. When intercom wires and AC lines feeding dimmers must cross, be sure to cross them at right angles.

Wire PA speakers with shielded conductors grounding shielding at one end only. Install PA systems using a line filter or filtered power supply.

Permanently installed microphone cable should be of high quality and should be run through metal conduit. Keep dimmer control load and line wires a minimum of 3′ from microphone cables when run in parallel, whether shielded or not.

Check to see if the microphone cables are tight. Cross the microphone cables with the dimmer load wires only at right angles. Ground the amplifier case. Also ground the microphone cable at the PA amplifier. If dimming systems still cause interference, connect a 0.1 MF capacitor rated at 600 VDC across the two wires of the wall-mounted dimmers.

Never "spark" wires together to check for power. Damage must not be recessed and the heat sink fins must be vertical for convective air flow cooling. Fig. 12-69. It is normal for the wall-mounted dimmer face plate to feel warm. Mount the dimmer where there is adequate air flow.

As the older transformer units wear out the newer electronic dimmers will be used both in commercial and industrial applications. This will call for more attention to the installation of the newer systems since they do have inherent interference generation characteristics. Both the small electronic dimmer and the larger units will have to be checked for their interference with other equipment in the building or on the same lines.

Fluorescent Lamps

Controls have been designed to permit fluorescent lamps to be dimmed. These electronic dimmers will control fluorescent lamps over a wide range of brightness. The lamps will start quickly at any intensity setting, and will operate without flicker even at low or high settings. Models rated for use with up to 120 lamps are available. Lamps must be 40-watt, rapid-start fluorescent and equipped with special dimming ballasts. White or colored lamps may be used. No dimming auxiliaries are needed. The dimmers incorporate filters for suppression of radio-frequency interference (RFI). They need no additional chokes, as was formerly the case (a choke is a coil of wire or inductor).

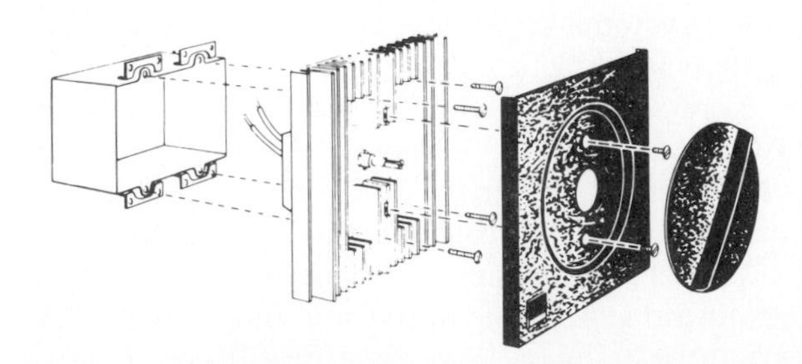

12–69. Dimmer mouth with fins vertical.

Power Consumption. Operation of fluorescent lamps at reduced intensity provides substantial cost savings because power consumption is also reduced. Fig. 12-70. The curve of light output, contrasted with input current and watts, shows the decrease in power supplied to the ballasts as the lighting is varied from full intensity to blackout.

Light-control models are offered in self-contained and remotely controlled varieties. Self-contained models mount on the wall. They are offered in maximum capacities of 10 to 30 lamps. Remotely controlled models can be mounted against any flat surface. They require a separate controller which can be located up to 200 feet from the dimmer. Maximum capacities of 30 to 120 lamps are available.

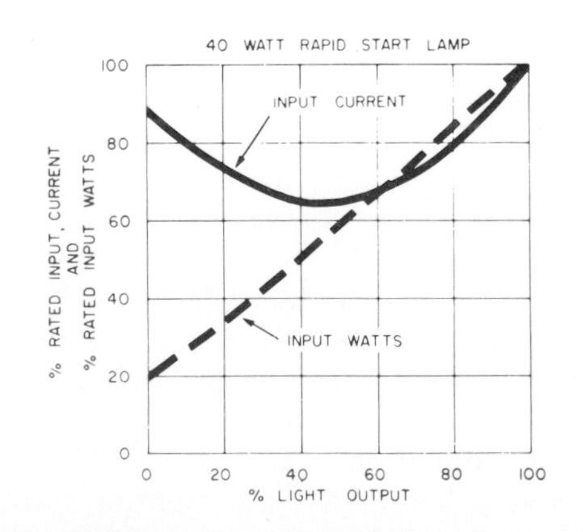

12–70. Power consumption of dimmer fluorescents.

Lamps. Multiple lamps for fluorescent fixtures should be the same size and age. Ideally, the lamps should have approximately 100 hours burn-in time to reach maximum efficiency. Lamps should not be installed in a cold draft, either natural or from air-conditioning outlets. Cold lamps become unstable and will not dim satisfactorily. Fig. 12-71.

Motor-controlled remote dimmers are available. A momentary switch is used for control, or a positioner controller may be used. Figs. 12-72 & 12-73.

Self-contained models can be combined with three-way switches to allow lights to be turned on and off from either of two locations. An added advantage of this method is that it is not necessary to disturb the intensity setting when the lights are turned off. The lights will then be at the same intensity when turned on again.

12–72. Momentary-contact switch for motor-driven, remote-controlled light controllers.

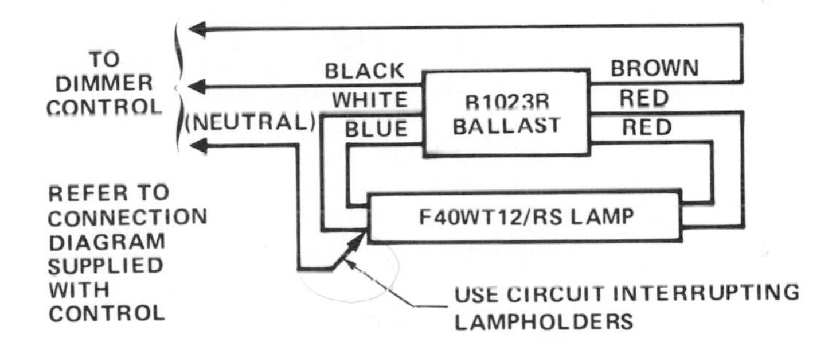

SINGLE-LAMP DIMMING BALLAST

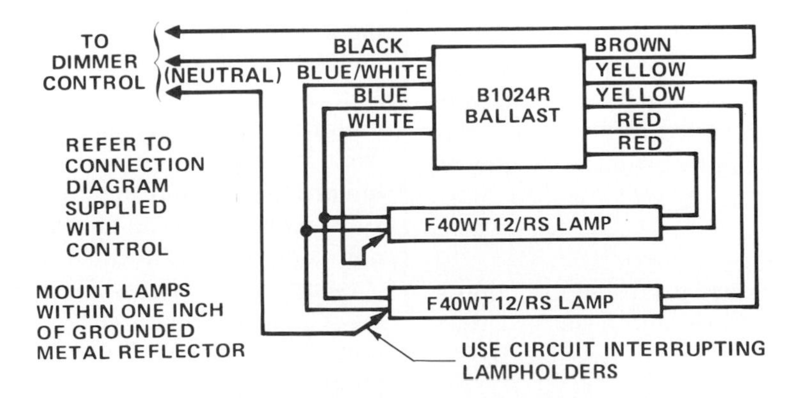

TWO-LAMP DIMMING BALLAST

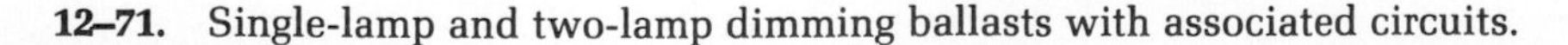
12–71. Single-lamp and two-lamp dimming ballasts with associated circuits.

12–73. Positioner controller.

A bypass switch allows the lighting load to be quickly transferred from the dimmer output to the line. A two-pole, double-throw switch is required. The switch can be placed at the control station or at a remote point. Fig. 12-74.

FARM LIGHTING†

Lighting for a farm should be carefully planned. Such a plan can be complex and technical, largely because there are many alternative ways to obtain good farm lighting. The suggestions here have been simplified to facilitate wiring and planning specifications.

Incandescent Lamps

Incandescent lamps are generally selected when light is needed for short periods of time and when lamps are turned on and off frequently.

Incandescent fixtures have their maximum wattage stamped on them. To increase the wattage of lamps for these fixtures may cause high temperatures, resulting in fire hazards and reduced lamp life.

Fluorescent Lamps

Fluorescent lamps are used mainly indoors. Fixtures include a ballast, and a starter circuit in some cases.

Better brightness control and efficiency of light output are obtained when "egg crate-type" metal or plastic louvers are part of the fluorescent fixture. These louvers keep the angle of direct light to not more than a specified angle, usually 45°. Fig. 12-75. Other transparent covers can be used to diffuse light.

Mercury Lamps

Mercury lamps combine the small size of incandescent lamps with the long life and high efficiency of fluorescent lamps. Like fluorescents, mercury lamps require ballasts to regulate the flow of current. Mercury lamps are available with partial color correction for use where the greenish color is objectionable.

Most mercury lamps operate with special fixtures and ballasts. Make sure of the type of mercury lamp being used before screwing it into the socket. Some mercury lamps do not require ballasts.

Planning Good Lighting

Begin a plan for good farm lighting by evaluating areas and activities where light is needed. Next, determine the foot-candle level (lumens) needed for each.

Then, select suitable lighting equipment and have it installed properly for good light distribution.

A *footcandle* is a unit of measurement. It is the amount of light provided on a surface one foot square. The surface is located one foot from a one-candela light source. The surface is uniformly curved so that all parts are one foot from the light source. The metric system uses lumens/meter^2, but the footcandle is used here since most American charts on this subject are not using SI units at this time.

Lamp requirements for lighting most work areas can be computed on a basis of fixtures mounted 7′ to 10′ above the floor. For higher mountings, the total lumens required to light the area must be increased. The finish of the walls and ceiling should have from medium to average reflectance.

†This information is provided through the courtesy of the United States Department of Agriculture.

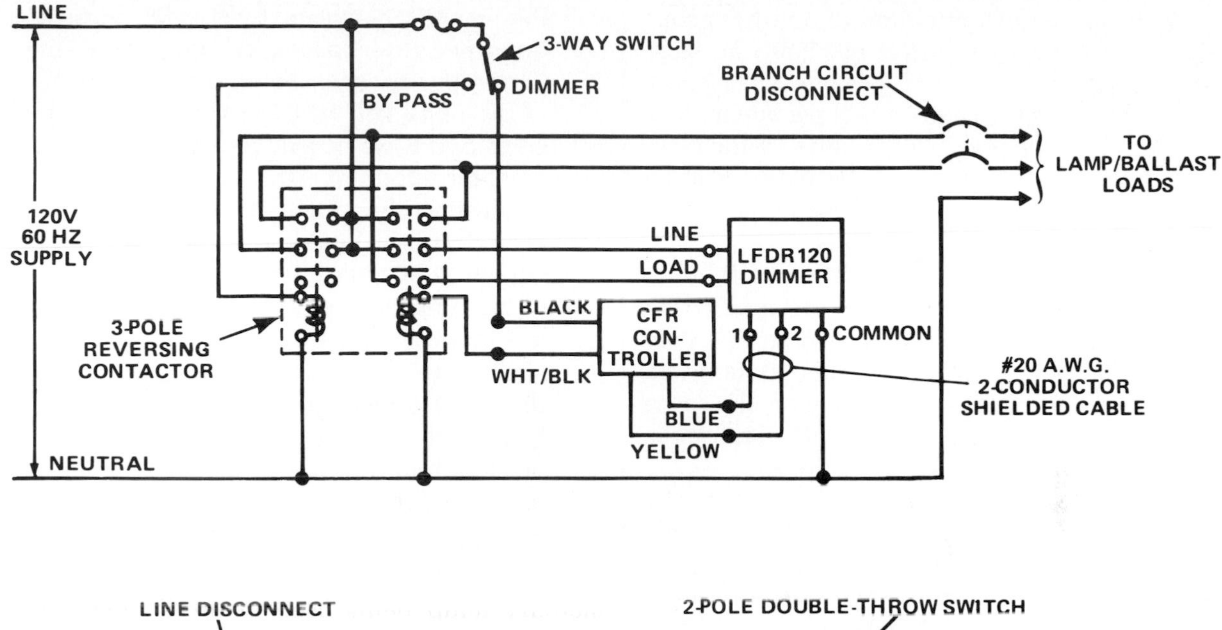

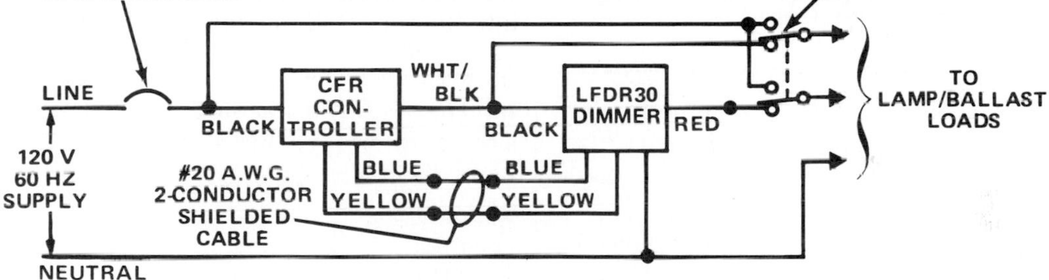

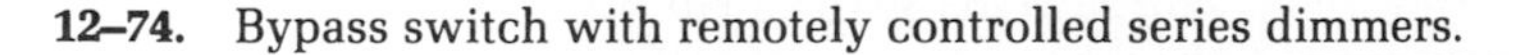

12–74. Bypass switch with remotely controlled series dimmers.

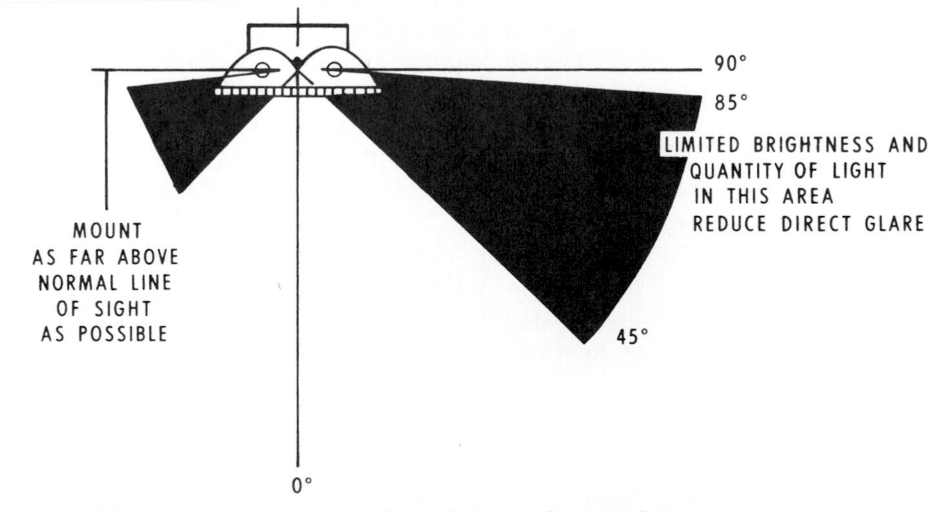

12–75. Louvers in fixtures keep the angle of direct light to about 45 degrees.

With the fixtures mounted under these conditions, 2 lumens of output per lamp for each square foot of floor area provide an average light level of about 1 footcandle per square foot at the working level. This ratio, which accounts for the efficiency of the entire lighting system, includes lamp output, the use of reflectors, and the reflectance of walls.

Normally, mercury lamp fixtures will be installed in rows the length of the area to be lighted. They should be arranged so that they provide uniform lighting. When no more than 5 to 10 footcandles of light are needed, the space between lamps should be from 1 to $1\frac{1}{2}$ times the distance from the lamps to the floor.

When more footcandles are needed, the spacing between lamps can be reduced to less than the distance from the lamps to the floor to give the required light. The distance from outside lamps to the walls should be no more than one-half the distance between lamps.

When spacing between lamps is $1\frac{1}{2}$ times the distance to the floor, use shallow-dome reflectors. If the spacing is equal to the distance from the lamp to the floor or less, install standard-dome or deep-bowl reflectors.

The following example shows how you can determine the number of fixtures and the size of lamps that should be used in most work or storage areas.

1. Suppose you need 5 footcandles of light in a machinery-storage area which is 24′ wide and 48′ long, with the lamps to be installed 10′ above the floor. The fixtures should be 10 to 15′ apart. If a row of fixtures is installed 6′ from each long wall, 12′ remain between the two rows. This provides an acceptable spacing across the 24′ width of the area.
2. Determine the number of fixtures needed in each row. Since the area is 48′ long and the spacing between rows is 12′, four fixtures per row, spaced 12′ apart, will give the proper spacing. The fixtures at the end of the rows will be 6′ from the walls. This gives two rows of four fixtures each in the area, for a total of eight fixtures.
3. Determine the size of lamps required for each fixture. You know the floor space equals 1152 square feet and that 2 lumens of lamp output per square foot are needed to obtain a light level of 1 footcandle per square foot. Multiplying 1152 by 2, you find that 2304 lumens are required to obtain a light level of 1 footcandle. The work area needs a light level of 5 footcandles. Multiply 2304 lumens by 5 and you get 11 520 lumens, the number you need for a light level of 5 footcandles.

To find the lamp size, divide the number of fixtures into the total lumens. This gives the lumens required per lamp. Select the correct size lamp from Table 12-F. If the lumens per lamp fall between lamp ratings, use the next larger size, or recalculate, using other lamp spacings.

In this example, divide 11 520 lumens by 8 lamps. This gives 1440 lumens per lamp. From the list of lamps, you find that either 100-watt incandescent lamps or 40-watt fluorescent lamps will give the required lumens for 5 footcandles at floor level.

In areas with dark walls and a dark ceiling, you will have to increase the size of the lamps. Check light levels with a light meter.

You can use this example to guide you in computing the lighting requirements for any of the areas discussed in this section except outdoor areas. For outdoor lighting, specific sizes and types of lamps are recommended.

12-F. Average Lumens per Lamp

INCANDESCENT, STANDARD		
WATTS		LUMENS
15		125
25		225
40		430
60		810
100		1600
150		2500
200		3500
300		5490
INCANDESCENT, TUNGSTEN HALOGEN		
WATTS		LUMENS
400		7000-7500
500		9000-10 000
1000		20 000-21 000
1500		30 000-33 000
FLUORESCENT		
WATTS	LENGTH, INCHES	LUMENS
15	18	500-700
20	24	800-1000
40	48	2000-3000
60	48	3000-3500
110	96	5000-6000
210	96	10 000-12 000
MERCURY		
WATTS		LUMENS
75-85		2000
100		2500-3000
175		5500-6500
250		9000-10 000
275 (high-pressure sodium-type)		25 000
400		17 000-19 000
400 (high-pressure sodium-type)		40 000

Poultry Lighting. One or two footcandles, for more than half of each day, are required to stimulate production in poultry houses. Twenty footcandles are recommended for good visibility when work is being done. You may find it convenient to have two lighting systems—one for egg and meat production and another for working in the poultry house. Table 12-G.

Dairy Lighting. Both incandescent and fluorescent lights are used in dairy barns. Fluorescent lamps may be used for general lighting. Incandescent spots or floods may be used for concentrated light in specific areas. In a dairy barn, locate lamps on the ceiling, both behind and in front of the cows. Use Table 12-H as a guide for lighting dairy buildings.

General Indoor Lighting. Storage areas, stairs, alleyways, as well as machine sheds and similar buildings, need a general lighting system for safety, convenience, and efficiency. Provide supplemental light as needed for specific tools or special locations. Use Table 12-I as a guide for general indoor lighting.

Outdoor Lighting. Outdoor lighting enables work to be completed more quickly, easily, and safely after dark. It helps protect buildings, machinery, and livestock from prowlers and reduces accidents.

Weatherproof wiring and fixtures are used in outdoor lighting. Reflector-type incandescent lamps are used for lighting small areas and for areas where lamps are used occasionally for short periods of time. Mercury, fluorescent, and similar lamps are used where light is needed for long periods of time.

Use flood or spotlight fixtures placed so that the light will be directed where it is needed. All outdoor lights attract insects, but yellow, orange, amber, or red lights attract fewer insects. Place outdoor lamps in areas where the insects attracted by the lamps will not be objectionable. (See "Insect Control," later in this chapter.)

Footcandle distribution data for flood and spot fixtures may be obtained from power suppliers and fixture manufacturers. Use Table 12-J as a guide for lighting outdoor areas.

12-G. Lighting Guide for Poultry Production

TYPE OF POULTRY	AGE, WEEKS	MINIMUM FOOT CANDLES
Chickens:		
Broilers	0-3	1.0
Broilers	3 and up	0.5
Pullets, layers and breeders	0 and up	1.0
Turkeys:		
Market stock	0-5	2.0
Broilers	5 and up	0.5
Breeder hens and breeder toms	0 and up	2.0

12-H. Lighting Guide for Dairy Buildings and Equipment

AREA OF ACTIVITY	RECOMMENDED FOOTCANDLES	TYPICAL INSTALLATION
Feeding area	20	Incandescent or fluorescent
Milking area	20	Incandescent or fluorescent
Cow's udder	50	Incandescent spot or high intensity fluorescent in milking parlors; fluorescent in stall barns
Milk-handling equipment:	20	Incandescent or fluorescent
Milk room, general		
Washing area	100	Sealed fluorescent fixtures, 2-lamp, 40-watt; adjust height above vat, not above tank
Bulk tank, inside	100	Incandescent spots directed at tank interior; portable lamps may be used

Placement of Fixtures. In a farmyard, locate lights so that they do not create shadows. Direct the lights away from the vision of travelers on public roads. Light the boundaries of yards and the entrances of driveways to discourage intruders. Use automatic controls to turn the lights on and off when you are away.

Inside buildings, place lighting fixtures where lamps are not likely to be broken or to create shadows. Keep lamps out of reach of livestock. Fixtures installed in dusty places, such as haylofts and feed rooms, should be dustproof.

A light source gives off light in all directions. Reflectors help control light. The proper reflector will direct the light where it is needed, reduce glare, and spread or concentrate light as desired.

For safety, direct lights downward as much as possible to avoid glare. Outdoors, recess fixtures in curbs or steps, or mount them to the side of the house. Provide convenient switches for all lighting both indoors and outdoors.

Because dust and dirt reduce light output, the lamps should be dusted regularly. The fixtures should be washed, and the walls and ceilings cleaned or repainted. Burned out or blackened lamps require replacement.

Light for Grading Farm Products. The lighting system used for grading agricultural products should provide well-diffused light of the correct color and amount. Graders must be able to see fine detail and determine the color of the products they grade.

12-I. General Indoor Lighting Guide

AREA OF ACTIVITY	RECOMMENDED FOOTCANDLES	TYPICAL INSTALLATION
Feed storage:		
Haymow, silo, grain bins	3	Incandescent in protected or dust-proof fixtures
Feed inspection area, silo room	20	Incandescent flood
Concentrate storage, feed processing area	10	Incandescent flood
Livestock housing	7	Incandescent or fluorescent
Livestock examination area	20	Incandescent spot or flood, or fluorescent
Stairways and ladders	20	Incandescent flood at top and bottom of stairs
Feeding areas	20	Incandescent or fluorescent
Machinery storage	5	Incandescent
Machinery repair area	30	Incandescent or fluorescent with supplemental portable flood
Farm shop:		
General	30	Incandescent or fluorescent; color-corrected mercury if ceiling over 12 feet high
Bench and machine work, sheet metal, painting	50	Incandescent flood or fluorescent; adjust height and group lamps as needed
Machine tool and detailed bench work	100	Incandescent flood or fluorescent; adjust height and group lamps as needed
Farm office	70	Incandescent or fluorescent
Rest rooms	30	Incandescent or fluorescent
Pump house	20	Incandescent

12-J. Lighting Guide for Outdoors

AREA OF ACTIVITY	RECOMMENDED FOOTCANDLES	TYPICAL INSTALLATION
Protective lighting	0.2	175-watt mercury refractor mounted 25 feet high for lighting 8000 sq. ft.; or incandescent floodlight
General work areas, driveway, walks, barn lots	1.0	400-watt mercury refractor mounted 25 feet high for lighting 8000 sq. ft.; or incandescent floodlight
Activities area, fuel storage, building entrance, electrical load center, feedlots and equipment, livestock loading, and recreation area	3.0	400-watt mercury refractor mounted 25 feet high for 2000 sq. ft.; or incandescent floodlight or spotlight

For grading tobacco, a combination of fluorescent and incandescent lamps has been used in the past. However, fluorescent lamps are now on the market which combine the necessary color qualities in one lamp.

Place the lamps 3′ to 4′ above and 12″ in from the working side of the grading table. Use two 40-watt fluorescent lamps in industrial-type reflector fixtures, placed end to end the length of the work area.

For grading potatoes, apples, and other produce, use fluorescent lamps. Provide at least 50 footcandles of light for general inspection of produce, 75 to 100 footcandles for grading, and 150 footcandles for extremely close inspection or grading. (Lighting recommendations for grading different products may vary from one state to another.)

Insect Control. Many insects that fly at night are attracted to light. In general, ultraviolet, blue, and green lights are much more attractive to insects than yellow, orange, amber, or red. Therefore using yellow lamps in patios, porches, and yards helps reduce nuisance insects.

Fluorescent black-light lamps produce radiation in the near ultraviolet region which is very attractive to many insects. Farm pests attracted to these lights include European corn borers, corn-ear worms, tomato and tobacco horn worms, pink bollworms, and cucumber beetles.

Light traps containing black-light lamps are used in insect surveys to follow migrations, detect new pests, and to monitor insect population levels so that insecticides may be applied at the proper time.

Wiring. Make certain that all wiring complies with the local and national electrical codes. All materials should carry the Underwriters' Laboratories (UL) seal of approval. Low-cost, temporary wiring is not recommended for farm lighting systems. Have all wiring inspected by someone other than yourself before applying power.

Both wiring and fixtures in most farm buildings are often exposed to corrosive fumes and moisture. In buildings that house livestock, use approved moisture-resistant materials such as porcelain-insulated fixtures, galvanized steel or molded plastic junction boxes, and branch-circuit cables with plastic outer coverings.

Light switches should be conveniently located on walls. Do not use pull chains; they are inconvenient, likely to break, and may be safety hazards. Switches should be installed at all entrances and exits where lighting must be controlled from more than one point.

Plan for present and future needs. Lighting is only one of several key considerations in determining the total load on a system. Check the requirements of other electrical equipment before determining correct wire sizes and the type of distribution box.

Efficient lighting should allow you to turn on the lights ahead of you as you go from place to place. It should be possible to:

- Turn walkway lights on and off from all key work areas.
- Turn on stairway lights before you start up or down the steps.
- Turn lights on and off from each entrance, in rooms or areas with two or more entrances.
- Control outdoor floodlights or yard lights from inside all frequently used buildings.

Portable Lighting

Lighting for special purposes or occupancies is covered by the National Electrical Code. The Code specifies the types of lighting that may be used in hazardous locations and other areas.

A frequently mentioned application is a trouble light for garages and service stations. Fig. 12-76. In the service areas of such locations, the entire area up to a level of 18″ above the floor is considered a Class I, Division 2 location (subject to a judgment by the enforcing agency). The actual gasoline dispensing area, and service pits or depressions below floor

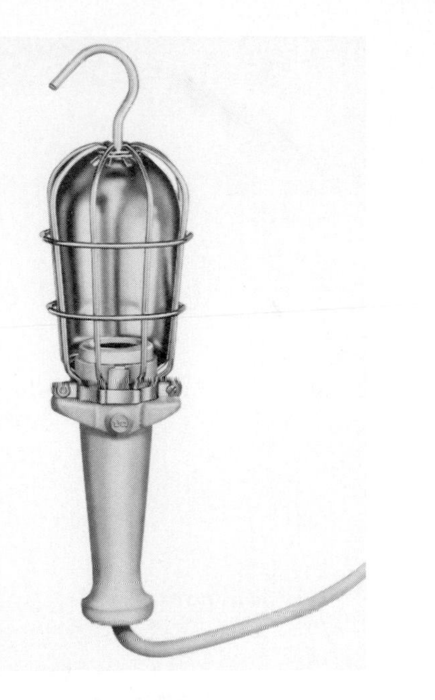

12–76. This type of "trouble" light is not intended for use in hazardous classified locations such as service stations.

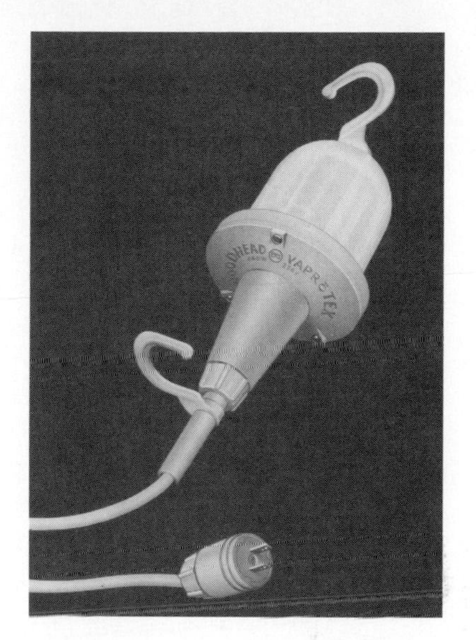

12–77. Vapor light, without an outlet, is suitable for use in hazardous classified locations.

level, are considered Class I Division 1 locations (again subject to the decision of the enforcing agency). For either of these classifications, hazardous location hand lamps are indicated. Fig. 12-77.

For use in service areas above the 18″ level, ordinary location hand lamps may be used, with certain restrictions. Exterior surfaces of the hand lamps should be of nonconducting material, or effectively protected with insulation. No switches or side-outlets are permitted in the hand lamps. Lamp and cord must be supported so that it cannot be used in the hazardous areas outlined above. See Fig. 12-78.

Fig. 12-79 shows the construction of a portable electric hand lamp for hazardous (classified) locations. Note the globe, guard assembly, handle and cord clamp. The tempered glass globe is explosion-, heat-, and impact-resistant. The idea is to guard against spark-ignited explosions. The raintight and weather resistant qualities also aid in its utilization. This lamp can be used in Class I, Group C & D, and Class II, Group G, atmospheres.

Keep in mind that Class I locations are those in which flammable gases or vapors are or may be present in the air in quantities sufficient to produce explosive or ignitible mixtures. Groups C and D include most commonly used chemicals.

Class II locations are those that are hazardous because of dust. Group G atmospheres are those containing flour, starch, or grain dust.

In still another requirement, OSHA states that portable electric lighting used in moist and/or other conductive locations such as drums, tanks, and vessels, shall be operated at a maximum of 12 volts. Low voltage lamps shown in Fig. 12-80 comply with this requirement.

In some situations a low-voltage extension light will serve the purpose. In some cases a hazardous location lamp is required. In other situations, both the low voltage and the hazardous location feature are indicated.

12–78. This lamp is enclosed but has an outlet, which makes it unsuitable for hazardous locations. However, it is usable in most other locations.

Low Voltage Lights. Ordinary portable lighting can prove hazardous when used under grounded or wet conditions. To follow good safety practice, OSHA specifies that portable electric lighting used in these conditions shall be at a maximum of 12 volts. Note how the step-down transformer is wired into the cord of the lamp so it is not accidentally left behind when the lamp is needed in a hurry. Fig. 12-81.

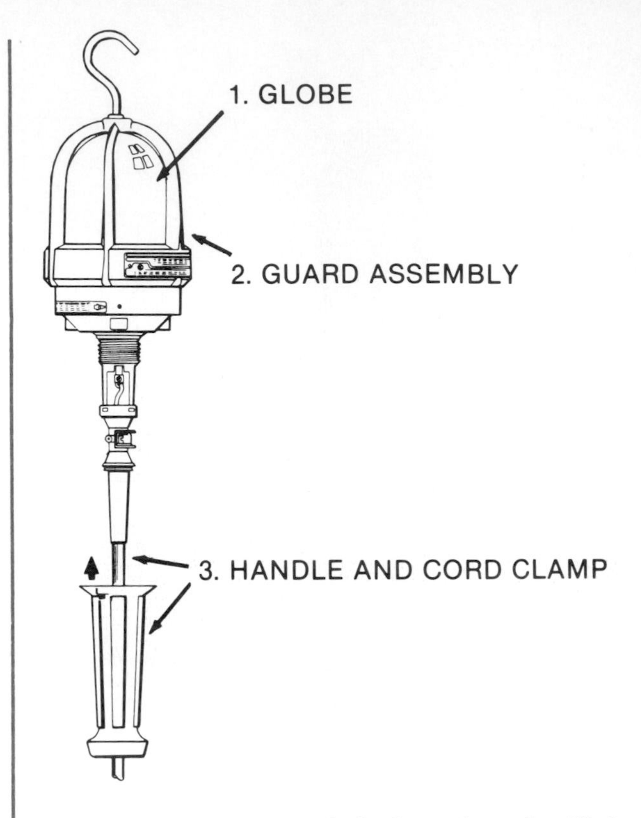

12–79. Portable lamp made for hazardous classified locations.

Isolation Transformers. The primary and secondary windings of the transformer are isolated from each other so that there is no complete electrical path between the convenience receptacle and the portable lamp. Therefore, no ground current flows even if the user comes into contact with one of the secondary conductors, because there is no conductive path.

Cord Lengths. Primary cord lengths were established to offer optimum safety. A short primary will keep the transformer close to the convenience receptacle, away from the possibly hazardous work area. Since there is no significant voltage drop, any practical length of cord can be supplied on the primary side without affecting the lamp performance. See Fig. 12-82.

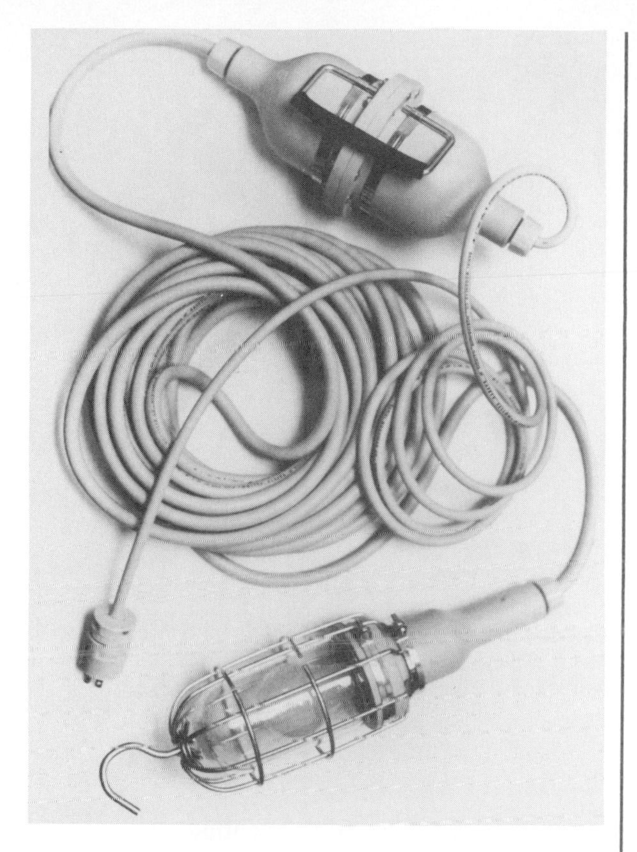

12–80. Low-voltage lamps such as this use either 6 volts or 12 volts. The transformer in the line isolates the lamp end of the cord from the 120-volt ground system.

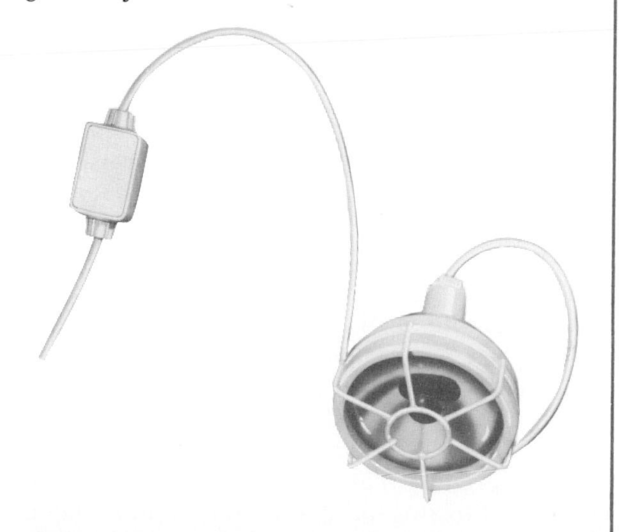

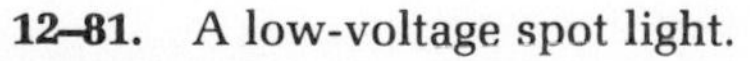

12–81. A low-voltage spot light.

12–82. Low-voltage lamp on the job.

The length of the secondary cord is limited by the permissible voltage drop. For example, the maximum secondary length for the standard 6-volt unit is 60′. For the standard 12-volt unit, it is 120′. With spot or floodlight units with the 30-watt transformer, the maximum secondary length is 50′. Fig. 12-83.

An additional safety factor can be achieved by using left-hand threaded lamps and sockets. This makes it impossible to accidentally insert the lamps into a standard voltage socket.

Inspection Lights. Inspection lights can be obtained in a number of sizes and shapes. One of the most popular is the fluorescent lamp in the 4-watt and 8-watt size. The 15-watt and 30-watt sizes are also available with the ballast located in the cord. Fig. 12-84. The fluorescent tube is enclosed in a shockproof, durable butyrate plastic shield to protect against shattering glass. Fig. 12-85.

12–83. Using the spot light on the job.

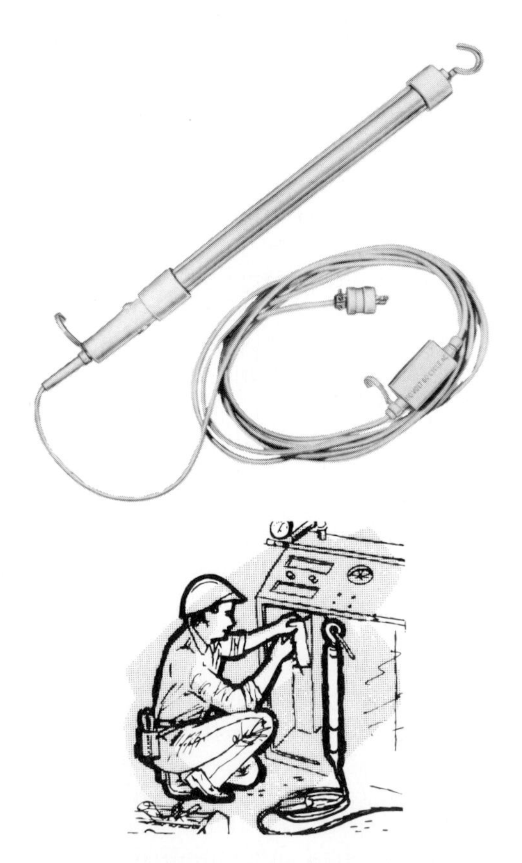

12–84. The small enclosed fluorescent lamp is suitable for use in service stations.

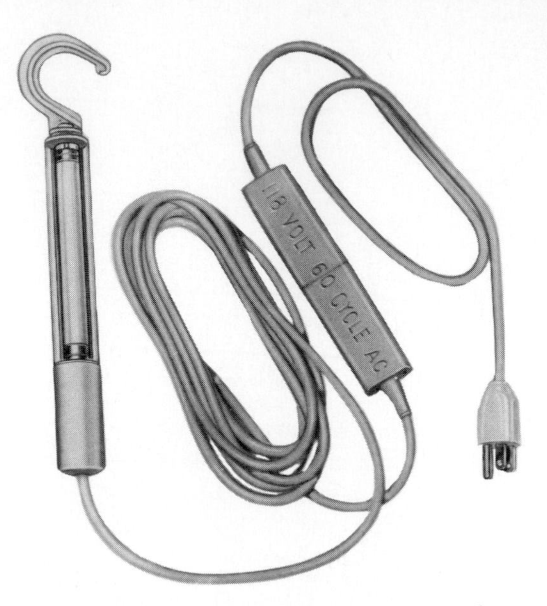

12–85. Industrial duty trouble or inspection lamp has a safety shield and is insulated. The ballast is rubber covered and molded into the line cord.

In locations where low voltage is not required, the standard type of enclosed "trouble" lamp can be used for inspection purposes. Fig. 12-77 shows a vapor-tight design that can be used in most locations. The plastic glove protects against breakage and damage from dropping.

String Lights. When more light is needed, a fluorescent twinlight can be used. These can also be connected with a base feed-thru unit to produce a string for more light where needed. Fig. 12-86.

Molded string lights have provision for effective grounding of the lamp guards as required by OSHA. The basic unit lampholder is molded to the cable with a threaded copper medium-base screw shell and ceramic disc insulator. The flexible rubber lip provides a weatherproof seal around the neck of the lamp. A brass bushing reinforces the support hole. Fig. 12-87.

12–86. For more light, the enclosed fluorescent twinlamp is available for use as a single or as part of a string.

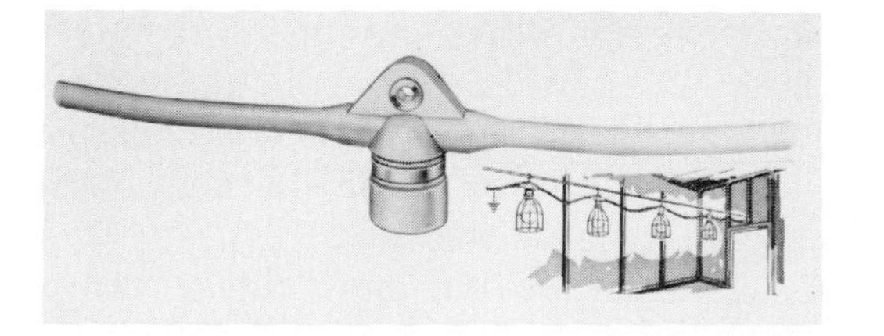

12–87. Watertight molded outlet for use in making a string of lights for outdoor use.

Then, if you need to make up a string of lamps to be used in the field, you can use the type shown in Fig. 12-88 with a guard attached to protect the bulb.

Construction Site Equipment Grounding. OSHA requires that the construction site have cord sets and receptacles that meet their requirements for personnel safety. To comply with requirements for construction sites, the employer has two alternatives. He or she must provide approved ground-fault circuit interrupters for personnel protection in all 120-volt, single-phase, 15- and 20-ampere receptacle outlets that are not a part of the permanent wiring of the building or structure, and which are in use by employees. Alternatively, he or she must establish and implement an Assured Equipment Grounding Conductor Program

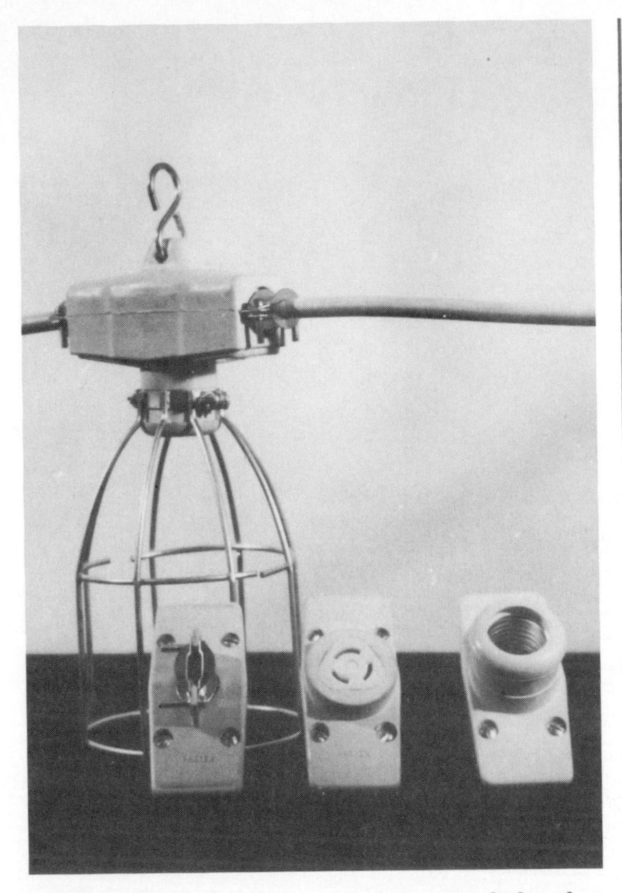

12–88. Another type of assembled string lights that have weatherproof sockets and a hook for suspension on a cable.

covering all cord sets, receptacles, and equipment connected by cord and plug that are available for use by employees.

Many employers have opted for the Assured Equipment Grounding Program, which entails a considerable amount of testing on a regular and frequent basis. It requires that all equipment grounding conductors must be tested for continuity, and each receptacle and attachment plug must be tested for correct attachment of the equipment grounding conductor. These tests must be performed before first use, before equipment is returned to service following repairs, before equipment is used after an incident that may have caused damage, and at intervals not to exceed three months.

All of these tests must be recorded. The records must be kept available at the jobsite for inspection by OSHA compliance officers. (Reference: OSHA Standard 1926.400(h).)

The ground continuity monitor provides a constantly monitored ground for extension cords. The ground continuity monitor is a simple device that is wired into an extension cord in place of a conventional connector. When the extension cord is plugged into the receptacle a light in the face of the GCM gives the following indications. (See Fig. 12-89 for the types available.)

1. If the equipment grounding conductor has continuity, the lamp glows.
2. If the ground wire of the connector is not connected, or the ground is not continuous, the lamp will not light.
3. The lamp also remains dark in the presence of the following conditions: reversed polarity, open hot, hot on neutral, hot unwired, open neutral, and hot and ground reversed.

By plugging the cord into the receptacle and leaving it in position to provide power for tools, lights, and other equipment, the glow of the lamp will afford an immediate check of the above conditions. If the light does not show, the employee knows he or she must call an electrician to check the cord.

Because this device maintains a constant check on the extension cords, the need for testing and record-keeping on these extension cords is eliminated.

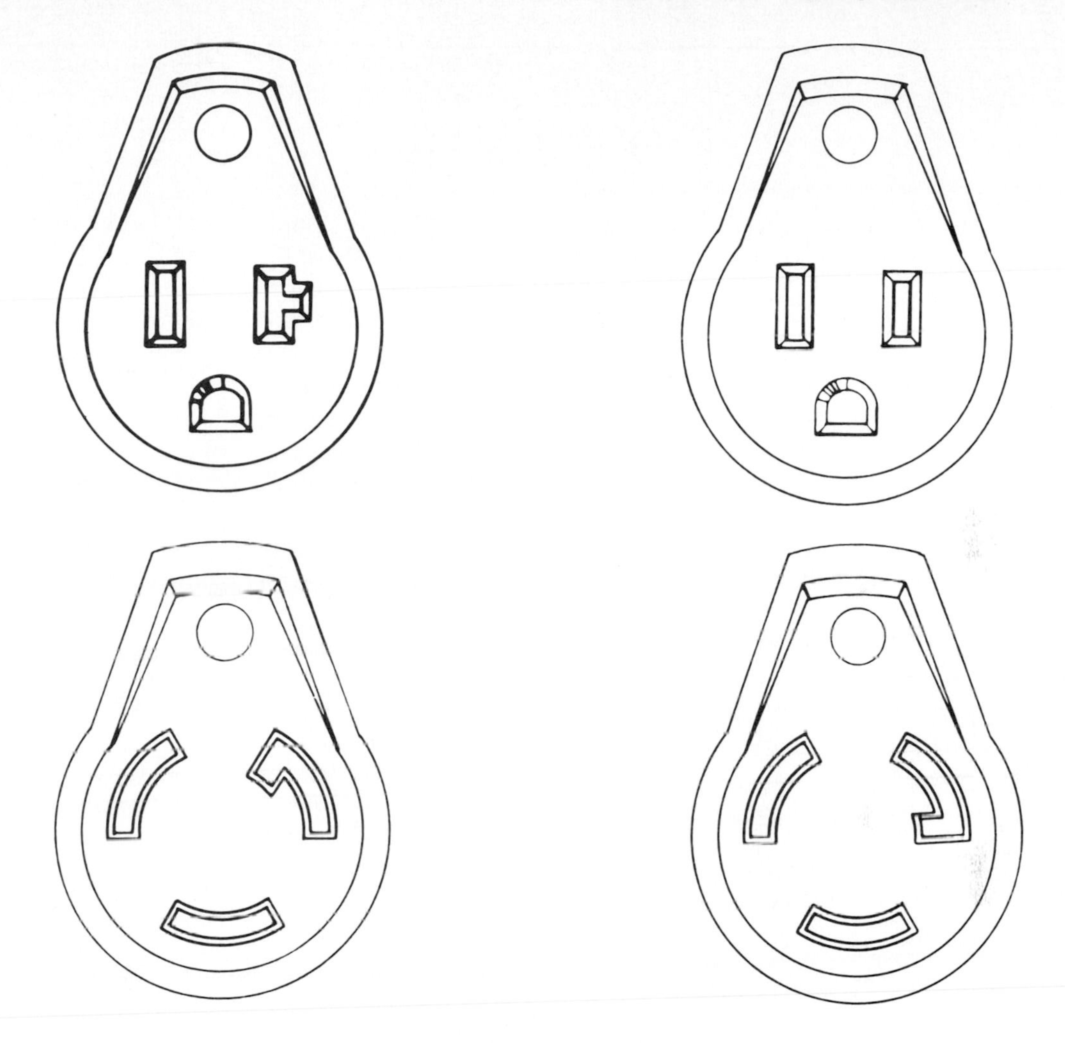

12–89. Ground continuity monitor extension cords light up at the end (top circle) to indicate that the correct wiring is present in the cord. It also indicates that power is available.

QUESTIONS

1. What are incandescent lamps?
2. How hot does an incandescent lamp get?
3. List eight types of incandescent lamps.
4. What is a fluorescent lamp's current-controlling device?
5. Describe the difference between a preheat and an instant-start fluorescent lamp.
6. What is meant by a trigger-start lamp?
7. How can ballast life be extended?
8. What is meant by duty cycle?
9. How is ballast hum reduced?
10. Can fluorescent lamps be used in outdoor installations?
11. How is a mercury-vapor lamp different from an incandescent lamp?
12. How is a self-ballasted mercury-vapor lamp different from the ballasted type?
13. What is black light? How is it obtained?
14. Why are some mercury-vapor lamps forced-air cooled?
15. Where is a pulsed-xenon lamp used? Why?
16. Where are neon lamps used? Why?
17. Name six methods of controlling light.
18. What is a lumen?
19. How are transformers used in lighting?
20. How does an electronic dimmer work?
21. How are fluorescent lamps dimmed?
22. What are the two primary conditions of the Assured Equipment Grounding Program?

KEY TERMS

ballast
electronic dimmer
fluorescent lamp
footcandle
incandescent lamp
lumen
mercury vapor lamps
neon lamp
photoelectric switch
solid-state dimming systems
spectral lamp
super actinic lamp
transformer

CHAPTER 13

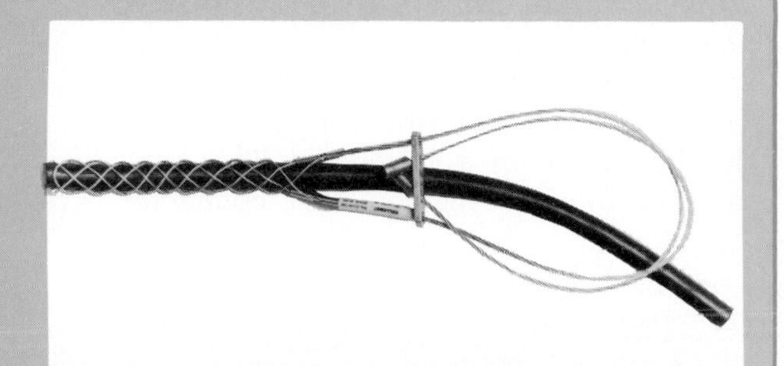

ENCLOSED AND EXPLOSION-PROOF WIRING SYSTEMS

Objectives

After studying this chapter, you will be able to:

- Identify the various types of connectors.
- Identify the various types of couplings.
- Identify the various types of clamps.
- Identify the various types of grips.
- Identify the commonly used hazardous substances.

CONDUIT

As you will recall from earlier chapters, conduit wiring, which has a metal covering, is not widely used for wiring houses. However, its use is common in industrial and commercial buildings. Here greater demands are placed on wiring. The metal guards the wire against crushing and shorting. It also helps prevent accidental damage to the insulation. It shields against arcing. Arcing can cause explosions in certain atmospheres.

In some locations, due to the nature of gases, it is necessary to totally enclose a switch or circuit breaker to prevent an explosion. This will be discussed later in the chapter.

BX CABLE

BX cable is the name applied to armored or metal-covered wiring. It is sometimes used in home applications. BX (the letters stand for metal-covered, flexible wire) sometimes meets the need in home applications for flexible wiring. It is used to connect an appliance, such as a garbage disposal unit, which vibrates or moves a great deal. For example, a garbage disposal unit is installed under a sink in the kitchen. The unit will vibrate when in normal use. This vibration calls for a flexible conduit or wire. The BX is attached to a junction box on the wall and through a special connector to the disposal unit.

BX, however, presents some rather unique problems. It must be cut properly. If it is not properly cut, the insulation covering the wires inside may be shorted to the protective metal covering. There is a fiber bushing designed to fit into the cut end of the BX. This bushing insulates the sharp end of the BX (where it was cut) and prevents shorts.

BX Cutter

A *BX cutter* has recently been designed to cut BX without the ragged edges produced by a hacksaw. Fig. 13-1. It is very efficient and can be carried on an electrician's belt with other tools.

Using the Cutter. Once the BX cable has been secured in the cutter, with the thumbscrew on the bottom, squeeze the handle. Squeezing the handle overcomes the force of the built-in spring. At the same time, it lowers the circular cutter against the BX casing. Do not apply too much pressure if you want a fast cut. At the point when the force required to turn the handle suddenly reduces, even when you increase the squeezing force, you have reached the pressure limit. Stop and release the tool. Release the hand pressure and turn the crank slightly. The tool should snap back to its original position for easy removal from the BX.

After the tool has been removed from the BX, hold the cable on each side of the slit and twist the casing counterclockwise until the casing separates. The insulated wires inside the BX casing are now exposed. If this separation does not happen easily, you have not cut the BX completely. You will therefore have to go back and insert the tool again to finish cutting the armored cable housing.

Once the BX has been cut, you must install it properly in a box for termination. There are a number of connectors for this type of job. Fig. 13-2.

FITTINGS

Connectors

■ *Box connectors* are designed to secure cable, metallic or nonmetallic, to junction enclosures. These connectors are not watertight and are used indoors. Fig. 13-2.

They are available in straight, 45°, and 90° angle styles. Some types have a screw-clamp method of gripping the cable; others utilize a squeezing principle. Figs. 13-3 & 13-4. Their purpose is to provide a clamping action without causing injury to the cable. The clamping action prevents connected wires from getting pulled apart by stress put on the cable from outside the enclosure. Fig. 13-5.

Box connectors used with metal-clad cable provide a continuous ground through the cable armor to the metal box. The cable is attached to the box with a connector. This assures a good ground connection. Box connectors are

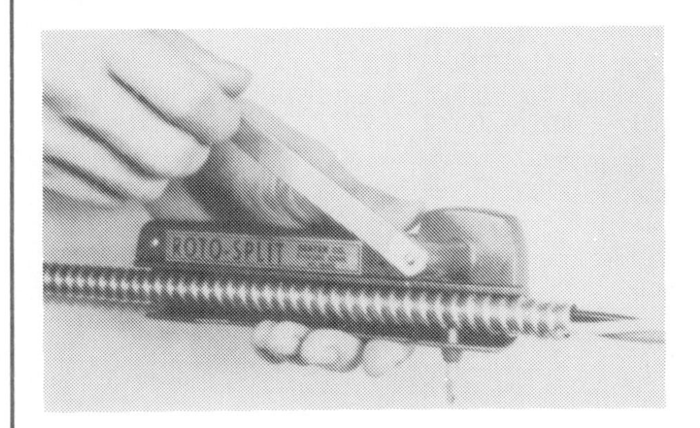

13–1. BX cable ripper.

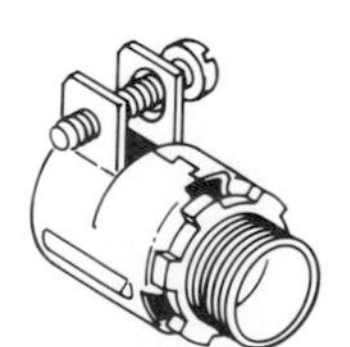

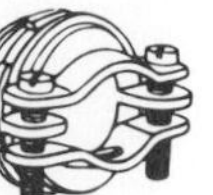

13–2. Box connector clamps.

easy to install. Insert them in a knockout or drilled hole and secure them in place with a locknut. Many outlet and switch boxes have built-in cable clamps. Fig. 13-6. When such clamps are provided, no box connector is needed.

Included in the box connector series is a straight duplex connector. This will take two metal-clad cables. All other connectors will take only one cable. The 45° and 90° types have removable hoods to allow easy insertion of the cable. Fig. 13-7.

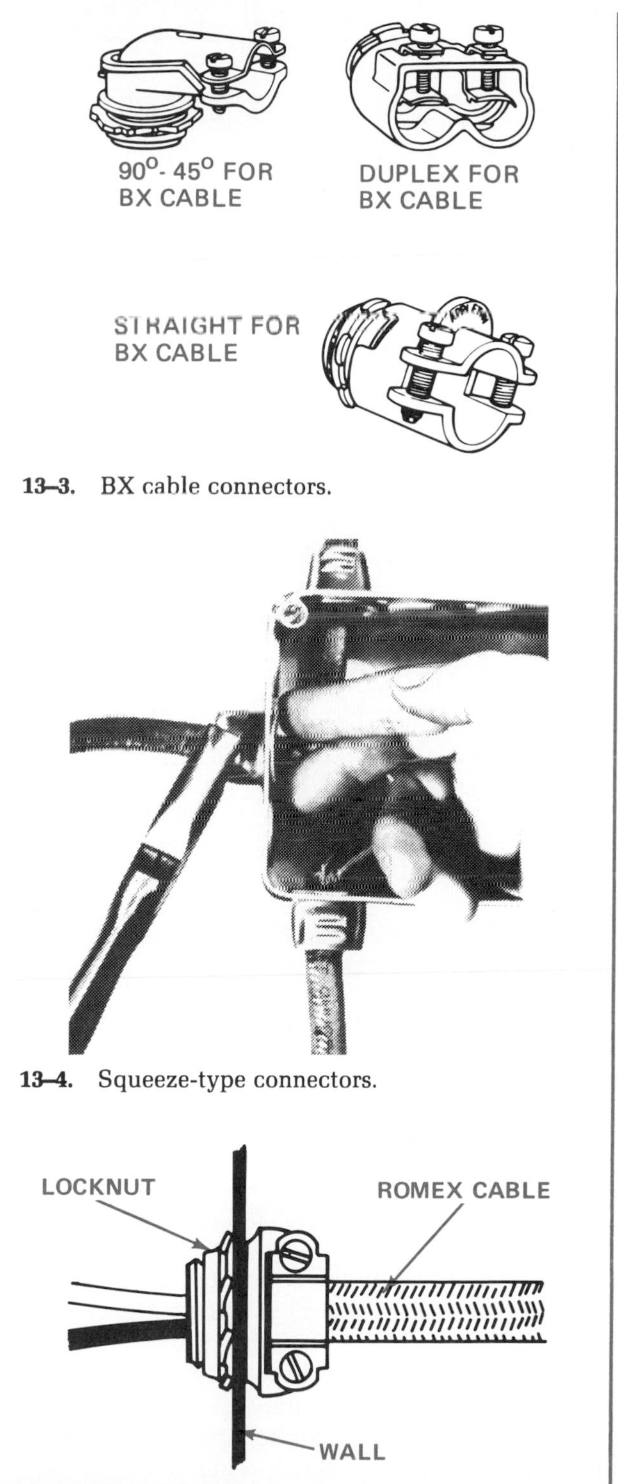

13–3. BX cable connectors.

13–4. Squeeze-type connectors.

13–5. Straight, 2-screw, clamp-type connectors.

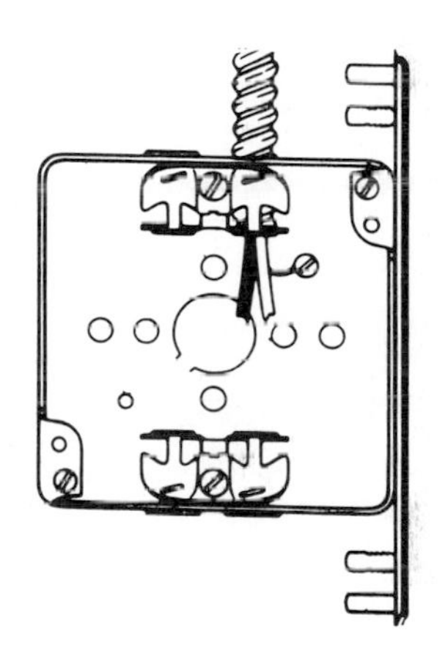

13–6. Outlet box with BX clamps built in.

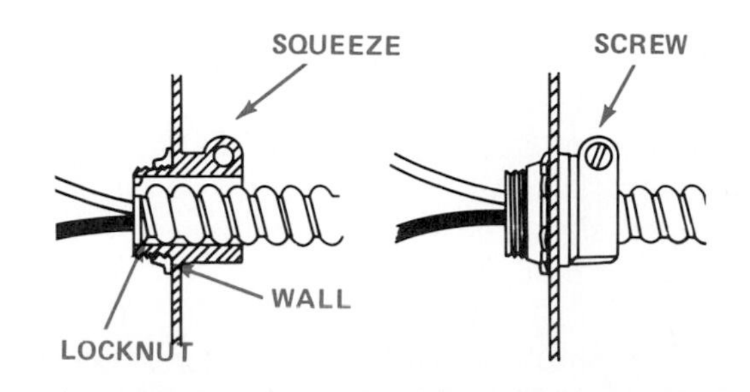

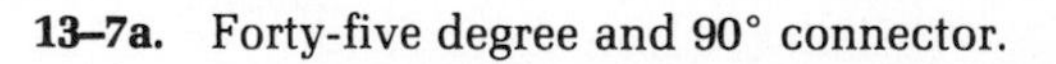

13–7a. Forty-five degree and 90° connector.

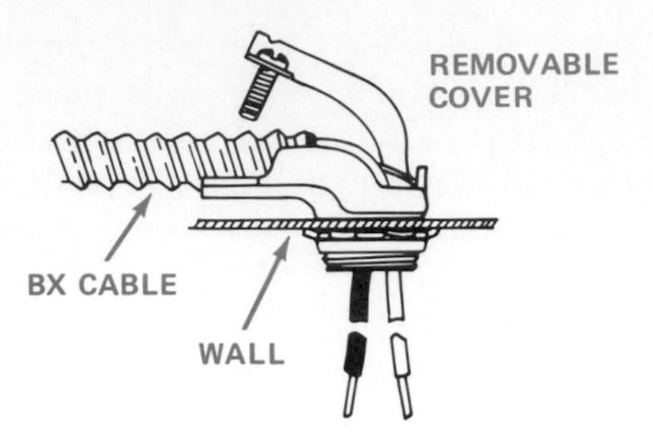

13–7b. Box connector with BX cable.

Box connectors are used in all types of construction where the use of metallic and nonmetallic cables and raceways is permitted. They are also applied to machines which have a light spray of oil or other liquids. Flexible cables are practical in most tight-area wiring. Fig. 13-8.

■ *Ninety-degree knockout box connectors* are designed for use in connecting conduit to outlet boxes. They may be used on other similar devices. Two types are available—threaded and no-thread. Fig. 13-9.

■ *Cable connectors (watertight)* are used for entrance of flexible cords and cables to electrical equipment. Fig. 13-10. They provide clamping action and reduce cable-insulation abrasion and wear. Watertights seal out vapor and dirt. They are both watertight and oiltight. Thus, they help prevent deterioration of valuable enclosed mechanisms. Their use can also prevent wire terminal strains. Strains result in dangerous broken connections. Use of watertights can prevent short circuits due to abrasion at the cable connector.

These connectors are made of lightweight aluminum. They consist of four parts: *body, grommet, ring,* and *cap*. An oil-resistant neoprene grommet provides a tight seal and firm grip on the flexible cord, or cable. The ring under the clamping cap allows the grommet to be compressed tightly around the flexible cord, or cable, without friction on the cap.

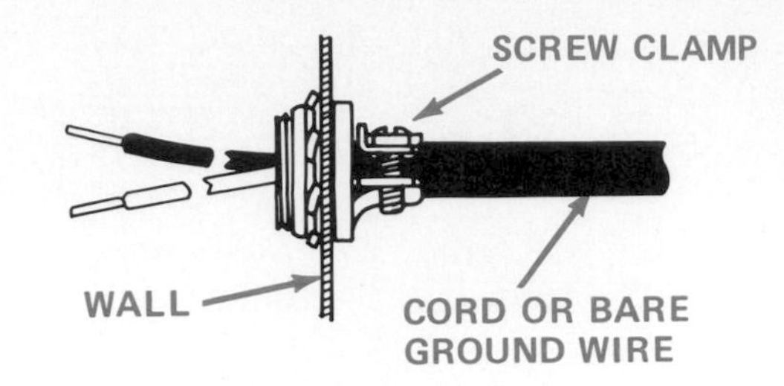

13–8. Connector holding a cord in a box.

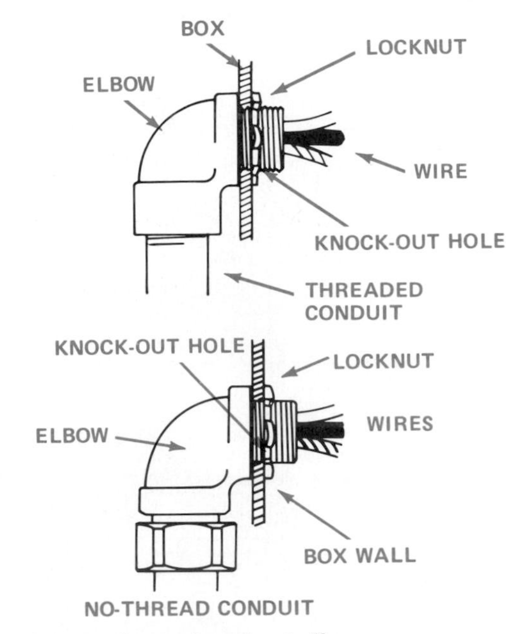

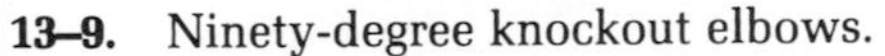

13–9. Ninety-degree knockout elbows.

Watertight connectors are available in straight and 90° angles. Hub sizes are $\frac{3}{8}''$, $\frac{1}{2}''$, $\frac{3}{4}''$, 1″ and $1\frac{1}{4}''$. A hub is an enlargement on the end of a piece of pipe or conduit. This means another pipe or conduit of the same size may be inserted and sealed with a satisfactory joint. In this case a hub means that the end of the connector is such as to accept a piece of conduit of the size specified on the connector. Installation of watertight connectors is easy. Fig. 13-11. These connectors have standard, tapered pipe threads for installation directly into a threaded cast fitting. A locknut may be added if the cable is to be used on an enclosure that is not threaded. This may occur when used on an outlet box, or a steel junction box.

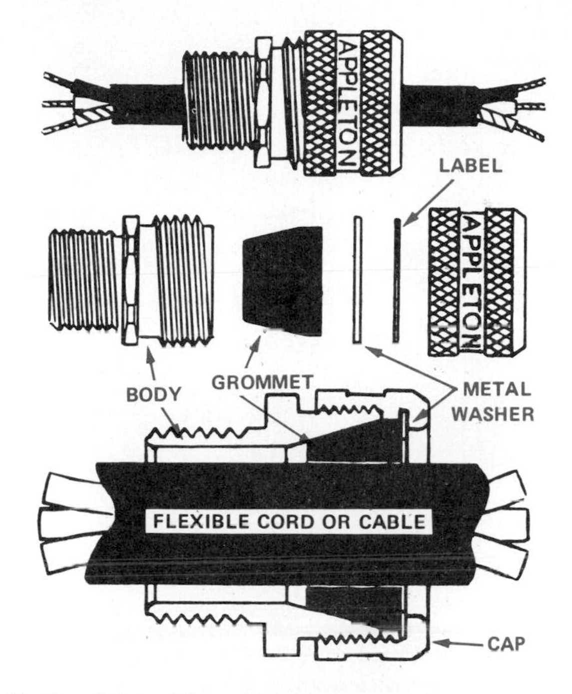

13–10. Watertight cable connectors.

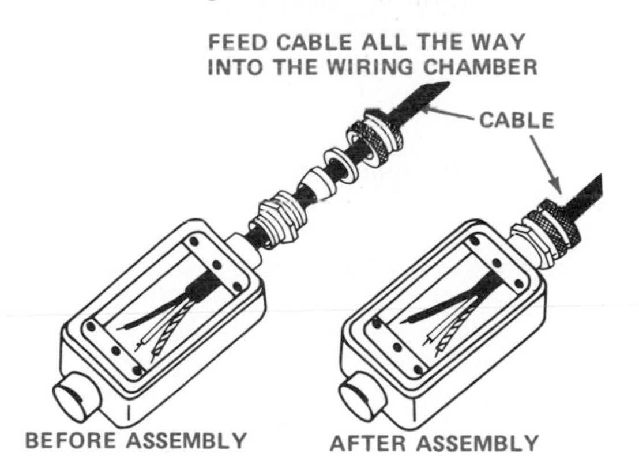

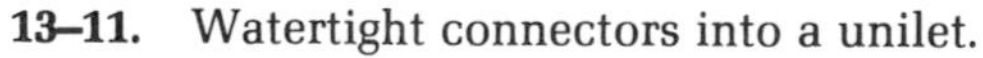

13–11. Watertight connectors into a unilet.

■ Rigid conduit hubs consist of two parts: the body and the hex-head wedge adapter. The threaded shank of the hex-head wedge adapter is placed through the knockout from the inside of the box. The hub body is screwed onto the wedge adapter until the fitting is reasonably tight with the box. Conduit is then installed to the hub and the fitting tightened securely. Fig. 13-12. The hex-head design permits tightening from either outside or inside the enclosure. Fig. 13-13.

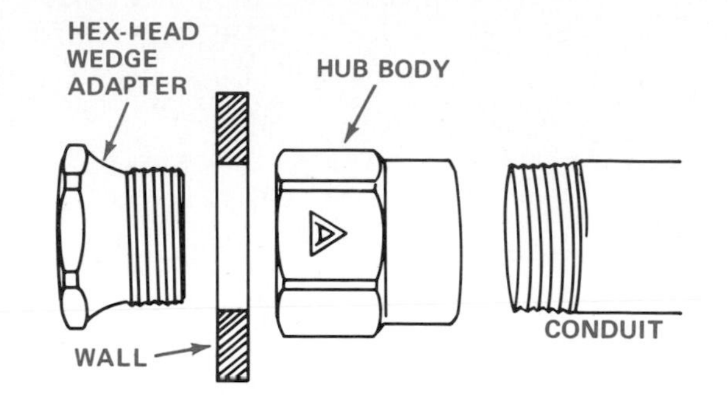

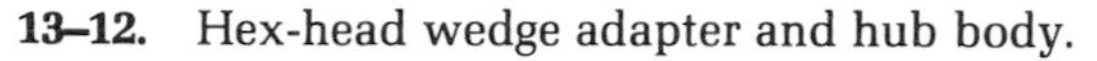

13–12. Hex-head wedge adapter and hub body.

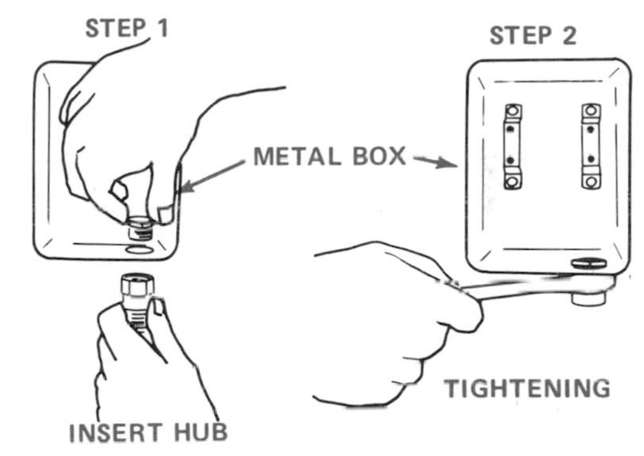

13–13. Attaching a hub to the box and adapter.

Often, when making an installation using threaded rigid conduit, junction boxes and device housings are used which do not provide suitable means for fastening the conduit to the enclosure. Knockouts, found in most enclosures, provide a location where rigid conduit hubs can be fastened. If there are no knockouts in the enclosure, a hole must be drilled. Fig. 13-14.

The self-locking, hex-head wedge adapter and the hub body exert a continuous, uniform 360-degree pressure on the inside and outside surface of the box wall. This eliminates the need for locknuts. All hubs have a built-in, recessed neoprene gasket, and a flame-resistant insulated throat. The latter eliminates the need for an end bushing. It protects the wire insulation and cable sheath from damage due to vibration. The insulated throat leaves more wiring room within the enclosure.

Connectors are made with a zinc finish over steel. Larger sizes are of malleable iron with a cadmium finish. Fig. 13-15.

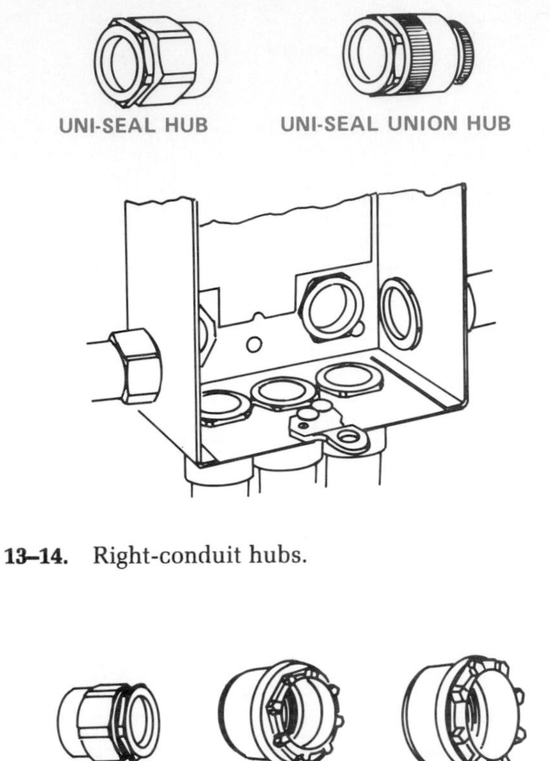

13–14. Right-conduit hubs.

13–15. Rigid-conduit connectors.

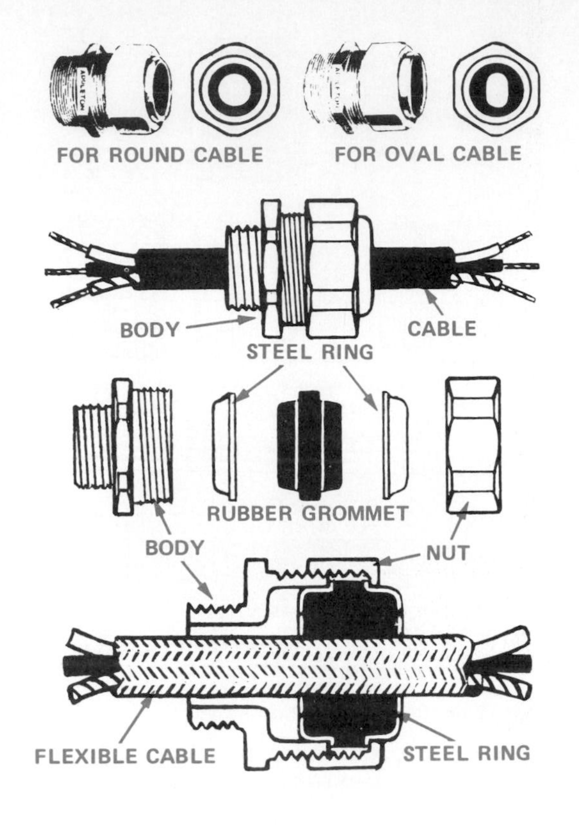

13–16. Watertight connectors and cable clamps.

■ *Watertight connectors and cable clamps* for service entrance cable are used for bringing in the cable and terminating into distribution panels, meter socket cabinets, or similar boxes. Fig. 13-16.

Connectors for this work use a soft-rubber grommet. The grommet is compressed when tightened. The compression makes a watertight seal around the cable. They can be used in knockouts, using a locknut, or threaded directly into a hub.

To install, disassemble the connector and slide the nut over the cable. Then, slide the rubber grommet onto the cable. Once this is done, slide the cable into the connector body. Place the rubber grommet in the connector and tighten securely with the nut.

Use a *sill plate* at the point where the service entrance cable enters the building. Fig. 13-17. The sill plate prevents water from following the cable into the building. Usually a soft, puttylike cement compound is used with the sill plate to seal any opening that might exist. Once inside, a number of liquid-tight connectors are available, if needed. Figs. 13-18 & 13-19.

■ *STCN connectors* are made for type CN nonmetallic sealtite conduit. STCN means *seal-tight connector, nonmetallic*. CN stands for *conduit, nonmetallic*. STCN connectors are made for use with Anaconda® Type CN sealtite conduit. Fig. 13-20. (*Sealtite* is the usual spelling for *seal-tight* conduit.) This type (CN) conduit differs from standard metallic sealtite conduit in that it does not utilize a flexible me-

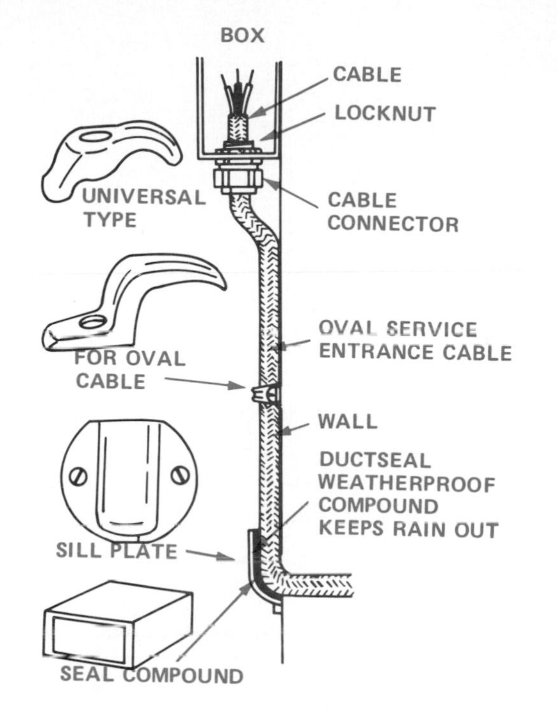

13–17. Cable clamps and waterproofing and entrance.

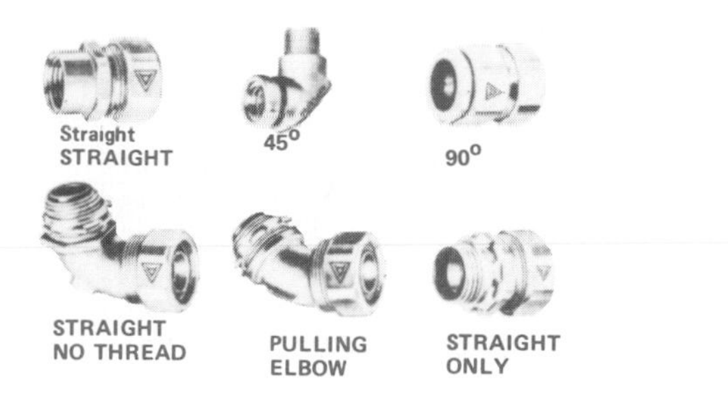

13–18. Liquid-tight conduit connectors.

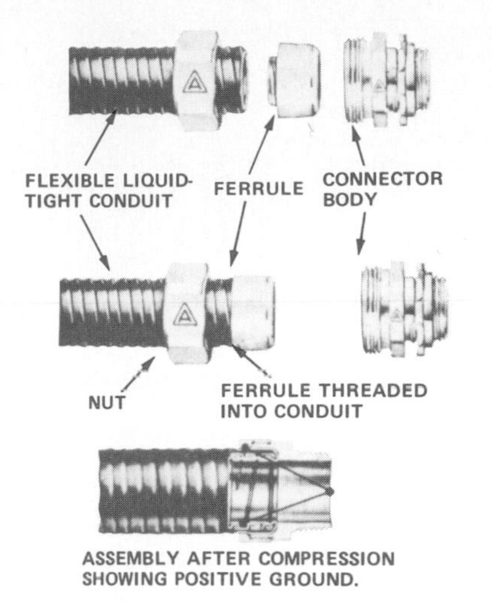

13–19. Flexible conduit and connectors.

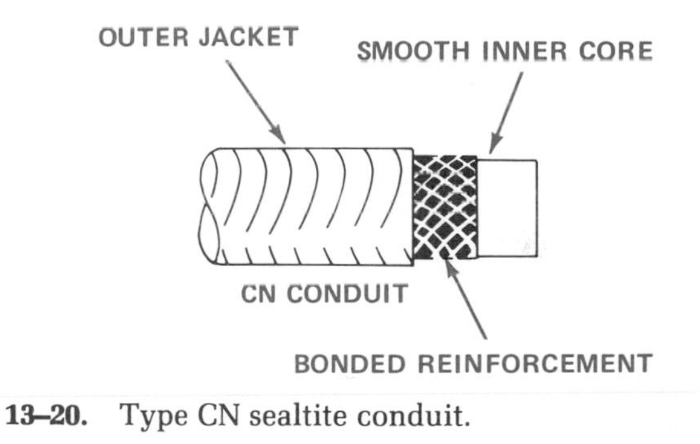

13–20. Type CN sealtite conduit.

tallic interior. Instead, CN conduit has a smooth plastic inner core, covered with a bonded reinforcing nylon cord. It also has a rugged plastic outer jacket for covering. The absence of any metallic elements in the conduit eliminates the problem of attack by corrosive atmospheres.

The STCN connector also has the advantage of making the conduit extra flexible and of providing an extremely smooth interior for protection of conductors. It is easily identified by its bright orange color. STCN connectors form a liquid-tight and vaportight connection of CN conduit to a circuit breaker, switch, junction box, and similar enclosures, as well as to boxes and threaded hubs.

The STCN connector has four simple, easy-to-install parts: a nylon ferrule, a steel body, a compression nut, and a tiger-grip locknut. Fig. 13-21. The steel body of the connector has a built-in, protective plastic-insulated throat. It also has a neoprene "O" ring.

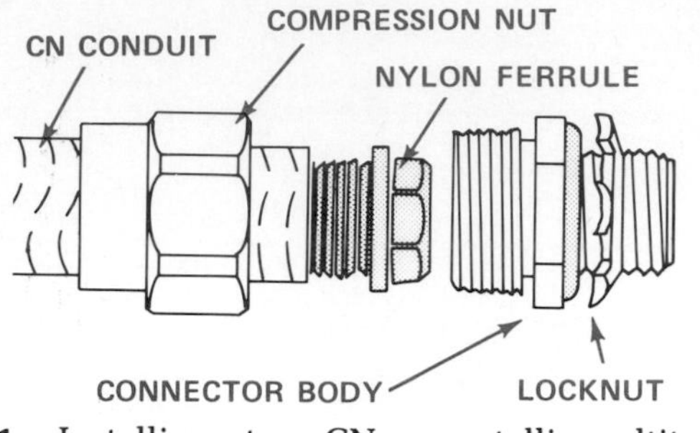

13-21. Installing a type CN nonmetallic sealtite conduit connector.

To install the connector:

1. Cut the CN connector to the desired length, being sure to make a very straight cut.
2. Feed the compression nut onto the conduit.
3. Press the threaded end of the nylon ferrule as far as possible into the end of the conduit with the palm of the hand. Use the connector body as a wrench to tighten the ferrule flange snugly to the end of the conduit. (The interlocking hexagonal-nut type design of the ferrule and body make this possible.)
4. Bring the compression nut up and tighten to a positive stop on the shoulder of the connector body.
5. Place the end of the body through the knockout and secure with the locknut provided. Fig. 13-22a. Check Fig. 13-22b for installation of conduit into a threaded unilet hub.

■ *Thinwall-conduit compression connectors and couplings* are used with a special EMT (thinwall) conduit. The EMT stands for *electrical metallic tubing.* Thinwall conduit connectors are employed in fastening (connecting) EMT to outlet boxes, switch boxes, panel, and other metal enclosures. Fig. 13-23. Thinwall conduit couplings are electrical fittings used to attach (couple)the length of one conduit to another. They are used to extend the overall length of a piece of conduit. Conduit usually comes in 10-foot lengths.

Throat openings of the thinwall connectors and couplings are chamfered to eliminate burrs. That means the ends have been slightly

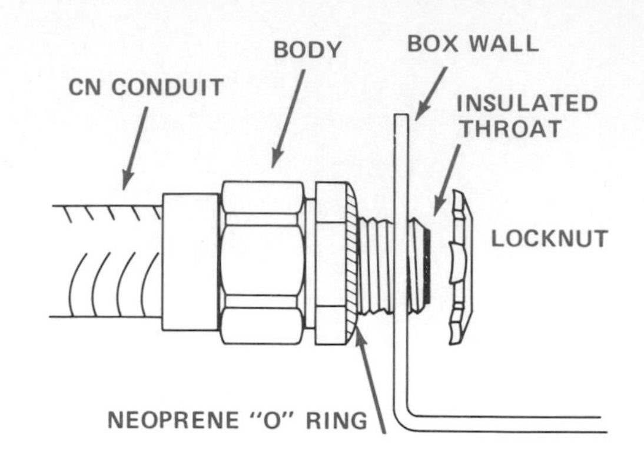

13-22a. Attaching CN conduit to a wall box.

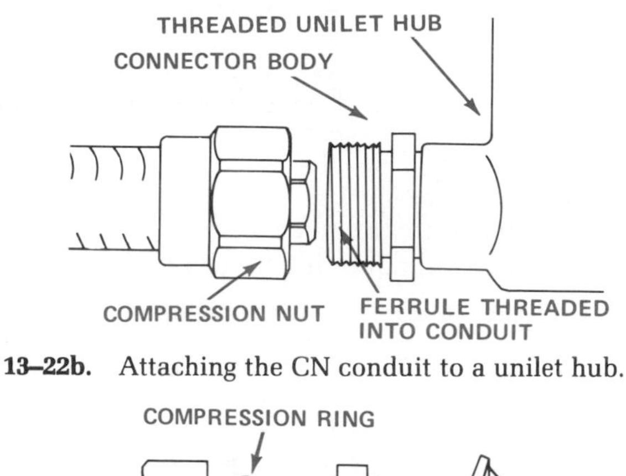

13-22b. Attaching the CN conduit to a unilet hub.

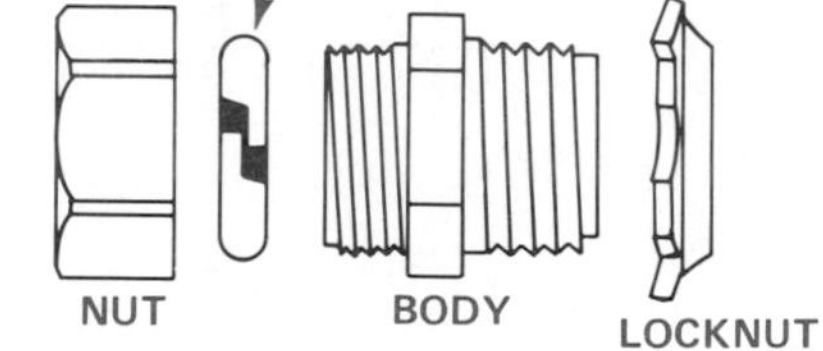

13-23. Compression connectors.

reamed out to eliminate small metal bits which can damage wire being pulled through the conduit. All sizes of these connectors and couplings have hexagonal nuts, and bodies which must be held securely. A wrench is used to hold the nuts when tightening. The body of the fitting has a conduit stop which allows the EMT to enter evenly for uniform strength.

The ends of the connector, or coupling, house a compression ring. When the nut is tightened, the inward motion of the nut forces the open compression ring into a closed position around the conduit. This action locks the fitting onto the conduit.

■ *Two-piece thinwall-conduit connectors* are used where thinwall conduit is to be connected to an outlet box, switch box, panel, or other metal enclosure. It has the advantage of increased wiring room inside the enclosure. Installation is fast with a single wrench. This means lower cost. Fig. 13-24.

This connector consists of a knurled, chamfered split-steel body with a hex nut. Fig. 13-25. To install, insert the body through the outlet box to the shoulder stop rim on the connector body. The conduit is then inserted through the nut into the body. Fig. 13-26. Tighten with a wrench placed over the hex nut. This draws the knurled chamfer of the body against the knockout hole of the box. At the same time it compresses around the conduit. Thus, slippage is prevented and there is no need for a locknut.

Two-piece, thinwall-conduit connectors may be used to install conduit between two stationary enclosures. If used here, simply cut the conduit to correct length. Slip the hex nuts on both ends. Insert the conduit through the knockout in one box far enough so that it may be backed up through the knockout in the second box. (Backing up is shown in Fig. 13-27.) Place the connector bodies over the ends of the conduit from the inside of the box. Tighten with a single wrench. These connectors are concrete-tight and UL approved.

■ *Indenter-type thinwall-conduit connectors* and *couplings* are used for attachment of thinwall conduit to metal enclosures. They are designed for joining two sections of thinwall. They provide a permanent, rigid connection or coupling. Install them with an indenter tool. Figs. 13-28 & 13-29. All indenter fittings feature chamfered edges to prevent damage to cable sheath and wire insulation. Some have insulated throats.

Install by inserting the EMT into the fitting until it rests against the conduit stop. This allows the conduit to enter only to the halfway point. Here it is held in check. This provides the strongest possible connection. With the

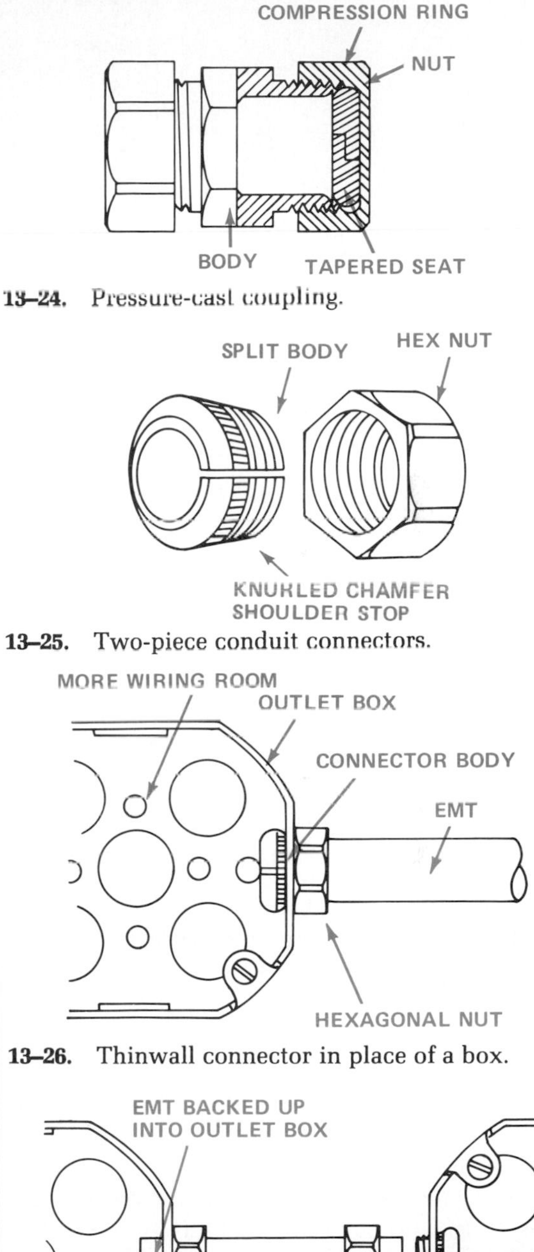

13-24. Pressure-cast coupling.

13-25. Two-piece conduit connectors.

13-26. Thinwall connector in place of a box.

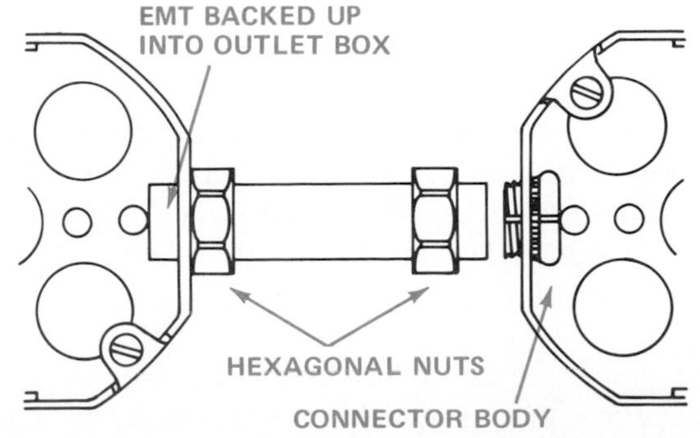

13-27. Two-piece connectors used between two boxes.

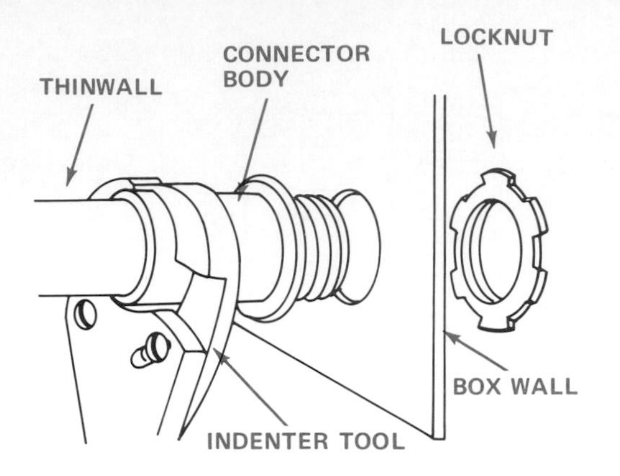

13–28. Indenter coupling.

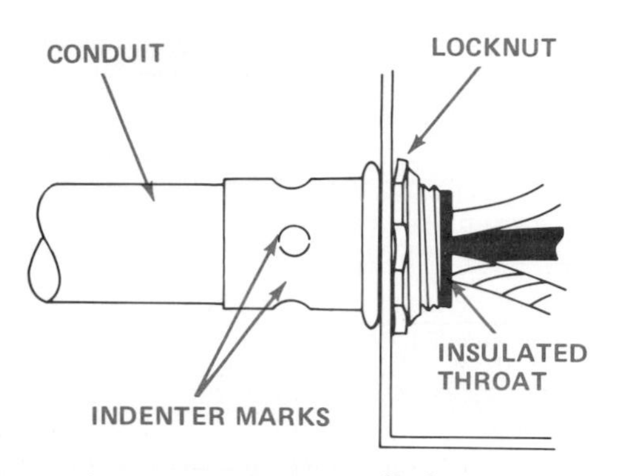

13–29. Indenter connector installed.

EMT inserted, the indenter-tool jaws are placed around the fitting and squeezed tightly. Fig. 13-30. The prongs make deep indentations in the fitting and conduit. The tool is then rotated 90° and another set of indentations is made.

Rotation and indentation are continued until a total of four have been made. The process is repeated for completion of coupling installation. Fig. 13-31. Indenter-type connectors have a tiger-grip locknut. The locknut is tightened on the inside of the enclosure for a slip-proof bond. Fig. 13-29. They are, of course, plated with cadmium to prevent rust or corrosion.

■ *Setscrew thinwall-conduit connectors and couplings* are used where EMT is to be fastened (connected) to a switch, outlet, or panel box. Setscrew couplings are applied where one

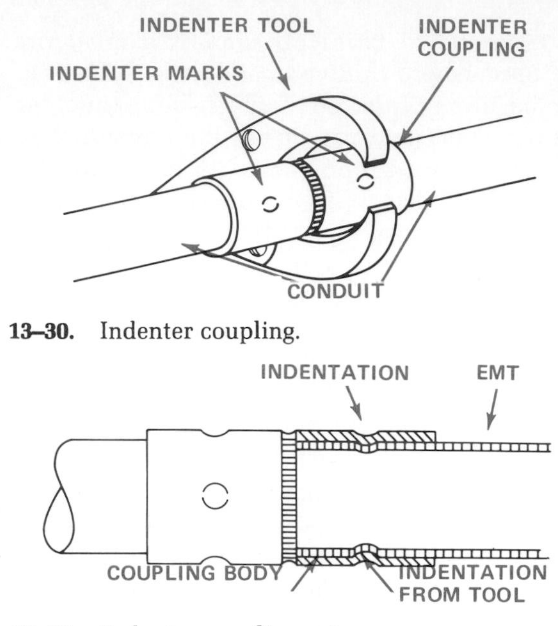

13–30. Indenter coupling.

13–31. Indenter coupling cutaway.

run of thinwall conduit is to be attached (coupled) to another. These connectors and couplings are designed for use on straight runs of conduit; that is, where two ends of the thinwall meet and line up properly. They may also be used where the two ends of the conduit line up with the knockouts of a box. Figs. 13-32 & 13-33 show a setscrew connector.

To install, simply loosen the setscrews on the fitting. Then insert the thinwall conduit until it falls against the built-in stop collar inside the fitting. This stop collar assures even holding pressure for EMT. Also, it provides a smooth surface over which wires may be pulled easily without damage to the insulation. All edges are chamfered for further protection.

After inserting the conduit, tighten the setscrews against the outer wall of the conduit. Deep-slotted, staked setscrews thread firmly into the embossed surface of the body of the fitting. Repeat the procedure to install the conduit in the other side of the coupling. To complete the connector installation, insert the conduit and the connector in the box knockout. Tighten the locknut to the interior of the box to prevent slippage. Fig. 13-34.

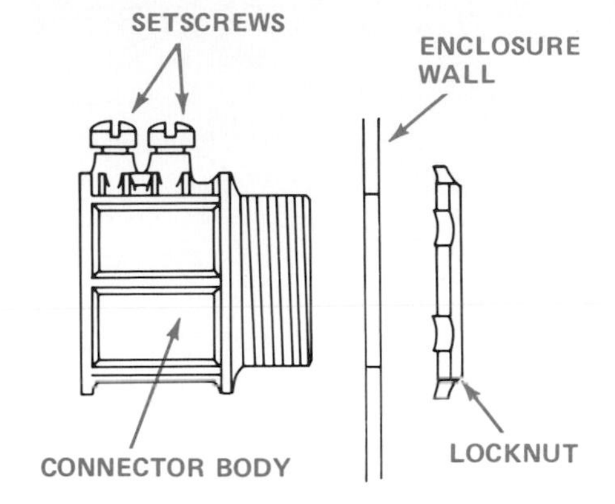

13–32. Setscrew connector.

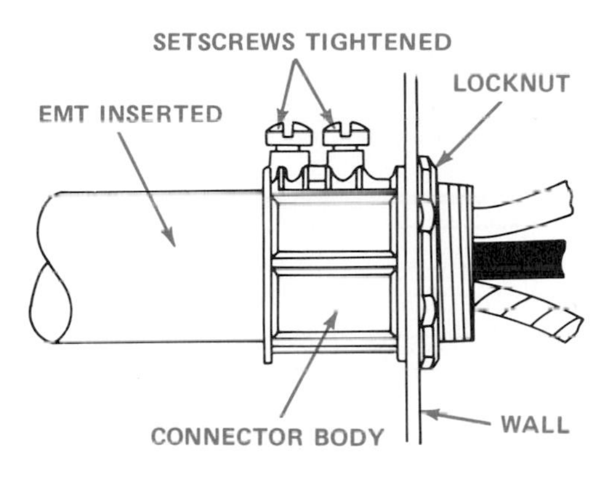

13–33. Installed setscrew connector.

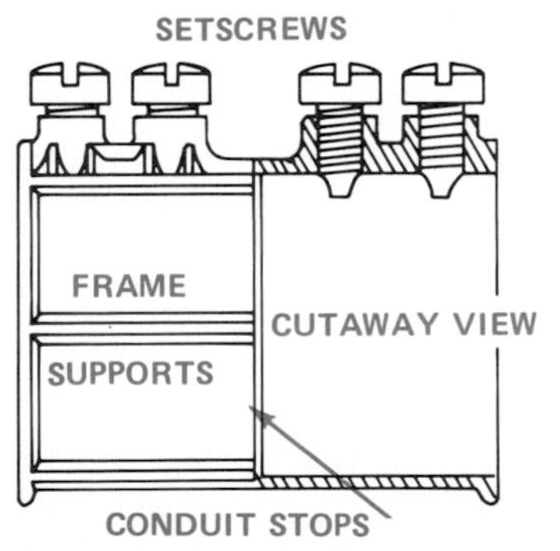

13–34. Cutaway view of a coupling, setscrew type.

■ *Rigid conduit no-thread connectors* are available in ½″ to 4″ sizes. The small sizes are steel; the larger sizes are malleable iron. This type of connector consists of a locknut, a body, compression ring, and nut. When the nut is tightened, it forces the ring into the tapered throat of the body, compressing the ring around the unthreaded conduit. Remember, this is unthreaded rigid conduit. It is different from EMT which is also unthreaded. Most rigid conduit is threaded, however. A tiger-grip locknut holds the connector in place to complete the installation. Fig. 13-35. Use rigid conduit, threaded, or no-thread couplings in installations where lengths, or runs, of conduit have to be joined (coupled) to attain the desired length.

■ *Pressure-cast spacing connectors* are used to connect two or more switches, panels, or outlet boxes. To do this, remove the knockouts on the side of the boxes to be ganged. Insert the spacing connector between them and tighten the two tiger-grip locknuts (furnished) onto the inside of the boxes. These fittings are made in the ½″ size. Fig. 13-36.

Couplings

■ *Expansion couplings* are installed at intervals in conduit raceways. These are installed to allow for the expansion of the metal conduit when it is used in buildings of considerable length. The exact length is determined by experience and observation. The couplings help prevent current interruptions that occur due to expansion and contraction.

These expansion couplings have the equivalent of weathertight joints for use with heavywall conduit. In addition, they contain a metallic packing ring that maintains the entire conduit system as a continuous electrical conductor.

To comply with safety standards prescribed by UL, an approved bonding jumper is available for use with expansion couplings. Fig. 13-37. When installing conduit with expansion couplings, it is imperative that the conduit on the right side be in perfect alignment with the conduit on the left side. This will allow freedom of movement of the conduit. Fig. 13-38.

To install couplings, remove the stationary end cap nut and thread it into the right-hand side of the raceway. Then remove the movable

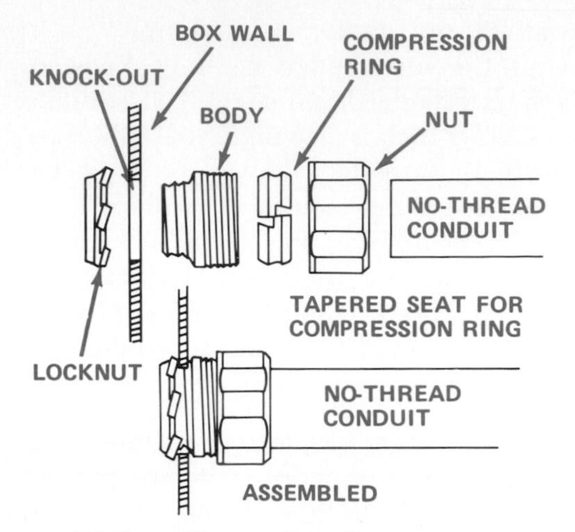

13–35. Rigid conduit no-thread connectors.

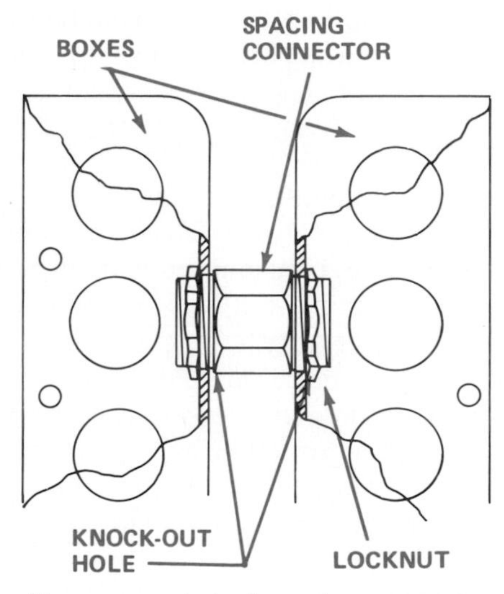

13–36. Pressure-cast steel spacing connectors.

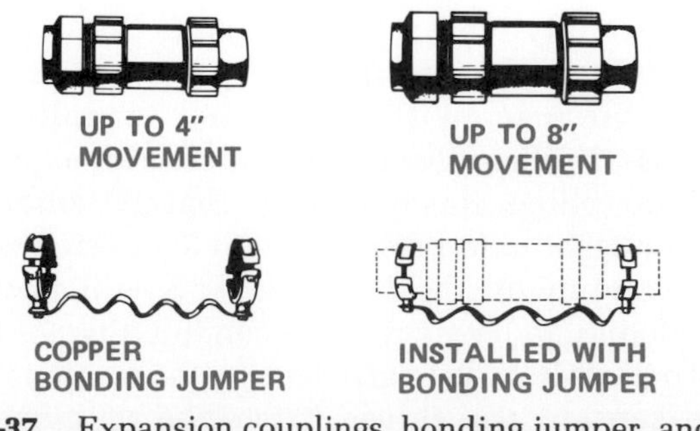

13–37. Expansion couplings, bonding jumper, and installed jumper.

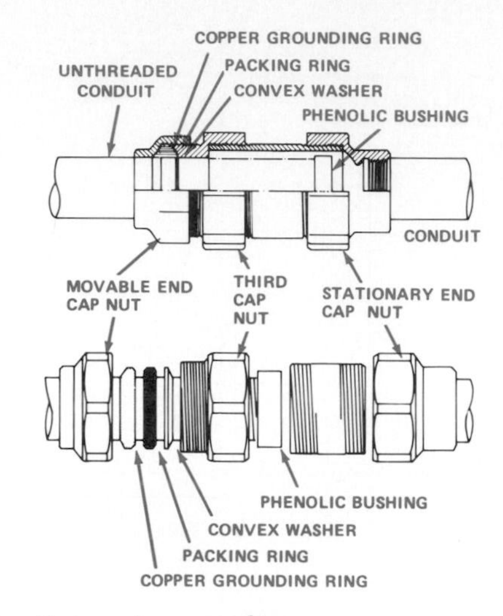

13–38. Expansion coupling.

end cap nut from the coupling. The end cap nut has a metallic copper grounding ring and convex washer. Place it on the left side of the raceway. Fig. 13-38. Now, remove the center cap nut and screw it onto the movable cap nut on the left. In the body of the coupling is a phenolic bushing. Remove it and mount it on the end of the left-hand side.

The depth to which the left-hand conduit run is inserted into the expansion chamber before final tightening of all parts depends on the expected expansion and contraction movement to which the conduit system will be subjected. Care must be taken that the movable cap is tightened properly. This provides a good grounding bond.

Expansion couplings provide for the natural movement which takes place on long runs of conduit. Buildings of greater length are the only ones concerned with such a problem. When in doubt, install an expansion coupling to make sure the movement of the conduit is taken into consideration. Fig. 13-39.

■ *Couplings for threaded conduit* are called Erickson couplings. They consist of three pieces. The three pieces are: a nut, a sleeve, and a threaded adapter. The body nut slides

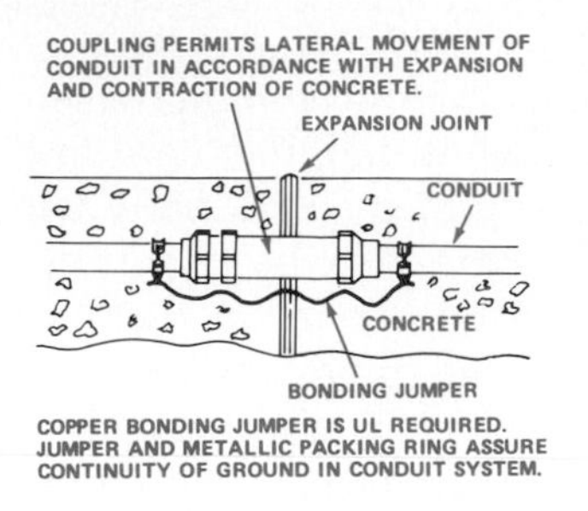

13-39. Copper-bonding jumper installed and embedded in concrete.

onto one piece of conduit. The sleeve threads on. The threaded adapter is turned onto the other piece of conduit. The body nut is then threaded onto the adapter retaining sleeve, completing the installation. Fig. 13-40. Sizes range from $\frac{1}{2}''$ to 5″ for threaded conduit couplings.

■ *No-thread couplings* have five pieces. They have two nuts, two compression rings, and a body. Tightening of the two nuts compresses the rings around the conduit. This makes a secure installation. Size range is from $\frac{1}{2}''$ to 4″. Fig. 13-41.

■ *Malleable-iron combination couplings* are used to connect flexible-steel conduit to rigid conduit. The use of malleable-iron combination couplings in installations gives durability and strength.

■ *The no-thread coupling* contains a compression ring in an easily tightened hex nut. This compression ring locks like a vise around unthreaded conduit and holds it intact. The BX, or similar armored cable, is slipped through the other end. It is fastened with a screw which clamps that end of the coupling together.

■ *The threaded malleable-iron coupling* is used on conduit which is attached to the threaded end of the coupling. Armored cable is installed the same way. Fig. 13-42.

■ *Combination thinwall to threaded rigid or flexible-steel conduit couplings* are designed for joining rigid conduit or flexible-steel conduit to thinwall conduit. Fig. 13-43. The combination couplings have one internally threaded end for threaded rigid conduit. The other end has a compression ring under a nut for thinwall conduit.

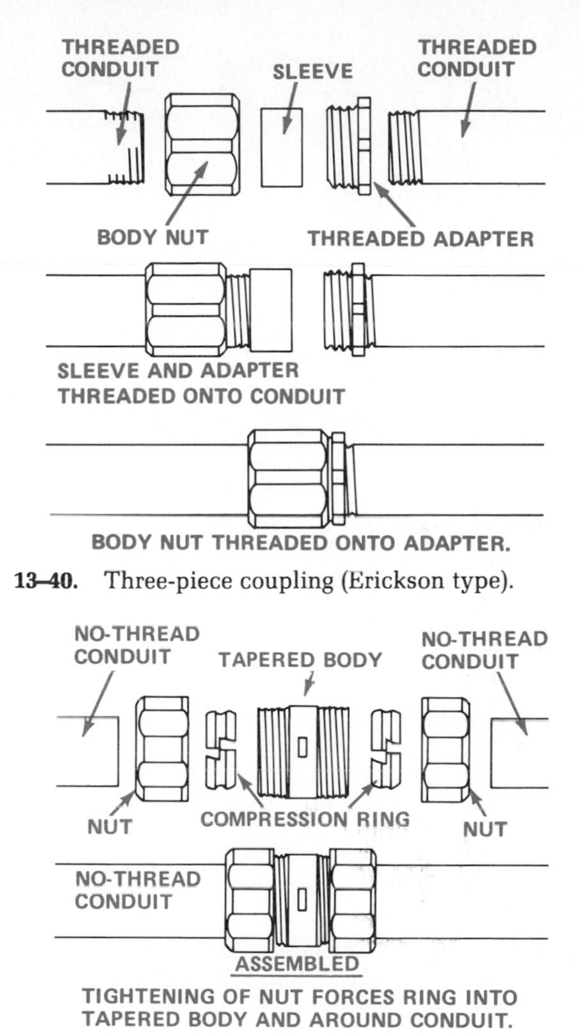

13-40. Three-piece coupling (Erickson type).

13-41. No-thread couplings.

To install the fitting, simply turn the threaded end onto the threaded rigid conduit and tighten the compression nut with a wrench. Then slide the EMT into the nonthreaded end and tighten the compression nut. As you tighten the compression nut, the open compression ring is forced into a closed position around the thinwall conduit. This forms a rigid coupling.

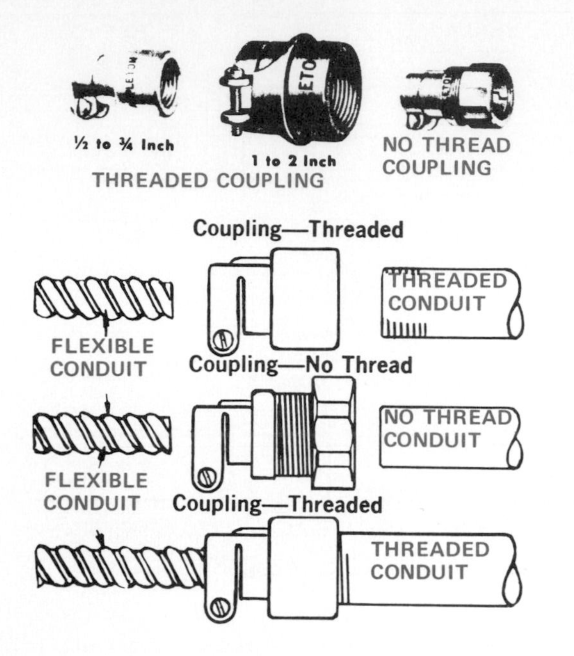

13–42. Malleable-iron combination couplings.

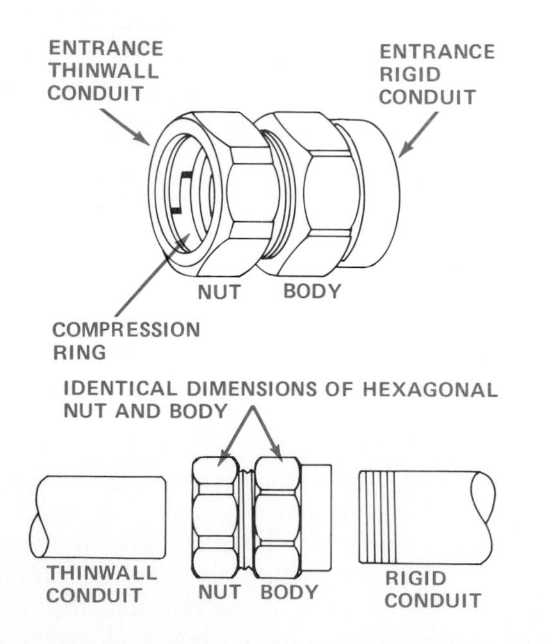

13–43. Combination thinwall to threaded rigid or flexible-steel conduit couplings.

The combination flexible-steel conduit to EMT coupling is manufactured in $\frac{1}{2}''$, $\frac{3}{4}''$, and 1″ sizes. One end has a split body with a tightening screw for flexible-steel conduit to a thinwall fitting. Place the end with the split body over the flexible conduit. Tighten the screw. Fig. 13-44. Then insert the thinwall conduit in the compression ring end and tighten the nut for a rigid bond.

A conduit stop inside the body of the fitting assures an even strength of bond. All edges are chamfered to prevent damage to cable sheath and wire insulation. Heavy-gage steel fittings are precision machined. They have a cadmium finish to resist rust and corrosion.

Clamps

■ *Beam clamps* are an accessory used to attach standard conduit hangers to I-beams and other structural members. Five sizes of beam clamps are available.

Each beam clamp has a tapped hole in the back and bottom surface. The sizes of these holes vary in the different sized clamps. The two tapped holes facilitate vertical and horizontal mounting by allowing hangers or

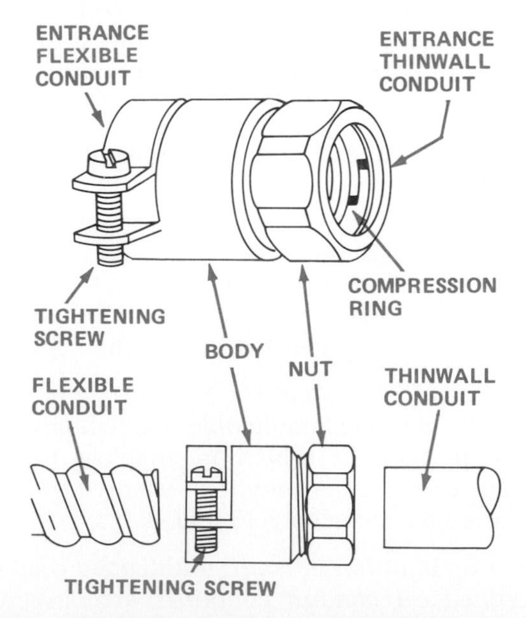

13–44. Thinwall to flexible-steel conduit fitting.

threaded rod to be mounted as desired. Take a close look at the clamp to see the function of the tapped holes more clearly. The square-head, angle-mounted setscrew gives maximum locking on any structural shape.

A beam clamp may be used in a single conduit run. One clamp is used for each conduit hanger. In some cases an obstacle makes it necessary to suspend a single run of conduit several inches from the clamp. A long mounting screw can be attached to the clamp under these conditions. Fig. 13-45. One beam clamp will support several conduit hangers. It will hold two, three, or more conduit lines when necessary. Groups of conduit can be suspended by using long mounting screws and two pieces of angle iron as shown.

■ *Sta-tite conduit hangers* are used to suspend rigid conduit or EMT from structural members. These structural members may be I-beams, angle irons, or braces. Fig. 13-46. The Sta-tite hanger has a square-head setscrew. The setscrew locks onto the beam. The square head allows for easy tightening. These hangers may be mounted either horizontally or vertically on the structural members. They provide safe, strong support for multiple-conduit raceways.

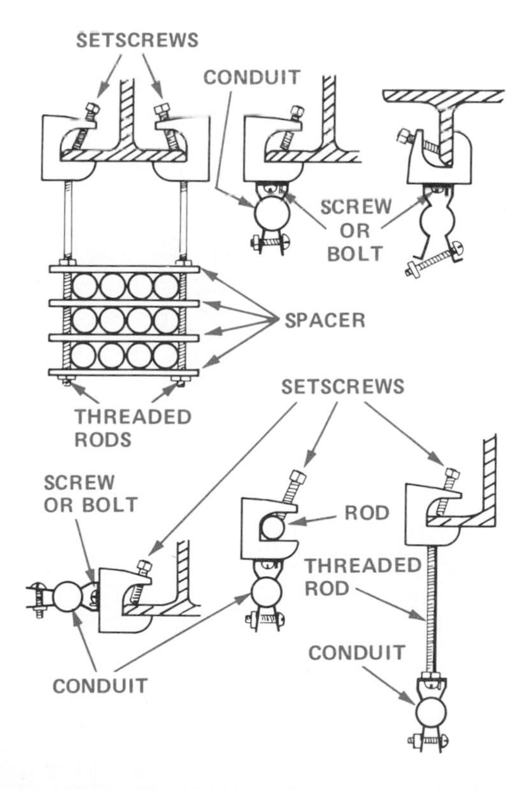

13–45. Beam clamps (insulator supports).

■ *Guy-wire conduit clamp hangers* are used mainly in industrial plants with irregular or high ceiling construction. It becomes necessary in these situations to suspend the conduit system at lower heights from guy wires. The wires are usually stretched the full length of the building. The hangers may then be mounted on guy wires up to $\frac{3}{8}''$ in diameter. The clamps are made of malleable iron for strength and durability. They are cadmium plated for rust resistance. Installation is easy, since the hanger is fastened to the guy wire and to the conduit by one bolt. Fig. 13-47.

■ *Conduit clamps* are used in fastening rigid conduit or EMT (thinwall) to a wall, ceiling, or floor. Similar devices, such as straps, are included in this series. A wide variety of types and sizes are available.

Use one-screw clamps to fasten rigid or EMT conduit to almost any surface. If you use them with clamp backs, installation time is cut because bending of conduit is not necessary to achieve perfect alignment. The offset bend (the distance between the conduit and the surface) is in perfect alignment with the knockouts of outlet boxes.

A one-screw, malleable-iron clamp is available in a large number of sizes and is cadmium plated. For use with EMT it is best to specify the clamp backs in one size smaller than the conduit size being used. Fig. 13-48.

One-screw, heavy, stamped-steel clamps are used in flush mounting rigid conduit or EMT to walls or ceilings. They are also available in sizes from $\frac{1}{4}''$ to 4''.

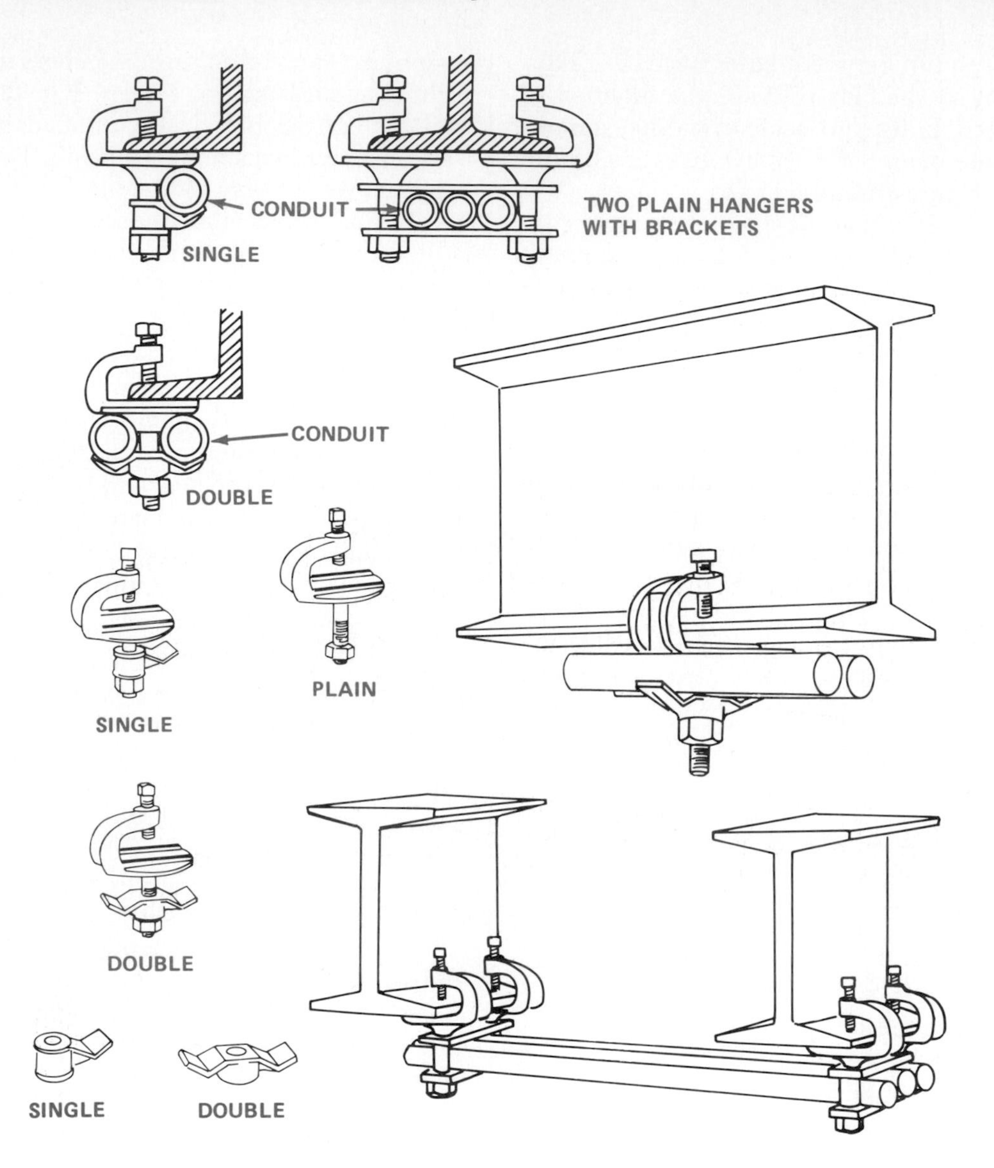

13–46. Sta-tite conduit hangers.

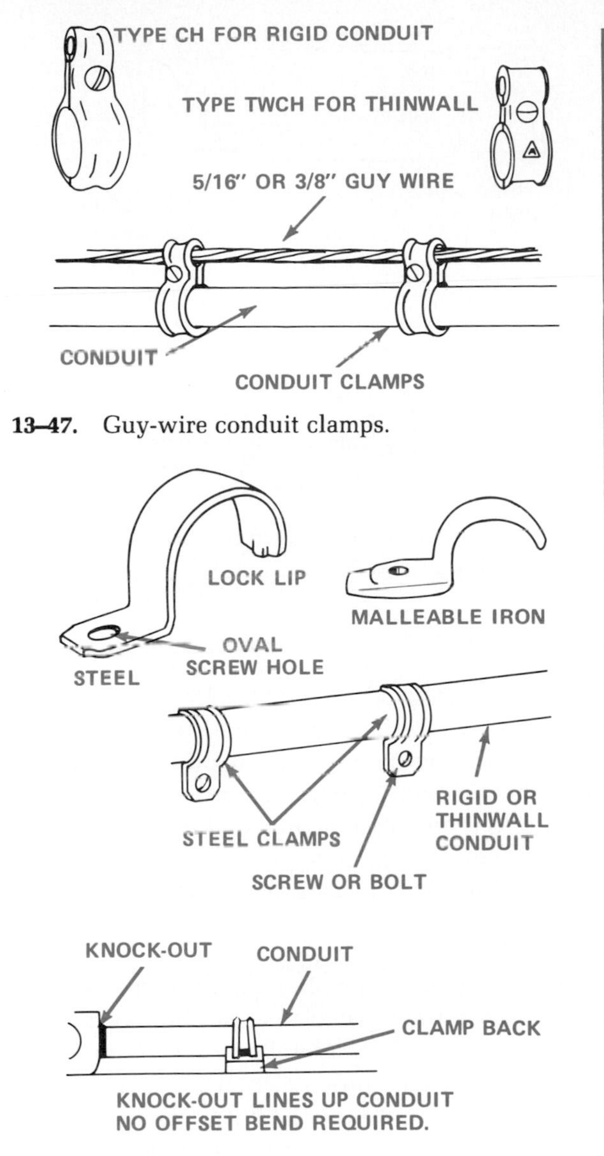

13–47. Guy-wire conduit clamps.

13–48. Conduit clamp, one-screw type.

At the open end of the clamp is a lip which locks conduit into place. An oval screw hole allows adjustment after installation. Sizes are stamped on the back of the clamp for easy identification.

■ *Two-hole pipe clamps, or straps,* are flush-mounted supports for fastening conduit to a wall or ceiling. They are light in weight, yet durable and versatile. On both sides of the arch of the 2-hole pipe strap are protruding punch marks that serve to lock the conduit, or cable, in place. Identification is easy. The size is marked on the back of each strap. Fig. 13-49.

■ *Bus-drop cable-suspension clamps* are especially designed to support power cable. The clamps may also be used to provide flexible connecting circuits between overhead, fixed-bus structures and machinery. Flexible air lines or lubrication lines are sometimes supported this way. Electric power cables are also supported by this type of clamp. Fig. 13-50.

Cable suspension clamps come in handy when installed with equipment that is likely to be shifted in location. One clamp covers a cable range from 0.400" to 1.187" in diameter. One clamp is required for a single 90° bend. By using two clamps back to back, a loop in the cable can be cradled. Thus, extra cable footage is acquired for possible future use when the machines are shifted to another location. (Most industrial machines are moved at one time or another to improve production.)

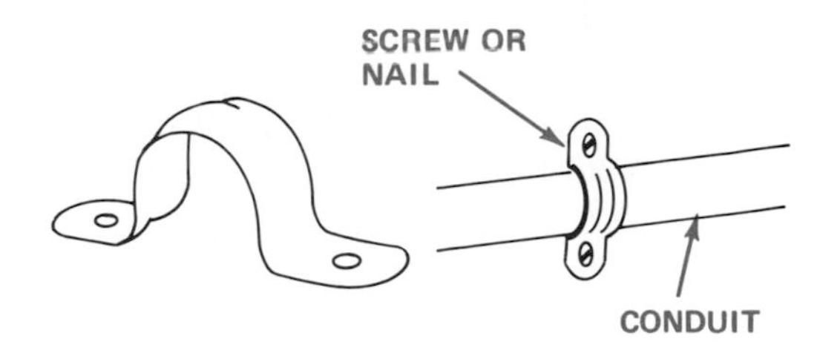

13–49. Conduit clamp, two-screw type.

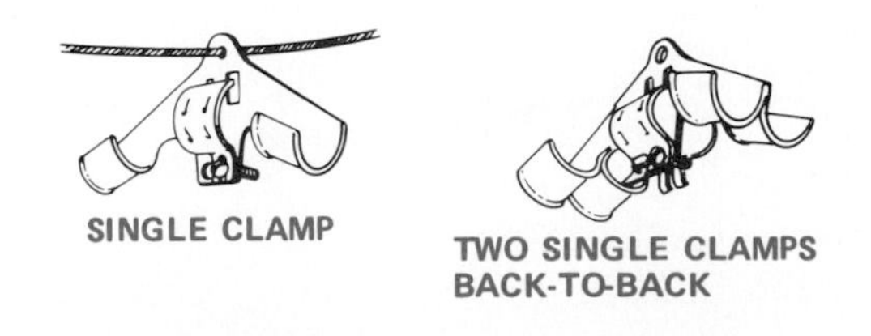

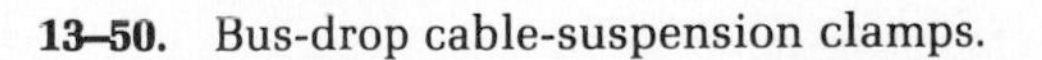

13–50. Bus-drop cable-suspension clamps.

The clamp is suspended above a machine with an aerial wire or cable. The support guy wire is usually fastened to a spring, which is, in turn, fastened to a rigid part of the building structure.

■ *Bus-drop mooring cable clamps* have an accessory to relieve the tension and vibrations of the suspended cable. This is the mooring cable clamp. It serves to relieve the tension and vibrations where the cable enters the bus structure. It is clamped onto the cable approximately one foot from the point where the cable enters the bus structure. The mooring cable clamp pulls the power cable from the bus-drop cable clamp by fastening the mooring clamp, with a guy wire, to a rigid structure. Fig. 13-51.

There are two sizes of mooring clamps available. One is for cable and has a 0.400″ to 0.875″ diameter. The other, for cables of 0.875″ to 1.187″ diameter, completes the range of sizes.

Guy wires, springs, and guy-wire cable must be furnished by the electrician. They are not included with the clamps.

■ *Adapters, clamps, and straps* for thinwall conduit are made for a number of jobs. The adapters are used for converted threaded, rigid-conduit fittings for use with thinwall conduit. Fig. 13-52.

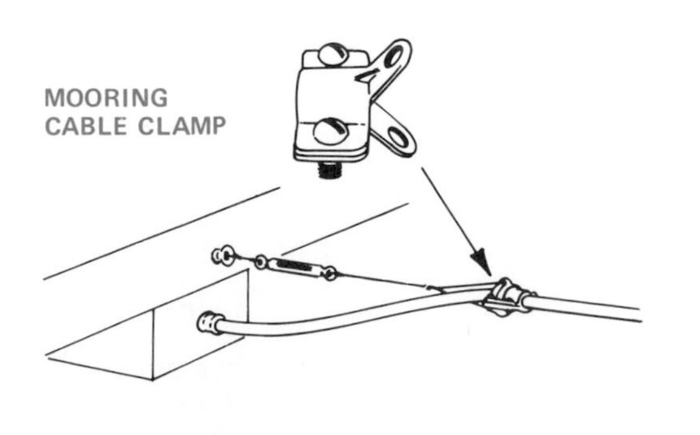

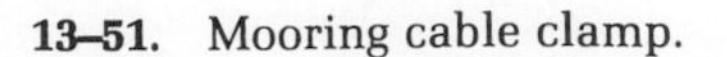
13–51. Mooring cable clamp.

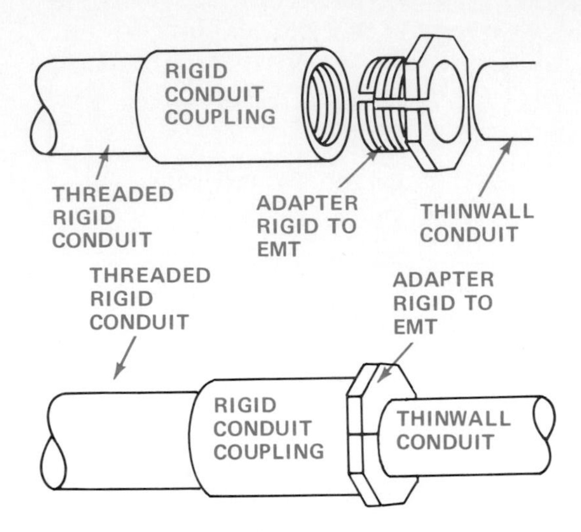

13–52. Installation of threaded rigid conduit to thinwall conduit adapters.

The adapter is a compression-type fitting with external threads. It has a split body. Thinwall conduit is placed inside the adapter and the assembly tightened into the threaded fitting. As you tighten the adapter, it is forced into a closed position. This closed position around the EMT forms a rigid bond.

Conduit Fittings

Thredmaker® connectors, couplings, and unions are used with rigid steel or aluminum conduit or intermediate metal conduit, indoors or outdoors. Thredmaker® fittings provide a means of putting standard NPT (National Pipe Threads) threads on a piece of in-place conduit or a cut conduit end without using hand dies or a thread cutter. Connectors provide NPT male threads. Couplings and unions provide standard electrical female threads. Fig. 13-53.

Conduit clamps and straps are used to fasten conduit to walls, floors, and ceilings.

■ *One-screw steel clamps* are used in flush-mounting thinwall. They are available in $\frac{1}{2}''$ to 2″ sizes. An oval-shaped screw hole allows for adjustment after installation. The size is stamped on the back of the clamp. Fig. 13-54.

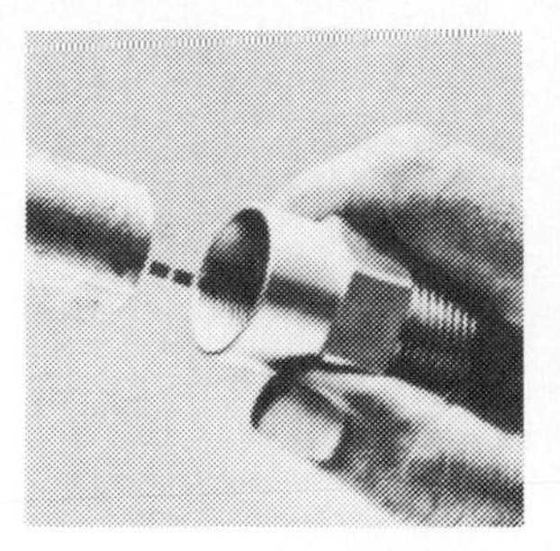

1. Slide the Thredmaker onto the conduit as far as it will go.

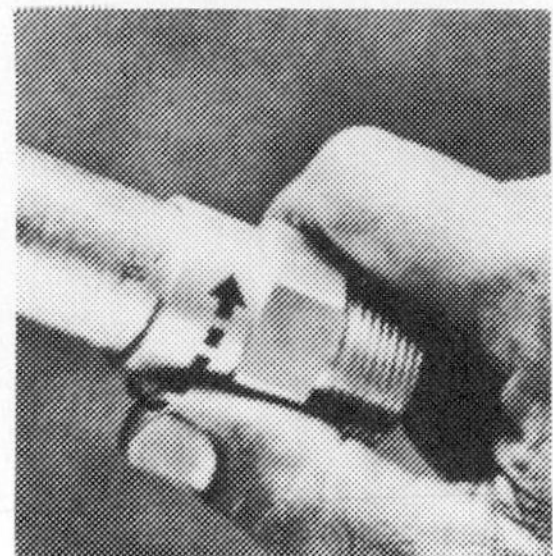

2. Turn the fitting by hand until it bites into the conduit—usually one turn is enough.

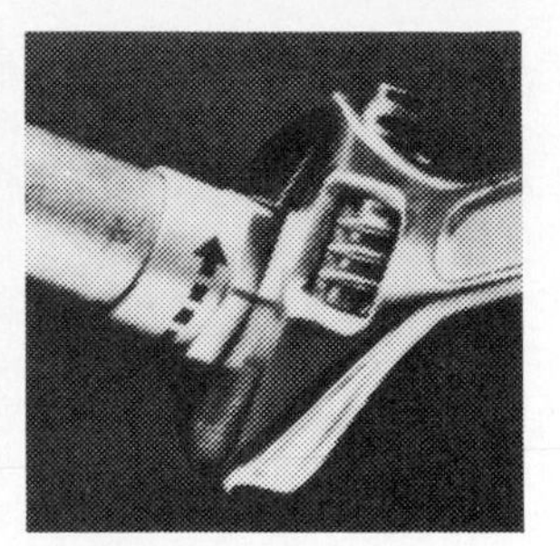

3. Wrench three to five turns using any available wrench.

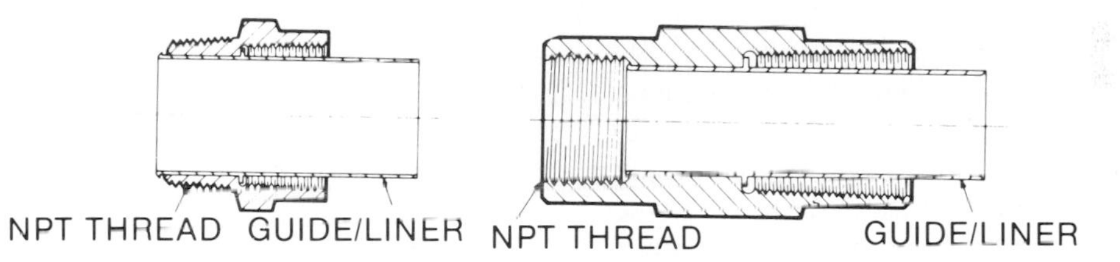

UNIONS

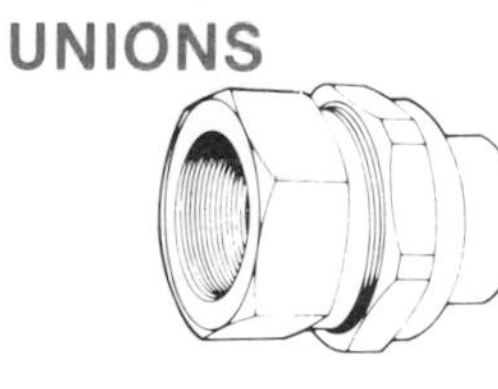

COUPLINGS—WITH GUIDE/LINER

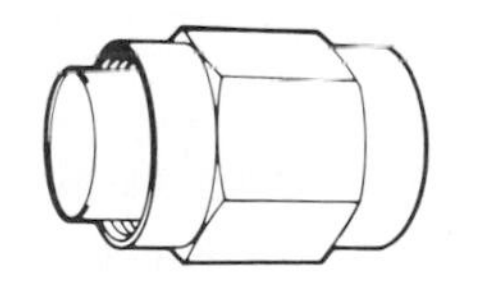

STRAIGHT CONNECTORS

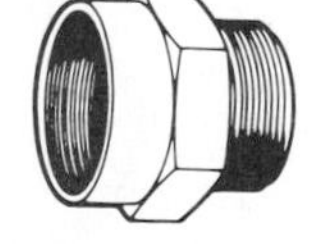

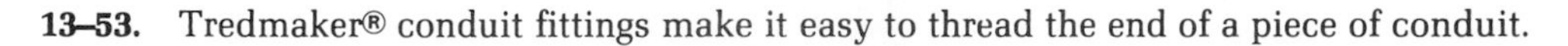

13–53. Tredmaker® conduit fittings make it easy to thread the end of a piece of conduit.

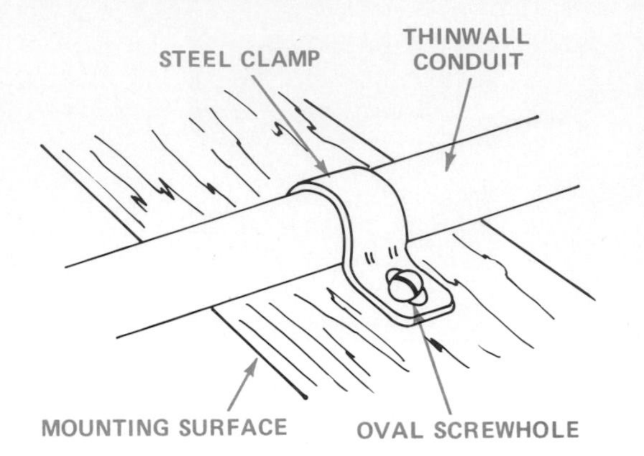

13–54. Mounting of steel EMT clamp.

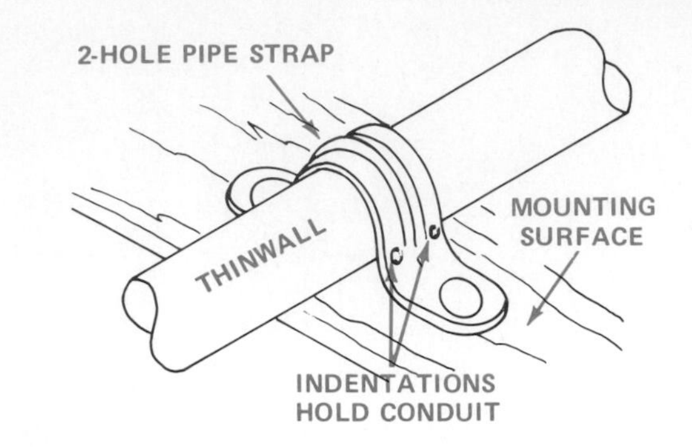

13–55. Two-hole pipe strap installed.

■ *Two-hole pipe straps* are flush-mounting supports for thinwall. They are inexpensive and are available in sizes up to $1\frac{1}{2}''$. Protruding marks on both sides (raised above the surface) of the strap serve to lock the conduit in place. The size is stamped on the back of the strap. Fig. 13-55.

Note that there is a distinct difference between clamp diameters for rigid conduit and those for thinwall conduit.

Support Grips

■ *Grips* are used to hold the weight of electrical cable as it hangs in a vertical, sloping, or horizontal position. Fig. 13-56. Electrical cable must be supported, or its dead weight can cause excessive strain or pullout at the connections. This will result in power failure. The National Electrical Code has a section that deals with supports. (See NEC Section 300-19 and 351-8.) Support grips also absorb additional strain from flexure, vibration, expansion, and contraction. Fig. 13-57. Supports are made of high-grade, nonmagnetic, tin coated bronze strand. Stainless steel grips are also available for severe service or unusual environmental conditions.

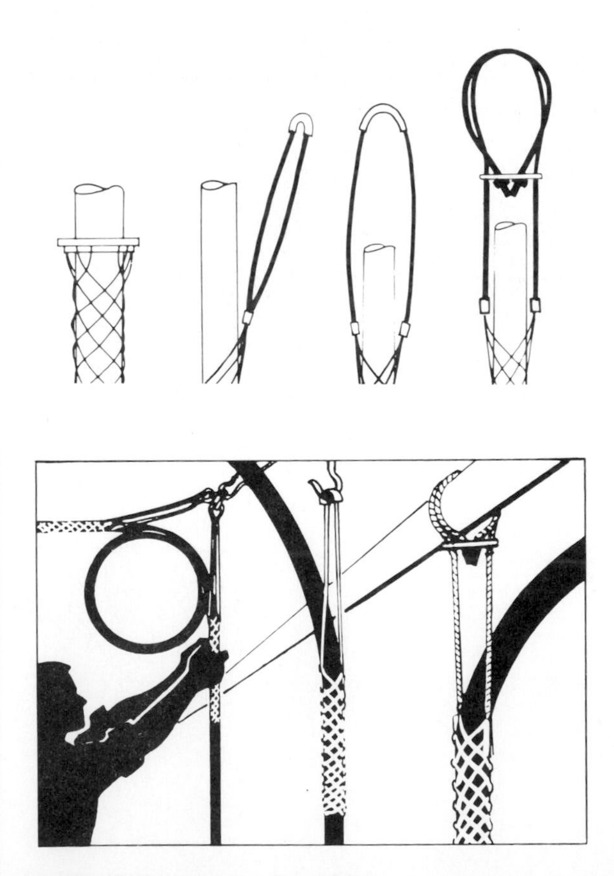

13–56. Support grips for cables.

Selecting the Correct Support Grip. Refer to Table 13-A to determine the grip style best suited for your application.

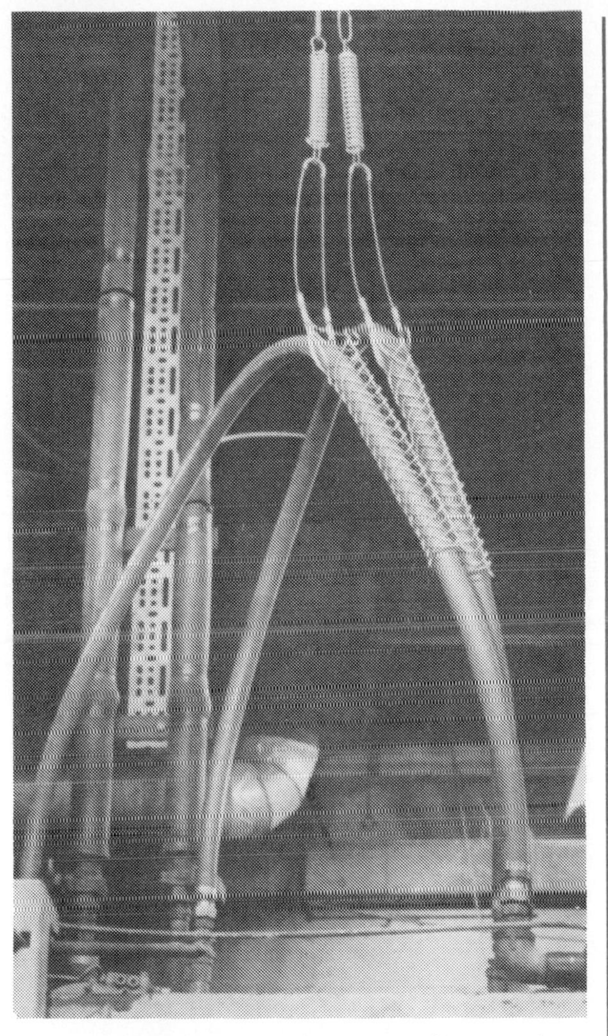

13–57. Cable support with grips to relieve strain.

13-A. Support Grip Selection

GRIP STYLES	APPLICATION
Closed mesh	Permanent support, cable end available
Split lace closing	Permanent support, cable end unavailable
Split rod closing	Temporary support, cable end unavailable
Material	Tin-coated bronze or stainless steel
Standard support grips	Support verticle runs to 99 ft. loads to 600 lbs.
Heavy-duty grips	Support verticle runs over 100 ft. loads over 600 lbs.
Service drop	Light duty to support service entrance cable
Conduit riser	Support cable runs in rigid conduit

(Kellems)

Determine the cable's outside diameter, then find the grip size that encompasses your cable diameter. Whenever possible, use a closed mesh that assembles over the cable end. If the cable end is not available, use a split mesh. Select an eye style that suits your needs. Fig. 13-58.

Select the proper material—tinned bronze or stainless steel. Estimate the tension to be put on the grip. Establish the working load you require. Compare this to the listed approximate breaking strength of the grip to ensure that the

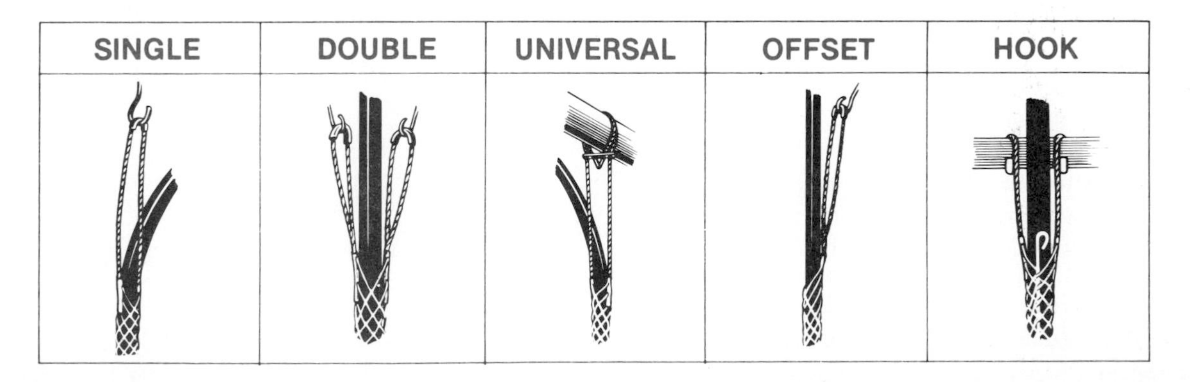

13–58. Various types of cable supports.

grip will be strong enough. Manufacturers list these factors in their catalogs or information may be obtained from the manufacturer. First, check the length and weight of the cable run to determine the miminum holding strength required. Be sure to leave a comfortable safety factor.

The *standard series* is used for vertical runs to 99′, continuous loads of 10 to 599 pounds, and cable diameters of 0.5″ to 4″.

The *heavy load series* is double weave and is used for vertical runs over 100′, continuous loads over 600 pounds, and cable diameters of 0.75″ to 4″.

■ *Split Lace.* Starting with the support end of the grip, this draws the lacing through one loop on either side of the split, drawing the lace through until the ends are evenly centered. Fig. 13-59. Criss-cross laces and thread through loops, down the length of the grip, so that the closure is spaced similarly to the mesh weave. When completely laced, twist the two stranded ends together, wrapping the twisted ends around the grip. Twist again to secure, and cut off the excess length.

■ *Split Rod.* Rod closing split grips are economical because they install quickly and the rod closure makes them reusable. Fig. 13-60. Wrap the grip around the cable. Using a twisting corkscrew motion, thread the rod through the loops. To remove, simply pull the rod out and the grip is ready to reuse. Next decide which of the attachments best fits your needs: double eye, single eye, offset eye, or locking bail. Fig. 13-61.

Selecting the Proper Cable Grip Attachments. Use an example to get a better idea of the proper selection of the grip attachment. For a 50′ run of three 1/0 AWG-THW wire (0.549″ nominal diameter), the cable grip diameter size to select is 100, which gives the cable grip range of 1″ to 1.25″. Since the end of the cable is available and it was determined that an offset eye would best suit your needs, a standard closed mesh support grip would be a correct selection.

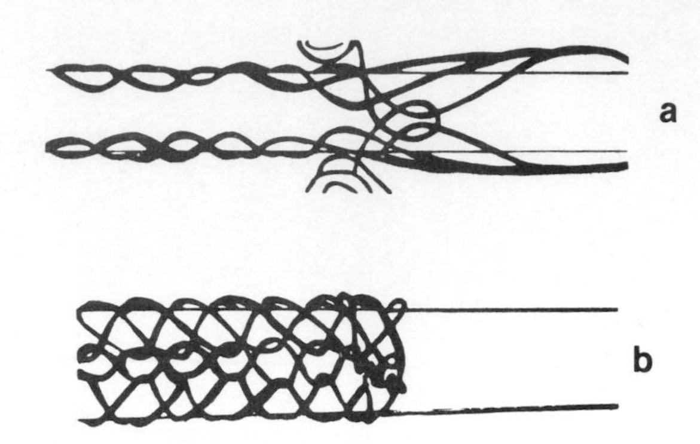

13–59. Lacing the grip on the cable.

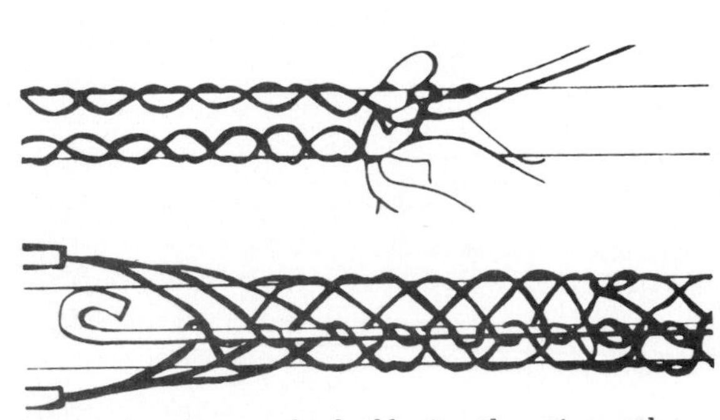

13–60. Another method of lacing the grip on the cable.

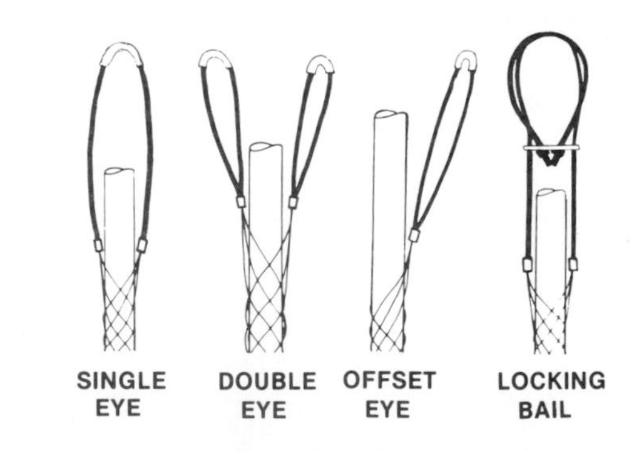

13–61. Various types of cable supports.

■ *Single eye.* This attachment is used for vertical cable and for applications where the cable bends, or where a single attachment is more advantageous for positioning.

■ *Double-U eye.* This attachment is used for a vertical cable and extends through the grip without bending. Eyes may be fastened to open hooks but should not be more than 15° from the axis of the vertical cable. When the eyes are supported equally, this attachment offers fully balanced load.

■ *Offset eye.* This attachment is similar to the single eye applications, but is intended for use when the offset positioning is required.

■ *Locking bail.* This attachment is adjustable and self-locking for use around a beam, pipe, or other continuous structural object.

Bushings

■ *End bushings* are used where threaded rigid conduit enters a metal enclosure. These enclosures may be boxes for controls, panel boards, or fuse boxes. A method to protect the wire from abrasion must be provided. This is necessary because the end of the conduit is very rough and is therefore capable of cutting the conductors. This cutting of the conductors causes a dead short or other damage. For this purpose, end bushings are designed to thread onto the end of the conduit so that the teeth will bear against the outlet box in which the conduit is fastened.

The end of the conduit is then slipped through the knockout. Fig. 13-62. There is a raised, rounded surface in the bushing over which the wire slides. The wire does not touch the conduit at this end. Fig. 13-63. After the bushing is installed, tighten the locknut on the outside of the box. The teeth of the locknut dig into the metal box, making a continuous ground.

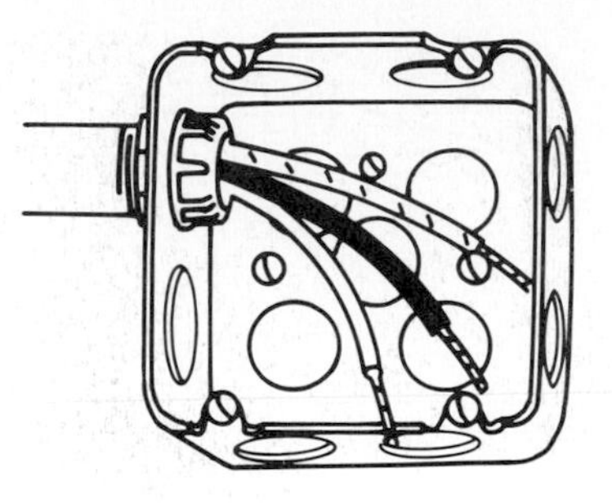

13-62. Locknut and bushing properly installed.

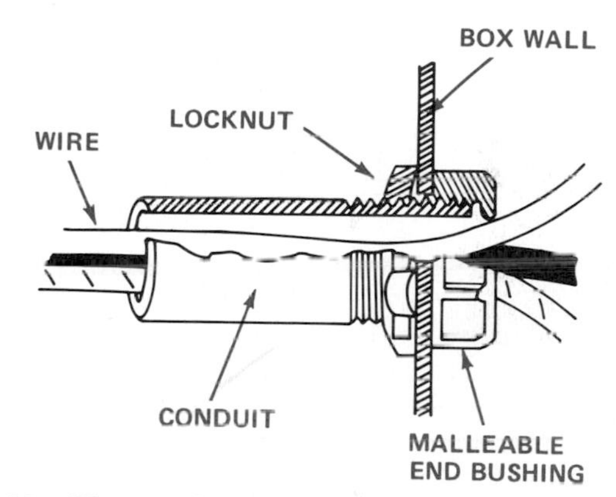

13-63. Thermoplastic bushings.

There are several types of end bushings available. For example, malleable-iron bushings secure the conduit in the same manner as a locknut. Plastic bushings protect the end of the conduit, but require the addition of locknuts for holding.

Wherever ungrounded conductor, No. 4 or larger, is used, insulating-type bushings are required. This means that either a wall and bushing with insulator or a thermoplastic bushing has to be used to comply with the NEC.

There is also a malleable-iron, capped bushing series which comes equipped with snap-in blanks. Such bushings are used to keep concrete, plaster, or other construction materials from getting into the conduit during installation. Fig. 13-64. These bushings may also be used to close off an unused piece of conduit which is found unnecessary. This is common when there is rework or an error. Fig. 13-65.

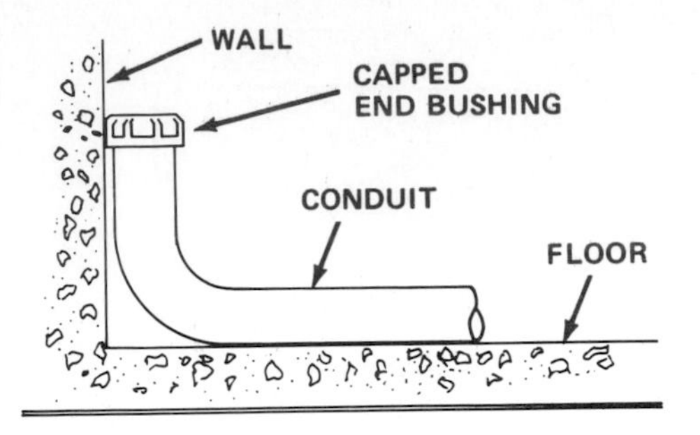

13–64. Capped end bushings.

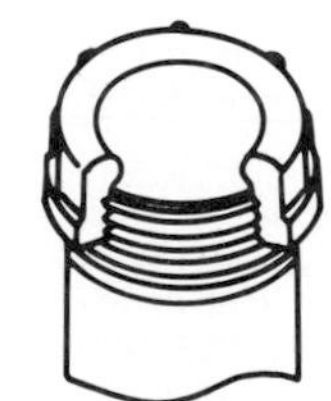

13–65. Heavy fiber stock with thick coat of wax makes this safety seal. Seals are used for protecting conduit during installation.

■ *Ground-type end bushings* come in a variety of sizes ranging from ½″ to 6″. They are insulated malleable-iron end bushings with a solderless lug terminal. The terminal is used to attach a ground conductor. They are available threaded or with a setscrew. They are installed by threading onto the conduit in the same manner as regular end bushings. Fig. 13-66.

■ *Reducing bushings* are used when conduits of different sizes are installed. Reducing bushings are a must in a case like this. They are threaded both inside and out. They are either threaded onto pieces of conduit to fit a hub larger than the conduit itself, or threaded into a hub to reduce the size to fit smaller conduit.

Reducing bushings come in a wide variety of sizes from ⅜″ to 4″ OD (outside diameter). Smaller sizes are made of steel, the larger sizes of cast malleable iron. There is an aluminum series as well. Some of the larger sizes have many different IDs (inside diameters). For example, a 2″ OD reducing bushing has a selection of tapped holes from ½″ through 1½″. Fig. 13-67.

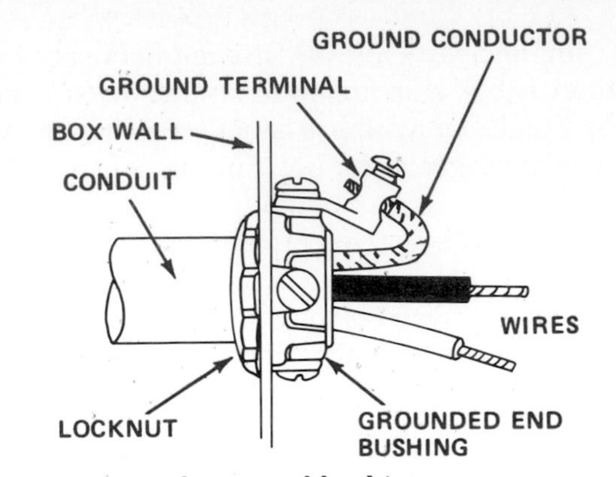

13–66. Ground-type end bushings.

■ *Steel bushing pennies* are used with bushings to temporarily seal an open end of conduit out of the ground in an unfinished installation.

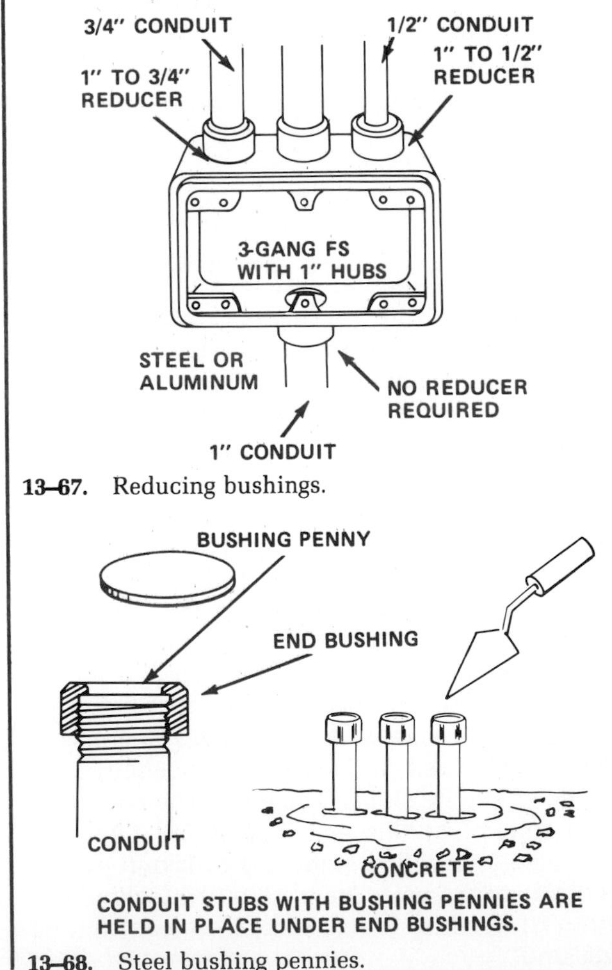

13–67. Reducing bushings.

13–68. Steel bushing pennies.

Figs. 13-68 & 13-65. The small discs resemble the cent, or penny, but are not made of copper. Pennies are placed over conduit ends to keep foreign matter out of the conduit while a building is under construction. To install, simply place a penny in the bushing and tighten the bushing onto the end of the conduit.

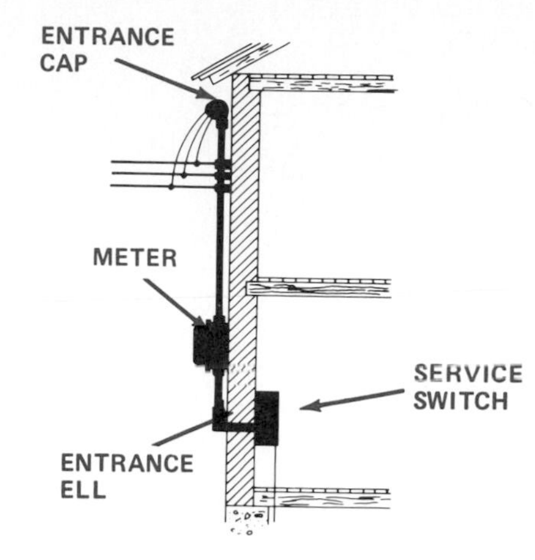

13-69. Entrance cap, meter, entrance ell, and service switch.

Other Fittings

■ *Entrance elbows* are used customarily where the conduit enters a building. The size of the conduit and fittings depends on the size of the entrance wires.

Entrance elbows must be raintight. Before pulling the wires through the conduit, remove the cover on the ell. (*Ell* is short for elbow.) Use a fish wire to pull the wires into place. After connection of the wires, replace the cover. Entrance ells come in almost any shape and size to fit almost any installation. Fig. 13-69.

■ *Ninety-degree female elbows* are designed for use with conduit-to-conduit, right-angle installations. Elbows are generally used instead of bending the conduit itself. Fig. 13-70.

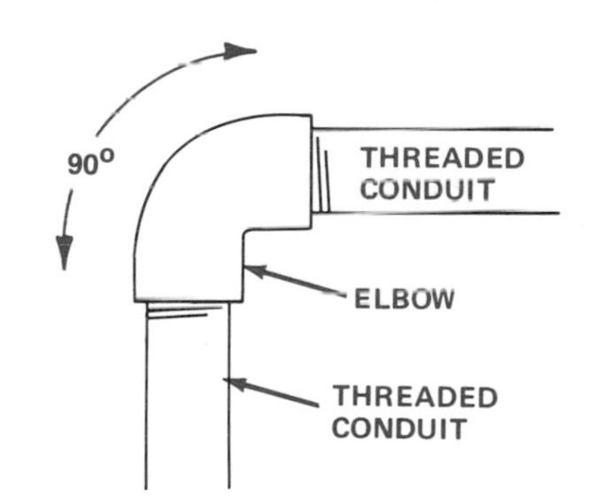

13-70. Ninety-degree female elbows.

■ *Ninety-degree bushed elbows* are used to make right-angle installations, either conduit-to-conduit, or conduit-to-box. Fig. 13-71.

■ *Pulling elbows* provide a simple yet effective means of pulling wires through raceways. They are used in cases where it is necessary to run the conduit around 90° corners. Fig. 13-72. The pulling elbows illustrated have removable covers which, when removed, allow you to pull the wires freely all the way through one side of the fitting, then through the other side. The cover is then replaced.

■ *The 45° female conduit elbows* are designed for use when a conduit-to-conduit, 45° angle installation is required. These elbows are generally used instead of bending the conduit. Fig. 13-73.

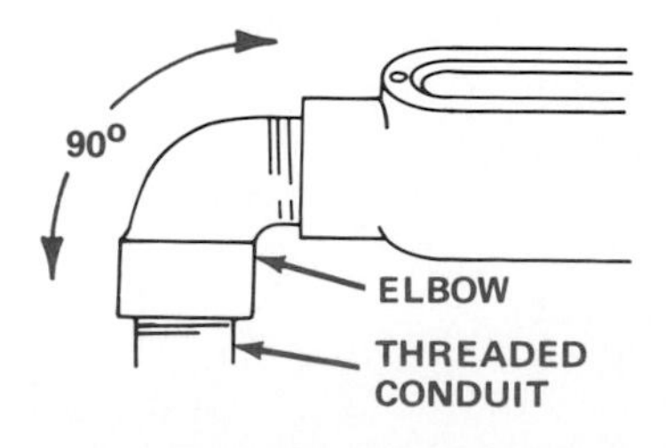

13-71. Ninety-degree bushed elbows.

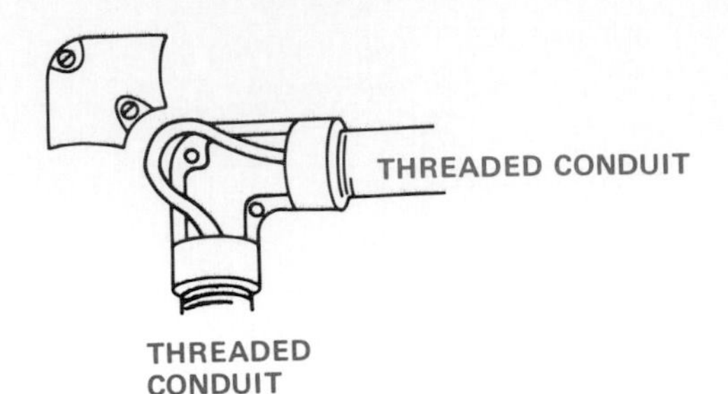

13–72a. Ninety-degree pulling elbows.

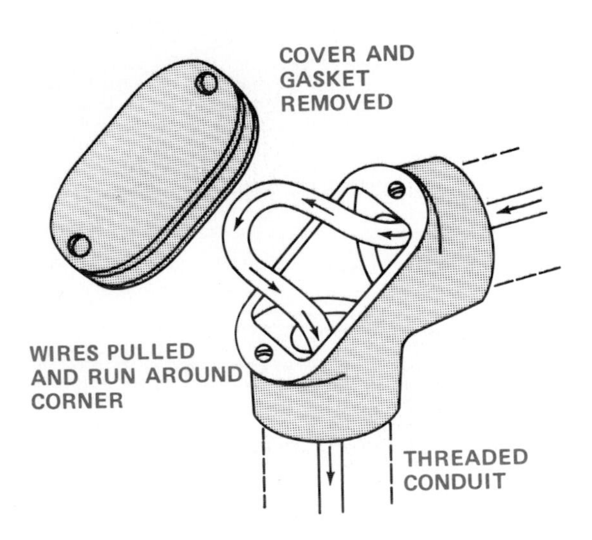

13–72b. Pulling elbows.

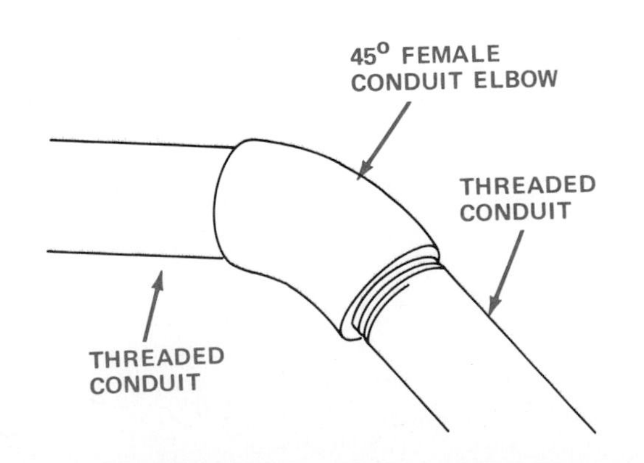

13–73. Forty-five-degree female conduit elbows.

■ *Steel-cupped reducing washers* are used with outlet boxes. They reduce the diameter of a knockout to a smaller opening. They are made of cadmium-plated steel and are used in pairs. One is placed inside the outlet box, the other outside. Conduit is placed through them with a locknut on either side. Fig. 13-74. Tighten the locknuts to complete the installation. A ledge on the edge of the washer prevents it from slipping after it has been installed.

■ *Entrance caps for use with threaded rigid conduit* are also known as service heads, entrance fittings, entrance heads, or weather heads. They are used to bring electrical power into a building. Current-carrying wires are brought from the outdoor lines to the entrance cap. The entrance cap is anchored on the building. Wires are then pulled from the service switch inside the building and through the conduit. The conduit passes from the inside wall to the outside wall of the building. The wires are pulled into an entrance elbow mounted on the exterior wall of the building. Fig. 13-75.

Conduit comes out of the building into an ell. Then conduit is connected from the ell (located at the bottom of the building or foundation) to the service entrance cap. Wires are pulled from the ell, through the conduit into the entrance cap, then threaded through the wire holes in the insulating block of the service

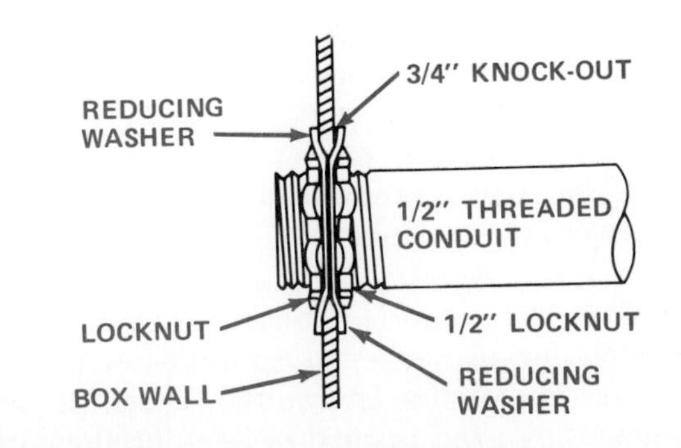

13–74. Steel-cupped reducing washers.

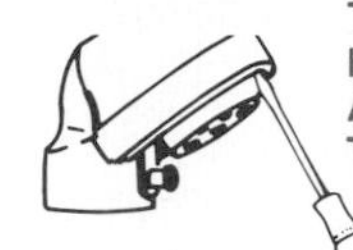

13–75. Entrance caps for use with threaded rigid conduit.

cap. The wires should be extended about 18″ and cut off. At this point, the interior circuit wires are spliced to the exterior wires.

Several types of service caps are available. All have removable covers, in which a detachable insulating block keeps the individual circuit wires separated. Some have snap-on covers that eliminate attaching screws on the cover.

In some cases, service-entrance cable is used instead of conduit. For these installations, the cable entrance caps have a clamp neck instead of threads or setscrews. The cap is fastened to the building instead of being supported by the conduit. The cable is secured to the building with clamps or straps. Fig. 13-76.

■ *Thinwall entrance caps and elbows* are like the threaded rigid conduit caps and elbows. However, they are made for thinwall instead of rigid conduit. They are used, as previously stated, to bring power into a building. The cables are brought from the overhead supply outdoors to the entrance cap anchored on the building. Wires from the inside of the building are then brought into an entrance el-

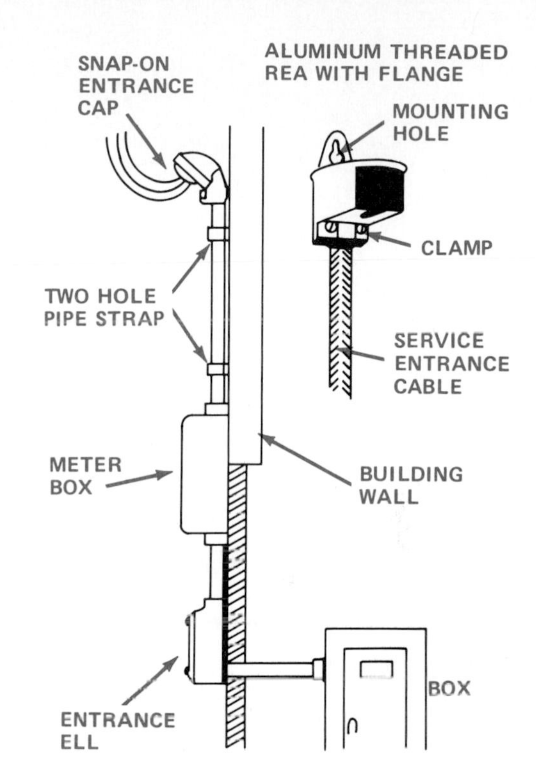

13–76. Entrance cap.

bow mounted on the exterior of the building. They are pulled through the ell, the conduit, and to the entrance cap. Connections are made at the entrance cap. Often a 90° elbow is needed where the conduit enters the service switch. Figs. 13-77 & 13-78.

Another type of entrance ell is shown in Fig. 13-79. It mounts to EMT with compression rings and nuts on either end. A removable cover facilitates the pulling of wires from one end and feeding them through the other.

Another 90° elbow is shown in Fig. 13-80. It is available in a long or short size. You use it to connect a run of thinwall conduit to a metal enclosure with a knockout at a 90° angle to the conduit run. The EMT is held in the female end with a compression ring and nut. The male end is secured to the entrance with a tiger-grip locknut. The edges of all these fittings are chamfered to prevent possible damage to wire insulation and cable sheath.

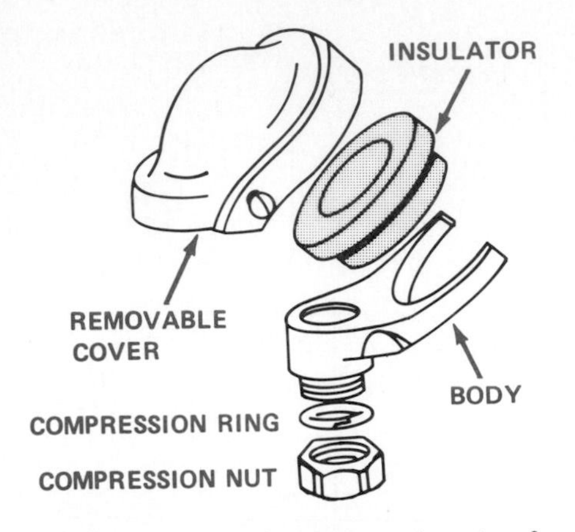

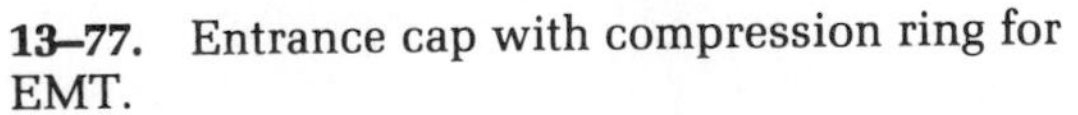
13–77. Entrance cap with compression ring for EMT.

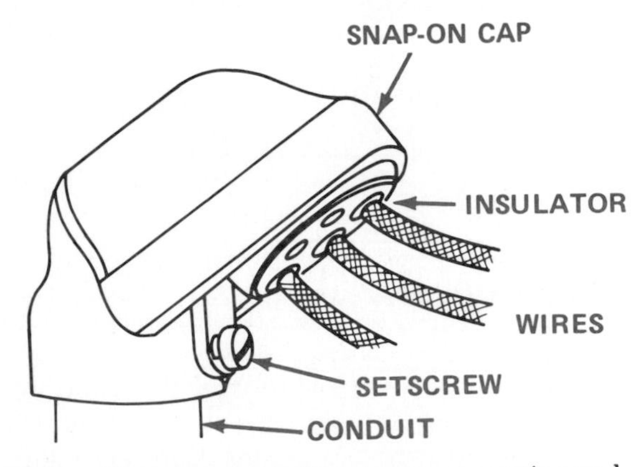

13–78. Entrance cap with setscrew mounting and snap-on cap.

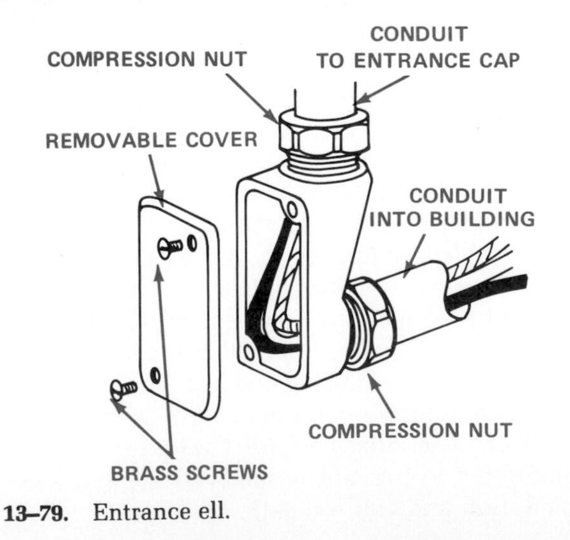

13–79. Entrance ell.

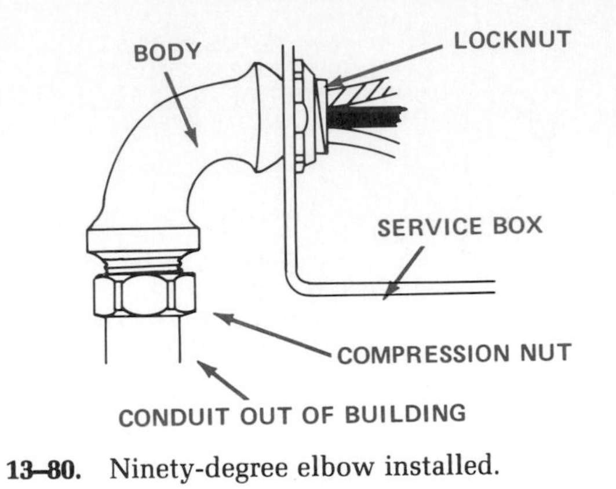

13–80. Ninety-degree elbow installed.

■ *Straight nipples or chase nipples* are used with conduit couplings to attach conduit to a box. Fig. 13-81. Place the nipple through the knockout from the inside of the box. Attach a coupling to the nipple, and the conduit to the coupling. To connect two boxes, place a chase nipple through each knockout from inside and connect with the conduit coupling. Chase nipples are available in sizes from $\frac{3}{8}''$ to 5″.

■ *Offset nipples* are designed to connect conduit to a box when the knockouts and conduit run do not line up. They are also used to connect two boxes side-by-side when their respective knockouts do not align. Fig. 13-82.

To connect conduit to a box using an offset nipple, secure the nipple to the box with a locknut. Connect the nipple to the conduit with a coupling. To connect two boxes, secure both ends with a locknut. Offset nipples are available in sizes from $\frac{1}{2}''$ to $1\frac{1}{4}''$. They are made of malleable iron.

■ *Conduit male enlargers* are vital fittings in conduit-to-box installations. They are used where the diameters of the conduit differ from those of the knockouts. They are made in three sizes, of malleable iron. Plated over the iron is a cadmium finish. The fitting is installed by threading over the end of rigid conduit, then inserting it through the knockout. Fig. 13-83.

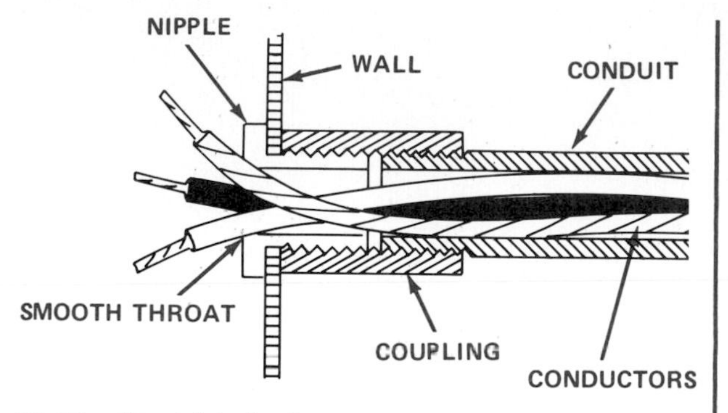

13-81. Straight nipple.

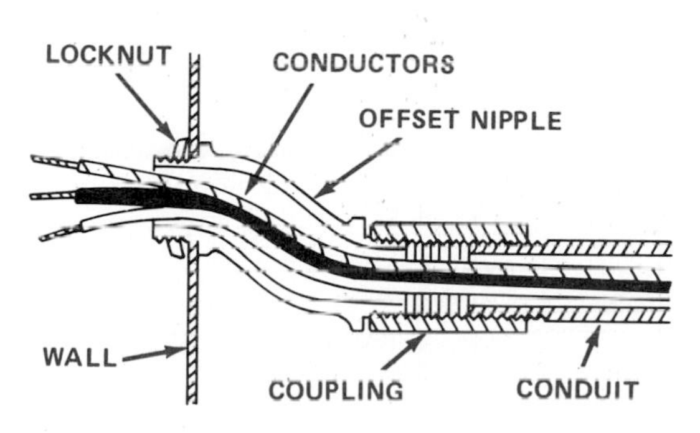

13-82. Offset nipples.

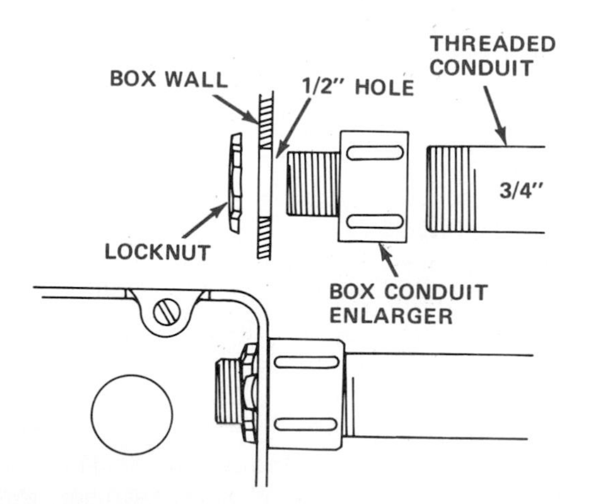

13-83. Conduit male enlargers.

■ *No-bolt fixture stems* are used with outlet boxes. They suspend fixtures from the ceiling without bolts. Male and female types are available. Both are threaded externally. The female stems also have internal threads. The shoulders have protrusions that match the knockouts on the outlet box. The fixture stem is placed through the base of the outlet box with the protrusions falling into place in the knockouts. A locknut is then tightened onto the stem. This holds the fixture stem in place. Fig. 13-84. Both types are made with a cadmium finish to prevent rust and corrosion.

■ *Hang-on conduit hangers* support rigid conduit or EMT. They come in handy to hang conduit. They can be mounted on beam clamps, screw, bolt, and drop-rod assemblies. They are available from manufacturers in three styles: swivel bolt, standard bolt, and without bolt. Fig. 13-85.

Hang-on hangers may be used for more than one run of conduit by stacking the hangers tandem. Fig. 13-85. This saves installation costs, since there is no need for independent suspension of each conduit run. Hangers may also be installed side-by-side, either vertically or horizontally, in conjunction with beam clamps, bolt, screw, and drop-rod assemblies. They can also serve as fixture hangers.

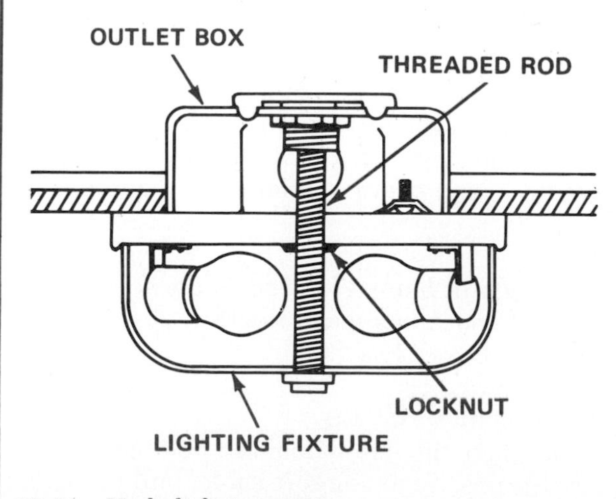

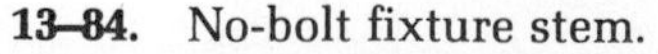

13-84. No-bolt fixture stem.

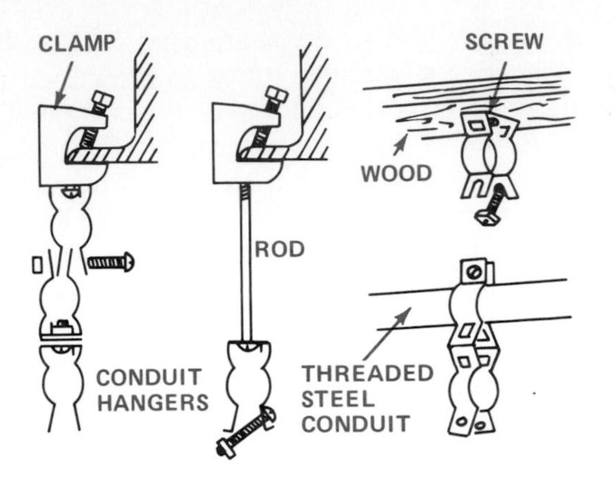

13-85. Hang-on conduit hangers.

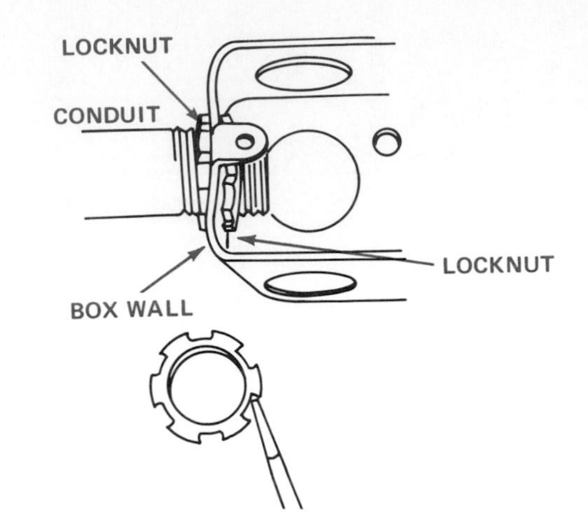

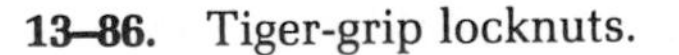
13-86. Tiger-grip locknuts.

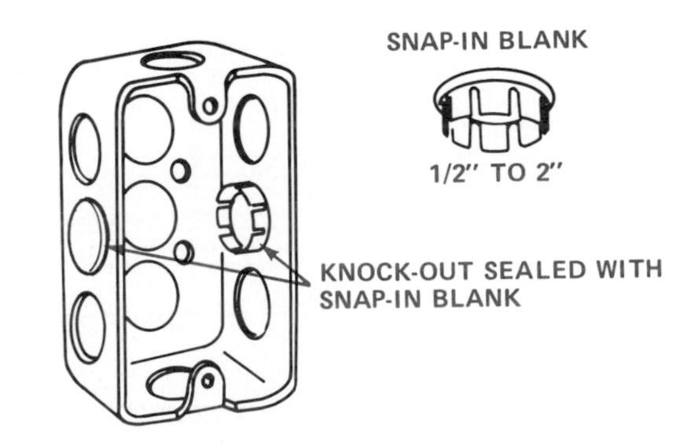

13-87. Snap-in blank.

■ *Tiger-grip locknuts* attach a piece of conduit or a connector into a knockout opening in a metal box. One inside locknut is used in connecting conduit to a box. Another locknut is used on the outside of the box. The pressure exerted by the two locknuts locks the conduit firmly in place through the wall. Fig. 13-86.

In tiger-grip locknuts there are non-slip notches around the perimeter of the locknut. These notches help prevent wrenches and screwdrivers from slipping. They also provide a perfect ground when tightened. You can be sure of the connection since their tilted edges bite into the metal wall of the box. The connection will not vibrate loose from excessive movement. Tiger-grip locknuts have been used where local codes demand an "approved" locknut.

■ *Snap in blanks for knockouts* are used to seal off knockouts made in error, or in rework in metal boxes. Fig. 13-87.

Strain Relief Grips

Grips are flexible wire mesh holding devices used primarily to support cable, pull cable, or relieve strain exerted upon cables. Figs. 13-88, 13-89, and 13-90. Wire mesh grips protect cables, relieve stress due to flexure of cables, prevent cable pullout from fittings, and aid in controlling the arc of bend at cable terminations. Fig. 13-91. Grips can hold all types of insulated and bare conductor, cable wire rope, flexible conduit, synthetic rope, pipe, tubing, and hose. Fig. 13-92.

Certain grips will help seal out contaminants and also extend the life of cable, flexible conduit, tubing, or hose. Fig. 13-93. The use of wire mesh grips ensures added safety to personnel and property.

The use of wire grips throughout an electrical system is common today. Fig. 13-94 shows how the various types of grips are used in a system.

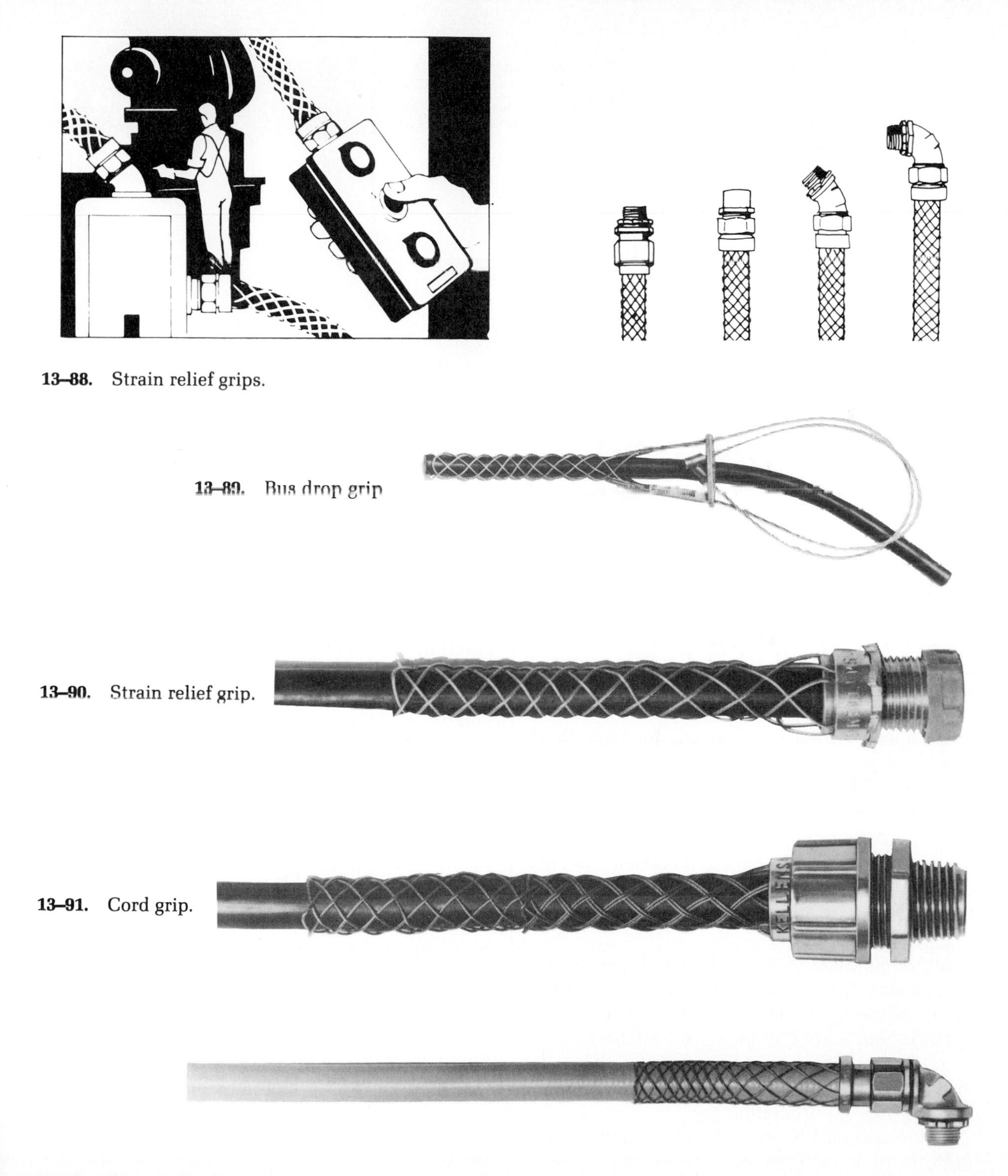

13–88. Strain relief grips.

13–89. Bus drop grip

13–90. Strain relief grip.

13–91. Cord grip.

13–92. Nonmetallic flexible liquid-tight conduit grip.

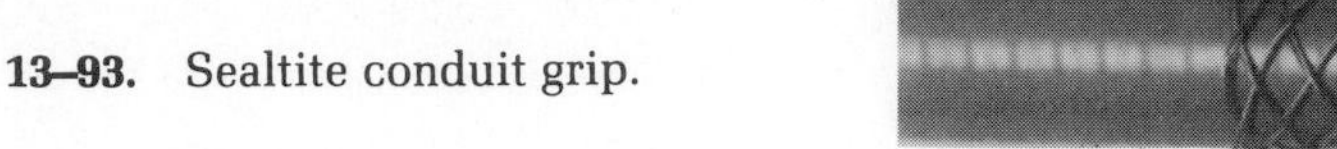

13–93. Sealtite conduit grip.

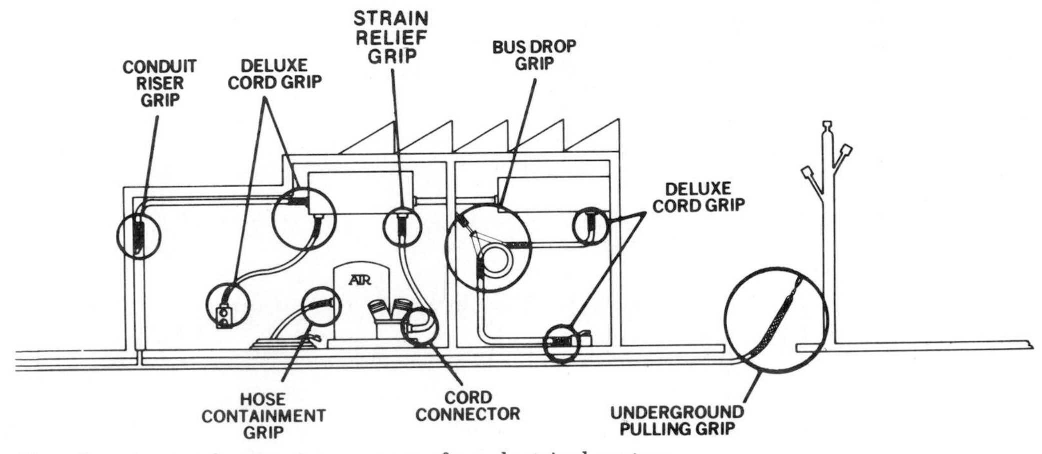

13–94. Uses for wire mesh grips in one part of an electrical system.

INSTALLATIONS FOR HAZARDOUS LOCATIONS

Electrical wiring installations for hazardous locations present complicated and perplexing problems. These problems may be confronted by engineers, architects, or contractors when they are called upon to install explosion-proof systems. A purchasing agent assigned the task of selecting, appraising, and specifying equipment to be used must also confront these wiring situations.

Many national, state, and local codes and regulations have been compiled for custom-made equipment that is manufactured for a specific job. State and local authorities as well as codes should always be consulted. This assures that the electrical systems conform to all installation requirements.

The National Electrical Code divides the locations of hazardous conditions into classes and divisions. If these classes and divisions are mentioned in your wiring specifications, it is best to consult your copy of the National Electrical Code. It will give you the exact requirements.

Hazardous Conditions

- Class I locations are those in which flammable gases or vapors are, or may be, present in the air in quantities sufficient to produce explosive or ignitable mixtures.

- Class II locations are those in which combustible dust is, or may be, present in the air in quantities sufficient to produce an explosive or ignitable atmosphere.

The phrase "quantities sufficient" is an attempt to define the amount of vapor or dust that will cause a dangerous atmosphere. This is difficult, since many factors are involved, some of which are variable. For instance, barometric pressure, humidity, air movement, and the amount and type of ventilation affect flammable conditions. Also, the ratio of room volume to the amount of vapor or dust, temperature, processes, and machinery all can contribute to hazardous situations.

It is vital that those who deal with dangerous materials such as gases and flammable liquids be aware of the hazards involved. Safety measures must be considered by persons who plan and install electrical equipment in hazardous locations. For exact ignition temperatures and explosive limits of commonly used hazardous substances, see Table 13-B.

Code Definitions. Articles of the National Electrical Code (NEC) deal specifically with problems associated with hazardous locations. These locations are defined according to classes and subdivided into divisions.

The three classes of locations and conditions are:

■ Class I locations: those in which flammable gases or vapors are, or may be, present in the air in quantities sufficient to produce explosive or ignitable mixture.

Class I, Division 1: those where such hazardous concentrations of flammable gases or vapors exist continuously, intermittently, or periodically under normal operation conditions.

Class I, Division 2: those where such hazardous concentrations of flammable gases or vapors are handled in closed containers or closed systems.

■ Class II locations: those where the presence of combustible dust presents a fire or explosion hazard.

Class II, Division 1: those where dust is suspended in the air continuously, intermittently, or periodically under normal operating conditions, in quantities sufficient to produce explosive or ignitable mixtures.

Class II, Division 2: those where such dust is not suspended in the air, but where deposits of it accumulating on the electrical equipment will interfere with the safe dissipation of heat, causing a fire hazard.

■ Class III locations: those where easily ignitable fibers or flyings are present but not likely to be suspended in the air in quantities sufficient to produce ignitable mixtures.

Class III, Division 1: those where ignitable fibers, or materials producing combustible airborne particles, are handled, manufactured, or used.

Class III, Division 2: those where easily ignited fibers are stored or handled (except in the process of manufacture).

Testing and approving conditions have advanced to a finer degree of classification with four separate designations: A, B, C, and D for Class I; and three categories for Class II: E, F, and G.

Equipment is tested by the Underwriters' Laboratories and the Canadian Standards Association.

Explosion-Proof Equipment

There is a common misconception that explosion-proof equipment is gastight. However, it would be impractical to make an entire wiring system gastight. Whenever an enclosure would be opened for servicing apparatus, for example, the explosive mixture would be trapped in the enclosure. The trapped atmosphere would explode the instant the apparatus was again operated.

The requirement, therefore, is not that enclosures be gastight, but that they be designed and manufactured strong enough to contain an explosion. That is, if an explosion occurs, the enclosures should prevent the escape of flame or heat that could ignite surrounding atmospheres. Fig. 13-95 (page 452).

Burned gases do escape from explosion-proof equipment. But their escape path has been engineered. It is controlled so that the temperature of the gas is well below the ignition point of the surrounding atmosphere.

Ground-Joint Construction. Ground-joint construction uses two carefully machined metal surfaces, bolted together. This keeps the hot, flaming gases caused by an explosion con-

13-B. Commonly Used Hazardous Substances

SUBSTANCE	FLASH POINT, DEGREES FAHRENHEIT	IGNITION TEMPERA-TURE, DEGREES FAHRENHEIT	EXPLOSIVE LIMITS PERCENT BY VOLUME LOWER	UPPER	VAPOR DENSITY AIR = ONE
Acetaldehyde	−36	365	4.0	57.0	1.5
Acetone	0	1000	2.6	12.8	2.00
Acetylcaloride	40	734	....	....	2.7
Acetylene	**Gas**	**571**	**2.5**	**80.0**	**0.91**
Acrolein	0 or less	532	2.8	31.0	1.9
Ammonia, anhydrous	Gas	1204	15.5	27.0	0.6
Amyl acetate (pure)	77	714	1.1	7.5	4.5
Amyl alcohol	94	650	1.19	....	3.0
Aniline	158	1143	1.3	....	3.2
Benzene	12	1044	1.4	7.1	2.8
Benzine*	0 or less	550	1.1	5.9	2.5
Benzyl chloride	153	1058	1.1	....	4.4
Butane, n	Gas	761	1.9	8.41	2.04
Butyl acetate, n	72	790	1.39	7.55	4.00
Butyl alcohol, n	84	650	1.45	11.25	2.55
Camphor	150	871	....	....	5.2
Carbon disulphide	−22	212	1.25	50.0	2.64
Carbon monoxide	Gas	1128	12.5	74.2	0.967
Coal gas	....	....	8.5	25.0	....
Collodion	0 or less	....	....	....	....
Crude petroleum	20 to 90	....	....	....	....
Cyclohexane	−4	500	1.3	7.8	2.9
Cyclopropane	Gas	928	2.4	10.4	1.5
Dibutyl ether, n	77	382	4.5	7.6	4.5
Dioxane	54	356	3.0	22.0	3.0
Ethane	Gas	959	3.0	12.5	1.035
Ethyl acetate	24	800	2.2	11.4	3.04
Ethyl alcohol	55	793	3.3	19.0	1.59
Ethyl chloride	−58	966	4.0	14.8	2.22
Ethyl ether	−49	356	1.85	36.5	2.56
Ethylene	Gas	842	2.75	32.0	1.0
Ethyl formate	−4	851	27.0	13.5	2.6
Ethyl methyl ether	−35	374	2.0	10.1	2.1
Ethyl methyl ketone	21	960	1.8	10.0	2.5
Ethyl nitrate	50	....	4.0	....	3.1
Formaldehyde	Gas	806	7.0	63.0	1.075
Furfural	140	600	2.1	....	3.3
Gasoline**	−45	536	1.4	7.6	3.4
Glycerine	320	739	....	....	....

13-B. Commonly Used Hazardous Substances (Continued)

SUBSTANCE	FLASH POINT, DEGREES FAHRENHEIT	IGNITION TEMPERATURE, DEGREES FAHRENHEIT	EXPLOSIVE LIMITS PERCENT BY VOLUME LOWER	UPPER	VAPOR DENSITY AIR-ONE
Hydrogen	**Gas**	**1085**	**4.0**	**74.2**	**0.1**
Hydrogen sulfide	Gas	500	4.3	45.0	1.2
Kerosene	100-165	445	0.7	5.0	4.5
Lacquer	0 to 80	....	....	....	....
Methane	Gas	999	5.0	15.0	0.554
Methyl acetate	14	935	3.2	15.6	2.6
Methyl alcohol	52	867	6.0	36.5	1.11
Methyl chloride	Gas	1170	8.3	18.7	1.78
Methyl ether	Gas	662	3.4	18.0	1.6
Methyl formate	−2	853	5.0	22.7	2.1
Naptha (high flash)	100-110	900-950	....	....	....
Naptha (solvent)	100-110	450-500	1.1	6.0	....
Nonane, n	88	403	0.8	2.9	4.4
Paint, liquid	0 to 80	....	....	....	....
Parting liquid	less than 80		Properties of parting liquids vary		
Phthalic anhydride	305	1083	1.7	10.5	....
Propane	Gas	871	2.1	9.4	1.56
Propyl acetate, n	58	842	2.0	8.0	3.5
Propylene	Gas	770	2.0	10.3	1.49
Pyridine	68	900	1.8	12.4	2.7
Rubber cement	50 or less		Hazard depends on solvent used		
Styrene	90	914	1.1	6.1	3.6
Tolual	40	997	1.4	6.7	3.1
Turpentine	95	464	0.8	....	....
Toulene	40	1026	1.27	7.1	3.14
Varnish	less than 80	....	....	....	....
Varnish shellac	40-70	....	....	....	....
Vinyl acetate	18	800	2.6	13.4	3.0
Vinyl chloride	Gas	882	4.0	21.7	2.2
O-Xylene	63	867	1.0	6.0	3.66

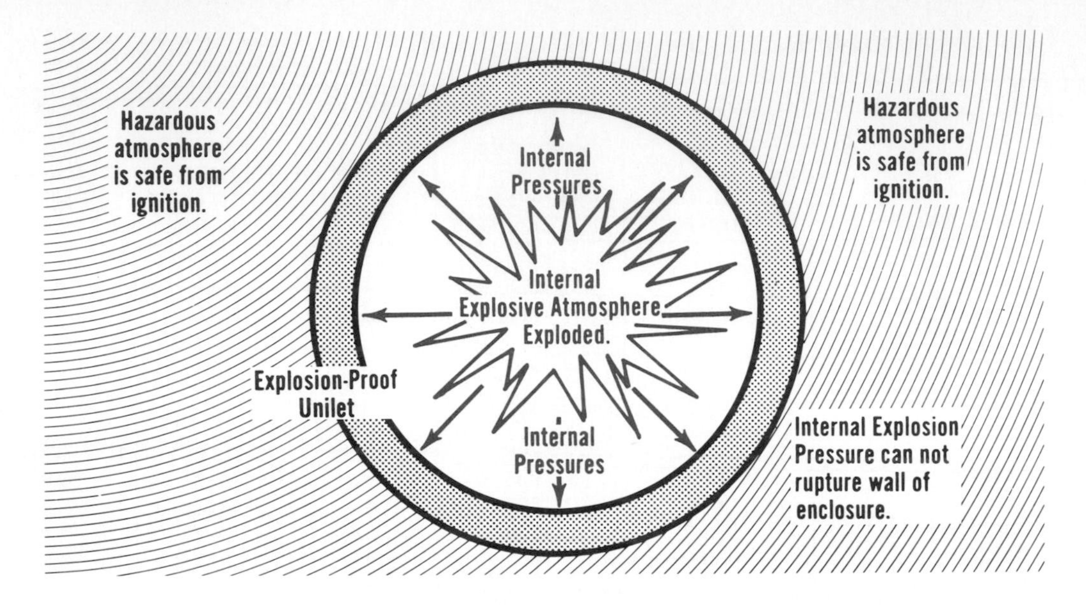

13–95. Explosion-proof equipment will contain an explosion.

tained. Internal pressures force the hot gases out between the ground surfaces. They are cooled in the process and therefore cannot ignite the hazardous atmosphere into which they are escaping. Fig. 13-96.

A *unilet* has both covers and bodies carefully machined. Their metal surfaces must fit so as to form a joint where they are bolted together. The joint must be kept clean and protected from damage. When assembling the unilet, be careful to remove foreign matter, such as paint or dirt, from matching surfaces. During the machining operation, the surfaces are quality controlled to meet UL requirements. Fig. 13-97.

Threaded-Joint Construction. In threaded-joint construction, one part of an enclosure threads into another to form an enclosed unit. This type of construction requires a minimum of five full threads. In the event of an explosion, the threaded surfaces will allow the internal gas pressures to be dissipated and cooled. However, hot flaming gases cannot escape to the surrounding hazardous atmosphere. Fig. 13-98.

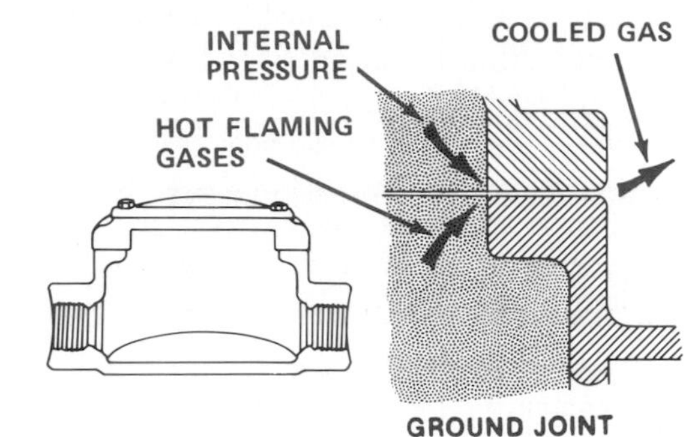

13–96. Ground-joint construction.

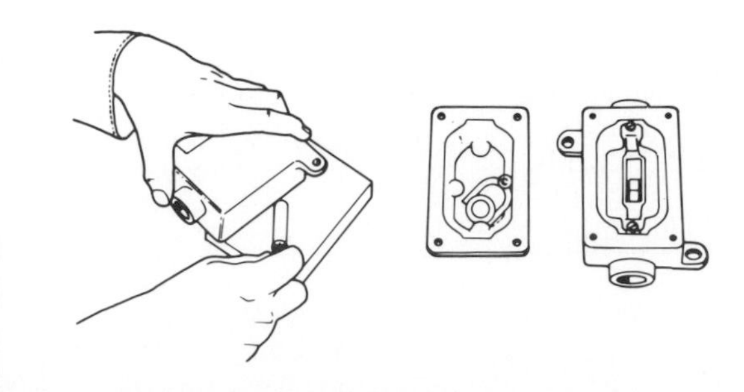

13–97. Precision-machined parts.

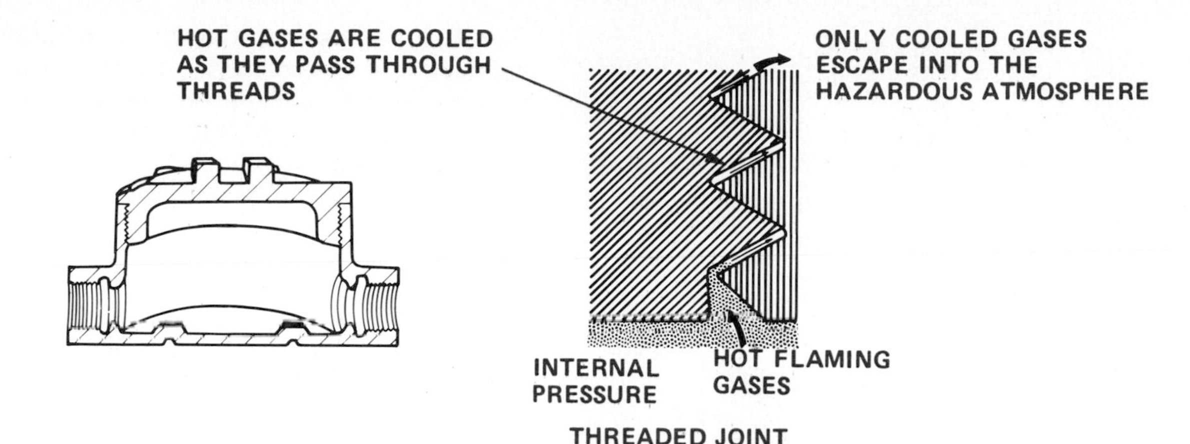

13–98. Threaded-joint construction.

Close-Tolerance Shaft Construction. Threaded-joint construction is normally used for covers of explosion-proof equipment. It is also utilized for construction of shafts having limited rotation. Fig. 13-99. Close-tolerance shaft construction is employed where threaded joints cannot be used, such as where two closely machined surfaces make contact over a prescribed distance. This arrangement allows for sufficient cooling and for the reduction of the internal pressures. Also, it keeps flames out of the surrounding hazardous atmosphere.

Examples of close-tolerance construction include push-button stations and similar equipment. Fig. 13-100. Factory-sealed pilot lights are supplied complete with red-jewel guard and 120-volt lamp. These are mounted on either single or two-gang unilets. They can also be supplied in combination with factory-sealed push buttons and selector switches. Fig. 13-100.

Seals. The drawing in Fig. 13-101 illustrates hazardous internal vapors. These vapors can be ignited by turning on a switch. Note that the explosion is contained between two seals. Control of this explosion is possible mainly because the volume of hazardous vapors is limited by the installation of sealing unilets. Seals must be correctly installed in each run of conduit entering or leaving any enclosure containing switches, fuses, circuit breakers, relays,

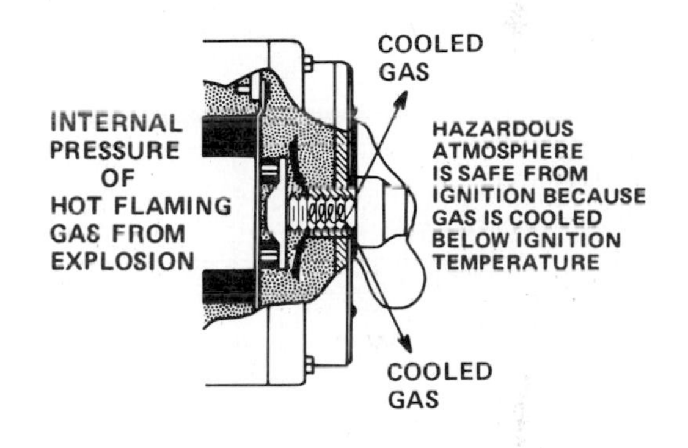

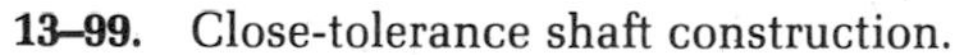
13–99. Close-tolerance shaft construction.

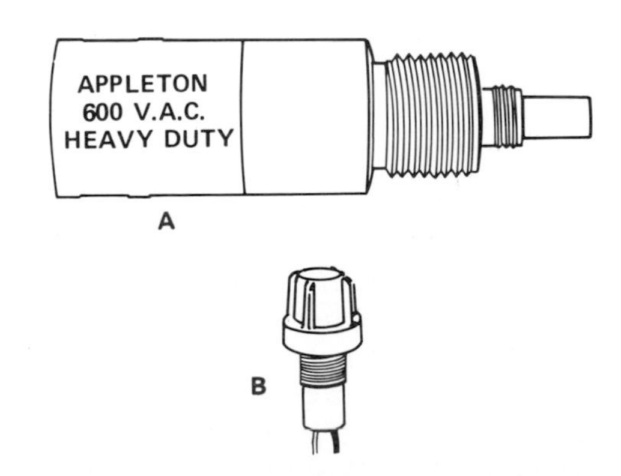

13–100. Factory-sealed push-button switch (A) and pilot light (B).

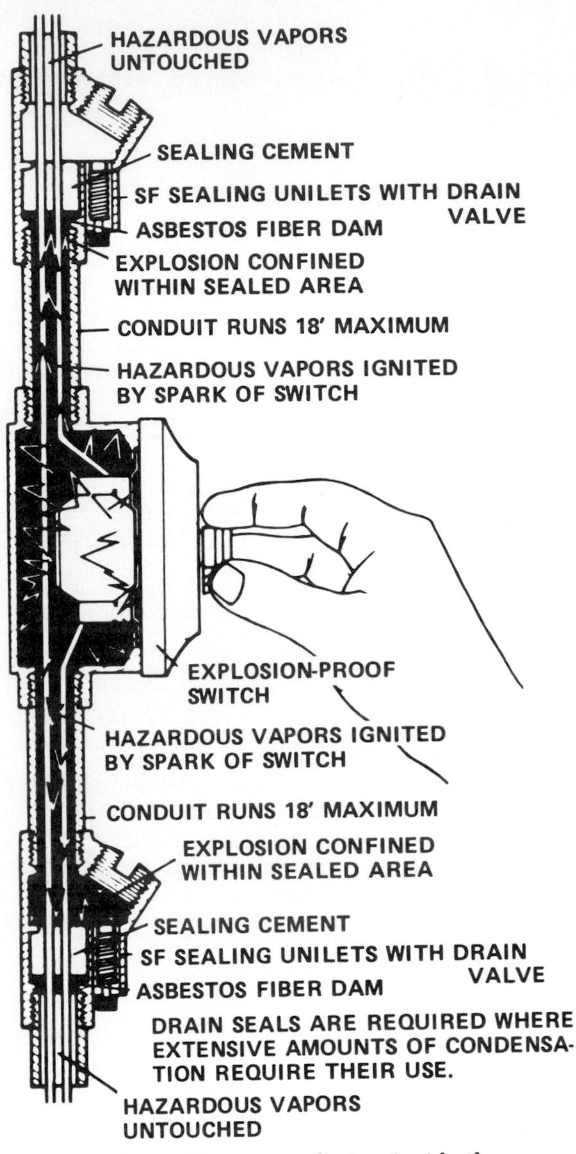

13-101. Controlling an explosion inside the equipment.

and devices which may produce arcs, sparking, or high temperatures. Seals are also required in conduit runs passing out of a hazardous area into a nonhazardous area.

Seals should not be placed more than 18″ from an arcing device such as a switch. This will contain the explosion if it occurs. Every conduit run that leaves a hazardous area should also be sealed.

For Code Divisions 1 and 2 installations, rigid conduit with threaded joints may be used between the seal and the explosion-proof enclosure. Threadless connections are not flame-tight and cannot be used. Fig. 13-102 shows how a seal is poured into a unilet.

Classifications of Hazardous Locations. This series of diagrams illustrates the different classes, divisions, and groups of hazardous locations.

- Fig. 13-103: a hazardous-area switch plug with receptacle, and a switch plug connected to prevent an arc that could cause an explosion.
- Fig. 13-104: a lighting diagram for a hazardous location. This is for Class I, Groups A and/or B in Division 1 lighting.
- Fig. 13-105: a power diagram for a hazardous location. Class I, Division 1.
- Fig. 13-106 (page 458): a lighting diagram for a hazardous location, Class I, Division 1.
- Fig. 13-107 (page 459): a power diagram for a hazardous location, Class I, Division 2.
- Fig. 13-108 (page 460): a lighting diagram for a hazardous location, Class I, Division 2.
- Fig. 13-109 (page 461): a power diagram for a hazardous location, Class II, Division 1.
- Fig. 13-110 (page 462): a lighting diagram for a Class II, Division 1 hazardous location.
- Fig. 13-111 (page 463): a power diagram for a hazardous location, Class II, Division 2.
- Fig. 13-112 (page 464): a lighting diagram, Class II, Division 2.

Electrical equipment in hazardous areas can be surface mounted or flush mounted. Fig. 13-113 (page 465). Note that the seal is placed at the 18″ level. This should prevent flammable gases or vapors from traveling through the entire conduit system. Seals may be boxed in by using a switch box cover. This method permits access to the seals later, after sealing is completed and the walls have been finished.

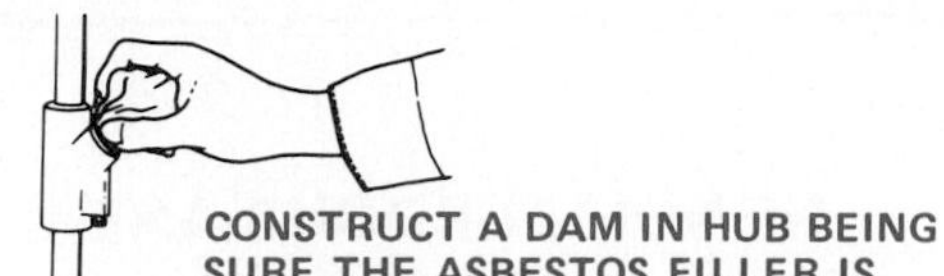

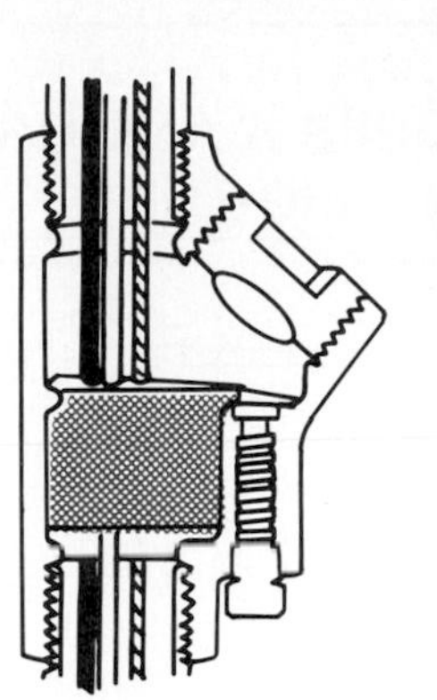

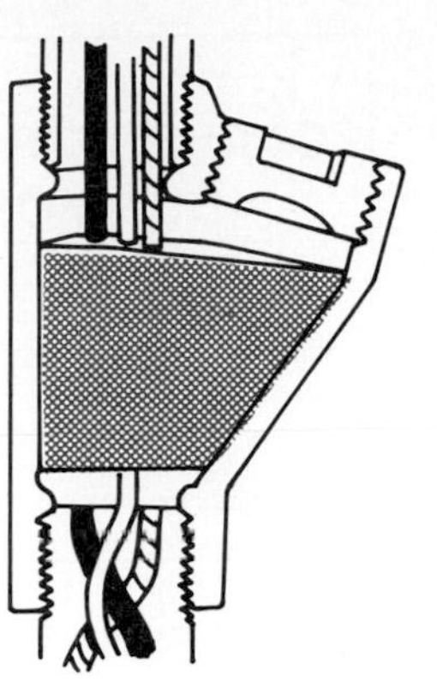

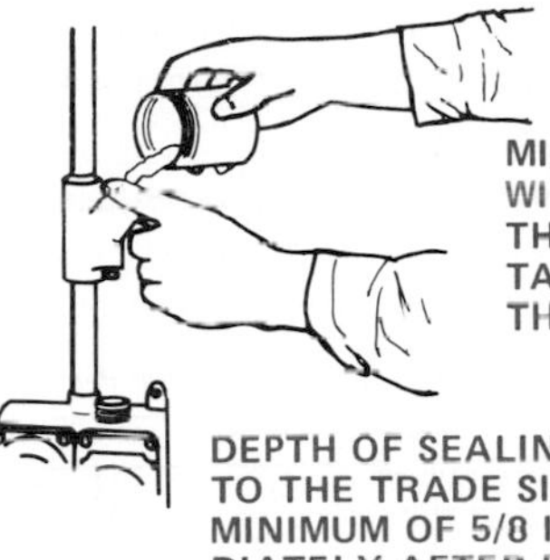

13–102. Sealing unilets for use as explosion-proof fixtures.

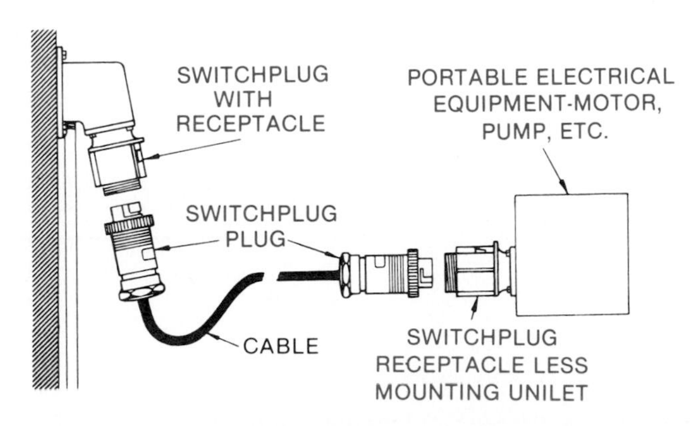

13–103. Hazardous-area plug and receptacle.

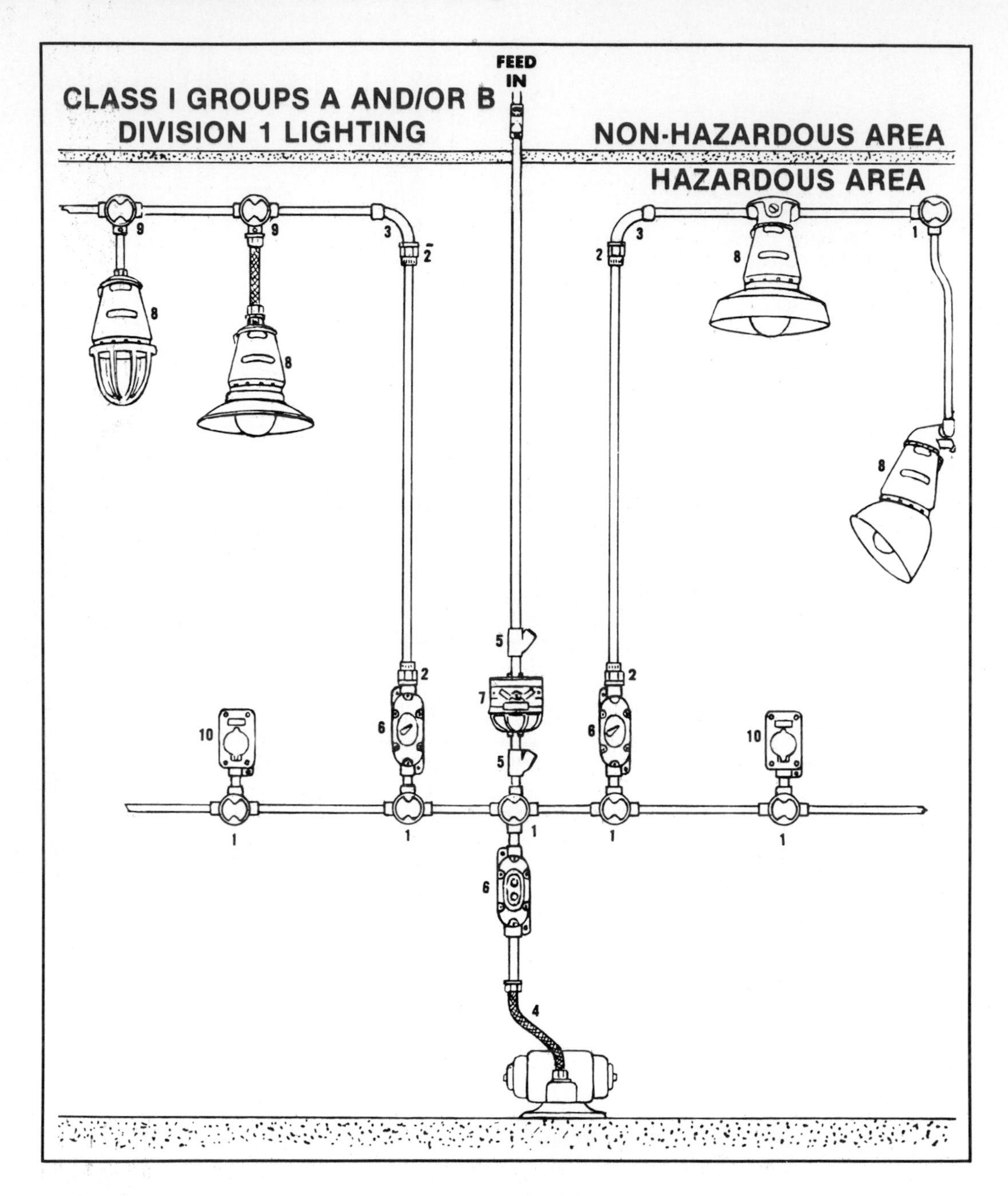

13–104. Hazardous location lighting diagram: (1) Junction unilets. (2) Unions. (3) Elbows. (4) Flexible couplings. (5) Vertical sealing unilets. (6) Factory-sealed control station and switch unilet. (7) Motor-starter unit. (8) Incandescent lamp. (9) Fixture hanger. (10) Plugs and receptacles.

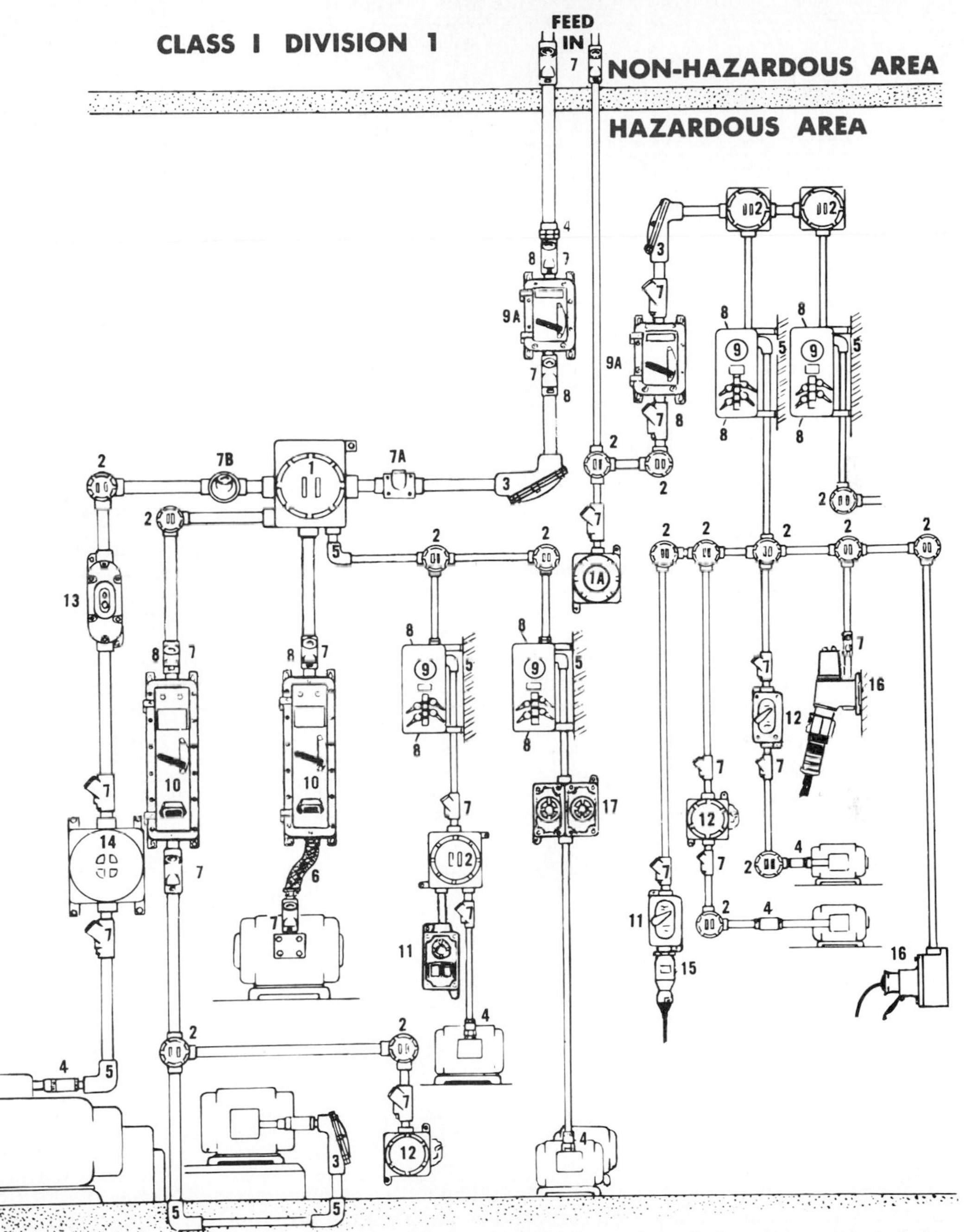

13–105. Power diagram: (1) Explosion-proof enclosure. (1A) Explosion-proof glass for viewing of the instrument inside. (2) Junction unilets. (3) Pulling unilet. (4) Unions. (5) Elbows, plugs, and miscellaneous fittings. (6) Flexible couplings. (7) Vertical sealing unilets. (7A) Horizontal sealing unilets. (7B) Junction unilets. (8) Drains (bottom) and breathers (top). (9) Circuit-breaker unilets. (9A) Explosion-proof switch which can be turned on and off with the elbow, if necessary. (10) Combination circuit breakers and line starters. (11) Switch unilets. (12) Motor-starter unilets. (13) Factory-sealed control station and switch unilets. (14) Explosion-proof unilet enclosure. (15) Cord connectors. (16) Plugs and receptacles. (17) Pilot lights.

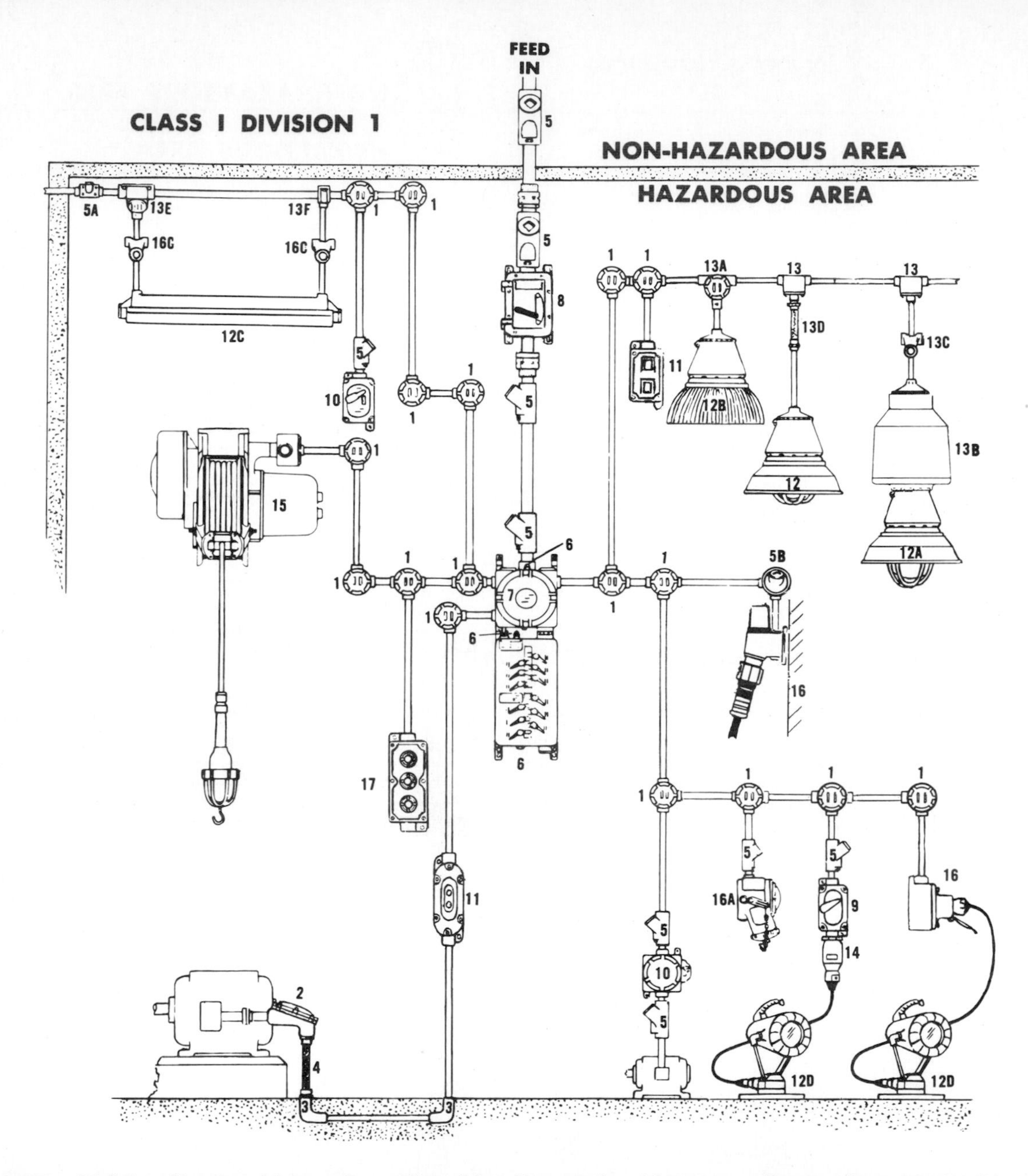

13–106. Lighting diagram: (1) Junction unilets. (2) Pulling unilet. (3) Unions, elbows, plugs. (4) Flexible couplings. (5) Vertical sealing unilets. (5A) Horizontal sealing unilets. (5B) Junction unilets. (6) Drains (bottom) and breathers (top). (7) Panel boards. (8) Circuit-breaker unilet. (9) Switch unilet. (10) Motor-starter unilet. (11) Factory-sealed control station and switch unilets. (12) Incandescent lamp. (12A) Mercury-vapor lamp. (12B) Prismatic incandescent. (12C) Explosion-proof fluorescent. (12D) Explosion-proof portable floodlight. (13) Fixture hanger. (13A) Fixture hanger. (13B) Mercury-vapor ballast. (13C) Swivels. (13D) Series flexible-hanger supports. (13E) Flanged junction unilet with canopy. (13F) Juncton unilet with hub cover. (14) Cord connectors. (15) Reelites. (16) Plugs and receptacles. (16A) Receptacle with interlocking safety switch. (16C) Dummy conduit support.

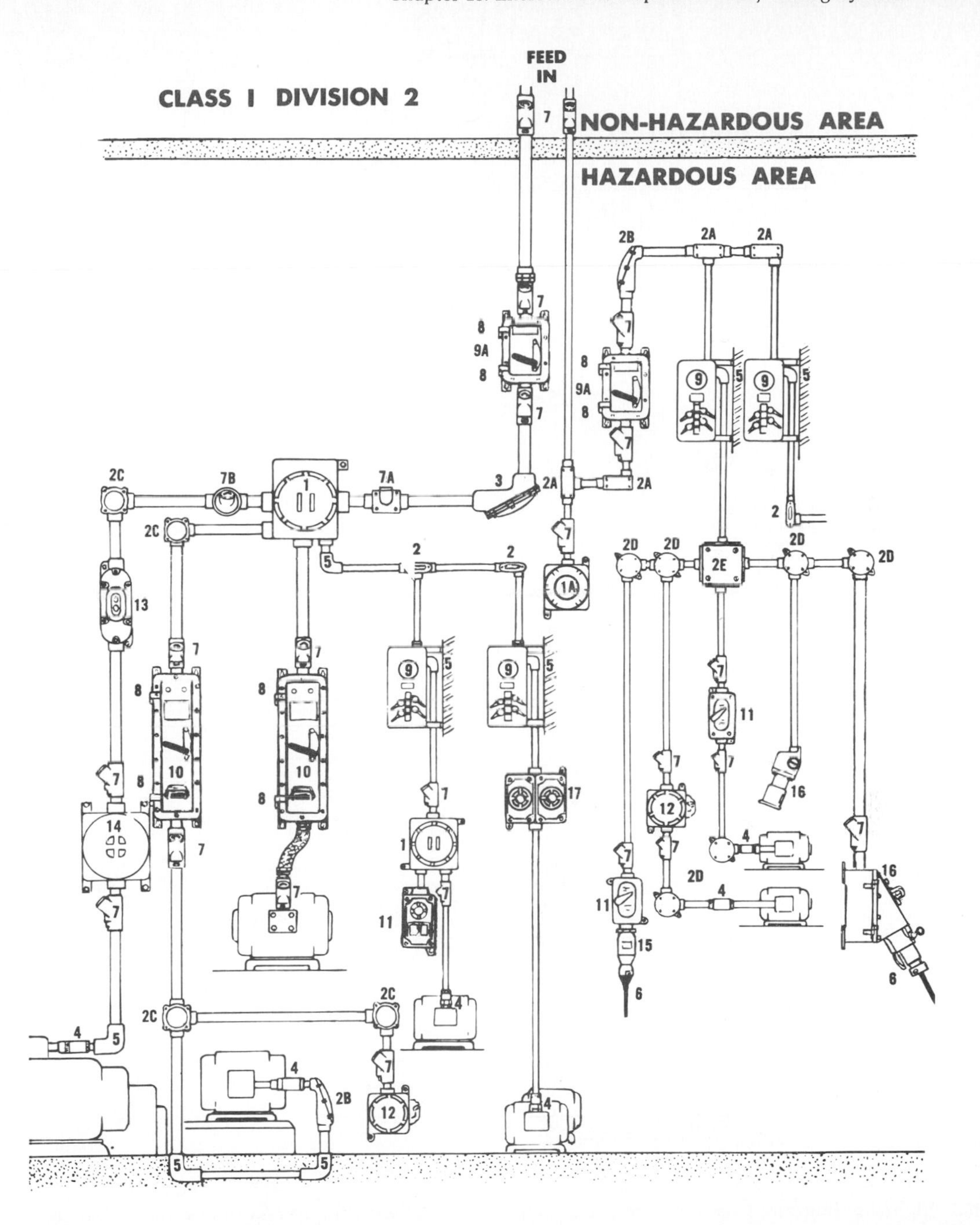

13–107. Power diagram: (1) Enclosures. (1A) Glass for viewing. (2) Junction unilets. (2A) Mogul junction unilets. (2B) Pulling unilets. (2C) Junction unilets. (2D) Junction unilets. (2E) Junction unilets. (3) Pulling unilet. (4) Unions. (5) Elbows, plugs, and miscellaneous fittings. (6) Flexible couplings. (7) Vertical sealing unilets. (7A) Horizontal sealing unilets. (7B) Junction unilets. (8) Drains (bottom) and breathers (top). (9) Circuit-breaker unilets. (9A) Circuit-breaker unilets. (10) Combinaton circuit breakers and line starters. (11) Switch unilets. (12) Motor-starter unilets. (13) Factory-sealed control station and switch unilets. (14) Explosion-proof unilet enclosure. (15) Cord connectors. (16) Plugs and receptacles. (17) Pilot lights.

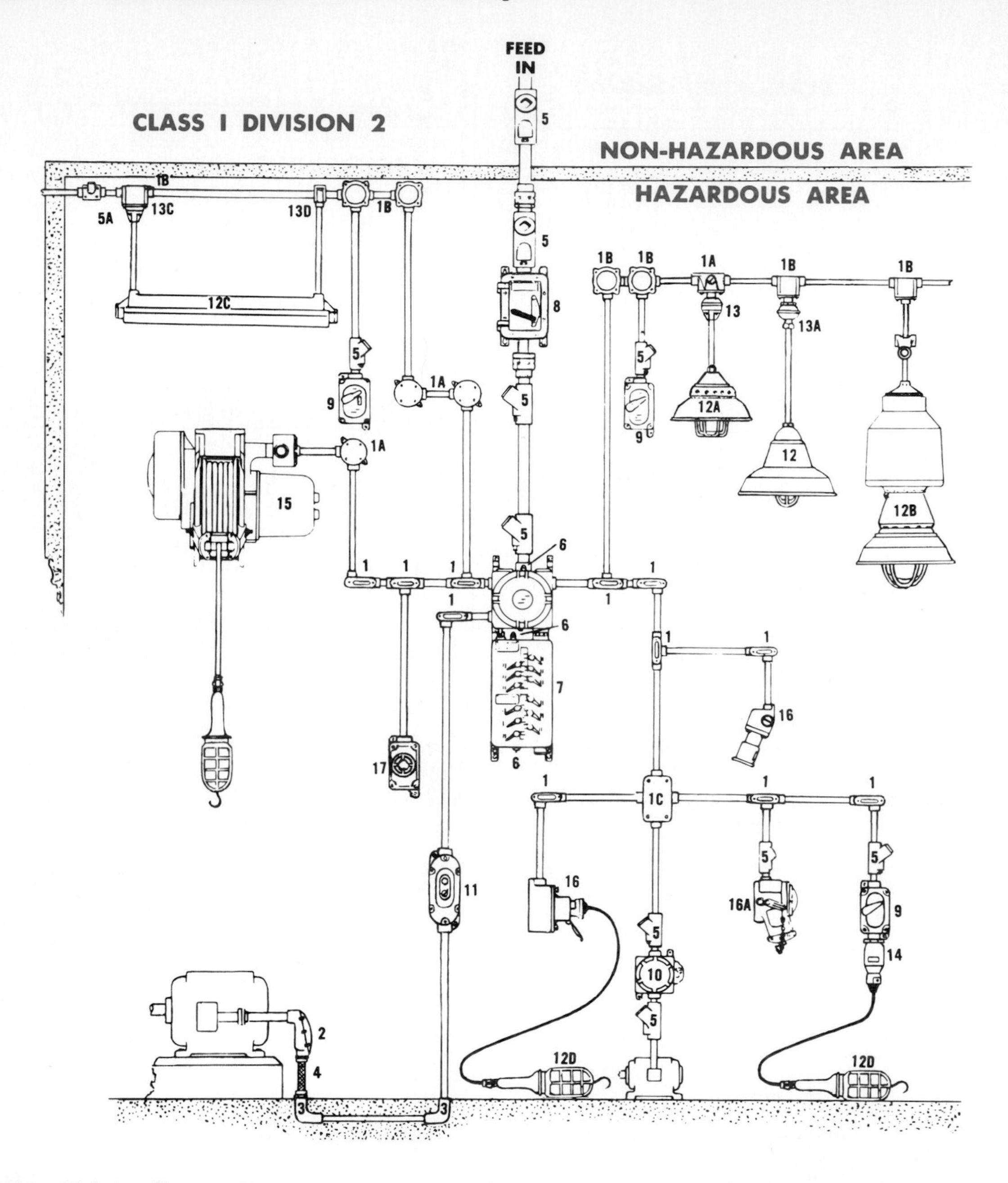

13–108. Lighting diagram: Nos. 1–17 are same as in Fig. 13–92, with the exception of 12D, vaportight handlamp.

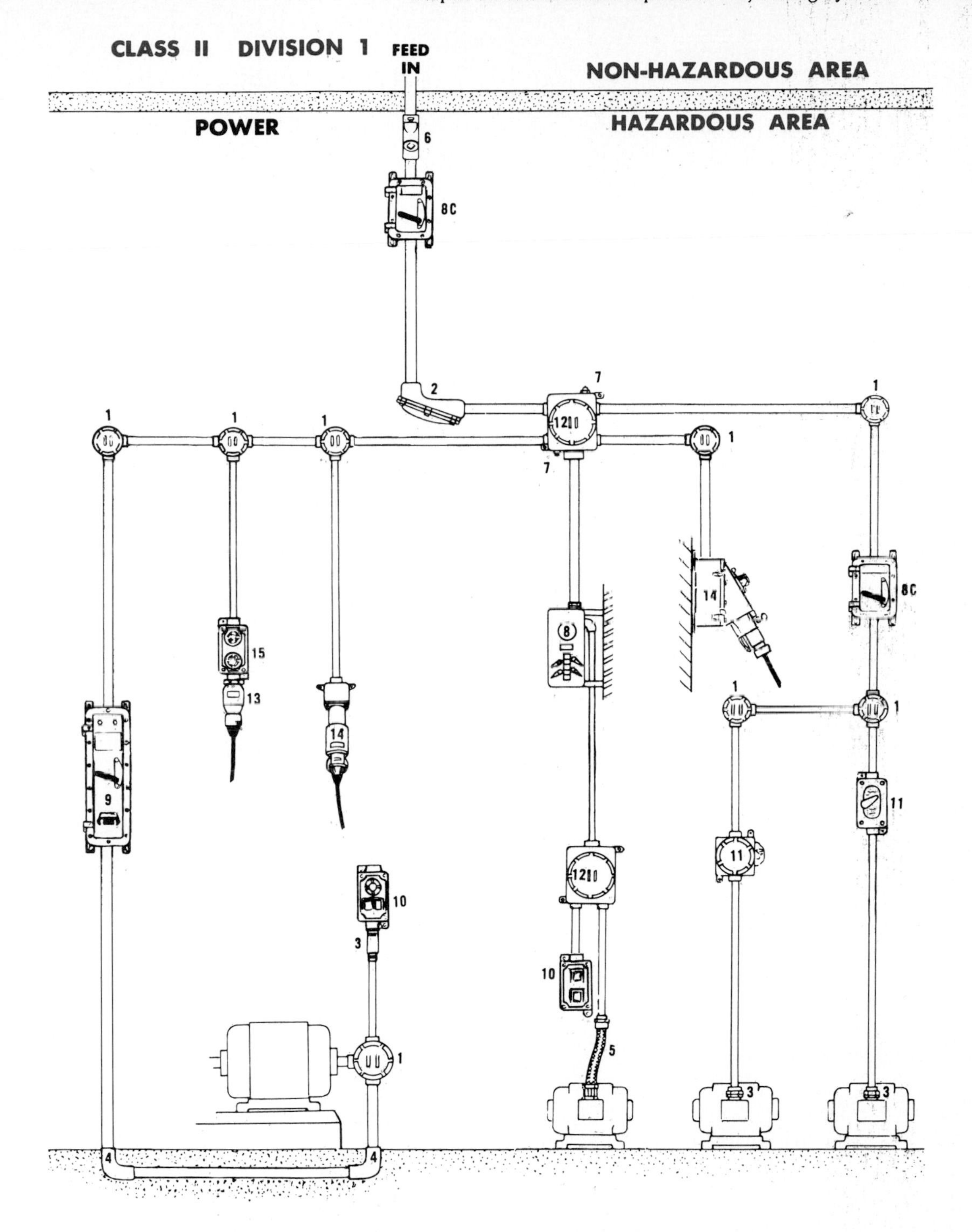

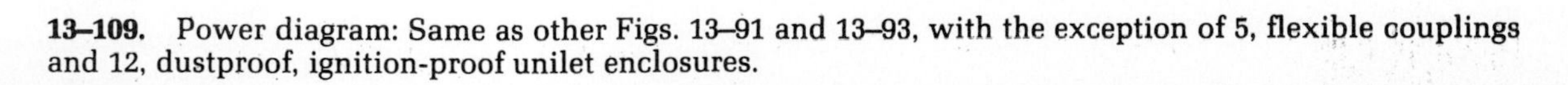

13–109. Power diagram: Same as other Figs. 13–91 and 13–93, with the exception of 5, flexible couplings and 12, dustproof, ignition-proof unilet enclosures.

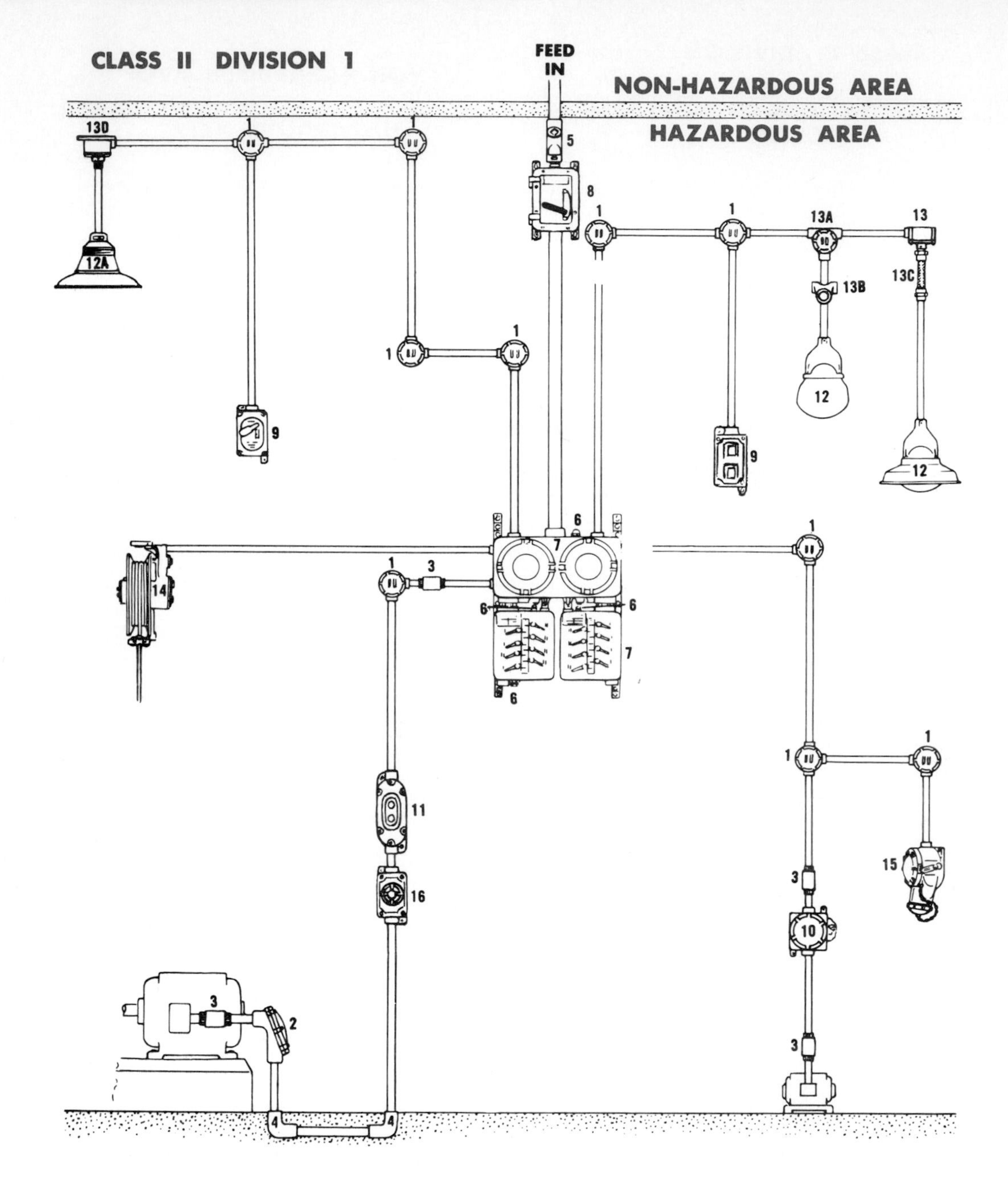

13–110. Same as other illustrations for lighting diagram.

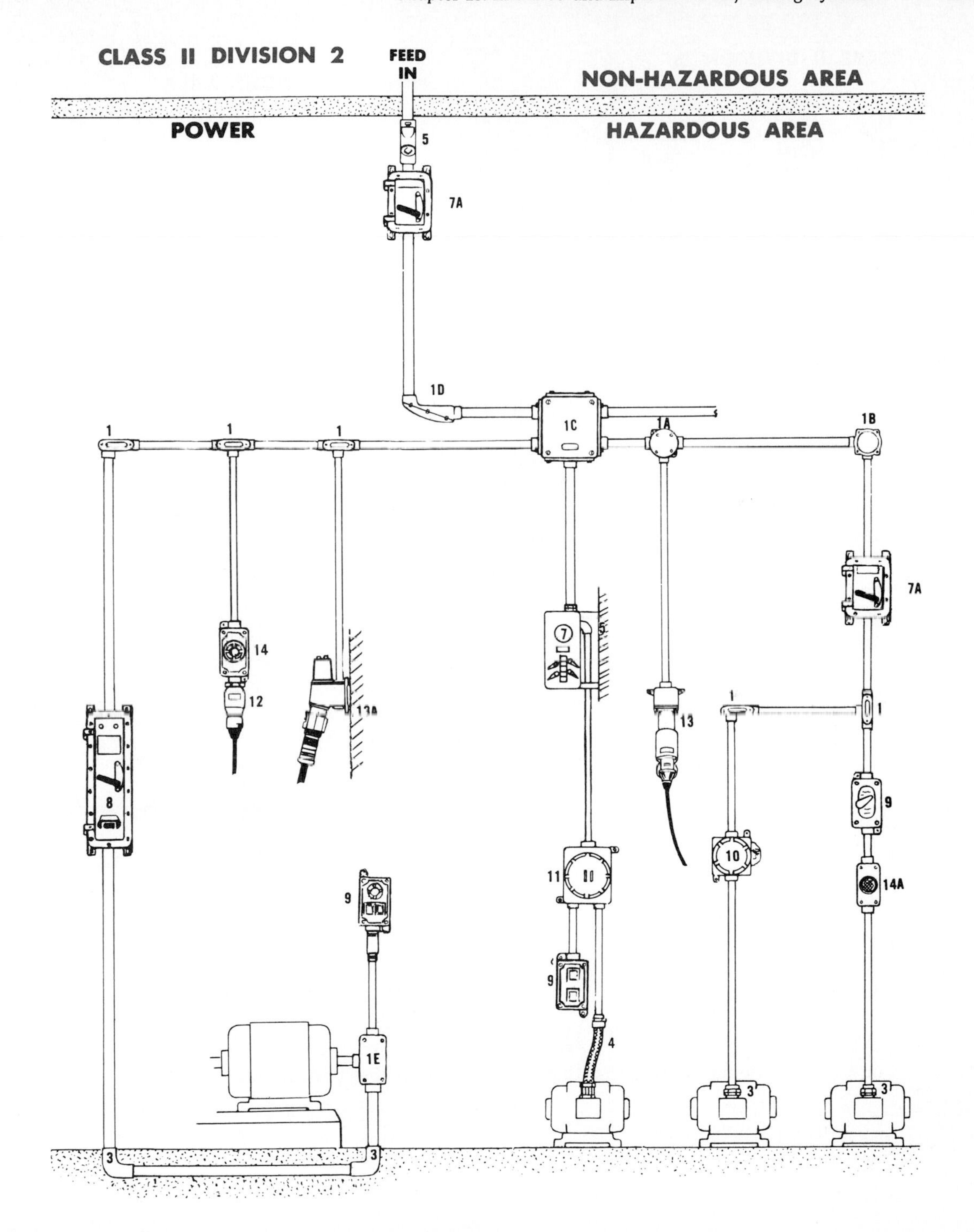

13–111. Same as other illustrations for power diagram.

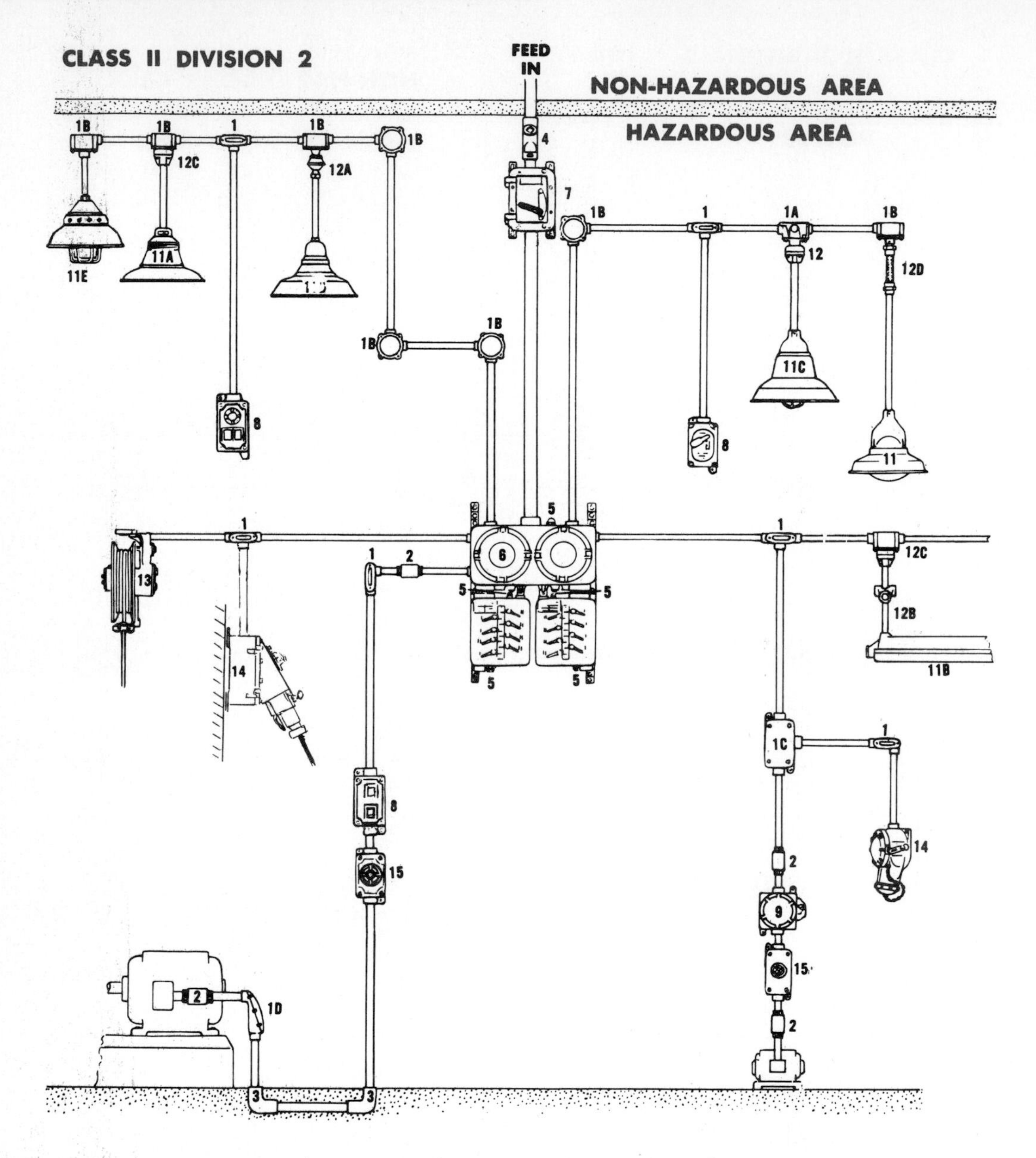

13–112. Same as other illustrations for lighting diagram, except that parts are dustproof.

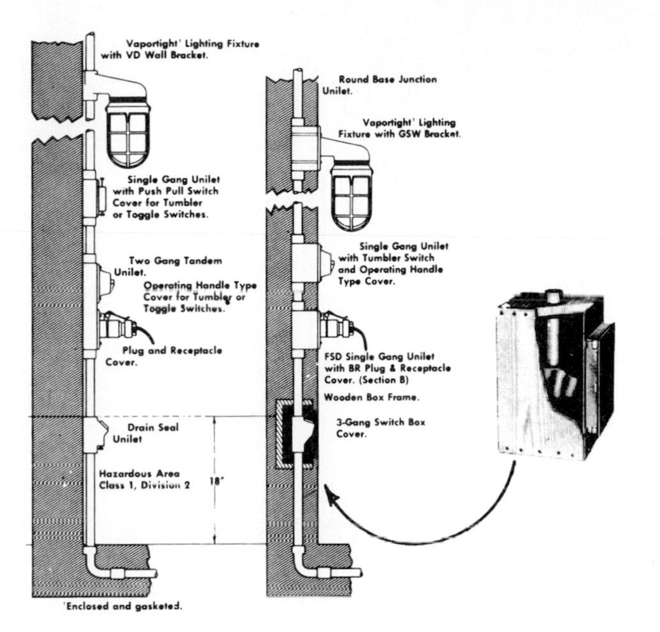

13–113. Electrical equipment can be surface mounted or flush mounted.

Other hazardous locations include:

- *Repair and storage areas of commercial garages.* Fig. 13-114.

- *Aircraft hangars.* Fig. 13-115. These have a Division 1 and 2 classification and related wiring requirements meet this class.

- *Service stations.* Such stations, especially those with gasoline pumps, require a particular wiring method to prevent explosions due to gasoline vapors. Figs. 13-116 & 13-117 (page 468). They are classified as Class I, Division 1 and 2 installations.

- *Bulk storage plant structures.* These require explosion-proof fixtures around the bulk storage area. It is a Class I, Division 1 and 2 classification. Fig. 13-118 (page 469).

People, vehicles, machinery, the flow of liquids and gases in pipes and containers—all are generators of electricity. Although static charges are usually minute, in enclosures they accumulate because the conducting object is insulated. Therefore the charge cannot escape.

For instance, the rubber tires on a gas truck (tanker) prevent an electrical charge from leaving the truck. A spark is possible when the nozzle touches the fill pipe or similar metal object. The static charge can ignite hazardous fumes surrounding the dock area. To avoid this danger, use a static grounding steel cable to preground the trucks and other objects. Several feet should be maintained between the truck and possible spark conditions.

Spray painting areas require special attention in the installation of electrical wiring. Before installing electrical equipment in spray

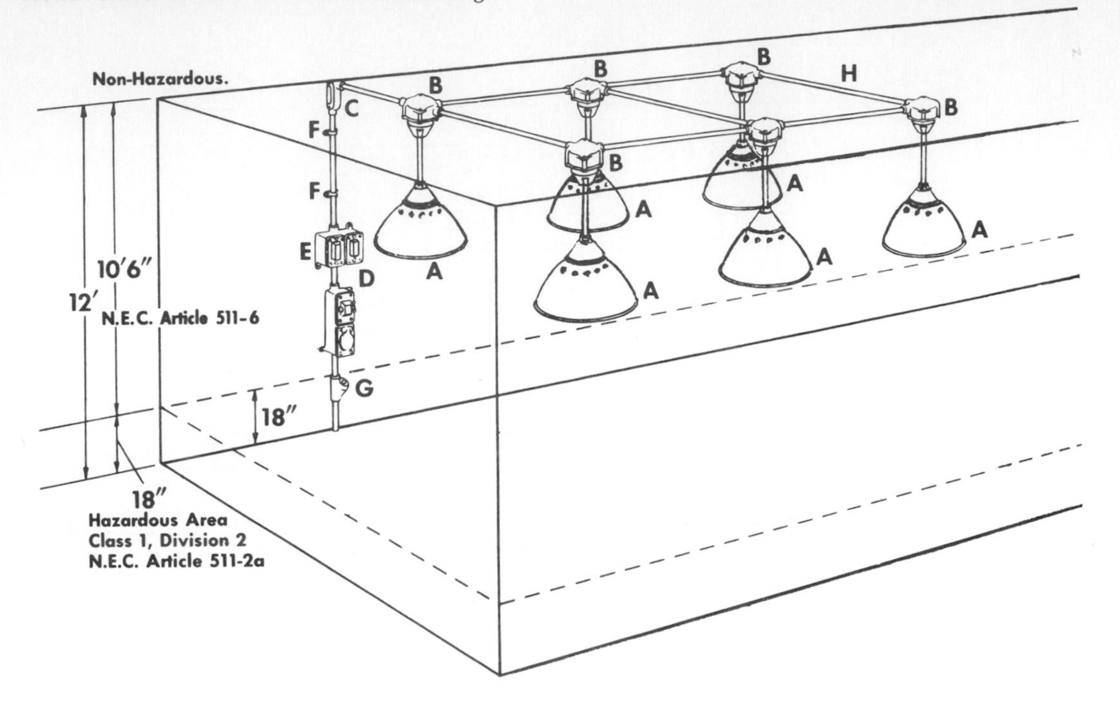

F—MALLEABLE IRON CLAMPS AND CLAMP BACKS for holding of conduit. Additional fittings such as Erickson couplings, meter connectors, elbows, reducers and pipe straps

G—EXPLOSION-PROOF SEALS are required in locations where service, storage and repairs of motor vehicles are in commercial garages. See N.E.C. Article 511 and 512.

D—FS AND FD MALLEABLE IRON UNILETS available in one, two, three, or four gang plus two gang tandem for use with general type switches, receptacles, pilot lights, etc.

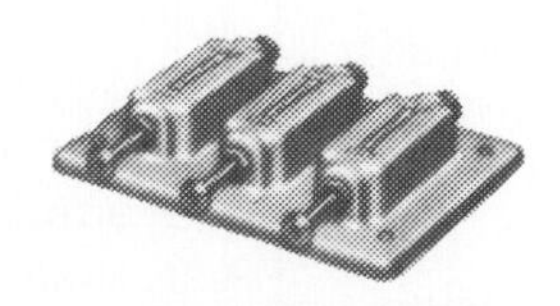

E—FS MALLEABLE IRON VAPORTIGHT* COVERS. Cast construction furnished with gasket for Vaportight* seal. Wide variety of covers are available.

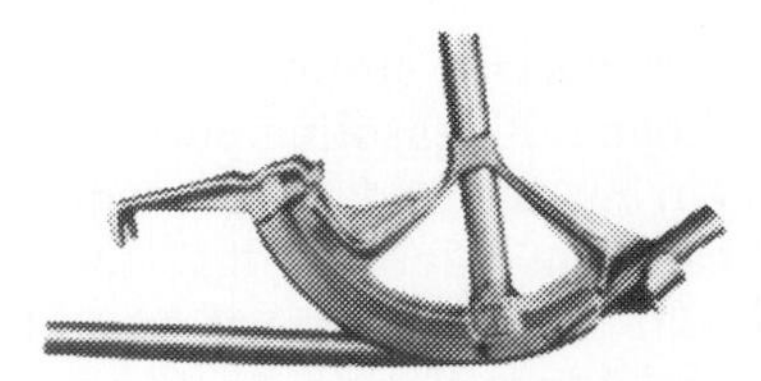

H—RIGID CONDUIT is required for all vaportight installations and can be bent easily

13–114. Commercial garages, repair and storage areas. See the picture below the diagram for the proper A-, B-, C-, or D-type fixture to go along with the diagram.

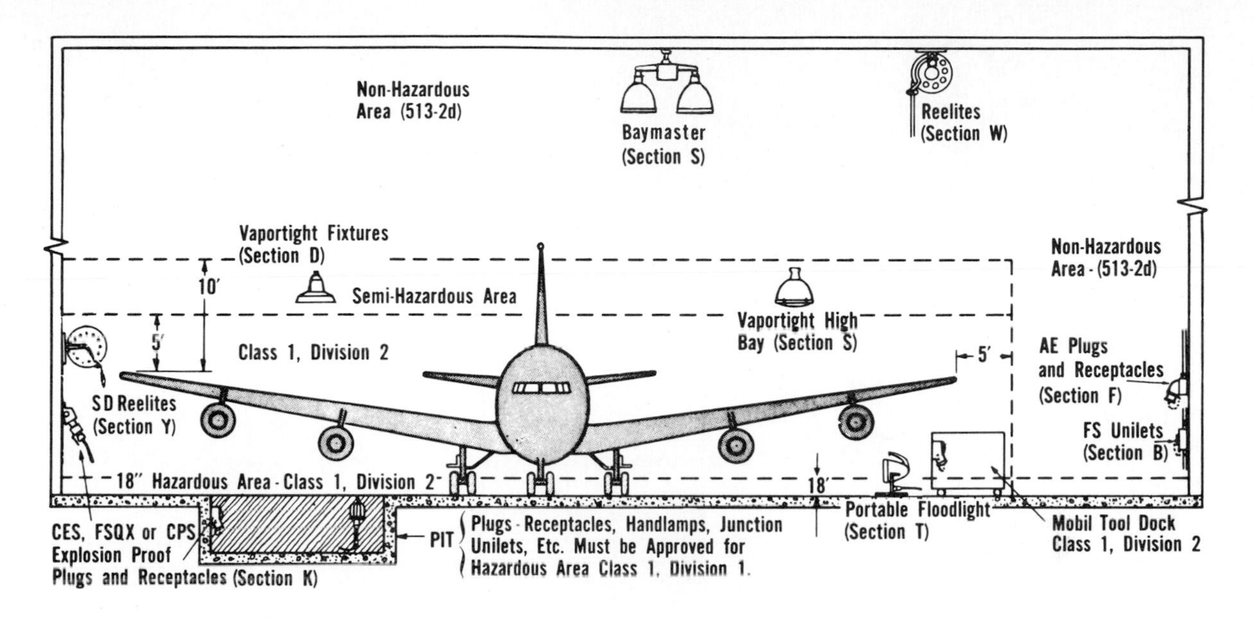

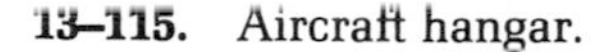
13–115. Aircraft hangar.

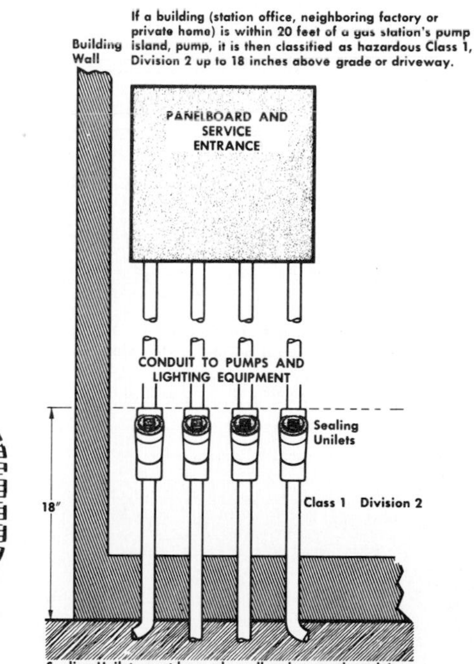

Ground Level
Installation.

Hazardous Area
Class 1, Division 1

Sealing Unilets

Junction
Unilets

Class 1,
Division 2

Ground Level Installation.

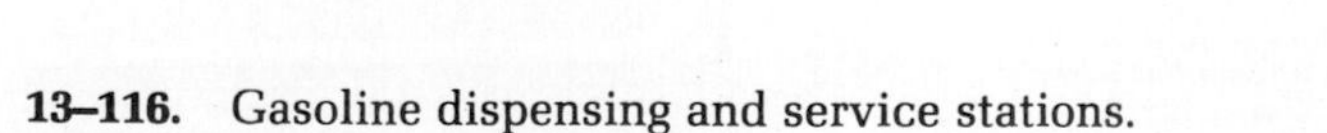

Gas Pump
Installed
on Ground
Surface

Area 18″ from all sides of the Pump extending 4′ High is Class 1, Division 1. The Area below ground is Class 1, Division 1.

13–116. Gasoline dispensing and service stations.

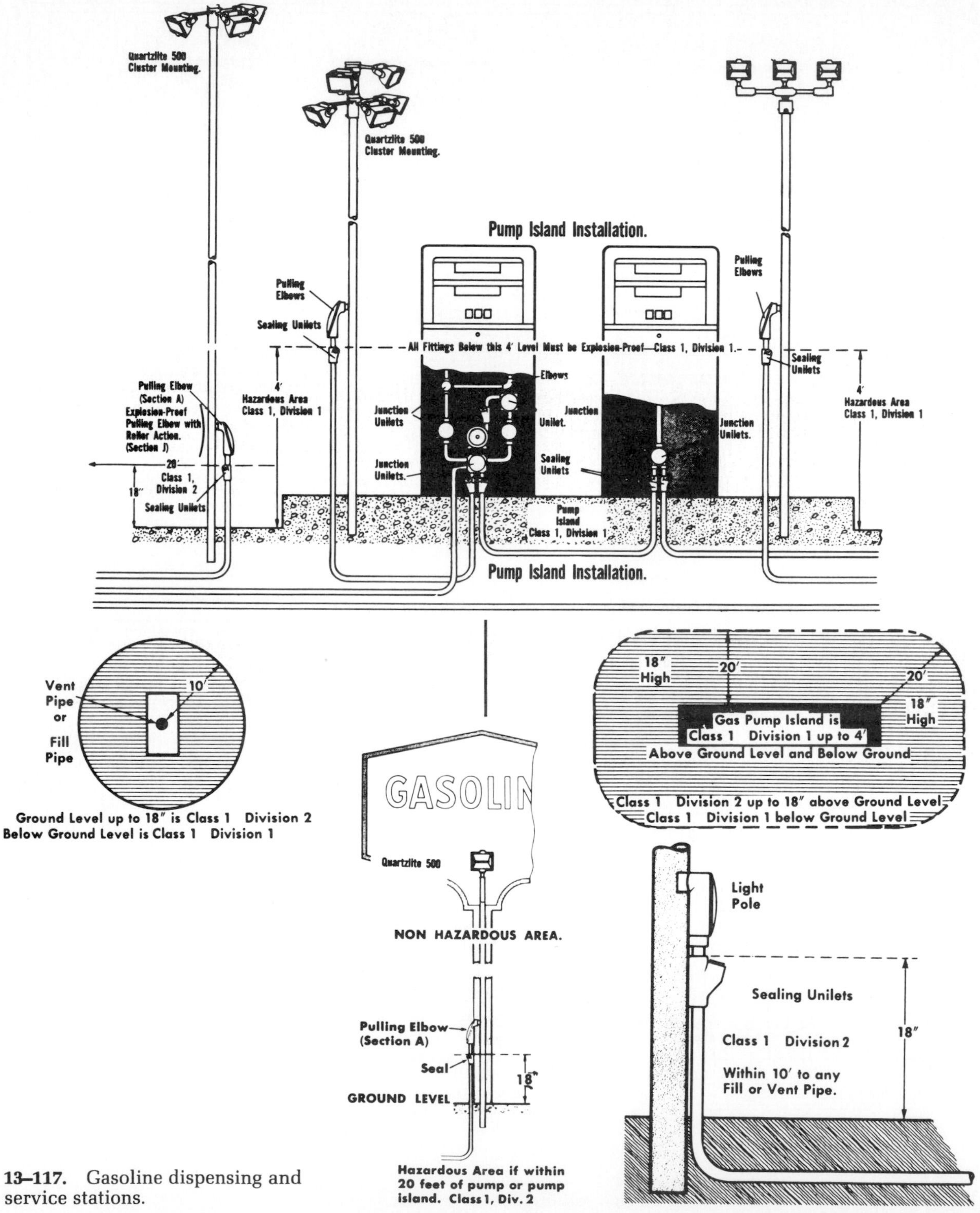

13–117. Gasoline dispensing and service stations.

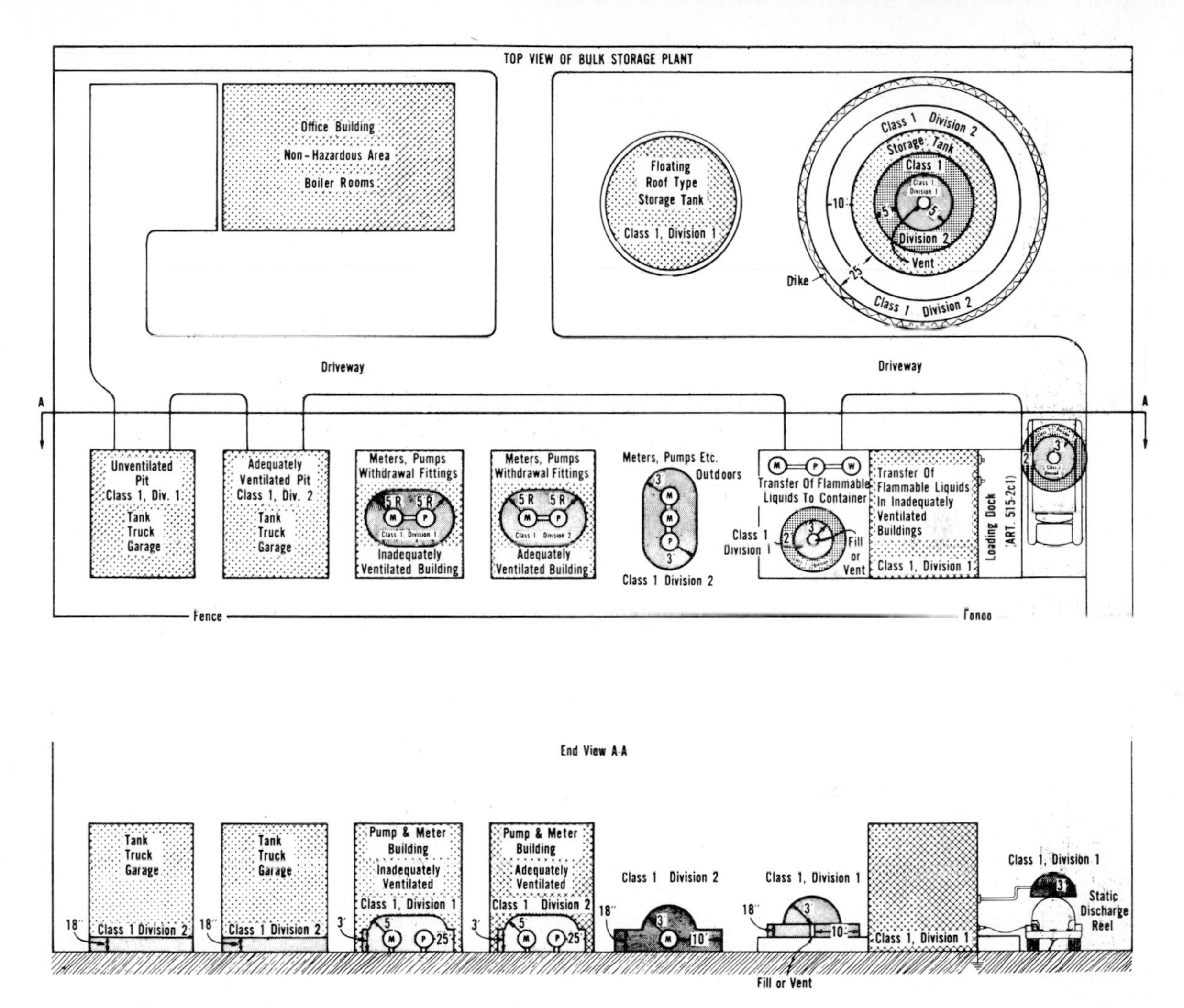

13–118. Bulk storage plants.

paint finishing areas, investigate the standards thoroughly. They would be classified Class I, Division 2. Fig. 13-119.

Operating rooms in hospitals require complex planning. You will need special types of plugs and outlets to wire a hospital operating room. Fig. 13-120. In the National Electrical Code, this comes under Class I, Division 1: flammable anesthetics. Fig. 13-121 illustrates the type of plug and receptacle needed to contain an explosion in case anesthetic vapors inside the receptacle become strong enough to ignite.

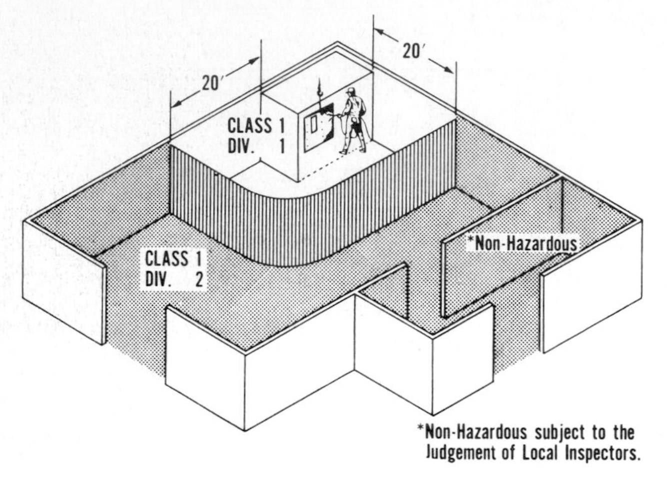

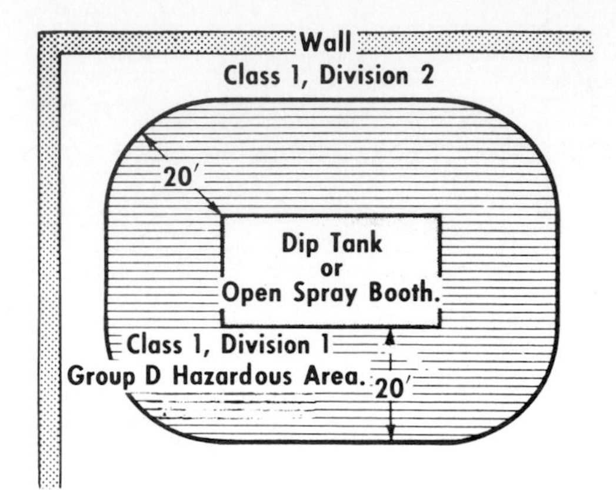

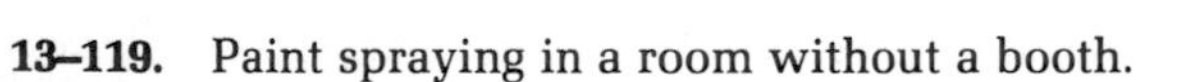

13–119. Paint spraying in a room without a booth.

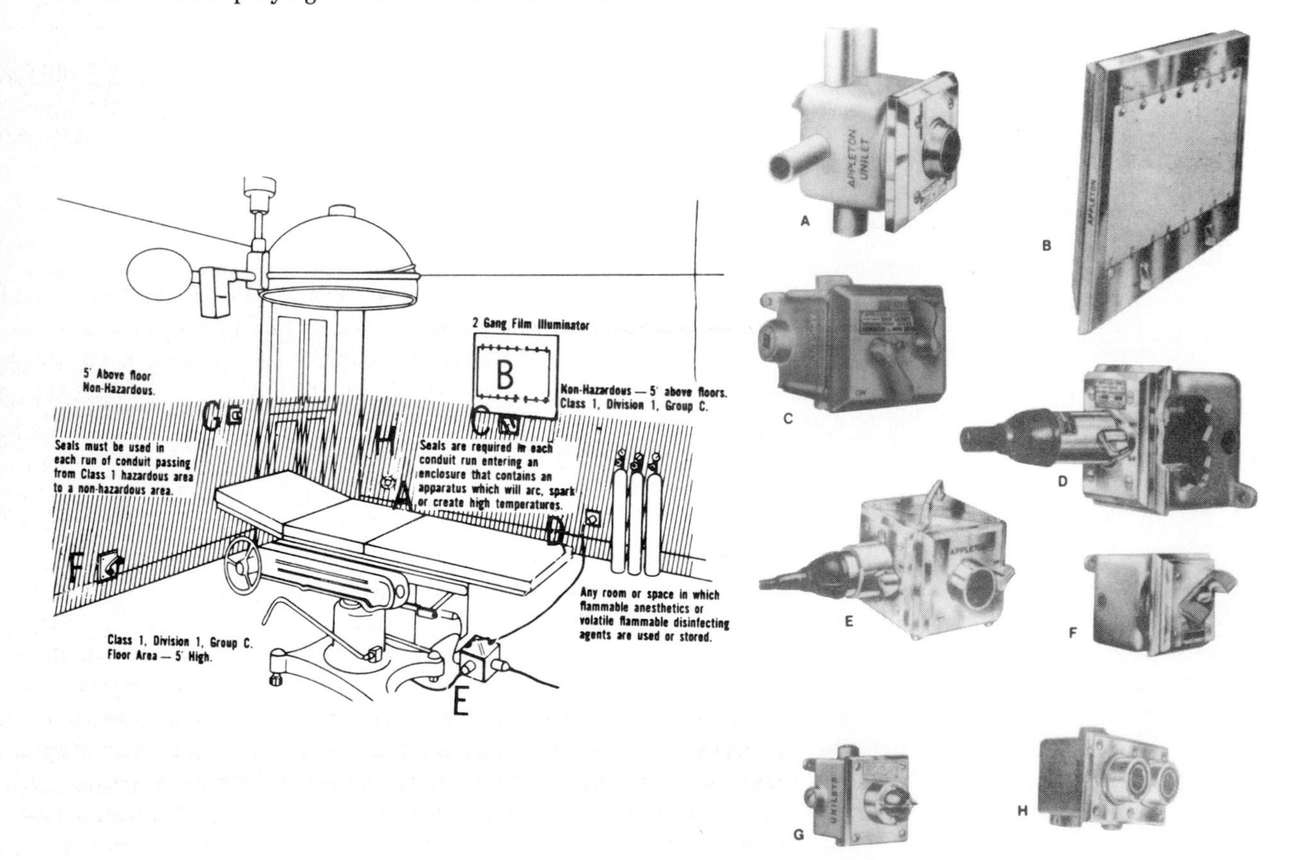

13–120. This operating room requires special fixtures as specified by the NEC. Where flammable anesthetics are used, explosion-proof fixtures must be used: (A) Receptacles are available with adjustable sleeves. (B) X-ray film illuminator. (C) Switch unilet. (D) Receptacles with interlocking switch. (E) Current tap unilet. (F) Foot switch unilet. (G) Pilot light unilet. (H) Plugs and receptacles.

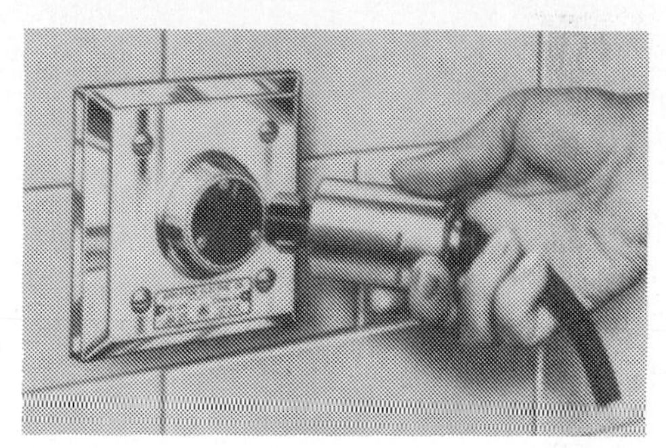

13–121. Receptacle and plug for use in a hospital operating room.

QUESTIONS

1. What is the advantage of using the BX cable ripper?
2. What is a box connector? Where is it used?
3. Where are expansion couplings used?
4. What are beam clamps?
5. Where are end bushings used?
6. What is the purpose of support grips?
7. List four types of cable grip attachments.
8. Why are reducing bushings sometimes needed?
9. Where are elbows used? How many types are available?
10. Where are offset nipples used?
11. Differentiate between Class I and Class II hazardous locations.
12. What is the minimum number of full threads required in threaded joint construction?
13. How do hospital outlets and plugs differ from those used at home?

KEY TERMS

BX cable
BX cutter
clamps
connectors
couplings
grips

SECTION FOUR PRODUCING and CONSERVING ELECTRICITY

CHAPTER 14

GENERATORS AND MOTORS

Objectives

After studying this chapter, you will be able to:

- Describe the principle of operation of a DC generator.
- List the three main types of DC generators.
- Describe the principle of operation of the DC motor.
- Demonstrate the right-hand rule for motors.
- Identify the primary components of an AC generator.
- Compare a synchronous motor with an induction motor.
- Troubleshoot three-phase motors.

DC GENERATORS

Generators are devices that generate, or produce, electricity. A generator may be used to produce either alternating current (AC) or direct current (DC). The AC generator is usually called an *alternator*.

The mechanical energy required for operation of a generator can be supplied by a gasoline engine, a diesel engine, an electric motor, a steam turbine, or some other kind of prime mover. The electrical energy developed by the generator can be either AC or DC depending on the construction features of the machine. Under certain circumstances, a generator can be used as a motor, in which case electrical energy is supplied to the terminals of the machine and mechanical energy is delivered to some kind of load.

Motors and generators are used extensively in the military services and are integral parts of many communications systems. For example, field radio and telephone apparatuses employ hand generators and dynamotors. Motors, ranging in size from less than 1 horsepower to over 25 horsepower, are used to furnish power for a variety of purposes.

The fundamental principle for the generation of an electromotive force (EMF) was explained in Chapter 1. The generation of an EMF depends on the relative motion between a conductor and a magnetic field. The magnetic field, or flux, may be stationary and the conductor moved, or the conductor may be stationary and the flux moved. In either case, the result is the same. The amount of EMF induced in a conductor moving in a magnetic field depends upon three factors:

- The strength of the field.
- The length of the conductor, or the size and number of turns in a coil (if the conductor is wound in the form of a coil).
- The speed at which the conductor sweeps through the magnetic field, or the speed of rotation of a coil in a magnetic field.

The basic AC generator, Fig. 14-1, is made up of a permanent magnet to produce a magnetic field; a single-turn loop which is rotated in the field and in which is induced an EMF; and a slip ring and brush assembly which makes possible the connection of generated EMF to an external load. This simple generator, although helpful in the understanding of basic theory, has little practical application because the amount of electrical energy generated is very small.

Practical generators, on the other hand, are capable of supplying electrical energy in almost any desired amount. The practical generator may use one or more electromagnets instead of the permanent magnet, an iron-core coil composed of many turns of wire instead of a single loop, and a superior (though similar) slip ring and brush assembly is replaced by a commutator and brush assembly. The housing and parts of practical generators are designed for a maximum speed. This speed is consistent with use and the mechanical limits of safety.

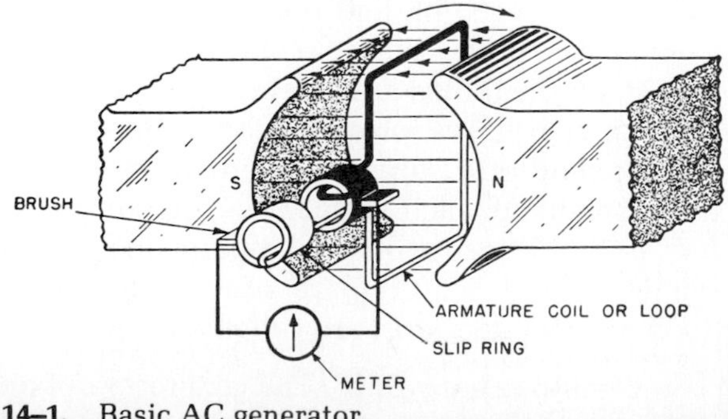

14–1. Basic AC generator.

Consider the four positions of the single-turn loop and slip rings as they rotate clockwise in a uniform magnetic field produced by the poles of a magnet. Fig. 14-2.

When the loop rotates through the position shown in #1 of Fig. 14-2, the black coil side is moving toward the north pole and the white side is moving toward the south pole. Because the coil sides are moving parallel to the direction of the magnetic field, no flux lines are cut and the EMF induced in the loop is zero. This condition may be verified by connecting a suitable meter across the output terminals as shown by the arrows in Fig. 14-2 (center-scale position of needle indicates zero volts). Note that the coil ends are connected to the slip rings. The ends rotate simultaneously with the loop. The stationary carbon brushes make contact with the slip rings. They are used to conduct the generated voltage to the external load—in this case the meter.

After the loop has rotated 90° from its initial position and is passing through the position shown in #2, the black coil side is moving downward and the white side upward. Both sides are cutting a maximum number of flux lines. The induced EMF, as indicated by the meter, is at a positive maximum.

As the loop passes through an angle of 180° in #3, the coil sides are again cutting no flux lines. The generated EMF is zero.

As the loop passes through an angle of 270°, as in #4, the coil sides are cutting a maximum number of flux lines. The generated EMF is at a negative maximum.

The next 90° turn of the loop completes a 360° revolution. The generated EMF falls to zero.

The complete rotation may therefore be summed up as follows: As the loop makes one complete revolution of 360°, the generated

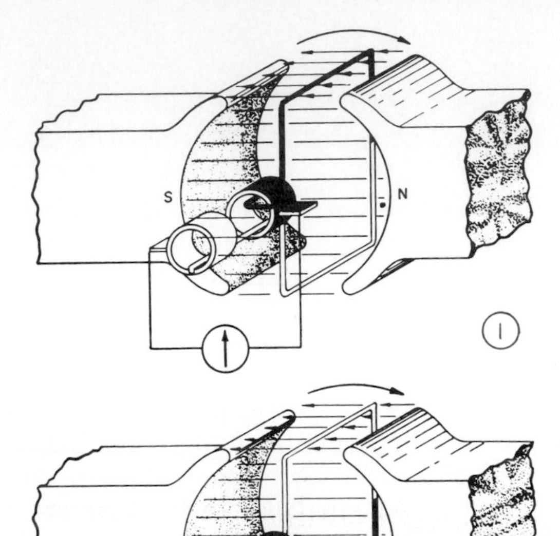

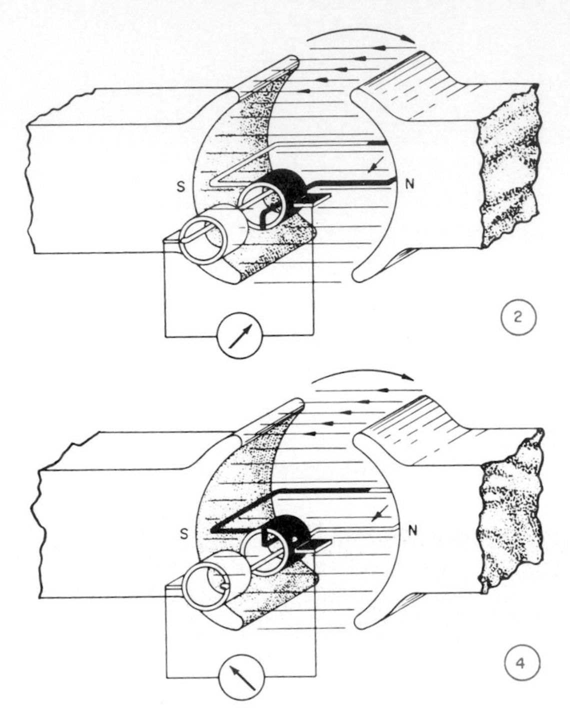

14–2. Generation of an EMF.

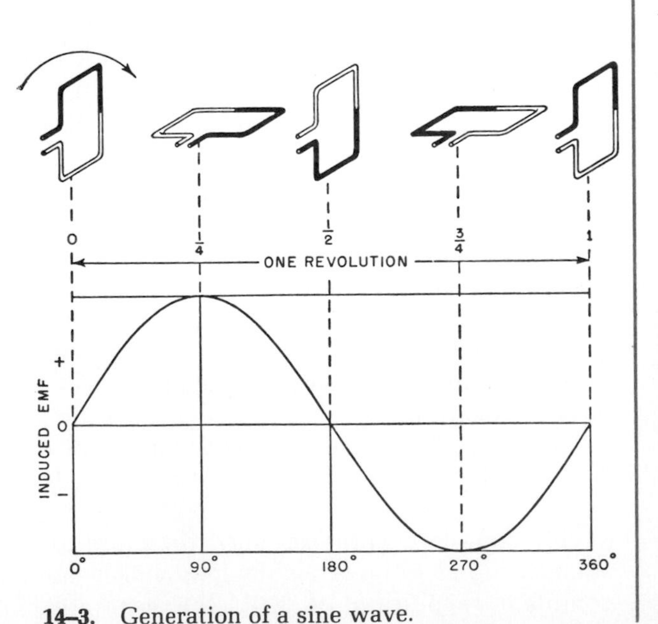

14–3. Generation of a sine wave.

EMF passes from zero to a positive maximum, to zero, to a negative maximum, and back to zero. If a constant speed of rotation is assumed, the EMF generated is a sine wave. Fig. 14-3.

Basic DC Generators.

By replacing the slip rings on the basic AC generator with two semicylindrical segments called a *commutator*, a basic DC generator is obtained. Fig. 14-4. In this figure note that the black coil side is connected to the black semicylindrical segment, and the white coil side is connected to the white segment. These segments are insulated from each other. The two stationary brushes, placed on opposite sides of the commutator, are so mounted that each brush contacts each segment of the commutator as the latter revolves simultaneously with the loop. The rotating parts of a DC generator (coil and commutator) are called an *armature*.

■ *Generating an EMF*. The generation of an EMF by the loop rotating in the magnetic field

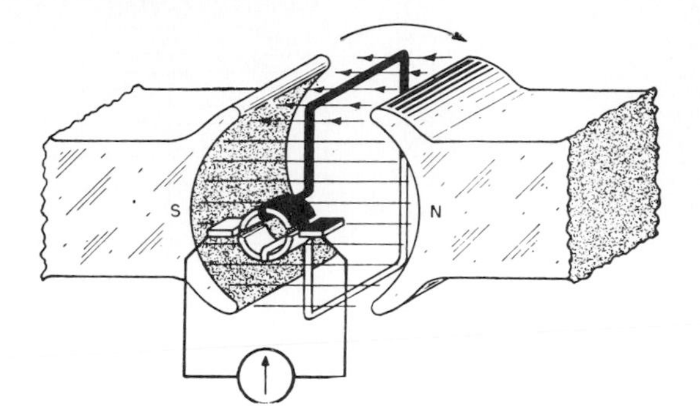

14-4. Basic DC generator.

is the same for both AC and DC generators, but the action of the commutator produces a DC output voltage.

Examine Fig. 14-5 to see how the output of the DC generator is produced, as compared to that of an AC generator. Note that the switching action of the commutator segments makes the output of the DC generator above the line, with no part of the output going in the reverse direction. At the instant each brush contacts two segments of the commutator, such as in A, C, and E of Fig. 14-5, a direct short circuit is produced. If an EMF is generated at this time, a high current will flow in the short circuit, causing an arc and thus damaging the commutator. For this reason, the brushes must be placed in the exact position where the short will occur when the generated EMF is zero. This position is called the neutral plane.

■ *Ripple.* Since the voltage of the basic DC generator varies from zero to a maximum each time the generator armature revolves 180°, there are two zero-to-maximum variations in each complete rotation of the armature. Fig. 14-6. This variation of the DC voltage, called ripple, may be reduced by using more loops or coils, as shown by comparing Figs. 14-6 & 14-7. As the number of loops is increased, the variation between maximum and minimum is decreased, and the output voltage of the generator approaches, a steady DC value.

Practical DC Generators. A practical generator uses electromagnets, instead of permanent magnets. This produces a much stronger magnetic field without increasing the physical size of the magnets.

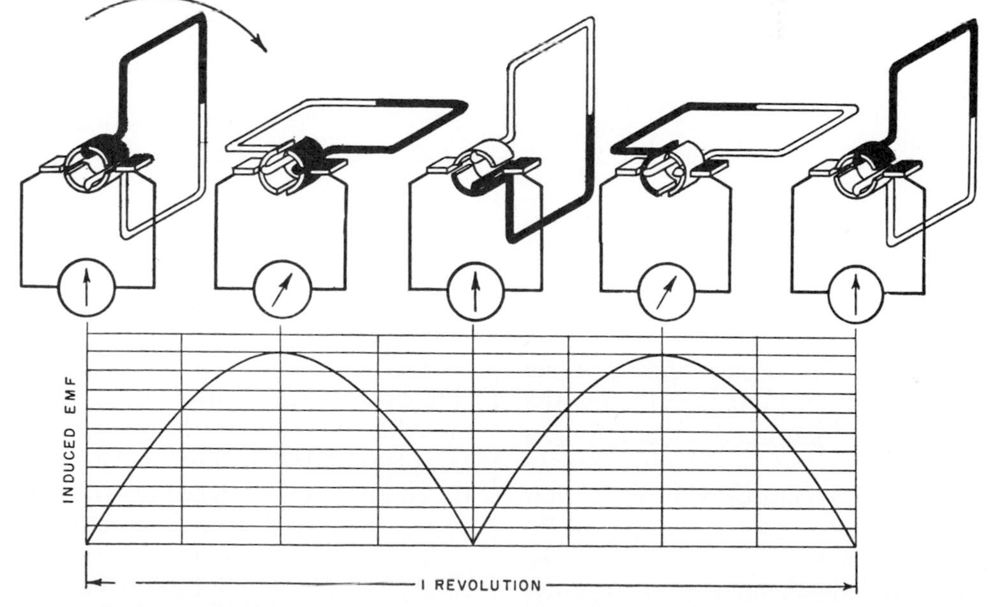

14-5. Operation of a basic DC generator.

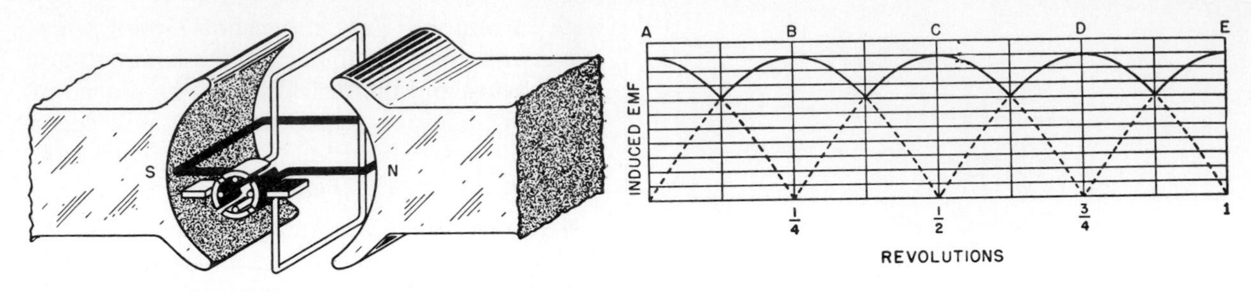

14–6. Output voltage from a 2-coil armature.

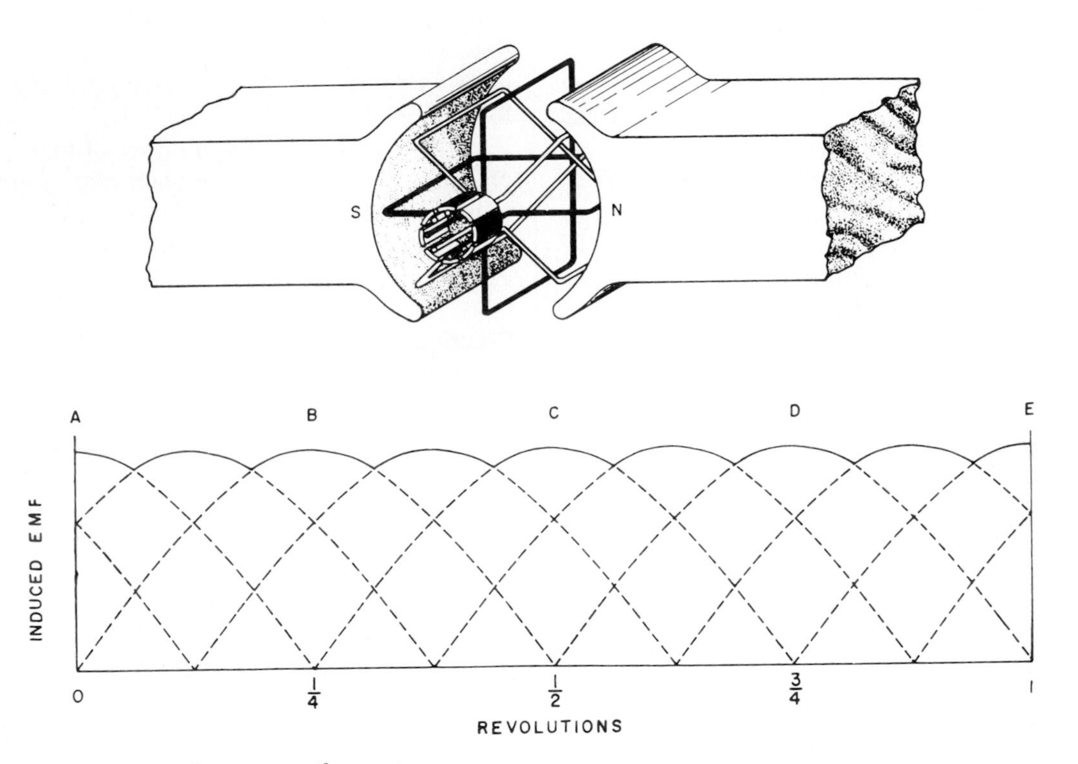

14–7. Output voltage from a 4-coil armature.

Fig. 14-8 shows the pole piece bolted to the yoke. The field winding is wrapped around the poles. Note the north-south poles and the lines of flux.

The same basic generator, but with four poles, is shown in Fig. 14-9. Note the north-south pole orientation here, and the way the field windings are connected. Fig. 14-10 shows a more complicated arrangement with an eight-section, ring-type armature.

Commutator segments are more numerous for a practical generator design. Fig. 14-11. Note the coils and slots in the drum-type armature. Mica is used to insulate the copper segments of the commutator from one another. Brushes ride on the surface of the commutator. The brushes form the electrical contact between the armature coil and the external circuit. Brushes, made of high-grade carbon, are held in place by brush holders. The holders are insulated

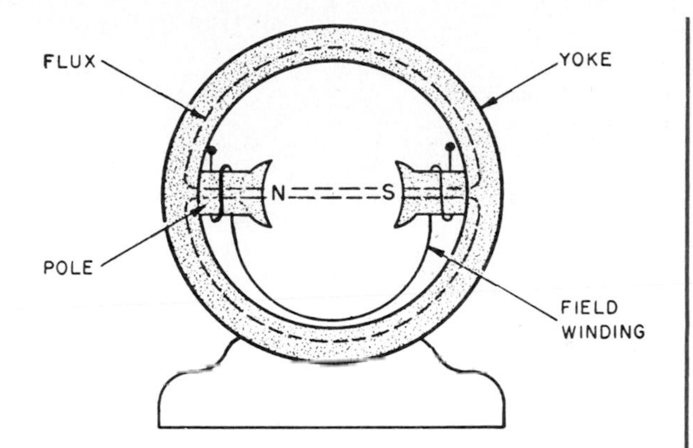

14–8. Two-pole assembly.

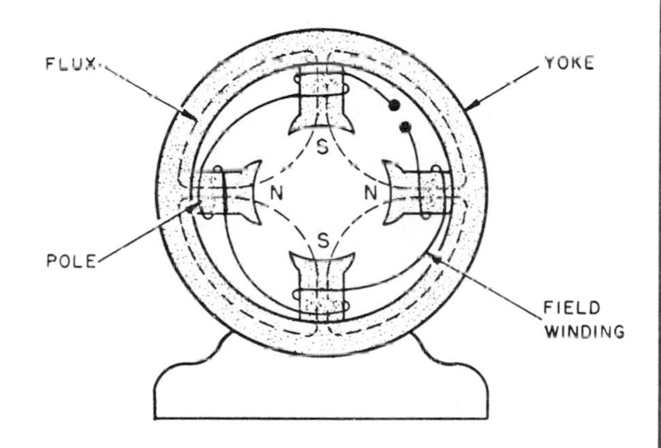

14–9. Four-pole assembly.

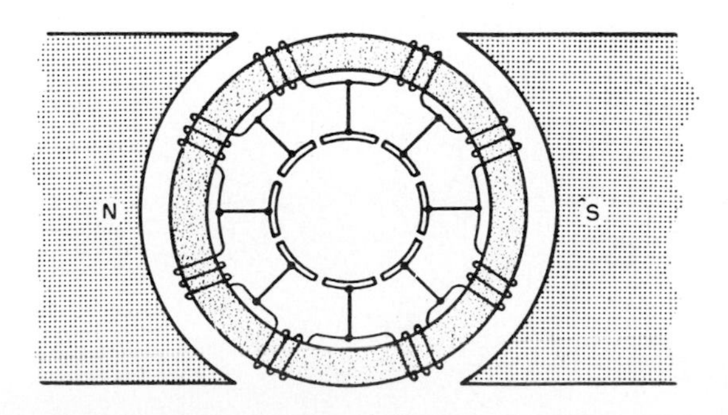

14–10. An eight-section, ring-type armature.

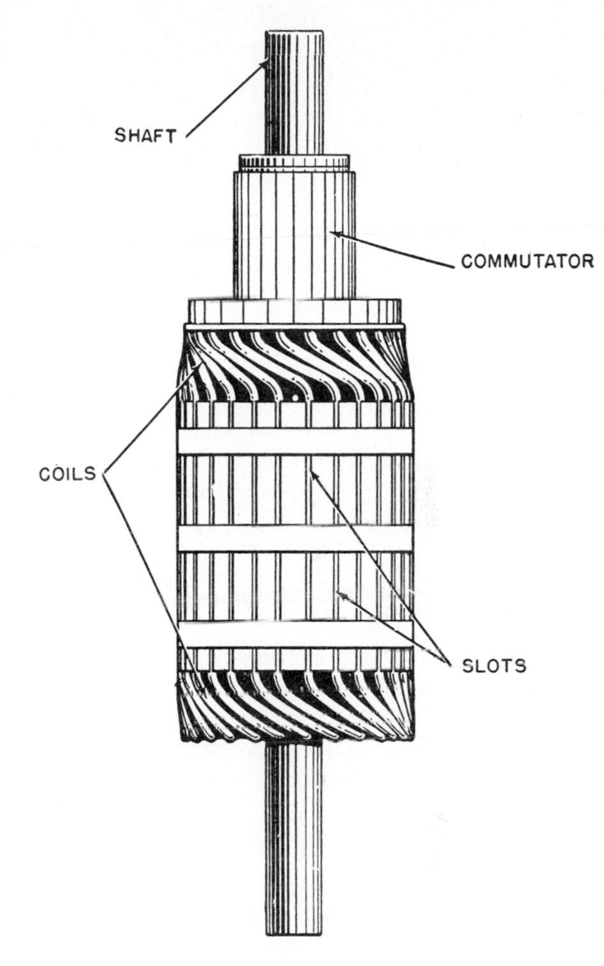

14–11. Drum-type armature (partly assembled).

from the frame. The brushes are free to slide up and down in their holders so that they can follow the irregularities in the surface of the commutator. The pressure of the brushes may be varied, their position on the commutator may also be adjusted.

Commutation. As an armature revolves in a DC generator, the armature coils, passing under the pole pieces, have a voltage induced in them which appears at the brushes of the machine. As the commutator segments (to which the coils are connected) pass under the brushes, two actions occur:

1. Current is drawn from the segments, as a result of the voltages induced when the coil passes under the poles.
2. Those coils that are in the interpole spaces are shorted momentarily, and the connections to the coils are reversed.

These actions can cause brush damage if the proper material is not used in the brushes. For this purpose, high-grade carbon is specified, due to its ability to glide over the polished commutator without creating scratches or too much wear. It is hard enough to provide long brush life.

To protect the armature, and especially the commutator, oil and grease must not be allowed to touch it.

■ *Armature reaction*. There is an EMF generated in a moving armature that opposes the magnetic field used to produce the electrical output of the generator. The neutral plane of the armature, Fig. 14-12, is perpendicular to the lines of force, or flux field, when there is no current in the armature. The line **ab** on Fig. 14-12 indicates the axis of the neutral plane.

Fig. 14-13 shows the armature excited and the field unexcited. The flux lines produced by current flowing through the armature coils alone are shown. No field is produced by the poles.

Both the field and armature are excited in Fig. 14-14. Note the shift of the neutral plane. This shift of the main field brings the armature coils, which are being shorted by the brushes, under the influence of an additional slight field, which induces a low voltage in the coils. This low voltage is short-circuited by the brushes, causing sparking, which results in the burning and pitting of the commutator. As the load current and resulting armature reaction increase, this effect becomes more pronounced.

The effect of armature reaction can be minimized or overcome by shifting the brush assembly, by using chamfered poles, by using

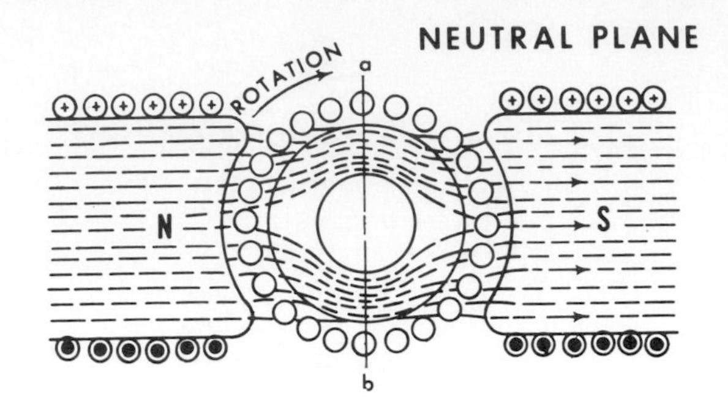

14–12. Field excited; armature unexcited.

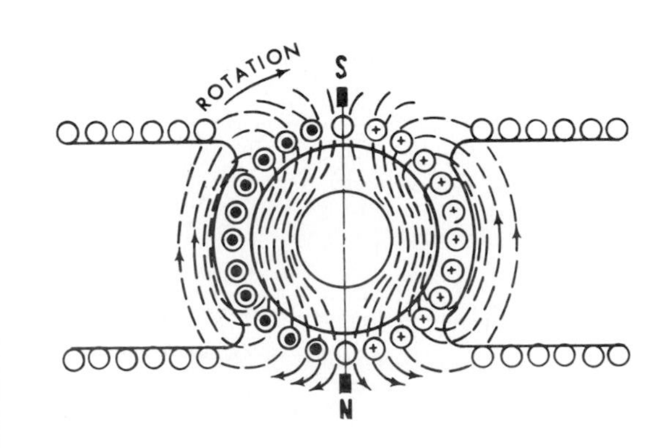

14–13. Armature excited; field unexcited.

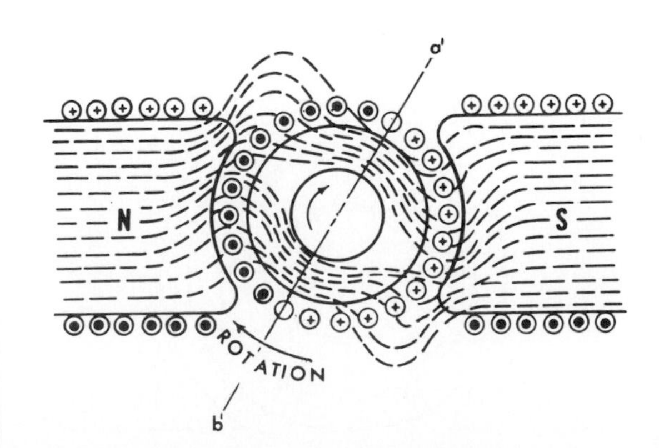

14–14. Both field and armature excited.

commutating poles, by the use of pole face windings, or by combining two or more of these methods.

Methods of eliminating the shift in the neutral plane result in actually nullifying the change.

1. The entire brush assembly can be adjusted to bring the brushes in line with the shifted neutral plane. Because the neutral plane shifts with the load, however, it would mean shifting the brushes every time the load changed. Therefore this is not a practical method of adjustment.
2. The poles can be chamfered slightly. That is, the radial distance between the pole face and the armature can be increased slightly at the edges of the poles. Fig. 14-15. This produces an increase in the air gap at the edges of the poles, offsetting to some extent the tendency of the field to shift because of armature reaction.
3. Commutating poles can be placed in the interpolar spaces. These poles are smaller and narrower than the main field poles. Their winding is in series with the armature, and so connected that the field set up by them opposes the field caused by the armature reaction.
4. The faces of the main field poles may be slotted longitudinally and windings placed in the slots. These windings are then connected in such a manner that the field produced by them opposes the field set up by the armature.

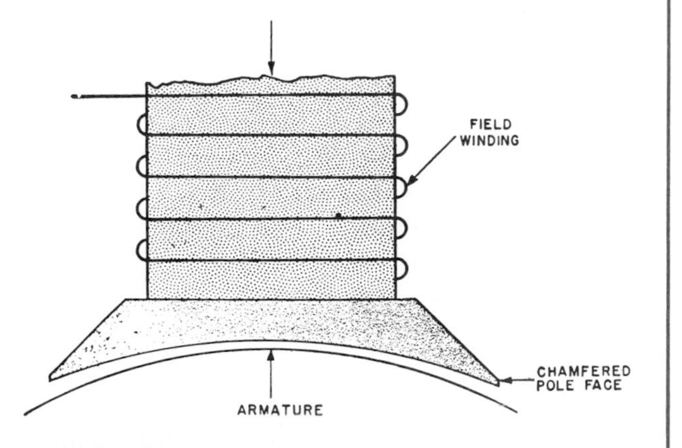

14–15. Chamfered pole.

Types of DC Generators

DC generators are classified according to the method used to supply the exciting current to the field windings. When the field current is obtained from a separate source, the generator is said to be separately excited. The self-excited generator has the excitation current supplied by the generator itself. The self-excited generator is further divided into at least three groups: *shunt, series,* and *compound.*

Shunt Generator. The poles of a generator retain some magnetism, known as *residual magnetism,* when not in operation.

Because residual magnetism produces a weak magnetic field, when the generator is started a small voltage is induced in the armature and appears at the output terminals. Then, because the armature output voltage is connected across the field windings, a small current flows in the windings. Figs. 14-16 & 14-17.

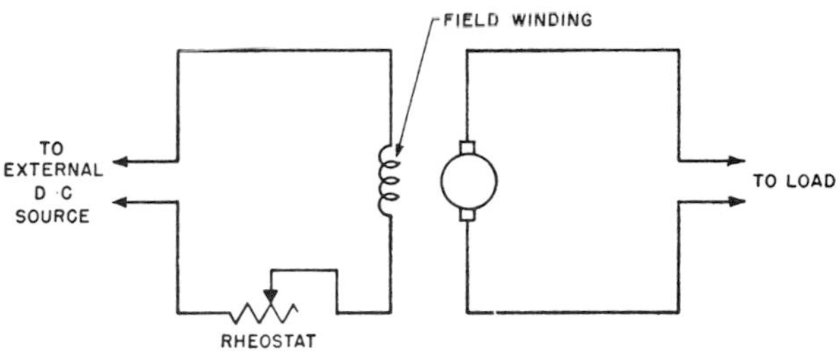

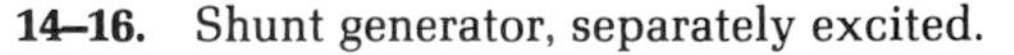

14–16. Shunt generator, separately excited.

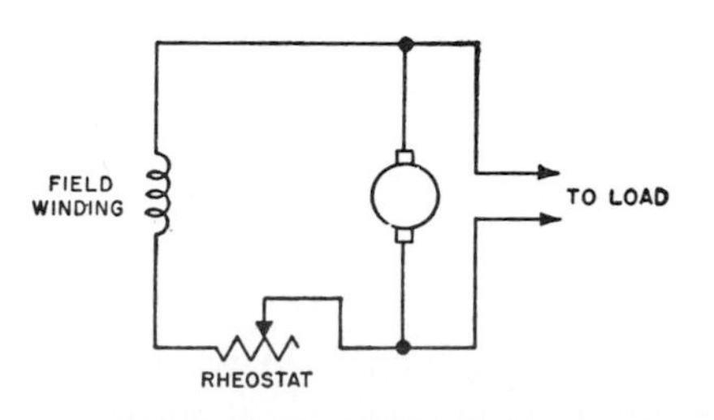

14–17. Shunt generator, self-excited.

This field current, in turn, strengthens the magnetic field, the output voltage increases accordingly, and a larger current flows in the windings. This action is cumulative, and the output voltage continues to rise to a point (called field saturation) where no further increase in output voltage occurs.

If the initial direction of the armature rotation is wrong, the cumulative action does not occur, since the small induced voltage opposes the residual field and there is no build-up of output voltage.

When the field windings of a shunt generator are excited, the machine is said to be *separately excited.* Fig. 14-16.

The terminal voltage of a shunt generator can be controlled by means of a rheostat inserted in series with the field windings. As resistance is increased, the field current is reduced, and consequently the generated voltage is reduced also. For a given setting of the field rheostat, either for a separately excited or self-excited machine, the terminal voltage at the armature brushes will be approximately equal to the generated voltage, minus the IR (voltage) drop produced by the load current in the armature. Consequently, the voltage available at the terminals of the generator will drop as the load is applied.

This voltage drop is greater in a self-excited generator than in a separately excited generator. Certain voltage-sensitive devices are available which automatically adjust the field rheostat to compensate for variations in load. When these devices are used, the terminal voltage remains essentially constant.

Shunt generators are used primarily in such an application as battery charging, which requires a constant voltage under varying current conditions. Separately excited shunt generators are often used in certain speed-control systems.

Series Generator. The field winding of a series generator is connected in series with the armature output voltage. Fig. 14-18.

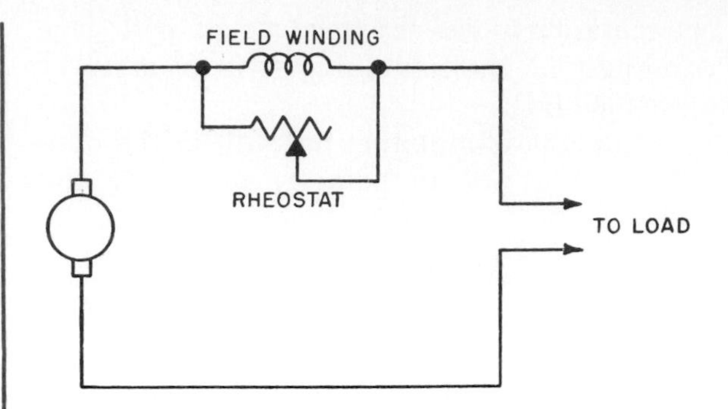

14–18. Series generator.

The field coils of the series generator are made up of a few turns of heavy wire. All current flowing through the field coil also flows through the armature. Series generators have very poor voltage regulation under changing load conditions. The greater the current through the field coils to the load, the greater will be the induced EMF, and the greater will be the output voltage of the generator. Therefore when load is increased, voltage will increase; when load is decreased, voltage will also decrease. Because the series generator has such poor regulation, however, only a few are in actual use.

Compound Generator. A compound generator has both a series and a shunt field, both windings being on the same pole structure. The series field may be connected to aid or oppose the shunt field. Figs. 14-19 & 14-20.

There are several types of compound generators:

■ A *cumulative compounded* generator has the series field aiding the shunt field. Fig. 14-19.

1. *Flat compounded*: voltage remains constant for all loads.
2. *Over compounded*: voltage rises with increased load.
3. *Under compounded*: voltage drops with increased load.

■ A *differentially compounded* generator has the series field *opposing* the shunt field. Fig. 14-20.

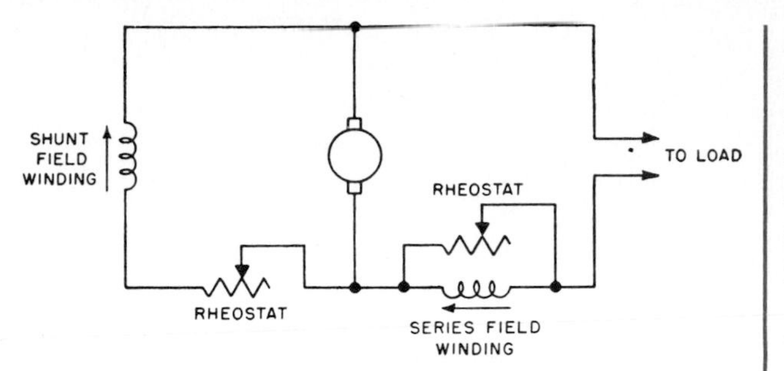

14–19. Cumulative compounded connections.

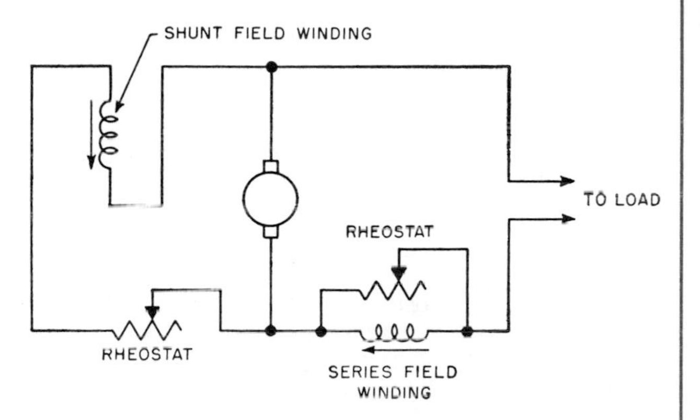

14–20. Differentially compounded connections.

Compound generators are usually designed to be over compounded. This feature permits the degree of compounding to be varied by connecting a variable shunt across the series field. Such a shunt is sometimes called a diverter. Compound generators are used where voltage regulation is of prime importance.

Differential generators have somewhat the same characteristics as series generators in that they are essentially constant-current generators. However, they generate rated voltage at no load, with the voltage dropping as the load increases. This constant-current characteristic makes them ideally suited as power sources for electric-arc welders which is the principal application for them.

A *long shunt* is produced by connecting the shunt field of a generator across both the armature and series field. If the shunt field is connected across the armature alone, it is called a *short shunt* connection. Short-shunt and long-shunt generator connections produce similar results.

Three-Wire Generators. Some DC generators are called three-wire machines. They are designed to deliver 240 volts with a neutral connection that provides 120 volts on either side of neutral. This is accomplished by connecting a reactance coil to opposite sides of the commutator. Fig. 14-21. The neutral is connected to the midpoint of the reactance coil. Such a reactance coil acts as a low-loss voltage divider. If resistors were used, the I^2R loss ($P = I^2R$, or power) would be prohibitive unless the two loads were matched perfectly.

The reactance coil may be built into the machine as part of the armature, the midpoint connected to a single slip ring with which the neutral makes contact by means of a brush. Or, the two connections to the commutator may be in turn connected to two slip rings, in which case the reactor is located outside the machine. In either case, the load unbalance on either side of the neutral must not be more than 25% of the rated current output of the generator. The three-wire generator permits simultaneous operation of 120-volt lighting circuits and 240-volt motors from the same generator.

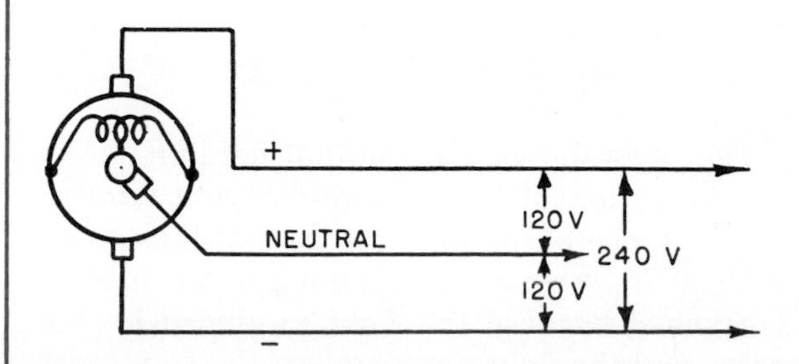

14–21. Three-wire generator.

DC MOTORS

Motors, like generators, are simply a means of transforming energy or power. Motors convert electrical power into mechanical power. The essential features and parts of a DC generator and DC motor are the same: both have field coils, armature coils, a commutator, and a brush assembly. It is also possible to use generators as motors. For example, when a voltage is applied to the terminals of a generator, the currents in the field coils and armature coils, respectively, set up magnetic fields which react to each other and cause the armature to revolve. When this happens, a DC motor is in operation.

Principle of Operation

An understanding of basic motor action may be obtained by considering the following simple facts concerning magnetic fields.

Fig. 14-22 shows the uniform magnetic field which exists between the poles of a magnet when its field coils are connected to a DC source. The lines of flux are directed from the north pole to the south pole.

Fig. 14-23a represents a cross section of a current-carrying conductor when the direction of current flow is away from the observer. By applying the left-hand rule for current-carrying conductors, it is found that the direction of the field is counterclockwise, as indicated by the arrows on the lines of flux. The left-hand rule states that if a current-carrying conductor were grasped in the left hand with the thumb pointing in the direction of current flow (negative to positive), the fingers would encircle the wire in the direction of the magnetic lines of force. Fig. 14-23b.

Fig. 14-24 shows the resultant field produced by the action of the current-carrying wire in the presence of the pole pieces and their magnetic field. Above the conductor, the field is weakened because the field produced by the poles and the field produced by the conductor are opposite in direction and tend to cancel

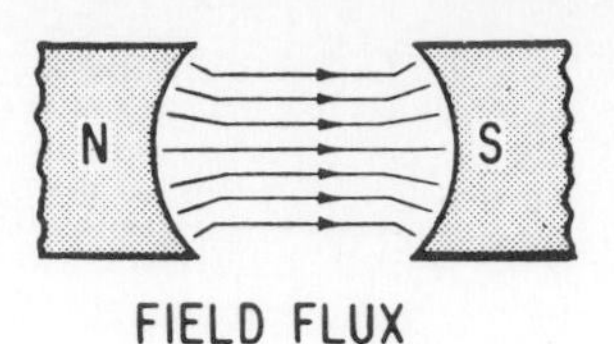

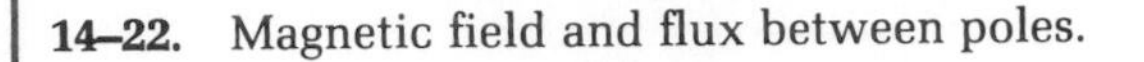

14–22. Magnetic field and flux between poles.

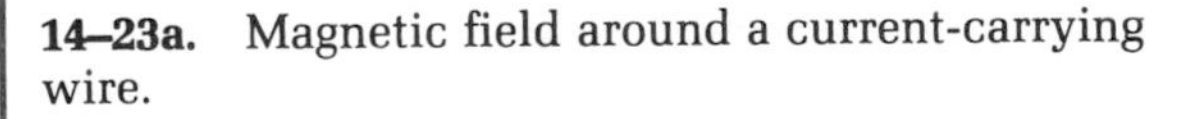

14–23a. Magnetic field around a current-carrying wire.

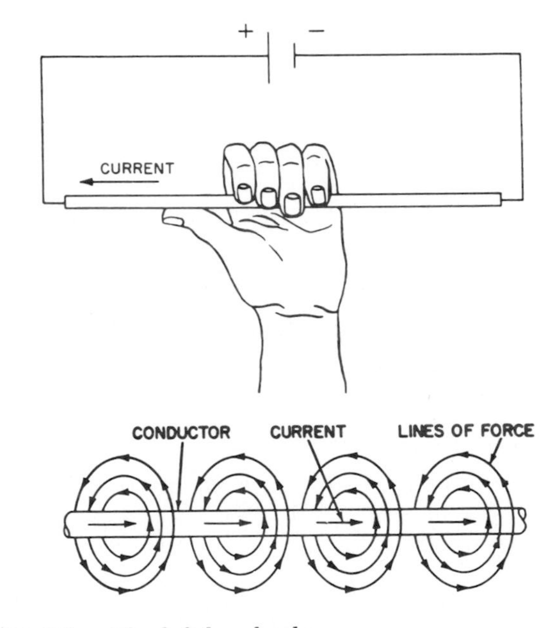

14–23b. The left-hand rule.

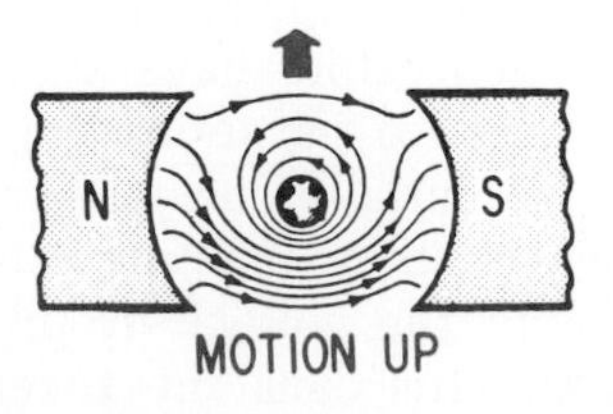

14–24. Note how wire is pushed upward when magnetic field is distorted by presence of current-carrying wire.

each other. Below the conductor, the two fields run in the same direction and the resultant field is strengthened. Because magnetic lines of force act to push each other apart, those lines below the conductor tend to push the conductor up, and those above the conductor tend to push the conductor down. However, because the field below has many more lines and is therefore much stronger, the push upward is greater, with the result that the conductor in Fig. 14-24 is moved upward.

Fig. 14-25 indicates the condition when the current flow has been reversed. The direction of the motion of the conductor will also be reversed because, in this case, the field above the conductor is strengthened and the field below is weakened.

Briefly, the principle on which motors operate may be stated as: *A conductor carrying current in a magnetic field tends to move at right angles to (across) the field.*

Basic DC Motors

A basic DC motor, like the DC generator, consists of poles which set up a magnetic field, an armature made of a single-turn loop and commutator, and a brush assembly. Each coil side of the loop lies in the magnetic field. Fig. 14-26.

■ *Motor operation.* When a DC voltage is applied to the brushes, a current flows around the loop. The magnetic field produced by the flow of current interacts with the magnetic field produced by the poles of the magnet. The resultant magnetic field is represented by flux lines, as shown in Fig. 14-26a.

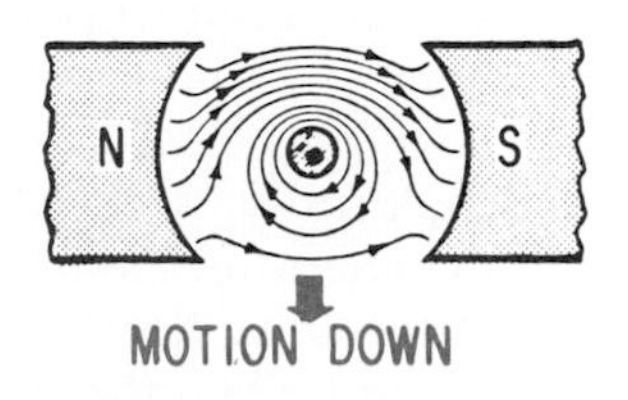

14–25. Note downward movement of wire with current flowing through it.

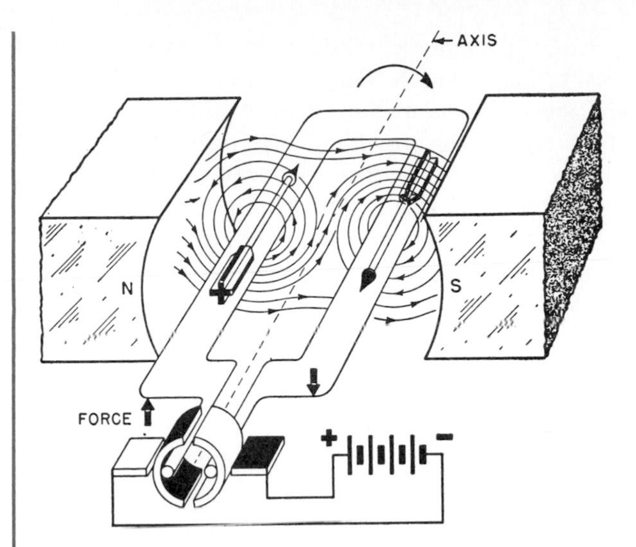

14–26a. Basic DC action.

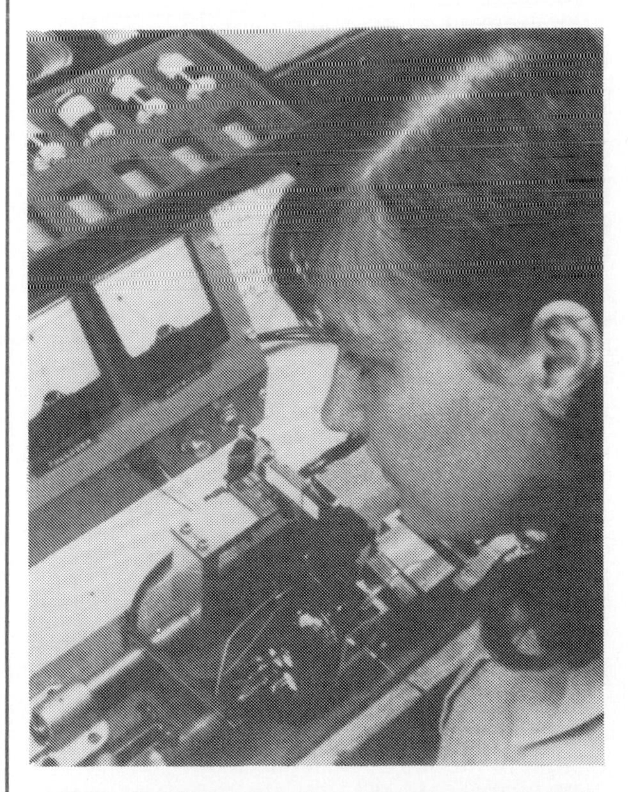

14–26b. Small permanent magnet motors are used in sound movie cameras. They have a permanent magnet field, commutator, armature, and brushes. Here, they are being checked for current drain.

Small permanent magnet motors are used to power toys, movie cameras, and many other devices. The permanent magnet limits its size. This means its power will be limited to fractional horsepower. It works only on DC. What would happen if AC were applied?

The permanent magnet motor has a permanent magnet for a fixed magnetic field. It has a commutator, brushes, and a wound armature. Not only does the permanent magnet motor show the basic principles of a DC motor in its simplest form, it also has practical applications.

The loop in the motor rotates clockwise under the influence of the magnetic field. It may be noted that the force acting in one direction (on one side of the loop) and the force acting in the other direction (on the other side of the loop) combine to cause the coil to turn on its axis. The loop thus acts as if it were a lever with a turning force, or *torque*, at each end. Because of this lever arrangement, the force at each end is magnified by the distance from the center. Thus, the torque is equal to the combined force on the two sides of the loop, multiplied by the distance of the conductors from the axis about which the loop rotates. The greater the torque, the more pull the motor has to drive a mechanical load.

Right-Hand Rule. There is a definite relationship between the direction of the magnetic field produced by the poles, the direction of current flow in the conductor, and the direction in which the conductor tends to move. This relationship is expressed in the *right-hand rule for motors*, which states: Place your right hand in a position so that the lines of force from the north pole of the magnet enter the palm of the hand. Fig. 14-27. Let the extended fingers point in the direction of the current in the conductor. Then the thumb, placed at right angles to the fingers, points in the direction of motion of the conductor.

From this rule, the direction of rotation of a DC motor can be determined, if the direction of the field flux and direction of current flow are known.

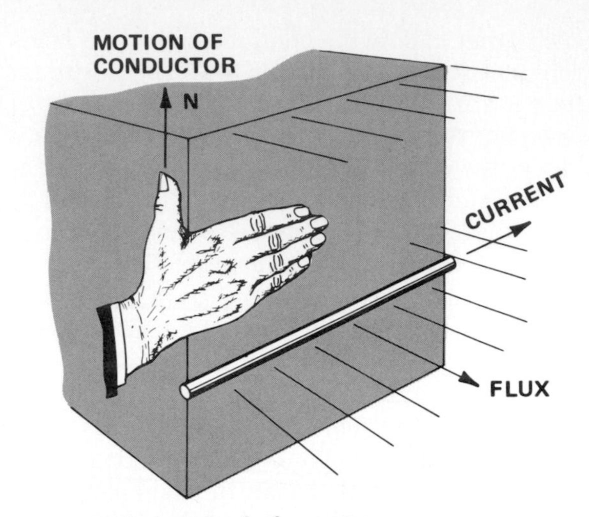

14–27. Right-hand rule for motors.

Basic DC Motor Theory

Figs. 14-28 through 14-30 show the continuous rotation of the armature of a DC motor.

Multi-Loop Motors. The torque developed by a single loop motor is too small for practical purposes. It is used only in explaining the theory of operation of an electric motor. Each time the loop passes through a neutral plane, the torque is zero, and a jerk is caused in the rotation of the armature. To develop more torque and to eliminate the effect of the zero-torque points on armature rotation, additional loops are added to the armature. Two loops can reduce the zero-torque condition. Fig. 14-31. More loops can reduce the problem even further and provide a more evenly distributed, smooth rotation of the armature. Both ring- and drum-type armatures are used on practical motors.

Shunt Motor. Structurally, a shunt motor is the same as a shunt generator. The field coils of a shunt motor are connected directly across the DC input terminals, and when the supply voltage remains constant, the current through the field coils remains constant. As a result, the

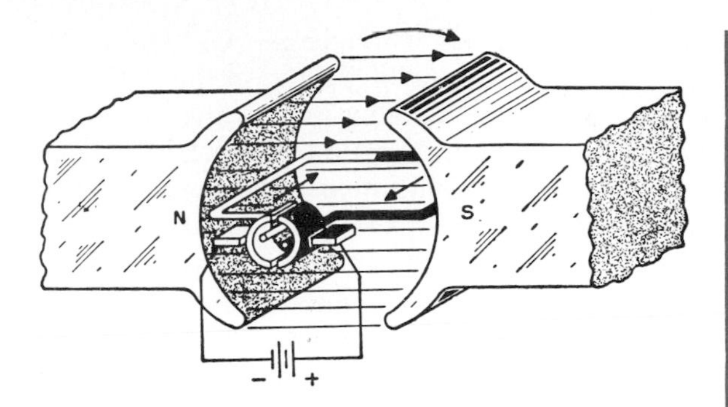

14–28. Note position of the armature loop and commutator.

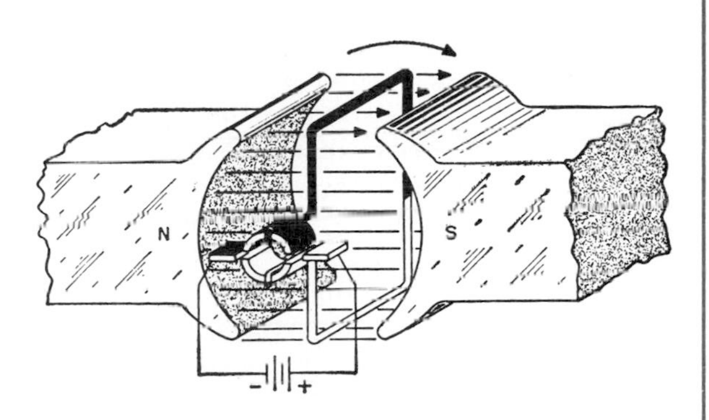

14–29. Loop has moved 90°. Note commutator location.

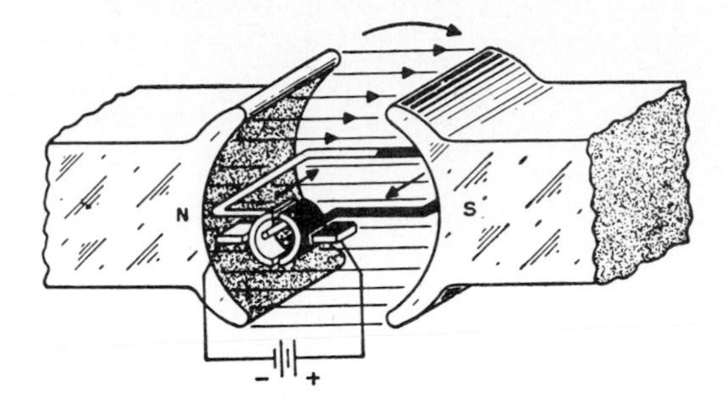

14–30. Loop has made a 180° rotation. Note commutator segments.

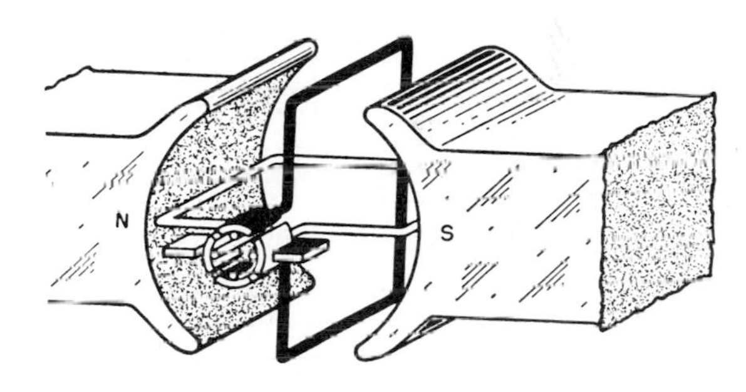

14–31. Two-loop motor.

magnetic field is also constant, and the torque of the shunt motor varies only with the current through the armature windings.

Whenever an armature rotates in a magnetic field, an EMF is induced in the armature coils. During rotation, the armature windings cut the lines of force and an EMF is induced in the windings. Therefore there is motor and generator action at the same time. The direction of the induced EMF opposes the applied EMF, and for this reason is known as a *counter electromotive force*, or CEMF.

In a shunt motor, the amount of current that flows through the armature depends upon the resistance between the applied voltage and the induced voltage (CEMF). The CEMF, in turn, depends solely upon the armature speed because the armature coils are fixed to the armature core and the field does not vary. Therefore the following formula is used for motor calculations:

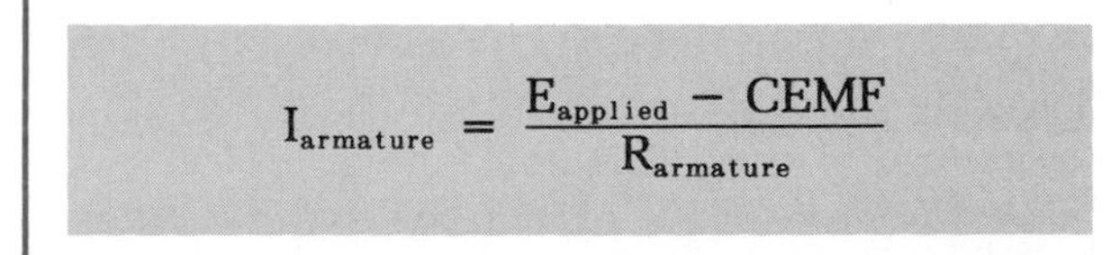

$$I_{armature} = \frac{E_{applied} - CEMF}{R_{armature}}$$

where $R_{armature}$ is the armature resistance, $I_{armature}$ is the current through the armature, and $E_{applied}$ is the voltage applied to the motor.

For example: Suppose that a motor develops a CEMF of 100 volts at 1000 RPM, the armature resistance is 1 ohm, and the applied voltage is 105 volts. Find the armature current at the speed of 1000 RPM.

$$I_{armature} = \frac{105 - 100}{1} = 5 \text{ amperes}$$

at 1000 RPM

Now, find the armature current if the speed is reduced to 900 RPM. In this case, the CEMF is reduced to $\frac{9}{10}$ of its value, or 90 volts. Therefore the current becomes:

$$I_{armature} = \frac{105 - 90}{1} = 15 \text{ amperes}$$

at 900 RPM

■ *Starting an electric motor.* The CEMF developed is proportionate to the speed of the armature. Consequently, no CEMF is developed when the armature is not turning, even though a voltage is applied to the input terminals. Under such a condition, the armature current is equal to the applied voltage divided by the ohmic resistance. For example, if the applied voltage were 100 volts and the ohmic resistance were 1 ohm, the armature current would be 100 divided by 1, or 100 amperes. Such a high current would burn out the armature windings. Also, if the field windings were not excited, no magnetic field would be produced by the poles, no flux lines would be cut when the armature was rotating, and the current again would be limited only by the value of applied voltage and the ohmic resistance of the armature windings. The current would be excessive since no CEMF would be produced.

To prevent such excessive current from flowing in the armature windings, a resistance is placed in series with the windings while the armature speed is building up to its proper value. This resistance is usually inserted by means of a starting box. In addition to protecting the motor as it starts, the starting box also has provisions for breaking the circuit if the field circuit should open or if the power supply should fail.

■ *Motor starters.* The starting box shown in Fig. 14-32 has the starting lever in the "off" position. To start the motor, the operator first moves the lever to the first contact of coil C, Fig. 14-32. This action closes the circuit to the field coil through the electromagnet, H, and also closes the circuit to the armature windings through high-resistance coils in the starter. As a result, maximum current flows through the field coil, and the magnetic field produced by the poles has maximum strength. Yet the current through the armature is limited by the series resistance. As the armature radually builds up speed, the operator moves the lever successively to other contacts in order to reduce the value of the series resistance.

Finally, the operator moves the lever to the extreme right position, in which it is held by the electromagnet. At the extreme right position, the series resistance is removed from the circuit. If the power fails, or if the field coils open for any reason, the electromagnet becomes deenergized and the lever is returned automatically to the "off" position by the spring, P.

■ *Armature speed.* The speed of a shunt motor is fairly constant under conditions of changing load. As a load is applied, the speed of the armature decreases, thus decreasing the CEMF and increasing the current. The increase

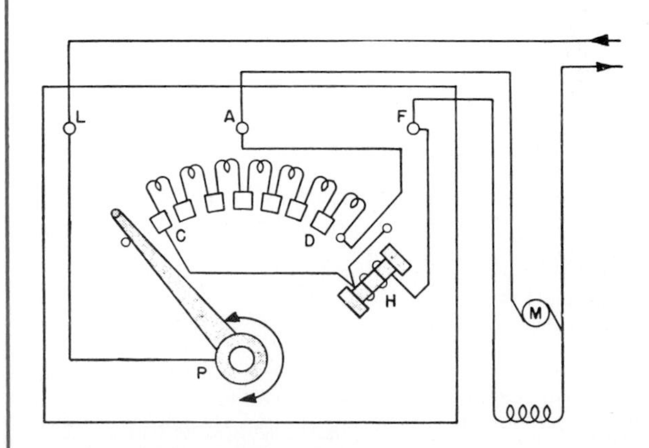

14–32. Motor-starting box.

in current boosts the coupling between field and armature and increases the torque, causing the motor to resume approximate running speed.

■ *Series rheostat.* The speed of a shunt motor may be controlled by means of a rheostat in series with the armature windings, or a rheostat in series with the field winding, or both. The rheostat in series with the field winding is the most commonly used method.

Adding resistance in series with the armature reduces the armature current and decreases the torque. This causes the motor to slow down until the current increases sufficiently to give the desired torque. When resistance is added in series, it also reduces the field strength, since the current through the windings is reduced. This in turn reduces the CEMF. The reduced CEMF makes the armature draw more current, a process which increases the speed of the motor due to the increased torque. The increasing CEMF produced by the increasing speed causes a current reduction to an amount sufficient to produce the torque required for the load.

A rheostat in series with the field winding is preferable to a rheostat in series with the armature. The field current is much lower than the armature current and the loss in the field rheostat is much less than the power loss in the armature rheostat.

■ *Armature reaction.* Because the current flow through a motor armature is opposite to the current flow through a generator armature, the armature reaction tends to shift the neutral plane back to the vertical plane, rather than forward as in the case of generators. Fig. 14-14. Therefore the brushes of a shunt motor are set back of the neutral plane instead of forward.

A shunt motor will operate only on direct current. An AC voltage applied accidentally to a DC shunt motor causes unpredictable motor operation.

Series Motor. A series motor is structurally the same as a series generator. (Fig. 14-18 shows a series generator.)

The series motor is adapted for giving a very high starting torque. Actually, the torque of this motor varies approximately as the square of the current. Remember, in a series motor, current flows through its series-connected armature and field coils. If the armature current is doubled, the flux is also doubled. Hence the torque, which is proportional to the current times the flux, is increased many times.

When an armature is at rest, the armature current (and therefore the torque) is at a maximum, because no CEMF is generated in the coils, and the current is limited only by the applied voltage and the ohmic resistance of the armature and field coils. As the armature gains speed, the CEMF increases, decreasing the armature current and torque. If additional load is applied to the motor, the armature slows down, the generated CEMF is decreased, a greater current flows, and a greater torque is produced. The speed of the motor is controlled by the load and, if the load is removed, the motor will race dangerously until centrifugal force causes the armature to disintegrate.

Series motors are generally used only where the load is constantly applied, and a good starting torque is required. An example of this is the automobile engine starter, where high torque is needed for cranking the engine. Hoists, streetcars, and other devices use a series motor.

One of the advantages of the series motor is its capability for operating on either DC or AC. On DC the brushes are set back of the neutral plane to compensate for armature action. For AC operation, both the field and the armature change polarity at the same time. The brushes are set in the vertical or neutral plane; the field core must be laminated to prevent eddy current losses. The theory of operation on AC will be explained later in this chapter.

A starting box is generally used with large series motors to limit current flow through the armature and field coils when starting the motor. This starter has a rheostat which can be connected in series with the motor windings. All the resistance is inserted in the circuit when the motor is being started, and the value of resistance is reduced gradually as the speed of the motor increases.

Compound Motor. A compound motor differs from the stabilized shunt type by its more predominant series field. Like compound generators, compound motors may be divided into two classes, differential and cumulative, depending upon the connection of the series field in relation to the shunt field.

■ *Differential compound motor.* A diagram of a differential compound motor is shown in Fig. 14-33. In this type, the series field opposes the connected shunt field. Therefore this motor operates at a practically constant speed. As the load increases, the armature current increases to provide more torque. The series magnetomotive force increases, thus weakening the shunt field and reducing the counter electromotive force in the armature, without causing a reduction in speed.

■ *Cumulative compound motor.* The cumulative compound motor diagrammed in Fig. 14-34 is connected so that its series and shunt fields aid each other. From this comes its name, cumulative compound motor. A motor

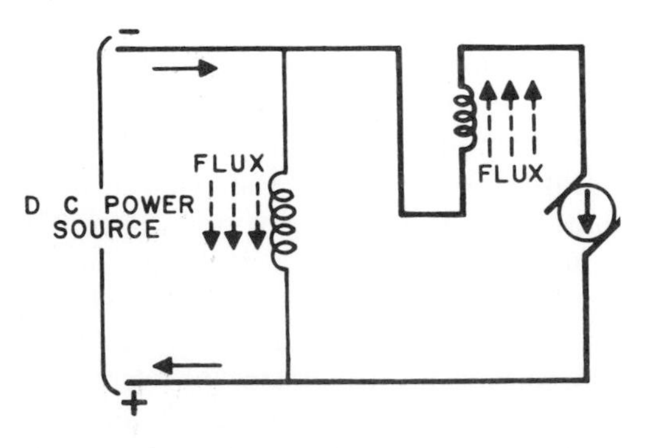

14–33. Connection of a differential compound motor.

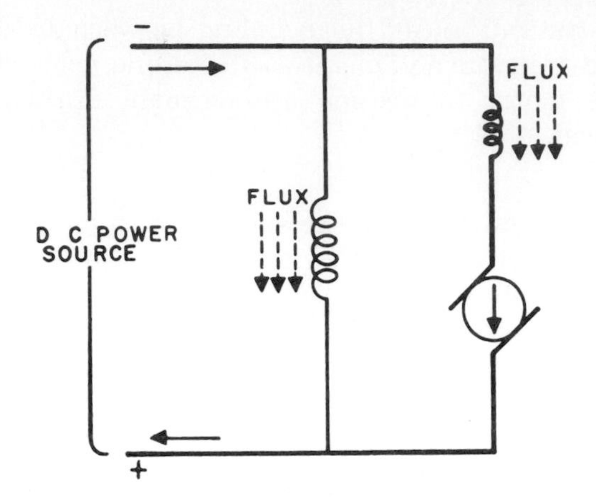

14–34. Connection of a cumulative compound motor.

thus connected will have a very strong starting torque, but it will also have poor speed regulation. Motors of this type are used for machinery where speed regulation is not necessary, but where great torque is desired to overcome sudden application of heavy loads. The operating characteristics of series, shunt, and compound motors are shown in the drawings and photographs in Fig. 14-35.

DC Motor Applications

Direct current motors have a decided advantage when it comes to reversing. In order to reverse a DC motor, simply change the polarity. The DC polarity can be changed with a switch or a relay. Usually, it is done with a relay. In the case of an elevator where the car moves up and down, the DC motor is capable of turning first in one direction and then the other. It serves this purpose very well.

In Fig. 14-35b the DC motor is designed for operation of an elevator. It can be seen in operation in Fig. 14-35c. Note the bullet-shaped black device near the control panel in Fig. 14-35c. This is the AC motor and DC generator connected together in one unit. The output of the DC generator is used to power the DC motor used to drive the elevator. This causes the elevator to run up or down on its track according to the polarity of the DC applied.

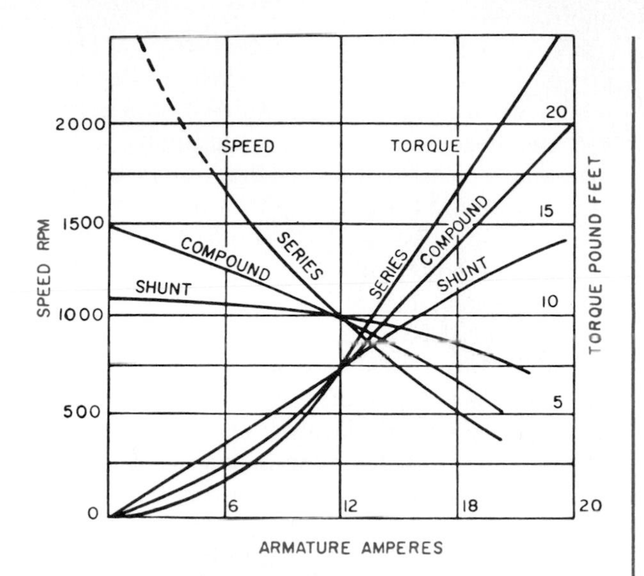

14–35a. Operating characteristics of series, shunt, and compound motors.

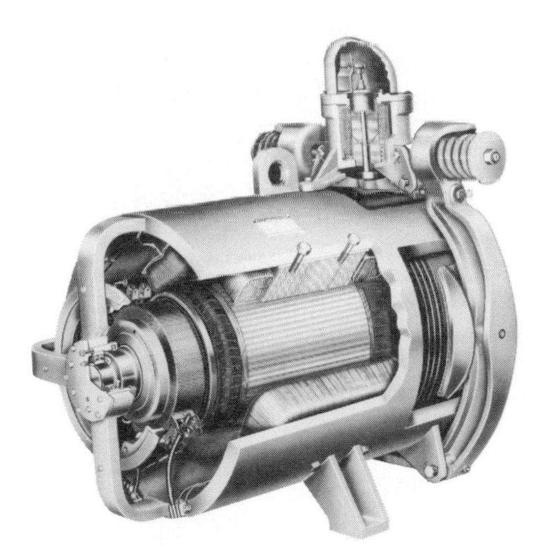

14–35b. Cutaway view of a DC motor used for elevator operations.

Note that the DC motor in Fig. 14-35b has brushes, commutator segments, and a wound rotor. The unit sticking up on the right end of the motor is a brake. This governs the speed of the motor during operation of the down cycle. It is also a safety device that will lock in place when the speed exceeds a certain rate of downward travel.

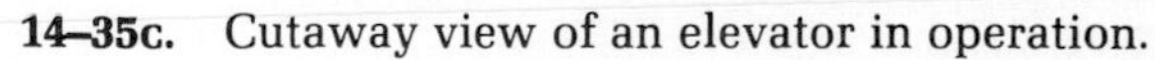

14–35c. Cutaway view of an elevator in operation.

AC GENERATORS

An elementary single-phase *alternator* (AC generator) was described in the first part of this chapter. This single-loop machine has little practical value except for laboratory demonstrations, inasmuch as it will generate very little electrical energy. The armature of a practical alternator is wound with many turns of wire, and its magnetic field is produced by electromagnetism. The EMF generated by this machine is limited only by the number of coils in the armature, the strength of the magnetic field, and the speed of rotation of the armature. In physical construction the alternator is similar to a DC generator. Their outputs do vary, however, Fig. 14-36.

Because the output of the generator is AC, and DC output is needed for the field windings, the generator voltage cannot be used as it is for field excitation. Therefore all AC generators are separately excited. Automobile alternators have the output rectified (changed to DC with diodes) before it leaves the generator, and some of the output is fed back to excite the field coils.

Slip rings operate satisfactorily at low power, but are a source of loss when large current is drawn from the armature. For this reason an alternator may use a stationary armature and a rotating field. In this case, the coils of the armature are placed about the circumference of the yoke, and the poles of the field (with their coils) rotate within the armature. Since the field current is low compared with the current delivered by the alternator, it is supplied through the slip rings. The load current usually is taken off the stationary conductors. A stationary armature makes it possible to generate higher voltages than can be done with generators using rotating armatures. This is because stationary conductors are not subject to the centrifugal force of rotation and may be more effectively insulated. The more insulation, the higher the voltage.

Components of an AC Generator

■ *Rotor*. As has been previously stated, the rotating field in a generator is called the *rotor*. When the field of an AC generator is placed upon the rotor, it is either of the *salient-pole* type or the *turbo* type, Figs. 14-37 & 14-38.

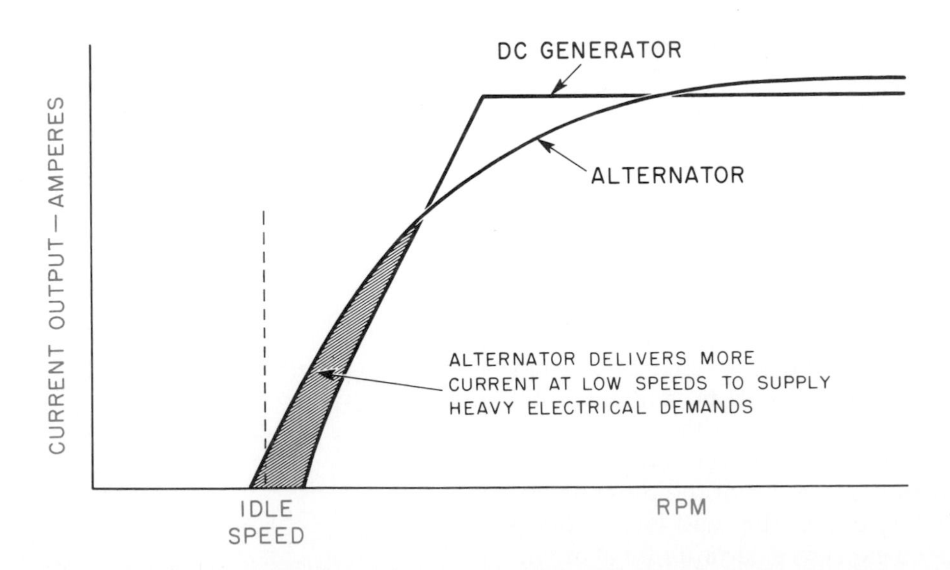

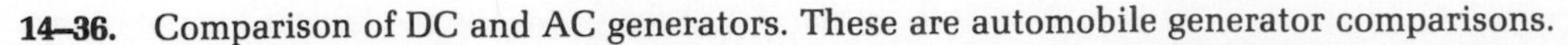

14-36. Comparison of DC and AC generators. These are automobile generator comparisons.

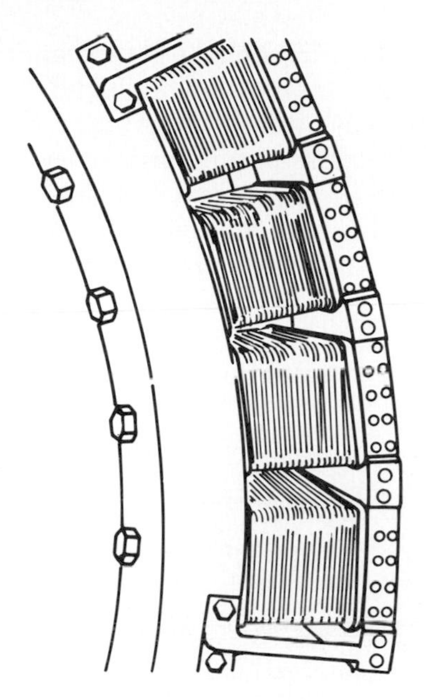

14–37. Salient-pole rotor.

When the AC generator, or alternator, is to be driven by a slow-speed diesel engine, or by a water turbine (from about 720 RPM), the *salient-pole*, or *projecting-pole* rotor, is used. Fig. 14-37. Field poles are formed by fastening a number of steel laminations to a spoked frame, or spider. The heavy-pole pieces produce a flywheel effect on the slow-speed rotor. This helps to keep the angular speed constant and reduce variation in voltage and frequency of the generator output.

In high-speed alternators (up to 3600 RPM), the smooth-surface, turbo-type rotor is used because it has less air-friction (heating) loss, and because the windings can be placed so that they can withstand the centrifugal forces developed at high speeds. Fig. 14-38. Turbo-type rotors are made from a solid-steel forging, or from a number of steel discs fastened together with the field coils locked in slots. These field coils are usually placed so that they distribute the field flux evenly around the rotor.

■ *Stators.* In a rotating-field AC generator, the armature windings are stationary, and are called the stators. The armature iron, being in a

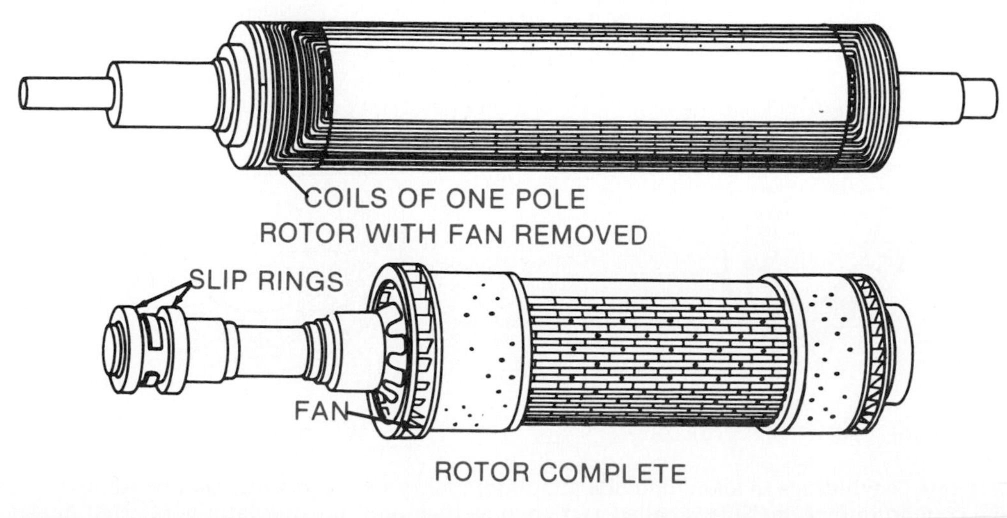

14–38. Turbo-type rotors.

moving magnetic field, is laminated in order to reduce eddy current losses. A typical AC generator stator is shown in Fig. 14-39.

In high-speed turbo generators, Fig. 14-40, the stator laminations are ribbed to provide sufficient ventilation because the high temperature developed in the windings cannot be dissipated in the small air gap between the rotor and the stator. In some larger installations, alternators are totally enclosed and cooled by hydrogen gas under pressure, which has greater heat-dissipating properties than air. Stator coils in high-speed alternators must be well braced to prevent their being pulled out of place when the alternator is operating at heavy load.

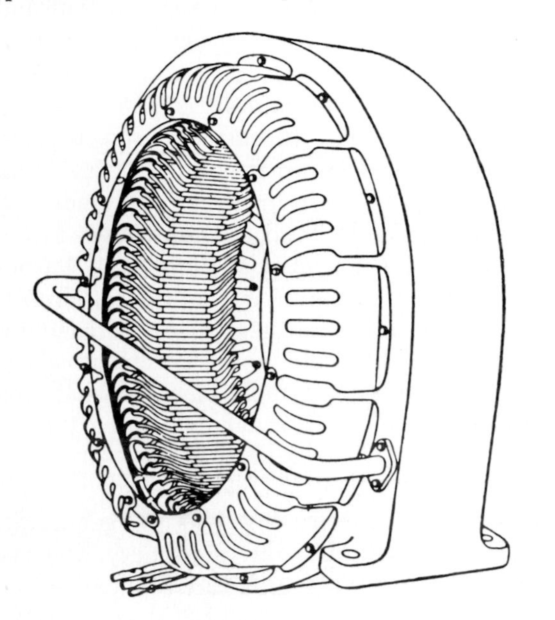

14–39a. An AC generator stator.

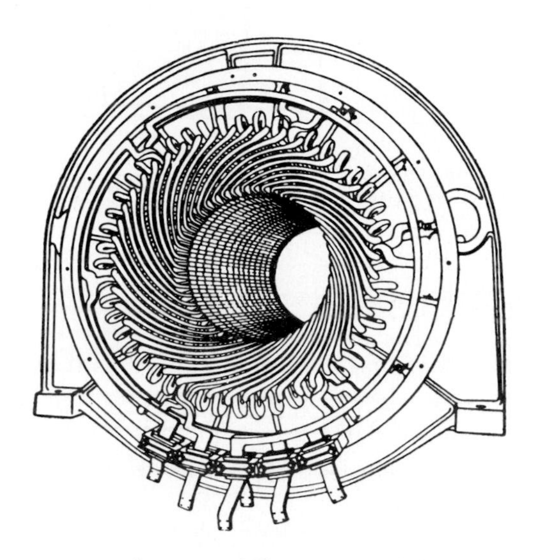

14–40. A turbo-type, AC generator stator.

EXTERNAL VIEW OF STATOR

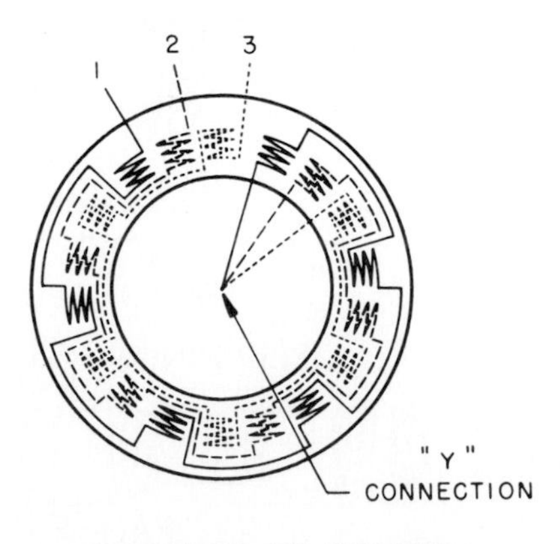

WINDINGS OF STATOR

14–39b. Three sets of windings in an automobile alternator stator. One end from each winding is connected to a common junction. This is called a "Y" connection, and the alternator is referred to as a 3-phase, "Y" connector stator. Output is taken off the other end of each winding.

■ *Exciters.* Like many DC generators, AC generators require a separate DC source for their fields. This DC field current must be obtained from an external source called an exciter. The exciter used to supply this current is often a flat, compound-wound DC generator, designed to furnish from 125 to 250 volts. The exciter armature may be mounted directly on the rotor shaft of the AC generator, or it may be belt driven. Fig. 14-41.

■ *Static exciters.* Another method of field excitation commonly in use is the static exciter, so-called because it contains no moving parts. In this method, a portion of the AC current from each phase of the generator output is fed back to the field windings (as DC excitation current). This is done through a system of transformers, rectifiers, and reactors. This system requires an external source of DC current to first excite the field windings. For engine-driven generators, the initial "field flash" may be obtained from the storage batteries, which are also used to start the drive engine.

■ *Frame and shaft.* The frame and shaft of the AC generator serve the same purpose as in DC generators. The frame completes the magnetic circuit of the field and supports the component parts and windings. The shaft, upon which the rotor turns, is supported on the end bells, or end frames.

Emergency AC Generators

A number of emergency power plants are available. Some are driven by gasoline engines, some by diesel engines, and some use natural gas as fuel. In all those shown in this chapter, the speed of rotor rotation is 3600 RPM. This is enough to classify them as high-speed alternators.

Emergency generators are built for use aboard ships and small craft, in hospitals, in industrial plants, and for home and farm use. They are available in many sizes and prices, but must be individually tailored for a specific job. Before purchasing one, it is best to know the power requirements and the frequency of use expected for the emergency power supply.

Fig. 14-42 uses a 5.5 HP gasoline engine at 3600 RPM to pull a generator which produces 2500 watts of electrical power at 120 VAC. The frequency output is 60 hertz. The generator uses a revolving armature and has inherent voltage regulation. The 1-cylinder, 4-cycle engine is capable of operating several hours on one tank of gasoline. It weighs only 185 pounds and is portable. It is a good emergency power unit for home use; for pumping flooded basements when the power is off; or for powering emergency equipment for electronics installations.

The generator in Fig. 14-43 is designed for marine use. It has a remote start-stop capability, a 13 HP gasoline engine with a 12-volt battery for starting. The speed of the armature is 1800 RPM. The generator can produce 4000 watts at 120 volts, 60 hertz. It has inherent voltage regulation and a revolving armature. This unit weighs 330 pounds.

The unit in Fig. 14-44 uses a diesel engine with 5.7 HP. The generator output is 3000 watts and can be used as 120 volts or 240 volts at 60 Hz. The generator, a 4-pole type with revolving armature and inherent voltage regulation, has sealed ball bearings. The engine is an air-cooled, 4-cycle, 1-cylinder, overhead-valve type, with a total weight of 348 pounds.

Fig. 14-45 is a design adaptable to a 5.5 HP, gas- or gasoline-powered, air-cooled engine. The generator turns at 3600 RPM and produces 120 or 240 volts AC. It weighs only 139 pounds and is very portable for use around a farm, home, or on a boat.

If additional power is needed for a hospital emergency room or a small industry, the 500 kW or 625 kVA power plant may be necessary. Fig. 14-46 shows a radiator-cooled diesel engine with 6 cylinders and a sealed ball-bearing generator. It can produce 277 or 480 volts AC at 60 Hz and 3-phase, 4-wire. In some cases the unit is equipped with a 12-cylinder Cummins diesel with 750 HP, which enables it to be used

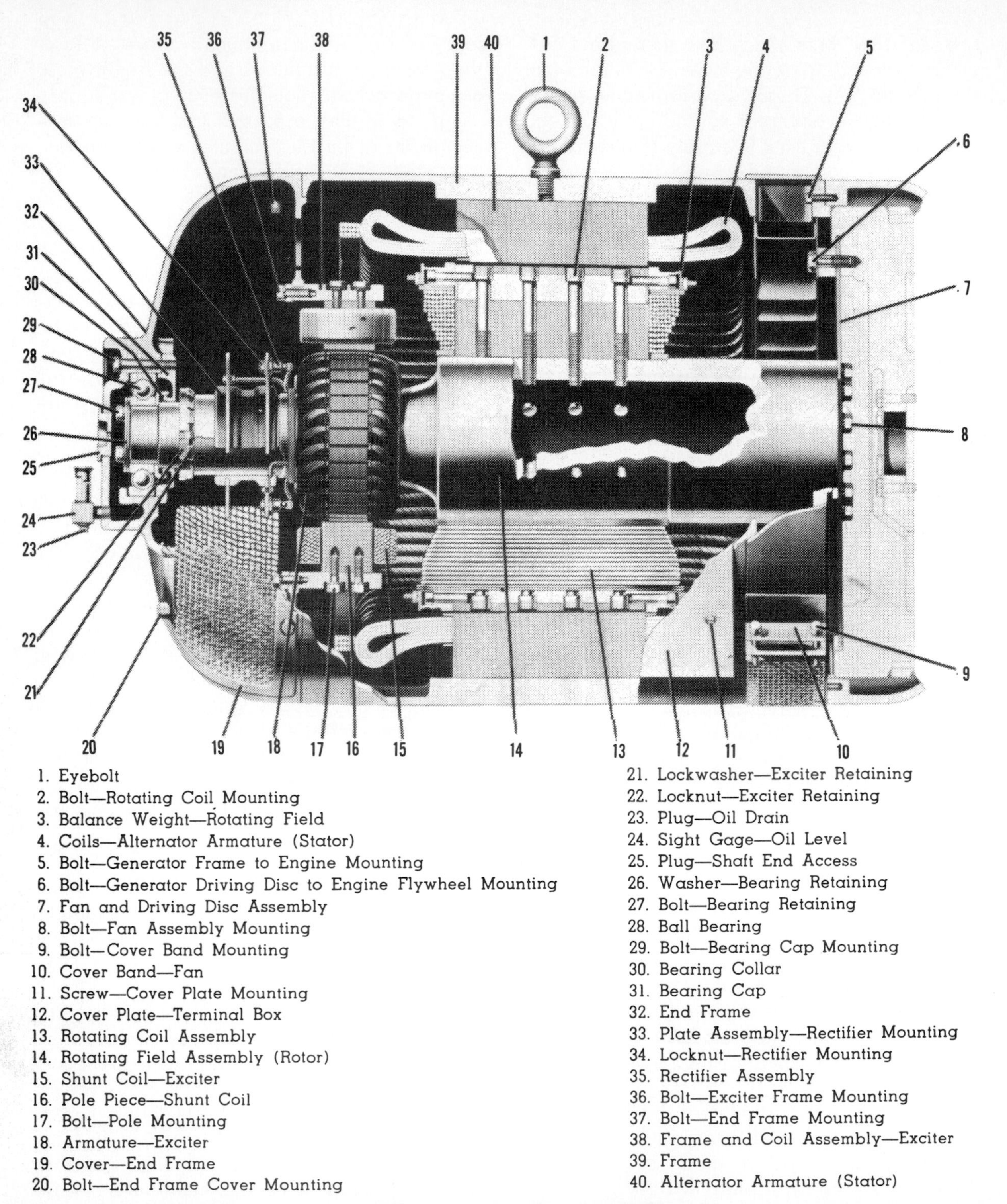

14-41. Exciter armature and alternator field mounted on the same shaft.

14–42. 2500-watt generator. Marine engine of 5.5 HP. It has remote starting, using a 12-volt DC starter, and rotates at 3600 RPM. Output is 120 volts AC.

14–44. 3000-watt generator. Diesel engine with 5.7 HP, air cooled. Puts out 120/240 volts AC, 60 Hz, single-phase, 4 wires. Weighs 348 lbs.

14–43. 4000-watt generator. Marine engine of 13 HP uses gasoline for fuel. It also has 12-volt, battery-operated starting. It puts out 120 volts AC, 60 Hz, single-phase, 4 wires, and weighs 3230 lbs.

14–45. 2500-watt generator. Single-cylinder engine with gas or gasoline capability. Generates 5.5 HP at 3600 RPM. Output is 120/240 volts AC, 60 Hz. Single-phase, 2 wires. Weighs only 149 lbs. and is aptly termed portable.

14–46. Full-time generator. Up to 12-cylinder engine, radiator cooled, diesel. It can put out a maximum of 625 KVA at 227/480 volts AC, three-phase, 4-wire. The engine can go as high as 750 HP. It weighs 12 020 lbs. The generator shown here is a 6-cylinder version, but can be used full time for oil field operations, farms, industry, and for hospitals during emergencies.

full time for oil fields, farms, or industry. The big unit weighs 12 020 pounds. It has remote-control capability, using a 24-volt DC battery-operated system for low-voltage controls. Automatic controls are available to start the engine in case of power failure of utility companies.

Polyphase Generators

The armature coils of a single-phase alternator are connected in series. At any given instant the voltage induced is the sum of the voltages induced separately in each coil. If the coils were not connected in series, and if their connections were brought out separately, the voltages induced would vary in phase with relation to each other. This is because each coil is in a different position in respect to the flux lines of the field.

A single-coil basic generator illustrates this principle. When windings, identical in size and of the proper phase relationships, are added to the single-phase alternator, a voltage output will be obtained from each coil added. These voltages will differ in phase but not in amplitude. The output from such a generator is known as a *polyphase-output system*. Systems of this type are generally 2- or 3-phase, with the 3-phase most widely used. In fact, the power generated by the large power utilities is 3-phase.

■ *Three-phase.* A 3-phase system is one in which the voltages have equal magnitudes and are displaced 120 electrical degrees from each other. Fig. 14-47. The three windings are placed on the armature 120° apart. Fig. 14-48. As the

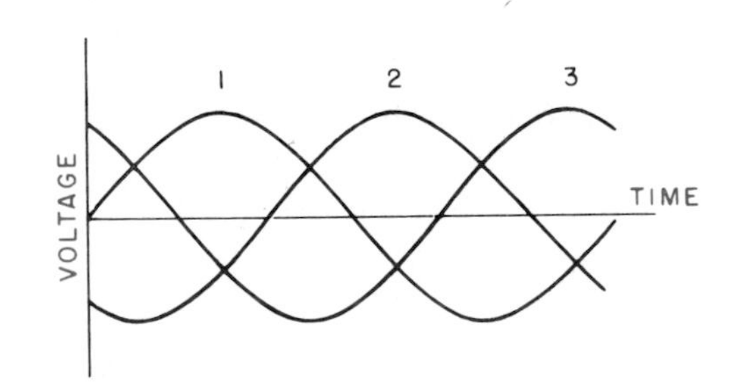

VOLTAGE INDUCED IN STATOR WINDINGS

14–47. Voltage induced in stator windings; 3-phase.

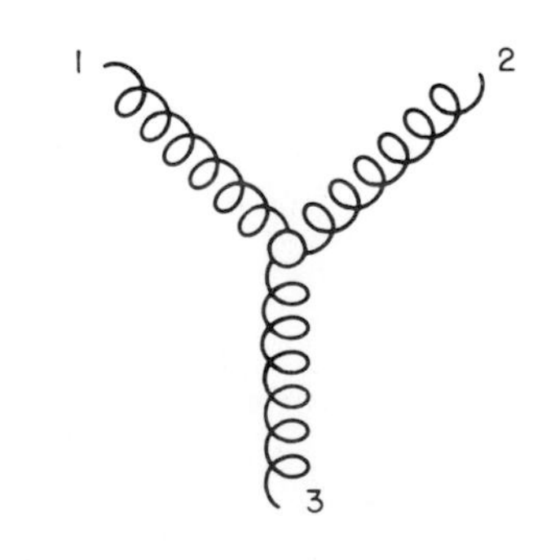

ELECTRICAL SCHEMATIC OF STATOR

14–48. Electrical schematic of a 3-phase stator connected in a "Y" configuration.

armature is rotated, the outputs of the three windings are equal but out of phase by 120°. Fig. 14-49.

Methods of Connecting Three-Phase Windings. Three-phase (3Ø) windings are usually connected in either a delta or wye configuration. Each of these connections has definite electrical characteristics from which the designations "delta" and "wye" are derived. Fig. 14-50 shows the delta-connected schematic and symbol. Fig. 14-51 shows the wye-connected schematic and symbol.

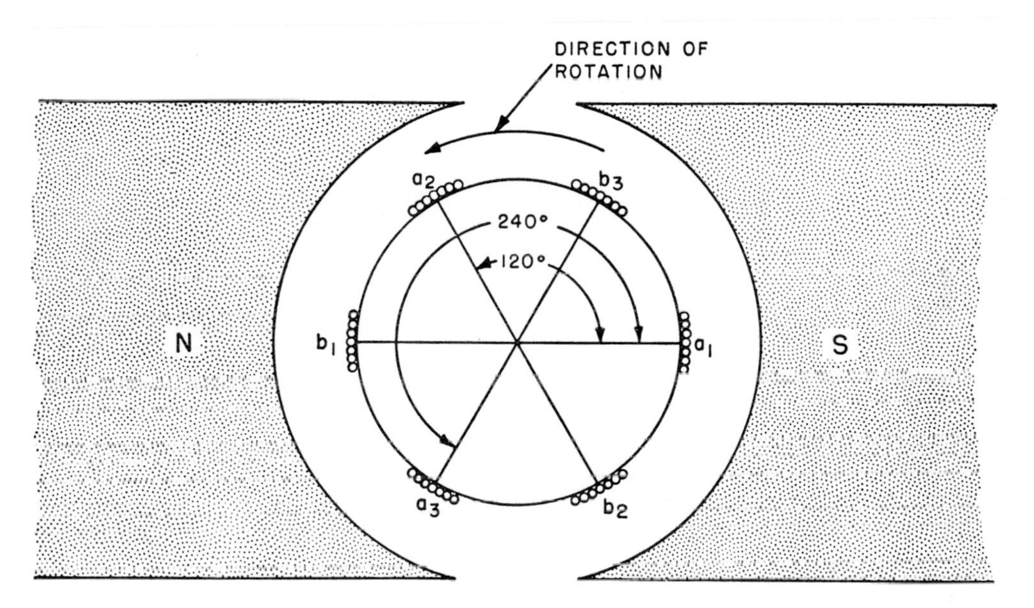

14–49. Elementary 3-phase alternator.

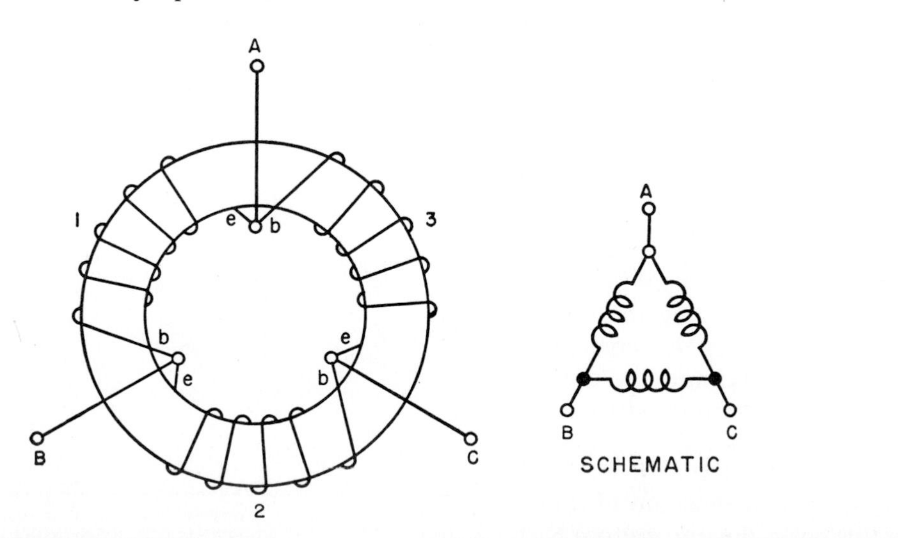

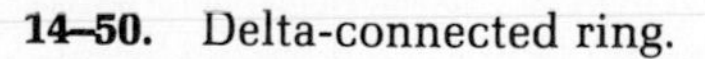

14–50. Delta-connected ring.

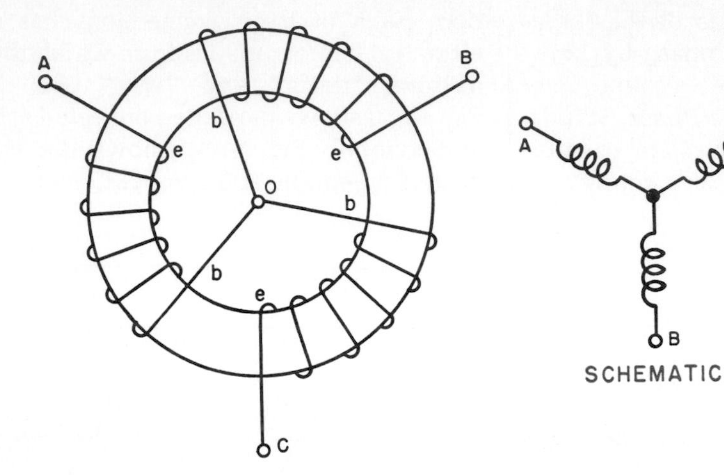

14–51. Wye-connected ring.

Electrical Properties of Delta and Wye Connections.

■ *Delta connection.* In a balanced circuit, when the generators are connected in delta, the voltage between any two lines is equal to that of a single phase. The line voltage and the voltage across any winding are in phase, but the line current is 30° or 150° out of phase with the current in any of the other windings. Fig. 14-52. In the delta-connected generator the line current from any one of the windings is found by multiplying the phase current by the square root of 3, or 1.73.

■ *Wye connection.* In the wye connection, the current in the line is in phase with the current in the winding. The voltage between any two lines is not equal to the voltage of a single phase, but is equal to the vector sum of the two windings between the lines. The current in line A of Fig. 14-53 is the current flowing through the winding L_1; that in line B is the current flowing through the winding L_2; and the current flowing in line C is that of the winding L_3. Therefore the current in any line is in phase with the current in the winding that it feeds. Since the line voltage is the vector sum of the voltages across any two coils, the line voltage E and the voltage across the windings E Ø are 30°

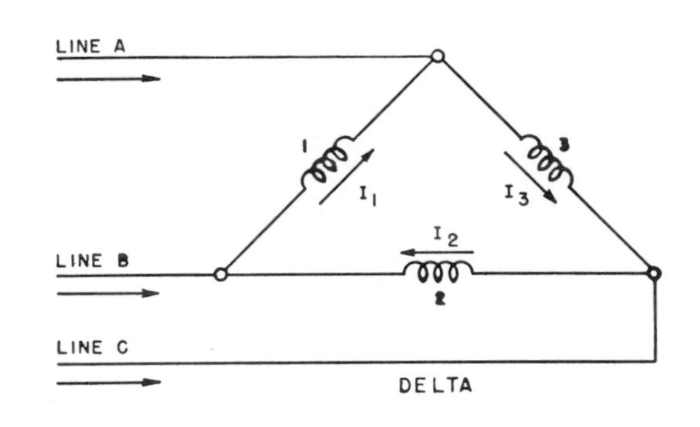

14–52. Delta-connected currents.

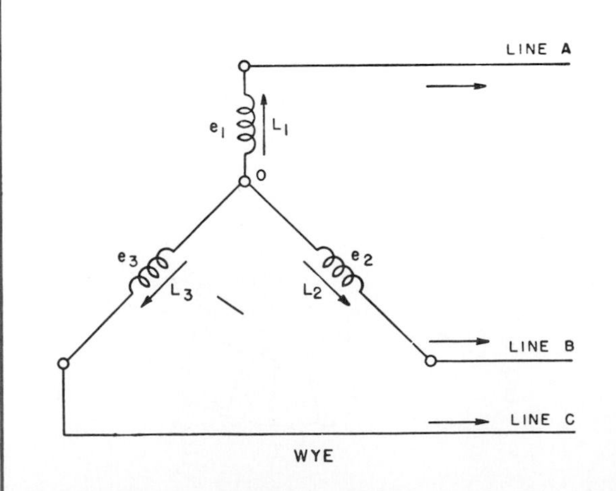

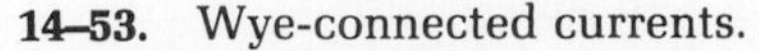

14–53. Wye-connected currents.

out of phase. The line voltage may be found by multiplying the voltage of any winding E Ø by 1.73.

■ *Summary of delta and wye connections.* The properties of delta connections may be summarized in this manner: The three windings of the delta connection form a closed loop. The sum of the three equal voltages, which are 120° out of phase, is zero. This means that the circulating current in the closed loop formed by the windings is zero. The magnitude of any line current is equal to the square root of 3, times the magnitude of any phase current.

The properties of the wye connection may be summarized: The three windings of the wye connection do not form a closed loop. The magnitude of the voltage between any two lines equals the magnitude of any phase voltage, times the square root of 3, or:

$$E_L = \sqrt{3} \times E\emptyset$$

The current in any winding equals the current in the line.

Transfer Connections for Three-Phase Power. The 3-phase transformer is used when large power outputs are required. Either a single transformer or three separate transformers may be used, but it is generally connected in delta or wye.

Practical 3-phase transformers use the shell-type construction shown in Fig. 14-54. This transformer is economical to construct and occupies less space than three single transformers. However, if one phase burns out, the entire unit must be replaced.

Commercial 3-phase voltage from power lines is usually 208 volts; the standard values of single-phase voltage can be supplied from the line, as shown in Fig. 14-55. The windings represent the wye-connected transformer, which is generally an outside installation. Fig. 14-56 illustrates the types of connections possible: A,

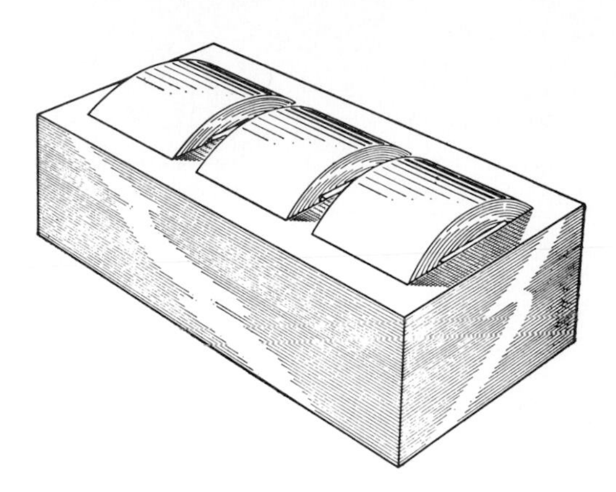

14-54. Three-phase, shell-type transformer. A shell-type transformer has the primary and secondary windings wound one on top of the other. These windings are placed, in single-phase units, on the center-leg of the core. Coupling reaches 100% in this type of transformer. Three-phase units use the same arrangements, but have the coils located on each of the core legs.

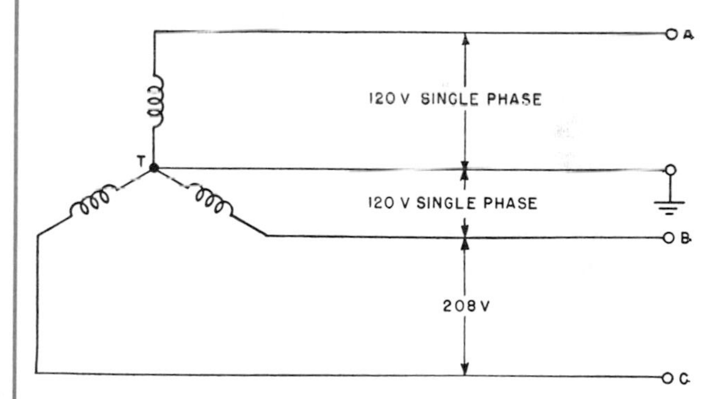

14-55. Commercial 3-phase voltage from power line.

has a specific purpose. Fig. 14-57 shows the voltages available from a delta-wye connection.

The ratio between line voltage and secondary voltage in the delta-wye transformer connection is not the transformer turns ratio, but the $\sqrt{3}$ times the turns ratio. The output voltage across any two windings is greater than that of a single winding. Therefore the single winding need be insulated for only $E_Y/\sqrt{3}$ volts. (E_Y is voltage across a single winding.)

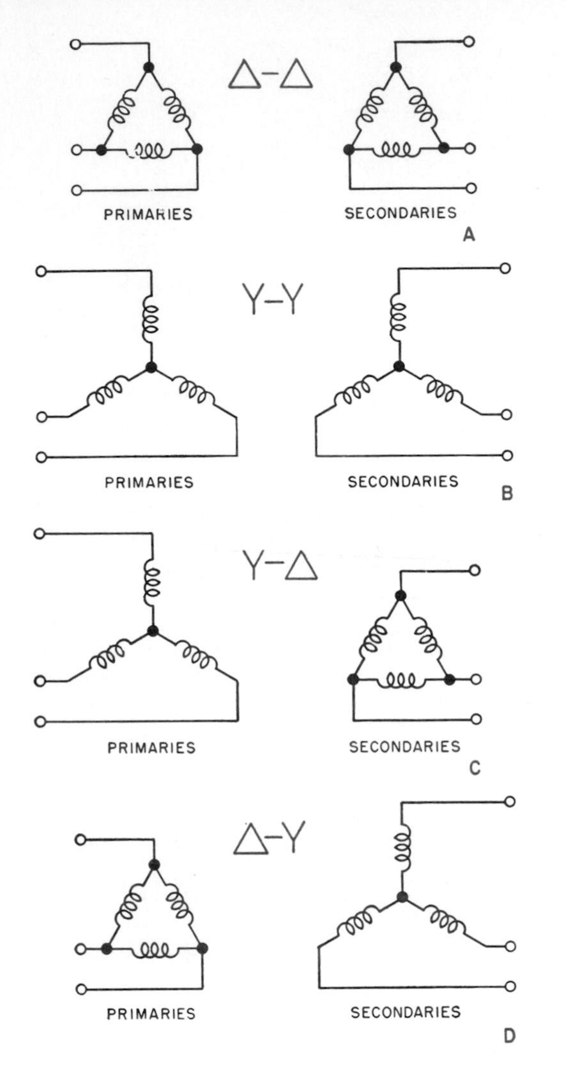

14–56. Methods of connecting 3-phase transformer.

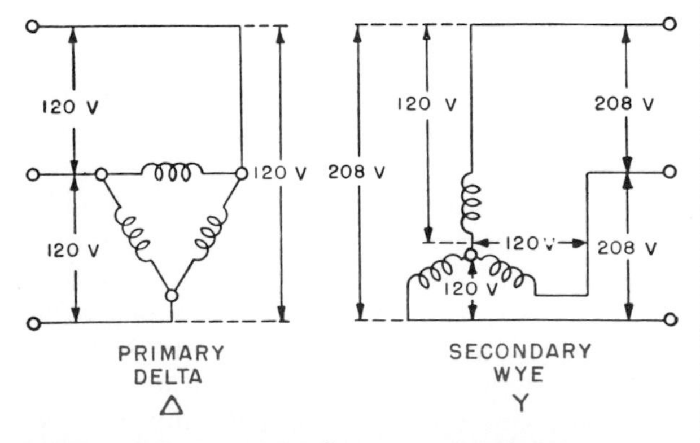

14–57. Delta-wye transformer connection.

For example, if the voltage across two windings of the secondary is 20 000 volts, the voltage across one phase is:

$$\frac{E_Y}{\sqrt{3}} = \frac{20\,000}{1.73}$$

$$= 11\,560 \text{ volts}$$

Because of this, the delta-wye connection permits cheaper construction of the transformer in high-voltage installations. A wye-delta connection, the opposite of the delta-wye connection, is often used for step-down voltages.

AC MOTORS

The induction motor is the most widely used of several available AC motors. Its design is simple, its construction rugged. Fig. 14-58. The induction motor is particularly well adapted for constant speed applications and, because it does not use a commutator, most of the troubles encountered in the operation of DC motors are eliminated. An induction motor can be either a single-phase or a polyphase machine. The operating principle, the same in either case, depends on a revolving, or rotating, magnetic field to produce torque. The key to understanding the induction motor is a thorough comprehension of a rotating magnetic field.

Rotating Magnetic Field

Consider the field structure of A in Fig. 14-59. The poles have windings which are energized by three AC voltages. Each phase can be represented by a letter: ØA, ØB, and ØC. These voltages have equal magnitude but differ in phase, as shown in B of Fig. 14-59.

At the instant of time shown as 0, the resultant magnetic field produced by the application of the three voltages has its greatest intensity in a direction extending from pole (1) to pole (4). (See A of Fig. 14-59.) Under this condition, pole (1) can be considered as a north pole, and pole (4) as a south pole.

14-58. An assembler attaches inner bearing cap and ball bearings on a rotor shaft for a 250-horsepower AC motor adapted for high-torque requirements.

At the instant of time shown as 1, the resultant magnetic field will have its greatest intensity in the direction extending from pole (2) to pole (5), and in this case, pole (2) can be considered as a north pole and pole (5) as a south pole. (See resultant poles shown inside pole (2) and pole (5) on B in Fig. 14-59.) Arrowhead shows resultant poles. Thus, between instant 0 and 1, the magnetic field has rotated clockwise. The magnet shown in A of Fig. 14-59 rotates counterclockwise. However, the magnetic field which caused it rotated clockwise.

At time 2, the resultant magnetic field has its greatest intensity in the direction from pole (3) to pole (6). It is apparent that the resultant magnetic field has continued to rotate clockwise.

At instant 3, poles (4) and (1) can be considered as north and south poles, respectively, and the field has rotated still farther.

At later instants of time, the resultant magnetic field rotates to other positions while traveling in a clockwise direction, a single revolution of the field occurring in one cycle. If the exciting voltages have a frequency of 60 hertz, the magnetic field makes 60 revolutions per second, or 3600 revolutions per minute. This speed is known as the synchronous speed of the rotating field.

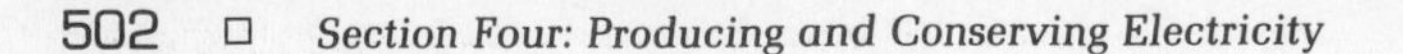

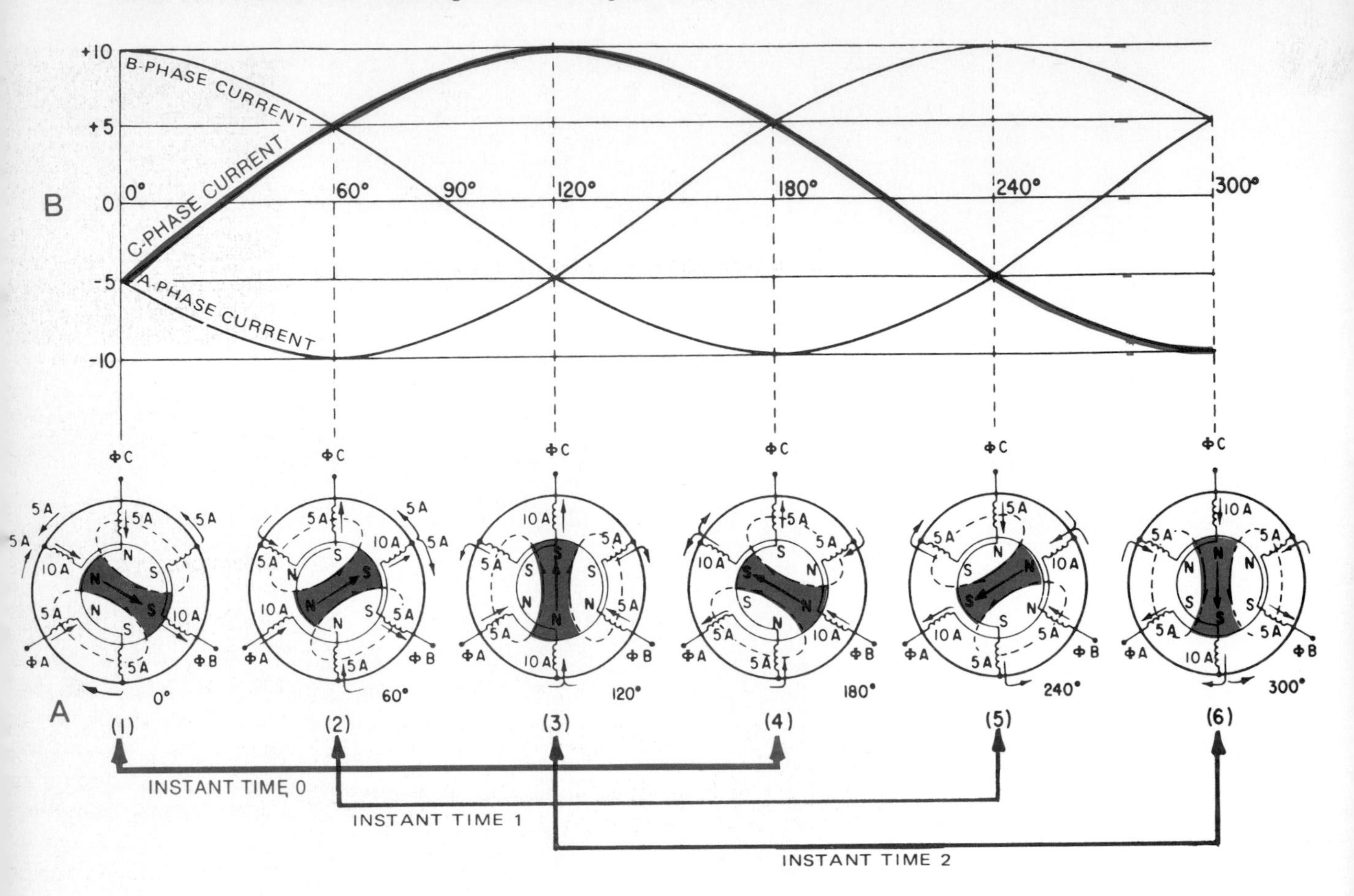

14-59. A rotating magnetic field developed by application of 3-phase voltages.

Construction of an Induction Motor

In an induction motor, the stationary portion of the machine is called a stator; the rotating member, a rotor. Instead of salient poles in the stator, as shown in A of Fig. 14-59, distributed windings are used. These are placed in slots around the periphery of the stator. Fig. 14-60.

It is not usually possible to determine the number of poles from visual inspection of an induction motor. A look at the nameplate will usually tell the number of poles. It also gives the RPM, the voltage required, and the current needed. This rated speed is usually less than the synchronous speed because of slip, to be discussed later in this chapter. To determine the number of poles per phase of the motor, divide 120 times the frequency by the rated speed:

$$P = \frac{120 \times f}{N}$$

P is the number of poles per phase, f is the frequency in hertz, N is the rated speed in RPM, and 120 is a constant.

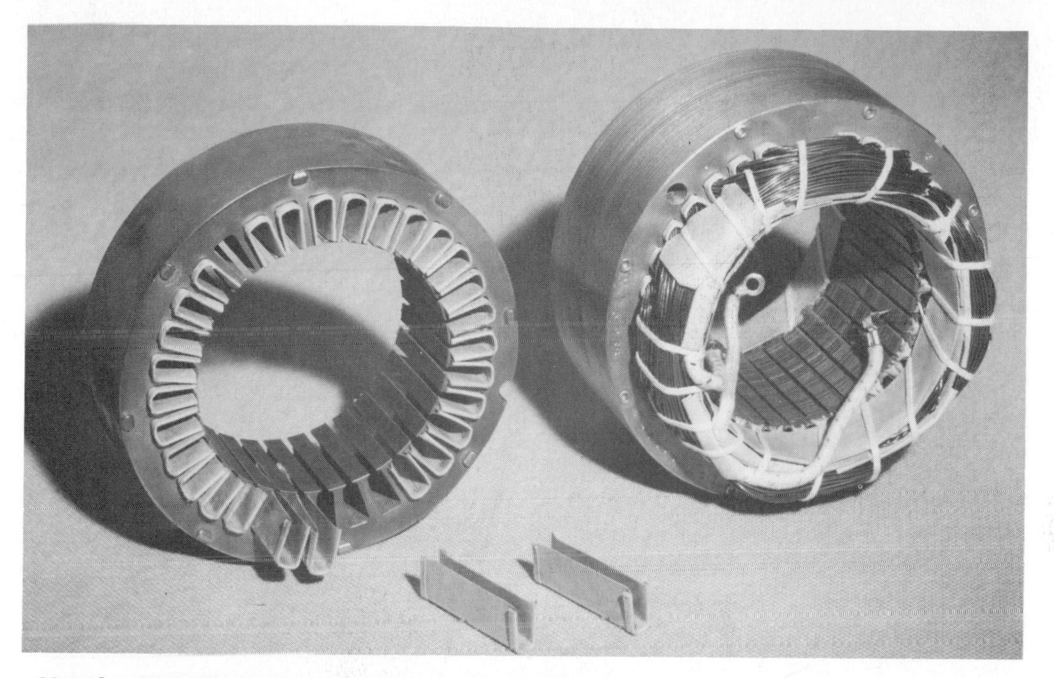

14–60a. Note how windings are inserted in a motor frame.

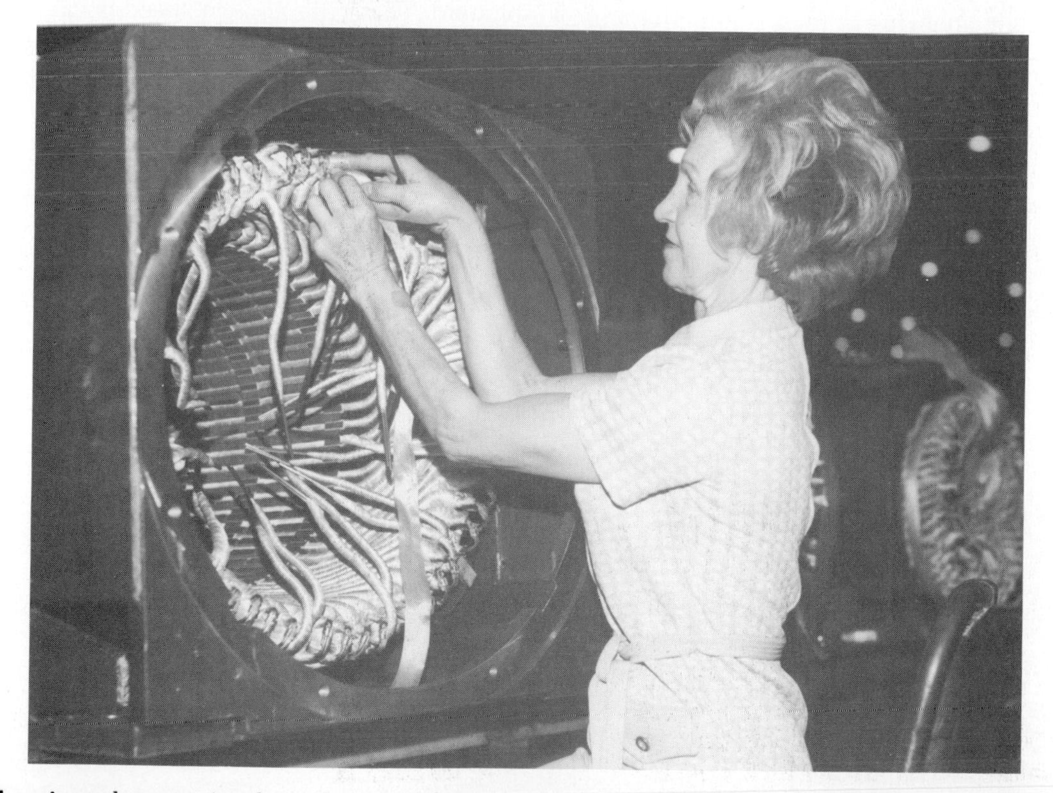

14–60b. A worker tape-insulates the AC coil connections on a stator for a 4160-volt four-pole motor.

The result is very nearly the number of poles per phase. For example, consider a 60-hertz, 3-phase machine with a rated speed of 1750 RPM. In this case:

$$P = \frac{120 \times 60}{1750} = \frac{7200}{1750} = 4.1$$

Therefore the motor has four poles per phase. If the number of poles per phase is given on the nameplate, the synchronous speed can be determined by dividing 120 times the frequency by the number of poles per phase. In the example just given, the synchronous speed is equal to 7200 divided by 4, or 1800 RPM.

The rotor of an induction motor consists of an iron core with longitudinal slots around its circumference, in which heavy copper or aluminum bars are embedded. These bars are welded to a heavy ring of high conductivity on either end. This composite structure is sometimes called a squirrel cage. Motors containing such a rotor are called squirrel-cage induction motors. Fig. 14-61.

Slip. *Slip* is the difference in RPM between a rotor and a rotating field. When the rotor of an induction motor is subjected to the revolving magnetic field produced by the stator windings, a voltage is induced in the longitudinal bars. Fig. 14-61. The induced voltage causes a current to flow through the bars. This current, in turn, produces its own magnetic field. The magnetic field combines with the revolving field in such a way as to cause the rotor to assume a position in which the induced voltage is minimized. As a result, the rotor revolves at very nearly the synchronous speed of the stator field. The difference in speed is just enough to induce enough current in the rotor to overcome the mechanical and electrical losses in the rotor. If the rotor were to turn at the same speed as the rotating field, the rotor conductors would not be cut by any magnetic lines of force. Therefore no EMF would be induced in them, no current could flow, and there would be no torque. The rotor would then slow down. For this reason there must always be a difference in speed between the rotor and the rotating field. This difference in speed is called *slip*. Slip is expressed as a percentage of the synchronous speed. For example, if the rotor turns at 1750 RPM and the synchronous speed is 1800 RPM, the difference in speed is 50 RPM. The slip is then equal to (50/1800), or 2.78%.

Single-Phase Motors

The field of a *single-phase motor*, instead of rotating, merely pulsates; and no rotation of the rotor takes place. A single-phase pulsating

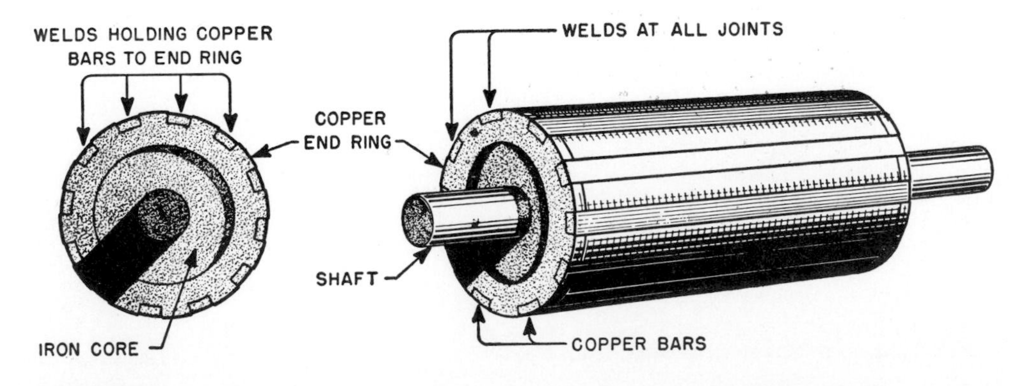

14-61. Squirrel-cage rotor.

field may be visualized as two rotating fields revolving at the same speed, but in opposite directions. It follows, therefore, that the rotor will revolve in either direction at nearly synchronous speed, provided it is given an initial impetus in either one direction or the other. The exact value of this initial rotational velocity varies widely with different machines, but a velocity higher than 15% of the synchronous speed is usually sufficient to cause the rotor to accelerate to rated, or running, speed. A single-phase motor can be made self-starting if means can be provided to give the effect of a rotating field.

■ *Shaded-pole motor.* The shaded-pole motor resulted from one of the first efforts to make a single-phase motor which would start without outside help. Fig. 14-62a. This motor has salient poles. A portion of each pole is encircled by a heavy copper ring. The presence of the ring causes the magnetic field through the ringed portion of the pole face to lag behind that through the other portion of the pole face. The effect is the production of a slight component of rotation of the field that is sufficient to cause the rotor to revolve. As the rotor accelerates, the torque increases until the rated speed is obtained. Such motors have low starting torque. Their greatest use is in small fans where the initial torque is low. They are also used in clocks, inexpensive record players, and some electric typewriters. Fig. 14-62b.

■ *Split-phase motors.* Many types of split-phase motors have been made. Such motors have a start winding that is displaced 90 electrical degrees from the main, or run, winding. In some types, the start winding has a fairly high resistance, which causes the current in it to be out of phase with the current in the run winding. This condition produces, in effect, a rotating field and the rotor revolves. A centrifugal switch is used to disconnect the start winding automatically after the rotor has attained approximately 75% of its rated speed. Fig. 14-62c.

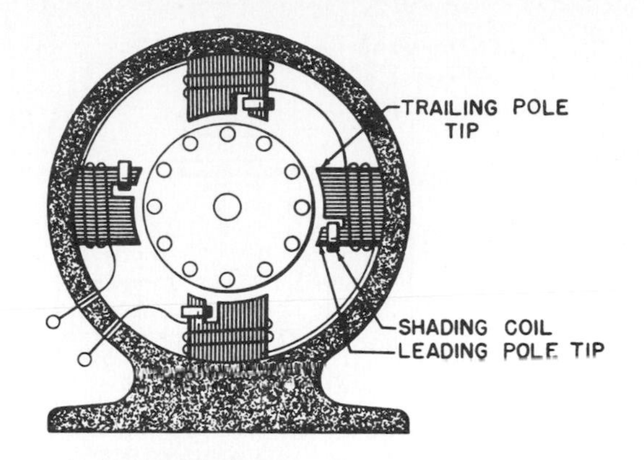

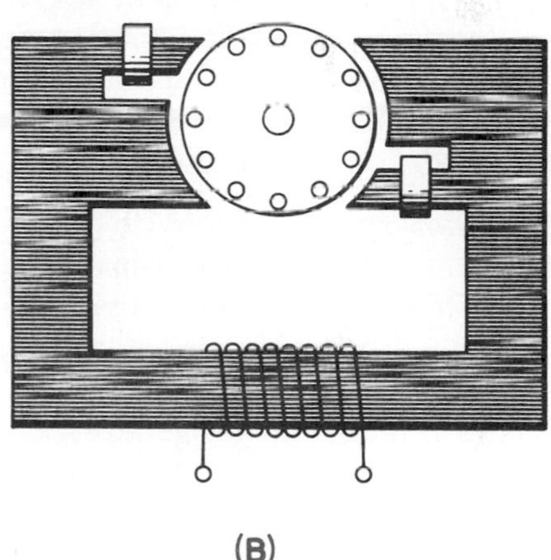

14-62a. Shaded-pole motor.

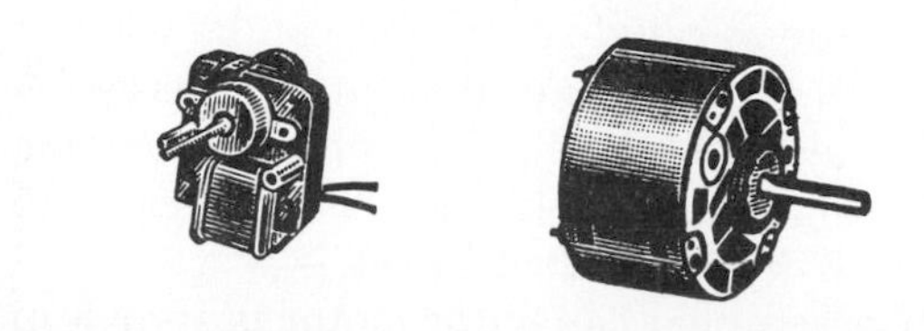

14-62b. Shaded-pole motors used for fans and clocks.

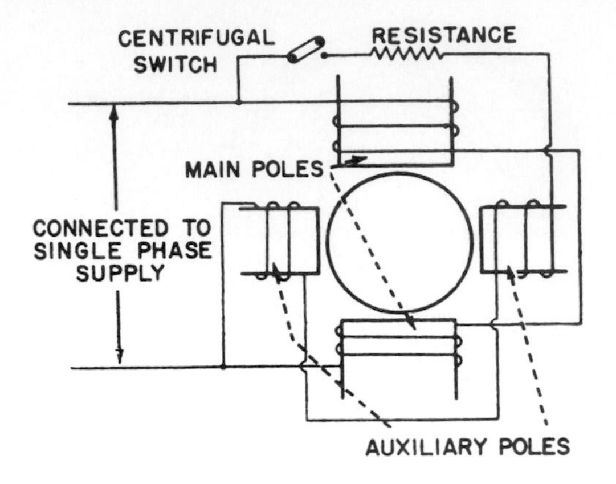

14-62c. Split-phase motor.

Split-phase motors are used where there is no need to start under load. They are used on grinders, buffers, and other similar devices. They are available in fractional-horsepower sizes with various speeds, and are wound to operate on 120 volts AC or 240 volts AC.

■ *Capacitor-start motors.* With the development of high-quality and high-capacity electrolytic capacitors, a variation of the split-phase motor, known as the capacitor-start motor, has been made. Almost all fractional-horsepower motors in use today on refrigerators, oil burners, washing machines, table saws, drill presses, and similar devices are capacitor-start. A capacitor motor has a high starting current and the ability to develop about four times its rated horsepower if it is suddenly overloaded. In this adaption of the split-phase motor, the start winding and the run winding have the same size and resistance value, the phase shift between currents of the two windings being obtained by means of capacitors connected in series with the start winding. Capacitor-start motors have a starting torque comparable to their torque at rated speed and can be used in places where the initial load is heavy. A centrifugal switch is required for disconnecting the starting winding when the rotor speed is up to about 25% of the rated speed.

A disassembled capacitor motor is shown in Fig. 14-63a. Note in Fig. 14-63b, also a capacitor-start motor, the centrifugal-switch arrangement with the governor mechanism. Fig. 14-64a shows the windings, the rotor, and the capacitor housing on top of the motor. Note that the windings overlap.

One of the advantages of the single-value, capacitor-start motor is its ability to be reversed easily and frequently. Figs. 14-64b & 14-64c. The motor is quiet and smooth running. If a 5 to 20 HP capacitor-start motor is called for, the two-value capacitor motor is used. Fig. 14-65. This motor has two sets of field windings in the stator—an auxiliary winding called a phase winding and the main winding. The phase winding is designed for continuous duty; a capacitor remains in series with the winding at all times. A start capacitor is added to the phase circuit to increase starting torque, but is disconnected by a centrifugal switch during acceleration.

In general, single-phase motors are more expensive to purchase and to maintain than 3-phase motors. They are less efficient, and their starting currents are relatively high. All run at essentially constant speed. Nonetheless, most machines using electric motors around the home, on the farm, or in small commercial plants are equipped with single-phase motors.

Those who select a single-phase motor usually do so because three-phase power is not available to them. Fig. 14-66 shows the simple methods used in the construction of a 3-phase motor. Note this is a half-etched, squirrel-cage rotor. The bearings are not sealed ball bearings, but are a sleeve-type with the oil caps placed so that oil may be added occasionally to keep the bearing lubricated.

A 3-phase motor is shown in the cutaway view in Fig. 14-67. Note the simple rotor and fan blades, windings, and the sealed ball bearings. This is an almost totally maintenance-free motor. Fig. 14-68 (page 498) is a polyphase motor which has been made explosion proof.

■ *Sizes of motors.* Some single-phase induction motors are rated as high as 2 HP. The major field of use is the 1 HP, or less, at 120 or 240 volts for the smaller sizes. For larger power rat-

14–63a. Disassembled single-phase, capacitor-start motor.

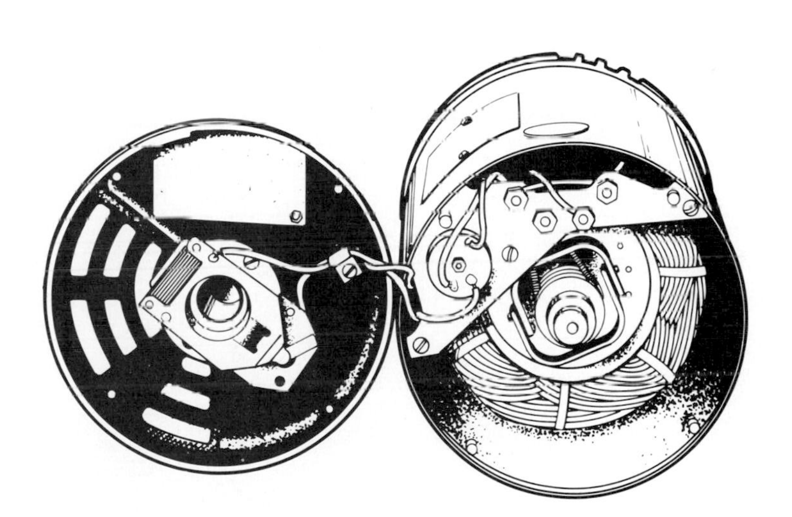

14–63b. Single-phase starting switch and governor mechanism.

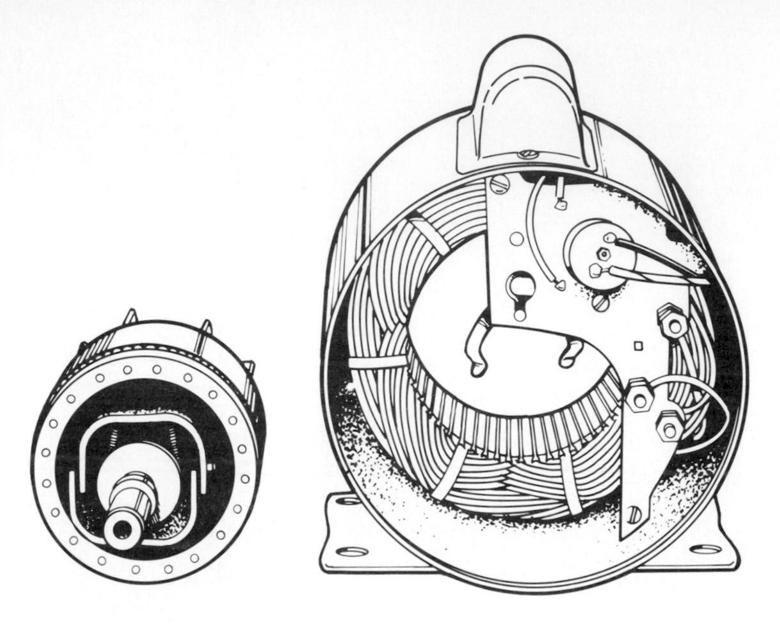

14–64a. Single-phase stator and rotor.

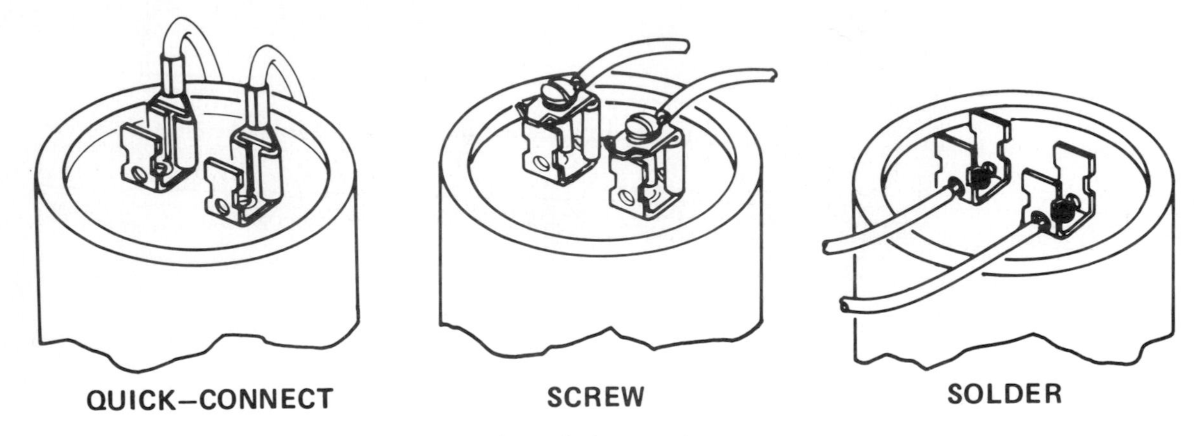

14–64b. Three methods of connection for a electrolytic capacitor.

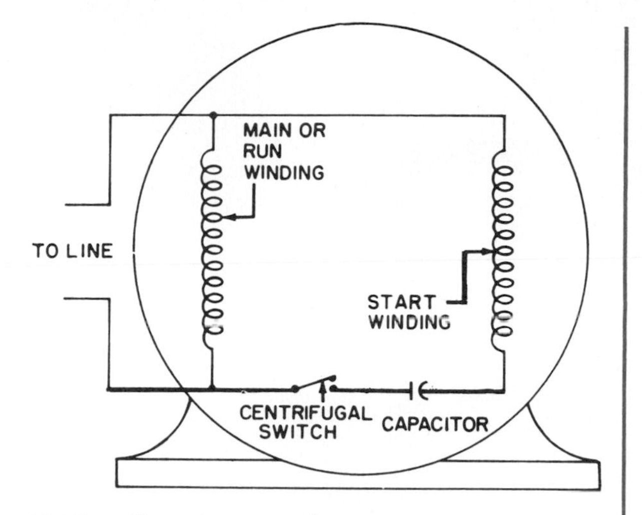

14–64c. Capacitor-start diagram.

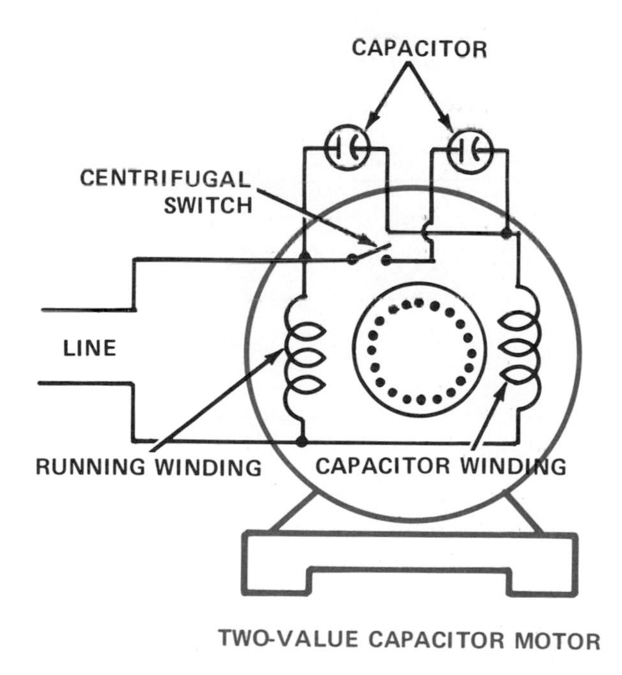

14–65. Two-value capacitor-start motor.

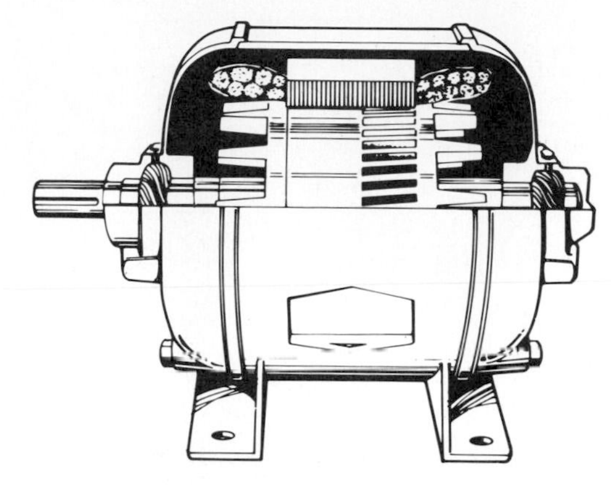

14–66. Cutaway view of a 3-phase motor with a half-etched, squirrel-cage rotor.

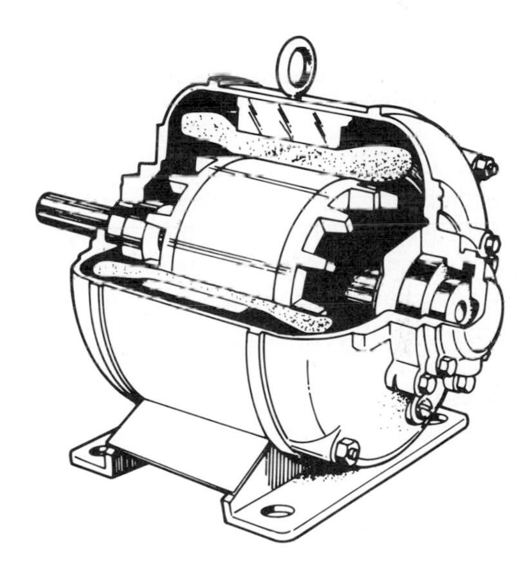

14–67. Cutaway view of a 3-phase motor showing the cast rotor.

14-68. This is a polyphase motor with explosion-proof construction.

ings, polyphase motors are generally specified, since they have excellent starting torque and are practically maintenance free.

Fig. 14-69 is a brush-lifting, repulsion-start, induction-run, single-phase motor. The following items should be noted: The rotor is wound, just like a DC motor. The brushes can be lifted by centrifugal force once the rotor comes up to speed; then the rotor can act as a squirrel-cage type. This type pulls a lot of current in starting, but is capable of starting under full load conditions.

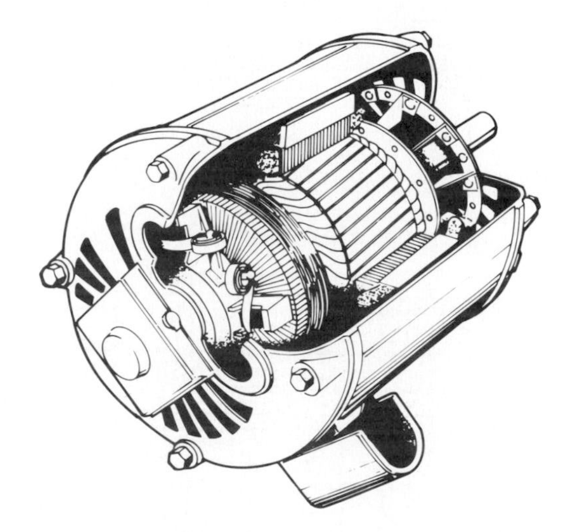

14-69. A brush-lifting, repulsion-start, induction-run, single-phase motor. Note the brushes and the wound rotor.

■ *Cooling and mounting motors.* Fig. 14-70 shows an improved motor ventilating system. A large volume of air is directed through the motor to reduce temperatures. The large blower on the right is located behind a baffle that controls air movement to the blower blades. The blower draws outside air in through the large drip-proof openings in the back end plate, then forces the cooling air around the back coil extension, through the rotor vent holes, the air gap, and through the passages between the stator core and the frame. A second blower on the front end of the rotor at left, cast as an integral part of the rotor, circulates the air around the inside of the front coil extensions and then speeds the flow of heated air out the motor through the drip-proof openings in the front end plate.

Fig. 14-71 shows the rigid base and the resilient base. Note that the resilient base has a mounting bracket attached to the ends of the motor, with some material used to make it more silent. However, the rigid base has its support mechanism welded to the frame of the motor. This can mean more of the hum or noise of the running motor can be transmitted to whatever it is attached to in operation.

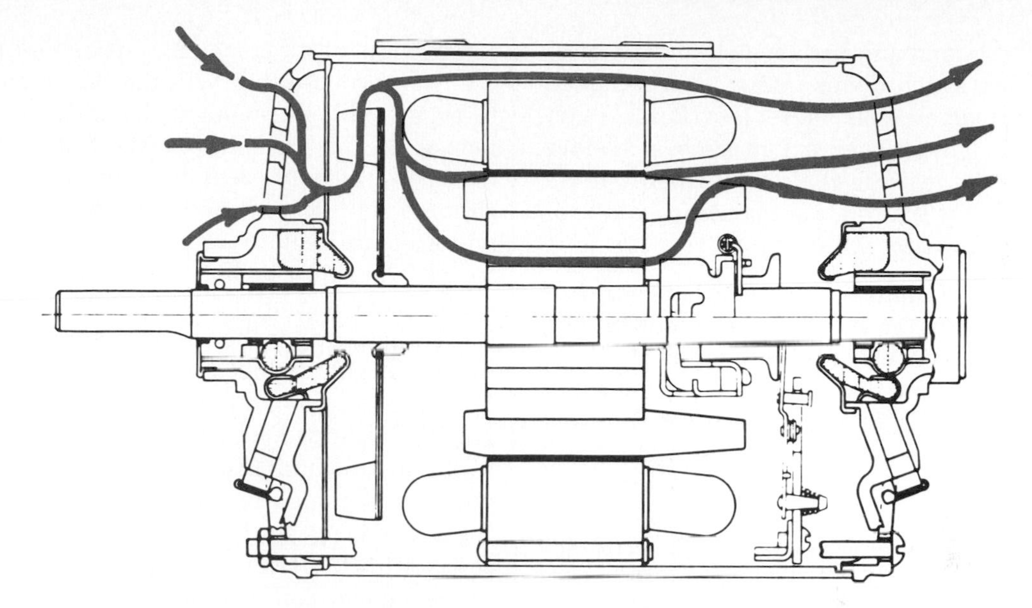

14–70. Cooling system using two fans to keep the air moving inside an electric motor.

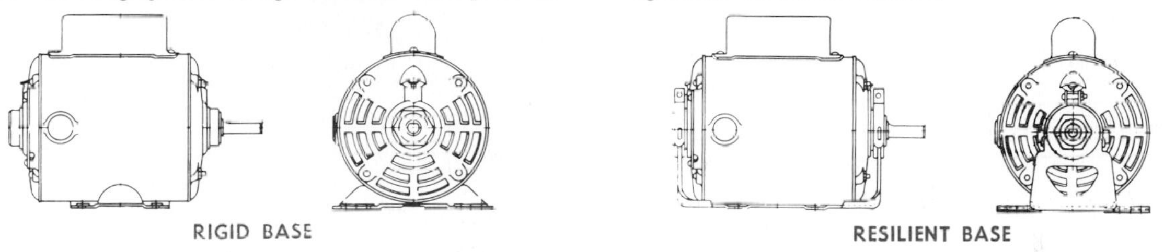

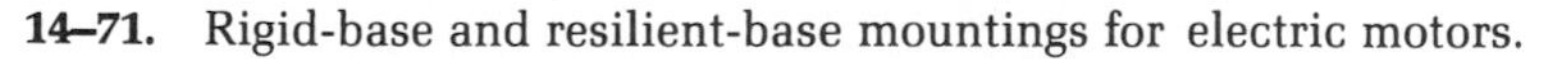
14–71. Rigid-base and resilient-base mountings for electric motors.

Direction of Rotation

The direction of rotation of a 3-phase induction motor can be changed by simply reversing two of the leads to the motor. The same effect can be obtained in a 2-phase motor by reversing connections to one phase. In a single-phase motor, reversing connections to the start winding will reverse the direction of rotation. Most single-phase motors designed for general use have provision for readily reversing connections to the start winding. Nothing can be done to a shaded-pole motor to reverse the direction of rotation. This direction is determined by the physical location of the copper shading ring.

If, after starting, one connection to a 3-phase motor is broken, the motor will continue to run but will deliver only one-third of the rated power. Also, a 2-phase motor will run at one-half its rated power if one phase is disconnected. Neither motor will start except by hand with one connection broken, but once started, the motors will run.

Synchronous Motor

A *synchronous motor* is one of the principal types of AC motors. Like the induction motor, the synchronous motor is designed to take advantage of a rotating magnetic field. Unlike the induction motor, however, the torque developed does not depend upon the induction of currents in the rotor. Briefly, the principle of operation of the synchronous motor is as follows:

A multiphase source of AC is applied to the stator windings and a rotating magnetic field is produced. A direct current is applied to the rotor windings and another magnetic field is produced. The synchronous motor is so designed and constructed that these two fields react upon each other. They act in such a manner that the rotor is dragged along. It rotates at the same speed as the rotating magnetic field produced by the stator windings.

■ *Theory of operation.* An understanding of the operation of the synchronous motor may be obtained by considering the simple motor in Fig. 14-72. Assume that poles A and B are being rotated clockwise by some mechanical means in order to produce a rotating magnetic field. The rotating poles induce poles of opposite polarity, as shown in the illustration of the soft iron rotor, and forces of attraction exist between corresponding north and south poles. Consequently, as poles A and B rotate, the rotor is dragged along at the same speed. However, if a load is applied to the rotor shaft, the rotor axis will momentarily fall behind that of the rotating field, but will thereafter continue to rotate with the field at the same speed, as long as the load remains constant. If the load is too large, the rotor will pull out of synchronization with the rotating field, and, as a result, will no longer rotate with the field at the same speed. The motor is then said to be overloaded.

Fig. 14-73 is similar to Fig. 14-72, except that it is 3-phase rather than single phase. The magnitude of the induced poles in the rotor in Fig. 14-72 is so small that sufficient torque cannot be developed for most practical loads. To avoid such a limitation on motor operation, a winding is placed on the rotor and this winding is energized with DC. A rheostat placed in series with the DC source provides the operator of the machine with a means of varying the strength of the rotor poles, thus placing the motor under control for varying loads.

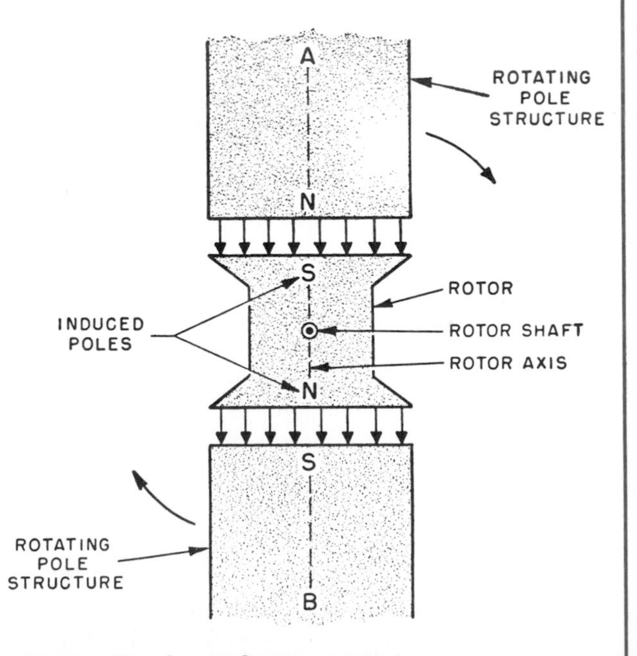

14–72. Simple synchronous motor.

■ *Synchronous motor as a capacitor or inductor.* If a synchronous motor is driven by an external power source and the excitation, or voltage applied to the rotor, is adjusted to a certain value called 100% excitation, no current will flow from, or to, the stator winding. In this case, the voltage generated in the stator windings by the rotor, or CEMF, exactly balances the applied voltage. If the excitation is reduced below the 100% level, however, the difference between the CEMF and the applied voltage produces a reactive component of current which lags the applied voltage. The motor then acts as an inductance.

If the excitation is increased above the 100% level, the reactive component leads the applied voltage and the motor acts as a capacitor. This feature of the synchronous motor permits its use as a power-factor correction device. When so used it is called a *synchronous* or *rotating capacitor*. Large steel mills use this device to correct the power factor and decrease their electric power costs. Other industrial uses have been found for the synchronous motor since it has a stable speed.

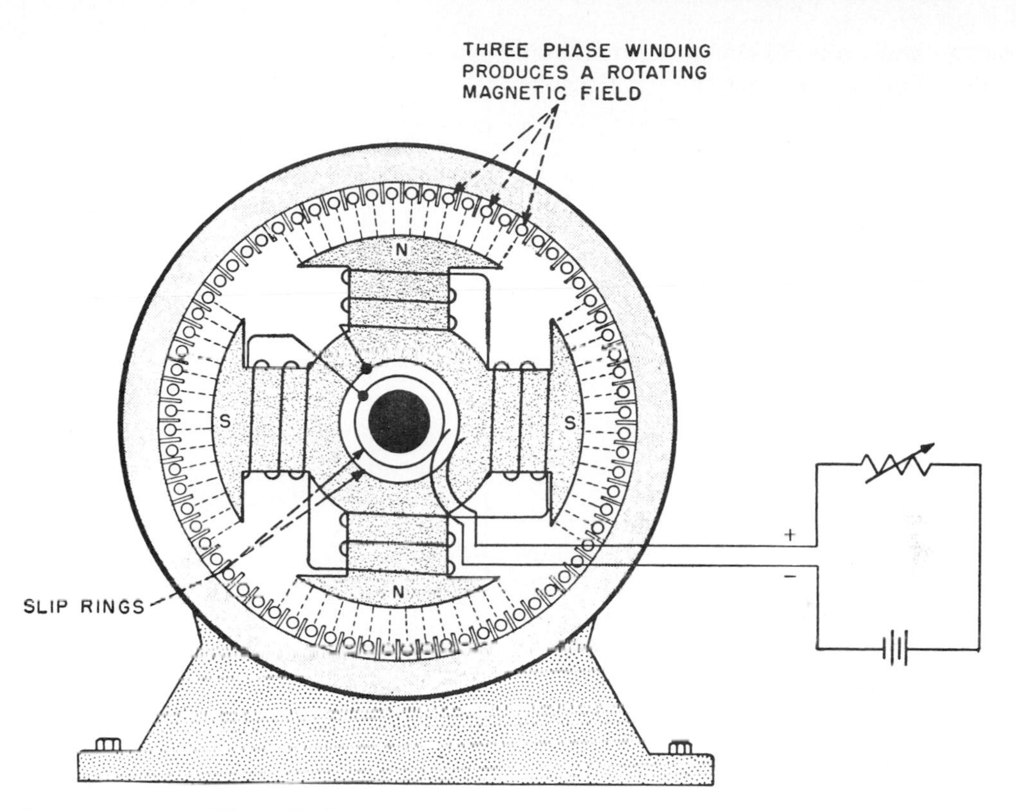

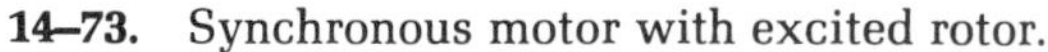
14–73. Synchronous motor with excited rotor.

Some advantages of the synchronous motor are:

1. When used as a synchronous capacitor, the motor is connected on the AC line in parallel with the other motors on the line. It is run either without load or with a very light load. The rotor field is overexcited just enough to produce a leading current which offsets the lagging current of the line with the motors operating. A unity power factor (1.00) can usually be achieved. This means the load on the generator is the same as though only resistance made up the load.
2. The synchronous motor can be made to produce as much as 80% leading power factor. However, because leading power factor on a line is just as detrimental as a lagging power factor, the synchronous motor is regulated to produce just enough leading current to compensate for lagging current in the line.

■ *Properties of the synchronous motor.* The synchronous motor is not a self-starting motor. The rotor is heavy, and from a dead stop it is impossible to bring it into magnetic lock with the rotating magnetic field. For this reason, all synchronous motors have some kind of starting device. Such a simple starter is another motor, either AC or DC, which can bring the rotor up to approximately 90% of the synchronous speed. The starting motor is then disconnected and the rotor locks in step with the rotating field.

Another starting method is a second winding of the squirrel-cage type on the rotor. This induction winding brings the rotor almost into synchronous speed. When the DC is connected to the rotor windings, the rotor pulls into step with the field. The latter method is the more commonly used.

Fig. 14-74 shows a small synchronous motor which has a number of applications. Because of their stable speed, synchronous motors are used for turntables in stereo equipment.

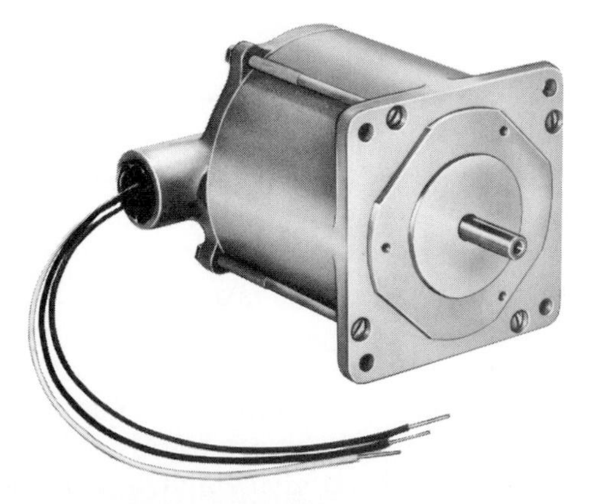

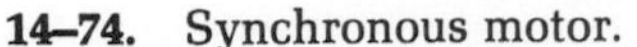

14–74. Synchronous motor.

Insulating Motor Terminals

Large electric motors require special consideration when their terminals are connected. To insulate the motor lead connections without spending excessive time in taping, you might use a heat-shrinkable motor connection kit.

Motor connection kits provide quick installation and dependable protection of the motor lead terminations up to 600 volts. Fig. 14-75. They are used for both new motor installations and periodic removal for routine motor maintenance. The heat-shrinkable caps fit easily into motor termination boxes. They are made from thick-wall, flame-retardant, cross-linked polyolefin. They are designed for fast shrinking with an electric heat gun or a broad torch flame, minimizing motor installation time. After shrinking, the thick-wall caps provide superior electrical insulation and resistance to abrasion caused by motor vibration. These labor-saving products deliver protection without the extra time normally needed for taping.

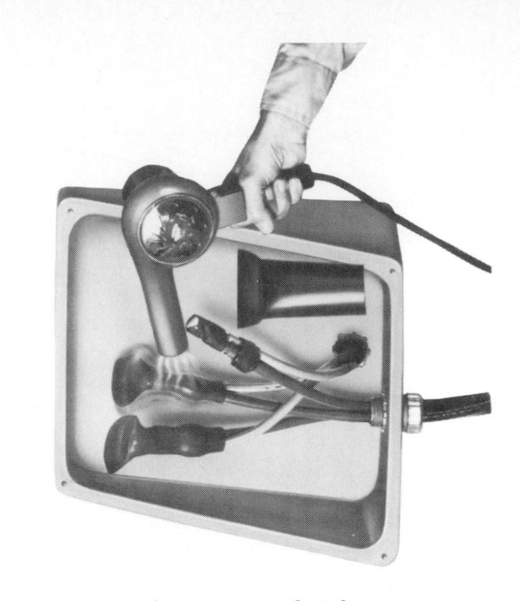

14–75. Using a heat gun to shrink caps on motor terminals.

The kit includes mastic sealant strips to wrap the cable jackets above the terminal lugs, leaving the bolt area clean. The shrinking of the caps squeezes the softened mastic around the cable jackets. This seals out moisture, acids, oils, and other contaminants. The mastic wrap also allows caps to fit several cable sizes by building up the jacket diameters.

A tough fabric tape is supplied to wrap the bolt and nut should there be any sharp edges after tightening. Fig. 14-76.

Caps can be removed in a matter of seconds when scored lightly with a sharp knife, warmed with a heat source, and peeled away with a pair of pliers. This leaves the feeder cable clean and ready to receive a new motor pigtail lead, keeping the motor replacement time to a minimum.

This kit is used for stub connections where limited terminal box size or bending space requires lugs to be joined back-to-back. These caps are designed for lugs with any length barrel. They can be cut to a convenient length. When cut shorter, caps should be long enough to just cover the mastic sealant wrapped around the cable jackets.

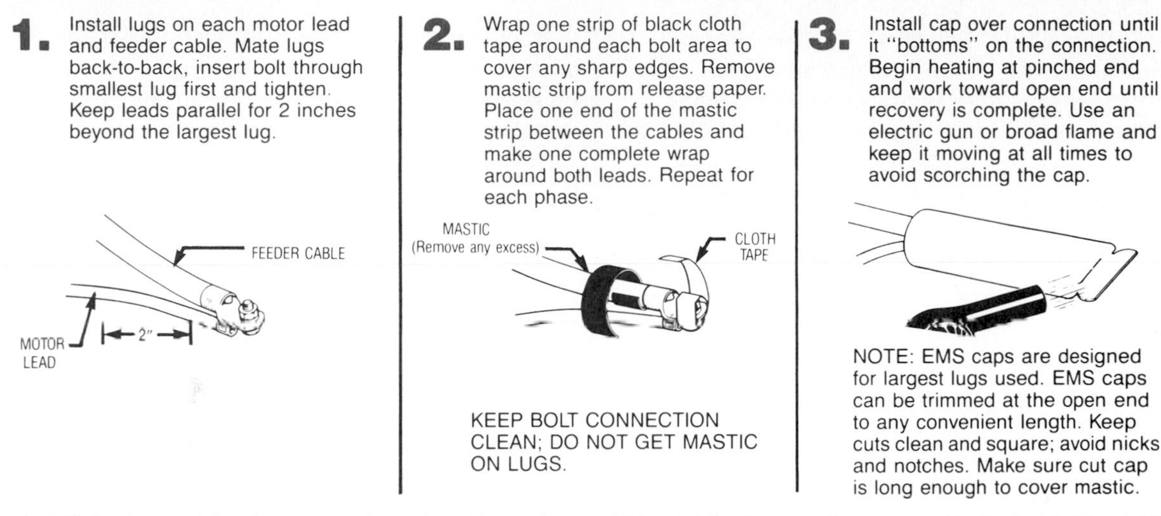

Installation is complete when cap conforms to cables and lugs, and mastic is squeezed out from open end. Do not use torch that produces pencil point flame.

Make certain all connections are cool to the touch (about 15-20 minutes) before replacing in motor terminal box. Pinched area can be trimmed with care when cool.

14–76. Installation of heat-shrinkable caps on motor terminals.

Troubleshooting Three-Phase Motors

Three-phase motors are used on all levels of commercial and industrial operations. They are rugged, long-lasting, and need very little maintenance. If abused or overloaded or supplied with under-voltage, they can develop problems.

The stator windings of integral horsepower AC motors are capable of full power operation for many years. However, winding life can be shortened by any combination of the following:

Mechanical damage produces weak spots in the insulation. It can occur during maintenance of the motor or result from such operating problems as severe vibration.

Excessive moisture encountered in service causes deterioration of the insulation.

High dielectric stress, such as voltage surges or excess input current, can cause overheating and insulation deterioration.

High temperature reduces the ability of the insulation system to withstand mechanical or electrical abuse. Over-temperature is usually a result of poor installation or misapplication of the motor.

Regardless of the reason for failures, the obvious result is thermal degradation of the insulation—or burnouts. The rate of insulation degradation is increased by higher temperature. In fact, insulation life is reduced by about one-half for each 10°C increase of winding temperature. Therefore, long winding life requires normal operating temperatures.

Stator temperature is affected by three factors—ventilation, ambient temperature, and input current.

Ventilation. Forced ventilation is generally an inherent design feature of induction motors. Decreased cooling air volume caused by blocked air passages, blower failure, or low air density at high altitudes leads to overheating and shortened winding life.

Ambient Temperature. Most motors are designed to operate at a maximum of 40°C ambient temperature. Any increase of ambient over 40°C requires derating the motor or its expected life will be shortened. Factors which raise input air temperature include placement of motors in discharge air streams from other equipment and high temperature locations.

Whenever a motor is derated or uprated, the starting, pull-up, and breakdown torques remain the same as nameplate rating. However, the bearings, shaft, and other components may be subjected to a life that is a function of the the "new" rating.

Input Current. An increase over the rated input current causes excessive winding heating proportional to the square of the current. Therefore, successful motor operation must be keyed around nameplate current. The common factors affecting input current are as follows:

■ *Starting.* The induction motor has a locked rotor current about six times nameplate. Therefore, at locked rotor condition the heating effect on the stator windings is 6^2 or 36 times greater than at rated load. A motor can sustain locked rotor currents for brief periods of only 10 to 15 seconds.

■ *Slow or repeated starts.* During acceleration approximately locked rotor current is present until the motor speeds beyond the breakdown RPM. Slow acceleration or jogging tends to increase motor temperature in the same way as repeated starting. For such applications the high heat input must be balanced by adequate cooling periods.

■ *Service factor.* Induction motors are designed for continuous operation at rated load. Many have a 1.15 service factor and can be safely loaded to 1.15 times their rated horsepower. They can be loaded beyond the service factor rating for only short periods of time without danger of overheating.

Motor failures resulting from loss of ventilation, high ambient temperature, or excessive motor loading generally exhibit a burnout pattern of uniform overheating. Secondary faults, such as phase-to-phase failures, may result, but the overall condition will usually be uniform.

■ *Voltage variation.* Voltage variation occurs when the voltage between the three phases remains equal, but increases above or decreases below the nameplate voltage.

Most motors operate successfully at rated load within a voltage variation of ±10%. The performance characteristics within this range change are provided by the manufacturer. These changed characteristics should be considered when selecting a motor. Voltage variation beyond ±10% may cause excessive heating. The effects of voltage variation on input current and heating differ with the frame sizes of the motors. Winding failures resulting from extreme voltage variation are identical to those overloads because the input current is uniformly excessive.

Ceiling Fans

Many makes of ceiling fans are available today. Fig. 14-77. Some have a single speed, others have multiple speeds. Most are reversible. They are used to keep the air moving in a room. They keep it cool during the summer and warm during the winter when the hot air tends to rise. The fan is reversed in the winter so it causes the air to be moved toward the ceiling and then follow the ceiling to the walls and down again, thereby eliminating a draft which may be desirable during the summer months. That means the fan rotates in the opposite direction during the winter than during the summer months.

14–77. Ceiling fan.

Reversing. A fan motor can be reversed by reversing the connections to the windings.

Speeds. A number of speed controls are available for fan motors. They may be the type with a switch that changes the number of windings by putting in or taking out windings to affect the speed. Or, the speed control may be electronic. It may use a semiconductor device in a circuit that will regulate the voltage and current. Different types of motors call for different types of controllers. In most instances the electrician is not responsible for wiring the motors for fans. They come pre-wired. The electrical installation of the fan usually requires the standard connection of two fan leads to two wires in the box where the fan is mounted.

14-A. Guide to Probable Causes of Motor Troubles

MOTOR TYPE	AC SINGLE PHASE				AC POLY-PHASE (2- OR 3-PHASE)	BRUSH TYPE (UNIVERSAL, SERIES, SHUNT OR COMPOUND)
	SPLIT PHASE	CAPACITOR START	CAPACITOR START & RUN	SHADED POLE		
TROUBLE						
Will not start	1, 2, 3, 5	1, 2, 3, 4, 5	1, 2, 4, 7, 17	1, 2, 7, 16, 17	1, 2, 9	1, 2, 12, 13
Will not always start, even with no load, but will run in either direction when started manually	3, 5	3, 4, 5	4, 9		9	
Starts but heats rapidly	6, 8	6, 8	4, 8	8	8	8
Starts but runs too hot	8	8	4, 8	8	8	8
Will not start but will run in either direction when started manually—overheats	3, 5, 8	3, 4, 5, 8	4, 8, 9		8, 9	
Sluggish—sparks severely at the brushes						10, 11, 12, 13, 14
Abnormally high speed—sparks severely at the brushes						15
Reduction in power—motor gets too hot	8, 16, 17	8, 16, 17	8, 16, 17	8, 16, 17	8, 16, 17	13, 16, 17
Motor blows fuse or will not stop when switch is turned to off position	8, 18	8, 18	8, 18	8, 18	8, 18	18, 19
Jerky operation—severe vibration						10, 11, 12, 13, 19

Probable Causes

1. Open in connection to line.
2. Open circuit in motor winding.
3. Contacts of centrifugal switch not closed.
4. Defective capacitor.
5. Starting winding open.
6. Centrifugal starting switch not opening.
7. Motor over-loaded.
8. Winding short-circuited or grounded.
9. One or more windings open.
10. High mica between commutator bars.
11. Dirty commutator or commutator is out of round.
12. Worn brushes and/or annealed brush springs.
13. Open circuit or short circuit in the armature winding.
14. Oil-soaked brushes.
15. Open circuit in the shunt winding.
16. Sticky or tight bearings.
17. Interference between stationary and rotating members.
18. Grounded near switch end of winding.
19. Shorted or grounded armature winding.

14-C. Mechanical and Electrical Characteristics

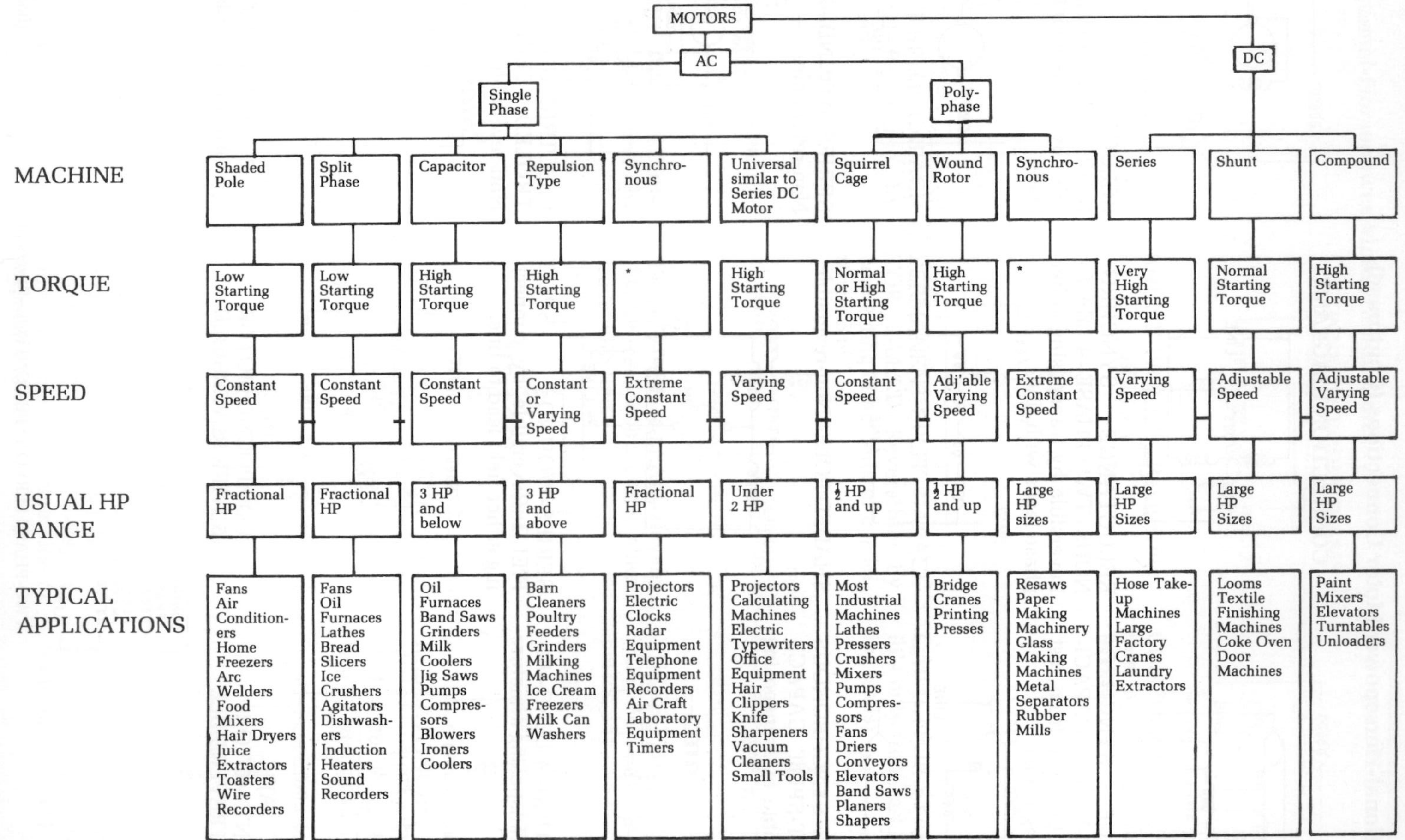

*Starting torque depends on starting method. Synchronous motors cannot start as synchronous motors, but must be started as one of the other types of AC motors.
(Lincoln Electric Co.)

QUESTIONS

1. What determines the amount of induced EMF in a conductor when it is in a magnetic field?
2. What is ripple?
3. Name three types of DC generators.
4. List the different classifications of DC compound generators.
5. Differentiate between long shunt and short shunt.
6. Briefly state the principle on which motors operate.
7. Why are multi-loop motors necessary?
8. List three types of DC motors, with advantages of each.
9. List the parts of an AC generator.
10. What is another name for an AC generator?
11. Where are emergency AC generators used?
12. What are two methods of connecting three-phase generator windings?
13. How does the synchronous motor differ from the induction motor?
14. Why is slip important in an AC motor.
15. Why do polyphase motors require less maintenance than single-phase motors?
16. How is the temperature rise in an AC motor held within safe operating limits?
17. How can a synchronous motor be used as in inductor or a capacitor?
18. What are some of the synchronous motor's properties?
19. What are some typical AC single-phase motor troubles?

KEY TERMS

alternator
armature
commutator
compound generator
counter electromotive force
residual magnetism
rotor
series generator
shunt generator
single-phase motor
slip
static excitor
stator
synchronous motor

CHAPTER 15

APPLIANCES AND CONSERVATION OF ENERGY

Objectives

After studying this chapter, you will be able to:

- Describe the wiring of major household appliances.
- List the important guidelines for conserving electrical energy.
- Explain the principle of programmable load control.

Today's household appliances are highly efficient devices when compared to those of years ago. Fig. 15-1. Electricity was first used to produce light. The first light sources were arc lamps. Electricity later became a source of energy for many kinds of equipment.

Many modern appliances are portable and use a standard 120-volt circuit with at least 15 amperes. However, some devices use 240 volts and need special plugs and wiring. Table 15-A.

■ *Electric ranges.* The largest appliance in the house is usually an electric range for cooking. It is wired with a permanent connection, or with a plug and pigtails. Fig. 15-2a shows the plug which connects to the range terminals provided by the manufacturer, and to the surface-mounted box in Fig. 15-2b. This plug and receptacle must be capable of handling at least 50 amperes and provide connections for the three wires used in such circuitry.

Fig. 15-3 is a flush-mounted outlet for 240 volts, which can be wired in a junction box with the wall plate to fit. Fig. 15-4.

■ *Dryers.* Electric dryers must have a special outlet for their connection to electrical power. Fig. 15-5. Note that the green wire is called the ground wire. The motor is run by 120 volts, and the heating element is across the 240-volt lines, red and black. These units, which draw approximately 30 amperes, should be wired with rigid conduit, thinwall conduit, or Romex with at least the proper wire to handle 30 amperes. A circuit breaker should be installed in the distribution box to handle the current. Old

15–1. Early styles of some electrical appliances.

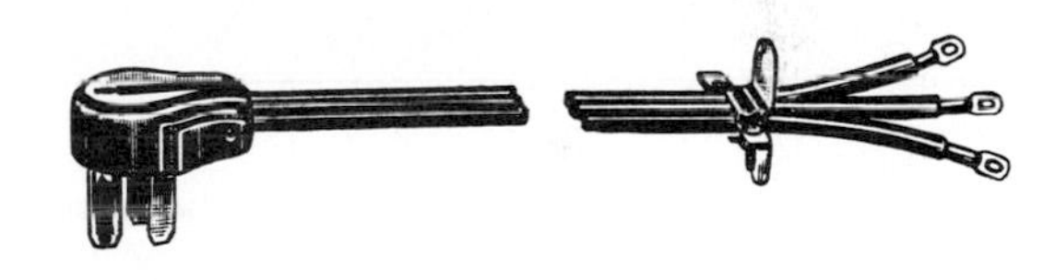

15–2a. Pigtail and plug.

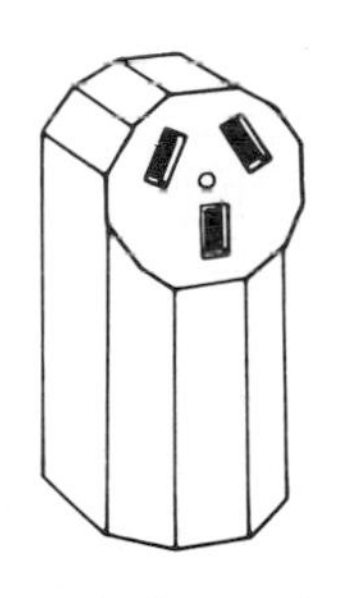

15–2b. Surface-mounted receptacle for 240 volts.

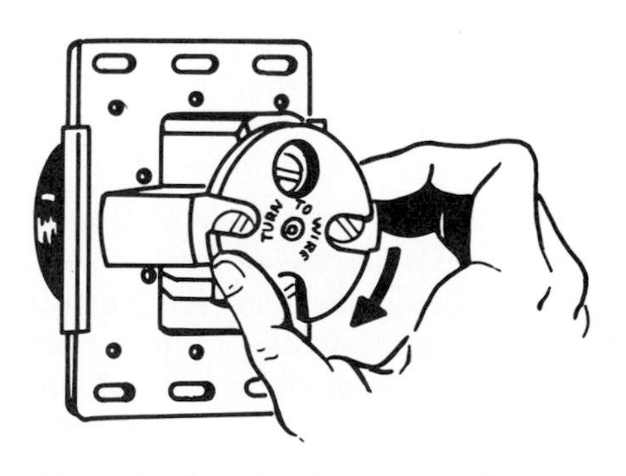

15–3. Note the slots for this receptacle.

15-A. Approximate Energy Consumption of Electrical Equipment

	AVERAGE kW-h PER MONTH
Air Conditioning (per ton on 12 hour day use)	360
Automatic Blanket	15
Clothes Dryer (Electric)	50
Dishwasher	3
Electric Fan (10-inch)	1
Garbage Disposal Unit	2
Home Freezer	40
Iron	12
Ironer	15
Lighting	60
Mixer	2
Oil Furnace with electric control	40
Radio	10
Range	90
Refrigerator	40
Sewing Machine	1
Television	40
Vacuum Cleaner	2
Washing Machine (Automatic)	10
Water Heater	350

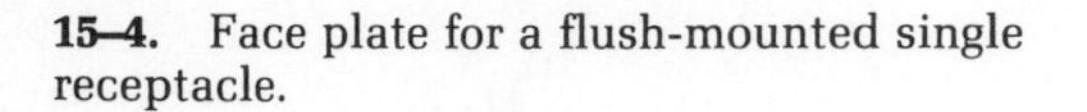

15–4. Face plate for a flush-mounted single receptacle.

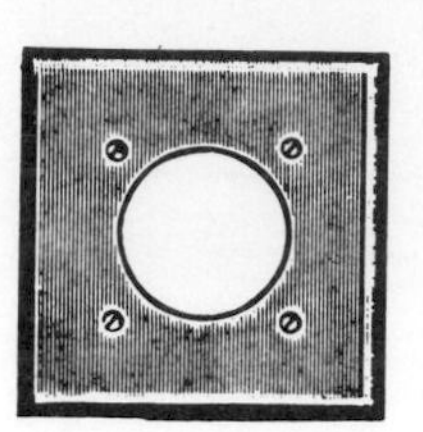

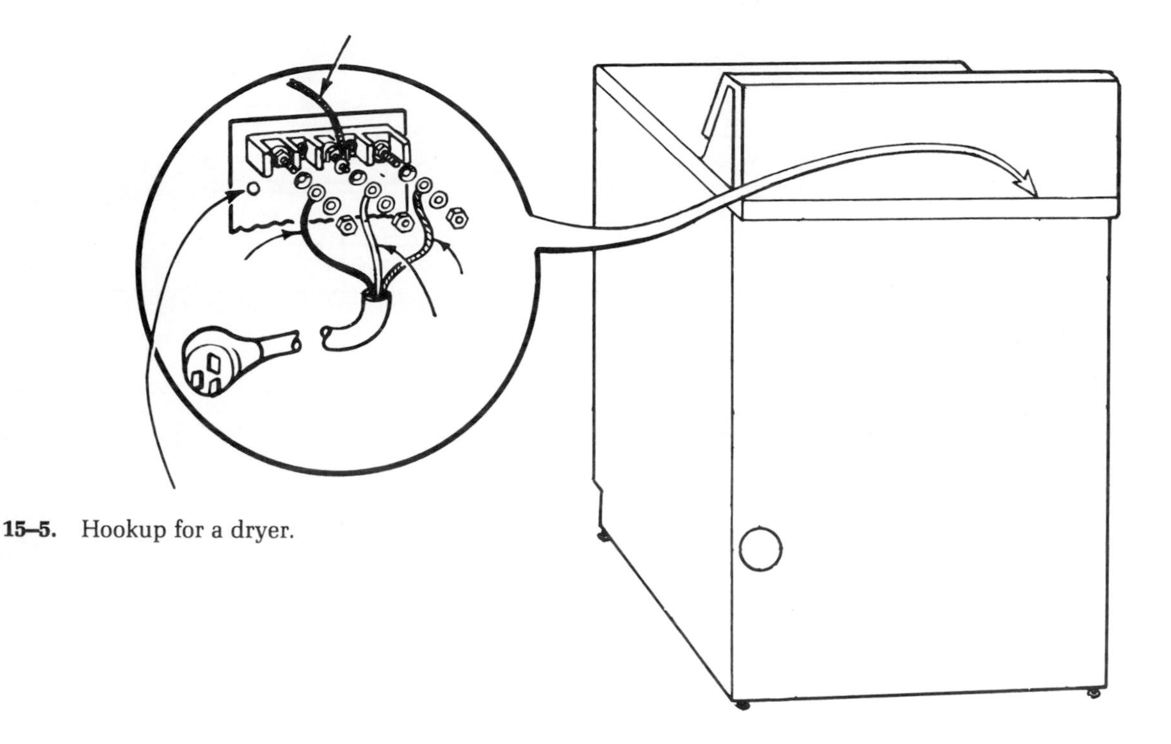

15–5. Hookup for a dryer.

homes, insufficiently wired to handle such a current, may require another distribution box to handle the load.

■ *Built-ins.* In some new or remodeled kitchens there are no separate ranges. Instead the ovens and cooking surfaces are built-in, or preset as units into walls or atop counters. These are usually permanently wired, with no plugs to bother with, since they seldom need removing for service.

Dishwashers, garbage disposal units, trash compactors, and microwave ovens are sometimes wired in, but in most cases a standard 120-volt outlet is all that is required. The kitchen area should be designed for built-ins, allowing extra 20-ampere outlets for appliances such as toasters, electric ovens (portable), and waffle irons.

■ *Refrigerators.* Refrigerators often look built-in when they are not, due to their planned fit into the space provided. They usually require a simple 120-volt outlet, yet some of the larger units should have their own circuit. The refrigerator in Fig. 15-6 has both a motor to drive a compressor and heating elements to keep butter at a different temperature

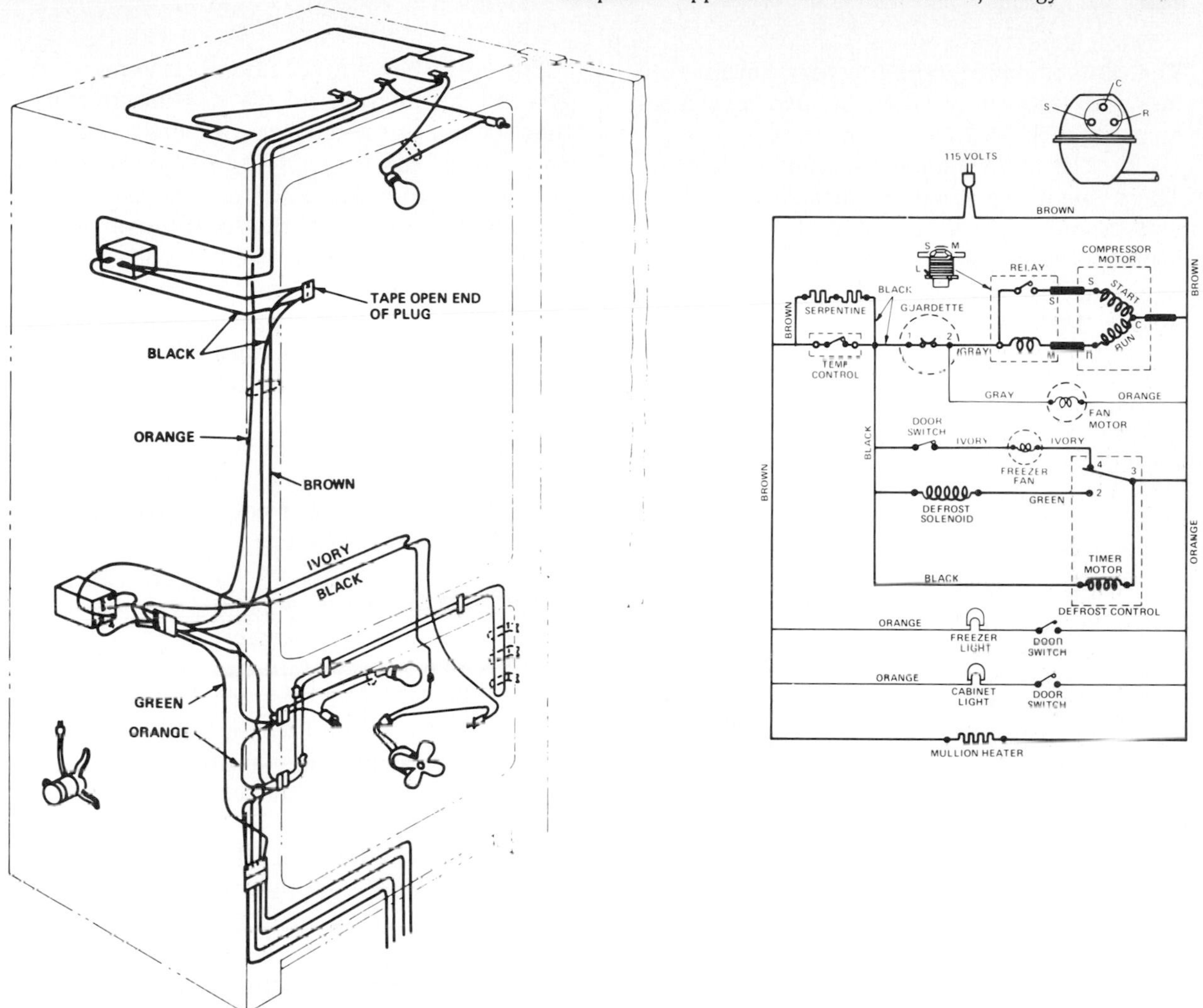

15–6. Wiring diagram and layout of parts for the electrical system of a refrigerator.

from that of the rest of the compartment. Automatic defrost, or frost-free, refrigerators use more electricity than the manual defrost type. Check the outlet and the circuit capacity before plugging it in. If a television picture shrinks every time the refrigerator turns on, then the circuit is overloaded and a separate circuit should be provided for the refrigerator.

HEATING AND AIR CONDITIONING

Houses that do not use electric heat still require electricity to provide warmth and air conditioning. Gas furnaces use electricity to control the on-off cycling and for thermostats.

The blower motor, which causes hot air to be pumped into every room, is also electrically operated. Fig. 15-7.

Fig. 15-8 shows some locations of furnaces. These locations must be furnished with 120 volts to operate the controls and the blower. In some cases where the unit has electric heating elements, it will be necessary to provide 3-wire, 240 volts.

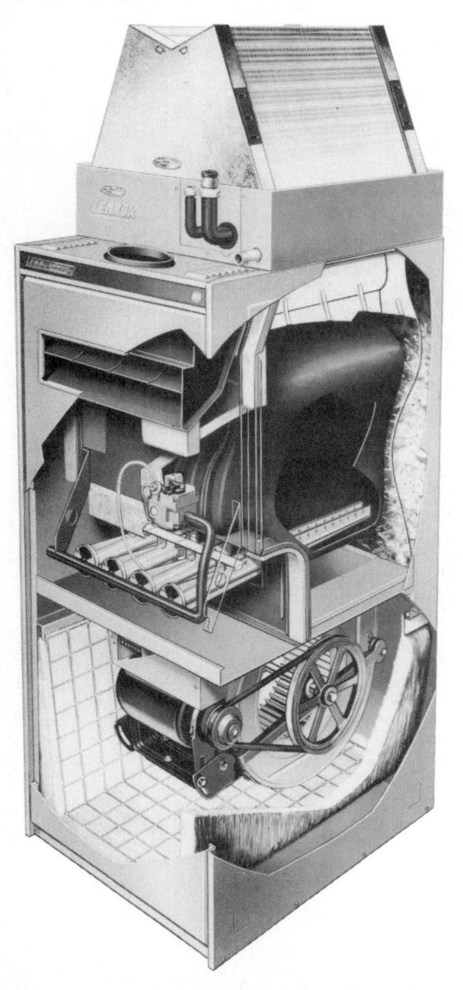

15–7. Typical hot-air, gas-fired furnace with air conditioning. Note the blower motor.

Fig. 15-9 shows a schematic drawing of the 240-volt installation where L_1 (line 1) and L_2 (line 2) are connected to the black and red wires in the service cable. Most furnaces come from the manufacturer with a box ready to accept the three wires of the local power source. Note in this schematic that 230 volts are supplied to the control transformer, and then stepped down to 24 volts to operate the control unit enclosed in dotted lines. This schematic also includes an air-conditioning unit contained inside the furnace, for providing cold air. Check with the manufacturer or the label on the unit to determine the voltage and the current. Provide the correct size wire and circuit breaker for the load.

Thermostats

The control unit (*thermostat*) is to be provided with a 5-wire conductor, as shown in the box to the left. Fig. 15-9. Wiring a combined furnace and air-conditioning unit is simply a matter of following the manufacturer's instructions, and connecting the wires to the proper terminals provided.

In Fig. 15-10, a 120-volt wiring diagram is furnished for study. Note that the thermostat control line calls for a 4-wire conductor cable, and L_1 is connected to the hot side of the 120-volt line. N is connected to the neutral, or white, wire.

If additional wiring information is needed, consult the factory-trained heating engineers who work with the equipment. Thermostats are to be checked closely for proper wiring procedure, since they are available for use on either 120, 208, 277, or 480 volts. Most of those used in the home are operated on 24 volts, AC. Make sure the thermostat is the one specified for the unit. The transformer is usually located inside the housing of the furnace and provides, in most cases, a stepped-down voltage of 24 volts. Make certain the thermostat is not mounted on an outside wall, or near a large glass area where the sunlight can strike it and influence the temperature reading. Proper

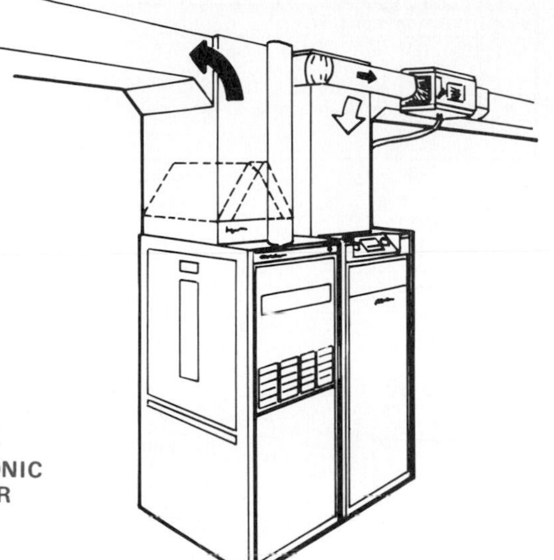

BASEMENT INSTALLATION
WITH COOLING COIL, ELECTRONIC
AIR CLEANER AND HUMIDIFIER

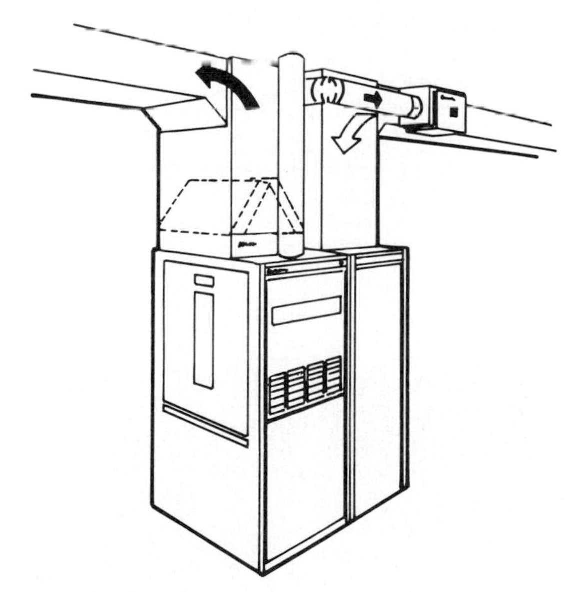

BASEMENT INSTALLATION
WITH COOLING COIL, RETURN AIR
CABINET AND POWER HUMIDIFIER

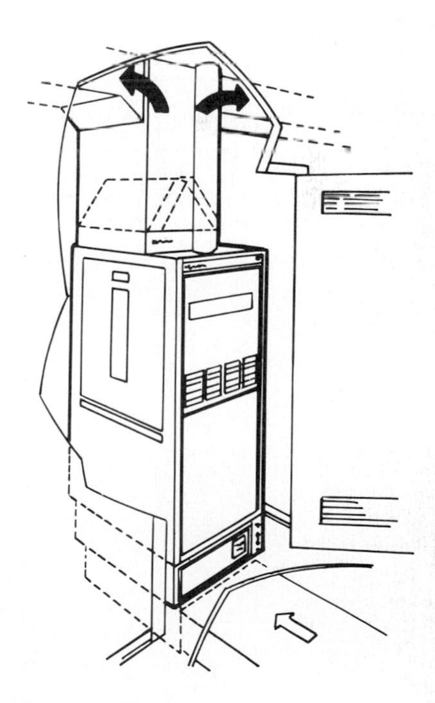

CLOSET INSTALLATION
WITH COOLING COIL AND
ELECTRONIC AIR CLEANER

15–8. Typical installations of hot-air furnaces.

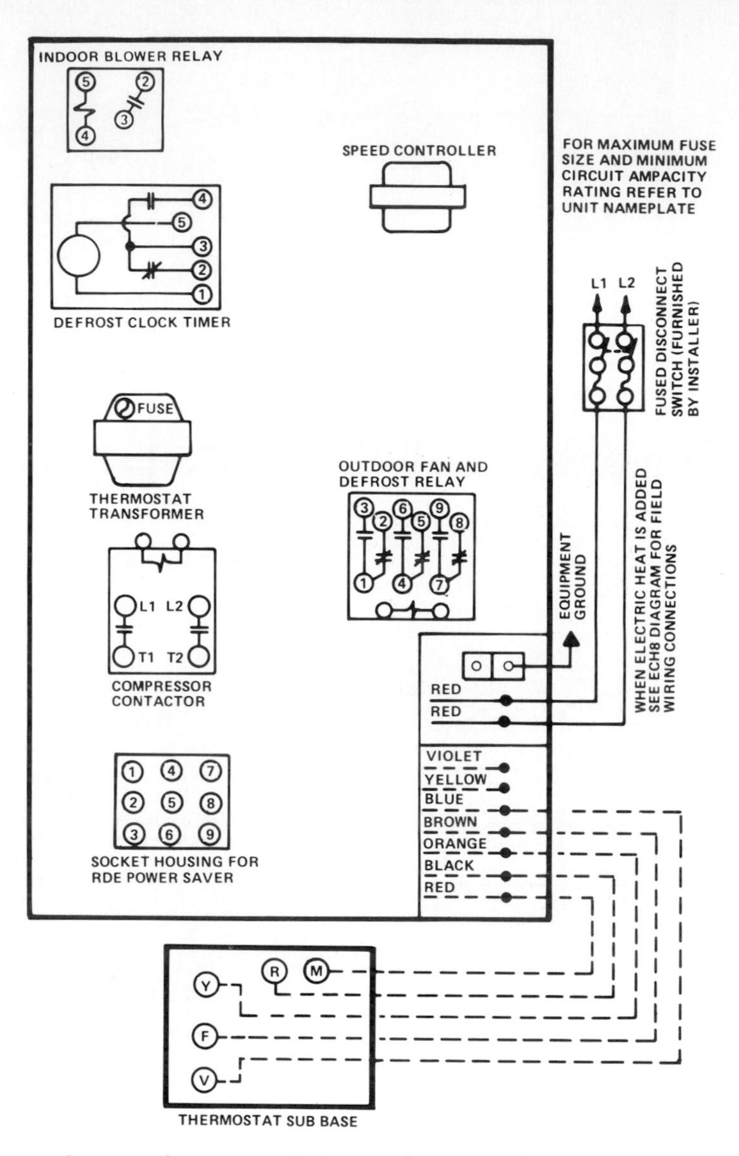

15–9. Wiring diagram and parts location for a 230-volt, hot-air furnace and air-conditioning unit.

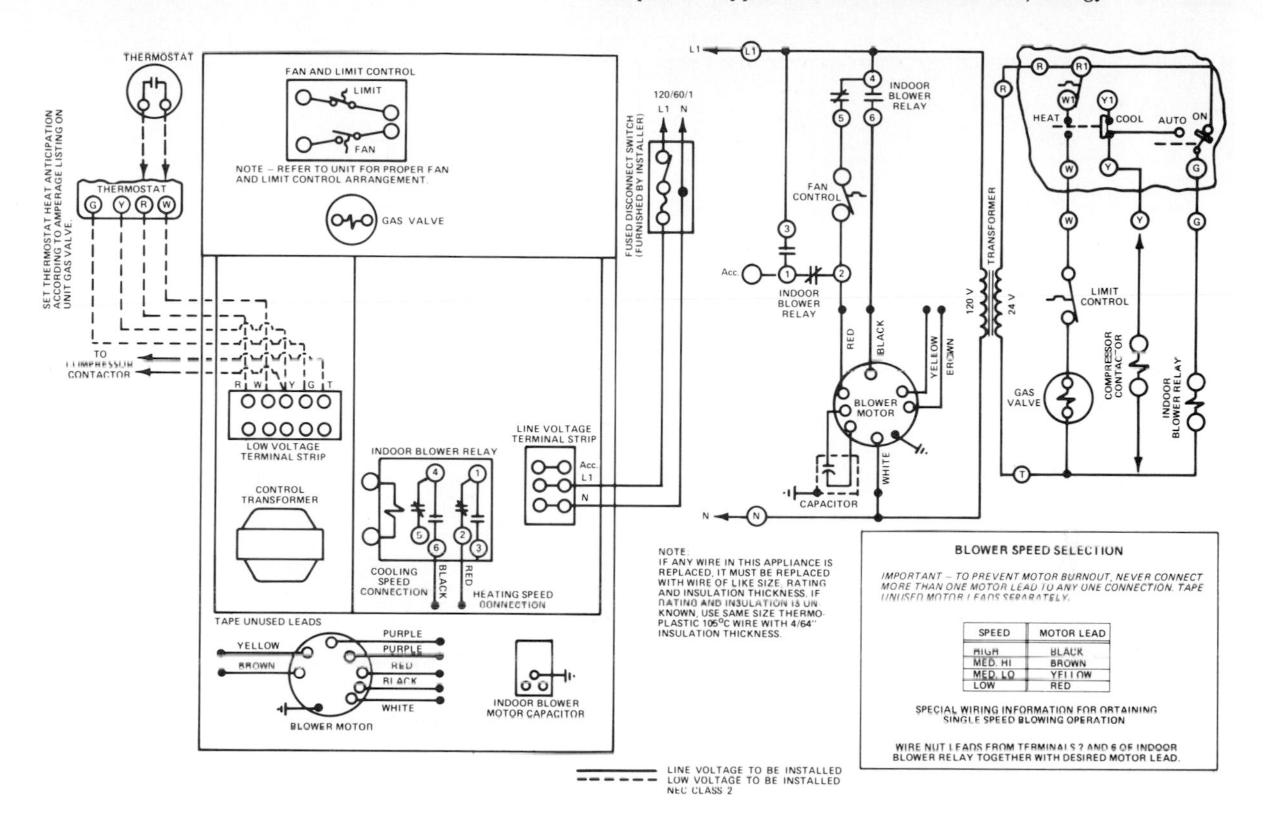

15–10. Wiring diagram and parts location for a 120-volt, hot-air furnace and air-conditoning, central heat-air unit.

placement of the thermostat is critical to proper operation of a furnace, and to providing uniform warmth throughout a room or house.

Although thermostats are made by several manufacturers, each has the common function of sensing temperature and controlling a heating or cooling unit. Thermostats are available in numerous voltage ranges.

The thermostat in Fig. 15-11 will register the actual temperature on the bottom scale, and should be rotated to set the desired temperature on the top scale. To prevent damage during construction, the thermostat is usually mounted in a house in the final building stage. However, a thermostat-wiring bracket is mounted on the wall earlier. Because most thermostats in homes operate on 24 volts AC, it is unneccessary to follow elaborate wiring methods for the low voltage. Fig. 15-12 is a single-pole switch that will operate on 120 to 277 volts and will handle up to 17 amperes. It is usually attached to an electric heater. Fig. 15-13 is a 2-stage thermostat, SPST, with a range of 40°F (4.4°C) to 80°F (27°C).

Fig. 15-14 is a unit with a sensor that can be mounted somewhere other than in the box mounting for the control unit. Designed for use in 1500- to 36 000-watt lines, it will operate on line voltage.

Thermostats are usually made of two dissimilar metals. The metals expand or contract to control a mercury vial. The vial has two terminals which are inserted into Mercury inside a glass enclosure. As the vial is tilted by the expanding metal spring, the mercury makes contact between the two terminals, or pieces of

15–11. Wall-mounted thermostat.

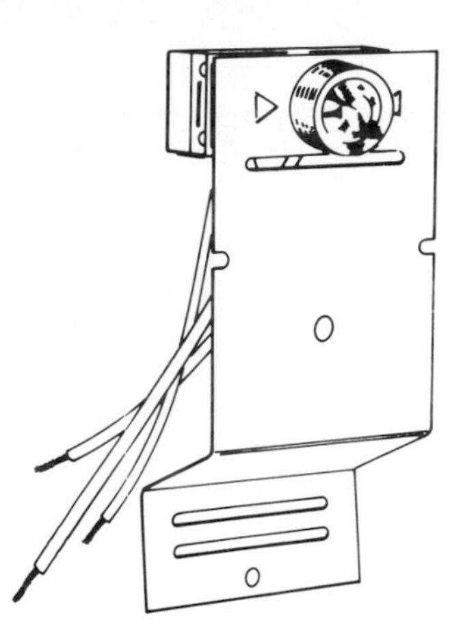

15–12. Thermostat. Usually mounted on a baseboard electrical heater.

wire, enclosed in the glass unit. At first the mercury may be vaporized slightly, but the vapor condenses and becomes usable again for the next cycle. Thermostats with mercury vials do not form deposits of copper oxide from arcing of the contact points, which does happen in switches using copper contact surfaces—even if the contacts are silver plated. An oxide buildup causes high resistance and draws

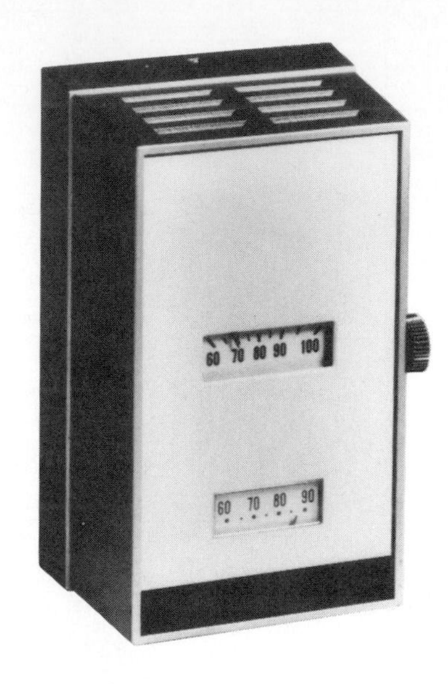

15–13. Two-stage thermostat.

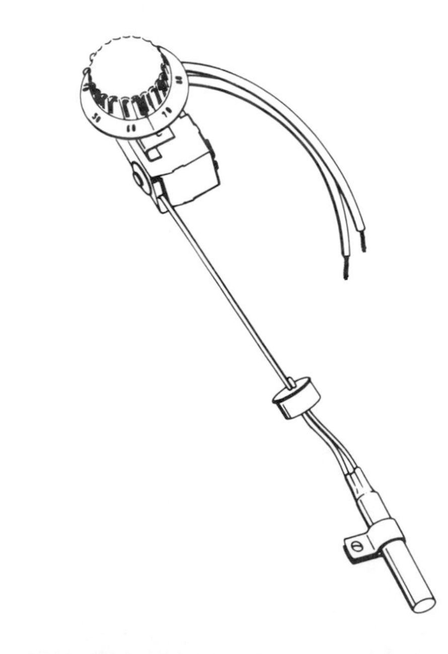

15–14. Thermostat with a sensor that can be mounted elsewhere.

more current each cycle, causing more oxide until the point is reached where no contact is possible and the switch fails. A mercury switch in the thermostat eliminates this difficulty.

Fig. 15-15 shows a protective cage for the thermostat. It should be used where the thermostat may be subjected to blows from heavy equipment or any movement of objects. Totally enclosed thermostats are available for locations where frequent adjustments are undesirable.

Fig. 15-16 is a 2-pole, 3-wire, 20-ampere, 250-volt single outlet for an air conditioner. It is mounted on a plate in a utility box. This summer-winter switch energizes either a heater or an air conditioner outlet.

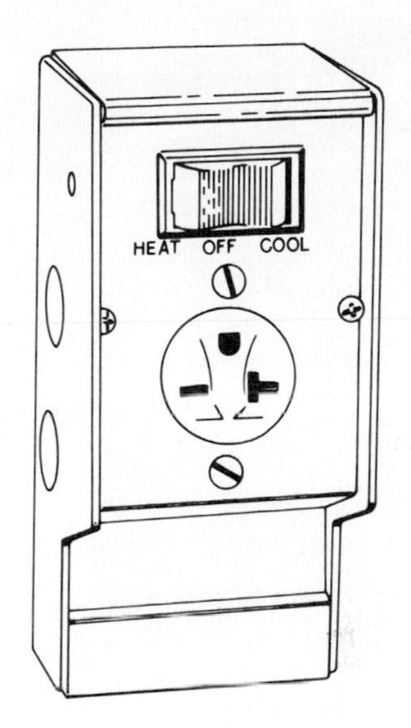

15–16. Summer-winter receptacle with switch.

Set-Back Thermostats. Energy savings are important today. To conserve energy, a number of newer types of thermostats are available.

These thermostats automatically raise and lower home temperatures to pre-set levels. *Set-back thermostats* can cut heating and cooling costs by 6% to 30%. Several models are available for use in the home. A microcomputer in the thermostat can program it to go on and off. You can program it for a week at a time without having to reset it. The thermostat also serves as a clock when it is not being programmed. Fig. 15-17.

Other types have a small shaded pole motor or other timer motor to drive the timing mechanism that attaches to the existing thermostat, such as that shown in Fig. 15-11. The timer is mechanically set and will turn the thermostat back or up. These attach directly to the existing unit. They turn it up or down according to a mechanical linkage.

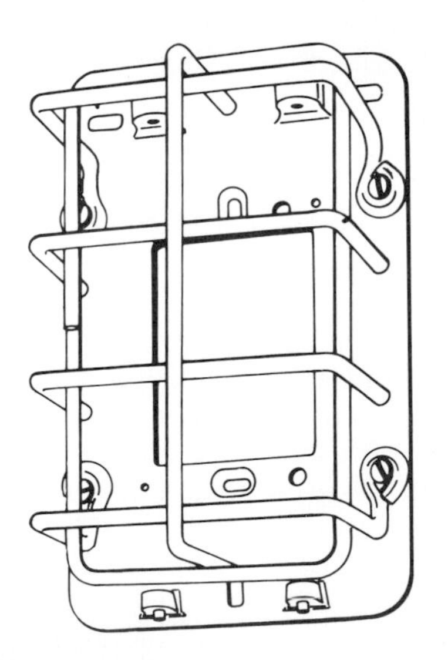

15–15. Thermostat cage.

Electric Heaters

Electric heaters come in a variety of configurations that may be flush mounted in walls, or in an open space, such as suspended from a ceiling.

Individual heaters are sometimes needed in such areas as bathrooms where extra heat is required. Most heating elements are made of a nickel-chromium alloy called nichrome. Some of the elements are enclosed in a ceramic material to keep them from being contacted by people and moving objects. Others are made with the springlike coils of wire exposed, but protected by an enclosure.

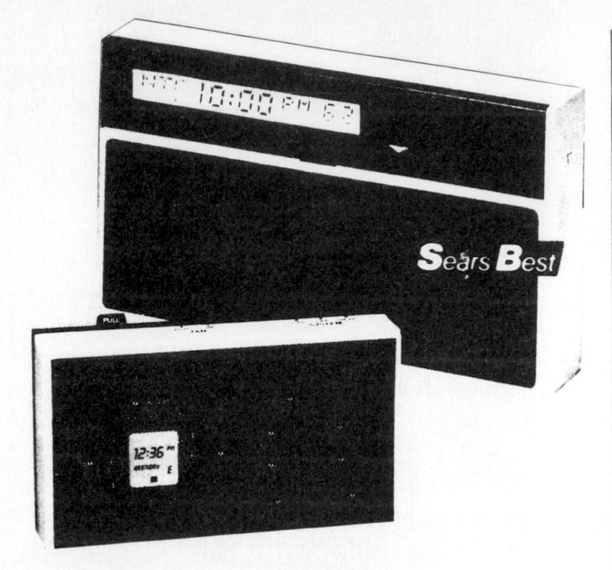

15–17. Digital thermostat. Used for programming temperature setbacks to conserve energy.

The following pictures show different types of heaters with varying wattages, Btu ratings, and voltage ratings. Control is usually by a line-voltage thermostat. Some units have individual thermostats.

The heater in Fig. 15-18 is used in offices, stores, and warehouses, and in remote locations such as garages, shipping rooms, storage areas, entryways, gymnasiums, portable classrooms, and greenhouses, where high-capacity heating is needed.

This heater is hung high in the air, away from congested areas. It may be mounted on a wall or ceiling bracket, or on rods, usually by one person. It is available in single-phase 120-, 208-, 240-, 277-, and 480-volt models. Three-phase units are available for 480 volts, delta or wye systems. No separate power source is needed for the motor or control circuits. Required contactors, relays, control-circuit transformers, and subdivided fusing are all factory installed. Wiring is color coded. It has an automatic-resetting thermal cutout that prevents overheating. Adjustable louvers direct the heat flow.

The heater in Fig. 15-19 has a cone diffuser hanging below the unit. This unit can produce

15–18. Unit heater.

15–19. Unit heater wiring.

122 868 Btu's on 480 volts, and will draw 43.9 amperes. It is available with 240-volt operation, but the current will be 89.1 amperes for the 122 868 Btu model. Control voltage is 120, which can be used on single- or three-phase power. The motor speed is 1725 RPM and is rated at $\frac{1}{4}$ horsepower to move 2150 cubic feet of air per minute. Special wiring and allowances must be made for the high current the unit draws.

Wiring diagrams are simple for electric heaters. Fig. 15-20 is a simple circuit with the 208- or 240-volt wires brought up to the unit and connected with wire nuts in a box provided. This unit is a 4000-watt heater for use on voltages up to 277, single-phase.

■ *Baseboard heaters.* Because all baseboard heaters are designed to fit near the floor and against the wall, they are similar in appearance. They have heating elements and radiator fins to dissipate the heat over the entire heating element length. They must meet standards for cleanliness since it is possible to cause fire if too much dust accumulates. In most cases the units do not have blowers, but depend upon the movement of the hot air to heat a given area. These units are referred to as low density and high density. Low-density units are made in about 8 different sizes starting with 376 watts for a 24-inch unit, and 1882 watts for a 117-inch unit. These operate on 208 or 240 volts.

High-density units range from 500 watts for the 24-inch model, to 2500 watts for the 117-inch unit. These can be obtained for use on 120, 208, 240, or 277 volts. Figs. 15-21 & 15-22 illustrate two designs for baseboard units.

■ *Wall heaters.* Wall units have the advantage of being out of the way and can be mounted in existing structures or in new construction. They are particularly handy in family rooms, garages, basements, vacation homes, and in commercial applications for foyers, laundry rooms, storage areas, and hallways. Industrial uses include foyers, lobbies, guard offices, storage rooms, hallways, and lavatories.

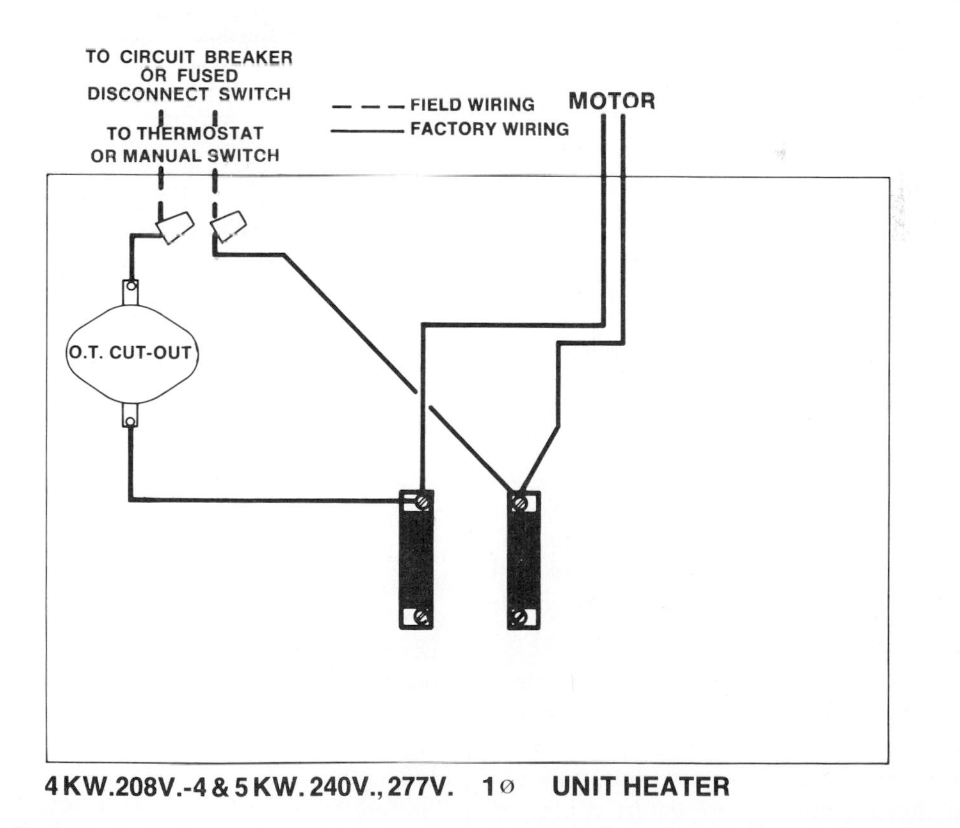

15–20. Unit heater wiring diagram.

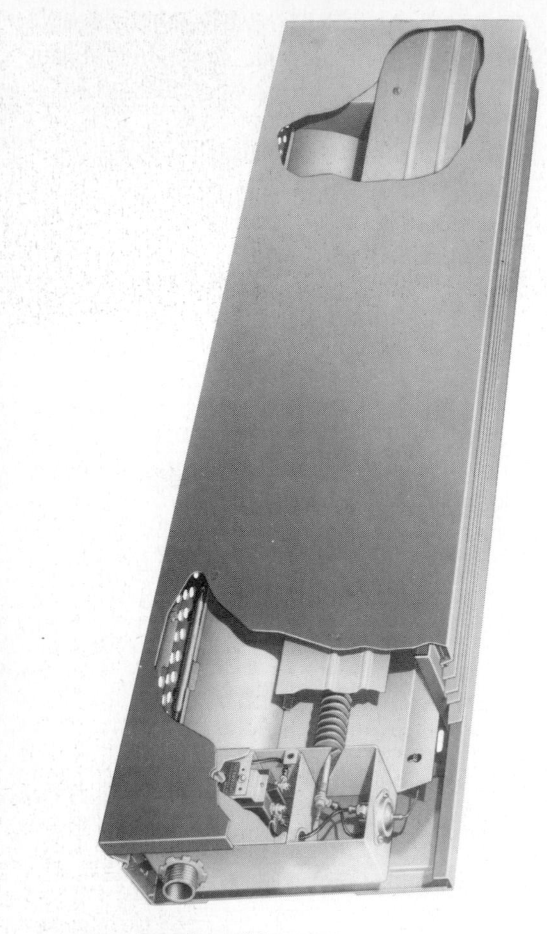

15–21. Baseboard heater.

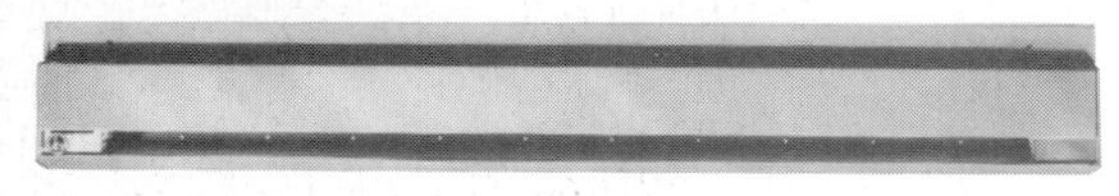

15–22. Baseboard heater.

These units have a fan to move the hot air. They come in 1500-watt to 4000-watt sizes, and may be obtained in 120-, 208-, or 240-volt models.

■ *Radiant bathroom heaters.* These heaters do not have fans. They depend upon the reflector in back of the heating element to move the heat. The nichrome-wire element is totally enclosed in tubular steel. The heaters are quiet and produce instant heat. Fig. 15-23.

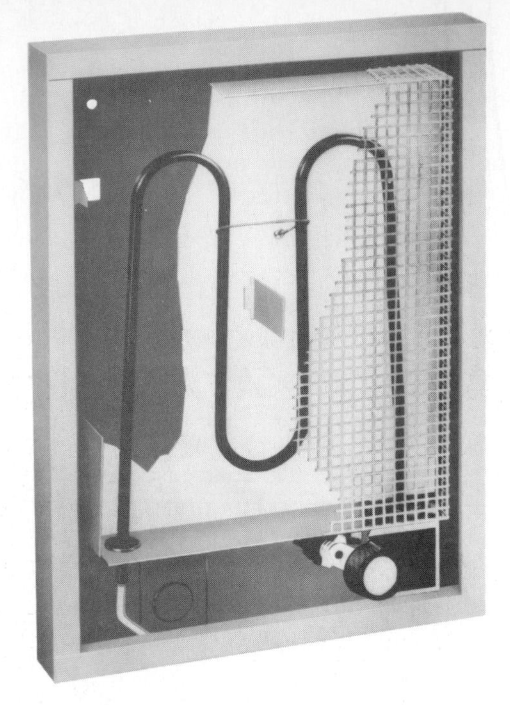

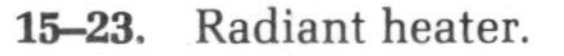

15–23. Radiant heater.

■ *Strip heaters.* This type of finned heater is ideal for fast heat dissipation for convection heating, process ovens, and dryers. Fig. 15-24. General uses for them are for clamp-on water heaters, warming ovens, ink drying, crane cabs, plastic-mold heating, core baking ovens, valve houses, pipelines, process welding, and process ovens and dryers.

Strip heaters have uniform heat transfer, low cost, and a long life. These qualities make them suitable for many commercial and industrial applications. They come in units ranging from 6″ to 43″ in length. They have both a high-temperature and a low-temperature capability. Each heater consists of coiled, nickel-chromium wire, centered and imbedded in a high-grade refractory material (zircon), which ensures good insulation and rapid transfer of heat. The fins may be slipped over the flat unit to help dissipate the heat more evenly and more rapidly, if needed.

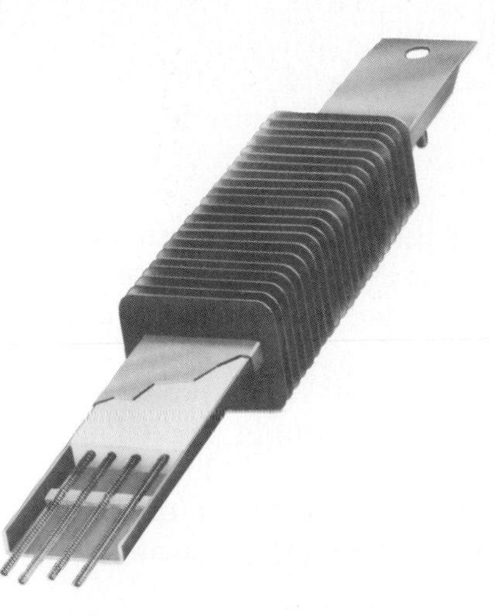

15–24. Strip heater.

■ *Infrared heat sources.* Infrared heat is available immediately at the flip of a switch. No warm-up period is needed. Energy is not wasted in heating an intervening conductor or convector. With infrared heating, an object is heated directly. Infrared is an extremely efficient source of heat energy. These units operate at a lower temperature than most heaters. Infrared is used for a number of applications in industry, among them the curing of paints, enamels, varnishes, and other applied finishes. It is a source of heat for drying or evaporating water in the processing of dehydrated food, and for taking the moisture out of cookies after they have been baked. This gives them a crisp quality desired by many people. Infrared heat can be used to soften plastic prior to forming, to expand machine parts before fitting, and to keep foods warm. It can dehydrate gumdrops or bake a finish on gun barrels.

Infrared radiation provides more comfortable working conditions than other heating methods because it raises the air temperature very little. It is also clean and soot-free. Because of its speed, efficiency, and adaptability, infrared heat has become standard in many industries that manufacture a wide variety of products. One model of an infrared heater is shown in Fig. 15-25.

15–25. Infrared heater.

CONSERVING ELECTRICITY†

Because of energy shortages, everyone should be aware of the efficient use of power. The electrician should be aware of the problems associated with conservation and the efficient utilization of electrical power.

Many suggestions for energy efficiency offered by electric power utility companies are practical only when considered during the planning stages of an installation. Some common-sense suggestions need to be brought to the attention of the householder or commercial establishment.

Because electricians are looked upon as experts in this field, their recommendations are likely to be taken seriously and applied. Therefore the electrician should become aware of ways to conserve energy.

Heating and Cooling

■ All electric appliances should be kept in good repair. When they don't operate efficiently, they require more power and you do not get full use from them.

†This information was supplied by the Niagara Mohawk Power Corporation.

■ Get the right size furnace, air conditioner, and water heater for your needs. A unit larger than necessary wastes fuel.

■ For heating systems, set the thermostat as low as you can without sacrificing comfort. Each degree over 70°F (21°C) uses $1\frac{1}{2}$% more energy.

■ Lower your furnace thermostat at night by several degrees. An extra blanket should enable you to do this. When you are away for several days or more, keep the temperature as low as you can without letting the water pipes freeze.

■ Make sure your thermostat is away from heat or cold sources such as windows, heating ducts, television sets, and lamps.

■ In summer, use an attic fan to help blow out hot air which may compete with your air-conditioning system. Keep your house at minimum comfort level. Each degree cooler than 75°F (24°C) uses more energy.

■ Clean thermostats yearly by removing the cover and carefully blowing away any dust.

■ Keep all air filters clean to make it easier for your furnace or air conditioner to do its job. Check them every 30 days.

■ Use insulated windows (combination or thermal pane) and doors to help prevent heat loss in winter. They also will help keep an air-conditioned home cooler in the summer.

■ Make sure you have effective weather stripping throughout your home to help avoid air leaks around all exterior windows and doors. If there is a $\frac{1}{4}$″ crack under the attic door, you lose several dollars worth of heat every winter.

■ Make sure your heat and/or cooling ducts are airtight. Make sure your home is well insulated. Installing insulation in the walls and ceiling will greatly reduce the energy needed to heat or cool your home. Storm windows or sheets of clear plastic can be helpful.

■ Close the damper on your fireplace chimney when not in use. Otherwise, heat from the furnace will escape through the chimney.

■ Make sure all sun-exposed windows have draperies. Let the sun in on cool days and keep it out on warm days to help your heating or cooling system. Insulated or lined draperies can help reduce energy consumption substantially.

During the summer, since many appliances give off heat, try to use them when demands on your cooling system are less—early mornings, evenings, and on cool days.

Laundry

■ When leaving your home vacant for some time, turn the temperature setting on the water heater to low.

■ Avoid wasting hot water while laundering. Use the cold and warm water settings on your washing machine as much as possible to cut down on the energy needed to heat water. Use the water-level control if your machine has one.

■ Follow detergent instructions carefully. Over-sudsing makes your washing machine motor work harder.

■ Clean the filter on your washer and dryer after each load.

■ Vent your dryer and make sure the vent stays free of lint.

■ Remove all garments from your dryer as soon as the cycle is finished. This will reduce the amount of ironing needed.

■ Use a dryer with durable-press settings. Such a dryer reduces the need for hand-ironing permanent press material.

■ If your dryer has an automatic sensing device on the cycle, use it. Overdrying wastes energy.

■ Iron large amounts of clothes at a time to avoid heating up an iron several times. If you are called away from your ironing, turn the iron off and unplug it.

■ Use the lowest setting on your iron that will do the job well.

Kitchen Appliances

■ Do not open refrigerator or freezer doors more than necessary, and close them quickly. Make sure door gaskets have a good seal. If your refrigerator-freezer isn't frostfree, defrost before the ice gets $\frac{1}{4}''$ thick. Ice acts as insulation and cuts the cooling power of the coils. A frostfree refrigerator-freezer requires more energy to operate than a manual defrost model.

■ Make sure the condenser coils on the back or bottom of your refrigerator-freezer are clean by dusting or vacuuming them.

■ Do not overfill your refrigerator. There should be enough space around each food container to allow for good air circulation. Keep air vents clear.

■ A chest-type freezer is more efficient than an upright freezer, since the cold air in a chest type will not escape as quickly as from an upright freezer when the door is opened.

■ Buy a refrigerator and freezer with separate doors, since less cold air is allowed to escape than with a single-door type.

Range.

■ Match pots and pans to the size of your range units. A pot too small for a unit will allow extra heat to escape.

■ Place pots and pans on the range before turning on the heat.

■ Use pans with flat bottoms, straight sides, and tight-fitting covers. Electric fry pans, portable broilers, and rotisseries generally use less electricity than an electric range when cooking the same food.

■ Thaw frozen foods before cooking, when practical. Be sure to cook them immediately after thawing.

■ Turn an electric range off just before your cooking is completed, thereby utilizing residual heat to finish the job and keep food warm for serving.

■ Microwave ovens reduce power consumption.

■ When possible, cook several foods in the oven at once.

■ When using glass or ceramic pots and pans in a conventional oven, gas or electric, lower the heat by 25 degrees Fahrenheit.

■ Preheat the oven only when necessary.

Other Energy Saving Tips

■ A shower usually uses less hot water than a tub bath. Don't leave the hot water running when shaving. Leaky faucets waste water, and leaking hot water is particularly wasteful and expensive.

■ Load your dishwasher completely to cut down on the number of loads. When handwashing dishes, don't leave the hot water running to wash or rinse each plate separately.

■ Change the bag and filters on your vacuum cleaner frequently. Keep permanent filters clean on tank-type vacuum cleaners.

■ Switch off radio and television sets when not in use. Solid-state sets (both color and black and white) consume less energy than tube sets. Larger screens consume more energy than smaller ones.

■ Keep lights off in the daytime except for safety, health, or comfort reasons. If a night-light is used, one of very low wattage will glow brightly in the dark.

■ Use fluorescent lighting instead of incandescent lighting when possible. Fluorescent lights are about four times as efficient as incandescent lights, and last seven to ten times longer.

■ Read your appliance manuals for specific information which will help you conserve energy.

These are but a few suggestions. They are not the complete answer to the energy problem, but following them can help conserve energy and keep the electric bill down.

Programmable Load Control

The advent of the computer has brought about the ability to monitor where power is being consumed and when. It also brought about a means for making sure that power is efficiently used. This is known as *programmable load control.* Industry has joined with the commercial area to obtain the benefits of a computerized load control.

The General Electric programmable load control shown in Fig. 15-26 is less complex and more practical than some others. This energy management system uses solid-state integrated circuitry and state-of-the-art microprocessors. The heart of the system is the central controller. It will control every electrical function in a building, turning needed energy on and uneeded energy off—automatically.

It takes about 4 hours to train a person to program or reprogram the system. Information is entered into the controller in much the same way that data is fed into a hand-held calculator.

15–26. This control center for an Energy Management System uses the latest in electronics to monitor and control energy usage.

The energy-saving program can be manually overridden by an occupant for his or her office or area. Lighting can be automatically "flicked" to warn the occupant that his or her area is about to be turned off. By connecting the building's telephone system to the controller, lights and air conditioning can then be manually overridden by any Touchtone® or equivalent push button phone in the building. Switches can be programmed to control any load from any location. Telephone or switch overides can also be centrally monitored to ensure savings.

The system has a troubleshooting capability built-in. It sounds an alarm whenever a facility "out-of-bounds" condition occurs, and reports the location of the problem, when it occured and when it was eliminated. The system even checks its own memory and operating functions to diagnose and alert the user to potential problems. With the telecommunications function, this alarm capacity is extended from remote sites back to a central station.

The system is capable of tracking load usage in every area of the building. It indicates this by time of day. Printouts of all system activity give a record of the facility's operation, building-by-building and area-by-area. Energy saving programs can be monitored and changed as needed to improve efficiency of energy usage.

QUESTIONS

1. How many amperes must the plug and receptacle of a kitchen range be capable of handling?
2. How much power does a 4-ton air conditioner use per month?
3. How much power does an electric water heater consume in a month?
4. What is the most common voltage used for a home furnace thermostat?
5. What is nichrome? Where is it used?
6. What is the difference between an electric unit heater and an infrared heater?
7. How are baseboard heaters rated?
8. What is a radiant heater?
9. Where are strip heaters used?
10. How can homeowners help conserve electrical energy?
11. What is programmable load control?

KEY TERMS

programmable load control

set-back thermostat

SECTION FIVE SECURITY SYSTEMS

CHAPTER 16

RESIDENTIAL SECURITY ALARM SYSTEMS†

Objectives

After studying this chapter, you will be able to:

- Plan an intruder detection system.
- List the primary components in a fire detection system.

†The author would like to thank the NuTone Division of Scovill for making it possible to obtain the illustrations and technical information necessary for this chapter. Mrs. Marian Finney, Mr. George Dorchak, and Ms. Linda Kobmann have been most helpful with their time and assistance.

Burglars strike every 12 seconds. Suburban homes and apartments have become their favorite targets. Over half of all residential burglaries take place in the daytime. Even amateurs usually have no difficulty getting in and stealing cash, television sets, stereo systems, jewelry, art . . . even clothes. Every once in a while, somebody gets in their way and the result is tragedy.

A fire starts every 20 seconds—not as often as burglary, but likely to do much more harm. Fires are unpredictable: a neglected cigarette, a short in an electrical appliance . . . and there's a fire. Fig. 16-1.

As an electrician you will be called upon to install the latest in burglary, fire, and smoke detection devices. It therefore becomes necessary for you to acquaint yourself with the latest available equipment.

A home needs a number of devices to be safe from fire, smoke, and burglary. In this chapter, several kinds of security systems are discussed. Which system is best for a particular home depends on the needs of the residents. Fig. 16-2.

BURGLAR ALARMS

Let's start by looking at a home as a space to be protected. The homeowner's first goal is to keep the burglar from getting in. There are only two practical ways a burglar can get in: through doors and through windows. Fortunately, both can be guarded with locks and detection devices.

Unfortunately, many homes are poorly protected at these entry points. Many doors, especially those leading to the basement, backyard, or garage, are hollow-core or panel construc-

16-1. Fires are a constant threat.

16-2. Newspapers tell the story of burglaries and fires.

tion. These doors are no barrier to a burglar armed with the proper tools. And it's important to remember that those tools can become weapons if the burglar is startled or frightened.

Basement and garage windows are often ignored. In many houses second story windows are as vulnerable as those at ground level. To add to the problem of protecting the perimeter, many homes have large areas of glass.

In most modern homes, in fact, the perimeter is relatively weak. The determined burglar has any number of chances to get in.

A well-designed, properly installed intruder alarm system, with well-placed sensors and a loud alarm bell, is very likely to scare a burglar off before any damage has been done. Burglaries can be prevented.

Planning an Intruder Detection System

There are two circuits that must be considered when planning an intruder detection system: the perimeter and the interior.

Let's look first at monitoring the perimeter—the windows and doors. There are two basic types of monitoring to detect intrusion past these entries: *contact switches* and *metallic window foil.*

Contact Switches. With very few exceptions, contact switches can be installed to monitor all the various types of openable doors and windows commonly used in homes. The choice of one device over another depends on the type of house construction. The necessity for concealment or the installer's personal work habits or customer preference also affect the choice of devices.

- *Magnetic detectors.* Surface-mounted magnetic detectors can be used on any wooden window or door. Fig. 16-3. With a plastic spacer, they can also be used on metal windows and doors.

Recessed magnetic detectors can be used when concealment is required. Fig. 16-4. Installed without nails in window sills or doorframes, they have a neater appearance than surface-mounted detectors. They fit very narrow areas. They are recommended to compensate for loose doors and windows because they will operate through a $\frac{3}{8}''$ (10 mm) gap. Use steel door mounting adapters for metal doors, windows, and casings.

- *Plunger detectors.* Recessed plunger detectors can be installed in wooden or metal doors or window frames. Fig. 16-5. This type of detec-

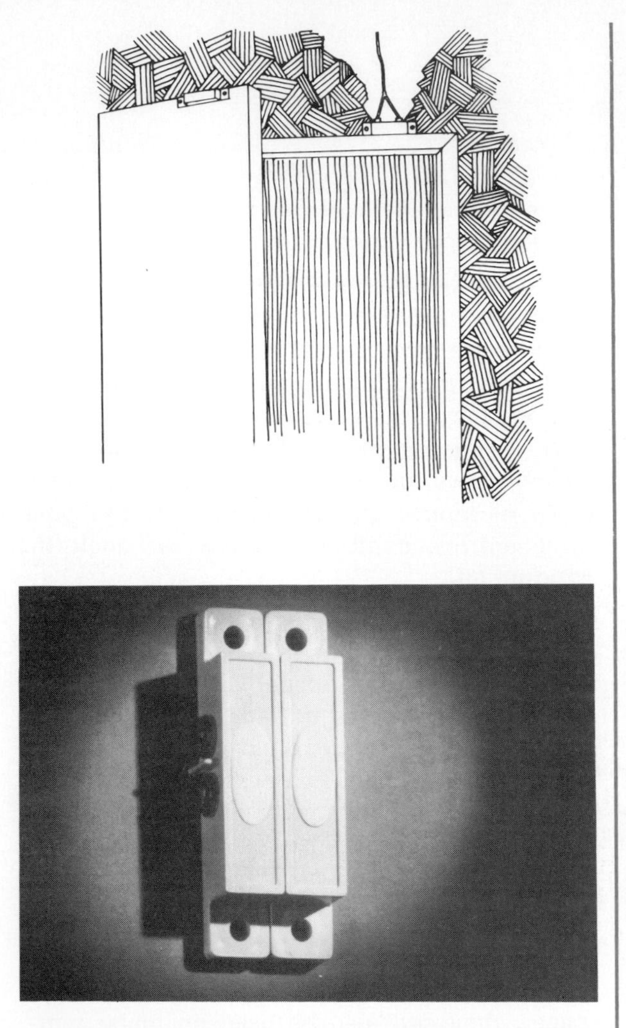

16–3. Surface-mounted magnetic detector. Can be mounted in almost any position.

tor is extremely sensitive. A special activator spring expands installation flexibility and improves performance on loose doors and windows.

Casement windows require a slightly different model of plunger detector. It has to be designed to work where screens or window construction prevents use of other detectors. Fig. 16-6.

Window Foil. For monitoring glass areas, metallic foil is the only reliable device now available. If the glass is broken, the foil tears and

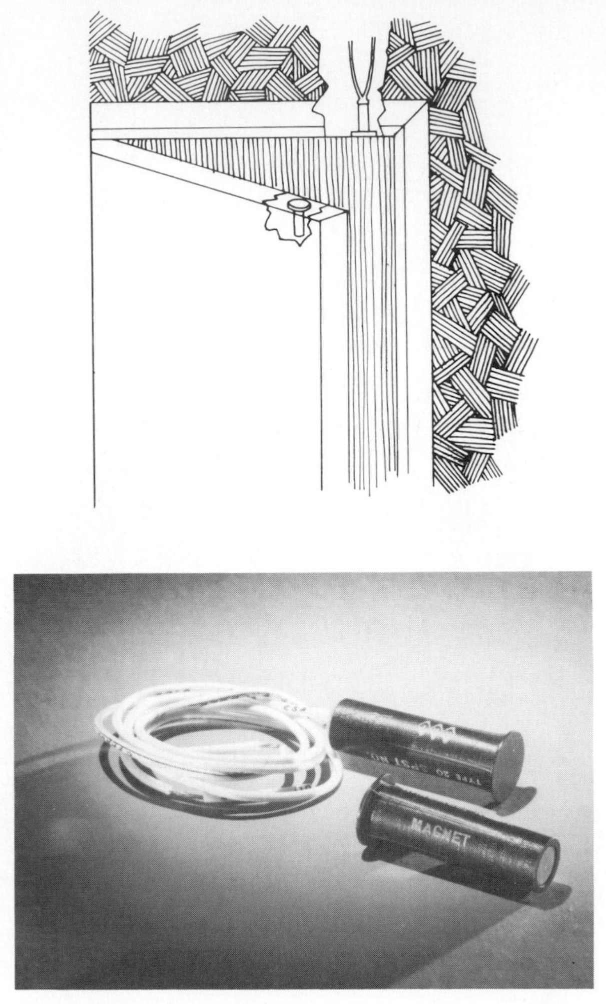

16–4. Recessed magnetic detector. Can be installed in window sills and doorframes.

triggers an alarm. Fig. 16-7. Other devices so far are troublesome in giving false alarms.

Aesthetics are often involved in the decision to use foil. A well-done job can actually add to the appearance of a home. But some homeowners simply don't want foil on their windows even though this decision reduces their perimeter protection.

Several devices are used to complete foiling jobs. Foil connector blocks are used to take foil tape circuits off glass so that soldering is unnecessary. Fig. 16-8. Foil contact switches permit opening of foiled windows when the sys-

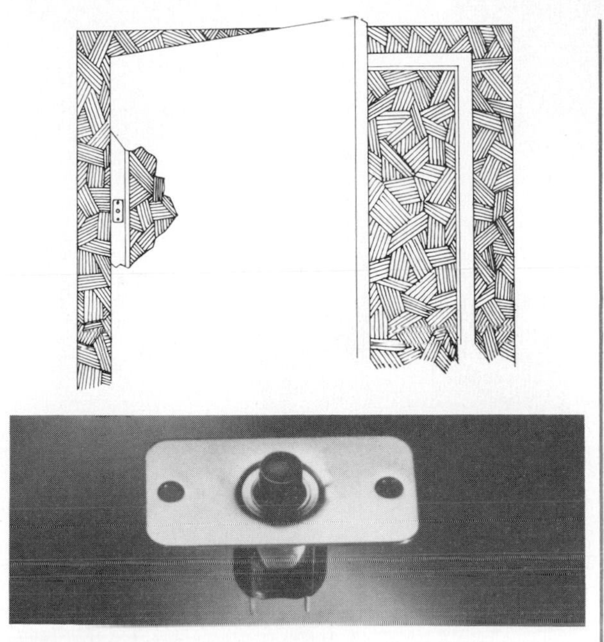

16–5. Recessed plunger detector. Installs in wooden or metal doors or window frames. Extremely sensitive.

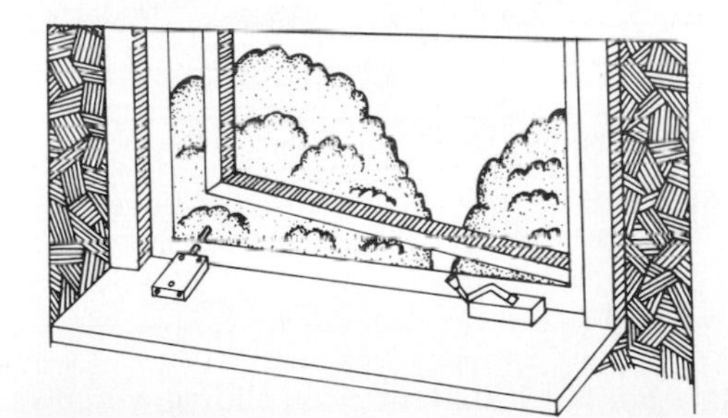

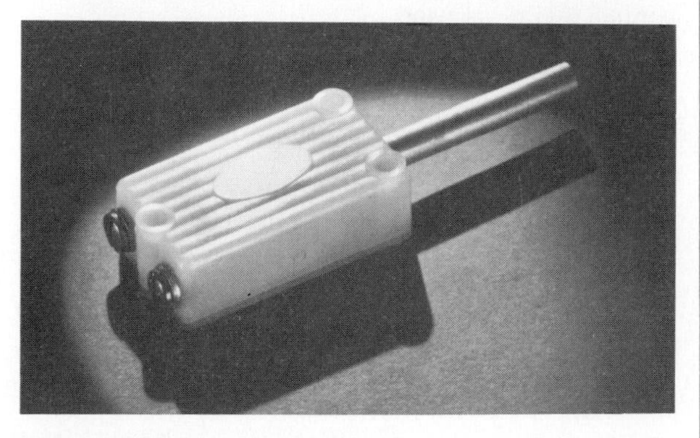

16–6. Casement window plunger detector. Designed for use with casement windows where screens or window construction prevents use of other detectors.

16–7. Window foil gives the best protection against entry by breaking a window.

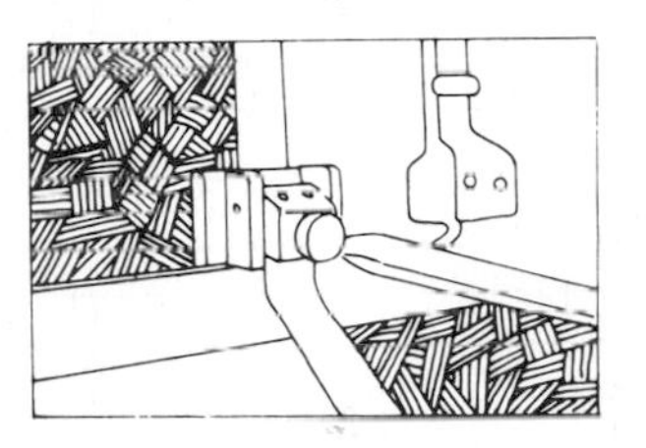

16–8. Foil connector block. This block is self-adhesive. Clear plastic blocks are for taking foil tape circuits off the glass so that no soldering is necessary. The block adheres directly to the glass surface.

tem is not armed. They detect opening of the windows when the system is armed. Fig. 16-9. Flexible cords connect foiled areas on movable doors and windows to the detection circuits. Fig. 16-10.

If all outside windows and doors are contacted and all vulnerable glass areas are foiled, then a good job of protection is assured.

Installation of Window Foil Tape. There are several things to remember when installing foil tape. Foil will not stick to wet, dirty, or cold windows. Windows must fit tightly in their frames. Severe bending and creasing will break the foil.

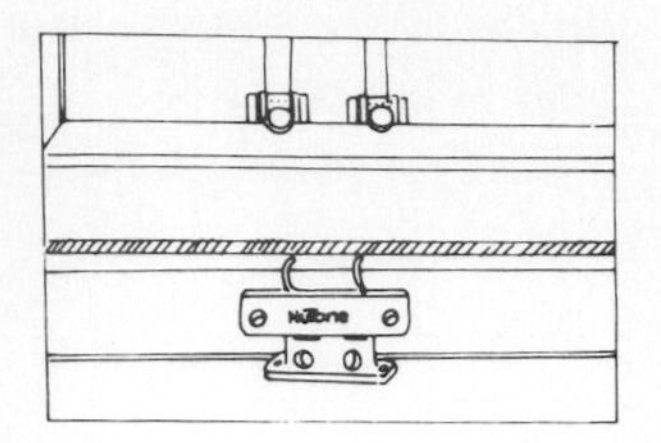

16–9. Foil contact switch. This is a two-part switch. One part connects to the foil on the window glass. The other part mounts on the sill and connects to the detection circuit. This allows the window to be opened without breaking the tape; that is, of course, if the shunt switch is used and the circuit is not armed.

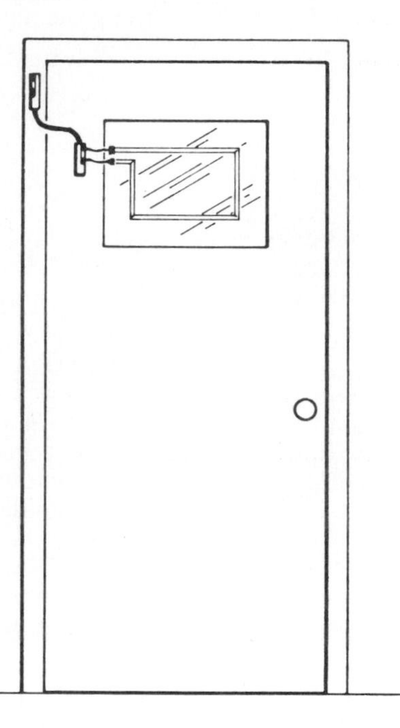

16–10. Foil is installed on door windows and crank-type windows with a flexible cord.

Foil must not touch any metal surfaces. Foil must be completely covered with foil *sealer* after installation to protect it from moisture and dirt.

Foil is used with connector blocks and foil sealer. Fig. 16-11. For foil installations on sliding windows, use contact switches. Fig. 16-12.

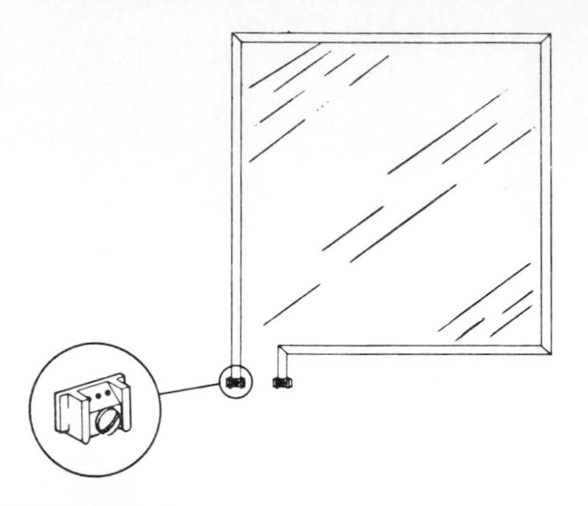

16–11. Window foil is used with connector blocks and foil sealer.

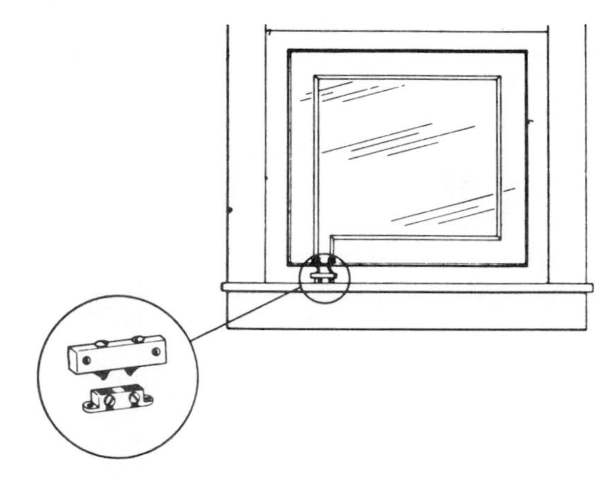

16–12. For foil installations on sliding windows, a contact switch is used.

Follow these steps for installation of foil:

1. Obtain a 3″ × 3 ″ (75 mm × 75 mm) block of wood (or plastic, Bakelite, etc.) and a grease pencil or a piece of tailor's chalk. Mark the outside of the window. This will be the guide line for the foil. Fig. 16-13. A block and a marking pencil are usually furnished by the manufacturer of the tape.
2. To help make installation easy, hang a foil reel above the window. Fig. 16-14. A foil holder-dispenser is usually supplied with the installation kit. If not, one can be made

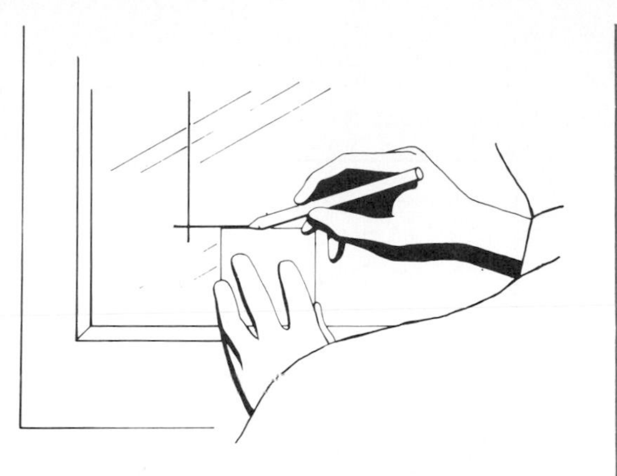

16–13. Using a 3″ × 3″ (75 mm × 75 mm) block of wood (of plastic, Bakelite, or similar material) and a grease pencil or tailor's chalk, mark the outside of the window. This will be the guide line for the foil.

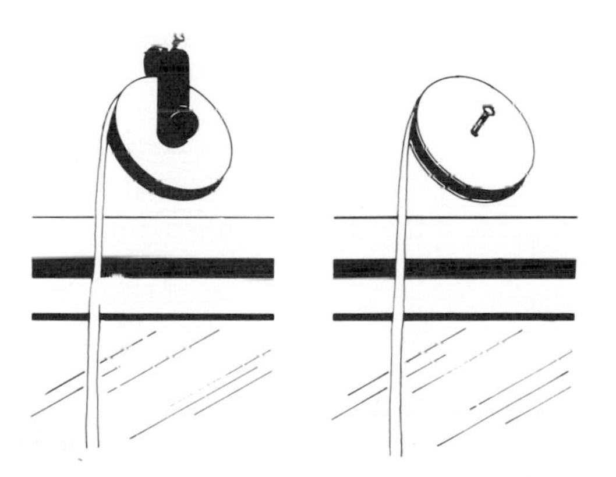

16–14. A foil reel (left). A foil holder-dispenser (right) can be made by using two pieces of cardboard and sticking a nail through the center of the coil of foil.

by sandwiching the foil reel between two pieces of cardboard and inserting a nail through the center.

3. Clean the inside of the window with a lint-free cloth and a solvent such as alcohol or benzine. Do not use normal window cleaner; it leaves an oily film on the glass. Store cloths carefully. Allow window to dry completely. Fig. 16-15.

16–15. Clean the window with a clean, lint-free cloth and a solvent such as alcohol or benzine.

4. Mark the location where the intruder detection circuit will meet the window. Mark the exact location for the contact switch or flexible cord. Fig. 16-16.
5. Important facts about foil installation:
   a. Foil must be stretched to assure that it will break easily if the window is broken.
   b. Foil must be smooth across the window. Use a small piece of stiff paper to smooth out the foil.
   c. Foil should be applied in one continuous strip to make it the best possible electrical circuit.

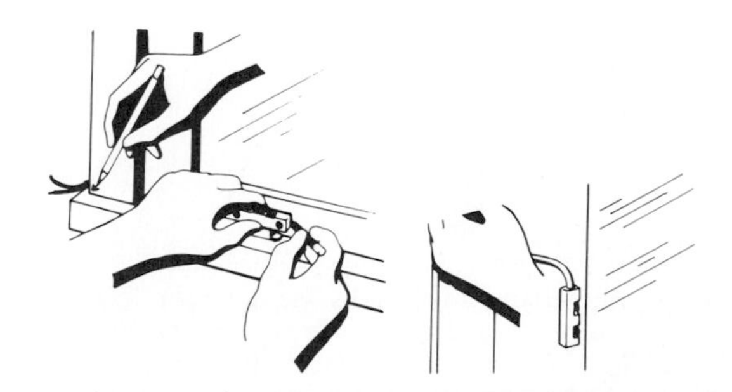

16–16. Mark the location where the intruder detection circuit will meet the window and the location of the contact switch or flexible cord.

- **d.** If foil breaks during installation, there are two recommended remedies: (1) If the area preceding the break covers a small or medium area, remove all of the foil completely from the window and begin a new strip. (2) If the area preceding the break covers a large area or an area with difficult turns or window frame crossings, splice the tape. Fig. 16-17.

6. Peel off the paper backing from a section of foil. Begin at the point where the intruder circuit will meet the window. Allow a 2″ (50 mm) piece of foil to extend over the window frame. Press and hold the foil at the edge of the glass. Stretch the foil away from the starting point and stick it to the glass. Fig. 16-18.

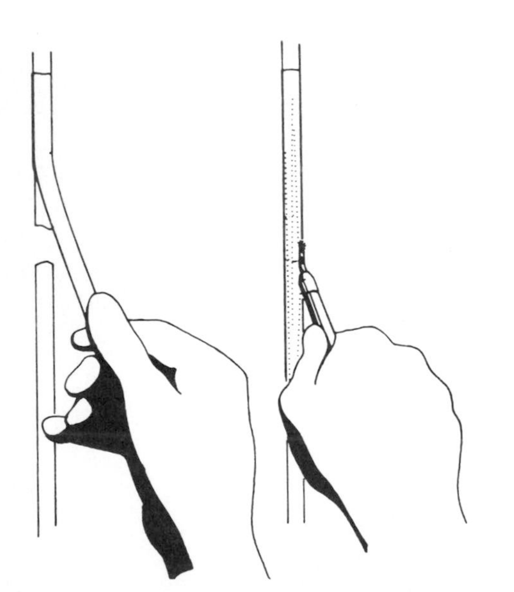

16–17. Repairing foil breaks. Remove the foil sealer (if already applied) with steel wool. Stick a new strip of foil over the existing foil, overlapping existing foil with foil about 3″ (75 mm). Use the foil splicing tool to prick holes over the overlapping foil. Check to be sure the circuit is complete through splicing. Cover with foil sealer.

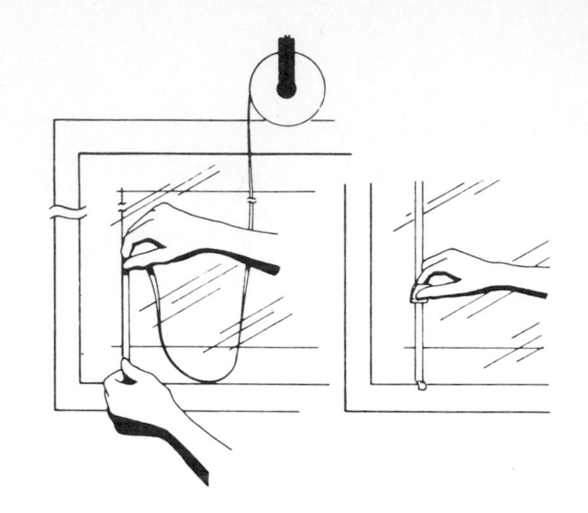

16–18. Press and hold the foil at the edge of the glass. Stretch the foil away from the starting point and stick it to the glass. Make the foil smooth on the glass by rubbing it with a small piece of stiff paper.

CAUTION: Foil must be stretched in order for it to break if the glass is broken. Smooth the foil on the glass by rubbing it with a piece of stiff paper.

7. Right-angle turns: Stick the foil to the glass up to the edge of the intersecting guide line. Remove at least 5″ (125 mm) of the paper backing from the foil. Fold the foil back over itself, gently creasing it. (Severe creasing and bending may break the foil.) Fold the foil at a 45° angle and press the corner down. Begin running the foil along the new guide line, holding it at the corner and stretching it as it is placed on the window. Smooth over the corner with a small piece of stiff paper. Fig. 16-19.

8. Continue to follow the guide lines around the window and work your way back to the starting area. Remember to stretch the foil as you stick it to the glass and to smooth the foil with a small piece of stiff paper. Fig. 16-20.

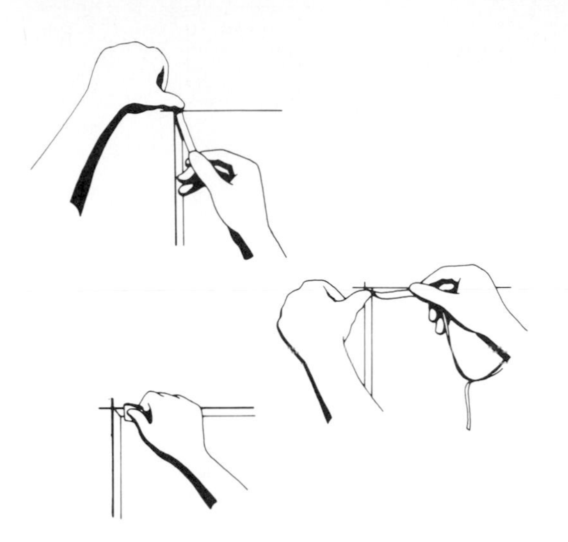

16–19. Installation at right-angle turns.

16–20. Follow the guide lines around the window, working your way back to the starting place. Be sure to stretch the foil as you stick it to the glass. Smooth the foil with a small piece of stiff paper.

9. Remove the paper backing from the bottom of the connector block. Pull the foil back away from the window frame and stick the connector block to the window. Overlap about $\frac{1}{8}''$ of the foil.
10. Remove the screw and clasp from the connector block. Lay foil over the connector block. Replace the clamp and screw to the block. Remove excess foil. Fig. 16-21.
11. Coat the foil with foil sealer. Sealer should overlap both edges of the foil. If the wick has dried, pull up the container and cut off the dried portion.
12. Cross metal frames.

CAUTION: Foil must not touch the metal crossbars or metal frames on windows. These windows may be grounded. A short circuit to ground could cause the alarm to sound. Most alarm grounds occur at poorly insulated foil connector blocks and crossbar locations.

To cross over a crossbar, install a connector block on each side of the crossbar and connect the blocks together with 18-gage insulated approved burglar wire. Fig. 16-22.

Protection by Foil. Even the best perimeter protection circuit will probably have some weak points. Part of the problem is money. Every opening that is contacted and every pane of glass that is foiled adds to the cost. It can reach a point where the increased protection for second story windows, for example, might not be worth what it costs.

Some kinds of building components make perimeter monitoring more difficult. There are windows, for instance, that simply cannot be foiled to provide security while maintaining an attractive appearance. Fig. 16-23. Some homeowners just don't want foil on their living room or dining room windows.

To allow for the inevitable weak points in the system, a second line of defense is available.

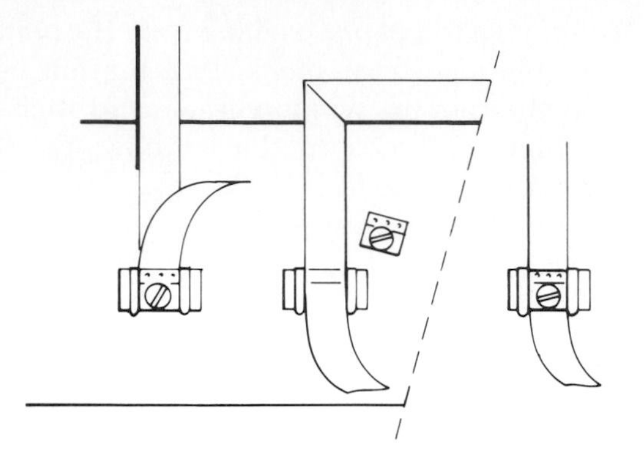

16–21. Attaching foil to the connector block.

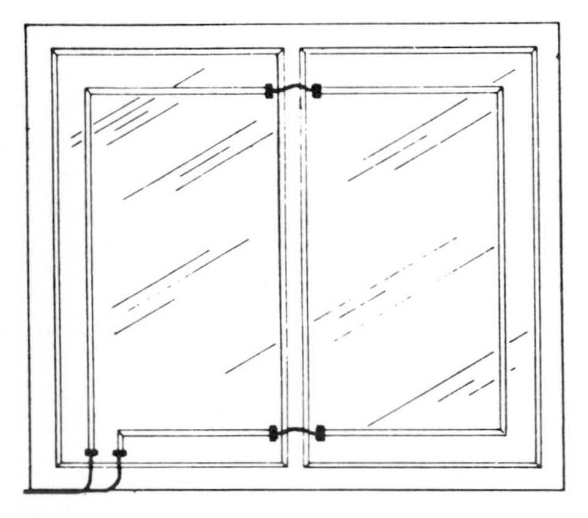

16–22a. Crossing metal frames. Install a connector block on each side of the crossbar. Connect the blocks together with 18-gauge insulated approved burglar wire.

16–22b. Completed installation.

16–23. This type of window is difficult to monitor. Foil would be very unsightly here, and most homeowners would not want it on a window of this type. Therefore, it is necessary to use other types of coverage here.

Planning the Interior Protection circuit. All homes have a recognizable traffic pattern. Determine the paths a burglar would have to take from any given weak point on the perimeter to an area of the house likely to be a burglar's target. Then monitor those paths.

The equipment for getting the job done is available. Contacts, for example, can be used to monitor interior doors. As in the perimeter circuit, these contacts can be magnetic detectors or plunger detectors.

Mats, placed under carpeting in hallways or stairs, can detect an intruder moving from one part of the house to another. Fig. 16-24. There is a special mat for homes with small household pets.

A sophisticated and particularly useful interior monitoring device is the infrared detector. The units are styled to look like wall sockets. The sockets are actually "windows" that are transparent to infrared light. Fig. 16-25. One window in the transmitter unit emits an invisible beam of infrared light. The beam is bounced back by mirrors behind the windows on the opposite wall socket reflector. This re-

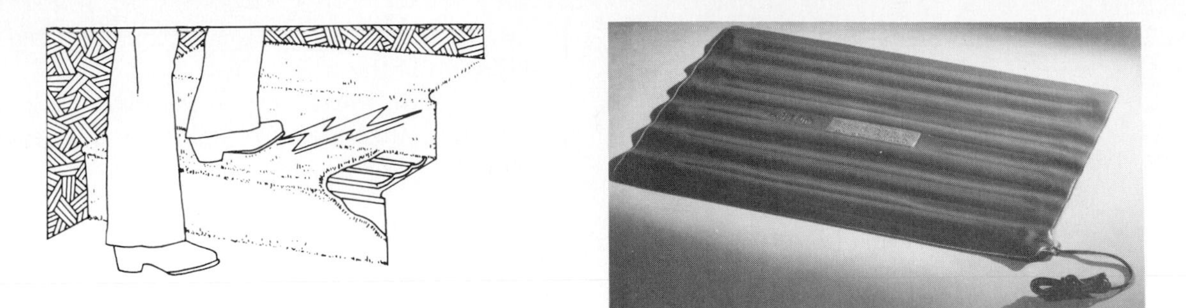

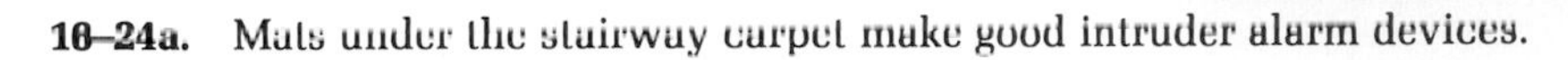

16–24a. Mats under the stairway carpet make good intruder alarm devices.

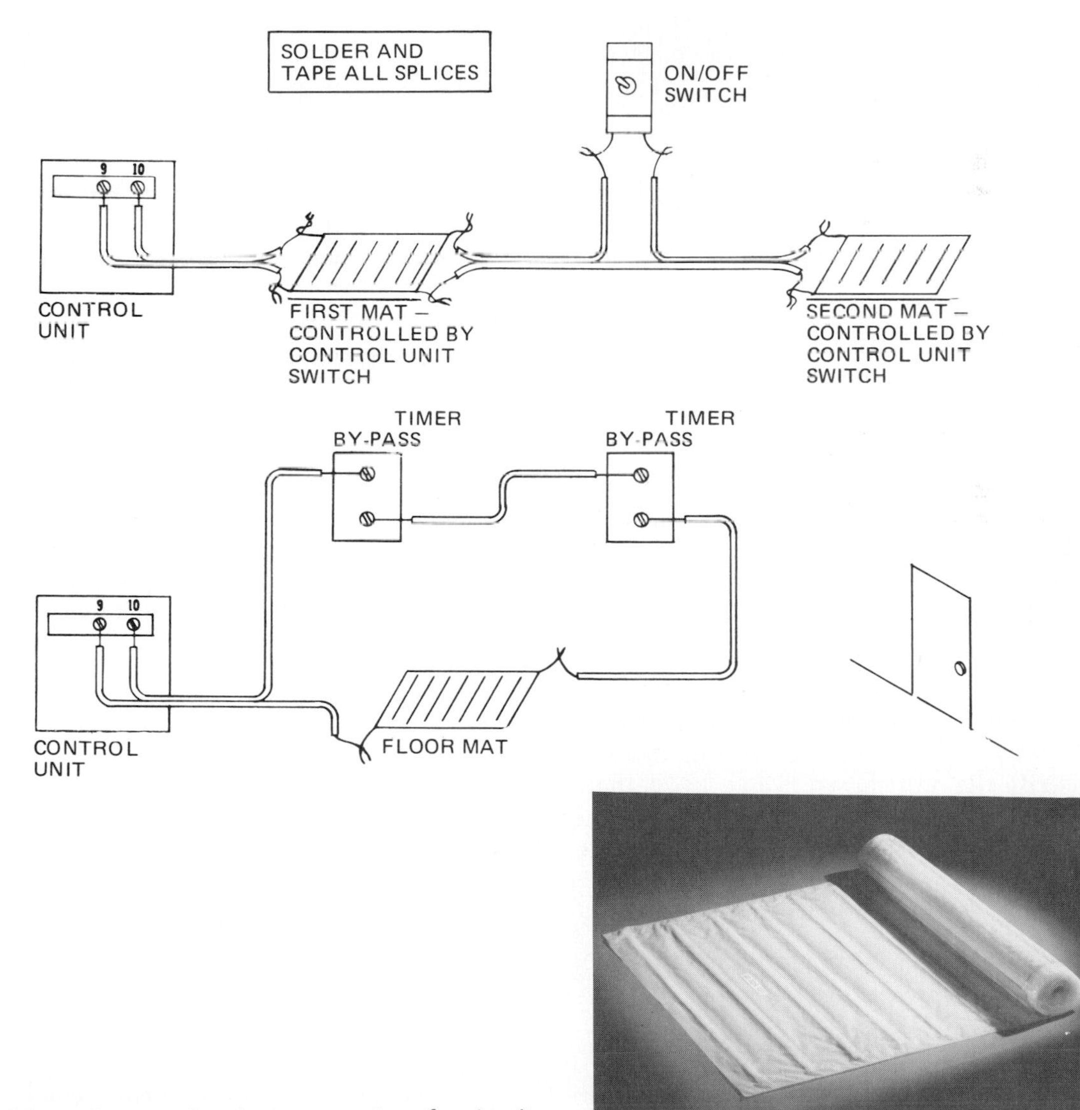

16–24b. Wiring diagram for placing mats into the circuit.

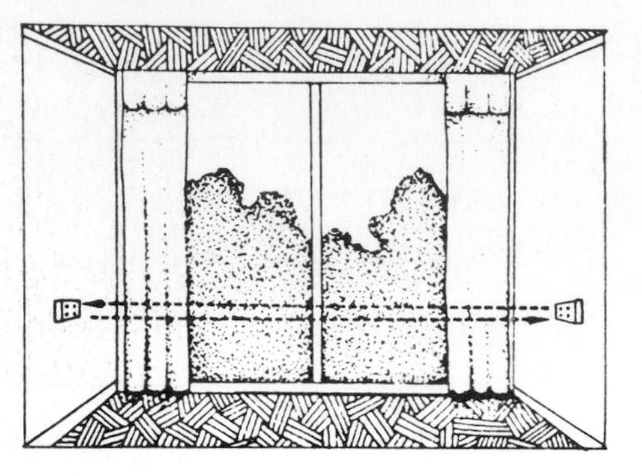

16–25. Infrared rays form an invisible line across the window area. If the line is broken, the rays set off an alarm.

flected light is picked up by the receiver behind another window on the transmitter unit. If any object breaks this beam of invisible light, the alarm sounds. In effect, the photoelectric cell is like a trip wire made of light. If that light is blocked, the beam is broken and the alarm sounds. The infrared beam is invisible to the naked eye, and it can cover any unobstructed opening from 3′ to 75′ wide (1 m to 25 m).

The infrared detector mounts inside a finished wall. The components needed are the power pack (LED to be connected), the transceiver, reflector, mounting plate, wall cover plate, and transformer. The plug-in transformer furnishes low-voltage power. The reflector plate can be mounted easily on the wall across the way from the transceiver. Fig. 16-26.

With the unit shown in Fig. 16-26, there is a minimum and a maximum distance (3′ and 75′) between the light source and the reflector. Other units may vary. Among the things to remember during installation are these:

- Opposing walls must be within the limits set by the manufacturer.
- There must be no obstacles (such as furniture, plants, etc.) between transceiver and reflector.
- Be sure the wall is deep enough for the installation of the transceiver.
- Do not locate the transceiver opposite windows, incandescent lights, or fluorescent lights.
- A 115-V AC electrical outlet should be near the transceiver and not controlled by a switch.

Fig. 16-27 indicates the possibilities of this type of system. Note that it can be used with other entry detectors. This way the infrared system is only part of a larger protection system. It has the ability to fill in some of the weak spots noted earlier.

The Emergency Circuit. A third detection circuit adds significantly to the homeowner's security in emergencies. It is the manual emergency push button. Fig. 16-28. If an intruder forces his way in or if there is a heart patient or an invalid alone in the house who has an emergency, a push button of this type will summon help. Fig. 16-29.

Excellent locations are by the front or rear doors, in bedrooms, basements, or kitchens. The manual emergency push button is connected to the control unit by an independent circuit. Even when the perimeter and interior circuits are not armed, pushing the panic button will sound the alarms.

Wireless Transmitters. For situations where wiring is difficult or impossible, the *wireless transmitter* is suggested. Fig. 16-30. While not recommended for use on fire detection components, wireless transmitters can be used quite reliably for intruder systems.

The wireless transmitter can be hooked up to any intrusion monitoring device (except window foil) on either the perimeter or the interior circuit.

When activated, the wireless transmitter sends a signal to a receiver unit hooked up to the master control panel. This signal causes the alarm to sound.

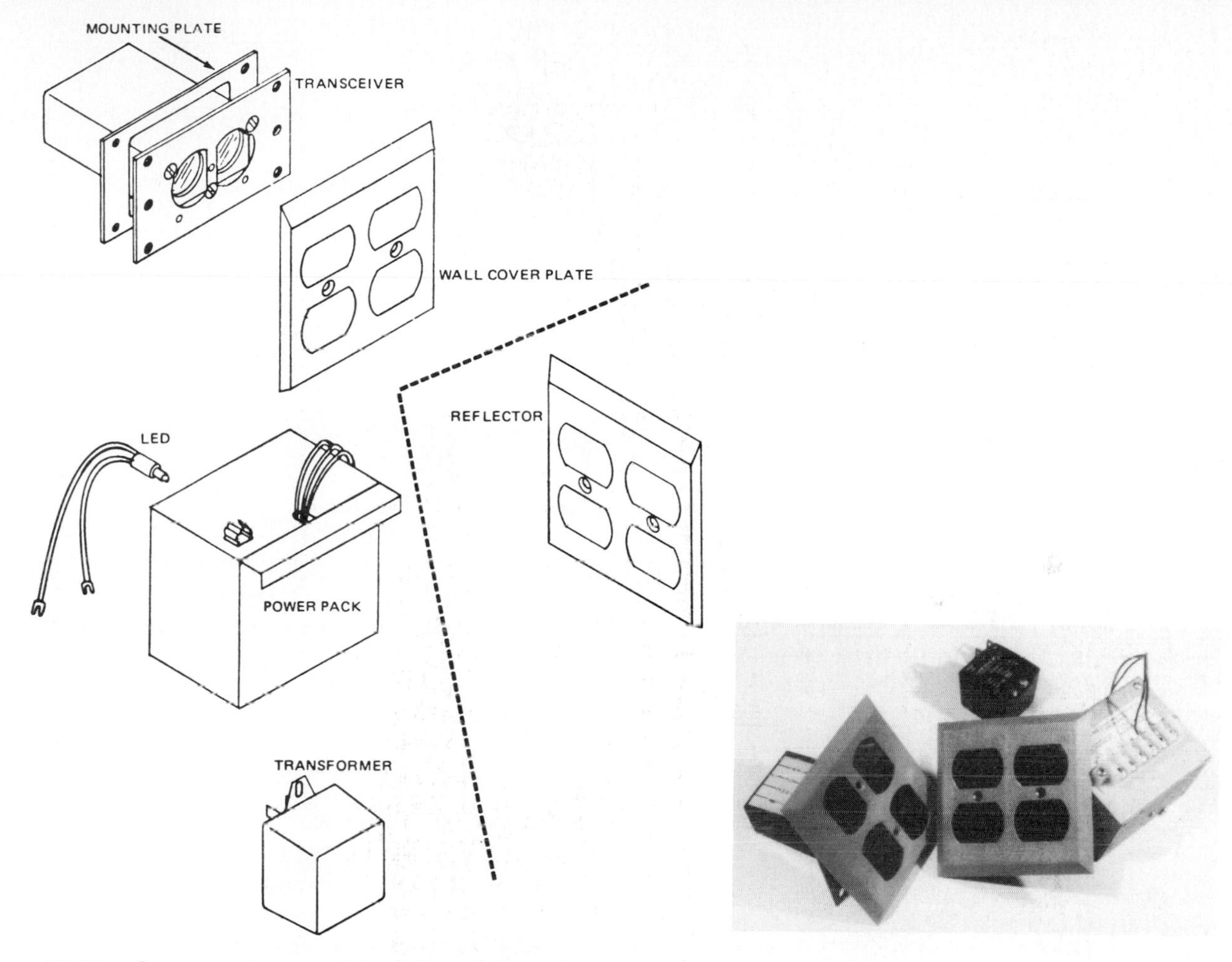

16–26. Component parts of the infrared photoelectric entry detector.

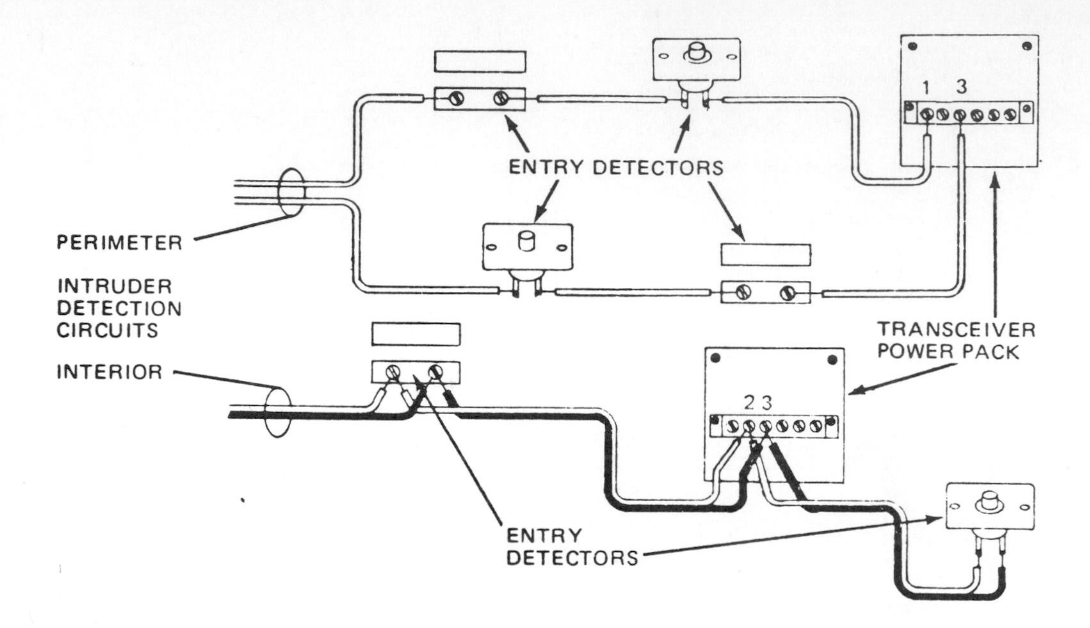

16–27. Wiring of the infrared detector into the interior circuit.

16–28. Manual emergency button mounted by the door.

16–29. Manual emergency button mounted by the bed.

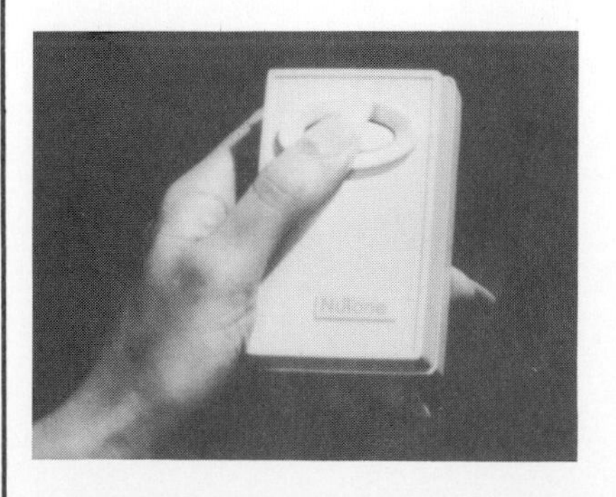

16–30a. Wireless transmitter in use.

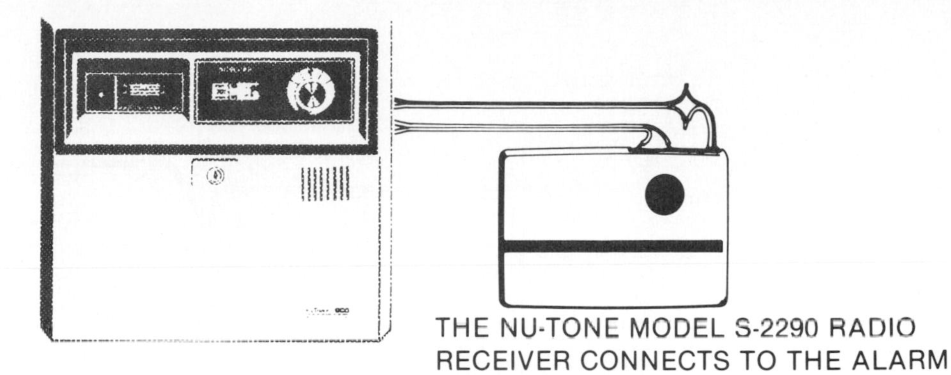

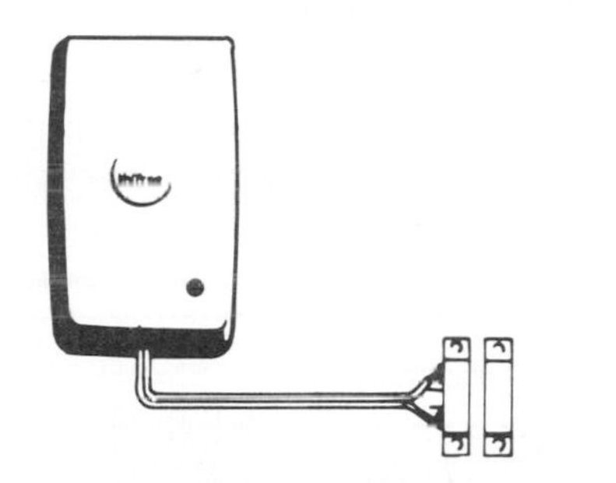

16–30b. Radio receiver connected to the alarm control unit.

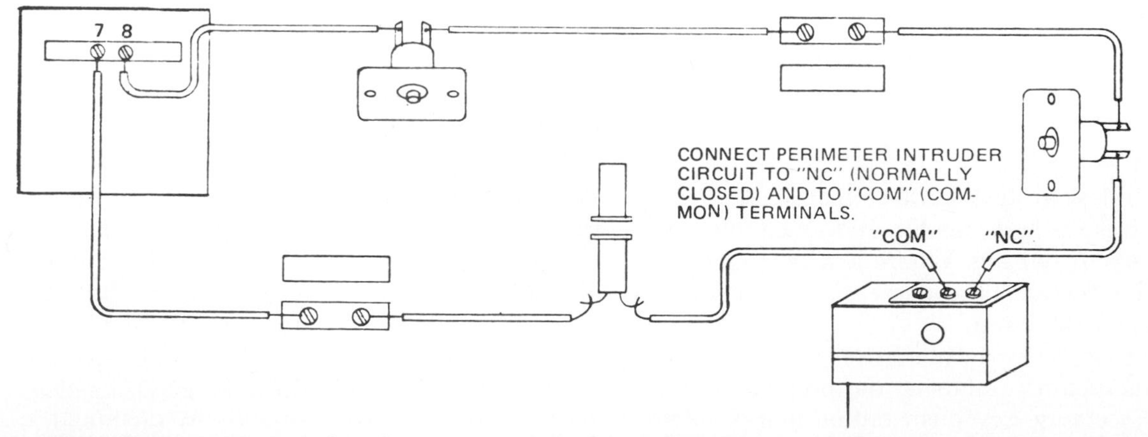

16–30c. Receiver connected in the perimeter intruder circuit.

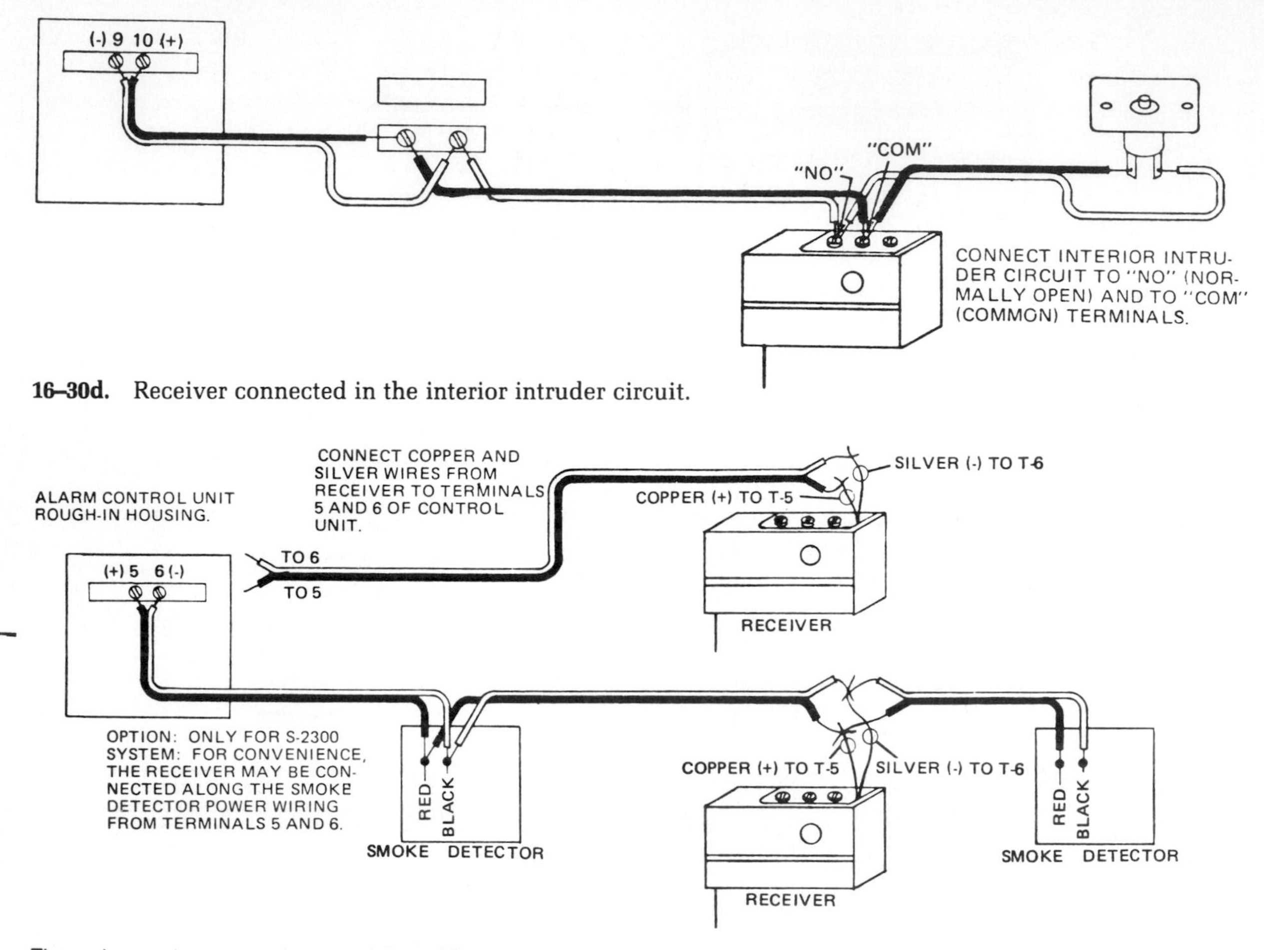

16–30d. Receiver connected in the interior intruder circuit.

The receiver can be connected to one of three different alarm activation circuits from the Control Unit: (1) the perimeter intruder circuit, (2) the interior intruder circuit, (3) the manual emergency alarm circuit.

IMPORTANT: Connect receiver to only one alarm activation circuit!

16–30e. Receiver connected to the power wiring.

A wireless system can be expensive. In most cases, it is cheaper, neater, and much more reliable to run wiring. Where a wireless system is used to replace hard wire, you sacrifice some degree of protection.

However, there is one wireless device that can significantly increase the protection. The wireless emergency push button is a portable panic button. It can be carried around the home. The unit has an effective range of more than 100′ (30 m). It can even protect residents coming from their car to the house late at night.

If there is a heart patient or an invalid who must frequently be left alone in the house, the portable panic button can give him or her the freedom to move around the house. If needed, help can be summoned quickly by pushing the button.

Home with Maximum Intruder Protection

A brief summary of what might be involved in obtaining maximum protection for a home is:

- Detectors at each door and window.
- Effective interior traps.
- An exit/entry system.
- Well-placed alarms.
- A manual emergency system.

Outside the House. An alarm should be mounted on a high point of the roof, gable, or outside wall. Fig. 16-31. Locate a siren where it will give maximum sound and not be easily accessible. The sirens usually have a weatherproof box and two wiring entrances: a threaded conduit opening and a wiring knockout in back. If the wiring knockout is used, insert the conduit plug in the conduit opening to seal the opening from the weather. Mount the box to the outside surface of the house with the conduit opening to the bottom. Connect the wires as shown in Fig. 16-3—observe polarity.

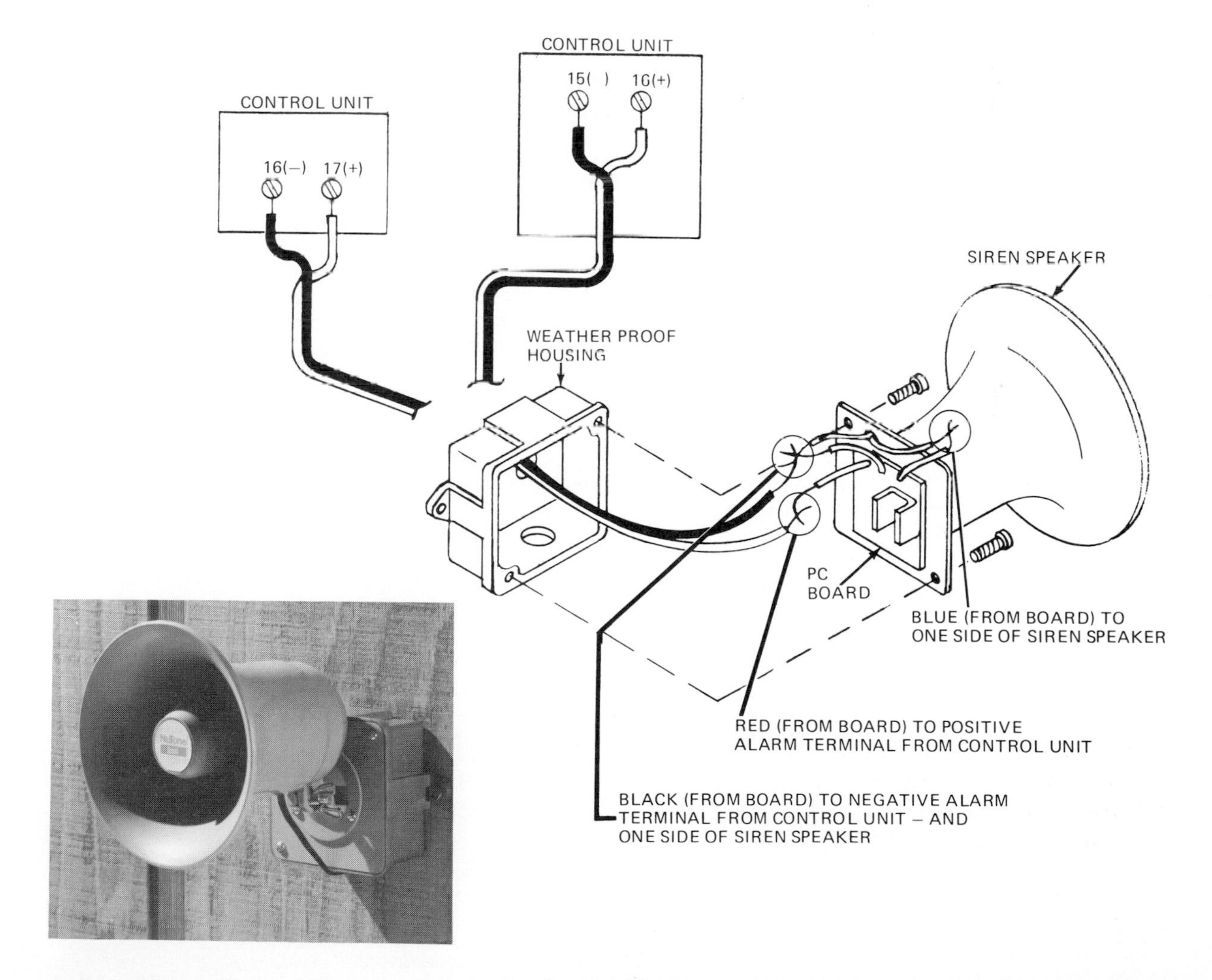

16–31. Electronic siren. Note mounting in weatherproof box.

An alarm bell can be mounted inside or outside. Fig. 16-32. An inside location can use a single-gang outlet box, 3″ or4″ (75 mm or 100 mm) octagonal or square box, or it can be attached to the wall. The exploded view in Fig. 16-32 indicates how the bell can be mounted.

In mounting the bell outside, use a weatherproof box. The box has two wiring entrances: a threaded conduit opening and a wiring knockout in back. If the wiring knockout is used, insert a conduit plug in the conduit opening to seal the opening against the weather. Mount the box to the outside surface of the house with the conduit opening to the bottom. Connect the two leads from the bell mechanism to the cable from the control unit. Refer to the exploded view in Fig. 16-32 when installing the bell. If a different type of bell is used, follow manufacturer's instructions.

If the bell does not ring properly, the mechanism may need adjusting. Remove the bell from the mechanism and the mechanism from the mounting. On the back of the mechanism, loosen the screw. Fig. 16-33a. On the bottom of the mechanism, turn the screw slightly to adjust the bell mechanism. Tighten the screw on the back of the mechanism and remount the unit. Fig. 16-33b.

Front Door.

- Use a plunger or magnetic detector.

- For exit/entry use the built-in module which plugs into the control unit.

- Next to the door locate a manual emergency button.

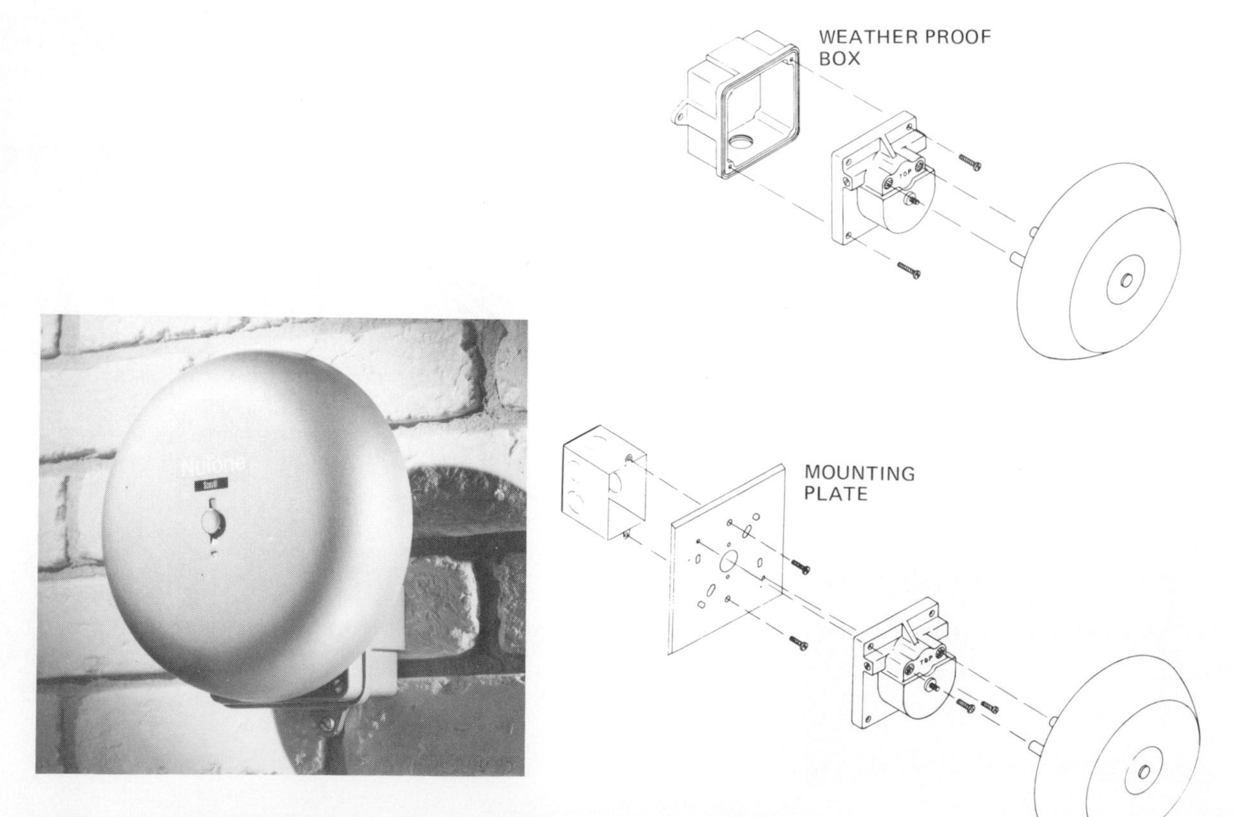

16–32. Alarm bell. Weatherproof box is used for outside mounting.

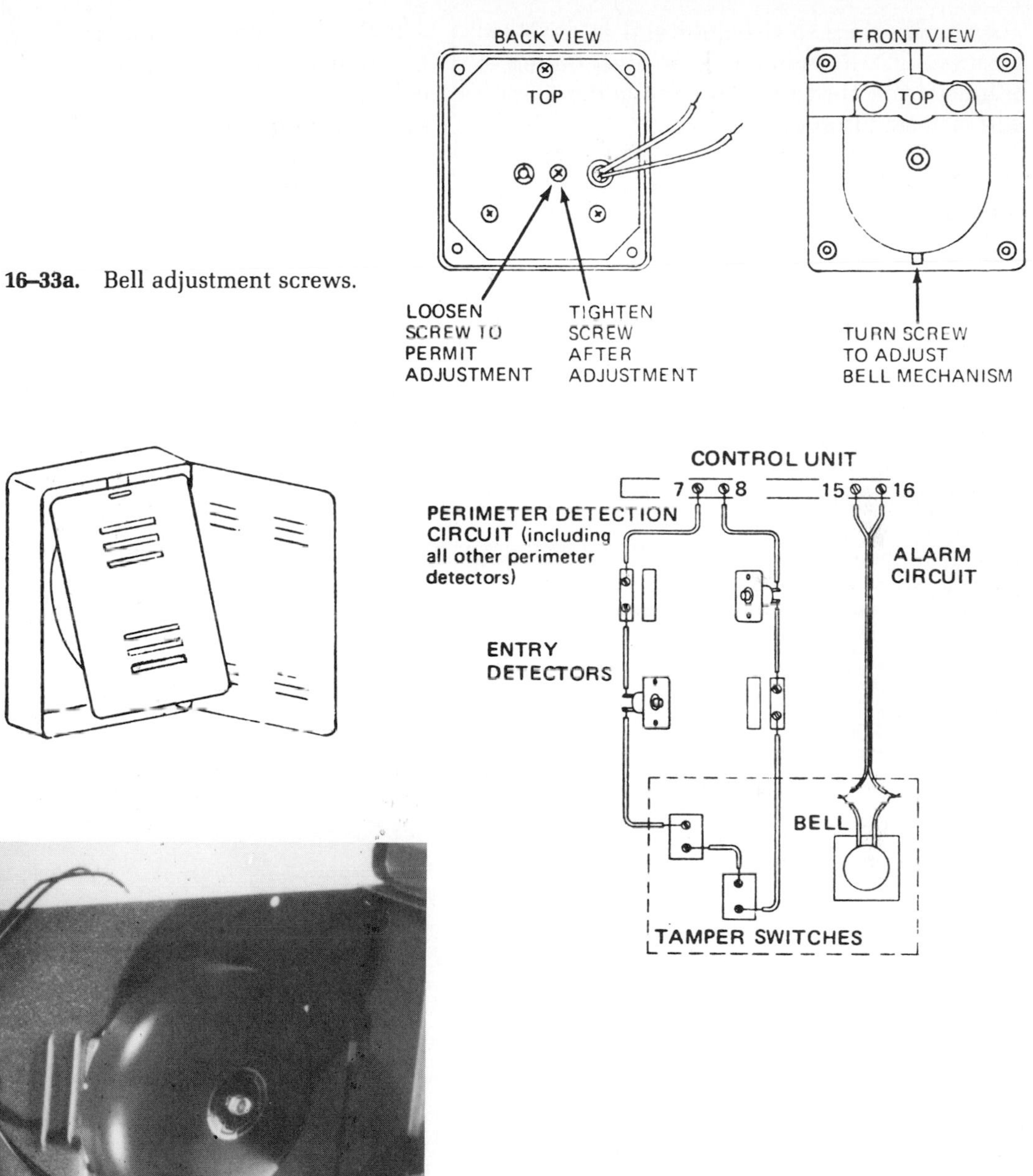

16–33a. Bell adjustment screws.

16–33b. Enclosed outside alarm bell with wiring circuit.

■ Inside, close to the door and in a hidden location, install a remote reset push button to eliminate the need to rush back to the control unit to reset the system.

Other Doors (Rear, Side, Garage-to-Kitchen).
■ Use a plunger or magnetic detector.

■ For exit/entry (because these doors are frequently used), equip them with the rotary exit timer switch. Fig. 16-34. Use it as a shunt switch during the day.

■ At the rear outside door, use an entry key switch to turn the alarm off before entering the door. Fig. 16-35.

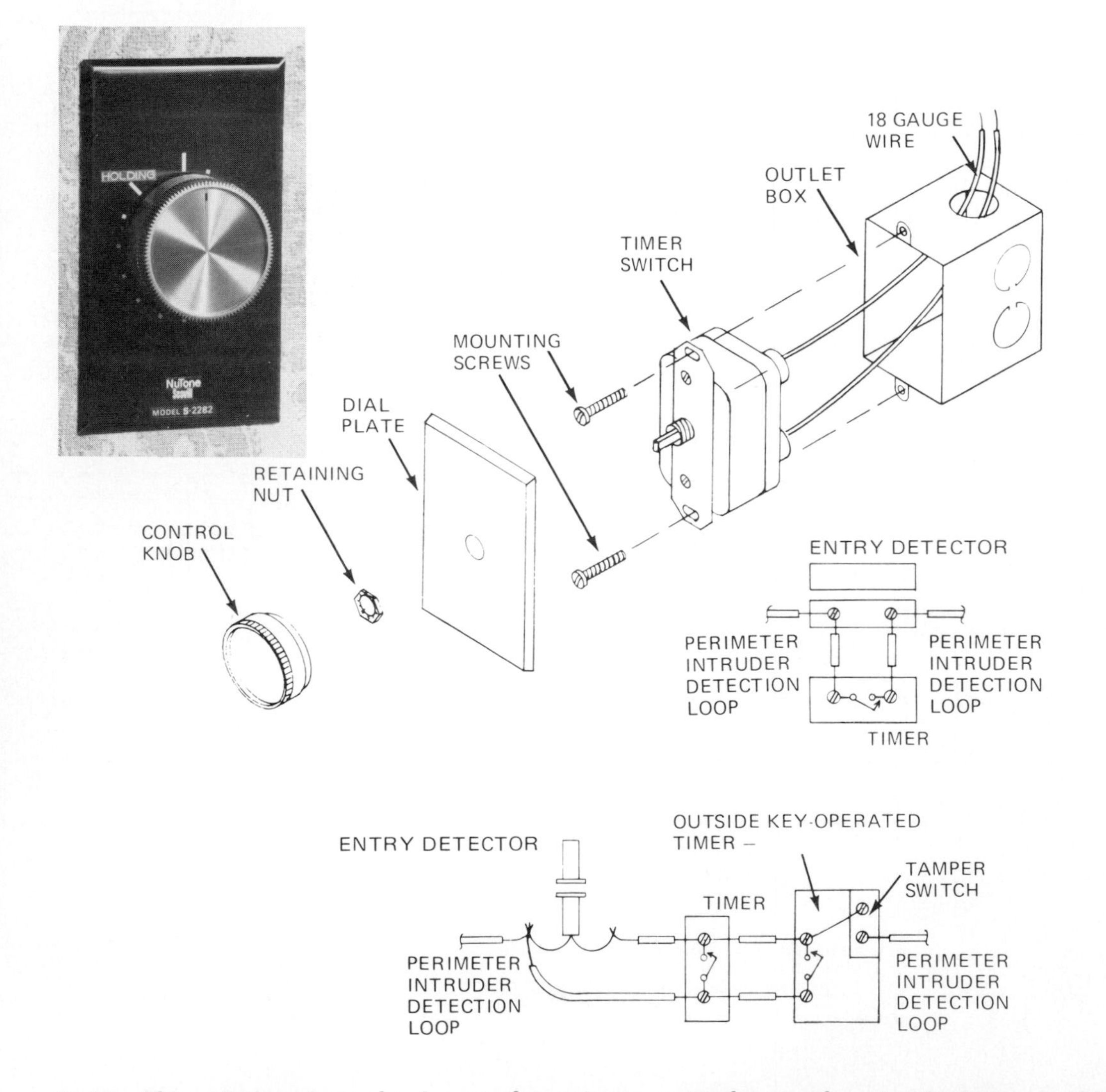

16–34. This exit timer is used to bypass the perimeter entry detector from inside the house. Often it is used in conjunction with an outside key-operated timer switch. It is usually located near an entry door at a convenience height, inside the house.

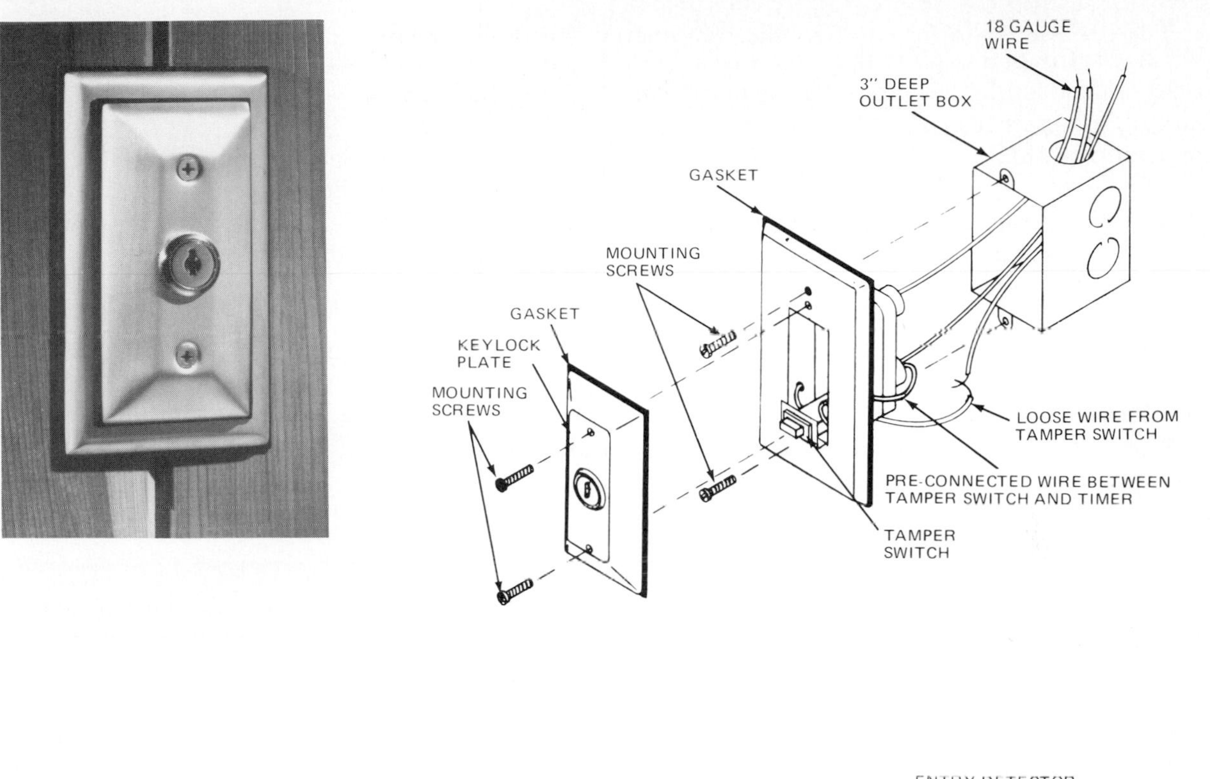

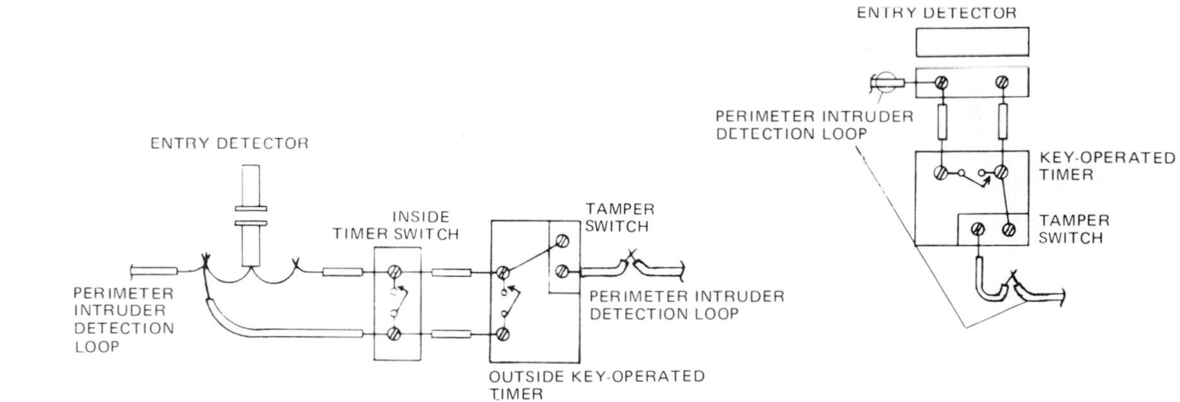

16–35. Key-operated timer.

■ On doors leading to the garage, foil any windows. Cover the foil with sealer. Use a door cord to connect the foil circuit to the perimeter circuit. To protect against forced entry, use a magnetic or plunger detector.

■ Sliding glass doors can be guarded with recessed magnetic detectors and a metal door adapter or with magnetic or plunger detectors. A long floor mat or an infrared detector just inside the door can improve security.

Windows.
■ Where possible, foil all glass areas. Depending on the type of window, use foil contact switches or magnetic or plunger detectors if the window needs to be opened.

■ For single-hung windows, use one plunger or magnetic detector.

■ For double-hung windows, use two plunger or magnetic detectors, one on each half of the window.

■ For casement windows, use the long plunger detector.

■ Use the shunt switch for windows that must be opened occasionally. This allows bypassing of these detectors when you want the window open without affecting the remainder of the perimeter circuit. One shunt switch may be used in the master bedroom and connected to all second floor windows. This allows the homeowner to shunt all bedroom windows from a single location.

Attic and Basement.
■ Don't overlook the attic windows. A large tree nearby could provide easy access.

■ Foil all basement windows.

■ Use interior plunger or magnetic detectors to guard the door leading from the basement to the living area.

Inside the Home.
■ Use a floor mat or infrared detector to guard the hallway to the sleeping areas.

■ Use plunger or magnetic detectors on all doors leading to closed areas where an intruder would enter. Plunger detectors may be preferable because they are hidden.

■ Install manual emergency push buttons next to the bed in the master bedroom, over the counter work area in the kitchen, next to the laundry area in the basement, and next to all outside doors.

■ Locate the control unit in the master bedroom.

Telephone Dialer. One way to make sure that the police check on an alarm is to install a *telephone dialer*. It will dial the police number and give the alarm—intruder, fire, smoke, or whatever you put on the tape. Once the dialer is installed it is possible to put on the tape your name, address, and any message necessary to alert the police to check on the house.

Since the dialer uses the telephone lines, the telephone company has to connect the coupler to its lines. Fig. 16-37. The unit operates on rechargeable batteries. It will need an AC power source and will need to be connected to the control unit. Fig. 16-37.

Determine the location for the dialer. Consider that it will be connected to the telephone connector block or telephone coupler. The dialer will also be connected to the control unit for the house alarm system. Dialer location should be convenient for the owner's operation and checkout—close to the alarm control unit. The dialer should be concealed from strangers. It mounts to a finished wall surface.

Local alarms do an excellent job of frightening burglars off. Under most residential conditions this is enough. The last thing a burglar wants to do is attract attention. The combination of light and sound draws attention to the break-in. The burglar thinks there will be

16–36. Ultrasonic motion detector.

someone on the scene immediately.

Local alarms also do an excellent job of alerting the family to danger. It's almost inconceivable that anyone could sleep through the racket that a well-placed set of local alarms will make.

If the burglar isn't scared off, if the family isn't home, or if the alarm has been triggered by someone pressing the panic button to get help in a medical emergency, the local alarm is not enough. There must be a way to summon help.

When local alarms need to be supplemented to summon help, the telephone dialer can send a prerecorded message to any other telephone.

There are two tracks on the dialer tape so that it can call different numbers and deliver different messages. The dialer tape can be programmed to call a single number three times in a row, or to call three different numbers. It can be programmed to give different messages to the different numbers, or to repeat the same message each time.

It can be programmed to delay a call for 20-25 seconds. That way, if there's a false alarm, the homeowner has a chance to stop the dialer before it sends the alarm over the telephone lines.

Before installing a dialer, be sure to check with local fire and police departments for any applicable regulations.

Motion Detectors. The *motion detector* operates on ultrasonic frequencies. Fig. 16-36. An ultrasonic frequency is just above the human hearing range. The transmitter is about the size of a twenty-five cent piece and so is the receiver transducer. The frequency is broadcast from the unit. Any motion within its range will cause a slightly different frequency to be reflected back to the receiving unit. This difference in frequency causes a relay to energize or a transistor circuit to energize and complete the circuit to the alarm control center.

An ultrasonic unit needs little or no maintenance if wired into a household circuit.

This type of detector is quite sensitive. It can be activated by the movement of a dog or cat. Even hot air furnaces can cause the curtains to move, setting off the alarm.

FIRE ALARMS

While fires are not always preventable, the amount of damage they cause can be reduced. The risk of injury or death can be minimized by a well-planned alarm system. Early warning is the key to fire protection. The sooner a fire is detected, the less likely it is to burn the house down and injure or kill the family.

Fires need three things to start: fuel, oxygen, and a source of heat. Oxygen, of course, is present in the air; so there's always enough of that. Most materials in a home can become fuel for a fire.

For a fire to start, though, a source of intense heat is needed. Again, we live with many heat sources. Any piece of equipment that produces flames or that contains electrical heating elements can start a fire, if conditions are right.

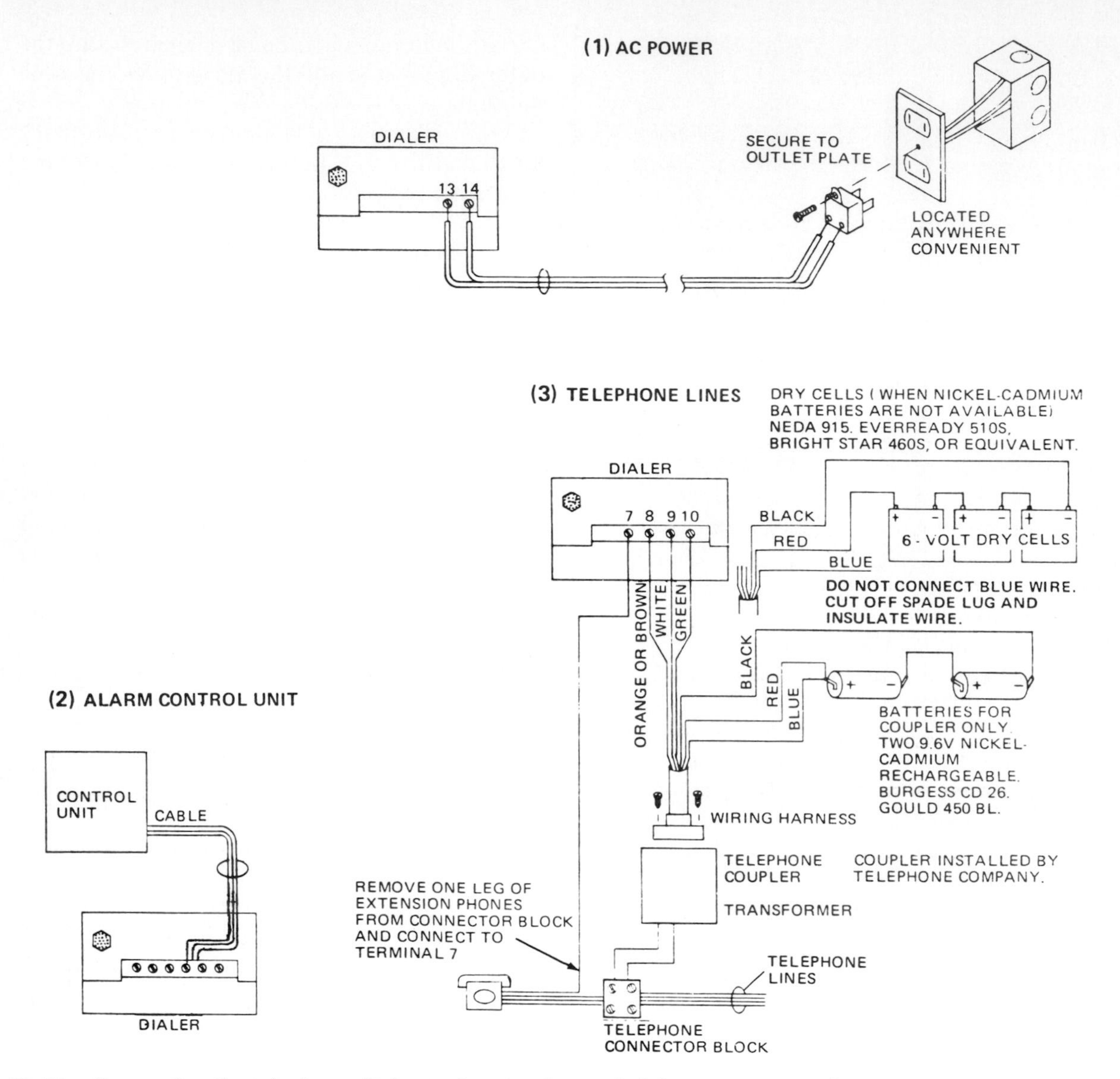

16–37. Connecting the telephone dialer to the circuits needed for proper operation.

Electricity can produce the heat to kindle a fire. Electrical appliances can fail. The protective circuits in a home's wiring system can break down or be bypassed. And, of course, there are cigarettes, a common cause of house fires.

All fires produce three things: heat, smoke, and the volatile products of combustion. These volatile products are mostly water vapor and carbon dioxide. The other gases vary from fire to fire, depending on what's burning.

All three products of fire can kill. Most of the deaths and injuries caused by fire are caused by breathing toxic gases, smoke, or superheated air. Very few fire victims actually burn to death. Fortunately, all three products of a fire can be monitored. The problem is to detect all three and provide time for the family to escape.

Fire Detection System

Smoke detectors sense the presence of abnormal smoke. Fixed-temperature heat detectors sense abnormal heat. One model triggers the alarm when the air reaches 135°F (57°C). Fig. 16-38. Another model sounds the alarm at 200°F (93°C). Rate-of-rise detectors sense abnormal, rapid rise in heat of 10-15°F (6-9°C) per minute.

Smoke Detectors. For the smoldering sort of fire that is usually the case in home fires, the photoelectric *smoke detector* is best. The ionization or radioactive type is usually better for detecting the smoke generated by business or commercial installations. The photoelectric model is also less susceptible to false alarms.

16–38. Fixed-temperature heat detector.

The photoelectric smoke detector shines a light into a tiny chamber inside the detector. A network of passageways brings a steady flow of room air into that chamber. If any of the air contains smoke, some light will be reflected. The reflected light is detected by a sensitive photoelectric cell. This reflected light triggers the alarm.

A concentration of less than 4% smoke in the air will sound the alarm. What is 4%? Examine a cigarette smoking in an ashtray. The plume of smoke rises and coils and spreads. It diffuses into the air. Somewhere above the cigarette is a region where the smoke is drawn out and wispy, just barely visible. That concentration is close to 4%. If that amount of smoke were concentrated in the chamber of a smoke detector, it would set off an alarm.

This doesn't mean that the alarm will go off whenever someone lights a cigarette. Even a roomful of people smoking won't trigger an alarm under normal conditions. The smoke spreads out so that the concentration is very low.

Smoke detectors should be located in halls between the sleeping quarters and the rest of the home. Or, they should be located at the top of stairways and in individual rooms. They can be located on ceilings at least 12″ (300 mm) from any corner or on walls between 6″ (150 mm) and 12″ from the ceiling.

Heat Detectors. *Heat detectors* come in two types: fixed-temperature and rate-of-rise. The fixed-temperature heat detectors work on the same principle as a furnace thermostat. Two different kinds of metal are bent into a disc shape. These two metals expand and contract at different rates as the temperature goes up or down. The curvature of the disc changes as the temperature goes up. When the temperature reaches a certain limit, the disc clicks down. This allows the contact to close and the alarm to sound.

The temperature limit of some fixed-temperature detectors is 135°F (57°C). This is the device installed in most living areas. A different type is used for boiler rooms, attics, and other storage areas where temperatures are usually high. It is set to trip the alarm when a temperature of 200°F (93°C) is reached.

Square of Protection. Most heat detectors on the market offer a certain area of coverage. This is called the *square of protection*. The square is determined by testing. For most models it is 10′ (3 m) in every direction from the detector. This figure applies to a smooth ceiling (one that is not interrupted with open joists or beams). The square of protection is thus 20′ (6 m) by 20′ with the detector in the center. If the detector is moved, its "square" moves with it.

If there's ever a question of whether to add another detector, keep in mind: walls, partitions, doorways or archways, ceiling beams, and joists will interrupt the flow of heat. Any of these creates new areas to be protected, whatever the size, especially if your goal is complete coverage.

Any area with at least one dimension exceeding 20′ (6 m) will require more than one heat detector for full coverage. Using the square of protection idea, arrange the detectors so that they are no more than 20′ (6 m) apart and so that no detector is more than 10′ (3 m) from a wall. In some instances, the "squares" may overlap. In others, they will meet at the center of the room. Fig. 16-39.

Rate-of-rise heat detectors monitor the rate at which the temperature changes. They are somewhat more sensitive than fixed-temperature detectors and can cover a 50′ × 50′ (15 m × 15 m) area if they are located in the center of that area. If the temperature goes up at the rate of 10-15°F (6-9°C) per minute, the alarm sounds.

The rate-of-rise units also incorporate fixed temperature sensors. Even when the temperature rises relatively slowly, if it reaches the limit temperature, it melts a special alloy and

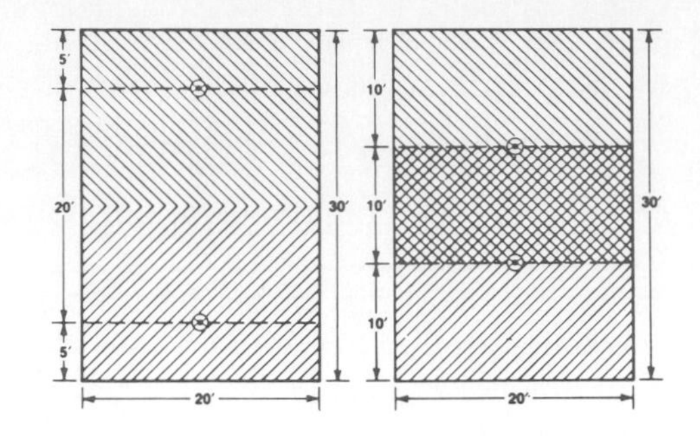

16–39. Use more than one heat detector if one dimension of the room exceeds 20′ (6 m).

closes a circuit. The same principle as that of an electric fuse is involved here. There are low-temperature units with a limit temperature of 135°F (57°C) and high-temperature models with a 200°F (93°C) limit. Fig. 16-40.

Planning for Fire Protection

The National Fire Protection Association defines several levels of fire protection. At the lowest level (Level 4), the Association recommends a minimum of one or more smoke detectors located so as to protect the sleeping quarters. At the top level (Level 1), the whole house is protected with an intelligently designed and carefully installed system of smoke and heat detectors in every confined living space.

Suggested Minimum Plan. The following minimum components provide protection against injury from smoke while the family is asleep. This plan consists of:

- Control unit.
- Alarm.
- Smoke/heat detector.

The great majority of fires start in the living room or kitchen between the hours of 9 p.m. and 6 a.m. This is a time when family members are asleep and helpless. Smoke inhalation, the most common cause of death in a fire, can kill before the air even gets warm, much less than 135°F (57°C). Therefore smoke detectors should be located between the most likely

16-40. Rate-of-rise heat detector.

source of fire and the sleeping areas. This means the detector should be outside the bedrooms, in the main hallway, for example, or at the top of the stairs leading to bedrooms. Even though fire/heat detectors are less expensive than smoke/heat detectors, one strategically located smoke/heat detector will sense smoke sooner and provide an earlier alert.

This minimum protection has a few disadvantages. It cannot be expanded into an intruder detection system later on. There is no provision for battery standby and no outside alarm.

Medium Protection Plan. The medium protection system offers greater protection against fire because it guards more areas of the home. It consists of:

- Control unit.
- Alarm.
- Smoke/heat detector.
- Heat detectors.

This medium system has the same three components as the minimum plan. Thus there is protection against smoke. The addition of heat detectors in strategic locations increases the area of coverage to more places where fire may start. This additional protection can usually be installed very economically because the devices are inexpensive and the wiring can usually be done quickly. The wiring from the attic areas over bedrooms and living areas can be done with small-gauge wire.

■ Install a rate-of-rise heat detector in the living/dining room. The greater sensitivity of the ROR detector would make it preferable to the fixed-temperature type because one detector could cover the entire area.

■ Install fixed-temperature detectors in all other rooms and confined areas like closets.

■ Install high-temperature heat detectors in areas that contain equipment such as furnaces, hot water heaters, or dryers.

The medium system plan locates a heat detector in all the places where a fire is most likely to start.

Maximum Protection Plan. The maximum protection plan increases the area of coverage by detection devices to the maximum degree possible. It would meet the NFPA's Level 1 requirements. It consists of:

- Control unit.
- Smoke/heat detector.
- Heat detectors (either fixed-temperature or rate-of-rise) in closed spaces and areas which harbor a possible fire hazard.
- Heat detector in each room (either type).
- Inside and outside alarms.

Fig. 16-41 illustrates the maximum protection system. Note the locations of the devices.

The kitchen presents its own fire hazards. Fig. 16-42 shows the location of fire/heat detectors. Any electrical device could cause a fire. Therefore it is a good idea to center a detector for maximum coverage.

If the ceiling is interrupted by beams or open joists, as in most basements (Fig. 16-43), detector placement must be tailored to suit the situation. Detectors should be mounted on the bottom of the joists or beams, not in the joist channels. The distance between the detector and the wall should be decreased to 5′ (1.5 m).

Fig. 16-44 illustrates the placement of heat detectors (200°F or 93°C) in the attic or crawl space. The attic temperature is usually higher

This microcomputer-based system's eight fully programmable zones (plus keypad fire, police, and medical emergency) provide unlimited options. Each installation may be custom-designed to meet your requirements. The system utilizes non-volatile Electrically Erasable Read Only Memory (EEROM) to store all data. This memory is retained even during complete power failures. Fig. 16-46.

All programming is accomplished by entering data through remote digital keypads. These keypads are available in three styles, the digital remote controls provide flexibility in system design. The deluxe remote control features a membrane keypad and a ten-watt speaker that may be used as in indoor siren and/or an intercom station. See Fig. 16-47.

16–46. Control center for the electronic alarm system.

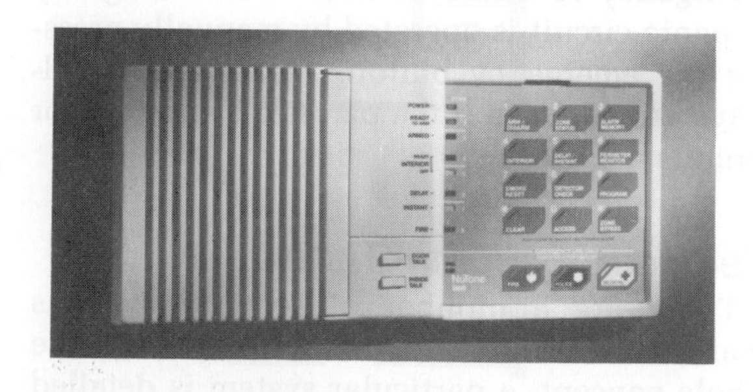

16–47. Remote keypad with a speaker.

The control panel houses the electronics for controlling the burglary/fire system and the digital communicator. The communicator itself is also keypad programmable. It may be programmed for most transmission formats. The control panel also contains a 12VDC, 1.5-amp power supply that will provide 900 mA of power for auxiliary devices. See Fig. 16-48.

■ *Entry delay times.* Each zone may be programmed for two entry times (from 000 to 255 seconds duration). After a time is programmed into each delay time's memory location, each separate zone may then be programmed for entry delay time #1 or #2. For example, a

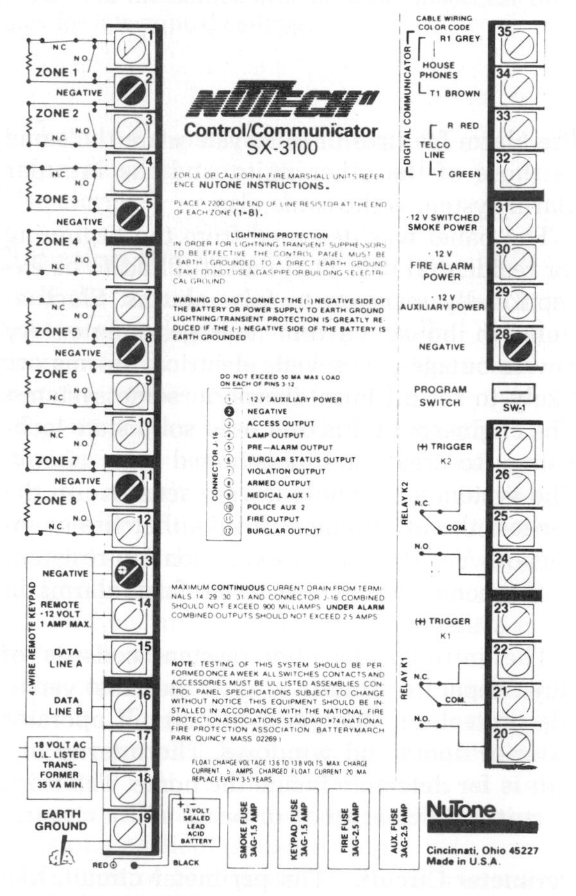

16–48. Terminal board for the control communicator.

long delay time might be used for a garage door, whereas a shorter delay time would be programmed for an entry door.

■ *Loop response time.* The system may also be programmed for fast or slow loop response times. These range from 40 milliseconds to 10 seconds in 40-millisecond increments. After the fast and slow loop response times are programmed, each separate zone may then be programmed for a slow or a fast loop response time.

■ *Keyswitch zones.* If the application calls for arm/disarm control from a keyswitch, the system may be programmed for a keyswitch zone. Through keypad programming, the installer assigns one zone as a keyswitch zone. This then can be controlled by a momentary normally open or a momentary normally closed keyswitch.

■ *Bypassing zones.* Zones may be individually bypassed (shunted) with a simple two-key operation at the remote control keypad. All manually bypassed zones are cleared when the system is disarmed. They must then be bypassed again before the system is armed. Zones cannot be bypassed once the system is armed. For special applications, fire zones may be programmed to be shuntable. The digital communicator can also be programmed to send a bypassed-zone-arming report to the central station.

■ *Fire zones and day zones.* Both fire zones and day zones sound an audible pre-alarm for a trouble condition. The user can easily silence a fire pre-alarm at the keypad by pressing the CLEAR key, while an LED will remain blinking to remind the user that a trouble condition exists.

■ *Alarm outputs.* A number of alarm outputs may be wired to operate two on-board auxiliary relays. The burglar alarm output can be programmed to be "Steady" or "Pulsing." The auxiliary relay outputs can be connected to a self-contained siren, a two-tone siren driver, or an eight-tone siren driver.

Alarm outputs can be programmed to provide a one-second siren/bell test upon arming the system. To extend the system's capability, the alarm outputs can be wired directly to a siren driver with "trigger inputs" (such as the eight-tone driver), freeing the on-board auxiliary relays for other tasks.

The system can also be programmed for silent or audible alarms. There are separate outputs for burglary, fire, police, and medical. Each can be individually programmed for a cutoff time of 001 to 255 minutes or for continuous operation.

The system has a "lamp output" terminal. This terminal is activated for two minutes each time a key is pressed, during entry and exit delay and during any alarm condition. It can be used as light control for a hall light or other applications.

■ *Arm/disarm codes.* The system has four arm/disarm codes and a master programming code which can be changed by the user. The codes can be one to five digits long; digits may be repeated. Each of the four arm/disarm codes can be programmed to be:

1. An arm/disarm code only.
2. An access code only. (The system includes an access control feature that can be used for an electric door release or other device. The access control output is timed. It is programmable for 1 to 254 seconds in duration.)
3. An arm/disarm code if "1" is entered first; an access code if "0" is entered first.
4. Arm/disarm and access simultaneously.
5. Same as 4, plus just "0" and code allows access.
6. A duress code.

The fourth code can be programmed to function for a pre-determined number of times (1 to 254). This code may be used to allow entry by house-keeping, service personnel, and the like. After the code is used the allowed number of times it will no longer work.

The system also includes a programmable "short arming" feature. This featue allows one- or two-digit arming. The full code must be entered to disarm the system. If the first digit of an arm/disarm code is "0," then one button arming is possible—just enter "1." This feature also allows all other command keys to functon simply by pressing the desired command key.

■ *Burglary alarms.* A burglary causes an audible pre-alarm. An alarm causes the ARMED LED to flash. Entering the arm/disarm code silences the audible pre-alarm, but the flashing LED remains on. Pressing the CLEAR key resets the flashing LED. During an alarm, entry of an arm/disarm code will abort the digital communicator unless the code is a duress code. A FIRE or MEDICAL alarm can be silenced by pressing the CLEAR key.

■ *Fire alarms.* Fire alarms can be silenced by entering an arm/disarm code or by pressing the CLEAR key. The fire zone LED remains on for an alarm or remains blinking for a trouble condition. If a trouble condition clears itself, the FIRE LED automatically resets. The FIRE alarm LED can be cleared by entering the arm/disarm code.

■ *Smoke reset.* When a fire alarm is activated, the system is silenced by pressing the CLEAR key. This does not reset the smoke detectors. After the smoke detector that caused the alarm has been identified (by viewing the alarm memory LEDs on the smoke detectors), pressing the SMOKE RESET key and the arm/disarm code interrupts power to smoke detectors for 5 seconds, resetting them.

SMOKE RESET also performs a battery test. It removes AC power from the control panel and dynamically tests the battery. If the battery fails the test, the POWER LED will flash. A "low battery" condition will be reported to the central station. The flashing POWER LED can only be reset by a subsequent test that recognizes a good battery.

Installation. The panel box is easily mounted and contains the terminal strips required to complete all the pre-wiring and terminal wiring connections. Fig. 16-49. The terminal strips themselves are marked for identification. All input connections are made on one terminal strip. All output connections are made on the other. This arrangement allows the installer to complete the pre-wiring, make the terminal connections, and then safely lock the box.

The main circuit board is a separate module. It is mounted with four hex screws, the board is easily mounted to the terminal strips and is simply plugged in with one connector. This arrangement allows for pre-wiring and protects the electronics from damage during wiring.

Once the installation and wiring are complete, power is supplied by a flip of the power switch. This aid eliminates the need to disconnect the power wiring during installation or service. Fig. 16-50.

When the system is powered up, the installer may program the system directly from the digital remote control keypads. Programming worksheets (complete with zone identification, programming memory locations, and factory program values) are provided to aid the installer in programming. A detailed programming manual is also provided.

After the programming is complete, the installation can be checked-out from the keypad. The installer simply uses the system status LEDs, the zone status command key, and the detector check key to verify the system installation. Then all user operations can be performed to verify the correctness of the program.

Troubleshooting. In this eight-zone system, the alarm memory function automatically reduces troubleshooting by one-eighth the time. Simply pressing the alarm memory com-

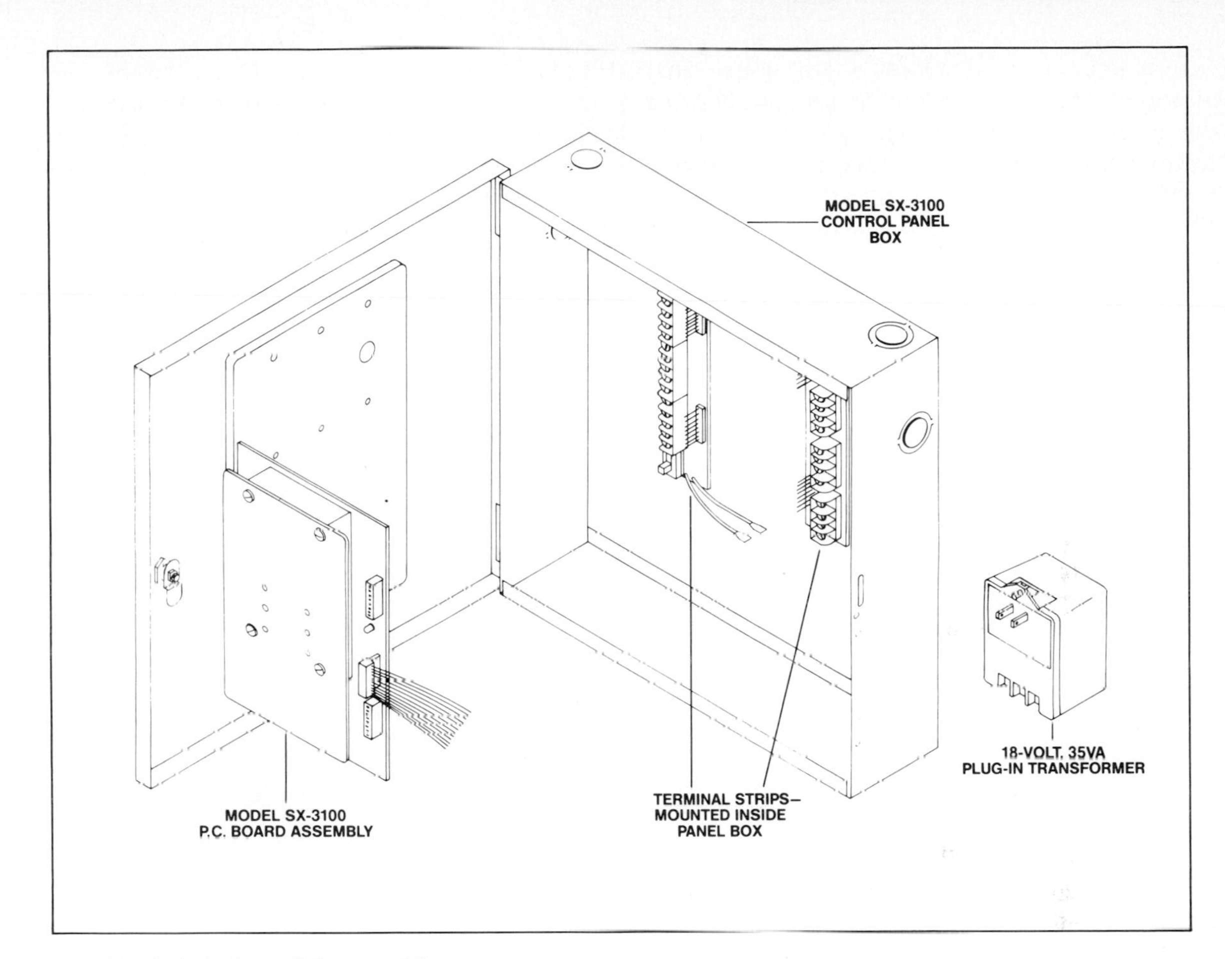

16–49. Exploded view of the panel box.

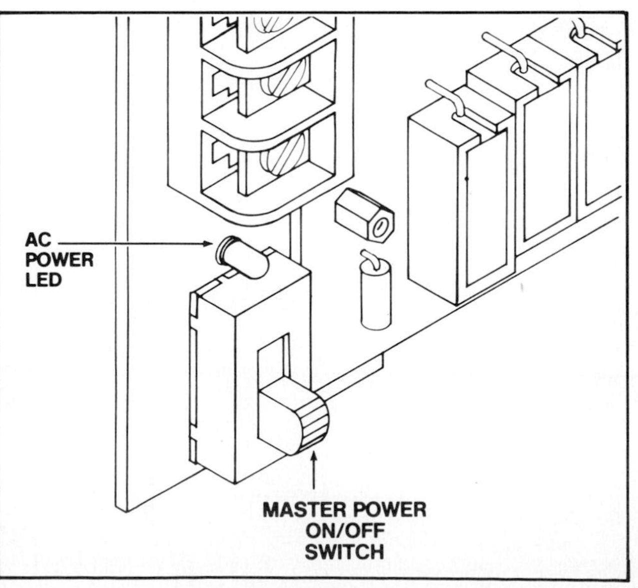

16–50. Master power on/off switch location on the terminal board.

mand key reveals which were zones last alarmed. Then, within that zone, the detector check command key may be used to troubleshoot specific detectors. For example, a typical system may contain forty-five contacts. With a conventional one-zone system, it would be necessary to troubleshoot most of the forty-five contacts by trial and error. With this system you may use the keypad to troubleshoot the entire installation. Fig. 16-51 shows a complete installation and its protection capabilities.

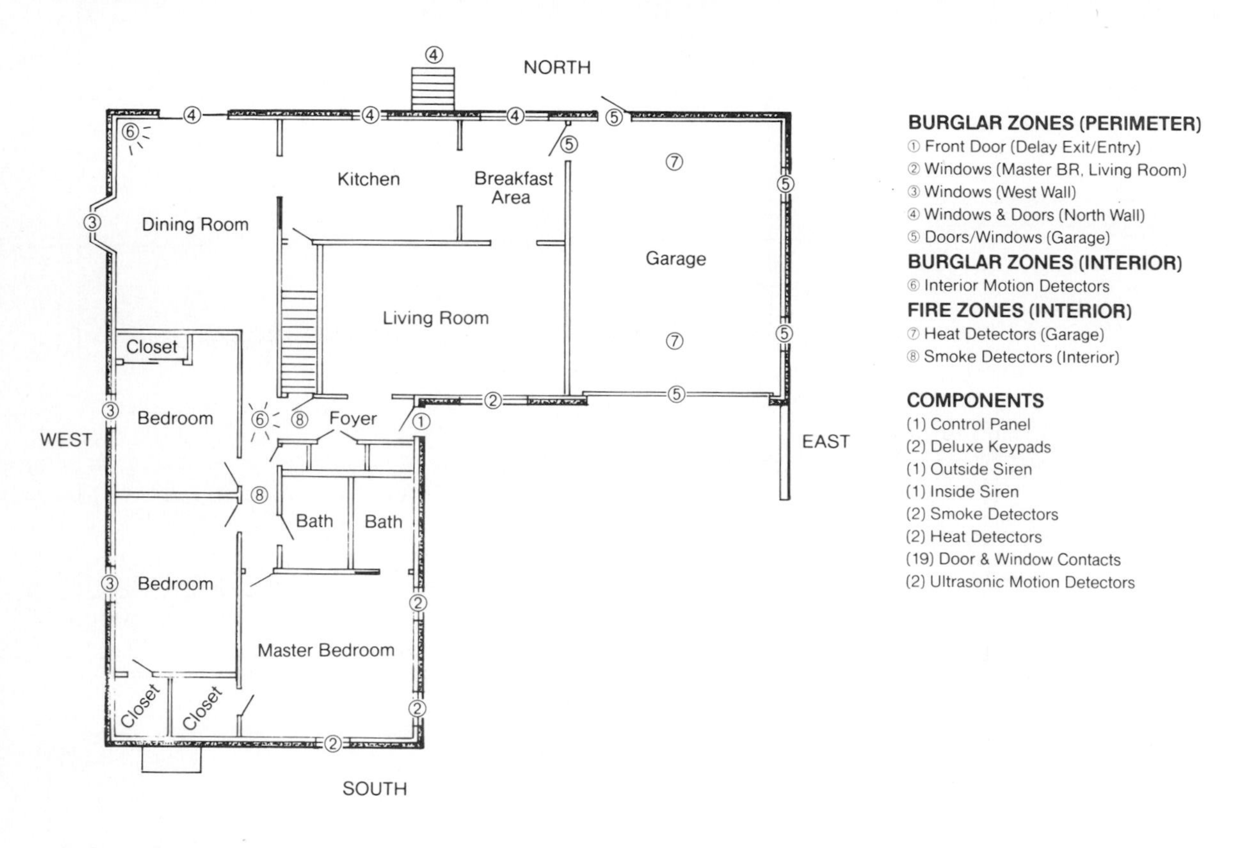

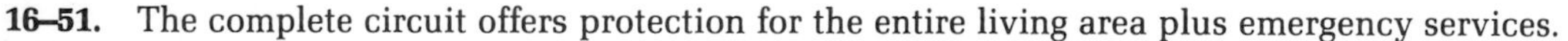
16–51. The complete circuit offers protection for the entire living area plus emergency services.

QUESTIONS

1. How often (on the average) do fires start?
2. Who installs burglar, fire, and smoke alarms?
3. How can burglaries be prevented?
4. What are the two circuits needed for an intruder detection system?
5. Where are magnetic detectors mounted?
6. How are plunger detectors mounted?
7. Where are mats placed in the home for detectors?
8. How is the infrared detector useful in an alarm system?
9. Is the infrared detector a complete security system, or should it be part of a larger system?
10. Why is an emergency circuit needed in an alarm system?
11. How does the wireless transmitter work?
12. Where should a siren be mounted on the outside of the house?
13. How are windows protected from entry?
14. What is a rate-of-rise detector?
15. What are the two types of smoke detectors?
16. What are the two types of heat detectors?
17. What is the square of protection?
18. List the units needed for a minimum fire protection plan.
19. List the units needed for a medium fire protection plan.
20. List the units needed for a maximum fire protection plan.
21. How is the emergency panic circuit operated?

KEY TERMS

contact switches
heat detector
magnetic detector
metallic window foil
motion detector
smoke detector
telephone dialer

SECTION SIX CAREERS

CHAPTER 17

Joyce R. Wilson/Photo Researchers, Inc.

CAREERS IN ELECTRICITY

Objectives

After studying this chapter, you will be able to:

- Identify the prominent career areas in electricity.
- List the educational requirements necessary to become an electrical engineer.
- Identify the employment qualifications for a construction electrician.
- Identify the job responsibilities of a maintenance electrician.
- Define *entrepreneur.*
- List the four similar problems faced by all entrepreneurs.
- Identify the qualities of an effective leader.

INDUSTRIES THAT EMPLOY ELECTRICIANS

The field of electricity offers a variety of career opportunities. Electricians are employed by many different industries, from steel plants to drug manufacturers, from coal mines to the merchant marine.

Many industries which rely heavily on electrical equipment hire full-time electricians to keep that equipment in good working order. For example, the aluminum industry needs electricians to install and repair electrical fixtures, apparatus, and control equipment. In the iron and steel industry, electricians install wiring and fixtures and hook up electrically operated equipment. Electrical repairers (motor inspectors) keep the wiring, motors, switches, and other electrical equipment in good working condition. The paper industry employs electricians to repair wiring, motors, control panels, and switches.

Railroad companies employ electrical workers to install and maintain the wiring and electrical equipment in locomotives, cars, and railroad buildings. Some workers also lay and maintain power lines.

Ships at sea need their own crew of maintenance workers on board. In the merchant marine industry, the ship's electrician repairs and maintains electrical equipment, such as generators and motors. The electrician also tests wiring for short circuits and removes and replaces fuses and defective lights.

In hospitals, the lives of patients often depend on equipment run by electricity. Proper maintenance and fast repair of electrical equipment are vital. Many hospitals therefore employ full-time electricians.

Factories that make medicines also rely on electricians to keep equipment running properly and prevent costly breakdowns. Electricians working for these companies also install and repair the various types of electrical equipment used.

Electricians are involved in the coal mining industry. They check and install electrical wiring in and around the mines. In addition, they help repair and maintain the machinery used for mining. Today's cars have a number of electrically operated devices and systems. In the auto industry, electrical engineers design the car's electrical systems, such as the ignition system, lights, and accessories.

In the construction industry, electricians assemble, install, and wire systems for heat, light, power, air conditioning, and refrigeration. Electrical inspectors then check these systems to make sure that they work properly and comply with electrical codes and standards. The inspectors visit worksites to inspect new and existing wiring, lighting, sound and security systems, and generating equipment. They also check the installation of the electrical wiring for heating and air-conditioning systems, kitchen appliances, and other components.

CAREERS IN ELECTRICITY

Although each industry has its own special requirements, careers in electricity can be divided into three general categories: Electrical engineer, construction electrician, and maintenance electrician.

Electrical Engineer

Electrical engineers design, develop, and supervise the manufacture of electrical and electronic equipment. The projects they work on include electric motors and generators; communications equipment; electronic equipment such as heart pacemakers, pollution-measuring devices, radar, computers, lasers, and missile guidance systems; and electrical applicances of all kinds. They also design and operate facilities for the generation and distribution of electric power. Fig. 17-1.

Usually, electrical engineers specialize in one major area, such as electronics, computers, electrical equipment, or power. Within these areas, there are still more specialized fields, such as missile guidance and tracking systems.

About 300 000 people are employed as electrical engineers. Most work for makers of electrical and electronic equipment, aircraft and parts, business machines, and professional and scientific equipment. Many work in the communications industry and for electric light and power companies. Electrical engineers also

17–1. Electrical engineers are responsible for the design and development of a wide variety of equipment. (Ray Ellis/Photo Researchers, Inc.)

work for government agencies, for construction firms, for engineering consultants, or as independent consulting engineers. Others teach in colleges and universities.

A bachelor's degree in engineering is required for most beginning engineering jobs. In a typical 4-year curriculum, the first 2 years are spent studying basic sciences (mathematics, physics, introductory engineering) and the humanities, social sciences, and English. The last two years are devoted mostly to specialized engineering courses.

Graduate training is now required for many jobs, particularly for teaching and for jobs in specialized areas. A number of colleges and universities now offer 5-year master's degree programs.

All 50 states and the District of Columbia require licensing for engineers whose work may affect life, health, or property, or who offer their services to the public. In order to obtain a license, an engineer usually must have a degree from an accredited engineering school, 4 years of related work experience, and must pass a state examination.

Engineering graduates usually begin work under the supervision of experienced engineers. With experience and proven ability, they advance to positions of greater responsibility. Electrical engineers should be able to work as part of a team and have creativity, an analytical mind, and a capacity for detail. They should be able to express their ideas well, both orally and in writing.

The employment outlook for electrical engineers is good. Increased demand for computers, for electrical and electronic consumer goods, for military electronics, and for communications and power generating equipment will increase the need for engineers in these fields.

Construction Electrician

As stated earlier, *construction electricians* assemble, install, and wire systems for heat, light, power, air conditioning, and refrigeration. They also install electrical machinery, electronic equipment, controls, and signal and communications systems. Fig. 17-2.

Construction electricians must be able to read blueprints and specifications. For safety reasons, they must follow the regulations in the National Electrical Code, and their work must also conform to local electrical codes.

There are about 250 000 construction electricians in the United States. Most work for electrical contractors or are self-employed as contractors. Some work for government agencies and businesses that do their own electrical work. Training for a career as a construction electrician usually involves a 4-year apprenticeship program. Apprenticeship programs are sponsored and supervised by local union-management committees. The programs include both classroom and on-the-job training. To qualify for an apprenticeship, an applicant usually must be a high school or vocational school graduate. Courses in electricity, electronics, mechanical drawing, science, and shop provide a good background.

Physical strength is not essential, but skill with one's hands, agility, and good health are

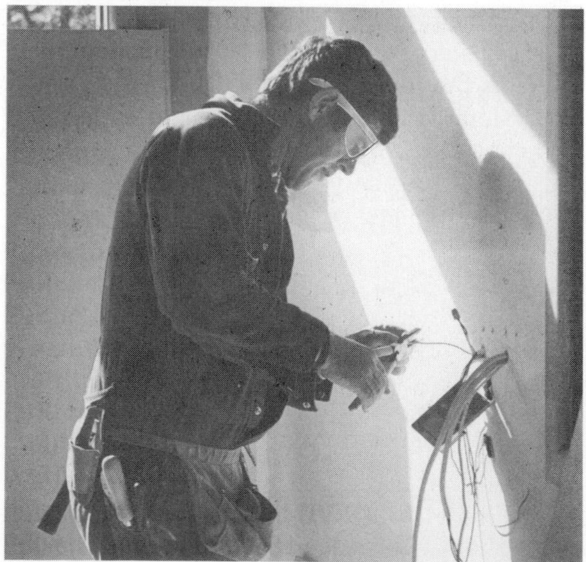

17–2. Construction electricians assemble, install, and wire a variety of electrical systems. (W. Backman/Photo Researchers, Inc.)

important. Good color vision is essential because electrical wires are often identified by color.

In most cities a license is required for employment. The electrician must pass a test which requires a thorough knowledge of the craft and of state and local building codes.

Employment outlook is good; the number of construction electricians is expected to grow faster than the average for all occupations. In any given year, the number employed depends on the amount of construction activity. When jobs are not available, however, construction electricians may be able to transfer to related jobs. For example, they may work as maintenance electricians.

Maintenance Electrician

Maintenance electricians keep lighting systems, transformers, generators, and other electrical equipment in good working order. They may also install new equipment.

The duties of maintenance electricians depend on where they are employed. Electricians working in large factories may repair particular items such as motors and welding machines. Those in office buildings and small plants usually fix all kinds of electrical equipment. Regardless of location, electricians spend much of their time doing preventive maintenance. They check equipment to locate and correct defects before breakdowns occur. Fig. 17-3.

Nearly 300 000 people work as maintenance electricians. More than half are employed by manufacturing industries. Many others work for utilities, mines, railroads, or the government.

Most maintenance electricians learn their trade on the job or through apprenticeship programs. The apprenticeship programs and the requirements for getting into the programs are the same as those for construction electricians. Hand skills, agility, and good health are important. Good color vision is required.

17–3. Maintenance electricians check equipment to locate defects before machine failure results. (R. Wood/Taurus Photos, Inc.)

Because of increased use of electrical and electronic equipment by industry, the demand for maintenance electricians will increase. Growth in the number of job openings is expected to be steady since the demand for maintenance electricians is not very sensitive to ups and downs in the economy.

Information Sources

■ For additional information about a career as an electrical engineer, write to:

Institute of Electrical and Electronic Engineers
345 East 47th Street
New York, New York 10017

■ For information about a career as a construction electrician, contact local electrical contractors, a local union of the International Brotherhood of Electrical Workers, a local union-management apprenticeship committee, or the nearest office of the state employment service or state apprenticeship agency. Some state employment service offices screen applicants and give aptitude tests.

■ Information about apprenticeships or other work opportunities for maintenance electricians is available from local firms that employ maintenance electricians and from local union-management apprenticeship committees. In addition, the local office of the state employment service may provide information about training opportunities. Some state employment service offices screen applicants and give aptitude tests.

■ For general information about the work of electricians, contact:

International Brotherhood of Electrical Workers
1125 15th Street NW
Washington, DC 20005

National Electrical Contractors Association
7315 Wisconsin Avenue
Washington, DC 20014

National Joint Apprenticeship and Training Committee for the Electrical Industry
1730 Rhode Island Avenue NW
Washington, DC 20036

STARTING YOUR OWN BUSINESS

In your town or city, there are many businesses—some large, others small. Some of the small businesses may be owned and managed by one person. Others may be medium-sized, employing several people. There may also be large businesses in your town or city. For example, there may be a large manufacturing plant. Or, the headquarters of a large company may be in your town. Either of these could employ hundreds—or even thousands—of people.

Most large businesses started as small businesses. This is true of some of the largest corporations. The building of a business requires skill and hard work. It can be difficult to imagine the energy needed to start a business. Anyone starting a small business faces a variety of tasks. He or she first must have the money needed. (This money may have to be borrowed.) He or she also must have a product or service that can be sold at a profit. He or she must have a way of distributing the product or service to the public. The smaller the business, the more of these jobs the owner-operator may have to do. In some small businesses, the owner-operator has all of these responsibilities. In larger businesses, such duties are divided. For example, one person (or one department) may be in charge of making the product. Another may be in charge of advertising it. A third may be in charge of delivering the item. Usually, the larger the business, the greater the number of people needed for each of these duties.

The person who starts a business is usually an *entrepreneur*. An entrepreneur is anyone who organizes and manages a business. This person also assumes the risks of the business. This means that the entrepreneur is responsible for paying the business expenses. All responsibility for the success of the business rests with the entrepreneur.

A person who is self-employed, or in business for himself or herself, is an entrepreneur. All entrepreneurs face four similar problems. These problems are:

- Identifying a *need*.
- Finding a *product* to satisfy the need.
- *Financing* the business.
- *Selling* the product.

Let's look at each of these concerns.

All successful businesses have one thing in common. They were started because someone noticed that people needed or wanted an item or service. Once an entrepreneur has noticed a need, he or she can then find a product to fill the need. This product might be an item (such as an electrical device). It might also be a service (such as an electrical product repair).

Starting a business is not easy. Succeeding in business is even harder. Anyone who starts a business is an entrepreneur. But not all entrepreneurs are successful. For an entrepreneur to be successful, his or her business must be successful.

THE DEVELOPMENT OF LEADERSHIP

Our society has always prized resourcefulness and initiative. These qualities were essential to the westward expansion of the American frontier and the early shaping of the values of the infant Republic. Some of the values of our American culture have been shaped by our history which, especially in its early days, required strong decision and firm enterprise. The events that are to us now simply history required fast commitments and unwavering resolution from those who took part in them.

Though the accidentals of historical influence may change, its basic forces remain much the same. This, by no surprise, results in problems and challenges for our society that are as puzzling and difficult as those posed for our ancestors. Because ours is a democratic society, every man and woman is prompted to develop leadership qualities.

Good leaders are characterized by effective communication, firmness of will, singleness of purpose, and moral integrity. While, it seems, these qualities are more readily apparent in some, all of us can develop, in some degree, the skills needed for effective leadership. Membership in student clubs—such as VICA or AIASA—can help develop leadership ability. A club, by its nature—bringing together as it does individuals with a single common interest—has perhaps less diversity than society generally. Still, it can offer you an opportunity to practice effective communication and learn the skills needed to work within a group, such as a committee. The valuable qualities of leadership are often exercised in the most quiet situations. Each of you will be given opportunities to exert the force of leadership to bring about a decision, resolve a crisis, or prompt an action. Leadership is essential to active participation in a democratic society.

QUESTIONS

1. What are the educational requirements for an electrical engineer?
2. What type of training is usually required for someone training for a career as a construction electrician?
3. What are the job responsibilities of a maintenance electrician?

KEY TERMS

construction electrician
electrical engineer
entrepreneur
maintenance electrician

APPENDIX A. Conversions

TO CONVERT-	INTO-	MULTIPLY BY-	CONVERSELY MULTIPLY BY-
Inches	Centimeters	2.54	0.3937
Inches	Mils	1000	0.001
Joules	Foot-pounds	0.7376	1.356
Joules	Ergs	10^{7}	10^{-7}
Kilogram-calories	Kilojoules	4.186	0.2389
Kilograms	Pounds (avoirdupois)	2.205	0.4536
Kg per sq meter	Pounds per sq foot	0.2048	4.882
Kilometers	Feet	3281	3.048×10^{-4}
Kilowatt-hours	Btu	3413	2.93×10^{-4}
Kilowatt-hours	Foot-pounds	2.655×10^{6}	3.766×10^{-7}
Kilowatt-hours	Joules	3.6×10^{6}	2.778×10^{-7}
Kilowatt-hours	Kilogram-calories	860	1.163×10^{-3}
Kilowatt-hours	Kilogram-meters	3.671×10^{5}	2.724×10^{-6}
Liters	Cubic meters	0.001	1000
Liters	Cubic inches	61.02	1.639×10^{-2}
Liters	Gallons (liq US)	0.2642	3.785
Liters	Pints (liq US)	2.113	0.4732
Meters	Yards	1.094	0.9144
Meters per min	Feet per min	3.281	0.3048
Meters per min	Kilometers per hr	0.06	16.67
Miles (nautical)	Kilometers	1.853	0.5396
Miles (statute)	Kilometers	1.609	0.6214
Miles per hr	Kilometers per min	2.682×10^{-2}	37.28
Miles per hr	Feet per minute	88	1.136×10^{-2}
Miles per hr	Kilometers per hr	1.609	0.6214
Poundals	Dynes	1.383×10^{4}	7.233×10^{-5}
Poundals	Pounds (avoirdupois)	3.108×10^{-2}	32.17
Sq inches	Circular mils	1.273×10^{6}	7.854×10^{-7}
Sq inches	Sq centimeters	6.452	0.155
Sq feet	Sq meters	$9.29 \times 10^{\times 2}$	10.76
Sq miles	Sq yards	3.098×10^{6}	3.228×10^{-7}
Sq miles	Sq kilometers	2.59	0.3861
Sq millimeters	Circular mils	1973	5.067×10^{-4}
Tons, short (avoir 2000 lb)	Tonnes (1000 kg)	0.9072	1.102
Tons, long (avoir 2240 lb)	Tonnes (1000 kg)	1.016	0.9842
Tons, long (avoir 2240 lb)	Tons, short (avoir 2000 lb)	1.120	0.8929
Watts	Btu per min	5.689×10^{-2}	17.58
Watts	Ergs per sec	10^{7}	10^{-7}
Watts	Ft-lb per minute	44.26	2.26×10^{-2}
Watts	Horsepower (550 ft-lb per sec)	1.341×10^{-3}	745.7
Watts	Horsepower (metric) (542.5 ft-lb per sec)	1.36×10^{-3}	735.5
Watts	Kg-calories per min	1.433×10^{-2}	69.77

APPENDIX E. Full-Load Currents in Amperes; Single-Phase Alternating-Current Motors

HP	115 V	230 V
$\frac{1}{6}$	4.4	2.2
$\frac{1}{4}$	5.8	2.9
$\frac{1}{3}$	7.2	3.6
$\frac{1}{2}$	9.8	4.9
$\frac{3}{4}$	13.8	6.9
1	16	8
$1\frac{1}{2}$	20	10
2	24	12
3	34	17
5	56	28
$7\frac{1}{2}$	80	40
10	100	50

(Sangamo Electric Co.)

The following values of full-load currents are for motors running at usual speeds and motors with normal torque characteristics. Motors built for especially low speeds or high torques may have higher full-load currents, and multispeed motors will have full load current varying with speed, in which case the nameplate current ratings shall be used.

To obtain full-load currents of 208- and 200-volt motors, increase corresponding 230-volt motor full-load currents by 10 and 15%, respectively.

The voltages listed are rated motor voltages. Corresponding nominal system voltages are 110 to 120, and 220 to 240.

These tables are based on the National Electrical Code 1987.

APPENDIX F. Full-Load Current* Three-Phase AC Motors

HP	INDUCTION TYPE SQUIRREL-CAGE AND WOUND ROTOR, AMPERES					SYNCHRONOUS TYPE† UNITY POWER FACTOR, AMPERES			
	115 V	230 V	460 V	575 V	2300 V	230 V	460 V	575 V	2300 V
$\frac{1}{2}$	4	2	1	.8		230	460	575	
$\frac{3}{4}$	5.6	2.8	1.4	1.1					
1	7.2	3.6	1.8	1.4					
$1\frac{1}{2}$	10.4	5.2	2.6	2.1					
2	13.6	6.8	3.4	2.7					
3		9.6	4.8	3.9					
5		15.2	7.6	6.1					
$7\frac{1}{2}$		22	11	9					
10		28	14	11					
15		42	21	17					
20		54	27	22					
25		68	34	27		53	26	21	
30		80	40	32		63	32	26	
40		194	52	41		83	41	33	
50		130	65	52		104	52	42	
60		154	77	62	16	123	61	49	12
75		192	96	77	20	155	78	62	15
100		248	124	99	26	202	101	81	20
125		312	156	125	31	253	126	101	25
150		360	180	144	37	302	151	121	30
200		480	240	192	49	400	201	161	40

(Sangamo Electric Co.)

For full-load currents of 208- and 200-volt motors, increase the corresponding 230-volt motor full-load current by 10 and 15%, respectively.

*These values of full-load current are for motors running at speeds for belted motors and motors with normal torque characteristics. Motors built for especially low speeds or high torques may require more running current, and multispeed motors will have full load current varying with speed, in which case the nameplate current rating shall be used.

†For 90 and 80% PF the above figures shall be multiplied by 1.1 and 1.25, respectively.

The voltages listed are rated motor voltages. Corresponding nominal voltages are 110 to 120, 220 to 240, 440 to 480, and 550 to 600 volts.

These tables are based on the National Electrical Code 1987.

APPENDIX G-A. Maximum Number of Fixture Wires in Trade Sizes of Conduit or Tubing

(40 Percent Fill Based on Individual Diameters)

Conduit Trade Size (Inches)	1/2					3/4					1					1¼					1½					2				
Wire Types	18	16	14	12	10	18	16	14	12	10	18	16	14	12	10	18	16	14	12	10	18	16	14	12	10	18	16	14	12	10
PTF, PTFF, PGFF, PGF, PFF, PF, PAF, PAFF, ZF, ZFF	23	18	14			40	31	24			65	50	39			115	90	70			157	122	95			257	200	156		
TFFN, TFN	19	15				34	26				55	43				97	76				132	104				216	169			
SF-1	16					29					47					83					114					186				
SFF-1, FFH-1	15					26					43					76					104					169				
CF	13	10	8	4	3	23	18	14	7	6	38	30	23	12	9	66	53	40	21	16	91	72	55	29	22	149	118	90	48	37
TF	11	10				20	18				32	30				57	53				79	72				129	118			
RFH-1	11					20					32					57					79					129				
TFF	11	10				20	17				32	27				56	49				77	66				126	109			
AF	11	9	7	4	3	19	16	12	7	5	31	26	20	11	8	55	46	36	19	15	75	63	49	27	20	123	104	81	44	34
SFF-2	9	7	6			16	12	10			27	20	17			47	36	30			65	49	42			106	81	68		
SF-2	9	8	6			16	14	11			27	23	18			47	40	32			65	55	43			106	90	71		
FFH-2	9	7				15	12				25	19				44	34				60	46				99	75			
RFH-2	7	5				12	10				20	16				36	28				49	38				80	62			
KF-1, KFF-1, KF-2, KFF-2	36	32	22	14	9	64	55	39	25	17	103	89	63	41	28	182	158	111	73	49	248	216	152	100	67	406	353	248	163	110

APPENDIX G-B. Maximum Number of Conductors in Trade Sizes of Conduit or Tubing

(Based on Table 1, Chapter 9)

Conduit Trade Size (Inches)		$\frac{1}{2}$	$\frac{3}{4}$	1	$1\frac{1}{4}$	$1\frac{1}{2}$	2	$2\frac{1}{2}$	3	$3\frac{1}{2}$	4	5	6
Type Letters	**Conductor Size AWG, MCM**												
TW, T, RUH,	14	9	15	25	44	60	99	142					
RUW,	12	7	12	19	35	47	78	111	171				
XHHW (14 thru 8)	10	5	9	15	26	36	60	85	131	176			
	8	2	4	7	12	17	28	40	62	84	108		
RHW and RHH	14	6	10	16	29	40	65	93	143	192			
(without outer	12	4	8	13	24	32	53	76	117	157			
covering),	10	4	6	11	19	26	43	61	95	127	163		
THW	8	1	3	5	10	13	22	32	49	66	85	133	
TW,	6	1	2	4	7	10	16	23	36	48	62	97	141
T,	4	1	1	3	5	7	12	17	27	36	47	73	106
THW,	3	1	1	2	4	6	10	15	23	31	40	63	91
RUH (6 thru 2),	2	1	1	2	4	6	9	13	20	27	34	54	78
RUW (6 thru 2),	1		1	1	3	4	6	9	14	19	25	39	57
FEPB (6 thru 2),	0		1	1	2	3	5	8	12	16	21	33	49
RHW and	00		1	1	1	3	5	7	10	14	18	29	41
RHH (with	000		1	1	1	2	4	6	9	12	15	24	35
out outer	0000			1	1	1	3	5	7	10	13	20	29
covering)	250			1	1	1	2	4	6	8	10	16	23
	300			1	1	1	2	3	5	7	9	14	20
	350				1	1	1	3	4	6	8	12	18
	400				1	1	1	2	4	5	7	11	16
	500				1	1	1	1	3	4	6	9	14
	600					1	1	1	3	4	5	7	11
	700					1	1	1	2	3	4	7	10
	750					1	1	1	2	3	4	6	9

APPENDIX H. Allowable Current-Carrying Capacity of Copper Conductors

SIZE		DC RESISTANCE	THREE CONDUCTORS† IN RACEWAY		SINGLE CONDUCTOR IN FREE AIR
GAGE AWG OR MCM	AREA IN CIRCULAR MILLS	OHMS PER 1000 ft 20°C, 68°F	RUBBER TYPE RUW THERMOPLASTIC TYPES T, TW	RUBBER TYPE RH	BARE AND COVERED CONDUCTOR
14	4107	2.525	15	15	30
12	6530	1.588	20	20	40
10	10 380	0.9989	30	30	55
8	16 510	0.6282	40	45	70
6	26 250	0.3951	55	65	100
4	41 740	0.2485	70	85	130
3	52 640	0.1970	80	100	150
2	66 370	0.1563	95	115	175
1	83 690	0.1239	110	130	205
0	105 500	0.09827	125	150	235
00	133 100	0.07793	145	175	275
000	167 800	0.06180	165	200	320
0000	211 600	0.04901	195	230	370
250	250 000	0.0415	215	255	410
300	300 000	0.0346	240	285	460
500	500 000	0.0207	320	380	630
750	750 000	0.0138	400	475	810
1000	1 000 000	0.0104	455	545	965

(Sangamo Electric Co.)

†For more than three conductors in raceway or cable: 4 to 6 conductors, derate to 80%; 7 to 9 conductors, derate to 70%.

These tables are based on temperature alone and do not take voltage drop into consideration.

Based on the National Electrical Code 1984. The current issue should be consulted for complete information.

APPENDIX I. Allowable Current-Carrying Capacities of Aluminum Conductors

(Based on ambient temperature of 30°C, 86°F)

SIZE		DC RESISTANCE	THREE CONDUCTORS† IN RACEWAY		SINGLE CONDUCTOR IN FREE AIR
GAGE AWG OR MCM	AREA IN CIRCULAR MILLS	OHMS PER 1000 ft 20°C, 68°F	RUBBER TYPE RUW THERMOPLASTIC TYPES T, TW	RUBBER TYPE RH	BARE AND COVERED CONDUCTOR
14	4110	4.140	....	....	....
12	6530	2.606	15	15	30
10	10 380	1.638	25	25	45
8	16 510	1.030	30	40	55
6	26 240	0.6482	40	50	80
4	41 740	0.4076	55	65	100
3	52 620	0.3233	65	75	115
2	66 360	0.2564	75	90	135
1	83 690	0.2033	85	100	160
0	105 600	0.1612	100	120	185
00	133 100	0.1278	115	135	215
000	167 800	0.1014	130	155	250
0000	211 600	0.08039	155	180	290
250	250 000	0.06804	170	205	320
300	300 000	0.05670	190	230	360
500	500 000	0.03402	260	310	490
750	750 000	0.02268	320	385	640
1000	1 000 000	0.01701	375	445	770

(Sangamo Electric Co.)

†For more than three conductors in raceway or cable: 4 to 6 conductors, derate to 80%; for 7 to 9 conductors, derate to 70%.

These tables are based on temperature alone and do not take voltage drop into consideration.

Based on the National Electric Code 1987. Consult current issue for more complete information.

Glossary

A

AC Generator. A generator that produces alternating current.

Alternator. Another name for an AC generator.

Ampere. The current produced when 6.28×10^{18} electrons flow past a given point in one second.

Ammeter. A meter capable of measuring amperes.

Armature. Rotating part of an electric motor or generator. May also be the moving part of a relay, buzzer, or speaker.

Atoms. The smallest particles of an element that retains all the properties of that element.

Automatic load transfer. A method of starting and switching to emergency power plants when the utility power is interrupted.

B

Ballast. A choke or inductor used in a gaseous discharge (fluorescent) lamp circuit.

Battery. Two or more cells connected together in series or parallel.

Bonding. Effective bonding means that the electrical continuity of the grounding circuit is assured by proper connections between service raceways, service cable armor, all service equipment enclosures containing service conductors, and any conduit or conductor that forms part of the grounding conductor to the service raceway.

Bonding jumpers. Are used to assure continuity around concentric or eccentric knockouts that are punched or otherwise formed in such a manner that they impair electrical current flow.

Branch circuit. A circuit that is used to feed small electrical loads such as lamps, small appliances, and kitchen equipment.

Brushes. Devices that make contact with the rotating connections to the armature of a motor or generator. Usually made of carbon or metal.

C

Canadian Standards Association (CSA). A parallel organization to UL in the United States. It is set up to develop voluntary national standards, provide certification services for national standards, and to represent Canada in international standards activities.

Capacitor. A device used to store an electrical charge. Consists of two plates and a dialectric.

Cell. A device used to generate an electrical current or emf.

Circuit. A path for electrons to flow or electricity to move.

Circular mil. A measurement unit for cross-sectional area of a round wire; equal to one-thousandth (0.001) of an inch in diameter.

Coil. A device made by turns of insulated wire wound around a core. A coil sometimes has a hollow center portion. Also called an inductor.

Color code. The carbon composition resistors are marked with their resistance with color bands. The bands are colored according to an established code as follows:

Black 0	Yellow 4	Gray 8
Brown 1	Green 5	White 9
Red 2	Blue 6	
Orange 3	Violet 7	

Commutator. A device made of segments of copper insulated by mica or some other material. Used to reverse direction of current flow from a generator or to a motor. Brushes are usually placed to make contact with the surface of the commutator. Segments of a commutator are connected to ends of the armature coils in a motor or generator.

Complete circuit. Is made up of a source of electricity, a conductor, and a consuming device.

Conductors. Materials through which electrons move.

Configuration. An arrangement of wires or components in a circuit.

Construction electrician. One who assembles, installs, and wires systems for heat, light, power, air conditioning, and refrigeration. They also install electrical machinery, electronic equipment, controls, and signal communications systems.

Convenience outlet. Outlets placed for the convenience of the home owner in everyday use of electricity.

Coulomb. The unit of measurement of 6.28×10^{18} electrons.

Current. The movement of electrons in a negative-to-positive direction along a conductor.

Current flow. The movement of free electrons in a given direction.

D

Dielectric. An insulating material used in a capacitor or other electrical device.

Diode. A semiconductor or vacuum tube device that is used to allow current to flow easily in one direction but retards or stops its flow in the other direction. Diodes are used in rectifier circuits and in switching circuits.

Direct current (DC). Current that flows in one direction only.

Draftsperson. One who draws plans and electrical schematics with the aid of mechanical devices.

E

Electrical engineer. One who designs, develops, and supervises the manufacture of electrical and electronic equipment.

Electrician. One who works with electrical equipment and wiring.

Electricity. The flow of electrons along a conductor.

Electrodynamometer. A type of meter that uses no permanent magnet, but two fixed coils to produce a magnetic field. The meter also uses two moving coils. This meter can be used as a voltmeter or an ammeter.

Electrolyte. A solution capable of conducting electric current, the liquid part of a battery.

Electrolytic. A capacitor that has parts separated by an electrolyte. Thin film formed on one plate provides the dielectric. The electrolytic capacitor has polarity (− and +).

Electromagmet. A magnet produced by current flow through a coil of wire. The core is usually used to concentrate the magnetic lines of force.

Electromagnetism. Magnetism produced by current flowing through a coil of wire or other conductor.

Electron. Smallest particle of an atom with a negative charge.

Elements. The most basic materials in the universe. Ninety-four elements, such as iron, copper, and nitrogen, have been found in nature. Every known substance is composed of elements.

Encapsulation. The embodiment of a component or assembly in a solid or semisolid medium such as tar, wax, or epoxy.

Energy. Ability to do work.

Entrance signals. Usually consist of a door bell, chime, or some other device used to alert the home occupant to someone outside wishing to see them.

Equipment ground. The grounding of exposed conductive materials such as conduit, switch boxes, or meter frames that enclose conductors and equipment. This is to prevent the equipment from exceeding ground potential.

Explosion proof. Term used to describe apparatuses that are enclosed in a case that is capable of withstanding an explosion within the case without igniting flammable materials outside it.

F

Farad (F). A unit of measurement for capacitance.

Fatal current. The amount of current needed to kill. Currents between 100 and 200 mA are lethal.

Feeder circuit. A circuit that is used to feed others or take electrical energy to where it can then be branched off to service other locations.

Filaments. Small coils of resistance wire in light bulbs and vacuum tubes that heat up to glow either red or white hot. The filament of a vacuum tube boils off electrons from the cathode. The filament in a light bulb glows to incandescence to produce light.

Fluorescence. A term that means *to glow* or *give off* light. Fluorescent lamps fluoresce, or produce light, by having ions of mercury collide and strike a fluorescent coating inside the tube.

Fluorescent lamp. A lamp that produces light through action of ultraviolet rays striking a fluorescent material on the inside of a glass tube.

Four-way switch. Used where it is necessary to turn a light or circuit on or of from three or more locations. It also needs a minimum of two three-way switches to be able to do its job.

Fuse. A safety device designed to open if excessive current flows through a circuit.

G

Generator. A device that turns mechanical energy into electrical energy.

Ground fault circuit interrupter (GFI or GFCI). A fast-operating circuit breaker that is sensitive to very low levels of current leakage to ground. It is designed to limit electric shock to a current and time duration value below that which can produce serious injury.

Grounding. Effective grounding means that the path to ground is permanent and continous; and it has a low impedance to permit all current carrying devices on the circuit to work properly.

Grounding conductor. A wire attached to the housing or other conductive ports of electrical equipment that are not normally energized. The conductor carries current for them to the ground.

H

Henry (H). Unit of measurement for inductance.

Hertz (Hz). Unit of measurement for frequency.

Horsepower. The unit of measurement that equates work done electrically with the work done by a horse. One horsepower is the energy consumed to produce the equivalent work done by a horse lifting 33 000 pounds for one foot in one minute. It takes 746 watts to equal one horsepower.

I

Impedance (Z). Total circuit opposition to alternating current. Measured in ohms.

Incandescent. A term that means to *glow white hot*. The filament in an incandescent lamp glows white hot to produce heat and light.

Incandescent lamp. A lamp or bulb that produces light by heating a filament to incandescence (white hot).

Induced current. Current produced by electromagnetic induction, usually from a coil or transformer.

Inductance (L). That property of a coil that opposes any change in circuit current. Measured in Henrys.

Inductive reactance. Opposition presented to alternating current by an inductor. The symbol is X_L. Measured in ohms.

Inerting. Consists of mixing a chemical inert, nonflammable gas with a flammable substance, displacing the oxygen until the percentage of oxygen in the mixture is too low to allow combustion.

Input. Term used to describe the energy applied to a circuit, device, or system.

Insulator. Nonconducting material lacking a sufficient supply of free electrons to allow the movement of electrons without exceptional force or high voltage.

Inverter. A device used to convert DC to AC.

J

Joule. Metric unit of electrical work done by 1 coulomb flowing with a potential difference of 1 volt.

K

Kilo. A prefix that means one thousand (1000).

Kilowatt. One thousand watts (1000 W), kW.

Kilowatthour meter. Used to measure the power used by a consumer for a month's period of time.

Kirchoff's law of voltages. The sum of all voltages across resistors or loads is equal to the applied voltage.

L

Load. Anything that may draw current from an electrical power source.

M

Magnet. A device which possesses a magnetic field.

Magnetism. A force produced by an electrical current in a wire or found in nature in certain materials.

Magnetohydrodynamic generator. A generator of electricity that uses hot plasma to produce electricity. Electrons in the gas (plasma) are deflected by a magnetic field. Between collisions with the particles in the gas, they make their way to one of the electrodes to produce electricity.

Magnet wire. Copper wire used to wind coils, solenoids, transformers, motors; usually coated with varnish or other insulation material.

Maintenance electrician. One who keeps lighting systems, transformers, generators, and other electrical equipment in good working order. He or she may also install new equipment.

Mega. A prefix that means one million (1 000 000).

Meter. Means *to measure.*

Mica. Insulating material that can withstand high voltages and elevated temperatures. Used in the manufacture of appliances and capacitors.

Micro. A prefix that means one-millionth (0.000001).

Microammeter. A meter limited to measuring microamperes.

Microampere. One-millionth (0.000001) of an ampere (μA).

Microvolt. One-millionth (0.000001) of a volt (μV).

Milli. A prefix that means one-thousandth (0.001).

Milliammeter. A meter limited to measuring milliamperes.

Milliampere. One-thousandth (0.001) of an ampere (mA).

Millivolt. One-thousandth (0.001) of a volt (mV).

Millwatt. One-thousandth (0.001) of a watt (mW).

Motor. A device used to change electrical to mechanical energy.

Multimeter. A meter capable of measuring (in most instances) volts, ohms, and milliamperes.

Multiplier. A resistor placed in series with a meter movement to handle the extra voltage applied to the movement. It serves to extend the range of the meter.

N

National Electrical Code (NEC). A book of standards produced by the National Fire Protection Association every three years. It describes the proper installation of various electrical machines and devices for safe operation.

Neutrons. Tiny particles that have no electrical charge.

O

Ohmmeter. A device used to measure resistance.

Ohm's law. Georg Ohm's law states that the circuit is equal to the voltage divided by the resistance.

Omega. The Greek symbol (Ω) is used to indicate the unit of resistance, the ohm.

OSHA. Occupational Safety and Health Act.

Outlet. Socket or receptacle that accepts a plug to make electrical contact, usually to provide power.

P

Parallel circuit. A circuit where each load (resistance) is connected directly across the voltage source.

Piezoelectrical effect. The process of using crystals under pressure to produce electrical energy.

Photosensitive. Sensitive to light.

Potentiometer. A variable resistor that has three contact points.

Power. Rate of doing work; abbreviated as P and measured in watts.

Proton. Smallest particle of an atom with a positive charge.

Pulsating direct current. Current produced when changed to DC and left unfiltered; abbreviated as PDC.

R

Rectifier. A device that changes AC to DC; allows current to flow in only one direction.

Relay. An electromatic device that is used for remote switching.

Remote control. The ability to start or control a device from a location other than that of the device being controlled.

Residential wiring. The electrical wiring used to supply a home with electrical outlets where needed.

Resistance. Opposition to the movement of electrons. Measured in ohms.

Resistor. A device that opposes current flow.

Rheostat. Variable resistor; usually has only two points connected into the circuit. Used to control voltage by increasing and decreasing resistance.

Root-mean-square (rms). A type of reading obtained by using a standard voltmeter or ammeter.

Rotor. A moving part of a motor or generator.

Rural grid. Consists of radial feeders that leave the substation as three-phase circuits with each of the phases fanning out to serve the countryside as single-phase circuits.

S

Series circuit. A circuit with one resistor or consuming device located in a string or one after another.

Series-parallel circuit. A combination of series and parallel circuits where a minimum of three resistors or devices are connected with at least one in series and with at least two in parallel.

Shock. A process whereby an outside source of electricity is such that it overrides the body's normal electrical system and causes the muscles to react involuntarily.

Short circuit. A circuit that has extremely low resistance.

Shunt. A resistor placed in parallel with a meter movement to handle or shunt most of the current around the movement. It serves to extend the range of the meter.

Slip ring. A ring of copper mounted on the shaft of a motor or generator through which a brush makes permanent (or constant) contact with the end of the rotor windings. Always used in pairs.

Solar cell. A cell that turns light energy into electrical energy. It is usually made of silicon.

Solenoid. A coil of wire wrapped around a hollow form, usually with some type of core material sucked into the hollow. The movement of the core is usually to move a switch or to open valves.

Splice. A form of electical connection in which wires are joined directly to each other.

Static electricity. A form of energy present when there are two charges of opposite polarity in close proximity. Static electricity is generated by friction.

System ground. The grounding of the neutral conductor, or ground leg, of the circuit to prevent lightning or other high voltages from exceeding the design limits of the circuit.

T

Terminal. A connecting point for wires. It is usually present on batteries, cells, switches, relay, motors, and electrical panels.

Thermocouple. A device made of two different kinds of metals joined at one end. When the junction is heated, an emf is generated across the open ends.

Thermostat. A device that acts as a switch which is operated when heat causes two metals to expand at different rates.

Three-way switch. Used where it is necessary to turn a light or circuit on or off from more than one location.

Toggle switches. Devices used to turn various circuits on and off or to switch from one device to another. They are made in a number of configurations.

Transformer. A device that can induce electrical energy from one coil to the other by using magnetic lines of force, stepping voltage up or down.

U

Underwriters' Laboratories. An organization which tests electrical devices, systems, and materials to see if they meet certain safe operation standards. The trademark is UL. It is a non-profit organization.

Universal motor. A motor that operates on AC or DC.

V

Volt (E). Unit of measurement of electromotive force.

Voltage. Electromotive force (emf) that causes electrons to move along a conductor.

Voltmeter. An instrument used to measure voltage.

W

Watt (W). Unit of electrical power.

Watthour meter. A meter that measures electrical power consumed in an hour.

X

X. Symbol for reactance.

Z

Z. Symbol for impedance.

Zener. A type of semiconductor diode that breaks down intentionally at a predetermined voltage. Usually used for voltage regulation circuits.

Index

D

H

N

O

P

T

U

Y

Z